D0753446

CONCRETE BRIDGES

*Inspection, Repair, Strengthening,
Testing and Load Capacity Evaluation*

CONCRETE BRIDGES

Inspection, Repair, Strengthening,
Testing and Load Capacity Evaluation

Dr V K RAINA
PhD, DIC, MICE, C. Eng. (London), P. Eng. (Ontario)
Civil Engineering Adviser

McGraw-Hill

New York   San Francisco   Washington, D.C.   Auckland   Bogotá
Caracas   Lisbon   London   Madrid   Mexico City   Milan
Montreal   New Delhi   San Juan   Singapore
Sydney   Tokyo   Toronto

ISBN 0-07-462349-4

Printed and bound by Quebecor/Book Press.

PREFACE

- Although a vast revenue resource is consumed in building bridges, 'managing' their maintenance and then actually 'executing their maintenance work' can prove even more exacting and costly if what has been built must remain operational for their intended long-term safe use. In the past, the maintenance work has traditionally been a mere ongoing process conducted many a time by those who long ceased to be technical — or almost! But never before has the need for maintenance been more pressing. This calls for a scientific assessment of the problem for a workman-like understanding and execution. This requires:

 (i) A thorough examination of the detailed inventories;

 (ii) Carrying out detailed condition-surveys and visual and 'hands-on' inspections;

 (iii) Analyzing the observations and structures in order to unfold the causes of the structural distresses;

 (iv) Carrying out the structural investigative computations, and, where called for, the appropriate *in situ tests* on material-samples and on existing structures; and

 (v) Ultimately writing the prescriptions for rehabilitation and repair or outright demolition and replacement, as may be necessary.

It can readily be realized that first the Maintenance Management (i.e. the rationalized methodology for assessment of the condition of what exists and how to repair and rehabilitate it to retain its operational status ... in order to enjoy its revenue return if not financial-return), and then the Execution of Maintenance (i.e. the physical execution of the maintenance decisions taken), form a whole new dimension of the society's rightful demand on the Engineer's Services if the existing bridges have to remain there for the purpose they were constructed in the first place! (The effective aggregate of the aforementioned two exercises is what is referred to as the Bridge Maintenance Management System, BMMS, in the present-day technical jargon.)

- The tendency of some bridge engineers to feel contended at involvement only with bridge-DESIGN is wrong, to say the least. Designing a bridge is only one (initial) activity, whereas what is required thereafter (for the next sixty years or more) is the understanding of its service-life 'tantrums' and the art and science involved in looking after these! The bridge rehabilitation is an on-going process of repair and strengthening. Its importance increases with the age of the bridge and it can be suicidal to overlook the possible potential problems and not be ready with their practical solutions. Bridge maintenance and rehabilitation presupposes the knowledge of analysis and retro-design.

- Bridge rehabilitation can be much more involved than designing and constructing a new bridge. It requires a great deal of effort to tie-in all the concerned disciplines together in-order to build-back a distressed bridge, cost effectively. While in the initial designing it is mandatory to design within the prescribed limits of Codal Specifications, it may not be possible to follow such regimentation when mending the distresses.

Alternates of even lesser load-carrying capacity may have to be seriously considered and even higher than permissible stresses accepted.

- As acknowledged in my other books, one of the prices a professional practitioner has to pay is that he, unlike those involved in research and laboratory work, classroom lectures, or staid office work, hardly has time to write. A chronic practitioner would rather spend the time in designing (and still more designing) and constructing (and still more constructing) and rehabilitating, than just writing! But of course it would be very useful if such a real-life-practising-professional, who has his fingers on the pulse of practice and, in fact, has a lot to write about, could squeeze time in order to 'also write' for the profession; however hard it might be for him to find that time! It would be even more meaningful if he, additionally, had a 'practical research background' that would help him sift the grain from the husk.

- Fired by this feeling, I took up writing the present book in the humble hope that it may provide an amalgam of 'practice' and 'theory', with the former subordinating the latter in order that the book is of 'gainful-use' to the practising engineers.

• Having worked with many Consultants and Contractors in many countries for many years on a multitude of projects, I assiduously sifted and stored these experiences over the long years with a view to sharing them with others through my various practice-oriented books, the present one being the sixth in this 'workman' series.

• Apart from drawing upon my own experience, and interaction with others, in preparing this book I have also drawn on some of the material published by the U.S. Transportation Research Board (TRB), the U.S. Federal Highway Authority (FHWA), the AASHTO and some of the states in the U.S., and also the practices followed by the *Ministry of Communications* in the Kingdom of Saudi Arabia and those proposed by the Road Directorate of Denmark. I am indebted to them and duly acknowledge them with grateful thanks.

• The reader will find this book a practitioner's delight, honed as it is with practical details whose treatment is fresh and compendious. The integrated coverage is comprehensive yet concise, even though somewhat encyclopaedic. Practical details cannot be short-changed or glibbed-over!

• The steps of the reader are guided in paths often trodden by and therefore familiar to the author, who, thereby, is perhaps able to layout a 'straight course' without the reader having to waste time in search of the 'route'. It has to be emphasized though that, in a given case, the choice and the application should be based on the concerned engineer's own judgment since he would know better all that is incumbent in his particular case, site and contract.

• Last but not least, I am indebted to Vinita, my dear wife, for putting-up with my rather rigorous work-calendar which never left me much free time for the social and domestic fronts. Despite her own stringent professional work-schedule (she is a London Ph.D. in bone pathology, and practices and teaches the subject), she is obviously more adept in evenly dividing attention between work and home. Be that as it may, it is not entirely un-understandable as to why she took to globe-trotting on her own each time my work-pitch soared!

DR V K RAINA

TALKING ALONE NEVER PULLED OUT A 'STUMP'!

Engineering is not just doing theoretical sums, nor is it a matter of blind adherence to graphs, formulae and 'files'. An approximate solution to an exact problem is more meaningful than an exact solution to an approximated problem. A useful book does not have to be graveyard of dead Ph.Ds. I am saddened if a book purports to be 'practical' when in fact it is packed with pages of iterative empirics, limited-use graphs, and bewildering morbid mathematical plethora that leaves one high and dry, wondering about its use. Worse still, if it is written by a non-practising pen-pusher or a mundane class-room lecturer, who has not 'stayed', 'survived' and 'surfaced' in the numbingly competitive, merciless, cut-throat, and cold commercial world of 'actual' construction and design. However exacting going through this might be, but this is where all the 'fun' lies. Here the 'doer' grapples with result-demanding survival-situations set to near-impossible and ever-shrinking dead-lines that can give ulcers. ('Talkers' are many , but 'doers' are few!) A non-practising engineer, be he the mundane class-room-lecturing type or the pen-pusher-gone-up-the-bureacratic-ranks-type, will find the heat unbearable if he has to face the cliff-hanging decision-demanding situations in the competitive private sector, because there he will be answerable to the dividend-hungry shareholders and the profit-demanding management! His month-end salary will depend on what he 'actually produced' during the month.

No amount of waving the academic certificates, producing non-productive publications or trumpeting about Committee-Memberships can foot this bill. Only the ones who have actually been moulded on the anvil of professional practice and have received long successful exposure to fiercely result-oriented and profit-motivated commercial experiences (where one is perpetually short on time and long on effort and cannot get away with waffling), are of real help. Only they qualify as practising professionals! Such practitioners have little time for self-congratulation, unlike the other category. They are like horses whose hooves are hardened on rough roads in 'actual' races and therefore can run on any road.

If circumstances force an engineer either to bureaucratize his engineering or to function as a staid class-room lecturer instead of practising as an active professional, the fault lies partly with the system he serves. But be that as it may, when faced with resolute and time-bound practical problems, such engineers tend to go into mile-wide and inch-deep monologues and never-ending meetings, ultimately only to fall back on the services of a Practitioner, because , all said, 'some one' has to do the actual job ultimately! (Closest that such engineers come to 'practice' is 'proof-read' the work done by that 'some one' and then . . . busily produce a Report! If they ever tried to do the work all by themselves, in open competition and in the limited time usually available, they would find it hard even to get started or decide upon the first-trial dimensions!) A good musician is far superior to the music-critic!

The classroom lecturers can 'teach' design-theories, but any attempt to teach the students the nuances and subtleties of practical design-detailing and construction by these learned lecturers, who may themselves lack their moorings in (fiercely) active commercially-oriented professional practice, could do more harm than good. Unlearning the lessons not based on active professional practice, may take longer than learning them first-hand from the professional practitioner who is closer to the 'scene of action'. Real-life design and construction practice lie far from the crowd of theories. Consequently Industry-University linkage is terribly important and cannot be over-emphasized for obvious benefits.

The fundamental truth is that 'practice' is mostly 'detailing and details' done on the drawing-board and at the job-site, not on the calculation-sheets and the computer alone. Engineering colleges and departments should open their doors wide-enough to bring more of the professional practitioner into their glasshouses if they really wish to improve the technical talent and reduce the time-loss. The practitioner has his fingers on the 'pulse of practice', and can also help sift the grain from the husk and make research work more use-oriented.

The developing countries, which must strive to achieve their targets faster, have the choice to learn from lessons already learnt in the developed countries over longer periods of time. They need more

of practice than theory and, fortunately, the option is available.

Practical engineers must be conceptual more than perceptual, creative more than analytical, and more visual than merely mathematical. They have to have a wide breadth of 'hands-on' experience gained by working with their own two hands from 'conception to delivery' in open competition, rather than merely proof-read others works and 'grin with glee at dashing a few ts'. Such experience is more useful than an 'isolated' narrow specialization alone.

Originality comes out of understanding, and understanding comes out of relentless result-oriented and profit-motivated practice, not from mere information. Good judgment comes out of experience and . . . experience often comes out of bad judgment. Preaching by those who themselves never actively practised, is like a ship that was never launched! A talented practical engineer has to have the ability of a trial-lawyer to punch a hole in a divergent argument and keep punching till he fixes the problem and wins.

Dr V K Raina

CONTENTS

Chapter 9　　***STRENGTHENING OF CONCRETE STRUCTURES BY***
　　　　　　　EXTERNALLY BONDED STEEL PLATES　　　　　**316**

CONCRETE BRIDGES

Inspection, Repair, Strengthening,
Testing and Load Capacity Evaluation

1

AN OVERVIEW

1.1 INTRODUCTION

Although a vast revenue resource is consumed in building bridges, 'managing' their maintenance and then actually 'executing the maintenance work' can prove even more exacting and costly if what has been built must remain operational for the intended long-term safe use. In the past the maintenance work has traditionally been a mere ongoing process conducted many a time by those who long ceased to be technical–or almost! But never before has the need for maintenance been more pressing. This calls for a scientific assessment of the problem for a workman-like understanding and execution. This requires:

   (i) A thorough examination of the detailed inventories;

   (ii) Carrying out detailed condition surveys and visual and 'hands-on' inspections;

   (iii) Analyzing the observations and structures in order to unfold the causes of structural distresses;

   (iv) Carrying out the structural investigative computations and, where called for, the appropriate *in situ* tests on material-samples and on existing structures; and

   (v) Ultimately writing the prescriptions for rehabilitation and repair or outright demolition and replacement, as necessary.

It can readily be realized that first the Maintenance Management (i.e., rationalised methodology for assessment of the condition of what exists and how to repair and rehabilitate it to retain its operational status . . . in order to enjoy its revenue return if not financial-return), and then Execution of Maintenance (i.e., the physical execution of the maintenance decisions taken), form a whole new dimension of the society's rightful demand on the Engineer's Services if the existing bridges have to remain there for the purpose they were constructed in the first place! (The effective aggregate of the aforementioned two exercises is what is referred to as the Bridge Maintenance Management System (BMMS) in the present-day technical jargon).

Bridge structures, like any other structure, deteriorate with time. Causes could lie in the inadequacy of design detailing, construction and quality of Maintenance, overloading, chemical attacks, atmospheric effects, abnormal floods and erosion, abnormal earthquakes, etc. Hence, they require looking after.

In this context it is therefore necessary to understand what is broadly meant by the following oft-used terms:

- Maintenance
- Repair and Rehabilitation
- Strengthening
- Replacement

Maintenance: Refers to the work needed to be done to preserve the intended load carrying capacity of the bridge and safety of the public using it.

Repair and Rehabilitation: Refers to the Maintenance work of larger 'Scope' and 'Cost' than simple routine maintenance. 'Rehabilitation' aims at rehabilitating (i.e., restoring) the bridge to the service level it originally had or was intended to have.

Strengthening: Refers to improving the existing load carrying capacity of the whole (or of the affected components) of the bridge to the value it originally had (but has lost now) or was intended to have (but never actually had). 'Widening' or 'raising' the deck may also be included here.

Replacement: Refers to reconstruction of the whole bridge or of its major components, since the cost and/or the extent of repair or strengthening may be beyond the acceptable economic or technical limits.

1.2 THE PURPOSE OF THIS BOOK

The purpose of the guidelines and prescriptions given in this book is to assist the engineer in a simple and practical workman-like manner to:

— Inspect various concrete bridges in a highway network and 'establish their comprehensive inventory',

— 'Rate' the condition of each structure's elements,

— Provide appropriate information in order to be able to determine as to which structures require routine maintenance—and what type, which ones require major rehabilitation—and what type, and, which ones must be replaced—in full or in part.

1.3   AN OUTLINE OF THIS BOOK

• *Chapter 2* broadly points out the 'Policy Principles' for individual decision-considerations once the structure-condition information has been made available. The decision may range from 'no action needed yet' to 'action needed immediately'. In the latter case the decision could range from 'analysis of causes of distress' through 'the structural investigative computations' to 'prescription for repairs' and 'design of strengthening'.

• *Chapters 3 and 4* outline the 'Types of Distress usually observed in various Bridge elements' and 'Cracking in concrete structures'.

• *Chapters 5, 6, 7, 8 and 9* provide rather detailed guidelines about 'Bridge-Inspection', 'Bridge-Repair' and 'Bridge-Strengthening' in a tool-kit approach, without mesmerizing the engineer with academic exotica.

• *Chapter 10* discusses 'Rehabilitation of Bridge Foundations' in quite some detail.

• *Chapter 11* summarizes the possible 'Structure Deficiencies, Remedies and Preventions' in a ready-reckoner tabular manner for convenience.

• *Chapter 12* touches on the aspects regarding Monitoring the service-performance of bridges.

• *Chapter 13* enumerates how to evaluate the Live Load Capacity of existing bridges.

• *Chapter 14* describes the details of a 'pre-engineered ready-to-assemble standard-panel-system' type of bridge. which can be very convenient for bridging in emergencies.

• *Chapters 15 and 16* deal with the details of certain Diagnostic Tests and Load Tests which may be required to be performed for ultimately deciding the load-rating of a bridge.

• *Chapter 17* reflects on some of the major bridge disasters on a somewhat historical basis, unfolding how some seemingly trivial but in fact important lessons were learnt in the course of development of bridges.

• *Appendix 1* pertains to Timber Bridges, giving some details regarding their maintenance.

• *Appendix 2* outlines specifications of some of the Repair-Items relevant to the subjects discussed in this book.

• *The remaining Appendices* are general-purpose reckoners for the convenience of the field engineer.

• A rather extensive list of *'References for Additional Reading'* has been provided towards the end of the book in order to lay open in front of the discerning engineer at least some of the topical literature, should he care to explore more.

1.4   A CONCLUSIVE NOTE

As pointed out earlier, in this book the steps of the reader are guided in paths often trodden by and therefore familiar to the Author, who, thereby, is perhaps able to recommend a straight course without the reader having to waste time on search through a longer route.

2
MAINTENANCE POLICY PRINCIPLES

GENERAL

2.1 The individual decision considerations will vary in accordance with the size and importance of the bridge in question. For rehabilitating or strengthening major bridges, elaborate analytical techniques for evaluation of various solutions may be applied, but an ordinary rehabilitation and strengthening work of lesser bridges may be carried out on the basis of simpler principles. In all cases, however, it should be ensured that available funds are allocated in accordance with the overall objectives and policies.

2.2 In principle, the decision (which may vary from no action, some degree of temporary action, full rehabilitation or strengthening, to an outright replacement) could be reached through a cost-benefit analysis. However, such an approach may prove to be laborious in day-to-day operations and perhaps unjustified in the majority of cases. A simple framework for decision-making, as indicated below, may therefore be advisable.

2.3 The bridges may be divided into elements, those with relatively 'short' life-spans (such as pavements, water proofing expansion joints, paint, etc.) and those with 'long' life-spans (such as structural elements in bridge-decks, columns, piers, foundations, etc.). Such division between 'short life' and 'long life' elements is useful because of the stronger technical motivation to rehabilitate and strengthen the 'long life' elements than the 'short life' elements. A division between ordinary bridges and major bridges is also warranted while considering difference in technologies to be applied and their relative economies. A division between bridges on rural roads, state highways, national highways and on expressways is also necessary, the basis being the difference in the volume of traffic.

The general policy for rehabilitation/strengthening of bridges must therefore be based on the following factors:
1. 'Short life' elements;

2. 'Long life' elements—'important' bridges;
3. 'Long life' elements—other bridges;
4. Bridges on expressways and national highways;
5. Bridges on state highways; and
6. Bridges on other roads, including village roads.

NOTE: 'Important' bridges are those bridges which are on *vital links* or on *links of strategic importance,* and all bridges on national highways and major arterial roads, even if they may be of simple structural design. A bridge is also important if there is little possibility of providing a detour in the event of its failure.

2.4 The best strategy can only be determined in the light of a thorough investigation and then the diagnosis of the causes of deterioration, faults and weaknesses, as also an assessment of the current condition of the bridge. Whenever possible, root causes should be eliminated before repairs are undertaken. Repair and strengthening work should be mechanically compatible with the original material properties of the concrete and steel and also with the basis of original structural concept.

2.5 ELEMENTS FOR A REPAIR-PLAN

The various elements that can help in forming a repair-plan for rehabilitation/strengthening, are:

(a) A documented data-base from inspections (i.e., from condition surveys).

(b) Locating the Damages / Defects / Distresses: The deterioration of a structure often appears in visible signs of damage. A detailed inspection provides information about the damages. An assessment of the structure may become necessary as a result.

Various tests may be conducted to complement the results of the visual inspection. Testing techniques and equipment should be determined relative to the extent and type of the deterioration and the importance of structure. To the extent possible, non-destructive test methods are preferable.

(c) Analysis of Causes of Damages / Defects / Distresses: The purpose of evaluating the structure-damage is not only to determine the effect of damage to its remaining service-life/load-carrying-capacity, but also (and perhaps more importantly) the determination of the causes of defects, so as to enable an intelligent determination of an effective retrofit. Before a repair-plan is implemented, either the causes of the damages have to be removed or else the repair-measures have to be so designed as to accommodate the causes, and build-in protection against them for future. Otherwise the risk of repetition of damage will continue to exist.

(d) Evaluation of Results of Structural Assessment: Conclusions from the structural investigation of a damaged structure form the basis for the decision as to what corrective action must be undertaken. This depends on the type and extent of the damage.

An initial concern should be as to whether or not there is a risk of failure to the damaged structure? If this risk is present, the first course of action must be to immediately provide an adequate auxiliary support mechanism to arrest or remove the risk. Where minor damage exists, a determination must be made as to whether the damage is stable or if the damage will propagate subsequently. This is often a difficult assessment and is made on the basis of a visual examination until such time as verification by calculations can be accomplished. The element of time and/or the harshness of the environment become important parameters when either the load carrying capacity is being diminished by deterioration or when the load carrying capacity has to be increased.

Sometimes an evaluation will be as to whether or not an economically effective repair-plan can limit or contain the damage and thus enhance the effective life of a structure. In some instances an evaluation will be concerned with the degree of urgency required to implement a repair-plan because of the advanced stage of damage.

(e) Design of Repairs for Rehabilitation: The first important step in the design of repair or strengthening is a careful inspection of the existing structure. The purpose of this assessment is to identify all defects and damages, diagnose their causes and evaluate the present adequacy of the structure.

Generally the repair 'design' shall conform to the relevant structural Design Specifications. However, it must be recognized that the repair for rehabilitation/strengthening is a special type of work and many a time accurate structural analysis, whether for assessment of the existing strength or for deciding the exact repairs, may not even be possible. Also the design in some cases may have to account for effects of 'secondary' stresses and 'forced' composite actions! When the structural system is too complex for accurate yet quick analysis, specifications more conservative than the regular specifications may have to be accepted. On the other hand, in certain special cases, consciously permitting

overstress may become unavoidable due to construction difficulties and hence 'calculated risk' may have to be taken. The designer of the rehabilitation/strengthening measures has, therefore, to be very judicious in his approach and objective in his understanding.

The techniques to be chosen will depend on the prevailing needs and access, allowable duration of lane-closures for traffic, atmospheric conditions, etc.

(f) Proposals and Estimation of Cost: The complexity and magnitude of the repair procedure will depend on whether:
— Only the cause of the damage has to be removed;
— the structure must be restored to original condition;
— the structure needs to be upgraded for its load carrying capacity and/or for its geometry.

There are several options to be considered in the evaluation of restoration of the true function of a damaged structure and these could be:
— total replacement of the structure;
— a combination of partial replacement and repair, based on the severity of damage in localized areas of the structure (e.g., in a multiple-girder bridge, only one or two girders may require replacement and others may be salvaged by repair)
— Undertaking extensive rehabilitation/strengthening measures.

2.6 The degree of restoration will depend on whether or not it is required to restore the original load carrying capacity. If for technical and/or economical reasons it is not feasible to achieve the desired restoration to original capacity and at the same time total replacement is not allowed by the budget, then a reduction in service-load capacity has to be accepted.

2.7 The owner, in reaching a decision as to the course of action, will have to evaluate not only the technically feasible options available, but also the cost of each option and the impact of various other considerations (e.g., economic impact on the communities served by the facility, life-expectancy associated with each option, any historical significance of the structure, risks that may be involved with change in safety level or reduction in load carrying capacity, etc.).

2.8 The rehabilitation and/or strengthening of major bridges is a complete task, requiring many a time inputs from several specialists. The bridge engineer, therefore, may have to consult the experts in various associated fields to work out the optimum proposals for repair.

2.9 ECONOMICS

Economics of various structure-schemes has been dealt with in detail in the author's book: *'Concrete Bridge Practice: Analysis, Design and Economics'* (Tata McGraw-Hill, New Delhi, India), to which reference may be made.

Purely as a matter of interest, the U.S. 1982: Bridge 'Construction Costs' and 'Reconstruction and Repair Costs' are reproduced in Table 2.1. These are only approximate and represent the average costs for usual highway and stream crossings, designed to AASHTO specifications.

For more realistic cost estimate, the quantities for major items should be computed, then multiplied by the current unit cost prices to obtain the 'raw' estimated cost. It is suggested to add 15 to 30 per cent over the raw cost for miscellaneous items, depending on what all items (e.g., overhead expenses, profits, promotional expenses, head office expenses, etc.) have been included in the unit cost prices.

Roughly, the estimated cost of the total bridge may be based on the average cost of the bridge per square foot of its deck area, plus the cost of the retaining walls per lineal foot basis.

Sometimes, an assumption is made that the superstructure cost will be about two-thirds of the total structure-cost on 'grade-separations' and about one-half of structure cost on 'stream-crossing' structures.

Economic analysis is a comparison of two or more alternatives, and deals with future expenditures and thus it has built-in elements of uncertainty. On bridge rehabilitation projects, all the deficiencies may not be revealed during inspections alone. The areas that need repairs may increase substantially during the actual repair or strengthening process.

Table 2.1   U.S.A. — 1982: Bridge Construction Costs (Dollars per square foot of deck area)

Construction Costs	Low	Av.	High
1. *Prestressed Girder*			
Stream Cross w/piles	40	50	60
Stream Cross w/o piles	35	40	48
Rdway Cross w/piles	36	45	55
Rdway Cross w/o piles	30	40	50
2. *Concrete T-Beam*			
Rdway Cross w/piles	50	60	70
Rdway Cross w/o piles	45	50	60
3. *Reinforced Concrete Box Girder*			
Stream Cross w/piles	60	68	75
Stream Cross w/o piles	57	60	67
Rdway Cross w/piles	40	50	60
Rdway Cross w/o piles	40	48	55
4. *Post tensioned Concrete Box Girder*			
Stream Cross w/piles	65	70	80
Stream Cross w/o piles	60	65	70
Rdway Cross w/piles	40	50	60
Rdway Cross w/o piles	40	48	55
5. *Reinforced Concrete Slab*	35	40	50
6. *Steel Structure (Rdway Crossing)*			
Rolled beam...	50	60	80
Plate girder...	50	65	90
Truss...	...	120	...
Arch...	...	130	...
7. *Treated Timber Str.*	30	35	45
Reconstruction and Repair Costs			
1. *Concrete deck Slab Replacement*			
Inc. removal... ...	20	30	50
2. *Concrete Deck Slab Repairs*	10	20	40
3. *Superstructure Replacement*			
• Steel grid deck slab	20	27	35
• Deck slab and steel girder	60	80	100
4. *Widening Exist Str.*	60	75	90
5. *LMC overlay (sq ft)*	3	4	7
6. *Asphalt overlay with water proofing (sq ft)*	1.5	2	3
7. *Silane Surface Treatment*	0.8	1	1.5
8. *Crack repairs (ft)*	15	20	30
9. *Repack joint (ft)*	8	10	15
10. *Conc. Repairs* (cu. yd.)*	300	600	900
11. *Steel Repairs (lbs)*	1	2	4

*price includes removal

2.10 ECONOMIC LIFE

In general the economic life of an alternative scheme is a period during which it provides service to the traveling public. Economic life starts only when the bridge opens to public- traffic. The estimated economic life may vary, depending on the type of structure and maintenance.

The following guidelines are to be used only in the absence of other specific information:

- Concrete bridge — 50 years
- Steel bridge — 50 years
- Timber bridge — 25 years

- • Rehabilitated bridge — 25 years
- • Major deck replacement — 35 years
- • Asphalt overlay with waterproofing membrane — 10 years
- • LMC overlay — 15 years

3

TYPES OF DISTRESS USUALLY OBSERVED IN VARIOUS TYPES OF BRIDGES

3.1 TYPES OF 'USUAL' BRIDGES

1. Masonry bridges—both in stone and brick
2. Reinforced concrete bridges
3. Steel bridges
4. Composite construction bridges
5. Prestressed concrete bridges
6. Timber bridges

These have the following forms:

1. Arches—in masonry and concrete (plain and R.C.)
2. Steel girders composite with concrete deck slab
3. Reinforced concrete slab-type, 'girders-and-slab' type, including box girders; which may be simply supported, continuous, balanced cantilevered etc.
4. R.C. rigid frame.
5. Prestressed concrete girders with R.C. or P.S.C. deckslab, as also box-girders; simply supported, continuous, balanced cantilevered with suspended spans, etc.

3.2 TYPES OF DISTRESS USUALLY OBSERVED IN DIFFERENT TYPES OF ELEMENTS

3.2.1 In Arch-Bridge Elements

(a) Changes in profile of the arch (any flattening of arch can weaken the arch).
(b) Loosening of mortar—this could be considered as ageing effect.
(c) Arch ring deformation—may be due to partial failure of the ring.
(d) Movement of the abutment or supporting pier—this is normally followed by the arch ring deformation, a hog or a sag.

(e) Longitudinal cracks—these could be due to varying subsidence along the length of the abutment or pier.
(f) Lateral and diagonal cracks indicate a dangerous state.
(g) Cracks between the arch ring spandrel and parapet wall.
(h) Old cracks which are no longer widening (which probably occurred immediately after the bridge was built).

3.2.2 In R. C. Elements

(a) *Cracking:* Cracks could be of different types. The significance of cracks depends on structure-type, location of crack, its origin and whether the width and length increase with time and loading. The cracks can be due to several reasons like (i) Plastic shrinkage and settlement, (ii) Drying shrinkage, (iii) Support settlement, (iv) Structural deficiency, (v) Reactive aggregates, (vi) Corrosion of reinforcement, (vii) Early thermal movement and restraint, (viii) Frost damage (ix) Sulphate attack, etc.

Plastic 'shrinkage' and 'settlement' cracks occur within the first hour of casting concrete (longer if retarders are used); they are caused by rapid surface drying and excessive bleeding, respectively. Former results in the disintegration of concrete while the latter causes loss of bond to bars and exposure of reinforcement. Early thermal movement cracks occur within first few weeks due to excessive temperature variation and restraint to movement. Drying shrinkage cracks occur in walls and slabs and take from a few weeks to a few years for development due to loss of moisture. High water-cement ratio, lack of curing and early exposure to wind and heat accentuate them. All these cracks create paths for seepage and leakage and ingress of deleterious materials. Cracks due to corrosion of reinforcement take several months to years to occur and lead to rapid deterioration of concrete. Alkali aggregate reaction can be the cause of extensive cracks on account of internal bursting forces set up by the expansive reaction of certain reactive aggregates (e.g. opal) in high alkali-content situations. Frost damage can occur at any age in porous concrete. Cracks due to sulphate attack may take from several months to years to develop on account of sulphate salts reacting in damp environment with tricalcium-aluminate in cement, creating expansive forces on account of the formation of higher-volume sulpho-aluminate crystals.

(b) Scaling: Scaling is the flaking away of surface matter of concrete. As the process continues, the coarse aggregate particles are exposed and eventually become loose and are dislodged. Such scaling is normally observed where repeated freeze-thaw action on concrete takes place.

(c) Delamination: Delamination is separation along a plane nearly parallel to the surface of the concrete. This can be caused by corrosion of reinforcement and differential shrinkage between over-vibrated and under-vibrated zones. Concrete surfaces and corners are particularly susceptible to delamination. Delamination ultimately can cause spalling of concrete.

(d) Spalling: Spalling of concrete can be a serious defect as it can cause local weakening and reinforcement exposure. It can also impair riding quality of deck and, with time, can cause structural failure. Spall is a depression caused by separation and disintegraion of concrete. Major causes of spalling are corrosion of reinforcement, overstresses, etc.

(e) Efflorescence: Leaching of salt-dissolved moisture can precipitate accumulation of salt deposits (white in colour) on the concrete-surface. These are noticed normally on the underside of concrete decks and along cracks on vertical faces of abutment walls, wing walls, etc. These indicate porous or cracked concrete. Calcium carbonate is formed by the reaction of atmospheric carbon dioxide with calcium hydroxide in hydrated cement, and this reaction is accompanied by reduction in pH, i.e., increase in acidity, in the moisture in concrete, which can cause rusting of steel.

(f) Stains: Most significant stains are those due to rust. This indicates presence of corrosion.

(g) 'Hollow' or 'Dead' Sound: If tapping with a regular hammer produces a dead or hollow sound, it is an indication of low quality concrete a hollow or a cavity, and/or delamination.

(h) Deformations: Swelling or expansion of concrete is usually an indication of reactive materials. However, localized swelling may be caused by compressive failure of the concrete within. Twisting of substructure or superstructure units may be evidence of settlement of foundation.

(i) Excessive Deflection: This could be due to deficiency in the structural capacity of the superstructure or due to passage of abnormal loads. Time dependent stresses also can cause such deflections if the estimated values of shrinkage and creep are different from the actual values. Deflection can be excessive if coarse aggregate of lower E-modulus (Limestone instead of e.g. Basalt) is used in making concrete.

3.2.3 In Prestressed Concrete Elements

In addition to the possible distresses mentioned in case of R. C. elements, the following are of special concern.

(a) Cracking: Cracking in prestressed concrete is an indication of a potentially serious problem. Cracks near the anchorage zones of prestressed members may indicate a deficiency of reinforcing steel against bursting stresses. Near-vertical cracking in the member could be due to serious understressing or loss of prestress. Cracks in the bottom of the unit, and near the supports, may be a result of constraint to movement of bearings. Cracks on precast members can be due to mishandling during transportation or erection. But these cracks may close subsequently.

(b) Stains: If rust stains in prestressed concrete are due to corrosion of prestressing cables, it is a very serious threat to structural integrity of the member.

(c) Spalling: Spalling in prestressed concrete can be a serious problem as it can result in loss of section properties and alternation of stresses.

(d) Excessive Deformations: In prestressed members, the abnormal deflection could occur also due to loss of prestress with time, which is a danger signal.

3.2.4 In Structural Steel Elements

3.2.4.1 *Following are the Chief Causes of Distress*
1. Corrosion
2. Possible excessive vibrations
3. Possible excessive deflections as also deformations due to buckling, kinking, and waving
4. Fractures
5. Distresses in 'connections'

3.2.4.2 *Deterioration of Steel Can Result From*
— *Deterioration of the protective paint system:* Accumulation of debris and moisture leads to cracks in the paint and even flaking of the paint.
— *Rust formation:* Resulting in section and strength loss
  • light rust (powder forms, pits the paint surface)
  • moderate rust (flakes develop)
  • severe rust (stratified delamination of scales)

For more detail see DIN 53210 and ISO 4628/I-1978.

— *Electrolytic action:* Other metals that are in contact with steel in contaminated moisture environment can set up electrolytic current, as in a cell, and this can cause rusting of steel.

— *Chemical or physical attack from:*
  • air and moisture
  • animal wastes
  • de-icing agents
  • industrial fumes, particularly hydrogensulphide
  • sea water
  • welds where the flux is not fully neutralized during welding.

3.2.4.3 Abnormal Deformation of Movements Can Result From

— excessive deflection
— long term deformation (e.g. due to creep)
— abnormal vibrations due to traffic and/or wind
— excessive wear in members accommodating movements such as pins (. . . Traffic effect),
— bucking, kinking, warping and waviness (due to overloading of members in compression)
— bent or twisted members (due to vehicular impact)

3.2.4.4 Fracture and Cracking

— Fracture due to:
  • over loading
  • brittleness
  • stress corrosion
  • fatigue
— Cracking
  • due to sudden change in the cross section of members (stress concentration)

  • in welds in adjacent metal because of fluctuations of stress-concentration.

3.2.4.5 Loosening of Bolts and Rivets due to

— overloads
— mechanical loosening
— excessive vibration

3.2.5 In Composite Construction

The distresses here are normally similar to those in concrete and/or steel bridges. However, it is usually observed that the distresses, such as 'cracks', are more common at the interface between the two materials—concrete and steel. It is due to horizontal shear, the shear connectors being either absent or being either absent or being of insufficient capacity.

3.2.6 In Timber Elements

1. Cracking and splitting of members due to overload, ageing, or under-design.
2. Abnormal deflection due to overload or under design or imperfect joints.
3. Insect attack and consequent decay, etc. (environmental).
4. Looseness in joints (due to lack of good workmanship, excessive ware, etc.).
5. Attack from marine environment (resulting in decay due to attack by marine borers).

NOTE: For details, refer to *Appendix 1 at the end of the book.*

4
CRACKS IN CONCRETE —
Types, Causes and Repairs

TYPES OF CRACKS

4.1 Concrete can crack in any or in each of the following three phases of its life, namely:
— in its plastic-phase while it has still not set (has just been placed)
— in its hardening-phase while it is still green (first three to four weeks)
— in its hardened-phase and in service (after first 28 days)

4.2 In its plastic-condition (i.e. before it has set), the concrete can crack due to
(i)   Plastic shrinkage
(ii) Plastic settlement
(iii) Differential settlement of staging 'supports'.

4.3 In its hardening-phase (i.e. during the first three to four weeks after setting), the concrete can crack due to:
(iv) Constraint to early thermal movement,
(v) Constraint to early drying shrinkage,
(vi) Differential settlement of 'supports'.

4.4 In its hardened-state and in service, the concrete can crack due to:
(vii) Overload,
(viii) Under-design,
(ix)   Inadequate construction,
(x)   Inadequate detailing,

(xi)   Differential settlement of 'foundations',
(xii) Sulphate attack on cement in concrete,
(xiii) Rusting of reinforcement due to
(a) — Chloride attack on reinforcement,
(b) — Carbonation effect on concrete,
(c) — Simple oxidation of reinforcement due to exposure to moisture.
(xiv) 'Alkali - aggregate' reaction,
(xv)   Fabrication, shipment and handling cracks in precast, prestressed or reinforced concrete members,
(xvi) Crazing,
(xvii) Weathering cracks,
(xviii) Long term drying-shrinkage cracks.

4.5 Cracks due to effects (i), (ii) (iv),(v) and (xii) – (xviii), are sometimes loosely referred to as 'Non-Structural' cracks and the remaining ones as 'Structural' cracks although the former too can lead to 'structural distress' and therefore are not non-structural in effect. This is only a loose terminology.

4.6 BRIEF DETAILS OF VARIOUS TYPES OF CRACKS

4.6.1 *Forms of different types of cracks, together with their pertinent details in summary form are indicated in Figs. 4.1 to 4.33 and Table 4.1, respectively.*

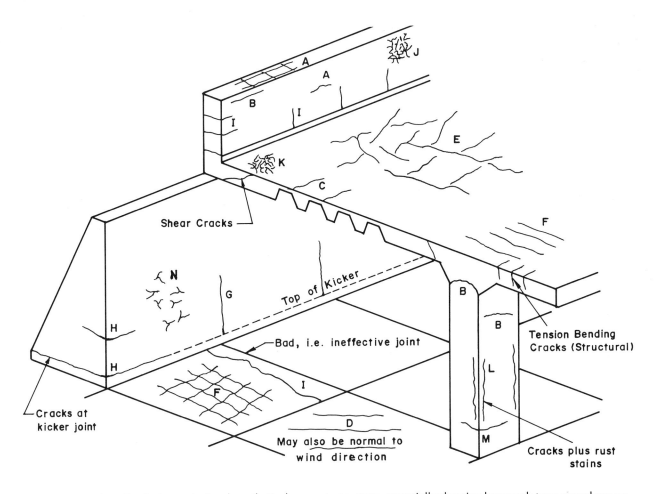

Fig. 4.1 Examples of intrinsic cracks in a hypothetical concrete structure, essentially showing how each type may show up. See details of A, B, C etc. in the following Table 4.1 (*Courtesy*: Concrete Society)

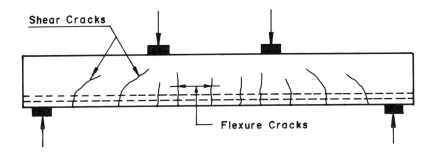

Fig. 4.2 Typical flexure and shear cracks in a simple beam (load induced)

Table 4.1

Type of Intrinsic Cracking (not caused by structural loading)	Letter Legend (see Fig. 4.1)	Subdivision	Most Common Location of Occurrence	Primary Cause (excluding restraint)	Secondary Causes/ Factors	Remedy, Assuming Basic Redesign is Impossible (In all cases reduce restraint)	Time of Appearance
Plastic Settlement cracks	A B C	Over the reinforcement Arching Change of depth	Deep sections Top of columns Trough and waffle slabs	Excess bleeding	Rapid early drying conditions	Reduce bleeding do air entrainment, or revibrate mildly	Ten minutes to three hours
Plastic shrinkage cracks	D E F	Diagonal (may be normal to wind direction) Random Over reinforcement (even mesh type)	Roads and slabs Reinforced concrete slabs Reinforced concrete slabs	Rapid early drying Ditto plus steel near surface	Low rate of bleeding and fast surface evaporation	Improve early curing and trowel	Thirty minutes to six hours
Early thermal contraction cracks	G H	External restraint Internal restraint	Thick walls Thick slabs	Excess heat generation Excess temperature gradients	Rapid cooling, curing by relatively cold water	Reduce heat and/or insulate	One day to two or three weeks
Long-term drying shrinkage cracks	I		Thin slabs (and walls)	Absence of movements joints, or, inefficient joints	Excess shrinkage and Inefficient curing	Reduce water content and Improve curing	Several weeks or months
Crazing cracks (occur only on surface)	J K	Against formwork Floated concrete	Fair faced concrete Slabs	Impermeable formwork Over trowelling	Rich mixes, poor curing	Improve curing and finishing	One to seven days sometimes much later
Cracks due to Corrosion of reinforcement (expansive reaction can lead to spalling of concrete)	L (and rust stains) M (and rust stains)	Natural and slow, or fast if excessive Calcium chloride present.	Columns and beams Precast concrete	Lack of cover, and dampness Excess calcium chloride and dampness	Poor quality concrete	See details ahead	More than about two years
Cracks due to Alkali-aggregate reaction (expansive reaction)	N (and may show gel type or dried resin type deposit in crack)		(Damp locations)	Reactive silicates and Carbonates in aggregates acting on alkali in Cement		See details ahead	More than five years

NOTE: 1. Through interconnected pores and permeations in concrete, improperly treated cold and construction joints, and in leaky patches in concrete, over a period of time, either white powdery salts (e.g. sulphates) appear after the solution containing them weeps out and dries up; or, where carbon dioxide content in the atmosphere is high, it reacts with calcium hydroxide (resulting from hydration of cement), leaving hard calcium carbonate formation in the pores and permeations and along the weeping cracks. This is *Efflorescence*. It can cause pH reduction in moisture in concrete, leading to corrosion of reinforcement. It can also cause expansion pressures from the deposits, and consequent cracking.

2. Under moist conditions sulphate salts from surrounding soil or water (or even from contaminated aggregates and the mix-water) react with C_3A in cement, leading to an expansive chemical reaction (sulphate attack) that causes splitting pressures and consequently cracking and disintegration of concrete. These cracks appear like those shown under N in Fig. 4.1 (but without the dried gel formation in the cracks). Remedy—use sulphate resisting cement (which is low in C_3A) or use Portland Blast Furnace Slag Cement. These cracks may show up after about 2 years, depending on the degree of pollution.

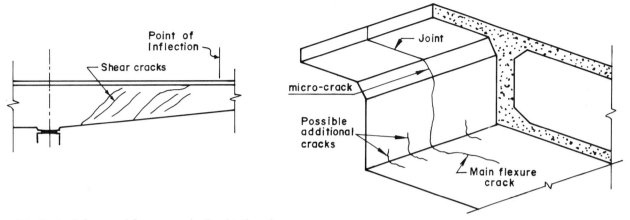

Fig. 4.3   Typical shear and flexure cracks (load induced)

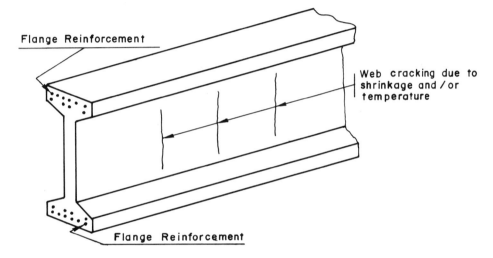

Fig. 4.4   Shrinkage and temperature cracks

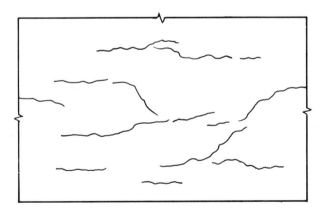

Fig. 4.5   Typical plastic shrinkage cracks on plan surface

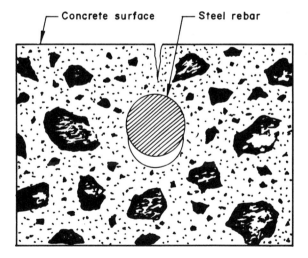

Fig. 4.6   Typical plastic settlement crack

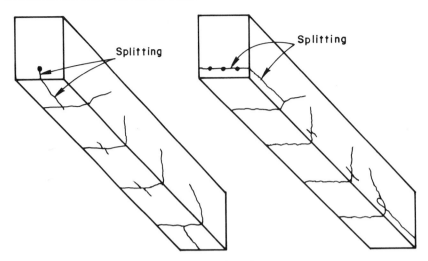

Fig. 4.7 Crack pattern due to corrosion of reinforcement

Fig. 4.8 Typical weathering damage of concrete

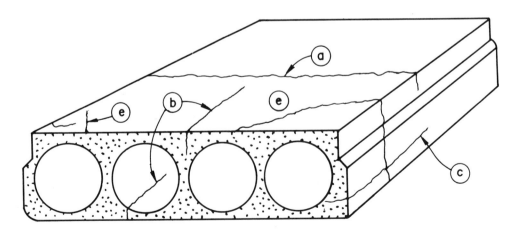

Fig. 4.9 (Contd)

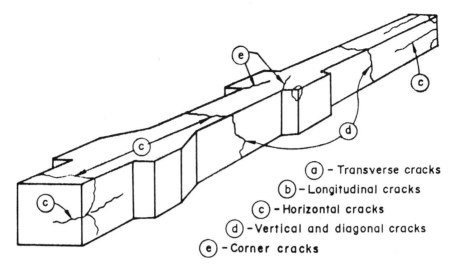

Fig. 4.9 Fabrication and shipment cracks in precast concrete member

Splitting Cracks

Splitting cracks are due to high concentrated loads, for example, at bearings and in the anchorage-zones for the prestressing cables.

Usually two types of splitting are possible. The first type of splitting is located very close to the acting concentrated force, which tries to split the concrete section locally. This is normally prevented by meshes of reinforcing bars.

The second type is caused by the trajectory distribution of the concentrated force in to the cross-section, which normally will take place over a certain distance depending on the geometrical conditions, (see Fig. 4.15).

4.6.2 *Details of various types of cracks indicated earlier are briefly explained in the following sections*

Plastic Shrinkage Cracks (i)

Plastic shrinkage cracks occur within about an hour (longer, if retarders are used) of placing concrete, although they may often not be noticed until long after. They commence from the exposed surface where the surface evaporation takes place which causes shrinking of that layer. Concrete in plastic state, having hardly any capacity to resist the consequent tension, relieves itself by cracking through paste and around aggregates and reinforcements and travels towards the opposite face. These cracks should not be confused with the long-term drying shrinkage cracks. Plastic shrinkage cracks are common in slabs, unless adequate care is taken.

Concrete slabs which are correctly trowelled should not exhibit plastic shrinkage cracks because the action of floating and trowelling is a form of recompaction that tends to close them as they form. (This trowelling can, however aggravate sedimentation of solids in the mix

and cause plastic settlement cracks.) Plastic shrinkage cracks usually occur in the following form:

— Diagonal cracks at approximately 45 degrees to the edges of the slab, the cracks being 0.2 to 2m apart. They also occur normal to the direction of wind since shrinkage would manifest in the direction of wind (Fig. 4.34, Plate 1.)

— A very large random map pattern (Fig. 4.35, Plate 1.)

— Cracks following the pattern of the reinforcement which could be of 'mesh' of 'grid' type ('mesh' formed by orthogonal reinforcement bars, placed in a grid-pattern)

Although cracks can be very wide at their start (up to 2 or 3mm), the width rapidly diminishes with depth. Nevertheless, in all but minor cases, they will usually pass through the full depth of a thin slab, in contrast with most types of plastic settlement cracks which do not propagate so much.

If not noticed in the soffit (of not easily accessible slab soffits) thorough wetting at the top of the slab may show them in case of full depth penetration. Taking cores can reveal them (Fig. 4.36, Plate 2.)

Plastic shrinkage cracks rarely reach the free ends of the slabs (e.g. the edges of a road slab) because these edges are free to move under plastic shrinkage. This is a very important way of differentiating them from long term drying shrinkage cracks if the time of formation is unknown. However, plastic shrinkage cracks will form up to the edge of a slab which has been cast against previous pour, especially if there is continuity of steel, because this acts as a restraint.

If the cracks follow the pattern of the top reinforcement it may be difficult at first to determine whether they are due to plastic shrinkage or plastic settlement! If

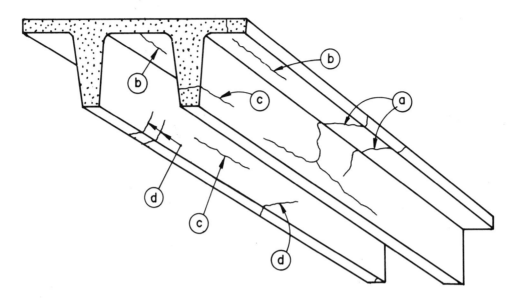

(a) – Transverse cracks

(b) – Longitudinal cracks

(c) – Horizontal cracks

(d) – Vertical and diagonal cracks

(e) – Corner cracks

(f) – Ledge cracks

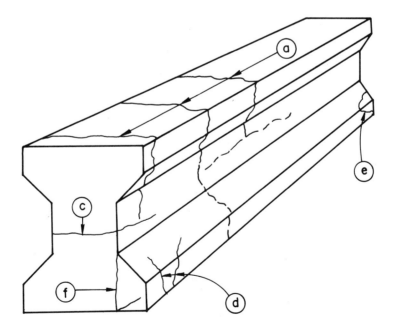

Fig. 4.10 Fabrication and shipment cracks in precast concrete member

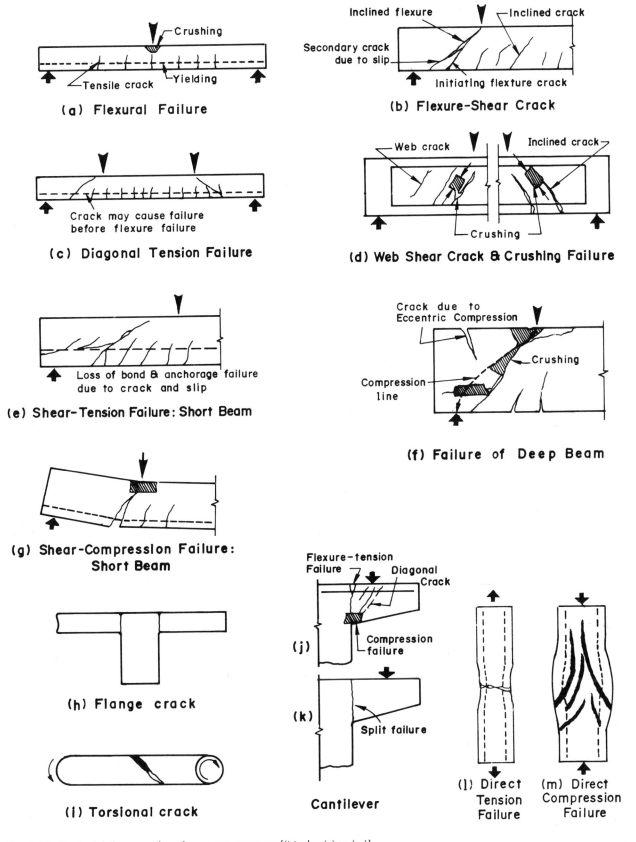

Fig. 4.11 Typical failure modes of concrete structure [(Modes (a) to (m)]

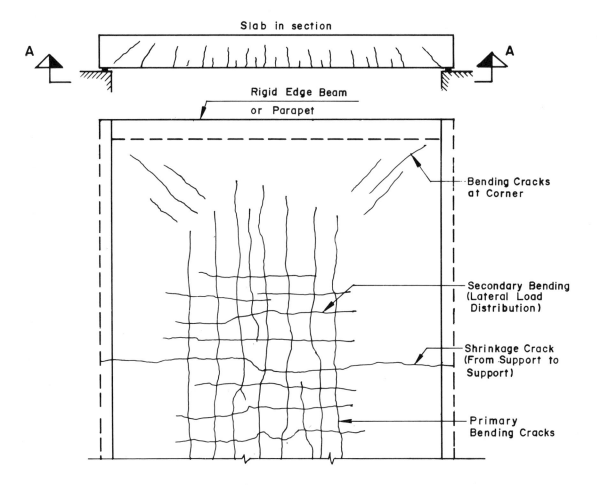

Fig. 4.12 Slab (A-A)

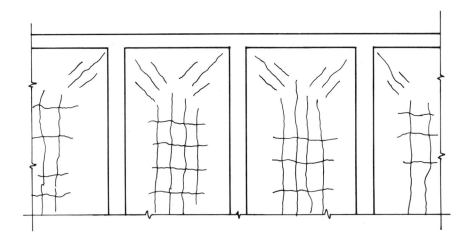

Fig. 4.13 Slab panels

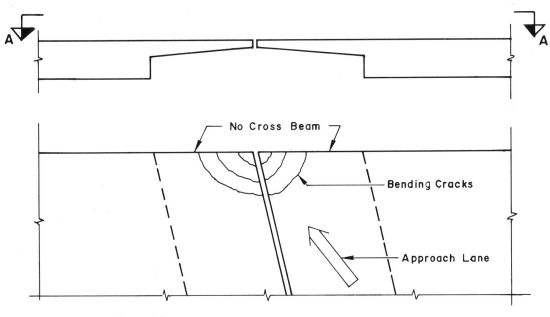

Fig. 4.14 Cantilever slab

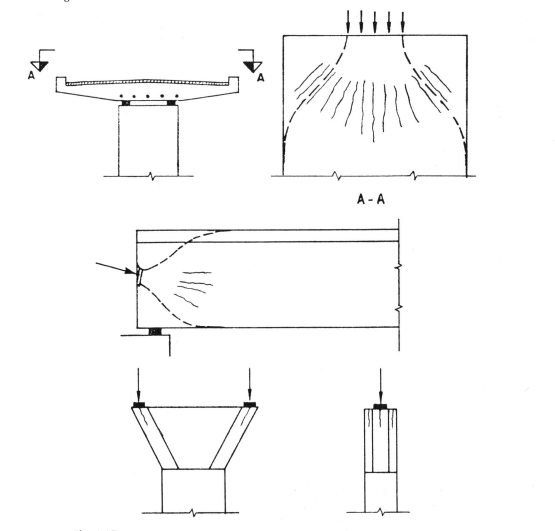

Fig. 4.15

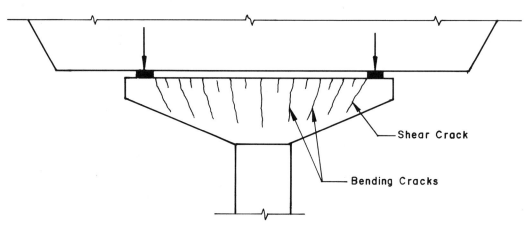

Fig. 4.16 Cap beam

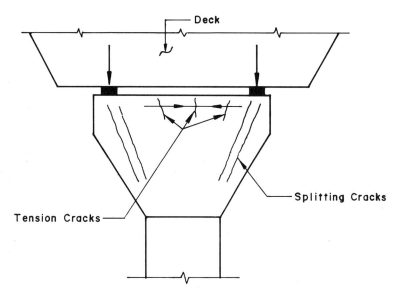

Fig. 4.17 Pier cap or pier as the case may be

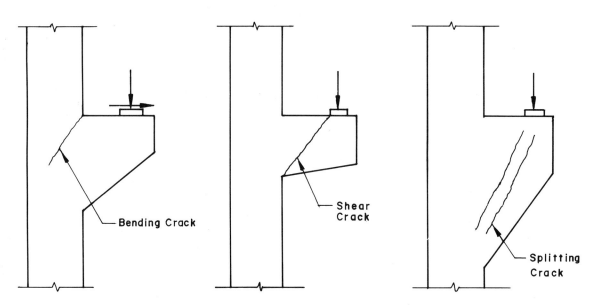

Fig. 4.18 Different types of cracks in corbels

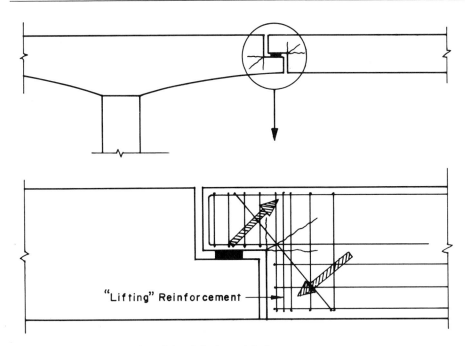

Fig. 4.19 Beam with reduced depth (halving joint)

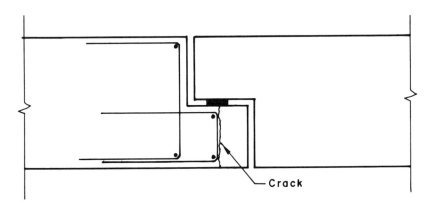

Fig. 4.20 Corbel with carelessly (wrongly) placed reinforcement

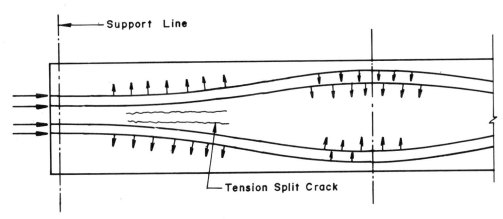

Fig. 4.21 Horizontally curved cables — plan view

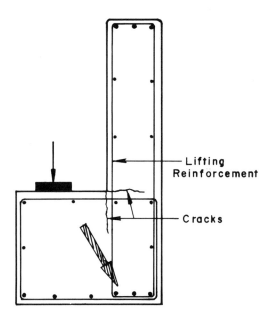

Fig. 4.22 Beam with corbel

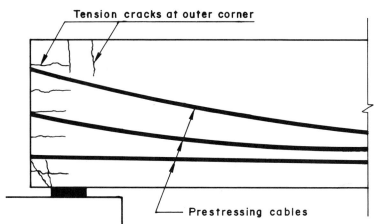

Fig. 4.23 Anchorage zone, prestressed beam. Unintended deflections and movements

Retaining walls

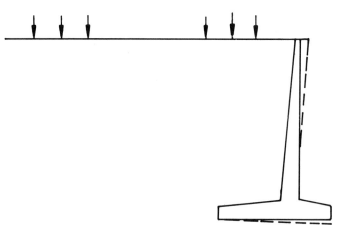

Fig. 4.24 Deflections caused by low stiffness and creep, compaction, or soil conditions

Girders

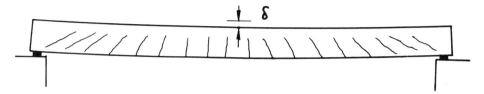

Fig. 4.25 Deflections caused by large span lengths in R.C. concrete (creep) and improper formwork

Bearings

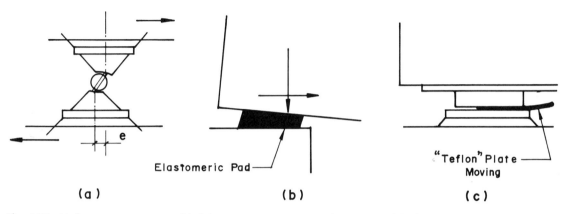

(a)   **(b)**   **(c)**

Fig. 4.26 Unforeseen movements (shrinkage, creep, temperature), wrong positioning

At Supports / Bearings

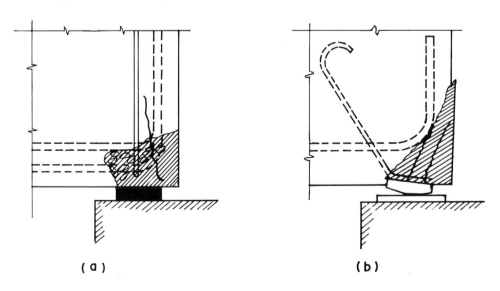

(a)   **(b)**

Fig. 4.27 Crushing as a result of honeycombs, wrong placement of bearings or bad workmanship

At Expansion joints

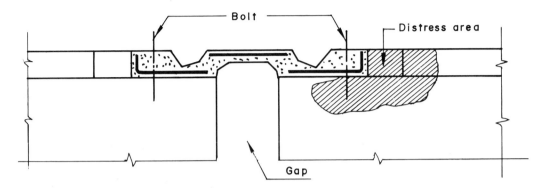

Fig. 4.28   Local distress (possible loose bolts or workmanship)

Vertical clearance shortage

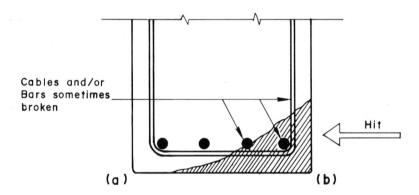

Fig. 4.29   Impact from vehicle

Plan View

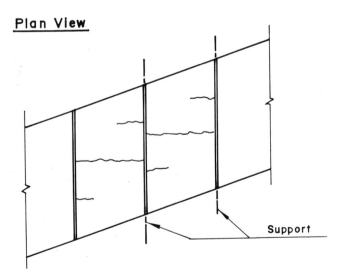

Fig. 4.30   Shrinkage cracks, slab

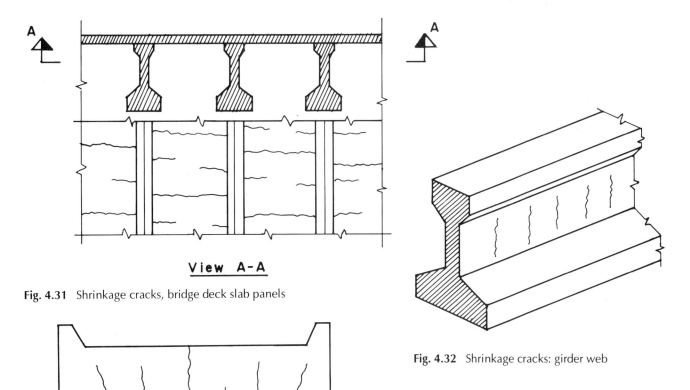

Fig. 4.31 Shrinkage cracks, bridge deck slab panels

View A-A

Fig. 4.32 Shrinkage cracks: girder web

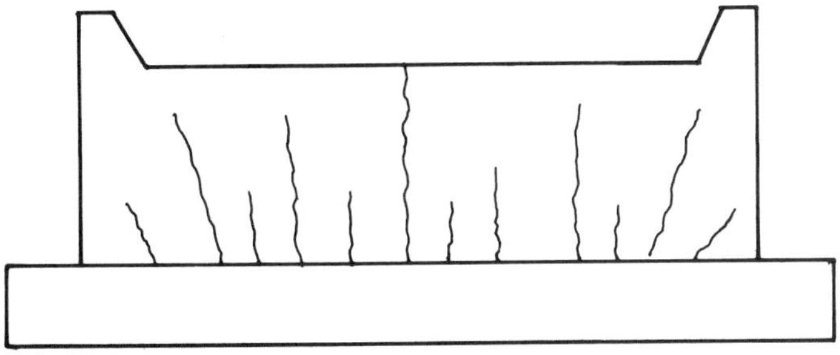

Fig. 4.33 Early drying shrinkage cracks, abutment footing (old concrete)
acting as a constraint on stem (new concrete)

NOTE: If such cracks traverse further down they could be due to differential settlement.

it can be shown that the cracks pass through the slab and follow the pattern of the steel then they are almost certainly plastic shrinkage cracks which have been orientated by the steel!

Another difficult case to diagnose with certainty is in trough and waffle floor construction. Generally, plastic shrinkage cracks will tend to form diagonal or random patterns but, because the formwork acts as a restraint, they may align themselves to follow the layout of the troughs. If, however, the cracks form only along the line of the troughs, then the mechanism is most likely that of plastic settlement, not plastic shrinkage.

The cause of plastic-shrinkage cracks is rapid evaporation of water from the surface of concrete. Immediately after concrete is placed the solid ingredients begin to settle. This process produces a layer of water at the surface since the water moves upwards as it is displaced by the settling solids. This process continues until the concrete sets. Under most weather conditions some of the water at the surface evaporates. As long as the rate of bleeding exceeds the rate of evaporation there

is continuous layer of water at the surface as evidenced by the appearance of a water sheen on the surface. If surface evaporation exceeds bleeding, the sheen disappears and the top surface of the slab is placed in tension. Since concrete in the early stages of setting has a negligible tensile strength, cracks form to relieve the tension. These are plastic shrinkage cracks.

The factors that determine rate of evaporation are the temperature of the concrete, the air temperature, relative humidity, and wind velocity of the air adjacent to the concrete. Table 4.2 shows how these factors affect drying. It may be seen that evaporation increases as the humidity decreases, as the wind velocity increases, as the air temperature decreases, and as the concrete temperature increases. Of particular interest is the fact that rapid evaporation is at least as big a problem in cold weather as in hot weather! Note in group 5 that even when the relative humidity is 100 per cent in cold weather, there will be a large amount of evaporation if the concrete is warm! Of all the factors listed in this table, only concrete temperature is subject to control. It

Table 4.2 **Effect of Variation in Concrete and Air Temperature, Relative Humidity and Wind-speed on Drying 'Tendency' of Concrete at Job Site (for a typical mix for ordinary reinforced concrete)**

Group-Case	Case No	Concrete Temp. (°F)	Air Temp. (°F)	Relative Humidity (%)	Dew Point (°F)	Wind Speed (mph)	Moisture Drying Tendency (lbs of water/ sq.ft./hr.)
1. Increase in wind speed	1	70	70	70	59	0	0.015
	2	70	70	70	59	5	0.038
	3	70	70	70	59	10	0.062
	4	70	70	70	59	15	0.085
	5	70	70	70	59	20	0.110
	6	70	70	70	59	25	0.135
2. Decrease in relative humidity	7	70	70	90	67	10	0.020
	8	70	70	70	59	10	0.062
	9	70	70	50	50	10	0.100
	10	70	70	30	37	10	0.135
	11	70	70	10	13	10	0.175
3. Increase in concrete temperature and air temperature	12	50	50	70	41	10	0.026
	13	60	60	70	50	10	0.043
	14	70	70	70	59	10	0.062
	15	80	80	70	70	10	0.077
	16	90	90	70	79	10	0.110
	17	100	100	70	88	10	0.180
4. Concrete at 70°F; decrease in air temperature	18	70	80	70	70	10	0.000
	19	70	70	70	59	10	0.062
	20	70	50	70	41	10	0.125
	21	70	30	70	21	10	0.165
5. Concrete at varied temperature; air at 40°F and 100% R.H.	22	80	40	100	40	10	0.205
	23	70	40	100	40	10	0.130
	24	60	40	100	40	10	0.075
6. Concrete at 70°F and air at 40°F; variable wind	25	70	40	50	23	0	0.035
	26	70	40	50	23	10	0.162
	27	70	40	50	23	25	0.357
7. Decrease in concrete temperature; air at 70°F	28	80	70	50	50	10	0.175
	29	70	70	50	50	10	0.100
	30	60	70	50	50	10	0.045
8. Concrete and air at high temperature; 10% R.H.; variable wind	31	90	90	10	26	0	0.070
	32	90	90	10	26	10	0.336
	33	90	90	10	26	25	0.740

may be seen that there is a definite advantage to cool the concrete! It should be placed as cool as possible in warm weather and should not be overheated in cold weather. Note in group 7 that if the concrete temperature is reduced from 80 to 60 °F, 75 per cent of the evaporation can be eliminated!

Since weather conditions cannot be controlled at will, and since construction personnel have only a lim-

ited control of the concrete temperature, primary reliance in preventing plastic shrinkage cracks must be placed in construction techniques. From group 8 it is clearly seen that in hot weather the rate of surface evaporation is the greatest in magnitude as the wind speed increases. Consequently plastic shrinkage cracking can be severe and extensive under hot and windy atmospheric conditions!

PLATE 1

Fig. 4.34 Plastic shrinkage cracks in R.C. wearing course (almost normal to wind)

Fig. 4.35 Random plastic shrinkage cracks in a deck slab

PLATE 2

Fig. 4.36 A concrete core drilled out of a deck-slab suffering from heavy plastic shrinkage crackings—the core shows the extent of penetration of such a crack

Fig. 4.38 A shear crack

Prevention of Plastic Shrinkage Cracks

The measures which have been found effective against plastic shrinkage cracking are:

1. Dampening the sub-grade and forms.
2. Dampening the aggregates if they are dry, absorptive and hot (but using them in surface -dry condition).
3. Starting curing immediately after initial setting of concrete but before the surface-sheen fully disappears.
4. Protecting the concrete with temporary coverings or applying a fog-spray during any appreciable delay between placing and finishing.
5. Erecting windbreaks to reduce wind velocity over the surface of the exposed plastic concrete.
6. Providing sunshades to reduce the temperature at the surface of the concrete.
7. Slight re-trowelling of laid plastic concrete before its initial setting.

On flatwork, not requiring hard steel trowelling, it is usually possible to adjust construction operations so that finishing is completed and 'membrane' or 'moist' curing begins before the water-sheen disappears. On particularly unfavourable days this requires finishing very soon after placing. Where trowelling is required, the problem is somewhat different since final trowelling and curing must be delayed. Sometimes cracks have occurred prior to final trowelling. These cracks might be closed by the trowelling. However, a more reliable method is the application of a fog spray or wet burlap to the surface before final trowelling.

Sometimes night construction has been resorted to for preventing plastic shrinkage cracks. This will be effective only if it produces significantly lower concrete temperatures or if the wind velocity is lower at night. The reduction of air temperature and not that of concrete (even with the increase in relative humidity which normally accompanies reduction of air temperature) will not significantly reduce the plastic shrinkage cracking! This must be understood carefully.

If it is not possible to eliminate the risk of plastic shrinkage cracks even by improved early curing, then changes to the concrete mix must be considered. First, check that the concrete does not contain an admixture with high retarding effects. If it does, try to replace it with the one that does not retard so much (rather than counter it by adding a compensating accelerator!). Second, consider the use of air-entrainment. Air-entrained concrete exhibits plastic shrinkage cracks less often than plain concrete. At first sight this might seem illogical because, as air entrainment permits reduction in water-content this would reduce the rate of bleeding, which should increase the risk of plastic shrinkage cracks occurring at a given rate of evaporation. However, most commercially available air-entraining agents are 'detergents' and therefore reduce the 'surface-tension' caused by drying, and consequently reduce the shrinkage cracking!

Plastic Settlement Cracks (ii)

Plastic settlement cracks (*see* Fig. 4.37) occur when there is a relatively high amount of bleeding and there is some form of obstruction (e.g. reinforcement bars) to the downward sedimentation of the solids. These obstructions 'break the back of concrete' above them as it were, and foment the formation of voids under their 'belly'. Thus we can have:

— Cracks directly over formwork-tie-bolts or over reinforcement near the top of a section,
— Cracks in narrow columns and walls where the said sedimentation is prevented by the resulting arching of the concrete due to narrow passage, which may be aggravated by the presence of horizontal bars
— Cracks at change of depth of section

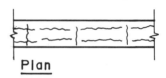

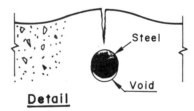

Fig. 4.37 Plastic settlement cracks

• *Prevention of Plastic Settlement Cracks*

There are three ways for this purpose
— Reduce the bleeding and hence the settlement
— Reduce the obstruction to sedimentation
— Applying light revibration

Air-entraining admixtures and plasticizers reduce water demand and thus are the most effective way of reducing bleeding and sedimentation and hence the plastic settlement cracks. These can also be eliminated by light revibration of the concrete if and immediately after they have formed. (They are formed within the first hour of concreting (later, if retarders are used) but before the concrete has set.) This light revibration should not be applied 'too soon' otherwise a second phase of bleeding can still cause settlement cracks to form! The correct time can easily be determined by simple site trials; It will be the last time that a vibrating poker can be inserted into the concrete and removed without leaving a significant trace. Revibration is often the only way to eliminate plastic settlement cracks, particularly in deep sections. Trowelling the surface can actually aggravate these cracks as the pressure may only cause further settlement of solids! Hence what relieves plastic shrinkage cracks actually aggravates the plastic settlement cracks and vice-versa.

- *Remedial Measures if Plastic Settlement and Plastic Shrinkage Cracks have Occurred*

Plastic cracks often form in the top face of a section e.g. plastic shrinkage cracks in slabs, plastic settlement cracks on top of deep beams and walls. Thus they can be accessible, and this coupled with the fact they form so early in the life of concrete, means that they may widen as thermal contraction and drying shrinkage take place. Consequently, it may not be wise to fill plastic cracks with 'rigid' materials until it is certain that these long-term effects have subsided.

The repair of plastic shrinkage cracks in road slabs is particularly critical. This is because the cracks are wide at the top and can rapidly take in pollutants which may cause subsequent spalling, and prevent the subsequent satisfactory application of sealing materials. Clearly, wide cracks in slabs are not likely to be self-healing at the top, and are likely to spall.

Plastic cracks, by their nature, pass through the cement matrix and around aggregate particles, therefore, they are very ragged and capable of transferring shear providing there is sufficient reinforcement to maintain aggregate interlock. Consequently full structural repairs (i.e. using epoxy or polyester resin) are not necessary, though preferable.

In many instances the simple expedient of brushing cement grout into the cracks will be satisfactory. This should be done as soon as possible after the concrete has nearly set.

However, it may be necessary to supplement this later with flexible penetrating sealant if thermal contraction or drying shrinkage causes further movement.

Plastic settlement cracks over steel must be efficiently 'sealed' if the concrete is in an exposed state, to eliminate the risk of corrosion of the steel. Reduced bond strengths due to gaps thus formed at the bar-bellies must be considered by the designer and suitable measures taken.

NOTE: These 'plastic' cracks are the most haphazard looking type of cracks. Structurally they may not be as damaging as they may look, but they can reduce durability and overall service-life of the structure and hence can be dangerous. Depending on their severity in terms of spread, crack-width and extent of penetration

(some intrinsic cracks may even appear on the soffit surface of some slabs where the load-induced cracks could in fact have descended only to part depth), the plastic cracks can jeopardise the monolithicity and overall integrity of the structure as a whole, even though the strength of concrete in portions bounded by these cracks may well be as per specification!

Support Settlement Cracks (iii), (vi), (xi)

In 'indeterminate' structural condition, significant differential settlement due to inadequate design of supports and foundations, introduces flexure and shear in the structure for which it may or may not have been designed. In the latter case, these additional stresses can lead to flexural and shear cracks. It may be noted that differential settlement of staging supports while the poured-concrete is still non-structural, can cause deformations, undulations, and cracks in the green concrete. As for the long-term differential settlement of foundations, it has to be noted that even a relatively large settlement (e.g. 30 to 60 mm) when the spans of the indeterminate structure are long (about 80 m or more) can have less detrimental effect than if the differential settlement is relatively small (e.g. 6 to 12 mm) but the spans are short (e.g. 15 to 20 m).

Cracks due to Constraint to (early) Thermal Movement (iv)

The reaction of cement with water, known as hydration, is a chemical reaction which produces heat. If an element of concrete is big enough and is insulated by adjacent materials including formwork, then the rate of heat development in the first 24 hours is likely to exceed the rate of heat loss to the atmosphere, and the concrete temperature will rise.

After a few days the rate of heat development falls to below the rate of heat loss and the concrete will cool. As with nearly all materials, this cooling will cause contraction of the element.

Hypothetically there will be no cracking if this contraction is unrestrained. In practice, however, there is bound to be some restraint and this can be considered as comprising of two components.

(a) *External Restraint*: If the concrete is cast onto a 'previously hardened' base, or if it is cast adjacent to or between two elements, without the provision of a movement joint, then it will be externally restrained.

(b) *Internal Restraint*: The 'surface' of an element of concrete is bound to cool quicker than its 'core'; It will also respond to daily temperature variations more than the core. Therefore, there will be differential thermal strains across the section and, where this differential is large, such as in thick sections, cracks may develop at the surface, at least. This internal restraint cannot be avoided, but the risk of cracking can be reduced, as discussed ahead.

In practice the net effect of these two components will vary according to several factors, the most important being the geometry of the element, the nature of the formwork and its striking time. Early removal of formwork can lead to relatively cooler surface and consequent thermal cracking.

The temperature of the core of the concrete in most sections cools to ambient in about 7 to 14 days. It is clear, therefore, that thermal movement cracks must occur during this period. Thus, one of the most important factors which assists in diagnosing if the crack is due to early thermal movement as opposed to long-term drying shrinkage is a knowledge of when the crack first forms. A crack which forms in the first two weeks is unlikely to be a drying shrinkage crack unless the element is a thin slab subjected to extreme drying conditions. Conversely, cracks which form after a period of several weeks or months cannot be early thermal movement cracks.

- *Factors Affecting Early Thermal Movement Cracks*

Type of Aggregate: Limestone and (to a lesser extent) granite aggregate concretes have lower coefficients of thermal expansion than other dense aggregate concretes and could, therefore, crack less.

Reinforcement Spacing: Crack 'widths' can be reduced by reducing the 'reinforcement spacing'. For a given percentage of reinforcement, crack widths are reduced by: (a) The use of small diameter bars, (b) The use of deformed rather than plain reinforcement, (c) Reduction of cover to the minimum allowable!

External Restraint: It can usually be dealt with by the provision of 'movement joints'. However, when two pours of concrete cannot be separated (e.g., a wall cast onto a hardened base), the effect can be minimized by reducing the time interval between the two pours.

Internal Restraint: This occurs when the outside face of a concrete section cools quicker than the core. It is most common in thick sections and can be minimized by delaying the removal of formwork, by the use of insulation, or by using chilled water and cooler concrete.

The traditional method of limiting early thermal movement cracks in large sections (e.g. with some dimensions exceeding 2 m) is to reduce cement content, impose strict limitations on depth of pour, profuse curing, providing movement joints in length direction, and using chilled water and cooler concrete.

If the maximum temperature difference between the core and the concrete surface is limited to about 20°C, then these cracks are less likely to form.

The temperature difference is limited either by keeping the full section of concrete cool, or otherwise by keeping the surface warm by insulation! Insulation will have little effect on the core temperature, but will significantly raise the temperatures and hence reduce temperature differences. It is usually necessary to monitor these temperatures by means of thermocouples (or other means) until sufficient experience is gained to know when the insulation can be removed.

More sophisticated techniques involving detailed knowledge of the properties of the concrete-mixes used are probably applicable to *major* civil engineering works only.

Exposed concrete slabs and webs tend to be subjected to severe changes in temperature-differentials as their 'exposed surfaces' rapidly respond to the daily changes in ambient temperature. This can lead to warping of thin paving slabs. Unless adequate 'jointing' is provided, this warping effect will produce cracks in a matter of few days. These will not be easily distinguishable from drying shrinkage cracks the latter mechanism will clearly add tensile strains to the top face upon drying. Thick external webs in box-sections have cracked at the turn of winter with sun beating on the outer surface.

- *Recommendation Against Early Thermal Movement Cracks*

There are no simple recommendations to minimize early thermal movement cracks because they can be controlled only through carefully coordinated planning by both the designer and the contractor.

The fundamental factors which have to be considered and specified are:

(a) *'Design and specification' wise*
  — restraint (overall size of pour, spacing of movement joints)
  — distribution steel (designed in relation to all other factors, particularly restraint)
  — heat development (section thickness, cement type and content)
  — aggregate type.

(b) *'Constraint' wise*
  — restraint (sequence and timing of and between pours, additional movement joints)
  — heat development (choice of concrete materials and formwork type)
  — cooling (striking off formwork, curing, insulation)

- *Remedial Measures Against Early Thermal Movement Cracks*

Thermal movement cracks in relatively thin sections will normally penetrate through the full concrete section. They should be regarded as 'live' cracks for at least a year and should be repaired accordingly (*see* ahead). Clearly the same applies to thermal movement cracks in mass concrete.

Cracks in large pours do not usually extend through the full concrete section and repairs may be concerned more with durability than with water retention for instance.

Warping cracks due to temperature differentials should be treated in exactly the same way as long-term drying shrinkage cracks. Temperature difference in concrete may be caused by cement hydration or changes in the ambient condition, or both.

Variation of temperature results in changes in shape and volume of the concrete. When free occurrence of such changes in volume and shape is restrained, resulting tensile stresses may exceed allowable limit, and hence, lead to cracking.

Cracks due to Constraint to Early-Drying-Shrinkage and Long-Term-Drying-Shrinkage (v), (xviii)

Volume changes induced by moisture-loss, commonly termed as shrinkage, are a 'characteristic' of concrete. When the shrinkage process is 'restrained' by another part of the structure or by subgrade, tensile stresses develop. When tensile strength of concrete is exceeded, it cracks. In massive concrete elements, tensile stresses are caused by differential shrinkage between the surface and the interior concrete. The larger shrinkage at the surface causes cracks to develop that may, with time, penetrate deeper into the concrete.

Drying shrinkage may be defined as the reduction in volume of concrete caused by the chemical and physical loss of water during the hardening process and exposure to unsaturated air. Shrinkage occurs as the cement-water-gel looses water partly through evaporation (physical loss) and partly through hydration of cement (chemical loss). The resulting reduction in volume can cause cracks only if the concrete is 'restrained' in some way, and its tensile strength exceeded.

Many concrete members which might be expected to crack according to theoretical data, either never do so, or the magnitude of the cracks is only a small fraction of that envisaged! This may be because much data still used by engineers derives from research work carried out on small unreinforced prisms using small maximum-sized aggregate! Also, the relative humidity and temperature to which the test specimens were subjected often differ considerably from those applicable to site conditions! Moreover member size has significant influence on rate of water-loss.

Also, the strain caused by the drying shrinkage occurs at a very slow rate in full size concrete members and therefore in it the relaxation due to creep can be of significant benefit!

Early thermal contraction strains far exceed long-term drying shrinkage strains and are primarily responsible for early cracking in retaining walls and similar reinforced concrete bridge structures. Long-term measurements on some large R. C. structures have shown that the strain due to drying shrinkage after 5 years was about 30×10^{-6}. As the tensile strain capacity of hardened concrete is in the range of 80 to 150×10^{-6}, it is clear that long-term drying shrinkage alone could not initiate the non-load induced cracks!

If adequate reinforcement and sufficient movement joints are provided against other forms of movement, the contribution of drying shrinkage movement to the incidence of cracking will often be too small to be of consequence. When unacceptable long-term drying shrinkage cracks do occur, they can usually be attributed to the design of mix, properties and proportions of its constituents, and the construction practice that was followed. The shrinkage of a particular concrete mix is also affected by additional factors such as temperature history, curing methods, relative humidity and ratio of volume to exposed surface. The more the water that is available to evaporate from the concrete, the higher the tendency to shrink on drying.

The type and the fineness of cement have little effect on drying shrinkage cracks. This may be because the tendency to shrink more is offset by earlier gain in strength and also higher tensile strain capacity to resist shrinkage better.

Sound, normal weight aggregates of concrete have low shrinkage values compared to paste. Apart from the obvious economic advantage of using aggregate, the more of it that can be included in mix, the more pronounced will be the reduction in the effect of the cement paste's very high drying shrinkage because, per unit volume, the quantity of fines will reduce (*it is the fines that shrink more*).

The water content of concrete, and consequently its long-term drying shrinkage, can be minimized (for a given workability) by using plasticizers and aggregates with the lowest possible specific surface area. (Rounded *aggregates* are better in this respect.)

An admixture affects drying shrinkage because for the same workability it reduces the water requirement of mix. A reduction in water content is usually possible but in the case of water-reducing and set-retarding admixtures a corresponding reduction in drying shrinkage will not necessarily ensue!

Calcium chloride as an accelerator additive increases drying shrinkage and greatly increases the risk of reinforcement corrosion (*chloride attack, see* ahead).

The materials present in admixtures are diverse and advice on the effect on a mix of a particular product should always be obtained from the manufacturers.

Loss of water by evaporation is the primary cause of drying shrinkage and hence the relative humidity of the air surrounding the concrete plays an important part in this shrinkage. As relative humidity decreases, the rate and amount of water loss from a concrete surface increase. Windy conditions increase the rate even more. As relative humidity increases, the mechanism is reversed until at about 95 per cent R. H. moisture movement ceases. If the relative humidity is maintained above 95 per cent R. H. the concrete may in fact absorb water and expand, not shrink!

Curing means preventing or at least delaying the drying up of the intrinsic moisture inside the capillaries of concrete which is needed both for hydration of cement (and hence gain of strength–compressive and tensile) as well as delaying shrinkage. The latter delays shrinkage-cracking and the former gives strength to stand its tension! It is hence that early and efficient curing is so very important! The rate of moisture-loss decreases rapidly as the distance from the surface of the concrete to its core increases. A large thick section will thus retain more water for a longer period than a thin slab.

- *Preventive Measures to Reduce Drying-Shrinkage Cracks*
  (a) Use minimum water content, using plasticizer for compensating the reduction in workability;
  (b) Use the highest possible aggregate content;
  (c) Use the largest possible maximum aggregate size;
  (d) Use concrete with a workability as low as is compatible with ease of placing and achieving full dense compaction;
  (e) Provide adequate and early curing to exposed surfaces, particularly on large flat areas;
  (f) Eliminate external restraints as much as possible, particularly by providing movement joints where applicable;
  (g) Provide sufficiently closely spaced reinforcement steel (generally 15 cm in slabs) in order to control crack width as it is better to have more but fine cracks than otherwise.

- *Remedial Measures Against Drying-Shrinkage Cracks*
To specify the type of repair to an unacceptable drying shrinkage crack, it is necessary to determine the basic cause, and the likelihood of its further widening. Repairs using 'rigid' materials (*see* ahead) should be limited to situations where a fundamental constructional defect has been discovered (e.g. misaligned bars, etc.). The cracks may be repaired as per 'Dormant Type' or 'Live Type' crack repair, as the case may appear to be (*see* ahead for details).

Cracks due to Overload, Under-Design, Inadequate Construction and Inadequate Detailing (ii), (iii), (ix), (x)

'Overload' or 'Under design' will obviously create excessive stresses (e.g. in flexure and shear), consequently leading to the corresponding cracking (*see* Fig. 4.38 to 4.40, Plates 2, 3).

Likewise, 'Inadequate Construction' (e.g. presence of cold joints, excessive honeycombs, porous concrete, etc.) or 'Inadequate Detailing' (e.g. shortfalls in overlaps of reinforcement bars, abrupt cutoffs and cut-outs, etc.) can similarly lead to excessive stress-peaks and possible cracking and even premature collapse.

Cracks due to Sulphate Attack on Cement in Concrete (xii)

Chemically Caused Intrinsic Cracks
The cracks resulting from corrosion of reinforcement (normally due to the oxidation under damp conditions resulting from inadequate concrete cover and porous concrete, or more severely, the chloride attack on reinforcement under damp conditions or from reduction in pH due to carbonation of concrete), or alkali (in cement) reacting with the reactive silicates and carbonates that may be present in certain aggregates (e.g. certain Dolomites, etc.) are caused essentially by the internal bursting pressures generated by the expansive nature of the chemical reaction (volume change action). Such cracks are consequently limited to where the surrounding atmosphere, water, soil and/or concrete constituents themselves can supply the pollutant, the concrete constituents can supply the reactant, and the porous or permeable nature of concrete-quality can allow the ingress through the vehicle of dampness. Consequently despite the normal care, corrosion-rust patches on concrete (followed by cracks roughly along the lines of reinforcements, finally followed by spalling of concrete unless remedied where possible) will manifest (seen more in the coastal areas and in structures built in the sea). 'Alkali-aggregate reaction' cracks will show (with a gelatinous or dried-up-honey-like-deposit in the cracks) where the aggregate contains the reactive silicates and carbonates which react with the alkali in the cement. Likewise, cracks resulting from the expansive chemical reaction between the sulphates (ingressing through damp and permeable concrete) supplied by the surrounding sulphate-laiden sea-water or ground-water or subsoil or by the contaminated constituents of concrete, and the cement (which is not low in its tricalcium aluminate component e.g. the ordinary Portland Type-I cement), will manifest where such conditions prevail. This reaction generally takes place where temperature to which the structure is subjected is not more than about $30°C$ for major part of the its life.

These 'sulphate-attack cracks' are also described under Note No. 2 in Table 4.1. These can be prevented by using Portland Cement with low C_3A (i.e. Sulphate Resisting Type V cement) or Portland Blast Furnace Slag Cement. (Where available, even Portland Fly Ash

Cement or Pozolona Cement can also be used. The central idea is to use such a cement as would give a very 'dense' mix so as to reduce the pores and permeability in concrete. This reduces the movement of dampness and hence inhibits the chemical reaction of sulphate attack.) These cracks can be repaired by epoxy-injection after they have occurred, treating them as 'dormant' cracks.

Cracks due to Rusting caused by Chloride Attack on Reinforcement in Concrete (xiii) (a)

When reinforcing steel is embedded in concrete it does not normally corrode because, in the inherent cement-alkaline environment, a 'protective passive layer' forms on its surface. However, if the depth of cover is insufficient or the concrete is permeable, then the passive layer will also be broken down in presence of excessive amounts of chloride ions. The chlorides can originate from sodium chloride (common salt) in marine locations or from de-icing applications, or from the use of a particular admixture e.g. calcium chloride (accelerator), or from the surrounding soil, from the contaminated unwashed aggregates themselves or even from the mix-water or curing-water.

When the protective passive layer breaks down, the steel is liable to rust or corrode and this is an expansive (bursting) process, which can cause the concrete to crack and spall. Cracking and spalling are particularly noticeable at corners of beams and columns and over the main steel, although the pattern of links and stirrups may also be visible. Such cracks will usually show signs of rust-stains along the lines of reinforcement.

However, other types of cracks which cross the lines of reinforcement may also pick up rust stains, and after sometime it will be difficult to separate the two!

Corrosion of a metal is an electro-chemical process that requires an oxidizing agent, moisture, and electron flow from one point in the metal to the other.

Corrosion of steel produces iron oxide and hydroxide, which have a volume much greater than the volume of the original metallic iron. This increase in volume causes high radial bursting stresses around the bars and results in local radial cracks. These splitting cracks can propagate along the bar, resulting in the formation of longitudinal cracks parallel to the bar, leading to spalling of the concrete. A broad crack may also form at a plane of bars parallel to a concrete surface, resulting in the delamination of the surface, a well-known problem in bridge decks. (For more details see ahead.)

• *Prevention of Corrosion*

The best way is to ensure that the concrete is dense and increase the cover of concrete on bars. Dense concrete can be made by increasing the minimum cement content (to 350 kg/m^3 of concrete generally but depends on situations and salinity), reducing water-cement ratio, using plasticizers for improving workability lost by the reduction of water content, thoroughly vibrating the concrete and then ensuring efficient and early curing.

Since Portland-blast-furnace-slag-cement gives a very dense cement-paste, it is advisable to use such a cement (containing nearly 65 to 70 per cent of blast-furnace slag by weight of cement, mixed together in the grinding process of cement manufacture). This denseness inhibits chloride-ion 'movement' because of almost impermeable concrete matrix (provided correct concrete mix and construction practices are carefully followed). Epoxy coating of bars is not very helpful! Chloride salts are soluble, and if any moisture movements take place in concrete, chloride ions may migrate through the moisture and lead to high concentrations in certain locations, This theory would certainly account for the extremely varied results obtained from chemical analyses of hardened concrete to which calcium chloride has been added as an accelerator.

See Figs 4.41 – 4.45, Plates 4, 5 and 6.

• *Remedial Measures Against Corrosion Effect*

Before repairs are carried out to concrete which has rust-cracked and spalled over reinforcement, it is essential that tests are carried out to establish whether excessive chlorides are present; this can be done by chemical analysis only. Guidance on sampling and testing for chlorides is given in ASTM Standards. Chloride levels are conveniently expressed as an equivalent percentage of calcium chloride by weight of cement.

(a) *Chloride Less Than 0.5 per cent (as CaCl$_2$/cement)*: If the results on such chloride level are fairly consistent then it is likely that these come from natural sources (e.g. the aggregates) and will not pose a problem. Cracking and spalling are then likely to have been caused by a combination of lack of cover and permeable concrete. These factors should then be investigated and, if confirmed, repairs should take form of remedying just these factors (*see* ahead).

(b) *Chloride Greater Than 0.5 per cent (as CaCl$_2$/cement)*: If the level of chlorides is extremely variable and generally in excess of 1.5 per cent (e.g. 2 to 4 per cent), then this is indicative of excessive-chloride. Structures such as bridge decks or structures which have high concentrations of chlorides near the surface (caused e.g. by de-icing salts) should be judged and dealt with accordingly (*see* ahead).

When excessive chlorides are proven, it must be remembered that the corrosive process cannot simply be chemically arrested by, for example, sealing. The chloride attack will continue because the moisture can never be eliminated.

PLATE 3

Fig. 4.39 Shear cracks in R.C. girders

Fig. 4.40 Initiation of flexural crack where the main tension reinforcement was inadequately stopped (poor detailing)

PLATE 4

Fig. 4.41 Bursting of kerb concrete due to expansive pressure from corrosion of reinforcement

Fig. 4.42 Bursting of concrete (in abutment cap) pursuant to corrosion of reinforcement

PLATE 5

Fig. 4.43 Delamination of concrete on the underside of the deck slab, due to hidden corrosion actively manifesting within

Fig. 4.44 Delaminated concrete has fallen off exposing the corroded bars (partial repair attempted)

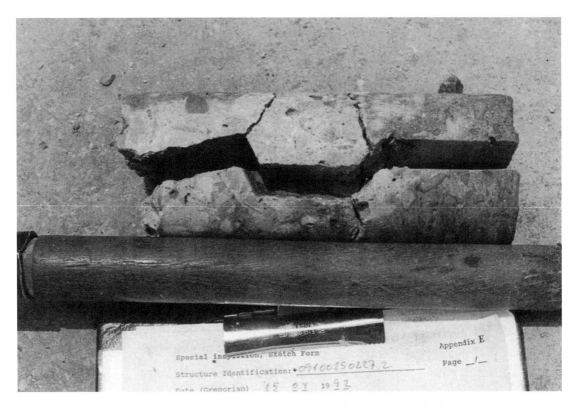

Fig. 4.45 A core drilled through a shear-key clearly showing delaminations at the key

Fig. 4.49 Carbonation front has penetrated to the reinforcement. Note that the phenolphthalene-spray has turned pink only beyond the bar, indicating that the pH has reduced to below 9 or 8 right through the concrete cover so that the passive alkaline environment has turned acidic, leading to corrosion, delamination, and bursting of concrete

All affected concrete adjacent to the rust surrounding the affected steel must be removed (all around the steel). The steel and exposed concrete, must be thoroughly cleaned, preferably by sand blasting, additional reinforcement added to replenish the loss, the exposed concrete saturated with water, and the concrete replaced with pneumatically applied layers of fine concrete or mortar. Saturation with water may not be allowed if epoxy-mortar or epoxy-concrete is used, unless the epoxy is water-insensitive. However, it is quite likely that some further areas of concrete will subsequently rust-crack; thus a decision will have to be made to determine how much 'apparently yet unaffected' concrete should be removed. This decision is usually governed by financial restraints. In some cases, periodic monitoring and repair is more economic than large removals.

NOTE: In order to better understand the phenomenon of steel-corrosion, a detailed appreciation is presented in Section 4.7 ahead.

Cracks due to Rusting of Reinforcement Caused by Carbonation of Concrete (xiii)(b)

Carbonation is a process in which carbon dioxide (in the atmosphere) reacts with dissolved calcium hydroxide in the pore water of the concrete structure, resulting in the formation of calcium carbonate and this reaction increases acidity as it reduces pH in the moisture in concrete. Consequently, the reinforcing steel corrodes, because the alkalinity of the concrete is reduced owing to the reduction of pH value through carbonation, reducing the passivity around steel.

The increased volume of the corrosion-product, (the ferrous oxide, which has a greater volume than the steel from which it is formed), causes radial bursting stresses around the reinforcing bar. This results in local radial cracks which initiate spalling of the concrete. These cracks further provide easy access for oxygen, moisture, and chlorides in the concrete, creating a condition whereby corrosion continues and causes further cracking, and finally significant spalling.

Here too 'dense' concrete helps as against any chemical attack, as described earlier.

Cracks due to Rusting of Reinforcement Caused by Exposure to Moisture and Atmosphere (xiii)(c)

Reinforcement, prestressing steel and most of the steel elements, if left exposed to moist atmosphere, will rust due to the chemical reaction of oxidation. This reaction will continue even if such steel is embedded in concrete.

Since rust occupies much more volume than the basic metal, the process creates radial expansive pressure on the surrounding concrete and bursts it, forming a line of crack initially. This can further lead to separation of sheets of concrete (delamination) and eventually the damage as associated with rust distress, described earlier.

'Dense' concrete is helpful against rust problems but rusted steel should not be used in the first instance, as pointed earlier.

Cracks due to Alkali (in cement) Reacting with Certain Reactive Types of Aggregates (xiv)

• *Alkali-Aggregate Reaction Cracks*

A rare form of expansion and cracking can occur under damp conditions following the reaction of some forms of silica and carbonates in certain aggregates with the alkalis in cement. The reaction between the silica and the alkali produces a gel which occupies more volume and hence, causes expansion and cracks, usually 'moving away' from the source of expansion. However, these cracks may join others and form a 'map' pattern not dissimilar to 'crazing'. Alkali-aggregate reaction, like chloride and sulphate attacks, can take place only under wet or damp conditions. Positive identification can be made only by means of laboratory tests on samples of concrete removed from the structure. Some of the symptoms which are often present, but not always seen, are:

(a) Presence of gel (like dried up resin or honey)-exuding from the cracks, usually clear but sometimes stained. On vertical surfaces it will be washed down as a stain; on horizontal surfaces it may stand out of the concrete. The gel may easily be confused with calcium carbonate deposits or efflorescence.

(b) Pop-outs: caused by popping out of pieces of reactive aggregate just below the concrete surface. The gel may be seen at the base of the pop-out, otherwise pop-outs look similar to frost damage (note the difference between the two cases).

(c) Other signs include persistent dampness, discolouration and discernible signs of expansion of the concrete. Alkali-aggregate reaction is a long term phenomenon, usually taking several years to manifest itself.

• *Prevention and Remedy of Cracks Due to Alkali-Aggregate Reaction*

If the aggregates do contain the undesirable reactive silicates and carbonates, the ensuing expansive chemical reaction with (high) alkali cement can lead to a considerable volume-increase, significant movements (causing closure of expansion joints and even misalignment of Bearings and other fixtures, etc.) and concrete cracking, about which not much can be done! (The movements are so significant simply because the major proportion out of the total concrete just happens to be the culprit 'coarse aggregate' itself.)

By Way of Prevention: Substitution to the extent possible by non-reactive aggregates is recommended even if uneconomical. Use of low-alkali cement, prevention of

dampness manifesting around concrete for any significant periods, use of Portland-blast-furnace-slag cement containing about 65 per cent slag by weight of cement (and more in case chloride and sulphate attacks on reinforcement and cement, respectively, are also expected, e.g. if the concrete structure is being constructed in the sea or close to it). Such a cement develops a very dense cement-paste giving impermeable concrete, which in turn inhibits the alkali-aggregate reaction physically and, since this cement has lesser alkali content, it reduces the reaction in the first place. (Lesser alkali content because 65 per cent or so is slag!)
By Way of Remedy: The cracks may be sealed by epoxy injection but only after 3 to 5 years, by which time, majority of this crack formation might take place. Until then these cracks may be surface-sealed (to prevent other pollutants from ingressing) by one of the suitable methods as described later for 'live' cracks. However, all this remedy is possible only if the damage due to these cracks is not extensive.

Fabrication, Shipment and Handling Cracks (xv)

Errors in fabrication, combined with stresses induced during shipment and handling, is one major cause for cracks in precast concrete members. These cracks may occur in variable shapes, sizes and locations of the member. Figures 4.9 and 4.10 show some typical patterns of these cracks.

Crazing Cracks (xvi)

Crazing is the cracking of the 'surface layer' of concrete into small irregularly shaped areas. These cracks do not affect the structural integrity of concrete and, in themselves, should not lead to subsequent deterioration of the concrete. The crazing cracks occur within the thickness of the laitance, where water-content is more, and hence shrinks more relative to inner layers of concrete. They are rarely more than a few millimetres deep and are caused by shrinkage of the surface layer! Crazing generally occurs:

    (a) In the over-floated or over trowelled surface layers of concrete slabs

    (b) In the 'formed' surfaces of concrete

Crazing is generally accepted as being the result of surface tensile stress caused by shrinkage of the laitance surface relative to the mass, which is due to differential moisture movement. Thus crazing is not necessarily an ageing phenomenon but occurs whenever these critical conditions are exceeded.

Avoid over-rich or excessively wet mixes. Mixes with high cement contents may result in surface-laitance which will allow differential moisture movement and hence be prone to crazing, particularly when over-vibrated at the forms or over-trowelled at the surface.

Compaction should remove trapped air and leave a concrete surface ready for the surface finish intended. If vibration continues for too long, the surface layer will be too rich and too wet and crazing may result.

Weathering Cracks (xvii)

The weathering processes that can cause cracking include freezing and thawing, wetting and drying, and heating and cooling.

Every type of concrete is porous, only that the porosity may be high or low and will, therefore, absorb moisture. When exposed to freezing temperature, the moisture will freeze and expand, resulting in hydraulic pressure which tends to cause the concrete surface to crack.

Other weathering processes that may cause cracking in concrete are alternate wetting and drying, heating and cooling. Both of these processes produce volume changes in concrete. If the volume changes are excessive, cracks may occur, similar to those discussed in drying shrinkage and thermal stresses. Typical weathering damage of concrete is shown in Fig. 4.8.

4.7 PHENOMENON OF CORROSION

Although the process of corrosion has been explained earlier, in view of its special importance, an attempt is made here to understand its mechanism in greater detail. It is discussed under the following heads (also *see* Chapter 15 for Tests).

— Initiation of Corrosion
— Carbonation
— Chlorides
— Carbonation and Chlorides
— Propagation of Corrosion and its Process
— Typical Corrosion Cases in a Bridge
— Combined attacks of Sulphates and Corrosion
— Corrosion Control
— Types of Corrosion

4.7.1 Initiation of Corrosion

Reinforcement is protected against corrosion when embedded in a concrete of a good quality and with a sufficient cover.

The protection is due to high alkalinity of the concrete (pH-value close to 13) leading to a thin grey 'passive' layer to be formed on the surface of the bar.

But, the protection against corrosion is not everlasting. The surroundings will always affect the concrete and finally lead to a breakdown of the passive layer. The breakdown of the passive layer may be caused by 'free chlorides at the reinforcement' of by 'carbonation of the concrete cover'. These mechanisms are described in the following sections.

The time until the passive layer breaks down is normally called the 'period of initiation'.

The duration of the initiation period depends on:

— The thickness of concrete cover—the thinner the cover the shorter the period of initiation
— The quality of the concrete cover (primarily water–cement ratio dependent). The initiation period decreases as the concrete quality reduces i.e. as the water/cement ratio increases. In special cases (honeycombs, cold joints, or too small a cover), poor workmanship can lead to early corrosion
— The aggressiveness of the environment e.g. the temperature, the humidity, and the level of contamination of sulphates, chlorides and penetration of carbon dioxide
— The kind of mechanism causing deterioration. (Carbonation and chloride penetration are by far the most aggressive mechanisms of deterioration. Chloride ions facilitate the corrosion process)

During the period of initiation, the corrosion process develops with no visible signs of deterioration, neither on the surface of the concrete nor on the reinforcement.

Therefore, the risk of future corrosion damage can only be assessed by performing special investigations (*see* Chapter 15).

4.7.2 Carbonation

Carbonation is caused by the carbon dioxide (CO_2) in the air. The CO_2 reacts with the calcium hydroxide, $Ca(OH)_2$, in the cement paste, eventually leading to a critical decrease of the alkalinity. The pH-value decreases to around 9, which normally is insufficient to protect the reinforcement against corrosion. Its surrounding passive layer is broken.

The carbonation-depth increases with time, roughly following the equation:

$$x = K\sqrt{T}$$

where:

   x = carbonation depth in mm
   K = coefficient
   T = time in years

The value of the coefficient K can be estimated as follows:

$$K = 72\left(\frac{1}{\sqrt{f_c}} - 0.126\right)$$

where f_c = compressive strength (MN/m^2) of standard concrete cylinder

The coefficient K is empirical and is based on experience. The formula is valid at a relative humidity of 50 per cent R. H. (assume that x is measured in mm and T in years). If the relative humidity in the pores of the concrete is different from 50 per cent, K must be multiplied by a factor < 1, depending on the humidity (Fig. 4.46).

Example

   Compressive strength of concrete: 20 MN/m^2
   Age of structure: 25 years
   Relative humidity in concrete :   60 % RH

$$K = 72\left(\frac{1}{\sqrt{20}} - 0.126\right) = 7.03$$

K must be multiplied by 0.95 (Fig. 4.46: R. H. being 60%)

An estimate of the carbonation depth will be
$$x = (0.95 \times 7.03)\sqrt{25}$$
$$= 33\text{ mm}$$

If the actual carbonation depth and the concrete cover are measured on site, a prediction of the time until initiation of corrosion can be made:

$$x = K\sqrt{T} \text{ so that } K = \frac{x}{\sqrt{T}}$$

If carbonation penetrates the cover C in a time of T years,

then: $C = K\sqrt{T_1}$, so that $T_1 = \left(\frac{C}{K}\right)^2 = \left(\frac{C}{x}\right)^2 T$

where x = the actual depth of carbonation in mm as measured at site
   C = cover (mm)
   K = coefficient
   T = age of concrete (years)
   T_1 = period of initiation (years) for the carbonation to penetrate the cover and reach the reinforcement.

Example

   Carbonation depth, measured: $x = 25$ (mm)
   Concrete cover, measured: $C = 35$ (mm)
   Age of concrete: $T = 20$ (years)

Initiation of corrosion is estimated to begin when the structure is T_1 years old:

$$T_1 = 20\,(35/25)^2$$
$$= 39\text{ (years)}$$

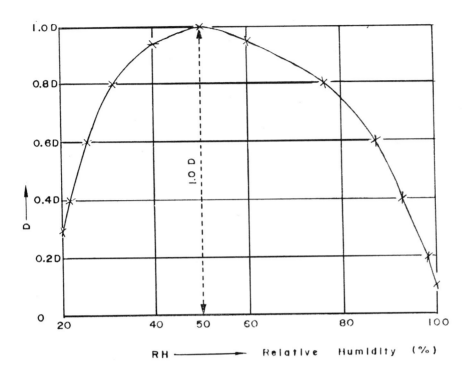

Fig. 4.46 Relative depth of carbonation w.r.t. that of 50 per cent R. H. (carbonation depth is D at 50 per cent R.H.)

4.7.3 Chlorides

Chlorides in the concrete may arrive from various sources, such as.
— mix water
— aggregates
— curing water
— surrounding soil (from where chlorides are washed out in wet periods)
— de-icing salt (in cold areas)
— in coastal areas from the seawater (reaching the concrete directly or airborne in windy periods)

In general, most of the chlorides already present in fresh concrete ('initial chlorides', i.e. chlorides from mix water, aggregates and some from the curing water) will become chemically 'bound' during the hardening of the concrete.

'Bounded' chlorides are not regarded as harmful if they remain bounded, since they then cannot break down the passive layer on the steel. As a rough estimate, 60 per cent of the initial chlorides are bounded, and 20–30 per cent of chlorides penetrating through hardened concrete become bounded. This is only a rule-of-thumb, and a repair strategy should not be based on this estimate alone.

Laboratory tests can measure the 'bounded' as well as the 'free' chlorides, because the concrete sample is dissolved in an acid that frees the bounded chlorides.

An assessment of the content of 'free' chlorides can be made by carrying out the measurements in distilled water instead of in the acid. The free chlorides will rapidly be dissolved in the water, while the bounded will remain bounded.

Diffusion of chlorides into concrete can be described roughly by the following two equations:

$$x = K_x \sqrt{T}$$

and

$$C_x = C_s - (C_s - C_i)\, \mathrm{erf} (0.5\, x/\sqrt{TD})$$

where
x = depth of critical chloride level (mm)
K_x = Coefficient
T = age of concrete (years)
C_x = content of chloride in the depth x
C_s = content of chloride in the surroundings
C_i = content of initial chloride
erf = Error function
D = diffusion coefficient

PLATE 7

Fig. 4.50 A classic example of corrosion through almost 70 mm penetration of carbonation front; The phenolphthalene-spray shows exposed concrete turning pink only at about 70 mm depth from surface

Fig. 4.56 Sulphate and chloride attack on a R.C. bridge-foundation in the sea in the splash zone

PLATE 8

Fig. 4.57 Sulphate and chloride attack on R.C. bridge foundation in the sea; Note heavy delamination in concrete and extensive rusting in the splash zone

Fig. 4.58 Classic case of combined attack of sulphate and chloride in the sea, made worse by alternate wetting and damp drying

The 'Error function' takes the following values:

x (mm)	$erf(x)$	x	$erf(x)$	(x)	$erf(x)$
0	0	0.6	0.604	1.4	0.952
0.1	0.112	0.7	0.678	1.6	0.976
0.2	0.223	0.8	0.742	2.0	0.995
0.3	0.329	0.9	0.797	2.4	0.999
0.4	0.428	1.0	0.843		
0.5	0.521	1.2	0.910		

Based on measurement of the 'chloride profiles' in the concrete, the different parameters can be calculated in order to predict the time until corrosion commences.

The calculations will be similar to the calculations regarding carbonation.

A typical chloride-content 'profile' is indicated in Fig. 4.47.

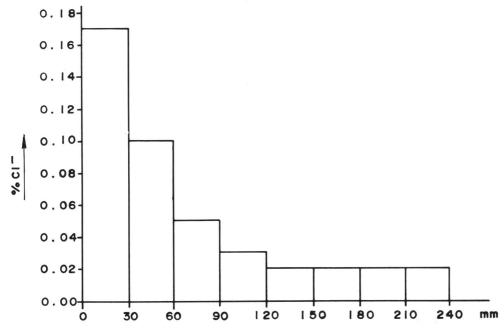

In this example, the initial chloride content is approximately 0.02%.

Fig. 4.47 Chloride profile

leading to accelerated corrosion and the critical limit of chloride content will be reached at the reinforcement faster!

A typical 'chloride profile' in a concrete with a carbonation depth of approximately 30 mm is shown in Fig. 4.48. Also *see* Figs 4.49 and 4.50, Plates 6 and 7.

If this mechanism takes place together with alternate wetting and drying in a chloride contaminated environment, the corrosion process can run very fast (as in tidal waters).

A simple method to evaluate the chloride penetration from outside in this case is to ignore the carbonated concrete layer. The thickness of the carbonated layer can be estimated approximately from the previously mentioned equations.

4.7.4 Carbonation and Chlorides

For concrete with a high initial content of chlorides, the chemically bounded chlorides in front of the carbonation front will get freed and accelerate the corrosion activity!

Since carbonated concrete provides almost no resistance against chloride penetration, the chlorides will breakthrough the carbonation-front (diffusion).

Thus, the chloride content will increase constantly when the carbonation front is moving into the concrete,

4.7.5 Propagation of Corrosion and its Process

When the chloride content at the reinforcement level reaches the critical limit or the carbonation front reaches the reinforcement, the passive layer is broken down and the corrosion process starts.

The corrosion process is an electrochemical process, where a low volt current runs between corroding areas (the anodes, where the passive layer has been broken down) and the non-corroding areas (the cathodes,

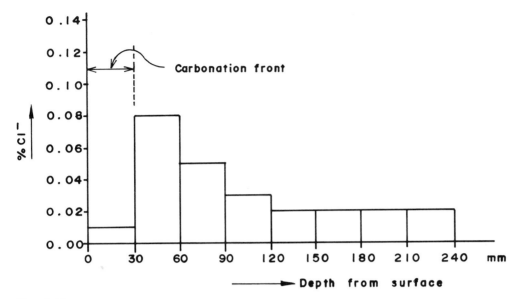

Fig. 4.48

where the passive layer is unbroken). These 'anodes' and 'cathodes', portions in strong and weak solutions of chloride, can be in the same steel bar.

The current is caused as a result of the natural behaviour of metals in liquids and concrete. If two metals with different electrochemical potentials are electrically connected, corrosion is likely to take place on the metal with the lower potential.

Therefore, the risk of corrosion can be evaluated by measuring the potentials.

When the passive layer breaks down 'locally', the area there changes its place in the electrochemical series, getting more acidic. This creates a potential-difference, causing the corrosion-current.

- Where the passive layer is broken, the following 'anodic- reaction' takes place.

Anodic Reaction

$Fe \longrightarrow Fe^{++} + 2\,e^-$ (producing electrons)

- If moisture and oxygen are present at the steel- surface, the following 'cathodic reaction' takes place:

Cathodic Reaction

$O_2 + 2H_2O + 4e^- \longrightarrow 4OH^-$ (consuming electrons)

The anode and the cathode may even be far apart, as long as there is an electrical connection between them! This connection is the medium of (acidified) moisture, the electrolyte, and the bar itself. The anodic reaction is producing electrons, and the cathodic reaction is consuming electrons. If there is an electrical connection between the anode and cathode, an electrical current will run between the two (the flow of electrons).

The corrosion process is illustrated in Fig. 4.51.

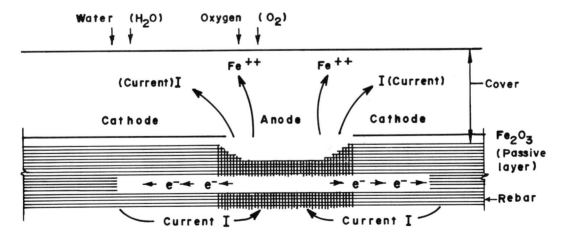

Fig. 4.51

- The Fe^{++} reacts with oxygen, OH^- and water, forming the corrosion-products! The 'type' of corrosion-product primarily depends on the 'available amount' of oxygen and water.

(i) If there is little available oxygen, the first products will be *white* $Fe(OH)_2$.

(ii) These white products may be transformed into *black* FeO and water. This type of corrosion -product is typical for local, chloride-initiated, corrosion.

(iii) If the oxygen is added to FeO, *black* Fe_3O_4 may be formed. Fe_3O_4 is not voluminous, so there are not necessarily any exterior signs of corrosion yet!

(iv) If water and oxygen become available to the white $Fe(OH)_2$, the products turn *via* a *green intermediate stage* into *brown* $Fe(OH)_3$.

(v) If additional water is available, voluminous *yellow/red/brown* $Fe(OH)_3 \cdot nH_2O$ (*rust*) is formed.

(vi) However, if plentiful oxygen is available, the voluminous *yellow/red/brown* $Fe(OH)_3 \cdot nH_2O$ (rust) is formed *without any of the stated intermediate stages.* This type of corrosion product is typical for carbonation-initiated corrosion in porous concrete (mark the difference).

The development of the corrosion-attack and the speed of the process primarily depend upon:
— temperature
— the ratio between corroding and non-corroding areas
— the content of moisture

As for most chemical reactions, the corrosion rate increases with increasing temperature.

If the area of the anode is small compared to the area of the cathode, the corrosion-rate will be high, because the corrosion is taking place in a small area (*local-corrosion*).

The process of 'local-corrosion' normally leads to corrosion-products which are *black, non-expansive, and liquid-like.* It means that local-corrosion cannot be expected to give visible signs of corrosion on the surface. Local-corrosion is hidden-corrosion and serious attacks can be developed without visible signs, increasing the risk of unexpected collapses!

In case of *chloride attack*, the chloride ions facilitate the formation of Fe^{++}, thus increasing the corrosion rate. Since chloride-initiated corrosion normally starts as local-corrosion, a considerable reduction of cross section of a steel bar may take place within a short time and without visible signs on the concrete surface.

When the anode-area and the cathode-area are nearly equal, we get what is called *general corrosion*. This type of corrosion leads to the well known yellow/red/brown corrosion-product, which quickly gives rise to visible signs of corrosion (spalling of the concrete and cracks due to bursting pressure—as rust occupies more volume, and the rust-stains).

The corrosion rate depends on the moisture content surrounding the bar. At R. H. less than 80 per cent, the corrosion rate is negligible. It has to be noted that the R.H. % refers to the moisture content in the pores of the concrete at the cathode area. The mean moisture content in the pores of concrete over a year can be different from the % R.H. in the air, specially in case of presence of chlorides in concrete.

Normally the corrosion-rate drops at R. H. above 95 per cent and is minimal at R. H. 100 per cent. See Fig. 4.52. But if there is electrical contact with areas with less moisture content, a cathode may be formed in this area and corrosion rate will then increase. This may be the case when a column is partly submerged in moist soil. Corrosion may take place at an anode area below

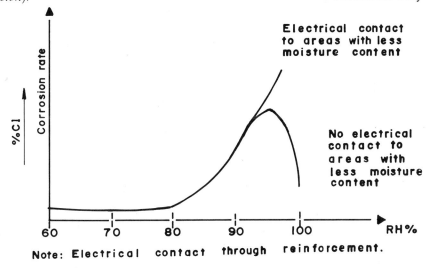

Fig. 4.52
NOTE: Electrical contact through reinforcement.

ground (R. H. high) because the cathode is formed above ground where R. H. is lower.

These two types of corrosion can normally be located at different places in the structure:

(i) *Local Corrosion:* Shows little visible signs on the surface (because rusting is localized and the expansive bursting pressure is low), is typically located:

— at 'restrained' cross-sections (max. negative bending moment), e.g. brackets and cantilever decks;

— at construction joints and cracks in concrete with high content of chloride and moisture;

— in concrete with high initial content of chloride.

(ii) *General Corrosion:* Shows visible signs on the surface (because rusting is profuse so that expansive bursting pressure is high), is typically located:

— in concrete of poor quality;

— in concrete with insufficient thickness of the cover;

— in concrete subject to alternate wetting and drying such as horizontal surfaces and vertical surfaces near water or ground level.

4.7.6 Typical Corrosion Manifestations in a Bridge

Three typical cases of corrosion: (i) in bridge deck, (ii) in piers, and (iii) in abutments and wing walls.

(i) *Bridge Deck (Fig. 4.53)*

in Zone A: Here the cracks are caused by negative bending moment—the corrosion will develop as local corrosion with a high corrosion rate and consequent reduction of cross-section.

in Zone B: Here the corrosion in the first phase will develop as local corrosion and in the later phase will lead to general corrosion, followed by spalling of the concrete.

The development of corrosion depends on the moisture content and the duration of wet and dry periods. In decks with no overlay/wearing course in hot climate and with occasional water-influence, the corrosion will quickly develop as general corrosion. The time, until the corrosion-attack becomes visible on the surface, is normally shorter but the damage to the reinforcement is less than in decks subject to longer presence of moisture (e.g. in case of an overlay which is not waterproofed, allowing water to penetrate but not quickly evaporate away).

In a deck with an overlay which is not waterproofed, the corrosion normally becomes visible late when the overlay cracks.

A high rate of corrosion may be seen even if there is a lack of the necessary oxygen at the shielded upper reinforcement layer. This is because the oxygen-consuming cathode-formation can take place on the lower reinforcement layer which relatively is not so well shielded because it only has concrete cover on it.

in Zone C: Here the risk of corrosion is limited to the soffit except if there is leakage especially with chloride contaminated water, in which case corrosion could attack on all surfaces. If larger areas are suffering from water (and chloride) influence, the mechanism of damage will be similar to that in Zone B, giving rise to spalling of the cover.

(I) In Bridge Deck

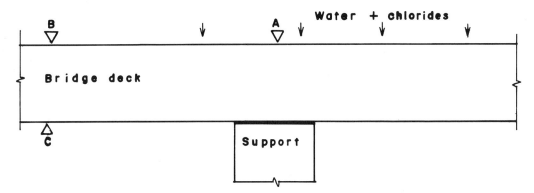

Fig. 4.53

(ii) Piers (Fig. 4.54)

(ii) In Piers

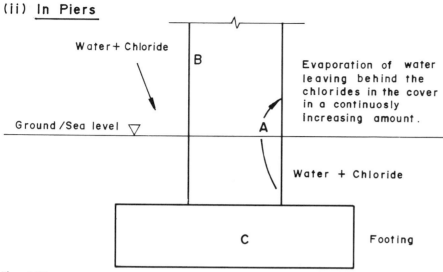

Fig. 4.54

in Zone A: Here is the risk of local corrosion, especially in constantly wet and chloride contaminated concrete. In concrete exposed to alternate wetting and drying, the corrosion product will expand, causing cracks and spalling of the cover.

in Zone B: Here general corrosion mainfests, mainly caused by carbonation. The corrosion will develop slowly, but visible damage will appear at an early stage.

in Zone C: Here is the risk of local corrosion. The lack of oxygen does not prevent corrosion because the cathode process will take place at the steel areas above ground level. The corrosion rate can be very high.

(iii) Abutments and Wing Walls (Fig. 4.55)

Abutments and wing walls may suffer from several types of corrosion attack:

in Zone A: At ground level and below, the conditions are similar to those in Zone A in a Pier.

in Zone B: Near the top, there is a risk of very high chloride content originating from surface water running from the overpassing road and the slopes (through chloride contaminated soil) to the top and the upper part of the back side of the wall. Due to alternate wetting and drying, visible signs of corrosion of the reinforcement are seen in 'splitting at the top'. This splitting causes easier access for water, which makes the corrosion process even faster.

in Zone C: On the back side of the wall there may be constant moisture and chloride. Even if there is a lack of oxygen, severe corrosion may occur in a manner similar to that in a pier footing explained earlier.

in Zone D: At the front of the wall, the chloride content may be very high, because the saline water, penetrating from the back, evaporates, leaving behind chlorides in the concrete.

NOTE: Any investigation of the corrosion problems on out-door concrete structures has to include an evaluation of:

— the actual corrosion activity

— the future risk of corrosion

Both in areas with evident corrosion and in areas without visible signs of corrosion (especially where corrosion may normally be expected to take place), the investigation should be carried out by breaking open the concrete and also by gentle hammer-test for detecting delaminations caused within (owing to expansive pressure resulting from rust formation).

4.7.7 Combined Attack of Sulphates and Corrosion

A special case of both sulphate attack and corrosion is seen on partly submerged structures, especially structures with large dimensions. *See* Figs 4.56 – 4.58, Plates 7 and 8.

The sulphate concentration will constantly increase in the evaporation zone, leading to a rapid deterioration. Capillary suction may also lead to a general sulphate attack on the concrete in the whole cross-section.

Partly submerged structures normally have the biggest problems in the splash zone. However, if the cover below sea level disintegrates due to sulphate attack, one big anode is formed under the sea level. (An

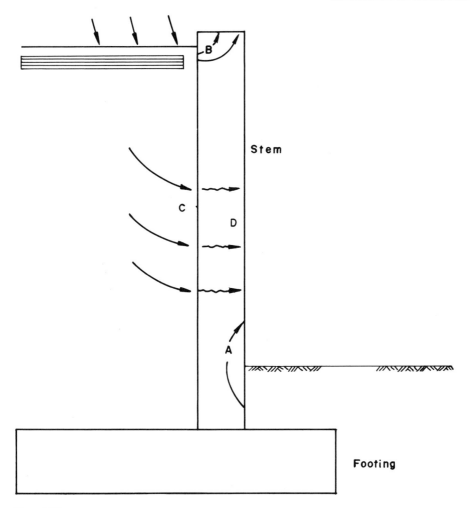

Fig. 4.55

'anode' is steel with chlorides on it, with its passive layer broken down. A 'cathode' is steel with moisture and air supply.)

The cathode will be the part above sea level. In structures with large dimensions, a large amount of current runs, leading to severe corrosion-attack on the submerged part.

NOTE: Also *see* Chapter 15 for relevant diagnostic tests.

4.7.8 Corrosion Control

To avoid future problems, improvements should be made in designs, specifications and materials for new and rehabilitation projects. If proper consideration is given in the choice of designs and materials during design and construction, many corrosion problems could be eliminated at little cost. The following are some of the factors to be considered.

(a) Avoid the Use of Dissimilar Metals
Avoid the use of dissimilar metals or burial of them in dissimilar soils, to avoid galvanic action. Where

dissimilar metals will be used, galvanize the steel or separate the contact surfaces of the two by stainless steel or organic coating materials.

(b) Select Only the Materials Best Suited for a Given Environment
In general, environmental characteristics may be grouped into following four categories, according to the nature of the atmosphere.

- *Rural:* Where no major harmful environment exists.
- *Marine:* Where ocean spray, salt, and humidity affect the structural corrosion (coastal zone).
- *Industrial:* Where high levels of carbon monoxide, sulphur oxide and nitrogen oxide exist (due to burning of high sulphur coal or oil), as also other pollution from the industries.
- *Arid:* Desert.

The above classification should be carefully examined and done according to individual locations instead of merely classifying by the geographic zones. For instance, a structure in a wet rural area would be more susceptible to corrosion than in a dry industrial area. The pH of rain (high humidity) is influenced by the level of sulfur dioxide, the time of wetness, geographic location, and maintenance. Therefore, it should be carefully considered in the selection of materials.

Construction materials are usually selected for reasons other than corrosion-resistance! The most common structural steel used in bridges is carbon steel which has no apparent corrosion resistance. However, low-alloyed i.e. weather resistant steel, which normally does not require painting, has also been used in some places. In addition to corrosion-characteristics, the following factors should also be considered in selection of steel members.

- Fatigue stress
- Notch-toughness for low temperature
- Weldability and through-thickness ductility to prevent lamellar tearing
- Maintenance repairs
- Life expectancy and life cycle cost
- Environmental restrictions

Where exotic and hostile environments are encountered, engineers currently prefer to use alloyed steel. In attempting to reduce detrimental corrosion, alloyed steel (containing chromium, molybdenum, nickel, copper and other metals) is used in place of the carbon steel normally used in structures.

The high chromium content retards corrosion and makes steel 'stainless'. Most commonly used stainless steel is the chromium-nickel grade for its weldability and corrosion-resistance.

The corrosion rates of steel in certain tropical areas can be much higher than in certain marine environments due to their relatively high temperatures and humidity.

Low alloy steel, the so called weather resistant steel, has been used effectively to resist atmospheric corrosion by the formation of a protective rust oxidate film. However, current research indicates that its corrosion rate will rapidly increase in industrial and marine environments and the use of weathering (ASTM-A-588) steel is not recommended in marine and polluted atmospheres and cost saving by using A-588 steel may not be realized in such areas. The useful life of the painted steel will be more in the absence of the rust film. (It is recommended that weathering A-588 steel should not be galvanized because of its silicon contents.)

(c) Care in Detailing

Details should be carefully examined to avoid water and soil penetration and not to create traps for water and debris. In corrosive environments, oversize sections may be used where the structural materials are placed underground or in the splash zone, or are inaccessible for proper routine maintenance. For steel bridge members in corrosive environments, an additional 1/8" thickness of steel would extend an average life of the structure about 30 per cent more. For concrete members, it is very helpful to have very dense concrete so as to inhibit movement of moisture, chloride and sulphate ions. (Details have been explained earlier.)

(d) Coating Systems

The use of coating system—metallic coating (hot dipping, clad steel or electroplating) on steel, organic coating on concrete (coal-tar or painting) and inorganic coating (e.g. cement coating) on steel and structure should be made to create a barrier against corrosion-attack.

(e) Cathodic Protection

The 'idea' behind the process of cathodic protection is to impose and establish a small current enough to overcome the 'corrosion current' created by the corrosion process. The corroding parts (anodes) are changed into non-corroding parts (cathodes) by the application of an external anode on the concrete surface! The system may, for instance, consist of a mesh (e.g. in titanium) fixed on the concrete surface and covered by mortar. The method requires information about the quantity of the existing reinforcements and investigation of the existing concrete. Mutual electric contact between the bars must be ensured during the installation.

The method requires instrumentation, continuous monitoring and adjustment of the current. If power supply is not directly available, solar panels may be used.

Cathodic protection methods have been used extensively for underground pipe installation and offshore drilling. However, cathodic protection has a limited use on the bridge. *There are two basic types of cathodic protection systems—the 'sacrificial-anode' and the 'impressed-current' systems.* The fundamental concept of protection is that all galvanic corrosion on the structure will halt when all points on the surface of the structure have been polarized to 'equal potential'. No difference in potentials means no flow of corrosion current and hence no corrosion!

(i) Sacrificial Anode System (Fig. 4.59): This galvanic type protection system requires no external power supply but it uses magnesium, zinc, or aluminum as the 'anode' and the metal structure (e.g. steel pile) becomes the cathode. For example, by zinc-coating the pipelines or by placing anode-zinc along the steel piles, the flow of current is induced by galvanic action. In this case the imposed anode i.e. the zinc-coating or the zinc-anode-piece on the piles is sacrificed in the process!

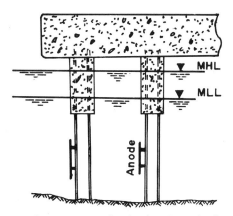

Fig. 4.59 Sacrificial galvanic method

Sacrificial anodes are zinc, aluminum, or magnesium. Zinc and aluminum are effective in salt water and magnesium in soil.

(ii) Impressed Current System (Fig. 4.60): In this applied potential system where an external current is applied, an external direct current power source (rectifier) is necessary to establish a potential of sufficient magnitude to offset and cancel out the 'corrosion current'. Since the current flow does not depend on the potential of the anode and metal of the structure, the selection of the anode is based on the capability of the conducting current and on transmitting current to the electrolyte with minimum corrosion to itself. This type of system uses graphite, carbon, silicon or cast-iron as the anode.

Since the current must pass from the anode through the soil or water to the structure to be protected, a cathodic protection system should be designed by considering such factors as conductivity, moisture content, soil and water characteristics, oxygen content, temperature, dissimilarity of metals, and environment. A cathodic system is used extensively to extend the life and reduce the maintenance of pipelines, water tanks, and marine structures.

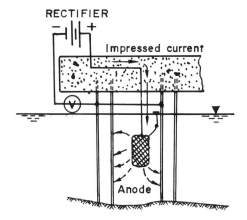

Fig. 4.60 Impressed current system

Anode will be selected based on its capability to transmit the current with minimum corrosion. Energized anodes are graphite or high silicone cast iron.

4.7.9 Types of Corrosion

(a) Corrosion of the metal due to *Carbonation* (explained earlier).
(b) Corrosion of metal due to *Chloride-ion attack* (explained earlier).
(c) Corrosion of metal due to *direct oxidation* (explained earlier).
(d) *Stress Corrosion:* Metals under tensile stress are susceptible to accelerated corrosion particularly in the presence of a 'pit' or a 'kink' where the tension will try to reduce the inter-fibural overlap between the molecules of the metal even faster. Corrosion is more active at the pit (or the kink) as it is, and the tendency of the grains to slip apart under tension, accentuates it.
(e) *Fretting Corrosion:* This type of corrosion generally occurs in the machinery of movable bridges on gears, on a shaft and metal fittings, due to surface contact of two metals under vibration. The contact surface becomes pitted and forms a red coloured deposit.
(f) *Chemical Corrosion:* Chemical corrosion may occur in industrial areas due to chlorides (marine atmosphere), carbon dioxide or sulphur dioxide pollution. These cause acid films and eventual corrosion.
(g) *Stray Current Corrosion:* Stray current corrosion, also known as electrolysis, occurs when direct current from outside source finds its way into, for example, a buried steel structure member. Corrosion occurs wherever the flow of direct current leaves the buried metal. Stray current corrosion may be much more rapid than other types of corrosion but it is also more localized, accruing only in the vicinity of direct current (DC) sources. Provable sources are pipeline protection rectifiers, protected pipelines, DC electric facilities including high voltage DC (HVDC) transmission lines, DC traction systems including DC powered electrified railways and apparatus in mines.

4.8 SOME DEFINITIONS CONCERNING PHYSICAL APPEARANCE OF CONCRETE

The following definitions, prepared in terms of physical appearance without assigning causes, may be noted for assistance while inspecting concrete elements, whether in superstructure or substructure.

Deterioration

Deterioration is any adverse change of the normal mechanical, physical and chemical properties, either on the surface or on the whole body of concrete, generally through separation of its components.

Disintegration

Disintegration into small fragments or particles due to any cause.

Scaling

Scaling is the loss of surface-mortar in the concrete. This procedure exposes the aggregate particles and, in its advanced stages, results in the loss of aggregate and loss of section. Scaling is often associated with cracking, and is principally located near gutters and kerb faces.

The degree of scaling is defined as —

Light Scale: Loss of surface mortar up to 6 mm in depth, and exposure of the top surface of coarse aggregate.

Medium Scale: Loss of surface mortar 6 mm to 12 mm in depth, with some loss of mortar between coarse aggregates.

Heavy Scale: Loss of surface mortar and mortar surrounding aggregate particles 12 mm to 25 mm in depth, so that aggregates are clearly exposed and stand out from the concrete.

Severe Scale: Loss of coarse aggregate particles as well as surface mortar and mortar surrounding aggregates, greater than 25 mm in depth.

Spalling

Spalling is a depression, resulting from the dislodgement surface of concrete. It differs from scaling, in that it refers to the removal of whole surface concrete portions, due to a horizontal or inclined fracture and not due to loss of surface mortar. Furthermore, it is typically found in old concrete structures and not in young concrete, as scaling does.

The major cause of spalling is the corrosion of reinforcing steel. In some cases overstresses may also be the cause.

Before developing into a depression, spalling may be present as a delamination below the concrete surface. Its detection on bridge decks is performed by hammer blows ('hollow sound') and other methods exciting the quality of sound. Cracks related to spalling are usually wide and long. Sometimes they are as deep as the reinforcing steel.

The degree of spalling is defined as

Small Spall: A rough 'circular' or 'oval' depression, generally not more than 25 mm deep, not more than about 150 mm in dimension, caused by separation and removal of a portion of the surface concrete, due to a roughly horizontal or inclined fracture. Generally a portion of the rim is vertical. In some cases it may be an elongated depression over a reinforcing bar.

Large Spall: May be a roughly circular or oval depression generally 25 mm or more in depth and 150 mm or more in other dimensions by separation and removal of a portion of the surface of the concrete, due to roughly horizontal or vertical fracture. Generally, a portion of the rim is vertical. In some cases it may be an elongated depression over a reinforcing bar.

Delamination

An area of concrete which, when struck with a hammer or a steel rod, gives off a hollow sound, indicating the existence of a nearby lamination-fracture near the surface (*see* Figs 4.43 and 4.44).

Cracks

These have been explained earlier in Sections 4.1 to 4.6. In addition, see the following definitions based purely on physical appearance:

Pattern Cracks or Map Cracks: Interconnected cracks, forming networks of any size, similar to those seen on dried mud-flats. They may vary in width from 'barely visible' to 'fine' to 'well defined' and 'open'.

'D' Cracks : Usually defined by dark coloured deposits and generally located near joints and edges.

Random Cracks : Meandering irregularly on the surface of the slab, having no particular form and not fitting other classifications.

Joint Spalls

An elongated depression along an expansion, contraction, or construction joint.

Popouts

Conical fragments that break out from the surface of the concrete leaving voids which may vary in size from 25 mm to as much as 300 mm in diameter at concrete surface. A shattered aggregate particle will usually be found at the bottom of the void, with part of the particle still adhering to the 'apex' of the popout 'cone'.

Pitting

Loss of thin coats of surface mortar directly over coarse aggregate particles, without apparent damage to the aggregate particles. The pits are usually not over 3 mm in depth. This is in contrast to popouts.

Mudballs

Small holes in the surface of concrete left by dissolution away from a concrete surface, either by deterioration or by adherence to forms.

Discolouration

Departure of colour from that which is normal or desired.

Efflorescence

A white (salt) deposit on concrete (or brick) caused by precipitation of dissolved salts brought to the surface by the capillary action (leaching) of moisture.

Exudation (stalactites)

A liquidy or viscous gel-like material discharged through a crack or opening in the concrete by the leaching water. (e.g. calcium carbonate formed by the reaction between the atmospheric carbon dioxide and the calcium hydroxide in concrete under damp conditions.

Encrustation

A crust or coating, generally hard, formed on the surface of concrete over a period of time by precipitation of minerals out of leaching water.

Leaching

Water seeping through concrete, dissolves water-soluble components (such as calcium hydroxide) in the concrete, which appear on the underside of the deck, as stains or efflorescence. In extreme cases stalactites are formed. In addition, sulphate attack on cement ingredients leads to formation of sulpho-aluminate crystals which cause cracks due to expansion forces caused by their larger volume. This phenomenon is prominent in Sabkha' regions. Corrosion of reinforcement is also probably in progress during this process (*see* Fig. 4.61, Plate 9.)

Examples of some of the defects in structural concrete are shown in Figs. 4.62 to 4.68.

4.9 CRACK WIDTH

Factors influencing crack-width in reinforced concrete in flexure

— Tensile stress in reinforcing steel
— Thickness of concrete cover
— Area of the concrete surrounding each reinforcing bar in tension

A number of methods have been developed on the basis of above factors to predict this crack-width. However, there is no universally accepted criterion which determines tolerable limits of crack-width in flexure, in shear or in torsion.

Engineers should use experience and judgment to determine these tolerable limits. As a guide to engineering-judgment, the following limits for flexural crack-width are generally adopted:

Fine — up to 0.30 mm
Medium — between 0.30 mm to 0.50 mm
Wide — more than 0.50 mm

Fine: Indicates no distress to structure. Immediate repair not necessary. Monitoring of the cracks is recommended.

Medium: Distress and loss of structural capacity is expected, but not in the range to cause immediate structural failure. Monitoring and eventual repair of the cracks is recommended.

Wide: Considerable loss of structural capacity. Immediate repair of the cracks is recommended.

NOTE: Generally permitted crack-width is 0.25 to 0.30 mm in flexure in reinforced concrete and in partially prestressed concrete.

4.10 EVALUATION OF CRACKS IN CONCRETE

Location and extent of cracks and general information about concrete in the structure should be determined by detailed visual inspection and then, if necessary, by non-destructive-testing and core-testing. Information about cracks should be checked from construction and maintenance records. All discrepancies of 'record' data with field-inspection and test data should be noted in the appropriate drawings. Stresses due to applicable loads and possible-causes of cracking may have to be determined. All noted data should be reviewed and analyzed by an experienced design engineer. The engineer will determine the effect of these cracks on the structure, and recommend remedial measures.

4.10.1 Visual Inspection

Cracks should be highlighted on the structure with a coloured highlighter, and their widths written across them along with inspection dates. Locations, widths and lengths of cracks should preferably also be noted on a sketch.

Crack widths can be measured by 'crack comparator' to an accuracy of 0.025 mm. Any change in crack-width and extent (crack movement) should be monitored regularly. Mechanical movement-indicator (or crack-monitor), which gives direct reading of crack-displacement and rotation, can be used to monitor cracks. Even simple plastic-templates, on which lines of different widths are printed in black, can also be used very conveniently to measure crack-widths.

4.10.2 Non Destructive Tests

These tests can give some idea about the pressure of internal cracks, voids, and depth of penetration of cracks visible at the surface. Tapping the surface with a light hammer is a simple technique to identify an internal crack near the surface. A hollow sound indicates a separation (delamination) or a crack below the surface.

The presence of reinforcement, its size, spacing, and cover can be approximately determined by using a pachometer (covermeter). In areas of congested reinforcement, it may be necessary to remove the concrete-cover to identify the bar size. Corrosion can be detected by electrical potential measurements using a suitable reference 'half-cell'.

These equipments should be operated by trained technicians, and the results evaluated cautiously by an

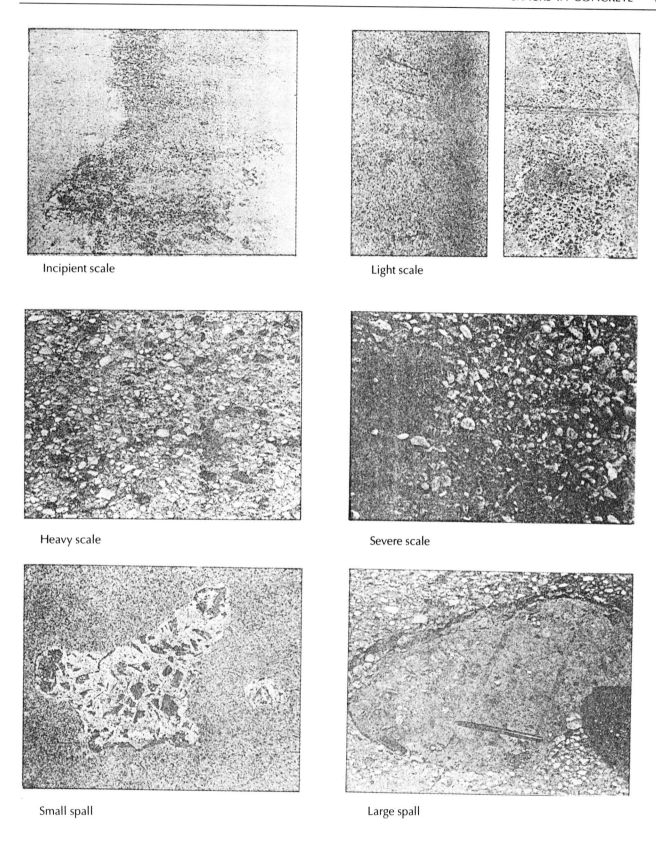

Incipient scale

Light scale

Heavy scale

Severe scale

Small spall

Large spall

Fig. 4.62

Small and large spalls covered with patches

Exposed reinforcing bars

Exposed re-bars and efflorescence

Exposed bars and spalls

Corings of delaminated deck

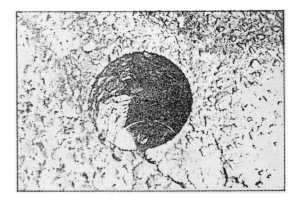

Delamination – subsurface fracture

Fig. 4.63

A large hole in deck slab

Steel plate covered hole in the deck slab (underneath)

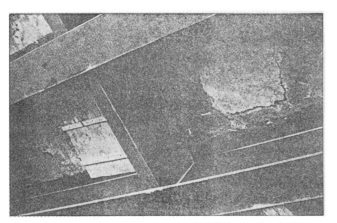

Concrete patched holes in the deck slab

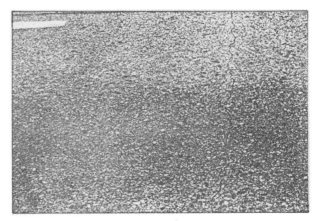

Transverse crack

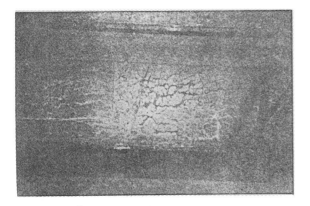

Map cracks

Map cracks, pitting and popouts

Fig. 4.64

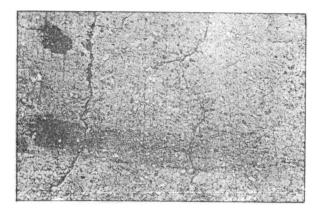

Random cracks

Broken steel grid grating

Exudation and stalactites

Joint spalls and 'D' cracks

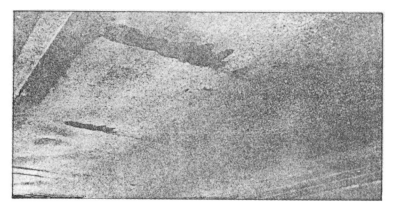

Transverse cracks and warping of deck slab

Fig. 4.65

Water penetrating concrete deck

Scaling deck

Fig. 4.66

Spalling deck

Severe deterioration at the bottom of the deck

Severe deterioration at the top of the deck

Water saturated concrete

Ponding of water on the deck caused by dirt in curb outlets

Ponding of water on the deck caused by snow

Ponding of water on the deck because inadequate deck drainage

Fig. 4.67

Deck cracks

Deck spalls

'Sealed' cracks

Fig. 4.68

experienced engineer. Radiography can also be used to detect internal discontinuities. Both X-ray and gamma-ray equipment are available.

4.10.3 Concrete Core Testing

Taking out 'cores' and crushing them like test-cylinders can indicate the quality of concrete as in compressive strength tests. Cores that contain cracks or pieces of steel, should not be used to determine concrete strength. Cores and core holes provide opportunity to see the extent of the width and depth of cracks. Core material and crack surfaces can be examined petrographically to determine the presence of alkali-silica reaction products or other deleterious substances. They can also be tested for the presence of excessive chlorides to detect possible corrosion of embedded reinforcement.

NOTE: *See* Chapter 15 for details of various diagnostic tests.

4.11 REPAIR OF CRACKS IN CONCRETE

4.11.1 Introduction

Before attempting to repair any cracks a full investigation should be made of the 'cause' of the cracks. Each case should be considered from first principles. Analysis of the cause may lead to the conclusion that the cracks taper or that there is a possibility of major faults within the structure which might need extensive repairs!

Before initiating remedial measures, consideration should be given to their necessity. Cracks in a floor slab which are not structurally significant and where the floor is to be finished, perhaps by a wearing course, can well be left alone.

All repairs will spoil the appearance of the concrete and additional cost may be incurred in applying finishes to reduce the effect of this deterioration! Repairs to 'live' cracks cannot be easily covered up. Autogenous healing, automatic closure of a crack either by subsequent hydration of cement of by deposition of calcium carbonate formed by atmospheric CO_2 reacting with calcium hydroxide released from hydration of cement, will sometimes occur when cracks are fine. It will generally take place in the first few days or weeks of the life of a structure. Such healing can be promoted by keeping the structure moist. (However, this needs extreme caution since moisture and calcium carbonate reaction can lead to reduction in pH value and consequent, corrosion of steel. It can also cause expansive cracks due to 'crystal' formation and carbonation shrinkage.) If repair is to be undertaken, full consideration should be given to consulting specialist companies in this field. After repairs have been completed, it may be prudent to ensure that

the remedial measures have been effective, either by a test or by taking confirmatory cores.

4.11.2 Classification of Cracks

Cracks may be separated into two classes for the purpose of deciding upon the type of repair:
(a) *Dormant* cracks which are unlikely to open, close or extend further. These are also called 'dead' cracks.
(b) *Live* cracks which may be subject to further movement (i.e. opening up).

4.11.3 Materials for Repair

4.11.3.1 Rigid Fillers (for Dormant Cracks)
To close against surface water penetration it may be sufficient to brush and push in a cement-grout.

The largest particle-size in the grout should be one-third or one-fourth of the crack-width. The grout can be modified adding a synthetic latex which will improve adhesion to the surface of the crack and reduce shrinkage of the grout. (Remember the grand mother's prescription of adding milk to the grout for filling the cracks!) During the last two decades epoxy resins have become the most commonly used materials for injection and sealing of dormant cracks and it is claimed that some formulations can penetrate cracks as fine as 0.01 mm! Epoxies are preferred materials because:

(a) Formulations are available which can harden even in wet conditions and will adhere even to moist concrete!
(b) Excellent adhesion to fresh and hardened concrete can be obtained.
(c) They have low curing-shrinkages.
(d) Some of them have good mechanical strength even in the presence of water and are resistant to a wide range of chemicals, including alkalis and aggressive ground waters, even at reasonably low and high atmospheric temperatures!

Polyester-resins and certain types of synthetic-latex have also been used for injection repairs. Polyester-resins cost less than epoxies, and have lower viscosities and thus achieve better penetration. However, hardening of polyester is adversely affected in wet conditions; they have higher curing shrinkage and are less resistant to alkali attack!

Synthetic-latex (such as styrene-butadiene acrylic, polyvinyl acetate, or a co-polymer of these) is cheaper than even polyester, but has considerably less strength than epoxy or polyester resins and may suffer loss of adhesion in conditions of constant immersion, e.g. in underground-structures or water-retaining structures.

4.11.3.2 Flexible Sealants (for Live Cracks)

A wide range of sealants for 'moving joints' and 'live cracks' is available. The most commonly used are bitumen-based compounds, polysulphides and polyurethanes. Rubber-modified bitumens and tar-modified polyurethanes are included in these groups. There is a much wider choice of coatings and membrane materials which can be used for the surface-sealing of fine live cracks.

4.11.4 Repair of 'Dormant' Cracks

4.11.4.1 Fine Dormant Cracks

Fine dormant cracks will most commonly be repaired by injecting with an epoxy resin formation. If the crack is structurally significant, it should always be repaired in this manner. Cracks which extend through a structure, are treated by fixing injection points (nipples) at intervals along the crack at the surface. The distance between the injection points is generally about one-and-a-half times the depth of penetration required. Between these points the 'surface' of the crack is effectively sealed (on both sides of the structure if the crack goes right through) by a thixotropic compound (epoxy and polyester resins and hot-melt thermoplastics have all been used for this surface sealing). Care must be taken when fixing the injection points to ensure that they intercept the crack and are not blocked by detritus.

A wide variety of devices is used to inject resin formations into cracks. These include modified grease-guns or sealant-guns, pressure-pots of the type used for spray-painting and special purpose twin-metering-pumps. The guns and enclosed pressure pots have the limitation that the resin will begin to 'set' soon after mixing and in relatively short time will harden so that it cannot be injected and may block and damage the equipment.

Twin-metering-pump delivers resin and hardener, in correct proportions, through separate delivery lines to a point close to the injection nipple. The materials are thoroughly mixed continuously in a mixing head, and pumped directly into the crack. Only a small amount of resin is retained in the mixing head and, when injection stops, this is flushed out with solvent from a tank which is a part of the equipment.

Epoxy resins start gaining in viscosity as soon as the resin and hardener are combined. When the components are first mixed and applied with a gun or pressure-pot system, the rate of injection will steadily fall as the viscosity increases. Alternately, the pressure must be increased to maintain the rate. A limitation of the process is that, while it is advisable to seal the reverse face of the cracked structure to prevent the resin flowing out, this may not always be possible. In some cases, e.g., a ground slab or a retaining wall, this is impractical.

It is important, when pressure injecting, to use the optimum pressure necessary to make the resin flow into the crack. Excessive pressure is likely to force the resin along the path of least resistance, leaving voids. Furthermore, where the surface area of the crack is relatively large, the injection pressure must be carefully monitored to avoid building-up large transverse forces that split open the crack!

Most epoxy resins which are currently marketed for injection, displace water from inside a crack and adhere well to damp surfaces, but whether they need a dry surface or can be used in damp or wet conditions should be ensured first, and the appropriate procedure followed.

4.11.4.2 Wide Dormant Cracks

Repair of wide dormant crack on a vertical surface will often be most practical by the injection method.

Cracks in horizontal surfaces can be repaired by injection but simpler methods can also be used. The crack can be treated by sealing the underside where this is accessible, opening out a slight 'V' on the upper surface and pouring the repair-material into the crack, either starting at the middle and working towards the ends or starting at one end. This technique will ensure that air is fully displaced from the crack. To avoid filling the crack with detritus, a small 'dam' can be formed on each side of the crack instead of cutting a 'V'. The materials used can be epoxy resins or cement grouts, the latter with or without a synthetic latex added to the mix. Cracks with a surface-width of up to about 1 mm will be easier to fill with an epoxy material because of the lower viscosity. Cement grouts will take about two to three weeks to develop compressive strength approaching that of the concrete structure while epoxies will cure in about two to seven days.

4.11.4.3 Dormant Fractures (Very Wide Dormant Cracks)

Economic factors will usually determine the choice of materials used to repair a dormant fracture. If it is relatively shallow, an epoxy resin mortar or grout may be the suitable material; the only preparation necessary will be to clean out the fracture. When the volume of filler material involved is significant, it may be cheaper to suitably open out the feature and use a fine concrete mix! In order to minimize shrinkage, the fine concrete should contain as little water as possible, consistent with workability, and precautions should be taken to prevent premature drying out and to ensure proper curing. Use of aluminum compound expansion agent can be considered gainfully.

4.11.4.4 Multiple Cracks (Dormant Type)

Where there is multiple dormant cracking, particularly where it is of a random character, the repair may prove to be slow and costly, a relatively recently patented repair-technique can be used to fill large number of cracks simultaneously. The method is the reverse of resin injection. Instead of displacing the air with resin under pressure, the air is first withdrawn with vacuum pumps and then a resin, or other material, is introduced to fill the voids and cracks! This is epoxy filling by vacuum process.

In order to evacuate the faults in the structure, it is necessary to be able to seal-off the area to be treated. This imposes limitations on the system, but it has obvious advantages where there is multiple random cracking and where the structure can be effectively sealed off.

4.11.4.5 A Summary of Repair Techniques for Dormant (i.e. dead) Cracks

By Epoxy Injection

Cracks as narrow as 0.05 mm can be bonded across by epoxy injection. This process has been successfully used in repair of cracks in buildings, bridges, dams and other types of structures.

This method can be applied to repair most of the structural and non-structural cracks with the limitation that the cracks are required to be dormant or causes of cracking shall be removed before application of this method. This method is not applicable for the cracks that are active or leaking and cannot be dried out. This method requires high degree of skill for satisfactory execution.

By Grouting and Sealing

This method is generally used on cracks that are wider than about 1 mm and of course dormant. This method involves enlarging the crack along its exposed face (recessing), and then cleaning, filling (grouting) and then sealing it with a suitable joint sealant if required. It is not suitable for active cracks and cracks subjected to hydrostatic pressure. This is the simplest of the repair methods and can be executed with relatively untrained labour. *See* Figs 4.69 – 4.71, Plates 9 and 10.

By Stitching

This method may be used when tensile strength must be re-established across major cracks. Stitching a crack tends to stiffen the structure, causing the crack to migrate elsewhere in the structure. Therefore, it may be necessary to investigate and strengthen the adjacent sections if necessary.

The procedure involves drilling holes on both sides of the crack, cleaning the holes, and anchoring the legs of the stitching bar in the holes, with either a non-shrink grout or epoxy resin based bonding system. The stitching bars will be varied in length and orientation and they should be staggered so that the tension transmitted across the crack is not evenly applied to a single plane within the section, but is spread over the area.

By Using Untensioned Reinforcement

This technique consists of suitably drilling holes normal to the crack plane, cleaning the crack and the holes, surface-sealing the crack, filling the hole and crack plane with epoxy (pumped under low pressure) and quickly placing reinforcing bars in the drilled holes. Typically 13 or 16 mm bars are used, extending at least 0.5 m on each side of the crack. The epoxy bonds the bars to the walls of the hole, fills the crack plane, bonds the cracked concrete surfaces back together in one monolithic form, and thus restores the section.

By Using Prestressing Steel

This method is often desirable when a major portion of a member must be strengthened or when the cracks that have formed must be 'closed'. This technique uses prestressing strands or bars to apply a predetermined compressive force. The effects of this additional compressive force on the structure should be carefully analyzed. Additional permanent prestressing can also be obtained by tensioning added external tendons placed normal to the cracks. This method shall be preceded by appropriately filling the cracks where possible so that some of the compression being applied is not lost in closing them; however, in post-tensioning this does not matter.

By Drilling and Plugging

Drilling and plugging a crack consists of drilling down the length of the crack and grouting it to form a key.

This technique is only applicable when the cracks run in reasonable straight lines and are accessible at one end. This method is most often used to repair vertical cracks in retaining walls.

By Grouting

Portland Cement Grouting: Wide cracks, particularly in gravity structures and thick concrete walls, may be repaired by filling with portland cement grout. The procedure consists of cleaning the concrete along the crack; installing built-up seats at intervals astride the crack; sealing the crack surface between the seats with a cement paint, sealant, or grout; flushing the crack to clean it and then grouting the crack. For small volumes, a manual injection-gun may be used; for larger volumes, a pump should be used. After the crack is filled, the pressure should be maintained for several minutes to ensure good penetration.

Chemical Grouting: Chemical grouts consist of solutions of two or more chemicals that combine to form a gel, a

PLATE 9

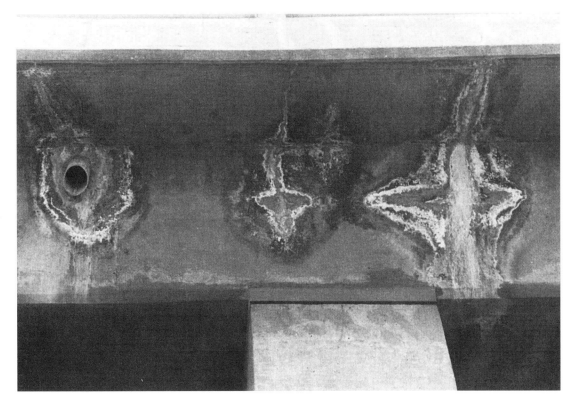

Fig. 4.61 Leakage of drainage water through the body of deck and around the down-spout. The water is contaminated with salts that have been wind-blown to the deck-surface; As this salt-laden water drains out the soluble salts precipitate, causing ugly staining apart from the sulphate and chloride attacks in concrete and steel touched by this leaked water

Fig. 4.69 Epoxy sealing of cracks in deck slab and girders in a reinforced concrete deck (cracks had been caused by temporary overload)

PLATE 10

Fig. 4.70 Another view of the sealed cracks referred to in Fig. 4.69

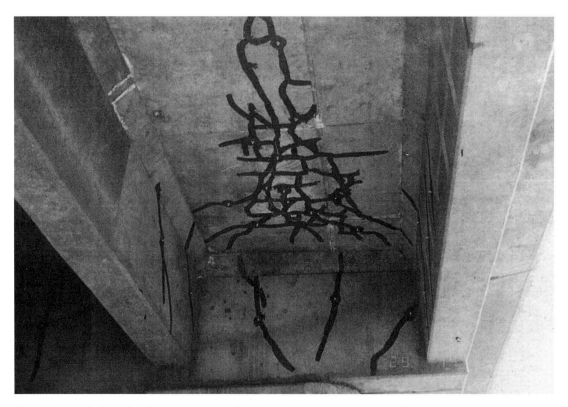

Fig. 4.71 Sealed overload cracks (structural) in R.C. slab, main beams and diaphragms

solid precipitate, or a foam, as opposed to cement grouts that consist of suspensions of solid particles in a fluid.

The advantages of chemical grouts include their applicability in moist environments, wide limits of control of gel-time, and their application in very fine fractures. Disadvantages are the high degree of skill needed for satisfactory use, their lack of strength, and the requirement that the grout does not shrink off in service.

By Drypacking

Drypacking is the hand placement of a low water-content-mortar followed by tamping or ramming of the mortar into place, producing intimate contact between the mortar and the existing concrete. Because of the low water/cement ratio of the material, there is little shrinkage, and the patch remains tight and is of good quality with respect to durability, strength and watertightness.

'Drypack' can be used for filling narrow slots (recesses) cut for the repair of dormant-cracks. The use of drypack is not advisable for filling or repairing active-cracks.

By Polymer Impregnation

'Monomer' systems may be used for effective repair of cracks. A monomer system is a liquid that consists of small organic molecules capable of combining to form a solid plastic. Monomers have varying degrees of volatility, toxicity, and flammability and do not mix with water. They are very fluid and will soak into dry concrete, filling the cracks, much the same way as water does. Monomer systems used for impregnation contain 'catalyst' or 'initiator' and the basic 'monomer'. When heated, the monomers join together, or 'polymerize', becoming a tough, strong, durable plastic that greatly enhances a number of concrete properties.

Polymer impregnation has not been used successfully to repair fine cracks, however.

By Overlays and Surface Treatments

Cracks in both structural and pavement slabs may be repaired using bonded overlays if the slabs are not subject to movement. Slabs with numerous fine cracks caused by plastic drying shrinkage or other one-time occurrences, can be effectively repaired by the use of overlays, unless the cracks are so extensive and deep that monolithicity cannot be achieved by repair.

Bridge slabs may be effectively 'coated' using a heavy coat of epoxy resin.

Slabs and decks containing fine dormant cracks can be repaired by applying an overlay of polymer-modified portland cement, concrete or mortar. In highway bridges, a minimum overlay thickness of 38 mm is recommended for successful application.

By Autogenous Healing

The inherent ability of concrete to heal cracks within itself is termed 'Autogenous Healing'. It is suitable for sealing dormant cracks in a moist environment as found in mass concrete structures.

The mechanism of healing occurs through the carbonation of calcium hydroxide in the cement paste by carbon dioxide in the surrounding air and water. The resulting calcium carbonate can fill the cracks. But such a chemical reaction results in reduction of pH in the moisture in concrete, which increases acidity, and thus can trigger rusting of steel.

4.11.5 Repair of Live Cracks

Live cracks should be treated as if they are 'moving joints'! Repair should cater for their anticipated 'potential movement'. A suitably dimensioned recess should be cut along the line of the crack and then sealed with an appropriate sealant. A surface seal made with a strip of pre-formed sheet material may be appropriate in certain circumstances.

The choice of a sealant will largely be determined by the amount of movement forecast, and the limitations imposed by the size of the recess which can be cut, together with the situation, i.e. whether vertical or horizontal crack. There are three types of sealant generally used:

(a) Mastics
(b) Thermoplastics
(c) Elastomers

(a) Mastics: These are generally viscous liquids, such as non-drying oils, or low melting-point asphalts, with added fillers or fibres. They are usually recommended where the total movement will not exceed 15 per cent of the width of the groove. The groove should be cut so that it has a depth to width ratio of 2:1. Mastics remain plastic and will not withstand heavy traffic or solvents. In hot weather the mastic will tend to be forced out by the expansion of the adjacent structures and the surplus will be flattened or knocked off by any traffic. Dirt and debris can become embedded in the material.

Mastics are the cheapest of the sealants but their use should be restricted to vertical situations or those which are protected from traffic.

(b) Thermoplastics: Thermoplastics become liquid or semi-viscous when heated. The pouring temperatures are usually above 100°C. They include asphalts, rubber-modified asphalts, pitches and coal tar. The groove depth to width ratio is of the order of 1:1 and the total design movement is of the order of 25 per cent of the groove width. Although these materials soften less than mastics, they may extrude at high ambient temperatures and debris may become embedded. Some of these

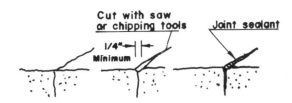

(i) Original   (ii) Routing   (iii) Sealing
     Crack

Figure (a) Repair of crack by routing and sealing is a method suitable for cracks that are dormant and not structurally significant. Routing and cleaning before installing the sealant add significantly to life of the repair.

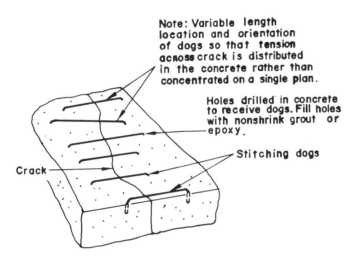

Note: Variable length location and orientation of dogs so that tension across crack is distributed in the concrete rather than concentrated on a single plan.

Holes drilled in concrete to receive dogs. Fill holes with nonshrink grout or epoxy.

Stitching dogs

Crack

Figure (b) Crack repair by stitching restores tensile strength across major cracks. Where there is a water problem, the crack should be made watertight first to protect the stitching dogs from corrosion.

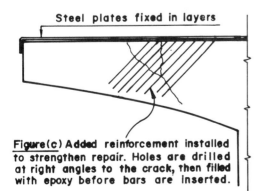

Steel plates fixed in layers

Figure (c) Added reinforcement installed to strengthen repair. Holes are drilled at right angles to the crack, then filled with epoxy before bars are inserted.

Fig. 4.72

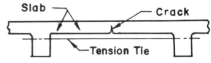

Slab   Crack

Tension Tie

(i) To Correct Cracking of Slab

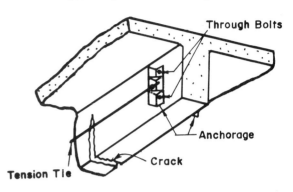

Through Bolts

Anchorage

Crack

Tension Tie

(ii) To Correct Cracking of Beam

Figure (d) External prestressing can close cracks and restore structural strength Careful analysis of the effects of the tensioning force must be made, or the crack may migrate to another position.

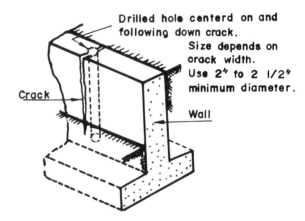

Drilled hole centerd on and following down crack.

Size depends on crack width. Use 2" to 2 1/2" minimum diameter.

Crack

Wall

Figure (e) Drilling and plugging is a repair method well suited to vertical cracks in retaining walls. The repair material becomes a structural key to resist loads and prevent leakage through the crack.

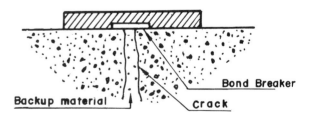

Backup material

Bond Breaker

Crack

Figure (f) Flexible surface sealant can be used over narrow cracks subject to movement, if appearance is not a consideration. Note bond breaker over the crack itself.

materials are degraded by ultra-violet light and may thus become hardened and lose elasticity after a few years exposure to direct sunlight.

(c) Elastomers: These cover a wide range of materials which include polysulphides, epoxy polysulphides, polyurethane, silicones and acrylics. Some are 'one part' and others are 'two part' materials. They can have considerable advantages over other types of sealants in that they do not have to be heated before application, they have excellent adhesion to concrete and are not susceptible to softening within the normal range of ambient temperature. Elastomers have a much higher degree of elongation than other sealants and many of them are capable of over 100 per cent extension but in practice this should be limited to 50 per cent (i.e., $\pm 25\%$). The groove depth to width ratio should be 1:2. The material should be prevented from adhering to the bottom that the crack remains free as a live crack.

A 'bond-breaker' should be provided to allow the sealant to change shape with a concentration of stress. The bond breaker may be a polyethylene strip, pressure-sensitive tape, or other material which will not bond to the sealant before or during cure.

Surface Sealing: Where movement is not all in one plane, where there is excessive movement beyond that accommodated by a recess of convenient size, or where there are factors which prohibit a recess being cut into a structure, 'surface-sealing' can be used. A strip of flexible 'preformed sheet' is bonded over the crack. Only the rims or edges are bonded to the concrete, the crack being covered with a strip of masking tape or other debonding agent. A typical surface sealer to a crack would be 100 mm wide with its central 20 to 25 mm width as unbounded part. Examples of this system include chlorosulphonated polyethylene, e.g., Hypalon, bonded with epoxy resin adhesive and butyl rubber bonded with a tar polyurethane.

4.12 CRACK ARREST

During construction of massive concrete structures, cracks due to surface cooling or other causes may develop and propagate into new concrete as construction progresses. Propagation of such cracks may be arrested by blocking the crack and spreading the tensile stress over a larger area.

A piece of bond-breaking membrane or a grid of steel-mat may be placed over the crack as concreting continues. A semicircular pipe placed over the crack may also be used.

4.13 DETERIORATION OF CONCRETE IN ARID AND SALINE ENVIRONMENT

The cracking of concrete and corrosion of steel in arid and saline environment can be more pronounced and hence of special importance. Reasons are the same as already explained but the chemical attack effects (i.e. rust distress due to direct chloride attack, carbonation of concrete and exposure of steel to moist atmosphere, distress due to sulphate attack and distress due to alkali-aggregate reaction) can be major problems.

Several preventives (e.g. epoxy-coating the reinforcement bars, perpetual cathodic-protection, etc.) and remedies have been tried. However, the best bet remains the one in which the concrete mix is very dense and construction supervision is accurate. The latter particularly in respect of ambient temperature and that of the concrete, and that during its pouring and curing, the wind speed and ambient humidity, the use of appropriate type of cement and admixtures, increased concrete cover, increased minimum cement content, reduced water cement ratio, appropriate compaction and early adequate curing.

The scope of all these suggestions centers around the effort to decrease concrete permeability. Also useful can be the impregnation of concrete with various polymers, such as methyl methacrylate. Also effective in combating the ingress of chlorides in to concrete is the use of latex-modified (water-styrene butadiene emulsion) concrete.

Although the coating of steel with organic or metallic coatings was heralded as a major breakthrough in combating the corrosion of steel in arid and saline environments, it was recently discovered that accidental breaks of this cover (during manufacture or construction), which cannot be detected, cause a concentration of the electrochemical phenomenon of corrosion at the breaks, leading to extreme localized corrosion of reinforcement at these locations.

A waterproofing membrane or a rubber-like paint has proved to be a viable protection against the penetration of chemicals in the concrete. To protect the concrete fully, these membranes or paints should be:
— elastomeric
— diffusion-resistant to carbon and sulphur dioxides
— waterproof
— capable of allowing the concrete to 'breath' (e.g. flex, etc.)
— resistant to degradation by ultraviolet light

Proprietary admixtures advertised to be corrosion-inhibitors have produced results in the laboratory environment, but have not proven yet in actual field conditions. The use of these admixtures should be done with discretion.

4.14 CAUTION

Repair of cracks in concrete is still only a 'repair', not necessarily strengthening!

If the cracks have caused sufficient damage to the member so that strengthening of the structure is required, full structural- investigative-analysis and stress-checks are essential in order to design suitable strengthening scheme for the structure.

4.15 CRACKING OF CONCRETE—*Some Reflections*

- Cracking and crack width-prediction-and-control have been much in the limelight in recent years. The problem has drawn growing attention of researchers, code-makers, designers and constructors, to such an extent that it has become one of the most discussed topics (in the last decade) among people dealing with concrete.

- All this talk about crack width-computation may lead one to believe that the panacea of all ills lies in accurate prediction of flexural-crack-width or shear-crack-width! Nothing could be farther from truth! Computing crack-width is an important design task but a designer with over-reliance on crack-width-computation for crack control may soon find it disappointing.

- Cracking of concrete occurs due to tensile stresses. But the occurrence of such tensile stresses may be traced sometimes to sources other than flexural tension! Had flexural stresses been the one and only reason for tensile stresses in concrete, full prestressing would have solved the problem. There are other kinds of tensile stresses! It is therefore to be expected that the concrete structure remains unimpressed by the most sophisticated prediction of flexural-crack-width, and may actually crack due to these other tensile stresses.

- One of the commonest causes of cracking in structures is the lack of reinforcing and detailing at locations where significant tensile stresses occur under predictable or unpredictable combinations of effects. Another kind comprises avoidable deficiencies in construction quality of concrete, resulting in so-called plastic cracking.

- In concrete, 'seeds' of the most cracking distress are 'sown' before hardening and application of loading! Deficiencies, and consequently damages that have

been done to concrete at its birth or at a very young age, cannot be undone after it matures and is put into service. If visible cracking has not appeared already in the concrete when poorly made, it remains very vulnerable due to 'unhealed scars'. Reinforcement does not help and computations go haywire.

- To guard against such cracking, one must understand one's concrete better. It is a multiphased, inherently heterogeneous, complex material. In engineering use, we interpret the behaviour of concrete at a simplistic level, which can be deemed as only little better than prescientific! The status of knowledge on microstructure of concrete and associated theories of crack-propagation and fracture of concrete have grown fast in recent years. It is now clear that micro-cracks, lying hidden from view, hold the key to the cracking behaviour of concrete like all other properties pertaining to its strength and deformation.

- Microcracks are minute flaws or discontinuities inherent in concrete due to its heterogeneous nature. They appear in concrete, not yet loaded, or while lying in the forms waiting to gain its mechanical strength. The cracks are too narrow to falsify either the use of elastic theory totally, or the assumption by the structural designer that the material is homogeneous and isotropic. Some microcracks may also get healed partly or completely with the crystals precipitating during the continued hydration of cement. But this autogenous healing does not solve all the problems. They may in fact create, through their geometry, zones of potential fracture. So it is but expected that any excessive microcracking, that occurs before hardening of concrete, may prove to be a major source of cracking-distress in the service of the structure.

- Restraint to free deformation while the concrete hardens, plastic shrinkage, plastic settlement, etc., are primary causes of random non-load-induced cracking. Cracking can occur in otherwise sound concrete, when still fresh, due to the tendency of concrete to contract, caused by loss of moisture or change in temperature. Other reasons are differential settlement of aggregates and paste over obstructions such as larger pieces of aggregates or steel (as the water bleeds upwards), leading to breaking the back of the plastic mass over the unyielding obstructions, etc. Segregation and faulty compaction can result in voids and honeycombs, in addition.

- Even when deficient quality of construction induces excessive microcracking (which soon develops into visible cracks), this may lie undetected for long, despite its early appearance. These cracks are usually obscured at an early age of concrete by water bleeding to the surface of concrete. All looks well until the curing is over or concrete is put into service. Unfavourable ambient conditions soon magnify the damages for all to see and ponder over. But postmortem analysis often ignores the original cause!

- The tensile strength of concrete develops very slowly and the concrete obviously remains vulnerable to cracking at young age. Deficiencies in the composition of concrete, it's placing and curing, etc., and more significantly the adverse environmental conditions (wind, heat and cold), are agents which magnify the distress. Even if visible cracking is not very evident, due to severe microcracking, the tensile strength of the hardened concrete may be much smaller than anticipated, and severe cracking distress under subsequent load-induced tensile stresses can occur. The cross-interaction between the already existing microcracks (randomly disposed) and the intended-cracks owing to structural loads applied, can subsequently develop a confusing crack-pattern, making it more difficult to pinpoint the cause of cracks. Concrete tries to fold at the 'existing' cracks before allowing development of new (discreet) cracks, and thus exhibits a crack-pattern that may not altogether allow the showing up of distinct structural cracks that were caused subsequently by the load-action.

- As mentioned earlier, cracking in concrete yet to harden, can be attributed basically to (i) temperature change (ii) plastic settlement and shrinkage. The mechanism of cracking due to temperature change is simple to appreciate. Cement liberates heat during hydration. While temperature rises, the plastic concrete hardens as it cools. A temperature differential can thus be set up between the outer and inner portions of the concrete-mass during this temperature change. This can be aggravated by environmental conditions. Restraint to free deformation results in tensile stress (not necessarily flexural).

- Similarly, drying shrinkage generates tensile stresses due to restraint to deformation. Drying is largest at the exposed surface which tends to shrink more than the interior, thus creating a differential-deformation relative to the interior layers.

- Conscious efforts to reduce these restraint-stresses and the consequent risks of cracking in plastic concrete may involve some elementary care in making good concrete. For reducing the cracking caused by temperature change, the temperature in the interior of concrete must be kept down and there are many ways of doing it. The most obvious one is to strike at the grass roots! Cement is the source of heat of hydration. So the lesser the cement, the lesser the temperature change. Of course, cement is also the primary source of strength. So an optimum quantity, just enough to gain the desired strength and durability, should be used. Not-optimal use of cement may make the concrete costlier and even stronger than desired, but never better! Merely richer mix for the same target strength can, in the limit, be poorer in quality. It has been reported from a survey of a number of mass concrete structures in U.S. that the only concretes relatively free from cracking were the 'leanest'. Perhaps temperature differences alone cannot explain the observed better cracking resistance of leaner mixes. Lower modulus of elasticity and drying shrinkage characteristics may also contribute equally.

- The above elementary rule of making better concrete with less cement may be too well known to be restated. Yet it remains a fact that misconceived reliance on liberal addition to cement content, to compensate for lack of controls on construction quality, continues. The accent should be on durability and bond, more than on extra strength alone. This calls for coarser gradation, cleaner dust-free aggregates of good quality, tight water-cement ratio, thorough compaction, and casting in low wind low heat low-cold weather with adequate curing. It is worse when people add extra cement hoping it will 'take care' of strength, when the worker quietly adds more water arbitrarily to permit him an easier workability locally. This, in fact, leads later to all sorts on non-load-induced random cracks that subsequently accentuate, which ultimately destroys the integrity of the structure.

  A designer doing voluminous computations of crack-width should perhaps first ensure that the mix of the concrete going into the structure is properly designed and followed, because excess of cement alone can lead to the loss of quality in respect of cracking behaviour.

- Caution against using too finely ground cement for not so high grades of concrete is also called for. Modern high strength cements are ground much finer than earlier. They liberate heat of hydration much faster, and increase the risk of setting up a thermal gradient in thick member. So, where cracking behaviour is a critical consideration, use of a very finely ground high-strength cement may not be favoured.

For preventing excessive drying shrinkage of concrete, it is important to protect the concrete-surface by continued curing. Members with large thickness, or those cast in adverse ambient condition, will require greater care. Premature exposure by removal of forms can also aggravate the risk. For the control of drying shrinkage, the effect of water content and mix proportions is necessary from the viewpoint of cracking resistance also.

- Plastic settlement cracking appears in concrete before hardening, as already pointed out. This type of cracking can be seen in poorly laid concrete on the upper surface of the slab following approximately the top reinforcing pattern. The obstruction to the downward movement of solid particles by the reinforcement grid is the obvious cause. The incompatibility of concrete mix, placing, and compacting techniques with the reinforcement arrangement, contribute to the distress. This can be controlled by slightly revibrating the concrete before its initial setting. Plastic shrinkage cracks (which start at the top and descend downwards, sometimes follow the obstruction-creating path of the reinforcement grid and sometimes appear normal to wind direction) can be controlled by robust timely trowelling. They are caused when the rate of surface evaporation is faster than the rate of internal bleeding (rising upward of water in fresh concrete). Trowelling aggravates plastic settlement cracks, while revibration aggravates the plastic shrinkage cracks. Bleeding reduces plastic shrinkage cracking, but increases plastic settlement cracking.

- The assumption of uniform stress distribution over the total 'effective' compression flange width only hides the considerable deviation from such stress distribution at the flange edges! But such inaccuracies do not spell any trouble, nor do they signify any gross error. Gross errors occur in detailing, when the designer fails to subjectively account for the possible significant impact of the inherent inaccuracies of his analysis and design. In spite of the sophistication in the tools for the mathematical analysis of a structure detailing of the structure cannot be done by computation alone! Unless design decisions are supplemented by a subjective engineering judgment of the actual structural behaviour, the finest efforts may end in unsightly cracks, if not disaster. This lesson has been learnt the hard way many times in the century. All the brilliant mathematics which went into the first Tacoma Narrows bridge design, did not save it from collapse due to aerodynamic instability! Small geometric inaccuracies ignored in analysis have brought about startling disasters of mighty steel bridges! Added to all of this, are the unavoidable differential temperature stresses whose prediction itself is inaccurate. A good robust workman-like detailing, combined with a good understanding of the structure-stress-flow, supplemented by some good common sense (which is uncommon), and first hand field experience, can be of immense assistance in carrying out practical and serviceable structural designs.

- In some of the recent concrete bridges, some unsightly cracking has appeared due to tensile stresses ignored in common design practice. The principal agent in these cases has been identified as temperature variation. Leonhardt has reported instances of severe unanticipated cracking of boxwebs due to non-uniform temperature distribution and consequent stresses. Priestley has talked about cracking of prestressed concrete box girders of a major urban motorway viaduct in New Zealand induced by thermal stress. All this cracking was found to be strongly correlated to the ambient temperature and solar radiation. Ignoring these stresses, as was often done in earlier design practice, cannot but be termed as a gross error. While these stresses may be precalculated, the cracking of concrete due to them cannot be avoided unless sections are thin. The best bet is to provide closely spaced reinforcement bars of small diameter along the skin in crack-critical zones. This will allow more cracks but of narrower width. Here again, quality control exercised in the making of concrete will greatly help in limiting the total cracking.

- A good deal of cracking problems in otherwise well-designed concrete structures would be avoided if correct construction practices were followed (e.g., washing aggregates; using rust-free reinforcement bars; proper batching of aggregates, cement and water; adequate mixing, transporting, placing compacting and curing of concrete; ensuring rigid forms and false-work; working at right temperature ranges, avoiding specified hot, cold and (above all)windy weather conditions, etc.) Also using aggregates that are dust and clay-free; not highly absorptive and dry; well graded; free from reactive elements that can chemically react with cement (create expansion, cracking and disintegration of concrete) or can attack reinforcement (chloride attack-leading to rusting, expansion, cracking and ultimately spalling of concrete), etc.; using adequate type of cement (e.g. ordinary type, sulphate resisting type, low alkali type, blast-furnace-slag type, etc., as appropriate to a given situation) combined with adequate plasticizer, retarder or accelerator and minimum water-cement ratio; using adequate minimum cement content (as appropriate for the prevailing conditions and impermeability requirements), and appropriate quality of water; is extremely important. Added to all this, if the constructor is aware of the precautions he should take against potential plastic shrinkage cracks (controling bleeding, lightly re-vibrating etc.), and early thermal cracks (avoiding thermal gradients across concrete thickness; cooling the aggregates, reinforcements, shutters, cement and the water, covering the structure, using low-heat and not-too-finely-ground-cement reducing body-constraints, etc.) and as to how to reduce constraints and restraints to concrete shrinkage, he will prevent more cracks than mere computations can control. He should also try not to space the main tension bars at more than about 15 cm in thin members.

5
BRIDGE INSPECTION

5.1 INTRODUCTION

The overall objective of the *bridge maintenance management system* is to identify the need for structural maintenance, rehabilitation and replacement, and, provide guidelines and methodologies to enable local engineers to reach rational, cost-effective decisions regarding maintenance and rehabilitation for bridges and other highway structures.

For decision-makers the decision of whether to rehabilitate or to replace a deficient structure, and its subsequent justification, has not always been easy. One principal cause for this has been the approach of piece-meal synthesis in decision-making that has been oriented toward emphasizing certain advantages of an alternative and underestimating its disadvantages.

A 'systems' approach will result in a coordinated step-by-step analysis, which, when applied to the maintenance of bridges and other structures, will integrate essential elements of reliable information, well-defined criteria, clearly perceived constraints, and uniform evaluation of all the available alternatives. Further, it will allow for and encourage the use of experience, judgement, and analysis of the impact of certainty and possible future decisions, ensuring the optimal or near optimal use of public funds.

Maintaining highway bridges and keeping them in fit condition to provide safe and uninterrupted traffic flow, is the primary function of a *bridge maintenance engineer*. Protection of the investment in the structure-facility through well programmed repairs and preventive maintenance, is second only to the safety of traffic itself. To achieve the desired result requires constant alertness and thorough inspection procedures.

5.2 QUALIFICATIONS OF INSPECTION PERSONNEL

The individual in charge of the organization unit that has been delegated the responsibility for bridge inspection, reporting, and inventory, should possess the following minimum qualifications, and should carry out the responsibilities listed here:

- He should be a qualified 'professional' engineer having a minimum of 10 years experience in bridge construction and inspection assignments in a responsible capacity.
- He shall be responsible for the thoroughness of the field-inspection, analysis of all findings ascertained by the inspection, and the subsequent recommendations for correction of defects, posting for restricted load and/or speed, or any other recommendations deemed necessary. The problems encountered in this work are numerous, variable, and often complex. Consequently, his judgment is frequently required for proper evaluation of the findings.
- He must be thoroughly familiar with the design and construction features of the bridge to properly interpret what is observed and reported. He or his supporting design office must be capable of determining the safe load carrying capability of the structure. He must be able to recognize any structural deficiency, assess its seriousness, and take appropriate action necessary to keep the bridge in a safe condition. He must also locate areas of the bridge where a problem is incipient so that preventive maintenance can be properly programmed.

Seldom will one individual have experience sufficient to qualify him as an expert in all the specialized fields of engineering which are a part of bridge science. Therefore, the engineer should be aware of any limitations imposed by his lack of experience in any area of the work. He should never hesitate to utilize the specialized knowledge and skills of associate engineers in such fields as structural design, construction, materials, maintenance, electrical equipment, machinery, hydrodynamics, soils, or emergency repairs.

5.3 CATEGORIES OF BRIDGE INSPECTION

All remedial and preventive maintenance or repair work, including replacement of components, should be planned in time, and economically, with minimum inconvenience to traffic. Original *completion reports* must be available for all bridges, and these should form the basis for detailed periodic bridge inspections. The data thus collected should be properly evaluated from time to time to assess the need for remedial measures required to be undertaken. Broadly, three categories of bridge inspection need to be conducted to collect the performance-data of bridges. These are discussed in this section.

Routine Inspection

These are broad general inspections, carried out quickly and frequently by highway maintenance engineers having reasonable practical knowledge of road structures, though not necessarily any specialized know-ledge in design details or special construction problems of any particular bridge or expertise in special problems of bridge inspection. The purpose of this routine inspection is to report fairly obvious deficiencies which could lead to accidents or future major repairs/maintenance problems. Such inspections should be carried out monthly.

Detailed Inspection

This type of inspection can be of two categories, viz., *general and major*, defined by the 'frequency' and 'intensity' of inspection, respectively. 'general inspection' could be made at yearly intervals, and it should cover all elements of the structures against a prepared checklist. It would be mainly a visual inspection supplemented by standard instrument aids. A written report must be made of the conditions of the bridge and of its various parts.

The 'major inspection' should be more intensive and would require detailed examination of all elements, even requiring setting up of special access-facilities where required. Such inspection, depending upon the importance of the structure, could be spaced between 2 and 3 years, and even smaller intervals for sensitive designs, or for bridges in aggressive environments.

Special Inspection

This could be undertaken to cover special circumstances such as occurrences of earthquakes, passage of high intensity loadings, unusual floods, etc. These inspections should be supplemented by testing as well as structural analysis, and hence the inspection team should have an experienced bridge design engineer available to them.

It is important that inspections are undertaken in those periods which offer the most critical evaluation of the performance of the structure. For example, items such as foundations, protective works, scour effects, flood levels, etc., should be inspected before, during and after the floods; bearing and joints should be inspected during temperature extremes; etc. The frequency of routine inspections could be determined by the importance of the structure, environmental conditions, and cost. The frequency indicated above may be considered as a guide. A comprehensive check-list of items related to the form, material, condition, and situation of the structure, should be drawn-up and followed by the inspecting team.

Besides being a qualified engineer, the inspection team leader must be familiar with design and construction 'features' or the bridge to be inspected, so that observations can be properly and accurately assessed for a meaningful report. His competency to recognize any structural distress or deficiency and assess its seriousness with complete recommendations for appropriate repairs, are important prerequisites for entrusting this assignment to him.

5.4 INSPECTION AND INVESTIGATIVE-STRUCTURAL-COMPUTATIONS

This comprises of Activities I and II as described below:

Activity I

This shall include the following:

(i) Detailed visual inspection of each element of the structure and its protection works from close range (not just looking for a mere overview using binoculars from distance).

Noting down any structural cracks more than 0.3 mm in width, their widths, and any signs of deterioration and distress. In a structure, this will range from the usual non-load-induced cracks (caused by drying shrinkage, early removal of shutters, plastic cracks, lack of curing, etc.) to serious structural cracks, and manifestation of distress zones. In the protection works and channel configuration, it will range from nothing of concern to serious undermining/scouring/dislodgement and choking of waterway and diversion in watercourse. All this has to be included in the report for the Activity I, giving necessary indicative sketches.

(ii) If the observed signs and manifestations of deteriorations and distress are such as can be adequately taken care of by routine type of repair and restoration work, then this report shall also give detailed methodology of the repair-work along with its specifications, quantities and cost estimate, together with workman-like sketches and notes for execution. In addition, the report shall list out the details of the causes that lead to the observed deterioration.

(iii) If the observed manifestations of distress in the structure are so serious as would require a detailed structural investigation (computations*, and possibly some tests‡) in order to enable to decide between 'repair' and 'part or complete demolition and replacement', then this report shall indicate so in detail (describing the likely causes of deterioration) and seek permission for taking up such work. Such work will then form Activity II, detailed ahead.

The present report, in the interim, shall propose various restorative measures to be taken to arrest furthering of the distress and maintaining the usability of the structure (for the time being) until the suggested investigative work is carried out and correct conclusions and recommendations drawn-up.

(iv) If the observed manifestations of distress in the structure are clearly such as would require its outright demolition (and replacement) then the report shall clearly say so, giving explicit supporting reasons and the details of the likely causes that obviously lead to such distress.

Activity II
Upon receiving the necessary permission to take up this activity with respect to a particular structure, the concerned shall then carry out all the necessary investigative structural computations (and tests, if needed), draw the necessary conclusions, detail out the relevant recommendations and submit all these in the report on *Activity II*.

For the work under this activity, use shall be made of the 'as-built' drawings and all other relevant information (as available). All necessary structural investigative calculations shall be carried out to investigate the structural stability and to estimate the material stresses at all critical sections under the operational conditions and possible load combinations. The current legal loading as well as the operating loading (if heavier) shall be used in the analysis. Appropriate conclusions shall then be drawn about the adequacy of the structure, complimenting the investigation with relevant tests if found necessary. Based on this, the report for *Activity II* shall clearly and unequivocally offer the most appropriate and technically sound recommendations for restoration of the structure, with full details of repairs, their specifications, quantities, and cost estimate.

However, if restoration is found either technically not feasible and/or economically not viable, then the report shall suggest the most appropriate method for demolition with possibility of re-use of any part/parts of the structure, describing the necessary procedure and precautions.

5.5   RATING THE CONDITION OF AN ELEMENT

In order to standardize the various condition-states of any bridge-element (so that uniformity of expression and understanding of the distress state prevails), it is necessary and convenient to rate each condition-state and give it a numerical designation. One of the convenient Numerical Rating Systems, ranging from '0' to '7' and 'N/A' and 'U', that may be followed, is as follows.

Rating	Guideline Definition
U	'Unknown' applies to components under inspection for which information is unavailable (such as footings below the ground line or the foundation piles under pilecaps). In certain cases these items can become exposed and be rated.
N/A	'Not applicable' applies to elements called for on the inspection forms (because these elements exist on other bridges) which do not exist on the bridge under inspection. For instance, some bridge piers and abutments have bearing pedestals and others support the bearings directly on the abutment seat or pier cap beam.
7	'New' or 'Like-New' Condition, no sign of distress or deterioration. No repairs necessary.
6	'Good condition' no repairs necessary.
5	'Functioning as Originally Designed'—insignificant deterioration or distress and does not reduce the capacity of the elements under inspection nor their ability to function. For example, a bridge expansion bearing which is corroded but has not lost any effective strength and still permits the required movements. Minor repairs can be made to alleviate distress or eliminate deterioration.
4	'Minimum Adequacy'—immediate rehabilitation of affected elements required to maintain design loading capacity.
3	'Not Functioning as Originally Designed'; serious deterioration (and/or distress), sufficient to reduce the element's structural capacity and/or its ability to function as designed. When this rating applies to primary elements, the bridge must have the maximum design loading reduced

---

* The structural *investigative computations* refer to all structural analysis needed to determine the governing bending moments, shears, torsions, reactions, etc., at critical sections of the existing structure, in order to establish the structural stability, material stresses, and other serviceability criteria (e.g. crack widths, deflections, vibrations, etc.) for ascertaining the acceptability of the structure. All investigative computations should be done for both substructure and superstructure, taking into account the current design live loads as well as other forces (e.g. flood, wind, etc.). It should be noted that these are only investigative computations, and are not the same as preparing a detailed design of a new structure.

‡e.g. concrete cores, **reinforce**ment tensile strength, various load tests (against bending, shear, etc.) of the slab, the longitudinals, etc.

accordingly. Immediate repairs must be made to return the structure to design capacity.

2   'Structurally Inadequate'—deterioration or distress so well advanced as to indicate the closing of the structure to all traffic pending immediate load rating analysis. This rating obviously applies to primary members only.

1   'Potentially Hazardous'—such a rating in primary members implies there is a danger of collapse under any further use of this structure and bridge should be closed to traffic immediately. When such rating applies to secondary elements, it can be the cause of vehicular or pedestrian accidents and should be corrected immediately.

0   'Dangerous' — bridge already closed, conditions beyond repair, imminent danger of collapse or already collapsed. Structure to be demolished.

NOTE: All items under inspection for which information cannot be obtained for any reason (denied access, etc.) will be marked U in the forms. Additionally, these items will be identified in a separate list including the reason for which such information was not obtained (unknown — but why?).

5.6 SAMPLE 'SPECIAL-INSPECTION-REPORT'

Note:
1.   Given below is a suggested 'format' of a SPECIAL INSPECTION REPORT. This is followed by a suggested LEGEND which 'numerises' a number of 'types of structures' and possible 'problems' in them. These numbers/abbreviations may be used in filling this Inspection Report, as appropriate.
2.   This is only a suggested 'Sample' FORM and does not include all the various Inspection Items detailed out in Section 5.9.1. The missing Items should be added appropriately to make this FORM complete.
3.   From the records and information about the Bridge in question, indicate the problems (if any) that were encountered or are likely to be encountered during its service-life since those may call for special efforts for Maintenance and Repair. The problems could be in Foundations, Abutments, Piers, Bearings, Expansion Joints, Superstructure, etc.

   Also, if possible, indicate those of the Construction-Stage problems, information about which may be important during Maintenance phase.
4.   Describe any other relevant information that may not have been covered otherwise.
5.   Purely for assistance some possible 'Structure Types' and possible 'Problems' have been indicated and numbered in the attached LEGEND. Please study these carefully and make use of these

numbers in this Inspection Report where details of problems in the Bridge can be appropriately expressed by these. Accordingly the appropriate Serial Numbers should be shown in the relevant places in this Report. This will save time and retain uniformity of expression and adequacy of detail. This will help in the overall evaluation task ultimately. Where the details are not covered by the descriptions given in the LEGEND, please enumerate.

(a)   Bridge Name . . . .
(b)   Stn . . . .
(c)   Location . . . .
(d)   Sketch the Longitudinal Spans and their Structural Arrangement; marking the Abutments as A_1 and A_2, and the Piers as P_1, P_2, P_3, etc. (Also indicate name of town or city on A_1-side and on A_2-side

   . . . .
   . . . .
   . . . .

(e)   Relevant sketch-details of Abutments and Piers and their foundations (give brief sketches)

   . . . .
   . . . .
   . . . .

(f)   Typical Cross Sections of Superstructure

   . . . .
   . . . .
   . . . .

(g)   Construction completed on . . . .
(h)   Construction contractor's name . . . .
(i)   Construction was supervised by . . . .
(j)   Design consultant was . . . .
(k)   Maintenance contractor's name . . . .
(l)   Date of this inspection . . . .

1. *Foundations*

	for A_1	for P_1	for P_2	for P_3	for P_4	for A_2
Type						
Problem						
Condition Rating						

Note:
(a)   Where a foundation for a support comprises of more than one element (e.g. piles and pile-cap), subdivide the appropriate column in to as many subcolumns, insert the element 'heading' and then indicate the structure-type, the Problem and the Condition-Rating as relevant for each.
(b)   Where helpful, convey the distress message by brief sketches with asterisks (and photographs).
(c)   Where the number of Piers is more than four, increase the number of Columns appropriately; in case of lesser number of Piers, strike off the unwanted columns.

2. Abutments and Piers

	A1					P1			P2			P3			P4			A2				
Type																						
Element	Dirt Wall	Wing Wall	Cap	Brg. Pedestals	Body	Cap	Brg. Pedestals	Body	Cap	Brg. Pedestals	Body	Cap	Brg. Pedestals	Body	Cap	Brg. Pedestals	Body	Dirt Wall	Wing Wall	Cap	Brg. Pedestals	Body
Problem																						
Condition Rating																						

NOTE: Same as Notes (b) and (c) mentioned in 'Foundations'.

3. Bearings

	on A1	on P1	on P2	on P3	on P4	on A2
Type						
Problem						
Condition Rating						

NOTE: Same as Notes (b) and (c) mentioned in 'Foundations'.

4. Expansion Joints

	above A1	above P1	above P2	above P3	above P4	above A2
Type						
Problem						
Condition Rating						

NOTE: Same as Notes (b) and (c) mentioned in 'Foundations'.

5. Superstructure

	Span A1 – P1							Span P1 – P2							Span P2 – P3							Span P3 – P4							Span P4 – A2						
	DECK							DECK							DECK							DECK							DECK						
	• Longitudinals	• Diaphragms	• Deck-slab	• Soffit-slab	• Footpaths	• Kerbs	• Parapets	• Longitudinals	• Diaphragms	• Deck-slab	• Soffit-slab	• Foot paths	• Kerbs	• Parapets	• Longiticlinals	• Diaphragms	• Deck-slab	• Soft-slab	• Footpaths	• Kerbs	• Parapets	• Longitudinals	• Diaphragms	• Deck-slab	• Soffit-slab	• Footpaths	• Kerbs	• Parapets	• Longitudinals	• Diaphragms	• Deck-slab	• Soffit-slab	• Footpaths	• Kerbs	• Parapets
Type																																			
Problem																																			
Condition Rating																																			

Condition of Approach Slabs and Recommended Action (if any)

NOTE: (i) 'Deck' refers to the longitudinal and transverse members of the superstructure, including deck-slab (and soffit slab if present).

(ii) 'Footpaths', if present, might be c.i.s. or p/c type, mark accordingly; if other type, then indicate appropriately.

(iii) 'Parapets' might be New Jersey Type (NJ), r.c. wall type (W), r.c. post and railing type (PR), metallic type (M), etc., mark accordingly.

(iv) Same as Notes (b) and (c) mentioned in 'Foundation'.

6. Drainage

Problem. . . . .

Location. . . . .

Condition Rating. . . . .

7. Corrosion in Structural-Steel Members (give member by member)

Member. . . . .

Problem. . . . .

Condition Rating . . . .

8. Any Special Remarks

. . . .

. . . .

9. Overall Condition Rating for the Bridge as a whole

. . . .

10. Name and Signature of Inspecting Engineer

. . . .

LEGEND . . . .

(Please refer to NOTE given at the beginning of the Sample Special Inspection Report)

1. Foundations
(use symbol *F*)

• *Types*

F.1 footings (open foundations)

F.2 piles

F.3 caissons

F.4 other (explain)

• *Some Problems may arise from*

F.a — undermining due to scour

F.b — dislodging of Gabions/protection works

F.c — exposure of piles

F.d — concrete cracking (from numerous causes*)

F.e — loosening of anchor-holds

F.f — settlement (differential, overall)

F.g — movement/tilting

F.h — overload/under-design**

F.i — Other (explain)

2. Abutments and Piers
(use symbols *A* and *P*)

• *Types*

A.1 — open-type (i.e. columnar) abutments

A.2 — closed-type (i.e. solid-type) abutments

A.3 — other (explain)

P.1 — solid wall-type pier

P.2 — frame (i.e. columnar)-type pier

P.3 — single shaft-type pier, solid (hammer headed)

P.4 — single shaft type pier, hollow (hammer headed)

P.5 — Other (explain)

• *Some problems may arise from*

AP.a — tendency to move/slide/rotate, which may also result in change in plumb and line and consequent misalignment/displacement in the bearings (will all the consequent effects),

AP.b — consequences of
— foundation scour
— foundation settlements

AP.c — impact hits from passing vehicles (and from barges and floating debris in case of bridges in water bodies)

AP.d — height more than about 10 m, so that unusual and special equipment may be needed for inspecting and carrying out the maintenance work in inaccessible zones

AP.e — Concrete cracking (from-numerous causes)

AP.f — overload/under-design

AP.g — other (explain)

3. Bearings
(use symbol *B*)

• *Types*

B.1 — *Elastomeric' type*
(generally used for relatively short-span and medium-span bridges)

B.2 — *'Pot' type*
(generally used for medium and long span bridges)

B.3 — *'Spherical' and 'Cylindrical' (teflon-coated) type*
(generally used where deck-rotation and/or loads are large)

B.4 — *Other*
(explain)

• *Some problems may arise from*

B.a — In the *Elastomeric Bearings* (which may sometimes be provided with a bonded PTFE layer and a stainless steel sliding plate when large movements are to be catered for), problems may arise from
(i) compression bulging of sides
(ii) unequal vertical deformation among a row of bearings in one line, owing to failure of one or more bearings due to excessive load
(iii) unequal vertical deformation owing to improper seating surfaces between the bearing and the deck-soffit and/or the pedestal underneath
(iv) 'surface' cracking
(v) 'radial' cracking in the bulges
(vi) excessive shear-deformation (due either to under-design or manufacturing defect)

---

* *These could be*

1. Plastic shrinkage of concrete, 2. Plastic settlement of concrete, 3. Early thermal and shrinkage movements and lack of adequately 'spaced' and 'functioning' movement-joints, 4. Drying shrinkage of concrete (long term), 5. Rusting of steel under moist conditions, due to: (a) oxidation of steel, (b) carbonation of concrete, (c) chloride-ion attack on steel (chloride salts from surrounding soil and water or from aggregates themselves if contaminated), causing rust-distress/delamination, 6. Sulphate attack on cement in concrete (sulphate salts may ingress from surrounding soil and water or may also be present as impurities in the aggregates), 7. Alkali-aggregate reaction (where aggregates contain reactive silicates and carbonates), 8. Differential settlement of supports, 9. Temperature: (a) variations, (b) gradients, 10. Weathering and disintegration of concrete, 11. Crazing in concrete, 12. Fire and/or explosion, 13. Overload/under-design, 14. Other (explain).

** *Resulting in distress due to*

(i) Excessive flexure, shear and/or torsion, (ii) Excessive direct force (axial compression/tension), (iii) Excessive bearing pressure, (iv) Restraint against designed deformation, (v) Inadequate soil-bearing conditions, (vi) Other (explain)

B.b — In the *Pot Bearings* (of 'fixed', 'free to slide in any direction' or 'guided to slide in a particular direction' types), problems may arise from
(i) Excessive rotation, causing excessive deformation in the potted elastomeric-disc and in the surrounding seal. (The temporary clamping bolts should either have been removed after installation or should have sheared subsequently.)
(ii) In the case of sliding-types, if the PTFE bonded to the top of the middle-plate is nakedly visible, then the top plate is not long enough in the direction of principle-movement. Such a bearing may need to be replaced (since such naked PTFE, open to ingress of dirt, may soon cease to function as designed).

B.c — In the *Spherical and Cylindrical Steel Bearings* (former capable of rotation in any direction and the latter capable of rotation in only one direction), with mating convex and concave (coated) steel surfaces, problems may arise that are somewhat similar to those in the case of Pot Bearings described earlier. The TEFLON sheet may inch-out.

B.d — In the case of *crude forms of Bearings* (such as felt or tar-paper layers), problems, may arise in the deck above and in the concrete under the bearings from cracking due to 'constraint against free rotation and sliding' of deck.

B.e — *Other*
(explain)

4. **Expansion Joints**
(use symbol *EJ*)
• *Types*

EJ.1 — *Buried-type*
(These are ancient type. In these a joint-filler is filled in the movement-accommodating-gap of the joint, and the wearing course is either carried across or butts with it from either side. If the (asphaltic) wearing course goes across, it is generally 6 to 12 mm saw-cut above the joint and the cut is filled with a mastic filler.)

EJ.2 — *Elastomeric-type*
(These are most commonly used in the modern-day bridges. They are waterproofed and 'modular' in build-up, making transportation handling and installation easy. Can take large movements of superstructure.)

EJ.3 — *Finger-type*
(Generally these are plates cut out into intermeshing finger arrangement, bolted to the edges of the adjacent decks).

EJ.4 — *Other*
(explain).

• *Some problems may arise from*

EJ.a — In the *Buried-type* joints
(i) swelling-up of wearing course
(ii) weeping out of the filler
(iii) consequent thudding by the passing wheels and riding-discomfort, further damage
(iv) cracking at edges

EJ.b — In the *Elastomeric-type* joints
(i) Tearing of elastomer (and consequent exposure of their reinforcing steel plates where such plates exist). This may be due to manufacturing defect or due to breaking of concrete in the supporting deck-edges
(ii) loosening of anchor bolts (and obvious consequences)

EJ.c — In the *Finger-type* joints
(i) loosening of anchor bolts
(ii) fracture of fingers (owing to combined effect of high-impact pounding from passing wheels and fatigue from highly repetitive nature of such loading, the effect becoming severer as the anchor bolts loosen.)

EJ.d — *Other* (explain)

5. **Superstructure**
(use symbol *S*)
• *Types*
The *span-arrangement* may be

S.1 — Simply supported spans

S.2 — Balanced-Cantilever and segmentally built Free-Cantilever Spans

S.3 — Semicontinuous spans (i.e. simply supported precast p.s.c. girders with R.C. deck slab continuous over piers, making them continuous for subsequent loads)

S.4 — Continuous spans (either cast as such, or built by stage- construction, or built by segmental cantilever method)

S.5 — Frame (single or multi-span)

S.6 — Cable-Stayed spans

S.7 — Other (explain)

• The *construction* may be in
— reinforced concrete
— reinforced and prestressed concrete, composite or non-composite structural steel and reinforced concrete
— structural steel (plate girder or truss)
— Other (explain)

• The *superstructure cross-section* may be
(i) 'R.C. solid slab' type (cast-in-situ, c.i.s.)
(ii) 'R.C. voided slab' type (c.i.s.)
(iii) 'R.C. box section' type (c.i.s.)
(iv) 'R.C. beams-and-slab' type (c.i.s.)
(v) 'Precast R.C. beams and cast-in-situ slab' type
(vi) 'Precast p.s.c. beams and cast-in-situ slab' type
(vii) 'p.s.c. voided slab' type (c.i.s.)
(viii) 'p.s.c. box' type (c.i.s. or precast segmental)
(ix) 'Precast box segments prestressed with precast concrete web, diaphragm and slab elements' type
(x) 'Steel beams and concrete slab' composite or non-composite type
(xi) Other (explain)

• *Some problems may arise from*

S.a — concrete cracking (from numerous causes*)
*See footnote (*) on previous page.

S.b   —   settlement of foundations (differential, overall)

S.c   —   movement/tilting of foundations

S.d   —   overload/under-design**

S.e   —   accidental hit from passing traffic

S.f   —   accidental fire

S.g   —   distress in Articulations (Halving-Joints) Shear lag effect (particularly during the construction stages) whereby the applied prestress is absorbed by only part of the assumed deck-section, leading to cracking in unprestressed regions

S.i   —   inaccessibility for inspection within.

S.j   —   Other (explain)

6. **Drainage**
   (use symbol D)

D – Inefficient drainage of water from and around the bridge can be hazardous for the traffic and can also lead to unsightly staining of concrete surfaces. When such ponded water gets mixed with any wind-blown sand and surrounding soil, containing sulphate and chloride salts, the resulting attacks on concrete and steel can lead to harmful cracking of concrete and rusting of steel and continuous disintegration of the effected reinforced (and prestressed) concrete components of the bridge. Footings, piles and pile-caps, buried and even the boxed portions of substructure, caps and seating beams, abutment-walls, boxed-deck, etc., can all fall prey to such distress and damage.

**See* footnote (**) on page 69.

7. **Corrosion in Structural-Steel Superstructure**
   (use symbol C)

C – Steel beams, boxes and trusses, used in superstructures, are susceptible to rusting and corrosion-distress due to oxidation and chloride-attack under moist conditions. Hence the anti-rust painting must always be ensured as a rust-protection means. Detection of paint-failure requires close hands-on examination. Where rust inhibiting steel (e.g. CORTEN) is used without any protective- coating, the effectiveness of the rust-inhibiting iron oxide coating should be checked for distress (e.g. signs of unevenness and flaking of the surface protecting layer). The most likely areas of rust-formation are: external faces of outer girders, all welds/rivets/bolts, site joints and gusset connections. Old steel-Bearings also need a special examination to ensure whether rust has not incapacitated them from performing the intended rocking/rolling/sliding, as the case may be.

5.7 SAMPLE 'GENERAL-INSPECTION-REPORT'

What has been described in Section 5.6 earlier, gives the details of a 'Special' Inspection. However, initially what is required is a 'General' Inspection Report. Only if this

BRIDGE GENERAL-INSPECTION-REPORT

Structure Number........................   STRUCTURE-ID........................

Inspection Date   _____.........................   Location........................
                        day   month   year

Element	Condition		X*	Stdd. Repairs			Y **	Photo Nos.
	Av.	Worst		Type	Qty.	Year		
1. Surface of Approaches								
2. Surface on the Bridge								
3. Expansion Joints								
4. Parapet and Railing								
5. Sidewalks/Median								
6. Slopes								
7. Wing Walls								
8. Abutments								
9. Bearings								
10. Edge Beams								
11. Deck Slab								
12. Primary Members of Deck								
13. Pier Caps								
14. Pier								
15. Waterway Channel								
16. Bridge, General								

X: *Is Routine Maintenance satisfactory?

Y: **Is Special Inspection needed?

- Remarks
    1. Does it need Special Inspection?.................................
        - if so, when necessary?.................................
    2. Other.................................

Name of Inspector

Signature

....

- Cleaning Condition

- Proposed next year of inspection if no special inspection reqd...................

'General Inspection' indicates that the condition of the bridge is of concern, a 'Special Inspection' is necessary.

The format of the 'General Inspection Report' is described on page 71.

5.8 ALTERNATIVE INSTRUCTION-CODE FOR THE PURPOSE OF PREPARING THE BRIDGE-INVENTORY

In Section 5.6 and its Legend, the description of the items required for preparing the bridge-inventory (e.g.

the 'types' of Abutments, Piers, Bearings, Expansion-Joints, Decks, etc.) are rather simply explained, merely indicatively, and are easy to apply for manual records. But where a whole databank of a very large number of bridges has to be prepared, so that computerization is necessary, then a digit-designated instruction-code is more convenient. For this purpose a simplified one-page Bridge Inventory Data Sheet (B.I.D.) and a Digit-Designated Instruction-Code (D.D.I.C.) (5.8.1 and 5.8.2) are presented below. These may be used both for the Special Inspection Report as well as for the General Inspection Report to suit convenience.

5.8.1 Bridge Inventory Data (B.I.D.)

(Simplified one-page Bridge Inventory Data Sheet per Bridge)

Structure-I.D. . . . .

- *Administrative Data*
  District Name . . . . Contractor Area Code . . . .
  Road No.. . . . . Station, Kilometer. . . . .
  Road Name . . . .
  Location . . . .
  . . . .
  Structure Number. . . . .
  Number of Crossing-Road. . . . .
  Year Built/Rehabilitated. . . . . / . . . .
  Over/Underpass . . . . Type of Service (On/Under) . . . . / . . . . . .
- *Technical Data*
  Type of Bridge (Material/Design). . . . . / . . . . . .
  No. of Spans/Sections. . . . . . / . . . . .
  Type of Abutment
  Type of Pier
  Type of Railings and Parapets
  Type of Deck Wearing Surface
  Type of Expansion Joint
  Type of Bearings on Piers/Abutments/in Primary Member
- *Geometric Data*
  Total No: of lanes on/under Structure
  Max./Min. Span Length (meter)
  Overall Length (meter)
  Vertical Clearance under Structure (meter)
  Approach Roadway Width (meter)
  Bridge Roadway Width Kerb-to-Kerb (meter)
  Bridge deck Width out-to-out (meter)
  Bridge Sidewalk Width (Left/Right) (meter)
  Vertical Kerb Height (meter)
  Bridge Median Width (meter)
  Bridge Curved Y/N
  Skewness* (approx. 00, 30, 45, 60 degrees)
- *Standard Photos*
  Approach   Elevation:
  1. . . . .   2. . . . .   1. . . . .   2. . . . .
  Abutment: Pier:
  1. . . . .   2. . . . .   1. . . . .   2. . . . .
- Data collected on: _____   Name of Inspector
                day   month   year   . . . .
- Now fill the General Inspection Report Form**   Signature

*Angle between centre-line of pier-length and perpendicular to longitudinal centre-line of roadway.

** *See* on page 71.

5.8.2 Digit Designated Instruction Code (D.D.I.C.) for Preparing the Bridge-Inventory

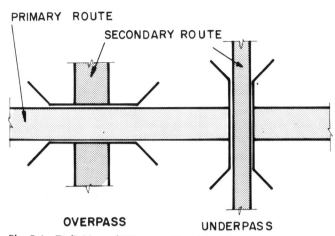

Fig. 5.1 Definition of "Overpass" and "Underpass"

TYPE OF SERVICE
Indicates the type of service of the bridge using 2 figures. 1st and 2nd figure refer respectively to the service on and under the bridge.

Code	Service
1	Highway
2	Street and Feeder Road
3	Second Level (interchange)
4	Third Level (interchange)
5	Pedestrian Exclusively
6	Railroad
7	Camel Crossing
9	Other

Type of Service on the Bridge

Code	Service
1	Highway
2	Street and Feeder Road
3	First Level (interchange)
4	Second Level (interchange)
5	Pedestrian Exclusively
6	Railroad
7	Camel Crossing
8	Wadi
9	Other

Type of Service Under the Bridge

Technical Data
Type of Bridge
Indicates the bridge's material and design characteristics. The code consists of 2 figures separated by a slash

("/") indicating material and design. (Material code (1 digit) / Design code (2 digits))

Code	'Material'
1	Reinforced Concrete Simply-Supported (S.S.)
2	Reinforced Concrete Continuous
3	Composite Steel / Concrete (S.S.)
4	Composite Steel / Concrete-(Continuous)
5	Prestressed Concrete (S.S.)
6	Prestressed Concrete (Continuous)
7	Prestressed Concrete-Continuous, Cast in Segments.
9	Other

'MATERIAL' CODE

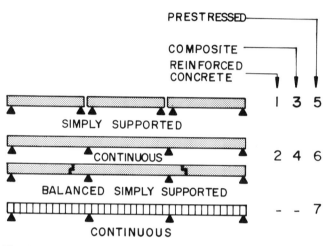

Fig. 5.2 "Material" code

If the superstructure is continuous over one or more piers, the structure is defined as continuous.

Code	'Design'
01	Slab
02	Stringer / precast I- or T-Beam
03	Girder and Crossbeam System (Grid syst.)
04	T-Beam System (Slab-Girder type, in situ cast beams and deck)
05	Single Box Beam
06	2 or more Box Beams
07	Frame
20	Mixture of above
99	Other

'DESIGN' CODE

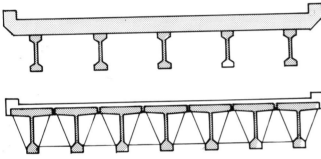

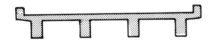

Fig. 5.3 Design type 01, slab

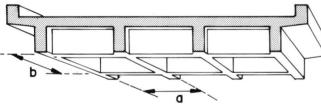

Fig. 5.4 Design type 02, stringer

b less than 2a

Fig. 5.5 Design type 03, girder and crossbeam system

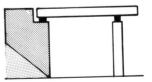

Fig. 5.6 Design type 04, T-beam system

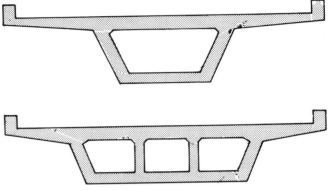

Fig. 5.7 Design type 05, single box beam

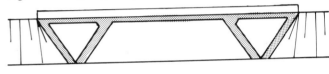

Fig. 5.8 Design Type 06, 2 or more box beams

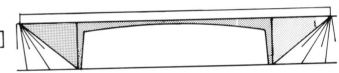

Fig. 5.9 Design type 07, frame

NUMBER OF SPANS/SECTIONS

Indicates the number of spans and sections, respectively. 'Sections' is only used for large structures, which have been divided into sections regarding the general inspections

TYPE OF ABUTMENT

Indicates type of abutment
1   Abutment Wall (advance position) Spill-through
2   Slope Protected Abutment, Spill-through or Solid
9   Other or more than one type
Abutment Wall, advance position type (code 1), is defined as a wall adjacent to the carriageway or other horizontal areas under

'Abutment' Type

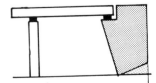

Fig. 5.10 Abutment type I, abutment wall

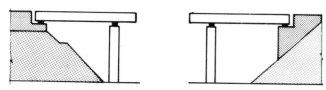

Fig. 5.11   Abutment type 2, slope protected

TYPE OF PIER

indicates the type of pier (1-digit)	
1	Single Column
2	2 or more Columns
3	Single Column with cap beam (hammerhead)
4	2 or more columns with continuous cap beam
5	Solid Pier
6	2 or more columns with separate cap beams (hammerheads)
	Other or more than one type

Type of Pier

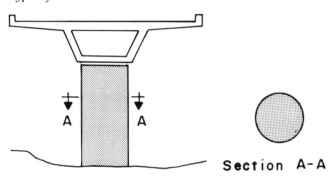

Fig. 5.12   Pier type 1, single column

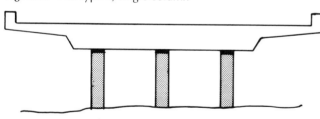

Fig. 5.13   Pier type 2, 2 or more columns

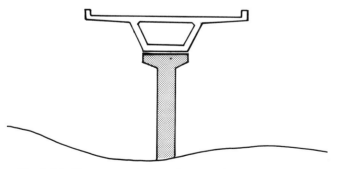

Fig. 5.14   Pier type 3, single column with cap beam (hammerhead)

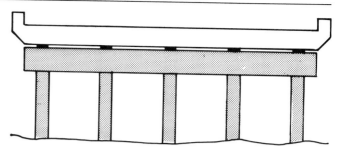

Fig. 5.15   Pier type 4, 2 or more columns with continuous cap beam

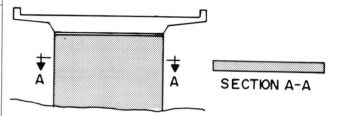

Fig. 5.16   Pier type 5, solid pier

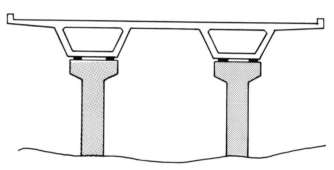

Fig. 5.17   Pier type 6, 2 or more columns with separate cap beams (hammerheads)

TYPE OF RAILINGS AND PARAPETS

Indicates the type of railing regarding material and construction at bridge edges and medians.

1	Concrete (with or without steel / aluminum top-railing)
2	Aluminum
3	Steel
	Other, or more than one type

TYPE OF WEARING SURFACE

Indicates the type of wearing surface on the bridge.

1	Concrete
2	Asphalt
3	Gravel
	Other

TYPE OF EXPANSION JOINT

1	Single Neoprene Strip Seal
2	Modular Expansion Joints
3	Single Neoprene Block
4	Multiple Neoprene Block
5	Finger Joint
6	Asphalt Joint
7	Covered Expansion Joint
8	No expansion joint device
9	Other, or more than one type
N/A	Not applicable

Type of Expansion Joint

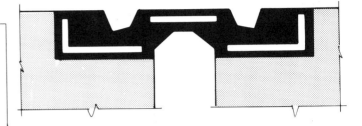

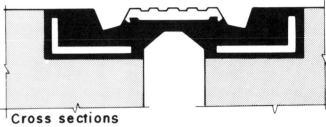

Cross sections

Fig. 5.20 Expansion joint type 3, single neoprene block

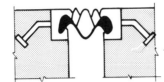

Cross sections

Fig. 5.18 Expansion joint type 1, single neoprene strip seal

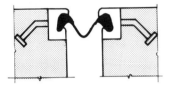

Cross section

Fig. 5.21 Expansion joint type 4, multiple neoprene block

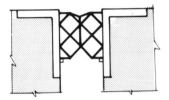

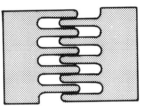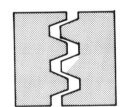

PLAN VIEW

Fig. 5.22 Expansion joint type 5, finger joint

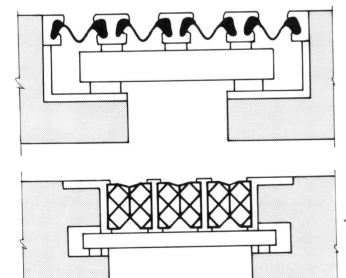

Cross sections

Fig. 5.19 Expansion joint type 2, modular expansion joints

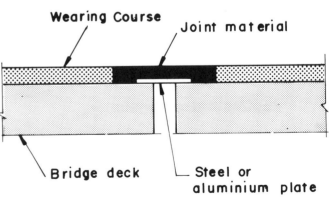

Fig. 5.23 Expansion joint type 6, asphalt joint

TYPE OF BEARINGS

Indicates by three 1-digit numbers the type of bearings on piers, on abutments and on primary members (in cantilever construction).

1	Construction joint with or without a layer of sliding material, such as a bituminous membrane (Tar-paper), etc.
2	Elastomeric Bearings
3	Pot-Bearings
4	Steel-Bearings
5	Roller Bearings
9	Other or more than one type

Type of Bearings

If there is rigid connection between primary member and pier/abutment (frame corner), the type of bearing is marked 'N/A' (not applicable).

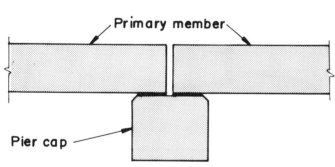

Fig. 5.24 Frame corner: bearing type N/A

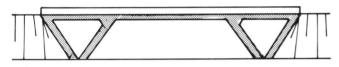

Fig. 5.25 Bearing type 1, construction joint with or without a layer of sliding material, such as a bituminous membrane

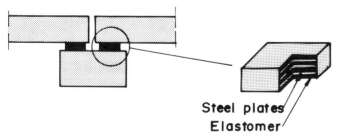

Fig. 5.26 Bearing type 2, elastomeric

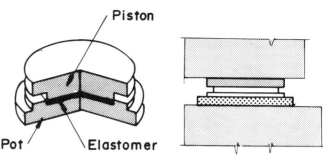

Fig. 5.27 Bearing type 3, pot-bearing

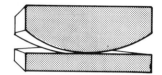

Linear Rocker Bearing (cylindrical rocker)

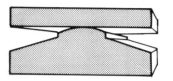

Point Rocker Bearing (spherical rocker)

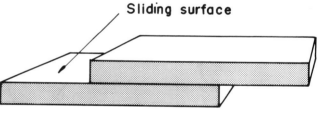

Plane Sliding Bearing

Fig. 5.28 Bearing type 4, steel bearing

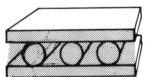 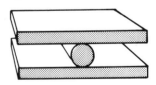

Multiple Roller Bearing   Single Roller Bearing

Fig. 5.29 Bearing type 5, roller bearing

GEOMETRICAL DATA

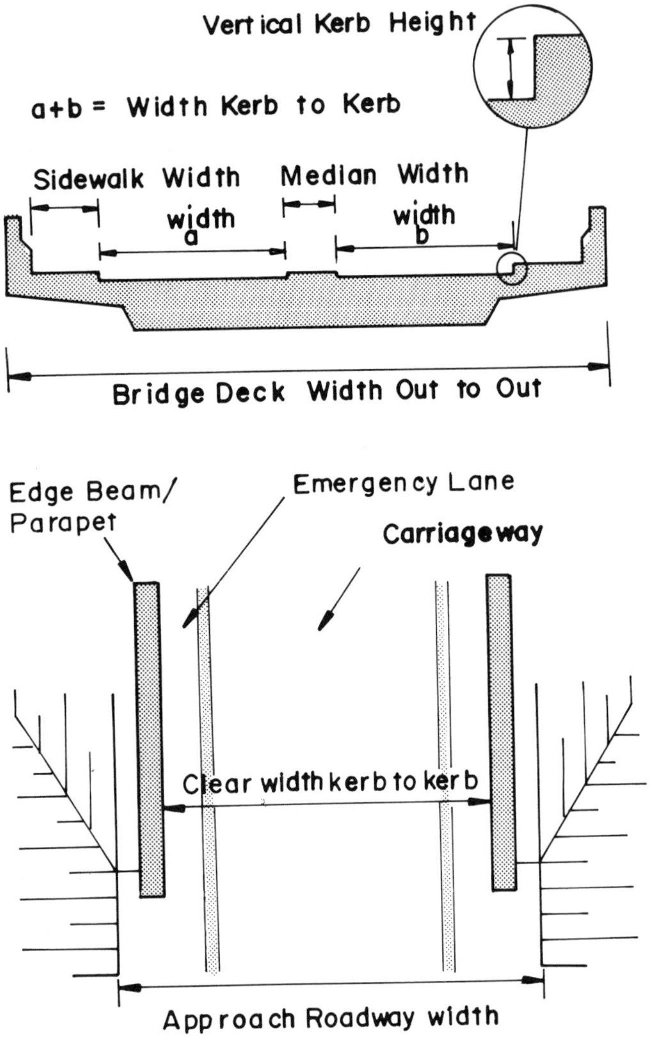

Fig. 5.30 Widths, kerb, height (geometrical data)

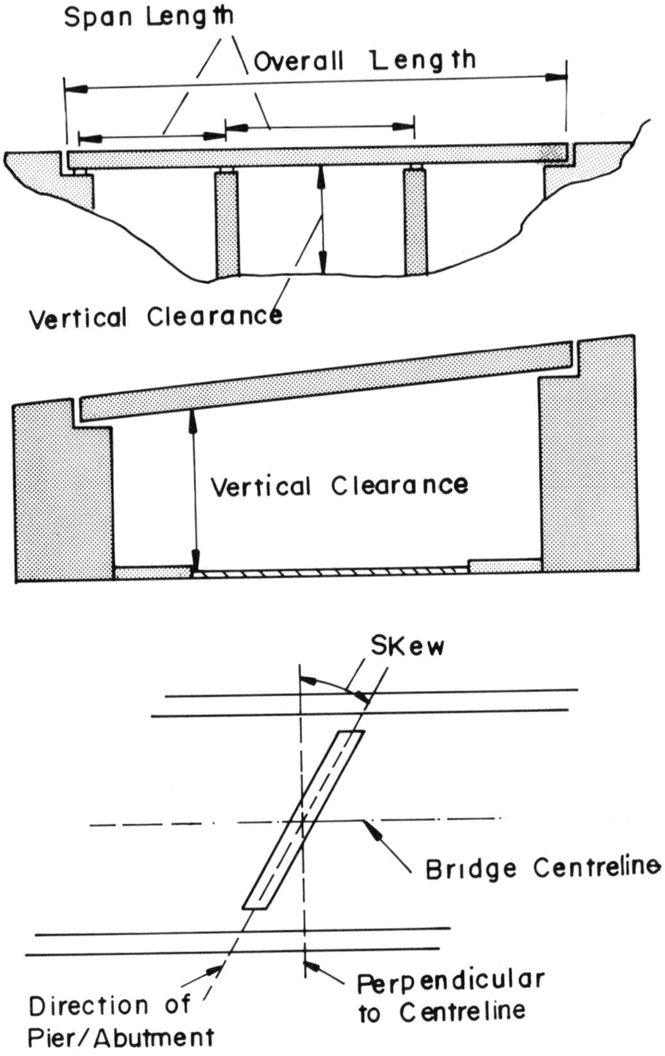

Fig. 5.31 Lengths, vertical clearance, skew

NO. OF LANES ON / UNDER BRIDGE
Indicates the number of through-lanes 'on' and 'under' the bridge (including possible exit- or entry-lanes, but exclusive of emergency lanes).
e.g. 06/04 (6 lanes on the bridge, 4 lanes under the bridge)

MAXIMUM/MINIMUM SPAN LENGTH
Indicates maximum and minimum span length to tenth of a meter between centre lines of piers/bearings on abutments, measured along the centre line of the bridge.
e.g. 31.3/29.3

OVERALL LENGTH
Indicates the overall bridge length to tenth of a meter, measured along the centre line of the bridge from

bridge-end to bridge-end (excluding approach slabs) – normally from joint to joint.

VERTICAL CLEARANCE
Indicates to a tenth of a meter the minimum vertical clearance under the structure. The clearance is measured over the carriageway or path if there is no carriageway. If the structure is an overpass over a river, indicate the average clearance in the main span.

APPROACH-ROADWAY WIDTH
Indicates to a tenth of a meter the nominal width of the approach-way including shoulders. For roads with medians, include width of both roadways, outside shoulders and median.

BRIDGE-ROADWAY WIDTH CLEAR KERB TO KERB
Indicates to a tenth of a meter the total roadway-width between kerbs (or parapets if there are no kerbs) (exclude possible medians), measured perpendicular to the centre line of the bridge.

BRIDGE DECK WIDTH OUT TO OUT
Indicates to a tenth of a meter the total out to out width of the bridge deck measured perpendicular to the centre line of the bridge.

SIDEWALK WIDTH
Indicates to a tenth of a meter the width of the left and right sidewalks (seen in the direction of chainage), measured perpendicular to the centre line of the bridge. The width must not include the width of the parapets.
e.g. 2.8 / 1.5 (left sidewalk 2.8 m, right sidewalk 1.5 m)

VERTICAL KERB HEIGHT
Indicates to hundredth of a meter the kerb height measured from the riding surface.

MEDIAN WIDTH
Indicates to tenth of a meter the width of the median measured perpendicular to the centre line of the bridge.

BRIDGE CURVED
Indicates whether the structure is on a curve.
Y — Yes
N — No (or minor curve)

SKEW
Indicates the approximate skewness of the structure. If the line through bearings on a pier/abutment is (approximately) perpendicular to the centre line of the bridge, the skewness is 00°. Use the following options: $00^\circ, 30^\circ, 45^\circ, 50^\circ, 60^\circ$, ($60^\circ$ is a very skew bridge!).

LOADS
The load information does not appear on the inventory form as it cannot be recorded on site.

DESIGN LOAD
Indicates the design live load of the structure.
The design live load code may assume the following values which express the live load the bridge was designed for.

1	HS 15 – 44 (HS 15)
2	H 20 – 44
3	HS 20 – 44 (=HS 20 =HS 20-S16-44
4	(HS 20 – 44) + 10%
5	40 Tonne Truck (5 axles), std
6	60 Tonne Truck (3 axles), std
7	any other
UN	Unknown

RATING FACTOR (R.F.)
The Rating Factor is a figure between 0.00 and 9.99 found as a result of Load Capacity Evaluation (see Chapter 12). The factor expresses the withstandable live load carrying capacity compared to the prevailing live load. It should normally be > 1.

REMARKS
'Remarks' may be attached to the Rating Factor, e.g. regarding the structural element determining the rating factor, max. crackwidth in the service limit state, or possible weight restrictions, etc.

STANDARD PHOTOS
The Inventory should include photos of the approach, bridge-elevation, abutments and piers. The approach photo is a photo showing the topside of the bridge including surface, kerbs, sidewalks, railings / parapets etc., seen from one end of the bridge.

If the approaches, abutments and/or piers are different, take two photos.

If the elevation cannot be shown in one photo, take two or take the photo at a skew angle.

All photos should be taken with the camera in horizontal position (landscape format). Take approach and elevation photos in the sunny side if possible (along the sun).

The abutment and/or the pier photos should also show the soffit of the superstructure.

Each photo must be indicated with the day of the month, hour and minute. This is normally done automatically by a 'data back' on the camera. The photos are stuck (and stapled) to photo forms (2 photos on each sheet), indicated with structure identification and date.

Photo pages are filed together with the inventory forms. Negatives must be filed in envelops marked with team number, structure identifications and dates. Each envelop should only contain negatives from one route number.

DATA COLLECTED
The 'date' on which the inventory data are collected.

TEAM NO.
Indicates the serial number of the inspection team, which has collected the data.

5.9 INSPECTION

5.9.1 Sequence of Inspection

In majority of (concrete) bridges, following is the sequence of the elements to be inspected:
1. Wearing surface (wearing coat),

2. Expansion joints,
3. Bearings,
4. All primary and secondary structural members forming the superstructure (Deck), e.g. deck-slab (and soffit-slab in box sections), longitudinal beams (stems or webs), transverse beams, etc.,
5. Piers, abutments (and wingwalls) and their foundations (e.g. footings, piles and pile-caps, caissons and their caps, etc.),
6. Effects of any foundation movements,
7. Waterways (stream/bed condition) and scour effects,
8. Dolphins and fenders
9. Underwater investigations
10. Submersible bridges
11. Movable bridges and suspension Spans
12. Miscellaneous items, e.g.
    (a) Light posts
    (b) Road kerbs
    (c) Drainage (scuppers, etc.)
    (d) Parapets and railings
    (e) Footpaths (walkways) and facias
    (f) Utilities
    (g) Medians
    (h) Paint
    (i) Signs
    (j) Rock slopes
    (k) Rock bolts/anchors
13. Encroachments
14. Aesthetics
15. Approaches
16. General

All these elements have to be thoroughly inspected (hands-on type close inspection). As to 'what to look for' and how to 'rate' the condition of the elements, is explained in the next section.

5.9.2 'What to Look For' in each Element and 'How to Rate' Condition

This is explained in Sections 5.9.2.1 to 5.9.2.16 below:

5.9.2.1 Wearing Surface (wearing coat)

5.9.2.1.1 Types
— concrete overlays, either cast on the deck-slab subsequently or cast monolithically with the deck-slab
— asphalt overlays

5.9.2.1.2 What to Look For
• *General:* The riding quality across the bridge is a major factor in determining the rating for a wearing surface. Beware that a wearing surface that appears good on a dry day may be hazardous when wet. A slippery surface should be rated low.

• *Concrete Overlays:* Look for scaling, spalling, cracking, rutting and exposed reinforcing. Determine whether the concrete surface is worn or polished. When softer limestone aggregates are used in the concrete, fine aggregates and cement paste will be worn away, exposing the surface of the coarse aggregates to the polishing action of rubber tires. The resulting deck surface becomes increasingly hazardous when wet.

• *Asphalt Overlays:* Asphalt overlays on bridge decks will not be rated by the bridge inspection crew, but rather by pavement crews. However, since the bridge inspection crew will be required to rate the condition of the deck below the asphalt surface and, since the condition of the surface may give indication of the condition of the deck, the following possible problems and causes of asphalt-deterioration are to be looked for and noted. See table on page 81.

Regarding bridge inspection, the bridge crews will be careful to note any reflective deck-surface cracks near or at the supports of continuous bridges. Such cracks are indicative of structural cracking at the top of the concrete-deck and probably of the underlying concrete members of the superstructure. This is particularly important in the coastal areas, where such defects are subject to the accelerating affects of a more corrosive environment.

5.9.2.1.3 Sample Ratings
• *What to Rate:* Rate the physical condition and the riding quality of the wearing course. When the wearing course is separate from the structural deck, the full thickness of the wearing course is considered. When the wearing surface is monolithic with the structural deck, only the surface that vehicles bear on is rated for the extent of spalling.

A rating of 7 indicates a surface in good condition with no spalls, delamination, or cracks.

A 5-rating is used to indicate the beginning of a spalling problem. No more than two or three isolated, moderate spalls or delaminations are present.

A rating of 3 indicates a more serious spalling problem, although large areas of the span are still unaffected. This could be one large area affected by spalls, but still less than half the span area, or a larger number of small, isolated spalls than indicated for a 5-rating.

A 1-rating is used where the area affected in any lane approaches half the total area of the lane.

NOTE: Ratings of 2, 4, and 6 are used for conditions in between their adjacent ratings, respectively.

Problem	Features	Probable Cause
Alligator or Map Cracking	Interconnected cracks forming a series of small blocks resembling an alligator's skin or chicken-wire mesh	Excessive deck deflection, drying of asphalt material
Edge Cracks	Longitudinal Cracks near the edge of deck	Lack of lateral support, drying out of asphalt, deterioration of concrete deck
Lane-Joint Cracks	Longitudinal separations along the seam between two paving lanes	Weak seam between adjoining asphalt-spreads
Reflection Cracks	No set pattern	Cracks in underlying concrete deck slab
Shrinkage Cracks	Interconnecting cracks forming large blocks	Volume change in concrete, cracking of underlaying deck slab
Slippage Cracks	Crescent shapes	Lack of bond between asphalt-coarse and deck slab beneath
Channels (Ruts)	Channelized depressions along wheel-lines of passing trucks	Consolidation or lateral-movement of surface under traffic
Corrugations	Ripples across surface	Lack of stability
Depressions	Dips in surface	Settlement of concrete deck
Potholes	Bowl shaped holes of various sizes	Localized disintegration of asphalt
Raveling	Separation of aggregate particles	Poor construction techniques, low asphalt content

5.9.2.2 Expansion Joints

5.9.2.2.1 Types

— The 'details' of different types of expansion Joints have been explained extensively in the book 'Concrete Bridge Practice — Analysis, Design and Economics' by Raina, V. K., Tata, McGraw-Hill, New Delhi.

— A few examples have also been indicated in the earlier pages here.

— Additionally, a refresher summary is given below:

• General: Expansion joints are necessary to allow the spans to expand and contract with temperature variations. All types of expansion joints must be free to move.

There are a number of types of expansion joints in use and some of these are indicated below in principle.

Span	Type of Joint
Short and Medium	Poured sealant, Compression-seal, Gland, Segmental (Modular).
Moderate	Gland, Segmental (Modular).
Long	Segmental (Modular), Finger Joint.

• The Gland-type expansion joint devices generally consist of continuous, flexible, sealing glands, held in place by Steel or Aluminium Extrusions which are anchored to either structural steel supports or directly to the concrete deck. The sealing gland must be positively secured by the extrusion and be intact to prevent water leakage. Examples of the various gland-type joints are shown in Fig. 5.32.

• The Compression Seal expansion joint device consists of a pair of continuous armour angles which reinforce the edges of the joint at the roadway surface. A continuous compressible labyrinth seal is compression-fitted in between the angles and may also be cemented in place for added water tightness.

• Modular expansion joint devices may be comprised of a series of continuous cells or glands secured to retainers. The intermediate retainers are supported by support bars, which in turn are supported by the major joint-elements that are embedded in the concrete on each side of the expansion joint. The spaces between the retainers (as shown by dimension 'w') should be equal and each space should be the same width throughout the entire length of the joint. See Fig. 5.34.

• The Segmental-type of expansion joint device is fabricated of steel and elastomeric material in short sections. Leakage between abutting sections and at kerb areas has been a common problem. Cross-sections of two common types of segmental joints are shown in Fig. 5.35.

A typical finger joint is shown in Fig. 5.36. The space between the fingers is designed to allow for expansion and contraction. As a consequence a trough is required beneath this type of joint to carry corrosive drainage water away and preclude deterioration of the bridge below. A cross section of a typical finger joint and the position of the drainage trough is indicated in Fig. 5.36

5.9.2.2.2 What to Look For

Poorly designed and maintained expansion joints are a constant source of nuisance and danger and should be examined carefully. Note if there is adequate space for thermal movement, if the joint is open an excessive amount, and if the joint is clear of all debris. Care must be taken in a sealed-type joint to see that the seal is in a condition which will prevent entry of stones, sand, or other noncompressible material.

Examine steel-finger joints and sliding plate joints for evidence of loose anchorages, cracking or breaking

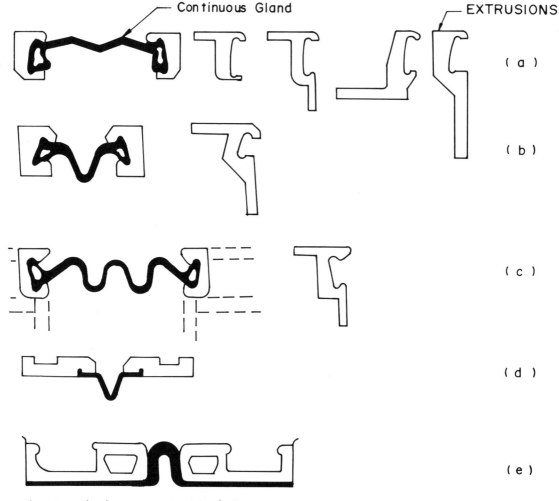

Fig. 5.32   Gland-type expansion joint devices

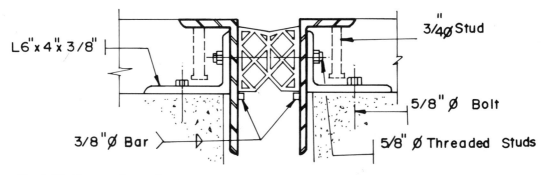

Fig. 5.33   Compression seal

of welds, or other defective details. Such defects may not only cause structural damage but may loosen-up and create a hazard to traffic. Sound the concrete deck adjacent to all expansion devices for voids or laminations in the deck.

Examine the underside of the expansion joints as far as possible to detect any impending problem. Lack of adequate room for expansion, especially in small areas of the joints, will create thermal deformation stresses, causing the concrete to shear and spall. This is a serious hazard in structures which cross over roadways, walkways, or any occupied areas. Besides inspection for proper functioning of the expansion joint, any deterioration in the material such as rusting, etc., shall be looked for along with remedial measures required.

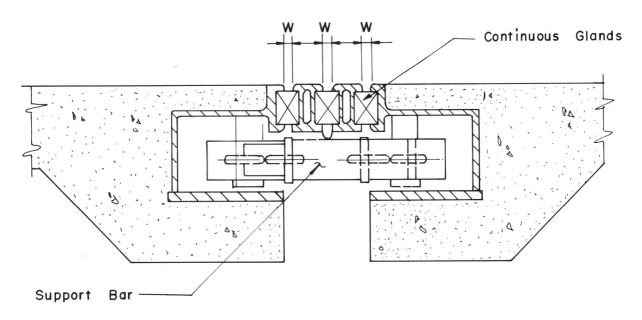

Fig. 5.34 Modular type

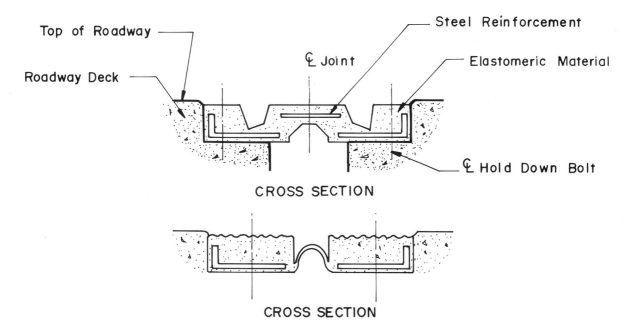

Fig. 5.35 Segmental expansion joint device

(i) Check all expansion joints for freedom of movement, proper clearance, and proper vertical alignment (Fig. 5.37). There should be sufficient room for expansion but the joint should not be unduly open. 'Closed' or 'widely-opened' joints, or a bump at the joint back-wall can result from substructure movements. Proper opening size depends on the season (temperature), the type of joint seals, the temperature range, and the length of the deck whose movement the joint must accommodate.

Normal temperature is usually assumed to be about 20°C (68°F). The Table 5.1 ahead lists some general data for various types of expansion joints. The expansion length in this table is the portion of deck or structure whose movement must be accommodated by the joint. This distance may extend from the end of the bridge, if that is the location of the nearest fixed bearing, or it may be the sum of the distances from the nearest fixed bearings from the two sides of the joint. Multiplying the

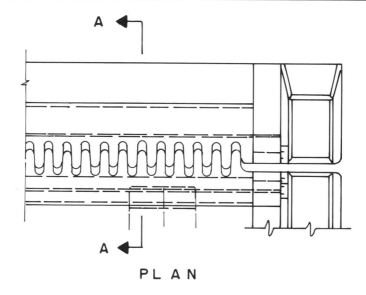

PLAN

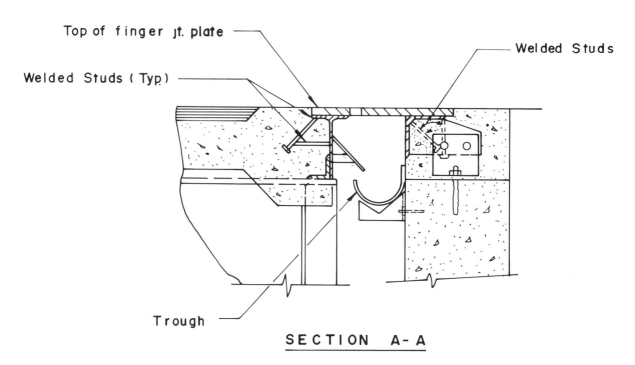

Top of finger jt. plate

Welded Studs

Welded Studs (Typ)

Welded Studs

Trough

SECTION A-A

Fig. 5.36 Finger joint

Table 5.1 Expansion joint data (General)		
Joints	Expansion Lengths	Joint Openings at 68°F (20°C)
Steel finger joints	60 m or more	80 mm (min.)
Steel expansion plates	60 m maximum	55 mm
Compression seals	40 m maximum	40 mm
Poured sealants and joint fillers	35 m maximum	37 mm

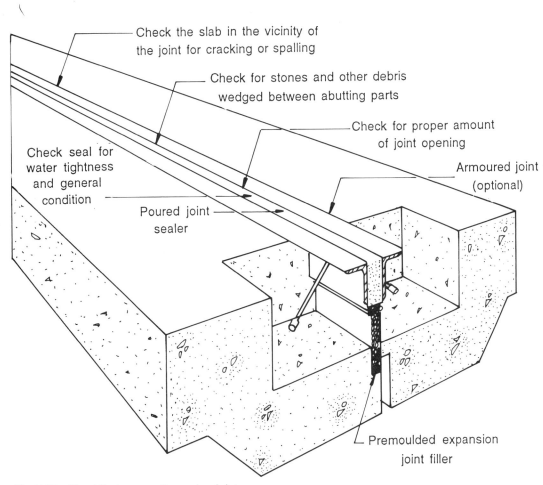

Check the slab in the vicinity of
the joint for cracking or spalling

Check for stones and other debris
wedged between abutting parts

Check for proper amount
of joint opening

Check seal for
water tightness
and general
condition

Armoured joint
(optional)

Poured joint
sealer

Premoulded expansion
joint filler

Fig. 5.37 Checklist items — Expansion joint

expansion length by the difference between the temperature at the particular moment and 68°F, and multiplying this product by 0.0000065, will give the approximate change in joint opening from the values listed below. (Very often, construction plans will give useful data concerning the setting of expansion devices.) However, this is only the thermal movement change.

(ii) Check seals for water-tightness and general condition.
Look For:
(a) Seal or sealant pulling away from the edges of the joint.
(b) Abrasion, shriveling, or other physical deterioration of the seal.
(c) Stains and other signs of leakage underneath the deck. Leakage seal permits water and brine to flow onto the bridge-seat and pier-cap, causing corrosion of Bearings, rusting, disintegration of concrete, and staining. Joints not properly sealed should be cleaned and resealed.

NOTE Above is true for a seal-type expansion joint. Other types can have their own problems peculiar to them. The inspector should study the relevant drawings.

(iii) Check to see the expansion joints are free of stones and other debris. Stones lodged in the joints can create localized stresses which may cause cracking and spalling of the deck. Large amounts of debris cause jamming, thus rendering the joints ineffective.

(iv) Examine steel finger-type joints and sliding plate joints for evidence of loose anchorages, cracking or breaking of welds, or other defective details. Sometimes the fingers may be damaged by traffic or by cracks which have developed at the base of the fingers. (Similar caution for neoprene finger joints too.)

(v) Verify that surfacing material has not jammed the finger joints on bridges that have been resurfaced.

(vi) Examine specifically the underside of the expansion joint (problem may be accessibility) to detect any existing or potential problem.

(vii) Look for deterioration of the joint materials.

When under the deck, check for deteriorated and broken joint-supports, troughs and baffles, and leakage joints which were intended to seal out water.

(viii) Check that rust of steel elements does not prevent movement for expansion and contraction.

(ix) Listen for the noise in the joints under traffic, which may indicate loose bolts or joint-components.

(x) Look for damaged or broken components, bolts, and welds.

(xi) Look for damage to supports, finger plates, glands, segmental components, and kerb or parapet plates caused by truck traffic.

(xii) Look for damaged, deteriorated or missing poured-type joint-sealants; ruptured or torn seals, glands and segmental components.

(xiii) Look for spalled concrete adjacent to the joint-supports or faces.

(xiv) Check seals and glands for water-tightness and general conditions. Joints designed for water-tightness that leak should be downrated one point. If the leak is causing damage again, reduce one point.

(xv) Sound the concrete deck adjacent to all expansion-devices for voids or laminations.

(xvi) Similarly, clogged joint-troughs and/or plumbing-systems should result in downrating of at least one point, depending upon the degree of harm the clogging is causing. All exposed parts of the joint-plumbing-system should be rated under the 'joint' item. *See* Figs 5.38 – 5.40, Plates 11 and 12.

5.9.2.2.3 Sample Ratings

7 — New condition.

5 — Good condition with some signs of minor deterioration. If the joint in Fig. 5.41 does not leak, then the filler material breaking away from the angle would qualify as minor deterioration.

4 — If leakage of the joint in Fig. 5.41 is causing moderate deterioration, then a 4-Rating is appropriate.

Examples of expansion joint ratings follow.

5.9.2.3 Bearings

5.9.2.3.1 Types

— The 'details' of the different types of Bearings have been explained extensively in the book: *'Concrete Bridge Practice – Analysis, Design and Economics'* by Raina, V. K., Tata McGraw-Hill, New Delhi.

— A few examples have also been indicated in the earlier pages here.

— Additionally, a refresher summary is given below

• *General:* Bearings transmit the superstructure loads to the substructure while permitting the superstructure to undergo necessary movements, without developing harmful stresses. Bearings are of two general types, 'fixed' and 'free'. The principal difference between these two types is that the fixed bearings resist translation but permit rotation, while the free Bearings permit both rotation and translation. Depending on structural requirements, the bearings may or may not be designed to resist vertical uplift. A number of different types of fixed and free bearing devices are used in bridge construction.

'Elastomeric' bearing pads are a type of free bearing device, made of rubber-like (elastomer) material, moulded into rectangular pads or strips. When placed beneath a beam, they permit the beam to undertake longitudinal movements and rotations. Two or more

Fig. 5.41 No leakage, Rate 5; Leakage causing moderate deterioration, Rate 4; Leakage causing serious deterioration, Rate 3.

Fig. 5.42 A vertically displaced finger-joint — the displacement and the poor riding quality result in a 4 rating.

PLATE 11

Fig. 5.38 Steel finger type expansion joint in a highly distressed condition

PLATE 12

Fig. 5.39 Detail of distress in the joint shown in Fig. 5.38

Fig. 5.40 Installed Waboflex SR expansion joint (working in good condition)

Fig. 5.43 A joint with deck that has been paved over, much of the asphalt has ridability over the joint. This joint is rated 3.

Fig. 5.45 If parts of the joint with deck are loose and protruding in a manner such that they might be snagged by vehicular traffic, or if the joint is broken out so that traffic must swerve to avoid a hazard, Rate 1.

Fig. 5.44 The trough of an open joint which is filled with asphalt concrete. This clogged trough is causing serious deterioration of the bridge seats and bearings (not shown) and is, therefore, low rated; Rate 3.

pads may be laminated with steel plates between them to allow for greater movement without excessive bulging. Longitudinal beam movement is accomplished by a shearing deformation of the elastomer.

The 'pot' bearing consists of a steel 'piston' acting upon a round elastomeric pad confined within a cylindrical steel base-pot. Under load, the elastomer allows tilting or rotation of the piston and the structural member that it supports.

To accommodate expansion and contraction, the upper surface of the piston is coated with a teflon-like material against which the stainless steel surface of the top plate slides. 'Keeper' plates act as 'guides' and limit specific movements.

Lessons From Some Actual Distress Experience with Bearings

(a) Bearing failure can result from a number of causes (e.g. damage or displacement following an accident, attack by chemicals, fire, and corrosion of contact surfaces) but probably the greatest cause of bearing malfunction, particularly of modern bearings, is

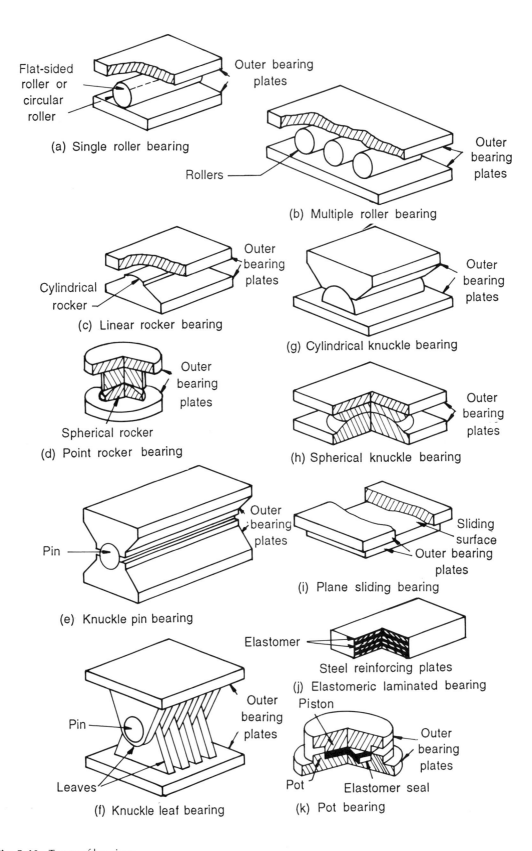

Fig. 5.46 Types of bearings

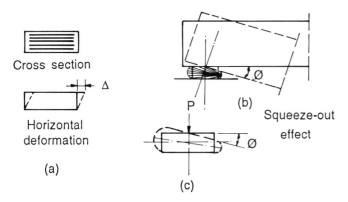

Cross section

Horizontal
deformation

(a)

(b)

Squeeze-out
effect

(c)

Fig. 5.47 Elastomeric bearing

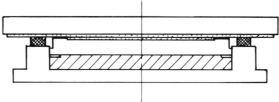

Components

- Tetron disc bases and rockers and all sliding plates are made of corrosion protected mild steel.
- Sliding plates are faced with a smooth surface of high quality stainless steel.
- Sliding surfaces are pure "dimpled" PTFE, incorporating grease pockets which allows a permanent reservoir of lubricant.
- Elastomer used for rotational purposes in Tetron disc bearings is high grade natural rubber to BS 1154.

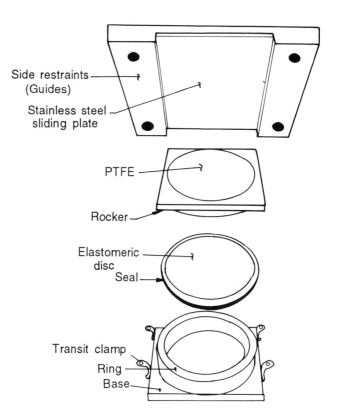

Side restraints (Guides)

Stainless steel sliding plate

PTFE

Rocker

Elastomeric disc

Seal

Transit clamp

Ring

Base

Fig. 5.48 Tetron disc (pot) bearing

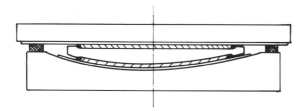

Fig. 5.49 Spherical bearing

due to inadequate or improper installation. It is not unknown for such simple looking bearings to be installed 90° out of phase or even upside down! It cannot be stressed too strongly that care in the installation of bearings is of the utmost importance.

(b) Bridges are usually designed with an expected life in excess of 50 to 70 years. Modern bearings and bearing materials have not been proved in service for this length of time so it is advisable to make provision in the design of bridges for bearing replacement, should this be found to be necessary. Facilities for correcting the effects of differential settlements, etc., should be provided unless the structure has been designed to accommodate such effects.

(c) Regular inspection of the bearings should be made so that any potential trouble is detected before serious damage is done to the structure. There should be adequate space around bearings to allow for inspection and maintenance in service. In certain circumstances, such as when piers or abutments are high or over water, it may be advisable to incorporate some form of travelling staging in the bridge design to facilitate inspection.

(d) Elastomeric bearings will take a considerable amount of maltreatment before failure unless grossly inferior materials are used. However, localized over loading due, for example, to uneven seating can cause breakdown of the bond between elastomer and steel reinforcing plates. Unreinforced elastomer strips can squeeze or work their way out under certain circumstances. Small seating plinths can disintegrate under shear forces generated by elastomer bearings (the seatings should extend at least 50 mm beyond the edge of the bearing, preferably 100 mm).

(e) Disintegration of poorly prepared bearing seatings (pedestals) is one of the most common cause of bearing failures. This problem has recently been highlighted at the Gravelly Hill motorway inter-change outside Birmingham, England. Here, the bearing seatings have disintegrated and allowed the deck-support beams to drop, causing tension cracks in the locally unsupported deck-slab above.

(f) At another project, it is thought that incorrectly proportioned constituents (too much hardener plus a small quantity of water in the aggregate) led to the failure of 2" high epoxy resin bearing plinths when the precast concrete beams were lowered onto the bearings.

(g) Incorrect installation procedures led to failure of bearings supporting a viaduct over a river estuary. Here large mechanical bearings were to be set on 12 mm thick pads of polyester resin-mortar with a sheet of polythene placed on top of the mortar bed to break the bond between the bearing and the mortar. The mortar was domed, the intention being that surplus material would squeeze out when the fixing bolts were tightened down. In practice, the large quantity of resin mortar needed for each bearing required that it be made up in a number of mixes, and consequently the material could not be considered as entirely homogeneous. On removing the damaged bearings it was found that the polythene sheeting had unevenly curled. Both these results led to a non-uniform support to the bearing, causing failure.

(h) Leakage expansion joints can lead to corrosion of metallic bearings. Unsuitable materials can give rise to problems. Many of the 18,500 sliding rocker bearings installed in the Midland Links viaduct (UK) are not functioning as they should. The bearings are made of three rolled steel plates, the middle one heavily chamfered to allow the top plate to rotate. The steel deck beams rest directly on the top plate with no special sliding medium at the steel to steel interface apart from an initial coating of molybdenum disulphide. Some of the bearings have seized and those that still slide do so very reluctantly! Attempts to introduce lubricant between the sliding surfaces have proved ineffective. This malfunction of rockers not rocking and rollers not rolling is a 'common' feature in such steel bearings.

(i) In a similar manner, the steel deck beams of Vauxhall Bridge over the River Thames in London, built in about 1906, rested directly on steel plates bedded on cill stones. Over the years these corroded and seized to the beams. Movement of the deck caused the front of the cill stones to break away. In 1976 the steel bearing plates were replaced by laminated rubber bearings set on new precast concrete bed stones.

(j) The abutment bearings of Wandsworth Bridge over the River Thames in London consisted of large knuckle leaf bearings, supported on a bank of four flat-sided forged steel (cut) rollers tied together with side bars bolted to each roller. The rollers ran on a bottom casting. As the bearings were subject to uplift, the lower casting of the leaf bearing was tied down to the bottom casting by four one-and-a-half inch-diameter bolts which passed through slotted holes in the middle, or lower leaf bearing, casting. The bottom casting in turn was bolted down to the concrete abutment bearing shelf. The bridge was built in the late thirties and inspection of the bearings in 1973 indicated that although the

PLATE 13

Fig. 5.50 Unevenly loaded neoprene bearing; bad detailing, poor construction; The bearing will eventually fail displacing even the load transfer area

Fig. 5.51 Neoprene bearing located too close to the girder-edge; bad detailing, evident distress and eventual malfunction

PLATE 14

Fig. 5.52 Neoprene bearing failure with accompanying distress in structure

Fig. 5.53 Another view of the failed bearing shown in Fig. 5.52

PLATE 15

Fig. 5.54 Steel plate bearing very poorly maintained; Locked-in debris hinders rocking; Even the quality of concrete very poor where it should in fact be dense and solid

Fig. 5.55 Another view of the bearing shown in Fig. 5.54

PLATE 16

Fig. 5.111 Sealed settlement cracks in abutment wall

Fig. 5.129 River bed very poorly maintained; growth of heavy vegetation will lead to waterway reduction and serious scour during floods

main casting and forged steel knuckle pins were in good condition, the forged steel rollers were badly corroded with no sign of any lubrication having been applied or any protection against the entry of dirt or moisture. Several bolts and a number of the tie-down bolts had broken or bent due to the heads binding on the intermediate casting. The bearings have subsequently been replaced by steel rocker bearings incorporating a PTFE/stainless steel sliding element. These have been set on new bearing plinths. No provision had been made for an expansion joint in the deck surfacing, which consequently cracked at the abutment, allowing water to penetrate down to the bearings.

(k) Other problems that have come to light include roller bearings which have overrun their design travel so that the gear pinions ran off the end of the guidance-rack and were sheared off when trying re-engage on their return; end-flanges sheared off rollers due to insufficient allowance for side thrust on these bearings. Compatibility of steel work fabrication with the drawings is necessary if the bearings are to function in accordance with the design.

(l) Replacing the bearings can be a very difficult operation unless suitable provision has been made in the design of the bridge structure for proper access to the bearings and for jacking-up the bridge-deck temporarily. Long has dealt with the problems of replacing bridge bearings.

5.9.2.3.2 What to Look For
Examine all bearing devices to ascertain that they are functioning properly. Keep in mind that small changes in other portions of the structure, such as pier or abutment settlement, may be reflected in the bearings. Bearings and lateral shear keys are subject to binding and damage from creep in bridges with a relatively high skew. Make a careful examination for any such defects.

Check anchor bolts for any damage and see that they are secure. See that anchor bolt nuts are properly set on the expansion bearings to allow movement as designed.

Expansion bearings must be checked to see that they can move freely and are clear of all foreign material. Rollers and rockers should bear evenly for their full lengths and should be in proper position relative to the temperature at the time of the inspection. Lubricated-type bearings should be checked to see that they are properly lubricated.

Note the physical condition of the elastomeric type bearing-pads and any abnormal flattening, bulging or splitting, which may indicate overloading or excessive unevenness of loading.

Examine grout pads and pedestals for cracks, spalls or deterioration.

Bearings must be examined carefully after unusual occurrences such as heavy traffic damage, earthquake, batterings from debris in flood periods, and temperature extremes.

Examine the concrete for cracks and spalls at abutment seats and pier caps. Check for shearing cracks in the ends of the beams and for edge cracks and spalling in the supporting member.

Rate the condition of the bearing pads, bearings, and anchor bolts at a pier or an abutment, as well as hangers at the end of a cantilever span. This rating should reflect the condition of the worst element rated. At a pier, bearings from two spans are included in the same rating.

• *Steel Bearings:* Look for heavy rust, lateral or vertical displacement (uplift), sheared bolts, cracked welds, rockers extended beyond their proper position for the temperature, and the presence of debris which may prevent free movement.

Where the bearing is subject to uplift, check for "slap" or "hammering" when a heavy vehicle crosses the bridge.

Rocker bearings, where slots are provided for anchor bolts, should be checked to ensure that the bolt is not rusted to the bearing.

Determine whether the bearings are in proper alignment and in complete contact across the bearing surface.

Check bearings that require lubricants to ensure adequate lubrication.

Where bronze sliding plates are used, look for electrolytic corrosion between bronze and steel.

Check anchor bolts for looseness, shear failure and missing nuts.

Measure the horizontal travel of the bearings to the nearest 3.0 mm (1/8 inch) from the reference point. The two punch holes are aligned vertically at the installation-temperature (usually 68°F/20°C). Record the temperature at the time of inspection.

On skewed bridges, bearings and lateral shear keys should be checked to determine if either are binding or if they have suffered damage from the creep effect of the bridge.

Check cantilever girder hanger connections and pin bearing connections for corrosion and improper alignment.

• *Elastomeric Pads:* See Figs 5.50 – 5.55, Plates 13 – 15. Look for delamination, cracking, deterioration, and excessive distortion. When the distortion of an elastomeric bearing exceeds 25 per cent of its height, it is considered excessive.

Check for splitting or tearing, either vertically or horizontally; bulging (caused by excessive compression); variable thickness (other than that which is due to the physical condition of the bearing pads) and any abnormal flattening (which may indicate overloading or excessive unevenness of loading).

• *Anchor Bolts:* Where the bearings must resist uplift forces, each anchor bolt should be struck with a heavy hammer to determine if it has been sheared off.

• *Hangers:* If available, examine bridge plans and shop drawings of the hanger details. Note any unusual features and, if discrepancies exist, analyze until consistency between the plans and the constructed connection can be established by field measurements. Observe the condition of the pin-and-hanger connection, noting any signs of distress such as dishing of the cups or washers; bowing of the hangers away from the web; cracked paint; any notches or cracks, particularly along the edges at the pins; and any noticeable movement or pounding under traffic, specially at the acute corner of skewed bridges.

Measure the thickness of each part in the cross section, sum the measurements including the spaces and compare with the planned dimensions.

Measure the center-to-center of pins and the out-to-out dimension of the hanger and compare with the planned dimensions.

Measure thickness of the hanger at the pins, the center, the end and at other locations where loss of section might have occurred. Compare these measurements with the planned thickness.

Hangers may be fracture-critical (where a single fracture would lead to catastrophic collapse) or redundant; depending on the number of the hangers supporting a member and the redundancy of the supported members.

All hangers are susceptible to both direct tensile and bending stresses. Hangers with only one pin (either top or bottom) are specially prone to cracking failure.

Hanger stresses are increased by corrosion at the pin/hanger interface, by stress-rises (such as deep corrosion pits, notches, and tack welds), and by section-loss from corrosion. These conditions should be observed and documented during the inspection.

Measure the alignment of the hanger plates as viewed along each edge and at the middle, to determine if any bowing has occurred.

Check wind-locks for excessive movement before engaging as well as for binding, jamming or improper-fit.

Measure the joint opening between the girder ends as well as at the roadway level. Record the temperature of the air and the steel at the same time of day, and also the weather condition.

Observe the vertical and the horizontal alignment of the expansion joints, kerbs, parapets and railings at the roadway level. (All these observations may have a correspondence with the bearings.) Each hanger should be subject to thorough hands-on inspection to verify its freedom from cracks and the problems enumerated above. The alignment of the suspended member should be checked to ensure that the hangers are not being subjected to racking forces and that any windlocks or guide-plates are functioning properly. All problems should be documented by photographs, sketches, and comments.

If the review of the information (obtained in the field) reveals non-conformity with dimensions within plus-or-minus 3 mm (1/8 inch); excessive corrosion; no apparent movement in the joints, or apparent distress in the assembly in the form of notches, cracks or distortion; notify the proper authorities immediately. Further cleaning, non-destructive testing and/or disassembly may be necessary.

• *Pot Bearings:* Check that bearing 'guides' are functioning properly and the 'keeper' plates are firmly in place.

There should be uniform clearance around the bearings between the 'keeper' plates and the bottom plate; if not, the steel is carrying some of the load instead of the elastomer.

Look for evidence that there is movement in the bearings in sliding type bearings. Look if seal is not burst or knocked out.

Check for brass chips or elastomer flakes around the base which may be extruded out of the bearing, if the bearing is failing!

• *Self-Lubricating Bronze Plate Expansion Bearings:* Check for evidence that there has been movement at the interface of the plates in the bearing. Expansion movement should occur at the flat surface of the bronze plate.

Look for cracks in the plates. Bronze plate may crack when the bearing is frozen. Check the alignment of the plates.

5.9.2.3.3 Sample Ratings

When rating the bearing-system, (bearings, pads, anchor bolts and hangers), consider the effect the condition of the bearing-system has on its ability to function (i.e. to support or tie-down the superstructure, and to provide for end-of-span movements) as well as consider the condition of the bearing it self. For instance, a bearing system in excellent condition but not operating would be rated lower than an bearing system in poor condition but operating as required.

Figure 5.56 shows a bearing in a good condition — the tilt of the bearing is reasonable for the ambient temperature and therefore rated 7.

Figure 5.57 shows the typical condition of the bearings on a pier. There is minor corrosion and the anchor bolt on the fixed bearing is bent, but the expansion bearing is in the proper position and both bearings are operable — rate 5.

Figure 5.58 shows a bearing which has a partial loss of support due to over-extension. This bearing is not functioning as designed, but, since the photograph was taken in cold season, there is no immediate danger of failure — rate 3.

Fig. 5.59 shows a fractured hanger. The temporary hanger, consisting of threaded rods, is disregarded when rating the bearings for this 'span', but should be noted in the comments — rate 1.

The bearing in Fig. 5.60 is in new condition and is in the proper position for the ambient temperature. For the bearings in this condition, rate the bearings as 7.

Figure 5.61 shows a steel expansion bearing with a roller nest. There is some rust, but the condition is not severe. Note the bent anchor bolts. This condition is also not severe since the bolts are still in place and the dead load of the truss is more than enough to prevent bearing-uplift. Rate bearings, anchor bolts, pads, as 5.

The bearing in Fig. 5.62 is tilted beyond its normal position. Since the photograph is taken when the air temperature was below 0°C, the proper position should be to the right of vertical. With summer temperatures, this bearing will probably tilt even more toward the backwall and possibly cause the bearing to be tilted beyond safe limits. This bearing should be downrated despite the fact that it is in good condition. The debris also should be noted. This bearing is rated 3.

An example of a 1-rating for a bearing is shown in Fig. 5.63. This bearing is almost disintegrated from rusting, and is inoperative. The rating of 1 is appropriate considering both condition and ability to function vis-a-vis original design.

Another example of 1-rating for a bearing is shown in Fig. 5.64. Although the physical condition of the bearing is excellent, the bearing is not functioning at all because it is completely off the pedestal. Such a situation is dangerous and should be corrected promptly. Maintenance forces should be contacted, immediately to make temporary repairs. Whenever such unusual condition is encountered, a description of the situation should be given in the "Remark".

Fig. 5.56 Rate 7

Fig. 5.57 Rate 5

Fig. 5.58 Rate 3

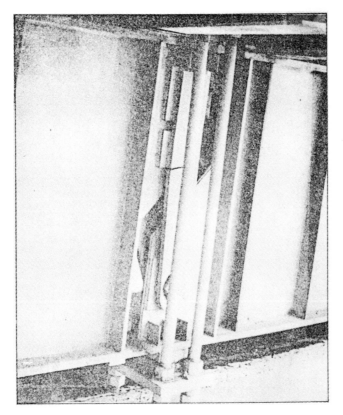

Fig. 5.59 Rate 1

Fig. 5.60   Rate 7

Fig. 5.61   Rate 5

Fig. 5.62 Rate 3

Fig. 5.63 Rate 1

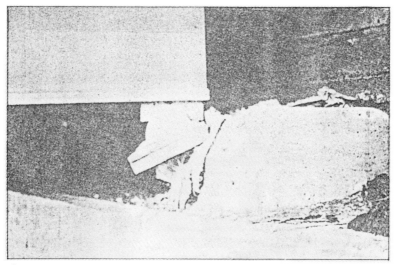

Fig. 5.64   Rate 1

Fig. 5.65   Cantilever girder hanger connection; this type of connection permits both translation and rotation, both windlocks and hangers operating "like-new" — Rate 7.

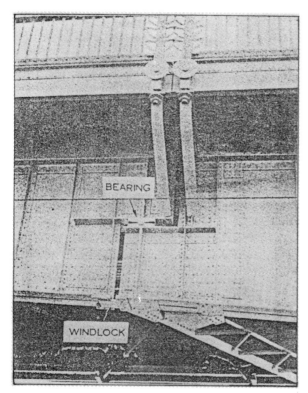

Fig. 5.66 Cantilever girder direct bearing connection, this type permits rotation only, both windlocks and bearing operating "like-new" — Rate 7..

5.9.2.4 Structural Elements of Deck
(i.e. of superstructure)

These are — deck slab (and soffit slab where it exists), longitudinal beams (stems, webs, etc.), transverse beams (diaphragms or cross beams), etc. Depending on their function in a given structure, some of the diaphragms and some slabs may not be primary members.

5.9.2.4.1 Types of Deck

- The *span arrangement* may be
- — simply supported spans
- — balanced-Cantilever and segmentally built Free-Cantilever Spans
- — semicontinuous spans (i.e. simply supported precast p.s.c. girders with R.C. deck-slab laid continuous over piers, thus making them continuous for subsequent loads)
- — continuous spans (either cast as such, or built by stage- construction, or built by segmental cantilever method)
- — frame (single or multi-span)
- — cable-stayed spans
- — etc.

- The *construction* may be in
- — reinforced concrete
- — reinforced and prestressed concrete

- — composite or non-composite structural steel and reinforced concrete
- — structural steel (plate girder or truss)
- The *superstructure cross-section* may be
  - (i) 'R.C. solid slab' type (cast-*in-situ*, c.i.s.)
  - (ii) 'R.C. voided slab' type (c.i.s.)
  - (iii) 'R.C. box section' type (c.i.s.)
  - (iv) 'R.C. beams-and-slab' type (c.i.s.)
  - (v) 'Precast r.c. beams and cast-*in-situ* slab' type,
  - (vi) 'Precast p.s.c. beams and cast-*in-situ* slab' type
  - (vii) 'p.s.c. voided slab' type (c.i.s.)
  - (viii) 'p.s.c. box' type (c.i.s. or precast segmental)
  - (ix) 'precast box segments prestressed with precast concrete web, diaphragm and slab elements' type
  - (x) 'steel beams and concrete slab' composite or non-composite type
  - (xi) etc.

- *Some problems may arise from*
- — concrete cracking (from numerous causes)
- — settlement of foundations (differential, overall)
- — movement/tilting of foundations
- — overload/under-design
- — accidental hit from the traffic
- — accidental fire
- — distress in articulations (Halving Joints)
- — shear lag effect (particularly during the construction stages) whereby the applied prestress is absorbed by only part of the assumed deck-section properly, leading to cracking in understressed regions
- — inaccessibility for inspection within
- — skew effect
- — etc.

Although the transmittal of loads and flow of stresses in a bridge-structure may be very complex, some general stress principles must be known to the bridge-inspector — at least the general types of stresses acting in a structure and their relationship to certain types of distress.

There are basically three types of stresses acting in a structure: normal, bending and shear. Normal stresses are stresses acting along the deck axis (tension or compression). The tension stresses tend to pull the member apart, whereas the compression stresses tend to compress it along the axis.

Bending stresses are developed when a deck is bent say downward, developing tension stresses on the bottom (or convex) side and compression stresses on the top (or concave) side. In continuous deck systems these

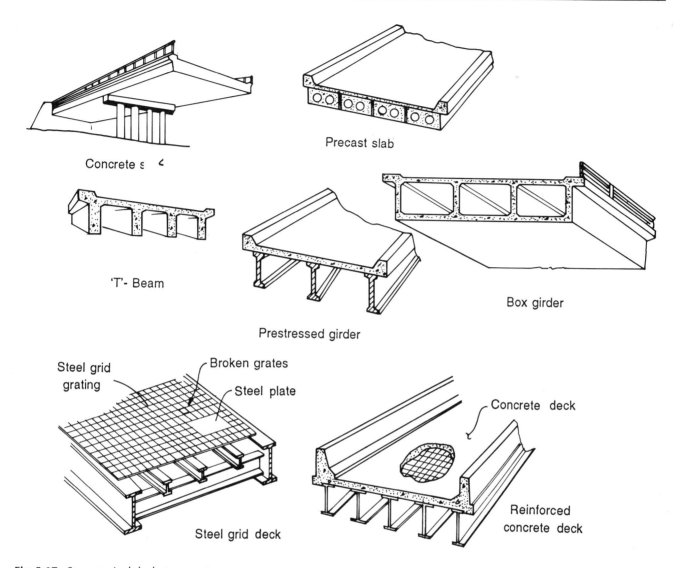

Precast slab

Concrete s ˂

'T'- Beam

Prestressed girder

Box girder

Steel grid grating

Broken grates

Steel plate

Concrete deck

Steel grid deck

Reinforced concrete deck

Fig. 5.67 Some typical deck structure types

stresses at the intermediate supports are reversed, i.e. the tension stresses are at the top and the compression stresses at the bottom.

Shear stresses act perpendicular to the normal stresses and are developed when the normal or bending stresses change in magnitude. These stresses tend to develop tension stress at an angle to the stress plane. These stresses are particularly important in concrete deck systems since tensile strength of concrete is very low.

Concrete is weak in tension. Therefore, it is reinforced with steel in the zones where tension stresses are expected to develop. This is also where potential cracks should be expected to develop. Figures 5.68 and 5.69 show the location of reinforcement steel in a typical concrete deck system and the potential load-induced cracks for a simply supported and a continuous beam

system, respectively. (Also *see* corresponding figures in Chapter 4, showing various types of crack patterns.)

Concrete is strong in compression, therefore in concrete members, subjected to compression, no cracks should generally be expected in compression — unless there is a special reason.

If the design and construction are right then generally distress due to deterioration alone may be expected. However, distress due to impact-damage, earthquake-damage, settlement of foundations, overload/under-design, etc. are possible nevertheless.

In prestressed concrete decks under normal conditions no stress cracks should be expected. Therefore, any cracks found in prestressed concrete decks, in locations where they would be expected in ordinary reinforced concrete, are dangerous, because they may indicate reduction in prestress or heavy over-stress.

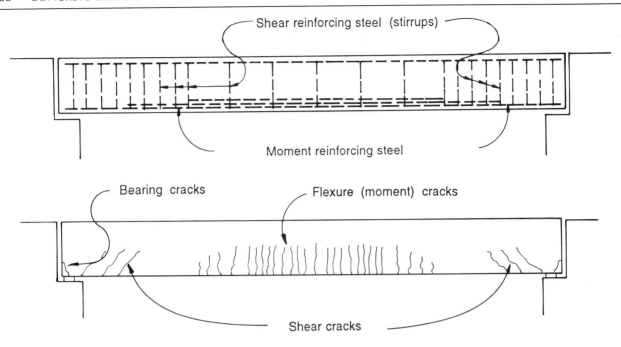

Fig. 5.68 Typical steel reinforcement and location of potential stress cracks (simply supported deck)

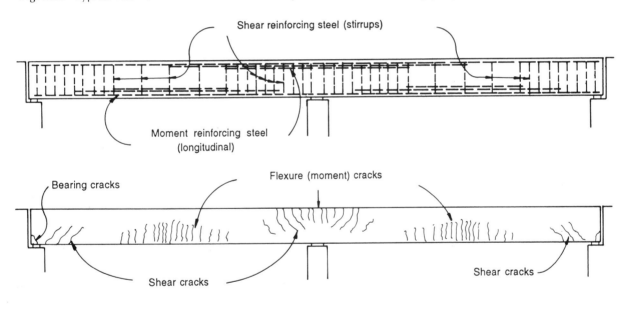

Fig. 5.69 Typical steel reinforcement and location of potential stress cracks (continuous deck)

Cracks around the prestressing anchors are due to inadequate details or poor construction. Other cracks, such as those due to shrinkage, settlement, corrosion, reactive-aggregates and loads are generally similar to those expected in ordinary reinforced concrete members, except that in prestressed concrete the effects due some factors may be more dangerous (e.g. tendon corrosion and misalignment).

Therefore, in prestressed concrete inspection, in addition to looking for ordinary concrete distress indicators, also particularly concentrate on possible potential loss of prestressing, damage and corrosion at anchorages (where accessible), and possible indications of 'tendon-corrosion' and 'increase in deflection'.

5.9.2.4.2 Damage Evaluation in Prestressed Concrete
The determination of degree of damage of a prestressed concrete girder is a matter of judgement based on detailed field inspection, monitoring of cracks, structural analysis and stress checks in the damaged girder, and then comparing with the original design.

The field inspection should include the following checking :

1. Damage to prestressed tendons — location of damage, the number of severed or yielded strands and their locations at cross section of the damage, and comparison with the original as-built plans.

2. Damages to concrete — location of damage, the size, pattern and severity of cracks and spalls, and determination of internal fractures by sounding or other methods, as in reinforced concrete case.

3. Damages to adjacent structural members (such as diaphragms, deck slab and parapets), as in reinforced concrete case.

   (i) *Minor Damage:* Consists of damages to concrete portions only. These can be either slight to extensive surface spalls without exposing prestressing strands, or fine cracks in concrete. Generally, minor damage does not require structural analysis (structural analysis and stress check preferred if the spall damage extensive).

   (ii) *Moderate Damage:* Moderate damage can consist of an extensive spall with exposed strands and fine to medium size cracks in the web and flange (bulb). Generally structural analysis and stress check is required for 'preloading' stress calculation for repairs, if necessary.

   (iii) *Severe Damage:* 'Severe damage' may include following damages to the strands and concrete:

(a) Substantial loss of flange (bulb) section
(b) Loss of strands and severely deformed strands
(c) Cracks extending to web
(d) Substantial loss of portion of web section including breaking of reinforcements
(e) Excessive horizontal and vertical misalignments of the girder; cracking at upper neck

It is apparent that some prestressing force will be lost in the damaged area due to reduced sections, loss of prestressing strands or strand shortening. Structural analysis must be made, based on the reduced section property of the remaining cross section of the damaged girder and number of effective strands, to determine whether the stresses can be accepted.

If the stress is slightly over the allowable value, then it may be considered as minor damage and the repair methods such as epoxy pressure-injection and concrete 'patching' may be adequate for the repairs. Another factor to be kept in mind is whether the damaged girder is a fascia-girder or a girder which is actually carrying live loads. Design specification by AASHTO requires full live load distribution to the fascia-girder but, in analysis of the damaged girder it is prudent to assume partial live load distribution by using simple-beam-actions between the fascia- girder and interior-girder or no

truck live load distribution if the fascia-girder only carries sidewalk area. But this depends on an individual case.

If the overstress is excessive, action should be taken to restrict live loads on the structure or use temporary supports .where possible, etc. until the repair is completed if the remaining strands cannot meet safe enough stress levels. Consideration should be given to strengthening by external post tensioning, preloading, etc., and if necessary even to replacing the girder.

(iv) *Critical Damage:* Includes the following defects (where damage to the girder is considered as not repairable):

(a) The loss of strands is extensive and the prestress force cannot be restored at a reasonable cost.
(b) Wide cracks extend across the bottom flange and well into the web.
(c) Abrupt (kinked) lateral deflection of the prestressed girder, and excessive vertical misalignment.
(d) Any development of wide cracks (this may indicate that the strands have yielded and permanent deformation has occurred).

5.9.2.4.3 Skew Effects

Numerous skew bridges are built in order to be aligned with the approach roadways. Field observations show that the long diagonal of the skewed bridge-deck has a tendency to lengthen during the service life of the bridge. A common problem is the lateral movement of the expansion bearings towards the acute corner of the deck.

Cracks near the fixed bearings (apparently resulting from unusual load-effects and temperature variations) and diagonal cracks at the acute-corner of the deck slab, are the possible problems.

The movements due to skew-effects are indicated in the Fig. 5.70

At the abutment, a lateral creep at the expansion bearing can cause damage to the wingwall. Cracks normally start at the edge of the bearing seat, where it is restrained from free movement.

In designing skewed bridges or curved bridges, problems from creep in deck should be considered so as not to permit the wingwalls to be overburdened by such thrusts.

5.9.2.4.4 What to Look For

The deck is the most exposed part of the bridge. It also provides the most extensive horizontal surface which comes closest to the direct loading from traffic! Hence, it is most susceptible to distress. Accordingly, its life span can be shorter and the likelihood of rehabilitation more imminent.

Concrete decks must be checked for cracking, scaling, pot-holing, spalling, and other evidence of

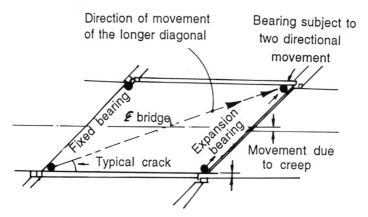

Fig. 5.70 Movements of skewed bridge deck

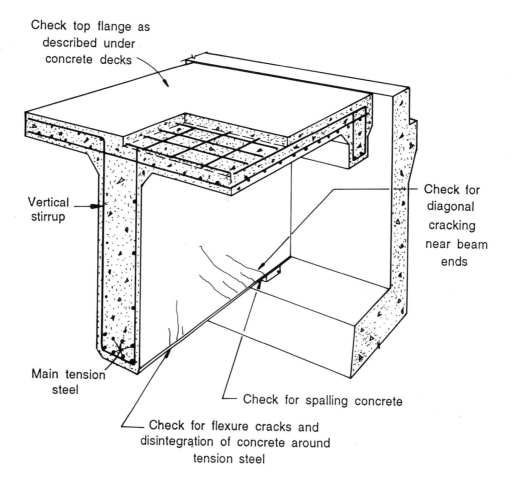

Fig. 5.71 Concrete 'T' beam system

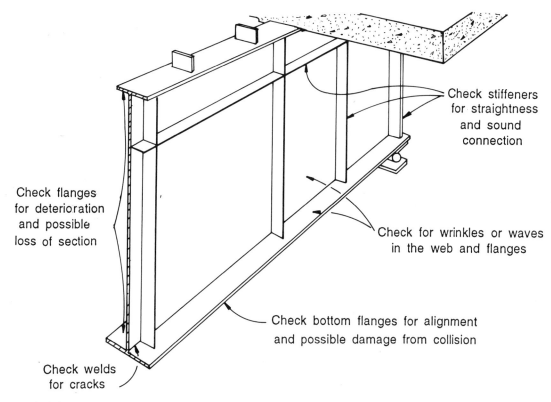

Check stiffeners
for straightness
and sound
connection

Check flanges
for deterioration
and possible
loss of section

Check for wrinkles or waves
in the web and flanges

Check bottom flanges for alignment
and possible damage from collision

Check welds
for cracks

Fig. 5.72 Steel girder system

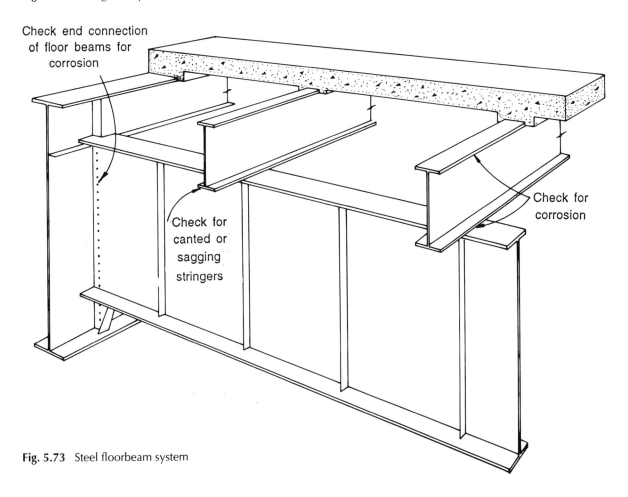

Check end connection
of floor beams for
corrosion

Check for
corrosion

Check for
canted or
sagging
stringers

Fig. 5.73 Steel floorbeam system

deterioration. Each item must be evaluated to determine its effect on the structure, and the work required to restore the loss of structural integrity and to maintain a smooth riding surface. Evidence of deterioration in the reinforcing steel must be examined closely to determine its extent. Decks which are treated with de-icing salts or are located in a marine environment, are specially apt to be affected.

Asphaltic or other type of wearing surface on a deck may hide defects in the deck until they are well advanced. The surfacing must be examined very carefully for evidence of deterioration in the deck. Such defects may show as cracking or breaking up of the surfacing, or, in excessive deflection. Areas, where deck deterioration is suspected, may require removal of small sections of the surfacing for a more thorough investigation. The underside of the deck slab should always be examined for indications of deterioration or distress. Note any evidence of water passing through cracks in the slab. When permanent forms have been used in construction of the deck, some such panels should be removed for a closer check of the portion on them.

Steel decks should be checked for corrosion and unsound welds. It is important to maintain an impervious 'surface' over a steel plate deck to protect against corrosion of the steel, especially in a marine environment and in areas where de-icing salts are used.

All decks should be examined for 'slipperiness' to determine if a potential hazard exists. Also, check drainage to see that the decks are well drained with no areas where water will 'pond' and produce a hazard to traffic. Check drains and scuppers to see that they are open and clear. In addition to being an immediate hazard to traffic, poor deck drainage will usually contribute to deck deterioration. Check to see that drain outlets do not discharge water where it may be determintal to other members of the structure, where they may cause fill and bank erosion, or spill onto a travelled-way below.

This exercise of 'what to look for', when inspecting a bridge-deck, may be summarized into the following, keeping-in view the various possible distresses in concrete described in Chapter 4 and indicated earlier.

1. General

- Examine the alignment and profile of members. Look for collision and fire damage as well as damage that may have occurred due to foundation or substructure movement.

- Observe the behaviour of members with the passage of live loads, and note any excess deflection, vibration or unusual noise.

- Inspect elements supporting water and/or sewer lines for corrosion damage from leakage.

2. Reinforced concrete

- Inspect concrete members for cracks, spalls, scaling, and efflorescence. Sound the concrete with a mason's hammer and if a hollow sound or dull thud is heard, the concrete could be delaminated or deterioration may have started.

- Check elements at points of bearing where friction from thermal movement and high bearing pressure may cause spalling.

- Check elements for diagonal cracks, especially near their supports because diagonal cracks on the web of a beam may indicate incipient shear failure. This is particularly important on older bridges.

- Carefully check cantilever box girder bridges, in which the suspended span rests upon bearings supported by anchor span 5. The re-entrant corners of the cantilevers should be inspected thoroughly for signs of cracking or other deterioration.

- Note any spalls where the reinforcing steel is exposed, check for stains indicating reinforcement is rusting.

- Describe the size and depth of spalling and scaling.

- Describe the location and width of opening of cracks > 0.25 mm.

- Inspect the condition of the overlay on the deck (including past maintenance of the overlay) and repairs to the deck itself that are visible. Cracks and patches in the overlay indicate that there may be problems with the deck slab. Patches of different colours are signs of a continuing problem and patches that do not hold, indicate deck deterioration.

- In open-steel-grating decks, look for broken welds, bolts and rivets. Check alignment and profile of open and filled grating decks. Look to see if the gratings are properly bearing on supporting members. Check the grating for cracks and listen for the sound of loose gratings as traffic crosses the bridge.

- Likewise, in steel components, check for cracked welds, leakage, corrosion, loss of section, secure fastenings and proper support. Check for vibration or banging as traffic passes.

- For concrete slabs, check for scaling, spalling, cracking, rust-staining, efflorescence, dampness, and leakage, and note the percentage of the deck area affected. Frequently, leakage appears on steel supporting members, indicating deck deterioration is taking place. Look especially close on the underside of the deck along curb lines, near joints, and other low areas of the deck where deck deterioration normally starts. When rating concrete deck slabs, remember that concrete deterioration normally starts at the top of the deck and along its periphery. From these locations, the deterioration progresses downward and inward until the entire slab is involved. Therefore, when minor deterioration is observed on the bottom of a slab, chances are that the deterioration is much above this point and the slab should be rated accordingly. Note the coincidence of bad areas between the top and bottom of the deck!

- The following aspects shall specially be looked for *spotting (rusting), cracking, leaching, spalling, and other signs of deterioration in critical areas of the structure, and in portions exposed to atmospheric attack on the windward side.* For the latter, the web of concrete girder and box girder, articulations, underside of decks, etc., shall particularly be inspected.

- Check for and estimate the percentage of exposed reinforcement. Where stay-in-place forms are used, inspect their condition. Deteriorating stay-in-place forms may present a hazard to pedestrians or motorist and are an indication of deck deterioration.

- Drains and scuppers also must be investigated. Extensive asphalt surface deterioration, requiring repeated patching or resurfacing, is an indication of deck deterioration below, especially where the surface repair coincides with bottom stains or other forms of deterioration.

- Look for ponding of water on the deck. If ponding occurs over an abutment or pier, it may indicate a settlement or foundation problem. Ponding between supports may indicate excess deflection in the deck stringers. It may be necessary to have a survey crew check the deck with a level periodically to determine if settlement is occurring.

- Hollow sounding areas may indicate delamination, the possible need for test should be noted.

- Check for flexural and diagonal tension (shear) cracks in the areas of high moment and shear. Discolouration of the concrete surface may be an indication of concrete deterioration or corrosion of reinforcing steel. In severe cases, the reinforcing steel may become exposed and the concrete may crack, delaminate and spall.

- Observe and sound the members exposed to roadway drainage and possible deterioration.

3. Prestressed concrete

- In addition to some of the inspection requirements mentioned in 5.9.2.4.1 for prestressed concrete, it shall also be inspected for all forms of deterioration possible in ordinary reinforced concrete members.

- Check for longitudinal cracks on all flange surfaces, especially on older prestressed bridges where insufficient stirrups were provided and perhaps enough care was not taken while removing the forms.

- Check for cracking and spalling in the areas around the bearings and at the interface between the cast-in-place diaphragms and the prestressed stringers where differential creep and humping of the beams may have some ill effects.

- On pretensioned deck units (either box beams or voided units), check the underside during the passage of traffic to see whether any unit is acting independently of the others.

- For prestressed concrete members, aspects like loss of camber, excessive deflection, and cracking–deterioration in concrete (viz., spalling, etc.), shall be looked for, the critical areas being anchorage zones, junctions of diaphragms, under-sides near the center of span, and near supports.

4. Box girders

- Where accessible, examine the inside of box girders for cracks and to see if the drains are open and functioning properly.

- Check the soffit of the lower slab and the outside and inside faces of the girders for cracking, rusting, delamination, etc.

- Check diaphragms for cracks and other deterioration.

- Examine the underside of the top slab and top flanges for scaling, spalling, and cracking, etc.

- Note any offset which might indicate problems with hinge bearings. An abnormal offset should be investigated.

- Stems of '*T' beams and box girders* are to be checked for abnormal cracking and any disintegration of the concrete, especially over bearings. Note any excessive vibration or deflection. Girders over a travelled-way must be checked for any damage resulting from being struck by overheight vehicles passing under the bridge.

5. Structural steel components

- In case of steel, let us first list out the **'factors' that can cause deterioration**, and then list out 'what to look for' in the steel components of a bridge-deck (same applies to steel-substructure elements also):

    (i) *Air and moisture:* Air and moisture, and chlorides and moisture, cause rusting of steel, especially in a marine climate.

    (ii) *Industrial fumes:* Industrial fumes in the atmosphere, particularly hydrogen sulphide, cause deterioration of steel. (Hydrogen Sulphide forms traces of sulphuric acid when mixed with moisture and atmospheric oxygen.)

(iii) De-icing agents: De-icing salts attack steel under damp conditions (chloride attack).

(iv) *Sea-water and mud:* Unprotected steel members, such as steel piles partly immersed in water and embedded in mud, can undergo serious deterioration and loss of section through chloride attack.

(v) *Thermal strains or overloads:* Where movement is restrained, or where members are overstressed, the steel may yield, buckle, or crack (welds, rivets and bolts may shear).

(vi) *Fatigue and stress concentrations:* Cracks may develop because of fatigue or poor details which produce high stress concentrations. Examples of such details are — re-entrant corners, abrupt changes in plate-widths or thickness, an insufficient bearing area for a support, etc.

(vii) *Fire:* Extreme heat will cause serious deformations of steel members.

(viii) *Collisions:* Trucks, over-height loads, etc., may strike steel beams or columns, damaging the bridge.

(ix) *Animal wastes:* These may cause rusting, and can be considered as a special type of direct chemical attack.

(x) *Welds:* Where the flux is not fully neutralized in welding, some rusting may occur. Welds may crack because of poor welding-techniques or poor-weldability of the steel, inadequate electrodes used, etc.

(xi) *Galvanic action:* Other metals that are in contact with steel may cause corrosion similar to rust action when two metals are connected by an electrolyte (anode-cathode reaction).

(xii) *Fracture and fatigue:* Cracks in structural steel may be grouped as follows.
— Tension-cracks due to overload
— Brittle-fracture cracks
— Stress-corrosion cracks
— Fatigue-cracks

• *Fracture and fatigue-cracks are far more critical in a welded structure than in a riveted structure.* A riveted structure is built up of several components and a crack stops at interface between components. In welded structures if a crack starts in a weld due to accidental impact, improper fabrication, or faulty details, the crack may progress until the member completely fails!

• *Critical bridge members* — any crack is considered undesirable but certain cracks are considered as being more critical than others. If the crack or failure occurs in a critical member, this can cause a sudden catastrophic failure without warning. The following members are considered as critical bridge members.

1. Main suspension cables and anchorages of a suspension bridge.
2. The main eye-bar of a suspension member. Eye-bar systems are more critical than wire or rope suspension members.
3. Hanger-bars and pins in suspended girder systems (e.g. in a span suspended from cantilevering spans). Cantilever-suspended span-system bridges are more critical than continuous multi-girder bridges.
4. A through-girder deck of two through-girders with floor beams, and a single or a double box girder deck. These types of decks are obviously more susceptible to total-failure than multi-girder system decks.
5. Tie-girders of tied-arch bridge deck.
6. Main tension chord of a truss.
7. Steel pier-cap of single or two-column steel piers.
8. Counterweight hanger of a Bascule bridge deck.
9. Anchors for cable stayed bridge decks.

• *Brittle fracture*

Brittle fracture usually occurs without prior plastic deformation or other warning signs. Brittle fracture causes catastrophic failure in the structural steel and generally there is no time available to take corrective action.

The following are the major causes of brittle-fracture :

1. Brittle-fracture occurs due to discontinuties, such as torch-cut edge, a stress-concentration, a mechanical-gouge, welding defects, quench-cracks, porosity, or lack of fusion. These discontinuities could grow to a critical size by fatigue or stress-corrosion.
2. When the metal has less ability to carry a load or to deform plastically in the presence of a notch, this lack of metal-toughness can cause brittle-fracture. For example, silicon steel is susceptible to poor notch toughness.
3. When tensile stresses are required to initiate the fracture, brittle fracture may occur.
4. Other factors affecting brittle fracture are as follows.
— temperature drop and icy conditions
— dynamic loading pulsations
— constraint
— redundancy
— stress-corrosion
— crack-growth
— plate thickness variation of abrupt nature

— fatigue crack growth
— residual stress
— flaws (e.g., in welding)

In 1973, the American Association of State Highway and Transportation Officials (AASHTO) adopted material toughness (Charpy V-Notch Impact) requirement for main load carrying members subject to 'tensile stress', and in 1978 issued the *Guide Specification for Fracture Critical Non-Redundant Steel Bridge Members'* to ensure responsibility of design, fabrication and non-destructive testing of Fracture-Critical Members (FCMs).

Fracture-Critical Non-Redundant bridge members are tension components with a single load path where a single fracture could lead to total failure. Examples of such fracture-critical non-redundant members are: the flange and web of the girders in a two-girder-system bridge, a hanger-plate, a steel cap of single or two-column pier, tie-girder of a tied arch bridge, counterweight hangers of Bascule bridges, and anchors for a cable-stayed bridge deck.

• *Stress-corrosion cracks*

Tensile stress (normal or residual) can cause the stress-corrosion cracks. The tensile force tends to expose more metal at its grain-boundaries (reduction of interfibural overlap), making it easier for the corrosion to bite. As the corrosion continues to penetrate the surface, it will start pitting and will cause a small intergranular stress corrosion crack along the grain boundaries and eventually through the center of the grains. However, not too many cases of stress-corrosion of bridge members are known except for a few cases with prestressed strands and the machinery of movable bridges. Stress-relieving heat-treatment can be given to steel to reduce the chances of stress-corrosion.

• *Fatigue cracks*

Steel may fracture at a stress less than its elastic limit value if the stress is repeated sufficient number of times giving a large range between values of maximum and minimum stress. There have been numerous incidents of fatigue-cracks in welded steel bridges under normal traffic conditions. The most notorious case of fatigue cracks is that which lead to the Silver Bridge failure in 1967. A broken eye-bar caused this catastrophic collapse which cost forty-six lives.

The fatigue and fracture problems should be considered during all phases of project such as study-phase, selection of structural type and materials, development of details, preparation of plans and specification, and erection of the bridge.

The minimum and maximum stresses affect the growth of fatigue cracks. These occur mostly at the ends of the cover plates, stiffeners and gusset plates of beam webs or flanges. **The following locations are also susceptible to fatigue damage.**

1. A loose connection (can cause unbalanced or excessive stresses).
2. Damaged, misaligned or bent member.
3. Areas of excessive vibration or torsion.
4. Corrosion-weakened areas under tension.
5. The toe of fillet welds on a welded steel cover plate on a rolled beam.
6. Welded stiffeners without sufficient gap between the flange plate and stiffener or the web and stiffener.
7. Stringer and floor-beam connection
   • Throat of connection angle
   • Damaged rivet head in the leg of connection angle attached to floor beam
   • Coped top flange of stringers connected to the upper portion of floor beam web
   • When end connection is rigid and it acts as fixed, the coped stringer web would be subject to relative movement between web and flanges and result in fatigue crack
8. Knee braces of floor beams and top connection plates of floor beam which cross over the main girders.
9. Chord splice, eye bars, hanger and diagonal rods, and tension chords of truss bridges.
10. Armoured expansion joints and hanger and pin assembly of suspended girders.
11. Torn or cut-out sections of web and flange.
12. Center and end gusset plate connections of lateral bracing.
13. Stress raisers such as rivet- or bolt-holes, craters in longitudinal fillet welds, intersections of longitudinal and vertical welds, sudden change of size in welded members without transition, severe porosity, discontinuity of back-up bars, lack of fusion, notch, flame cut edge or mechanical indentations.
14. Open-grating welded to top flanges of stringers and floor beams.
15. Previous repairs which used indiscriminate welding or cutting.

Also, certain details in welding design have caused fatigue damages. These could be excessive welding, intersecting weldings, unnecessary members (stiffener perpendicular to applied loads, double-diagonal members, etc.), neglect of strain compatibility, discontinued backing-bars for welds, welding an attachment in tension area instead of bolting, and other over-lookings of the structural behaviour.

Structural members should have as few attachments or connections as possible and the details should be simple and buildable, with enough room for welding. (If suitable welding details cannot used, bolting is preferable.)

What to Look for During Inspection of Steel Components

(i) *Rust:* Rusted steel varies in colour from dark-red to dark-brown. Initially, rust is a fine powder, but as it progresses, it becomes flaky or scaly in character. Eventually rust causes pitting in the member. The inspector should note the location, characteristics, and the extent of the rusted areas. The depth of heavy pitting should be measured and the size of any perforation caused by rusting should be recorded. Rust may be classified as follows:

(a) *Light* — a light and loose rust formation, pitting only the paint surface.

(b) *Moderate* — a looser rust formation with scales or flakes forming. Definite areas of rust are discernible.

(c) *Severe* — A heavy stratified rust or rust scale, with pitting of the metal surface. This rust condition eventually culminates in the perforation of steel section itself, and an eventual collapse.

(ii) *Cracks:* Crack in the steel may vary from hairline thickness to sufficient width to permit light through the crack. Any type of crack is obviously serious, and should be reported at once. Record the location and length of all cracks and indicate whether the cracks are open or closed, dormant or live.

(iii) *Buckles and kinks:* These conditions develop mostly because of damage arising from thermal strain, overload, or added load conditions. The kinks are caused by the failure of the yielding of adjacent members or components. Collision-damage may also cause buckles, kinks and cuts. Look for cracks radiating from cuts or notches. Note the members damaged, the type, location, and extent of the damage, and measure the amount of deformation.

(iv) *Stress concentrations:* Observe the paint around the connections at joints for fine cracks which are indications of large strains due to stress concentrations. Be alert for sheared or deformed bolts and rivets.

(v) *Structural steel:* Inspect structural steel (particularly where it is partially encased in substructure-concrete) at the face of exposure for deterioration and for movement.

(vi) *Galvanic corrosion:* This condition will appear essentially as in rust.

- *In particular, in case of steel frames/trusses*
  - Check connections, connection hardware and fasteners carefully.
  - Examine pins and eyebars on pinned eyebar trusses; check pins for corrosion, cracks and tightness of pin nuts, etc.
  - Inspect steel for corrosion and deterioration especially at the following places
    — along the upper flanges
    — around bolts, nuts and rivet heads
    — at gusset, diaphragm, and bracing connections
    — at cantilever 'hanger and pin' connections
    — under the deck joints and at any other points that may be exposed to roadway-drainage and debris build-up
    — at any point where two plates are in face-to-face contact and water can enter (such as between a cover plate and a flange) if rusting occurs at this interface, the expansive force created can be strong enough to prise the plates apart!
    — at the fitted end of bearing stiffeners
  - If rusting and deterioration is evident, use a chipping hammer to remove loose rust and check the members for possible reduced cross sectional area, using calipers, rulers, corrosion meters, or section-templates.
  - Examine along the seams of 'built-up' members and splices for signs of slippage.
  - Check for wrinkles, waves, cracks, or damage in the web and flanges of steel beams, particularly near points of bearing. This condition may indicate overstressing. Check the stiffeners for straightness and determine whether their connections are broken, buckled, or pulled from the web.
  - On cantilevered bridges, check hinges and hangers to see that they are functioning freely and without restraint due to scoring, jamming, dirt, or corrosion. If a hanger link is out-of-plumb beyond the limits expected for normal temperature variations, a further investigation should be made. Where a hanger has one end fixed rigidly to a web by welding or bolting, so as to develop an eccentric hinge, the hanger will develop both bending and axial stresses. Examine the web and the hanger adjacent to its fixed end for cracking.
  - Check the wind locks for excessive movement before engaging and binding, jamming, or improper fit.

- Where accessible, examine the inside of steel box girders for evidence of corrosion and deterioration.
- Examine welds, weld terminations, and adjacent base metal for cracks, particularly at connections transmitting heavy torsional or in-plane moments to the members. Typical connections of this type are
  — floor beam to girder connection, where either the floor beam or girder is continuous through the other
  — Brackets cantilevered from the fascia beams (or any cantilever connections from a beam).
  — Movement splices and connections.
  — Joint in rigid frame structures or in Vierendeel-type bracing.
- *Also check welds at*
  — sudden changes in cross-section or configuration and other locations subject to stress concentration or fatigue loading
  — Areas where vibration and movement could produce fatigue stress
  — Longitudinal stiffener connections — whether these longitudinal stiffener are also welded to the vertical stiffeners or are purely ornamental, the possibility of web and weld cracking is high
  — At unusual connections, curved sections, reentrant corners and copes
- *Rust and deterioration:* Check through-truss lower chord and framing members, adjacent faces of eye bar heads, pin plates, etc. On riveted trusses, check the horizontal surfaces and connections of lower chord members. Note any deformation caused by expanding rust on the inside of laminated or overlapping plates.
- *Alignment of truss members:* Check end posts and interior members which are vulnerable to collision damage. Buckled, torn, or misaligned members reduce the load carrying capacity. Misalignment can be detected by sighting along the truss chord members. Investigate and report any abnormal deviations.
- *Overstressed members:* Wrinkles in the flanges, webs, or cover plates indicate overstress of a compression member. Overstress of a tension member could result in localized narrowing in the cross section, usually accompanied by flaking of the paint.
- *Loose connections:* Cracked or displaced paint around joints and connections may indicate loose or slipped joints.

- *Pins:* Inspect pins for scoring and other signs of wear. Be sure the spacers, nuts retaining caps, and keys are in place.
- *Noise:* Note clashing of metal with the passage of live loads.
- *Tension members:* Check looped rods for cracking where the loop is formed. Examine eyebars for cracks in the eyes. Determine whether the spacers on the pins are holding the eyebars and looped rods in their proper position. Check the physical condition of threaded members such as truss-rods at turnbuckles.

5.9.2.4.5 Sample Ratings

The inspection ratings are closed tied into the degree of material-deterioration apparent in a Primary Member, as well as the extent to which the primary member retains its original design structural capacity. A primary member with no evidence of material decay at all, and performing at full-design capacity, is given the highest inspection rating of 7 — new condition. Where the primary member exhibits isolated areas of minor types of material decay, as defined herein, but still not to the degree where there is any significant effect on the member's ability to perform at full original design capacity, an inspection rating of 5 — minor deterioration, and functioning as designed, is appropriate. When the primary members have extensive, serious material deterioration, or the primary member system can no longer achieve its full original design capacity, although still able to react elasticity to loadings, thus retaining some degree of its original load-carrying capacity, the inspection rating should be 3, i.e. 'not functioning as originally designed'. Should the primary member system loose practically all capacity to sustain any loadings, and there is an apparent danger of collapse under any further use of this structure, the inspection rating of 1 — potentially hazardous, is applicable.

An important consideration in rating a primary item is how the material-deterioration or capacity-reduction in individual structural elements relates to the performance of the bridge as a whole, and this is a function of the structural type and the nature and extent of deterioration. This means that, although an individual primary member may warrant a low inspection rating, say 2, the primary member item on the inspection form may receive a higher rating, say 3 or 4 if the individual member, which is deficient, is not critical for the structure as a whole to essentially retain its original flexural capacity (although the isolated fascia stringer may have suffered significant structural damage for instance).

The opposite situation can also occur where a small deficiency in size can be so critical as to require a low rating of the primary member item for the whole span. For example, the primary member item on a truss bridge with 12.5 mm diameter hanger rods, each having 6.25 mm diameter loss at the connection points, should be rated very low, say 2 or 1, even if all of the other primary members of the bridge are in good condition.

It is important that the primary member item be downrated only for distress exhibited in the primary member itself. Do not downrate a primary member which is in 'like-new' condition if its structural effectiveness is reduced by problems not appearing in the primary member itself, such as frozen bearings, deteriorated pedestals, or failed substructure columns. Other inspection items and notes will record these deficiencies.

A Rating of U, (*unknown*) should rarely, if ever, be used for primary members. If a metal primary member is partially or completely encased in concrete, the primary member item should still be rated something other than U (*unknown*). In these cases, the condition of the encasement and the condition of the portion of the primary members not encased, if any, will be the basis for the primary member rating. When rating encased members, a note should be included in the inspection report indicating the presence of encasement and stating the limitations of and the basis for the ratings. Further investigation should be requested if the inspector believes there may be a *serious deficiency* that could only be determined by removing encasement.

EXAMPLE RATINGS — CONCRETE MEMBERS

The stem of a concrete T-beam bridge is being rated as primary member in Fig. 5.74. These stems are *rated 7*. Note the original form-lines do not affect the rating.

Fig. 5.74 Rate 7

Fig. 5.75 Rate 7

A reinforced concrete arch and concrete floorbeams are being rated in Fig. 5.75.

These elements are free from deterioration. They are rated 7. A concrete arch is being rated in Fig. 5.76. The arch is generally free from deterioration, although there is some dampness at the springing. There is no cracking, although original form lines are evident. This arch is *rated 5* — minor deterioration.

A *3-rated* concrete arch is shown in Fig. 5.77. Note the extensive leakage, efflorescence and dampness. The entire underside of the arch is deteriorated in this example. If only isolated areas were involved, the rating could be higher.

The stem of a reinforced concrete T-Beam bridge is rated in Fig. 5.78. The concrete stem shows signs of considerable efflorescence, random cracking, and dampness. Even though there are no apparent spalls, serious deterioration has already occurred. This primary member is *rated 3*.

The bottom of a stem of a concrete T-beam span is shown in Fig. 5.79. The reinforcing is completely exposed. The remaining concrete is not sound and shows signs of extensive cracking and dampness. The reinforcing has serious section loss. The primary member of this span is *rated 1*.

Fig. 5.76 Rate 5

Fig. 5.77 Rate 3

Fig. 5.78 Rate 3

Fig. 5.79 Rate 1

A concrete end diaphragm has diagonal tension cracks caused by differential settlement between adjacent beam stems. This diaphragm can still function fairly well and is rated 5, indicating a minor problem. In a case like this, the inspector should ascertain the cause of the stress cracks. If the cause cannot be determined, or if there is any uncertainty as to the seriousness of the situation, further investigation should be called for. Sketches and notes should be made of any problem of this nature.

Figures 5.81 and 5.82 are of the bottom surface of the same deck. The center portion (Fig. 5.81) appears free from deterioration. Near the kerb lines (Fig. 5.82) some dampness with a little cracking and minor efflorescence appears. This is to be expected as bridge deck deterioration usually starts at the low areas (normally along kerb or joint lines). While this deck could remain serviceable for some time, deterioration has started and the deck is *rated 5* indicating minor deterioration.

A concrete deck with transverse cracks in approximately a quarter of the bays is shown in Fig. 5.83. These cracks appear narrow but some efflorescence is coming through. The underside of the deck does not appear damp and does not have alligator-type cracking. There are no signs of leakage on the stringers. This is most commonly observed on decks 10 to 20 years old. This deck is *rated 4*. If the underside of the deck

Fig. 5.80   Rate 5

Fig. 5.81   Rate 5

Fig. 5.82 Rate 5 (isolated deteriorated area of the deck shown in Fig. 5.81)

Fig. 5.83 Rate 4

appeared damp or if leakage was apparent on the stringers, the deck would be rated lower.

The deck in Figure 5.84 is cracked throughout and shows signs of efflorescence. Considerable leakage is obviously coming through the deck, as evidenced by the staining on the floorbeams and the stringers. The deck in Fig. 5.85, although having fewer cracks, exhibits extensive leakage causing heavy staining of flooring members. Both of these decks are extensively deteriorated and are *rated 3.*

The deck slab in Fig. 5.86 is seriously deteriorated as evidenced by loss of concrete cover, cracking, leakage, and spalling. Localized slab failures may occur at any time with a deck in this condition. It represents a potential hazard and is *rated 1.*

A bridge deck constructed with stay-in-place forms is shown in Fig. 5.87. This deck is *rated 7.* There are no signs of leakage on the supporting flooring system and the forms are free from rust. This is a 'like new' condition. Considerably more judgment is necessary in rating decks constructed with stay-in-place forms because the bottom of the deck is covered. The inspector should be aware that rust can occur on the forms due to moisture condensing on the outside of the form. This is most commonly observed on the older jack arch type bridges with low clearances over streams.

EXAMPLE RATINGS — METAL STRINGER AND GIRDER BRIDGES

Figures 5.88 and 5.89 are examples of 7-rated primary member systems. There is no section-loss or cracking, and all primary members can function as originally designed. The diaphragms on the stringer-span are rated as secondary members (*see* Fig. 5.88). The stringer-floorbeam and floorbeam-girder connections are included in the rating of the primary member on the girder span (Fig. 5.89). Note that paint loss should not affect the primary member rating. Paint loss is rated under the paint item.

Figure 5.90 is of the end of a deck girder. Localized deterioration has occurred near connection plates. About 5 per cent metal loss has occurred in this isolated area and the remaining portion of the girder has no section-loss. This girder is *rated 5.*

Fig. 5.84 Rate 3

Fig. 5.85 Rate 3

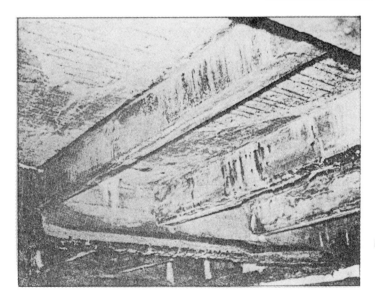

Fig. 5.86 Rate 1

Fig. 5.87 Rate 7— No visible defects; like new condition

Fig. 5.88 Rate 7

A stringer is shown in Fig. 5.91. On close inspection it is seen that some metal has been lost, but the magnitude is small and the stringers are still basically capable of functioning as originally designed. These primary members are *rated 5*. The paint system has failed, but the lack of paint is rated under 'Paint' and does not influence the primary member rating.

A stringer span with minor to moderate deterioration of the bottom flanges and webs, but severe deterioration of the top flanges is shown in Fig. 5.92. The top flange has deteriorated to varying degrees with a few isolated locations where the edge of the flange is extremely thin. The primary member is *rated 3 or 2*.

Figure 5.93 shows the web and bottom flange of a plate girder. After hitting with a hammer, it became apparent that approximately 60 per cent of the web has been lost due to corrosion. Also, numerous rivet heads are severely deteriorated. This is a serious deterioration problem and is *rated 2*.

Figure 5.94 shows a floorbeam whose top and bottom flanges have been badly damaged at numerous locations near the center of the span. This is obviously potentially hazardous and is rated 1. A condition this serious should be brought to the attention of the appropriate authorities immediately.

Fig. 5.89 Rate 7

Fig. 5.90 Rate 5

Fig. 5.91 Rate 5

Fig. 5.92 Rate 3 or 2

Fig. 5.93 Rate 2

Fig. 5.94 Rate 1

Fig. 5.95 Rate 7

Figure 5.95 shows the lateral bracing on the underside of a through-truss bridge. There is no section-loss due to corrosion, connections are sound and the alignment of the bracing is good. This bracing is rated 7 as no deficiencies are observed.

5.9.2.5 Piers, Abutments (and Wing Walls), and their Foundations (e.g. footings, piles and pilecaps, caissons and their caps, etc.)

5.9.2.5.1 These are generally built in reinforced concrete and occasionally partly in prestressed concrete or steel.

FOUNDATIONS
- *Types*
  - footings (open foundations)
  - piles
  - caissons
  - etc.
- *Some problems may arise from*
  - undermining due to scour
  - dislodging of Gabions/protection work
  - exposure of piles
  - concrete cracking (from numerous causes)
  - loosening of anchor-holds
  - settlement (differential, overall)
  - movement/tilting
  - overload/under-design
  - etc.

ABUTMENTS and PIERS (and WING WALLS)
- *Types*
  - open-type (i.e. columnar) abutments
  - closed-type abutments (solid type)
  - solid wall-type pier
  - frame (i.e. columnar)-type pier
  - single shaft-type pier, solid
  - single shaft-type pier, hollow (single or multi-cellular)
  - etc.

Some problems may arise from
- tendency to move/slide/rotate, which may also result in change in plumb and line, and consequent misalignment/displacement in the bearings (with all the consequent effects)
- consequences of
  - foundation scour
  - foundation settlements
- impact-hit from passing vehicles (and from barges and floating debris in case of bridges in water bodies)
- height more than about 10 m_t so that unusual and special equipment may be needed for inspecting and carrying out the maintenance work in inaccessible zones
- concrete cracking (from numerous causes)
- overload under-design
- etc.

5.9.2.5.2 *What to Look For*
- In case of concrete, look for cracking, scaling, spalling, delamination, leaching, rust-stains, impact-damage, etc., all as discussed in Chapter 4 and indicated elsewhere in this chapter.
- In case of steel, look for the various possible distresses indicated in the previous section in this chapter.

(i) In Pedestals: Look for signs of deterioration. In the case of concrete, look for cracking, scaling, spalling, delamination, and leaching (specially near the edges). This is particularly critical where concrete beams bear directly on the abutment. Check bearing-seats for the presence of debris and standing water. For masonry, look for cracks and deterioration of the mortar joints as well as cracks in masonry-units (stones or bricks) and lost masonry units. For steel, look for loss of section and bowing or buckling. Look for deterioration in the concrete making up the pedestal (usually the distress as a result of poor deck-drainage containing deleterious chemical compounds). Also investigate structural distress (pedestal-failure) in the form of larger cracks as a consequence of bearing-overload, structure-movement ('drag' cracks), and cracks caused by 'pavement-shove'

(ii) In Abutment Dirt-Walls (back-walls): Check for signs of tipping any effect from off-centered bearings and other movements. Check particularly the construction joint between the backwall and the cap beam wings. Abnormal gaps (clearances) between the backwall and the end beams are indications of probable movement. A common problem is a 'pavement-shove' damage, which usually shows up in the form of large crack in concrete backwalls, opening of the joints with adjacent wingwalls, or bowing of steel backwalls.

(iii) In Abutment Breast-Wall (i.e. abutment stem, wall or columns): Check for evidence of tilting, splitting, vertical and lateral movement. Also look for signs of material-deterioration specially at areas exposed to roadway-drainage. For concrete, check for mapcracking, leaching, spalling and hollowness. Check for opening of joints with adjacent wingwalls. Determine whether drains and weepholes are clear and functioning properly. Seepage of water through joints and cracks may indicate accumulation of water behind the abutment. Report any plugged weep holes. For masonry: check the stones and the joints for vegetation and deterioration. Note any missing stones.

(iv) In Wing Walls: Check for seepage, tilting, vertical and lateral movements. Check joints with abutment-stem for variation in opening, indicating movement :

In concrete type: Look for signs of deterioration, such as spalling, mapcracking and enfflorescence; sound for hollowness. Describe the type, location and width of the opening of cracks; note the percentage of cracking. Describe the size and depth of spalling and scaling.

• In masonry type: Look for loose joints, loss of joint material, loose or shifting stones, deterioration of the stones (splitting, cracking), and growth of vegetation in joints.

• In sheet piling type: For Steel, look for section-loss from corrosion, and bulging or buckling. Seepage of water through steel sheeting itself is not a deficiency unless the seepage is causing a problem.

• In cribbing type: Check the tightness of the joints and for any loose cribbing members. Look for splitting and bulging, loss of stone fill, settlement of soil by leaching into stone fill, etc.

• In proprietary designs: With the advent of new and more cost-effective construction products in the market, the inspector may encounter wingwalls and abutments constructed of *proprietary materials* such as 'Doublewall' or 'Reinforced Earth'. 'Doublewall' consists of precast interlocking reinforced concrete modules backfilled with earth to form a gravity-type retaining wall. 'Reinforced earth' consists of precast interlocking concrete facing panels that are used at the face of a reinforced soil volume — a composite material formed by the association of special backfill with steel reinforcing strips which together form an integral structure.

• In any type: Check for settlement and misalignment. Take any measurements necessary to describe and evaluate the problem. Measurements are particularly helpful over several inspections to evaluate whether conditions are deteriorating!

In the inspection also check against any uneven or unusual settlement; rotation of the modules or panels off the vertical; change in the batter of the front face (check the batter against the plans); concrete spalls that may indicate wall movement, concrete deterioration, or cracked and displaced units; loss of backfill material through the joints; and erosion or undermining of foundation material at the base of the wall.

(v) Erosion: Disturbance or loss of the embankment covering material is usually obvious from a casual perusal of the embankment. In its early stages, however, loss of embankment-material sometimes is not so obvious unless the inspector is diligent in his inspection. Unevenness on the surface of block-paving is sometimes an indicator of loss of embankment-material. Other signs include soil-marks on the face of the abutment or wingwall, irregularities in the embankment slopes and water seepage from indeterminate sources.

(vi) Scour: (Also *see* Section 5.9.2.7 ahead.) In clear shallow water the evidence of the scour can sometimes be observed, but in most cases it will be necessary for the inspector to probe or sound for scour evidence. If the streambed is near the bottom of footing elevation, the inspector should probe to make that the footing has not been undermined. When some of the underlying streambed material has been lost, it is necessary to further document the extent of the problem.

Whenever it is necessary to take channel profiles near structure units, the inspector shall also check for substructure loss of section by probing for the existence of subaqueous deficiencies. Using a bent rod or a similar probing tool, the inspector may check the structure for erosion of concrete, missing masonry units, broken or rotten timber logging, holes in sheet piling, and other forms of deterioration. The inspector should document the extent of deterioration with a sketch. Where deterioration extends above and below the water surface, the entire area of deterioration should be documented.

Where a bridge spans over an open channel, the latter shall be observed for obstructions, erosion, and the need for cleaning and stabilization. If the channel warrants any of the above actions, it should be indicated and described in the remark section of the inspection form.

The stream channel should be inspected to determine whether conditions exist that could cause damage to the bridge foundations. The two primary items to observe are scour and accumulation of sediment and debris.

Scour can result in fairly uniform degradation (i.e. lowering of the stream channel) and abrupt drops in the channel that move upstream during peak flows. This type of scour is referred to as 'head-cutting', and may be a serious problem if it is occurring in the channel downstream. It may threaten the bridge foundations as it moves upstream. Such items as undermining of trees or production of sediment that could block or reduce the bridge-opening, should be noted.

Deposits of debris or sediment that could block the bridge-opening, causing local scour in the stream-channel, should be noted. Accumulation of debris sediment in the stream may cause scour of the streambanks and roadway embankment, or could cause changes in the channel alignment. Debris and sediment accumulations upstream of the bridge will increase the flow velocity and may result in excessive ponding at and around the bridge.

In case of major bridges built across alluvial rivers, owing to concentration of flow (for various reasons) deeper scour may occur near some of the piers even with discharges smaller than the design discharge. Such deep scour may result in tilting of foundations and consequent damage to various components of the bridge. In order to avoid such situations, the following aspects need special attention for the safety and proper functioning of bridge structure

(a) As far as possible, take soundings before, during and after each flood at all foundation locations for all bridges built across major rivers in alluvial beds with foundations seated in sol-

ids, particularly where the rivers show the tendency to meander and give rise to the concentrated flow. Maintain a permanent record of the same.

(b) Observe the HFL, discharge, obliquity of flow, erosion of banks, functioning of bridge waterways, and changes in flow-pattern, etc.

(c) In cases where such records reveal that scours, as observed, have a tendency to approach/exceed the anticipated/design scour depths, appropriate protective steps (e.g. dumping of boulders around the foundation or full-fledged 'garlanding' of the foundation, laid at suitable levels), which will not cause adverse or deteriorating flow condition of the river, may be resorted to after obtaining necessary approval of the competent authority.

(d) In some cases it may even be found necessary to 'train' the river and guide the flow more uniformly through the various openings by means of a proper training works, such as guide bunds or spurs, etc.

(e) Since even small settlement of foundations may adversely affect the safety of the superstructure, in certain cases it is suggested that suitable non-bearing concrete blocks/wooden packings be suitably provided under the beams very near the bearings so that, should the superstructure get dislodged from bearings, it would ultimately rest on these blocks until rehabilitation is done, thus averting the risk of a total collapse. Also, observation on the movements and tilts of the rollers should be periodically made and permanent record of the same maintained. In case the bearings are found to have moved or tilted to critical condition, immediate action has to be taken for lifting the spans and resetting the bearings.

(vii) In All Protective Works: The following points may be kept in view for inspection and maintenance.

(a) Most careful patrolling and watch is necessary during each flood season, specially the first flood season, to detect the weakness in construction and to take a corrective action promptly.

(b) The engineer-in-charge should acquaint himself with the past history of the protective works and behaviour of the river, because it is only when he possesses this knowledge that he can deal effectively with any problem that may arise.

(c) It is advisable to have a reserve quantity of stone boulders (available at site) which can be used in case of an emergency. A part of the stone may be stocked on the guide bund itself, and a part in the nearest store where it can be loaded and transported quickly to the site. The quantity of reserve boulders would depend upon site conditions. However, 2 per cent of the total quantity of the boulders used in apron and slope-pitching may be kept as a reserve in stock for emergency use.

(d) It is necessary that during flood season the field engineers remain vigilant and keep a careful watch on the behaviour of the river as it affects the 'training' works. During flood season it is advisable to have regular patrolling of the guide bunds and the approach banks, and proper action taken when any abnormal swirls, eddies, or scour are apprehended. Any small rain cuts or displacement by waves along the guide bund or approach bank must be repaired immediately. If not attended to urgently, there is always the danger of a small cut developing into a major disaster.

(e) Any settlement in the bank or bridge, or a slip in the slope, needs immediate attention.

(f) During winter or dry weather, a survey of the river course has to be carried out to a sufficient distance on the upstream and downstream of bridges with guide bunds.

(g) Soundings, preferably with the help of an echo-sounder, should be taken near the guide bund when the river is in flood.

(viii) In case of Foundations, investigate them for evidence of any significant scour or undercutting. Making the inspection at the season of lowest water elevation will facilitate this work. Probing and/or Diving will be necessary at many piers. This will normally be required at approximately five-year intervals, except under unusual conditions. Particular attention should be given to foundations on spread footings where scour or erosion can be much more critical than in a foundation on piles. However, be aware that scour and undercutting of a pier or abutment on piles can also be quite serious. The vertical support capacity normally will not be greatly affected unless the scour is excessively severe, but the horizontal stability may be jeopardized. This condition becomes particularly unstable when erosion has occurred only on one face of a pier, leaving solid material on the opposite face. Horizontal loads may also have been produced by earth or rock fills piled against or adjacent to substructure units and such loads were obviously not provided for in the original design. Such unbalanced loading can produce an unstable condition which must be corrected.

Examine all exposed concrete for the existence and severity of cracks and deterioration of the concrete itself. The latter is becoming an increasing problem in cold weather countries in the areas where de-icing salts from the roadway are carried down by drainage through joints and cracks in the superstructure. The horizontal surfaces of the tops of the piers and abutments are particularly vulnerable to this attack.

Structural steel partially encased in substructure concrete should be inspected at the face of exposure for deterioration and for movement. Stone masonry should be checked for cracking in the mortar joints and to see that the pointing is in good condition. Check for erosion, cavities, cracking and other deterioration of the stones.

Any suspected movement or settlement should be checked with an engineer's level and compared with previous records. Action should be taken as necessary and complete records should be made for future reference.

Timber piles must be checked for decay, specially in areas where they are alternately wet and dry. The most likely place for this condition to be found is at the ground-line. Such determination will require boring at periodic intervals. Holes made for testing, which might promote decay, should be filled with treated wooden plugs. The timing of such borings will vary greatly from area to area because of climatic variations, species of wood used for piling, and the preservative treatment that had been given to the timber. Although piles may appear sound on the outer surface, some may contain advance interior decay. Creosoted piles, for example, may become decayed in the core-area where the treatment has not penetrated, even though the outside surface shows no evidence of deterioration. Hammer testing will many times reveal and unsound pile.

All timber piles in salt water must be checked for damage by marine borers which will attack timber in the area at and below the tide line down to mud line. Footing piles, which had been exposed by scour below the mud line, are highly vulnerable to attack. Attack may also occur in treated piles where 'checks' in the wood, bolt-holes, or other connections, provide an entrance to the untreated heart-wood area.

Special attention should be given to the contact surfaces of timbers when exploring for decay, and also to areas where earth or other extraneous material may have accumulated. Areas such as the top of piles where the cap bears, and where the bracing members are fastened, and also the checked or split areas, are sections very susceptible to decay. Caps must be examined for decay, cracks, and other evidence of over-stress.

Observe the caps under heavy loads to detect any excessive deflection. Steel caps should be observed for any rotational movement resulting from eccentric connections. Check condition of the web stiffeners.

Examine concrete and steel piles both in the splash zone and below water surface for corrosion and deterioration.

Bracing members must be checked to see that they are adequate, sound, and securely fastened. Observations should be made during passage of heavy loads to see if unusual movement occurs in any of the members.

(ix) Footings, Caissons and Piles: These are usually visible only after erosion of embankment material. When visible, look for signs of deterioration such as mapcracking, leaching, spalling and hollow sounding concrete. Also look for signs of distress in the form of large cracks or splitting. Look for loss of steel- pile section due to rust, and any pile-buckling. Look for concrete-pile deterioration, such as cracking, spalling and efflorescence. (Look for the various possible distresses in concrete and steel, indicated earlier. Look for erosion and scour as indicated earlier.)

Protruding parts of footings may be subject to damage from collisions by vehicles or water traffic. Check and record any damage found. Any footings in water are subject to deterioration at the water line; inspect this area closely.

(x) In the Stems (i.e. columns or shafts or walls) of Piers and Abutments: Look for signs of concrete deterioration, mapcracking, efflorescene, spalling and deterioration, and tap with hammer to ascertain limits of unsound concrete. Check for any distresses or damage from any impacts/accidents, or from any chemical effects. Take measurements to describe movements over several inspections and evaluate the cumulative condition. In *masonry* stems, look for loose, cracked and missing stones and vegetation growth in them. Check (and record) any damage resulting from collision by vehicles or waterborne traffic. Piers in water are subject to deterioration in the splash zone. Stems should be inspected down to ground line. Piers in water should also be inspected around low-water level or low-tide level, as the case may be.

Look for earth or rock fills placed against solid stems not provided for in the design.

All piers and abutments should be examined for any tilting and settlement and impact-damage from vehicles, ice, or debris. Check for deterioration, specially near the water-line, at the ground-line, and where the structure body is exposed to roadway-drainage either through a deck-joint or by vehicle-splash. Whenever possible, underwater portions should be visually inspected. Therefore these inspections should be conducted during periods of low-water or -tide. When the water depth is greater than 1 meter and the structure is not visible at the ground-line (stream bed), it should be noted by the crew leader that the inspection could not be accomplished and that a follow-up inspection is necessary. (*See* 'Erosion' and 'Scour' described earlier.)

In case of steel: look for loss of section of the main members, bracing and connections. Check for local and column buckling. Also look for various other distresses that are possible in steel, as described earlier.

(xi) Pier-caps and Abutment-caps: These usually have a relatively large amount of reinforcing steel in the top face or bottom face, or both. This steel is protected only by concrete-cover and is very susceptible to corrosion-damage induced by drain-water containing dissolved corrosive elements (and any debris) coming through deck joints. The early stages of corrosion-damage are cracking and delamination followed by spalling of the concrete-cover and loss of steel section.

Concrete cap-beams must be sounded with a hammer to find areas of delamination and/or unsound concrete. Delamination will produce a hollow sound when struck; unsound concrete will produce a dull thud and only 'sound' concrete will have a resounding ring when struck. The concrete must also be checked for map cracks, stress cracks, and efflorescence, etc., all as described earlier.

A heavy build-up of debris, combined with inefficient deck-drainage, creates a corrosive environment for the cap. Such a condition, while not serious in itself, should be a cause for low rating because of the potential for rapid cap deterioration.

For instance, a pier or abutment cap with a heavy build-up of debris and moisture, but otherwise in good condition, should be rated 4.

Differential settlements and other movements of foundations can cause serious distress, culminating in large cracks in them and/or in the pier and abutment stems. Look for any unusual movements during the passage of heavy loads.

5.9.2.5.3 Sample Ratings
Extensive deterioration of the pedestal and/or bridge-seat with noticeable loss of bearing-area under the bearings. *See* Fig. 5.96.

Potentially dangerous situation; either due to extreme deterioration of distress.

Figure 5.97 shows an example of a pedestal that has failed due to movement of the superstructure. This pedestal has lost virtually all of its ability to support the beam. This pedestal is rated 1. A condition this severe should be brought to the attention of appropriate authority immediately.

Figure 5.98 shows the deterioration area at the end of a solid pier. The spall areas and the extensive cracking are evidence of serious deterioration and the stem

Fig. 5.96 Rate 2

Fig. 5.97 Rate 1

would be rated 3, and possibly lower. Concrete such as this may be in worse condition than is visually apparent; sounding is essential to determine a proper rating. The location of the deterioration in relation to the bearing area is a consideration in rating this pier.

Figure 5.99 shows a pier which has obviously failed. Rate 0 — dangerous condition.

Figure 5.100 shows a railroad-type pier, with built-up and cross- bracing. There is some minor loss of section on all members so that the pier column item is rate 5. Note that the top member between the columns is included in the column rating since it is a bracing mem-

ber; if it carried direct load, it would be rated under the capbeam item.

Figure 5.101 shows a stem of a solid masonry pier that has settled and is severely deteriorated. The loss of stones, shown better in Fig. 5.102 endanger the bearing areas since progressive upward loss of stones is likely. Either the settlement problem, which has caused some buckling of the floor bracing in the near span, or the loss of stones, is a serious enough to warrant a 1 rating.

In Fig. 5.103, the deterioration of the concrete at the splash zone has exposed the main rebars causing them to rust. There has been a substantial but critical loss of section. This condition should be rated 3.

Fig. 5.98 Rate 3

Fig. 5.99 Rate 0

Fig. 5.100 Rate 5

Fig. 5.101 Rate 1

Figure 5.104 shows an individual column with a shear crack. This column has undergone shear failure. The temporary bracing is ignored for the purposes of rating the condition of the column. Therefore, the inspector must conclude that the conditon of the pier column is dangerous and should be rated 0.

Figure 5.105 shows considerable cracking throughout the cap beam. This type of cracking is a sure clue that spalling is imminent. Note also considerable efflorescence. Rate this 3.

Figure 5.106 is an example of pier cap that could be rated 2. Note shear cracks, extensive spalling and mapcracking; exposed and corroded reinforcement, deteriorated concrete core.

Figure 5.107 shows the entire bottom row of rebars exposed and heavily rusting. There is no rebar bond to the concrete. The concrete is soft and spalling heavily. The temporary column is ignored for the purpose of rating. The cap beam would be rated 1.

Figures 5.108 and 5.109 show two views of a failed cap beam. The crack between the bearings could cause a loss of support for the fascia girder. This situation should be flagged and immediate reported to the appropriate person. Rate 1.

Fig. 5.102 Rate 1

Fig. 5.103 Rate 3

Fig. 5.104   Rate 0

Fig. 5.105   Rate 3

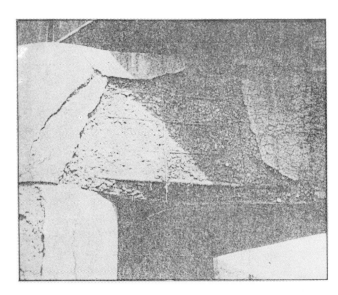

Fig. 5.106 Rate 2

Fig. 5.107 Pier cap Rate 1

Fig. 5.108 Rate 1

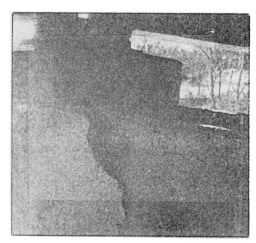

Fig. 5.109 Rate 1

5.9.2.6 Foundation Movements

Most foundation movements are cuased by movement of the supporting strata. For this reason, it is desirable to give a brief description of these movements. Soil deformations are cuase by volume changes and settlement or subsidence in the soil or by a *shear failure*. Slope-side and bearing-failures are good examples of shear failures. Where loads are not large enough to cause shear failure, settlement may still occur as a result of volume change. The length of time and magnitude of settlement depend upon the composition of the soil. Granular soil such as sand will usually undergo a relatively small volume change in a short period of time. However, cohesive soils, such as clay, can undergo large deformations, or volume changes, which may continue for years. This latter process, called *consolidation,* is usually confined to clays and clayey silts. Substructures that are supported directly by a cohesive soil, may continue to settle for a long period of time. Consolidation usually produces vertical settlement.

- *Types of Movements*

For convenience, foundation movements may also be classified, somewhat arbitrarily, in to the following.

(i) Lateral Movements: Earth-retaining structures, such as abutments and retaining walls, are susceptible to lateral movements, although piers too sometimes undergo such displacement.

(ii) Vertical Movements (settlements): Any type of structure not founded on solid rook may be subject to settlement (Figs 5.110 and 5.111 – Plate 16).

(iii) Pile Displacement: Pile settlement could be listed under lateral or vertical movements. Some of the causes of failure are peculiar to piled foundations (e.g., over-estimation of pile group capacity by ignoring reduction owing to overlapped pressure bulbs).

(iv) Rotational Movement (tipping): Rotational movement of substructures can be considered to be the result of asymmetrical settlements or lateral movements.

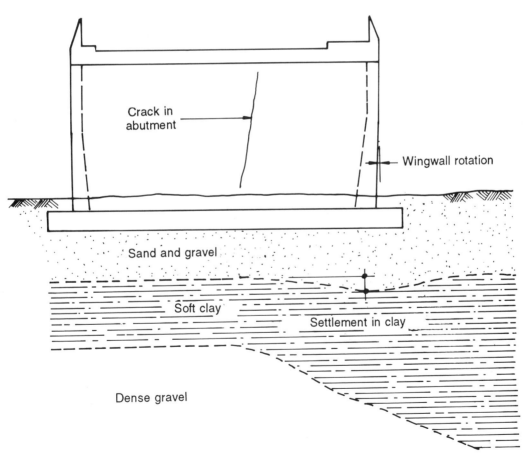

Fig. 5.110 Differential settlement under an abutment

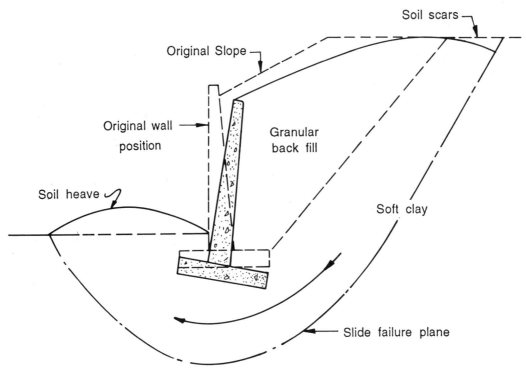

Fig. 5.112 Slide failure

- *Causes of Foundation Movements*

The following causes of foundation movements, except as specifically noted, can produce lateral and/or vertical movements, depending on the characteristic of the loads or substructures.

(i) Slope Failure (embankment slides): These are the shear failures manifested as lateral movements of hillsides, cutslopes, or embankment. Footing or embankment loads, imposing shear stresses greater than the soil shear strength, are common causes of slides (Fig. 5.112).

(ii) Bearing Failures: Bearing failures are settlements or rotations of footings due to a shear failure in the soil beneath (Fig. 5.113). When bearing failures or slope failures take place on an older structure, it usually indicates a change in the subsurface conditions. This may endanger the security of nearby structures and foundations.

(iii) Consolidation: Serious settlement can result from consolidation action in cohesive-soils. Settlement of bridge foundations may be caused by changes in the ground water conditions, placement of additional embankments near the structure, or increase in the heights of existing embankment.

(iv) Seepage: The flow of water from a point of higher head (elevation, or pressure) through the soil to a point of lower head is seepage (Fig. 5.114). Seepage develops a force which acts on the soil through which the water is passing. Seepage results in lateral movement of

the retaining walls by

- (a) An increase in weight (and lateral pressure) of the backfill because of full or partial saturation.
- (b) A reduction in resistance provided by the soil in front of the structure.

(v) Water Table Variations: Large cyclic variations in the elevation of the water table in loose granular soils may lead to a compaction of the upper strata. The effects of non-cyclic changes in the water table, such as consolidation, slides, and seepage, have been described above. Changes in the water table may also change the characteristics of the soil which supports the foundation. Changes in soil characteristics may, in turn, result in the lateral movement or the settlement of foundation.

(vi) Frost Action: Frost-heave in soil is caused by the growth of ice lenses between the soil particles. Footings located above the frost line may suffer from the effects of frost-heave and a loss in the bearing capacity due to the subsequent softening of the soil. The vertical lifting by frost and ice actions are not unheard of.

(vii) Expansive Soils: Some clays, when wet, absorb water and expand, placing large horizontal pressures on any wall retaining such soil. Structures founded on expansive clay may also experience vertical soil movements (reverse settlement).

(viii) Ice: Ice can cause lateral movement in two ways. Where fine-grained backfill is used in the retaining structures, and the water is above the frost line, the

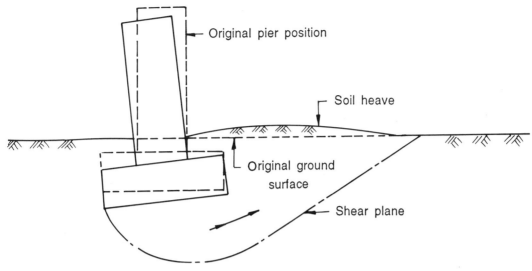

Fig. 5.113 Bearing failure

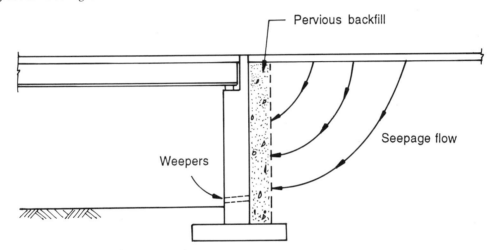

Fig. 5.114 Seepage forces at an abutment

expansion due to freezing of water exert a very large force against the wall. The piers of river bridges are also subject to tremendous lateral loads when an ice jam occurs at the bridge.

(ix) Thermal Forces from Superstructures: On structures without sliding bearings, or where the sliding bearings fail to operate, thermal forces may tip the substructure units. Any horizontal thrust is another force that will have the same effect.

(x) Drag Forces: Additional embankment load or slow consolidation of a subsurface compressible stratum will exert vertical drag down forces on the bearing piles. This may cause yielding or failure of the piles.

(xi) Deterioration, Insect Attack and Construction Defects: Piles may develop weakness, leading to foundation settlements from one or more of the following causes:

(a) Timber, steel and concrete piles are subject to loss of section because of decay, rusting and deterioration.

(b) Timber piles are vulnerable to marine borers and shipworms.

(c) Construction defects include over-driven piles, under-driven piles, failure to fill pile-shells completely with concrete, or imperfect casting of a cast insitu pile. Any of these defects will produce a weaker pile.

Settlement will probably be gradual in improperly driven piles or in piles with weak or voided concrete. Piles suffering severe loss of section due to rust, spalling, chemical action, or insect infestation, may fail suddenly under an unusually heavy load.

(xii) Scour (Erosion): Scour can cause extensive settlement which may also be uneven. Since water will carry

off particles of soil in suspension, a considerable hole can be formed around piers or other similar structural objects. This condition results in a greater turbulence of the water and an increased amount of soil particles can be displaced.

- *Effects on Structures*

The effects of foundation movements upon a structure will vary according to the following factors.

(i) Magnitude of Movements: All foundations undergo some settlement, even if only elastic compression. All sizable footings probably will experience at least a minute differential settlement. However, very small foundation movements may have no effect. Simple span structures, and those with enough joints, will tolerate even moderate differential displacements with little difficulty other then minor cracking. Movements of large magnitudes, especially when differential, can cause distress in structure. Large movement will cause deck joints to jam, slabs to crack, bearings to shift, substructures to crack, rotate, or slide, and supperstructures to crack, buckle, and possibly even to collapse. The larger the settlement to the accommodated within a given span, the more the anticipated structural damage.

(ii) Types of Settlement: These may be of the following types.

(a) *Uniform settlement:* A uniform and simultaneous settlement of all the foundations of a bridge will have little effect upon the structure. However, this rarely happens. If the spans are statically determinate, there is hardly any structural distress even if all foundations did not settle simultaneously, except for the formation of the humps and sags.

(b) *Differential settlement:* Differential settlements can produce serious distress. Where the differential settlement occurs between different substructure units, the magnitude of the damage depends on its span length, and whetherr or not the bridge type is statically indeterminate. Should a differential settlement take place beneath the foundation of the same substructure, damage can vary from an opening of the vertical expansion joints (if present), or formation of near vertical cracks, to severe tipping and cracking of walls or other members. Unaccounted scour can cause failure.

(iii) Types of Structure: These may be either determinate or indeterminate.

(a) *Simple spans i.e. determinate structure:* As mentioned above, the strength of a simple span or determinate structure usually is not affected by movements unless they are too large for the geometry. There are usually enough joints to permit the movements with out a major damage to the basic integrity of the structure. At most, some finger joints or bearings may require resetting, or beam support may need shimming. However, pile bent or trestle bridge are very vulnerable since a large settlement or movement of a bent could cause the superstructure to fall off a narrow bridge seat, leading to the loss of spans.

(b) *Statically indeterminate structure:* A statically indeterminate bridge superstructure can be seriously affected by differential movements, since such movements at supports will redistribute the loads, possibly causing large overstresses. For example, a fixed-ended arch could be severely damaged if a foundation rotated or settled. Continuous bridges have fewer joints than simple span bridges. Such bridges are very likely to be damaged if subjected to displacements which are greater in magnitude, or different in direction, from those that were considered in the original design. However, longer the span, lesser the damage due to a 'given' differential settlement because the bending moment caused by it is inversely proportional to the square of the span length.

- *What to Look for During the Inspection*

Foundation movements may often be detected by first looking for deviations from the proper geometry of the bridge. With the exception of curved structures, haunch members, and steeply inclined bridges, members and lines should usually be either parallel or perpendicular to each other. While not always practical, especially for bridges spanning large bodies of water or for those located in urban areas, careful observation of the overall structure for lines that seem incongruous with the rest of the bridge is a good starting point. For a more detailed inspection, the following methods are often useful :

(i) Check the Alignment: Any abrupt change or kink in the alignment of the bridge may indicate a movement of a foundation or bearings. Older bridges are particularly vulnerable to unequal pressures which can cause structural misalignment.

(ii) Sight Along Railings: A sudden dip in the rail-line is often the result of settlement of a pier or abutment.

(iii) Run Profile Levels Along the Centerline and/or the Gutter Lines: This inspection technique will not only help to establish the existence of any settlement, but will also identify any differential settlements across the roadway. Normally this kind of inspection technique will be employed only for large bridges or where

information concerning the extent and character of differential settlement movement is required.

(iv) Check Piers, Pile Bents, and Abutment Faces for Plumbness with a Transit: This inspection method provides an excellent check for the simpler techniques of plumbness determination. An out-of-plumb pier in either direction usually signifies foundation movement; it may also indicate a superstructure displacement. For small bridges and preliminary checks, the use of plumb bob is an adequate means for determining plumbness.

(v) Observe the Inclination of Rockers and the Roller Movements: Rocker inclinations inconsistent with seasonal weather condition may be a sign of foundation or superstructure movement. Of course this condition may also indicate that the rockers were set improperly. Undue roller movements and distortions in elastomeric bearings can also be indicators.

(vi) Observe Expansion Joints: Observe the expansion joints for signs of opening or rotating. These conditions may indicate the movement of subsurface soils or a bearing failure under some of the footings. Abnormally large or small opening elevation differential, or jamming of the finger joints can be caused by structure movements. Soil movements under the approach fills are also frequent occurrences.

(vii) Observe Slabs, Walls, and Members: Cracks, buckling, and other serious distortions should be looked for. Bracings as well as the main supporting sections should be scrutinized for distortion.

(viii) Check Backwalls and Beam Ends: Check the backwalls for cracking which may be caused by either abutment rotation, sliding, or pavement thrust. Check for beam-ends if bearing against the backwall. This condition is a sign of horizontal movement of the abutment.

(ix) Observe Fill and Excavation Slopes: Slide scarps, fresh sloughs, and seepage are indication of past or imminent soil movement.

(x) Scour: (Please *see* ahead under the section 'Waterways').

(xi) Unbalanced Post-Construction Embankment or Fill: Embankments or fills should be checked for balance and positioning. Unbalanced embankments or fills, not provided for in design, can cause a variety of soil movements which may impair the structural integrity of the bridge.

5.9.2.7 Waterways

Waterways should be inspected in order to determine whether any condition exists that could cause damage to the bridge or the area surrounding the bridge. In addition to inspecting the channel's present condition, a record should be made of any significant changes that may have taken place in the channel, attributable either to natural or artificial causes. When significant changes

have occurred, an investigation must be made in to the probable or potential effects on the bridge structure. Events which tend to produce local scour, channel degradation, or bank erosion are of primary importance.

* *Scour:* (Also *see* Section 5.9.2.5.2 earlier.) Scour is defined as the removal and transportation of material from the bed and banks of rivers and streams as a result of the erosive action of running water. Some general scouring takes place in all stream beds, particularly at the flood stage. The characteristic of the channel influences the amount and nature of scour. Accelerated local scouring occurs where there is an interference with the stream flow, e.g. approach embankments extended into the river or piers and abutments constructed in the river flow. The amount of scour in such cases depends on the degree to which stream flow is disturbed and obstructed by the bridge and on the susceptibility of the river bottom to scour action. Scour depth may range from zero in hard rock to about 10 m or much more in very unstable river bottoms. In determining the depth of local scour, it is necessary to differentiate between true scour and apparent scour. As the water level subsides after flooding, the scour holes that where produced, tend to refill with sediment. Elevations taken of the stream bed at this time will not usually reveal true scour depth. However, since material borne and deposited by water will usually be somewhat different in character from the material in the substrata, it is often possible to determine the scour depth on this basis. If, for example, a strata of loose sand is found overlaying a hard till substratum, it is reasonable to assume that the scour extends down to the depth of the till. This can often be compared by sounding or probing, provided the scour depth is limited to a few metres. Where coarse deposits or clays are encountered, sounding will probably be unsuccessful.

* *Stream Bed Degradation*

Stream bed degradation is usually due to artificial or natural alteration in the width, alignment, or profile of the channel. These alterations which may take place at the bridge site or some distance upstream for downstream, upset the equilibrium, or regime, of the channel. A channel is said to be *stable* or *in regime* if the rate of flow is such that it neither picks up material from the bed nor deposits it. In the course of years, the channel will gradually readjust itself to the changed conditions and will tend to return to a stable, i.e. regime conditions. Stream bed degradation and scour seriously endanger bridges whose foundations are located in an erodible river-bed-deposit and where the foundation does not extend to a reasonable depth below that of the anticipated scour. Removal of material adjacent to the foundation may produce lateral slope instability, causing damage to the bridge. Either concrete slope protection or riprap is often provided to prevent bank erosion or to 'streamline' the flow at obstructions. It is particularly important where flow velocities are higher or where considerable turbulence is likely. It may also be necessary where there is a change of direction of the waterway. Slope cones around abutments are very susceptible to erosion and should usually be protected.

● *Waterway Adequacy*

Scour and stream bed degradation are theoretically the result of inadequate waterway under the bridge. The geometry of the channel, the amount of the debris carried during high water periods, and the adequacy of freeboard should be considered in determining waterway adequacy. Where large quantities of debris are expected, sufficient freeboard is of the greatest importance.

● *What to Look for During Inspection*

(i) Maximum Water Level: Ideally, waterways should be inspected during and immediately following floods, since the effects of high water will be most apparent at these times. Since this is not always possible, a knowledge of the heights of past major floods from stream-gauging record, or from other sources, together with observations made during or immediately following high water, are helpful in determining the adequacy of the waterway openings. Other sources are:
  (a) High water marks or scars left on neighbouring trees
  (b) Water marks on structures
  (c) Debris wedged beneath the deck of the bridge or on the bridge seats
  (d) Information from established residents in the area

(ii) Insufficient Freeboard: This is a prime characteristic of inadequate waterways. In addition to the signs mentioned previously, lateral displacement of old superstructures is a prime indication of insufficient freeboard.

(iii) Debris: Debris compounds the problems of scanty freeboard. Check for debris deposits along the banks, upstream and around the bridge superstructure.

(iv) Obstruction: Debris or vegetation in the waterways, both upstream and downstream, may reduce the width of the waterways, and increase scour, and even become a fire hazard. Sand and gravel bars formed in the channel may increase velocity and lead to more scour near piers and abutments.

(v) Scour:
  (a) *Channel profile:* In stream beds susceptible to scour and degradation, a channel profile should be taken periodically, generally at 30 m intervals, extending to a few hundred metres upstream and downstream.

This information when compared with the past records, will often reveal such problems as scour tendencies, shifts in the channel, and degradation.
  (b) *Soundings:* Soundings for scour should be taken in a radial pattern around the large river piers.
  (c) *Shore and bank protection:*
    — examine the condition and adequacy of the existing bank and slope protection works,
    — check for bank erosion caused by improper location or skew of the bridge piers or abutments,
    — note whether channel changes are impairing or decreasing the effectiveness of the present protection works,
    — determine whether it is advisable to add more channel protection or to revise the existing protection.

(vi) High Backwater: Be particularly alert for locations where high fills and inadequate or debris-jammed culverts may create a very high backwater. The fill acts as a dam, and if it is drawn down by being washed out as is highly possible during heavy rainfall, a disastrous failure could result.

(vii) Observe the effect of wave action on the bridge and on its approaches.

(viii) Observe the areas surrounding the bridge and its approaches for any existing or potential problems, such as *debris jams*.

(ix) Observe the condition and functioning of existing *spur dikes* in the vicinity.

Figures 5.115 to 5.128 show some of the more common waterway problems. Also *see* Figs. 5.129 and 5.130, Plates 16 and 17.

● *Waterway — What to Look For*

— Observe the adequacy of the waterway-opening under the structure. The amount of debris carried during flood period must be considered in this determination. Inadequate free-board for (ice and) debris present a serious threat to the structure during high-water, the high-ice marks and the debris-marks, with the date of its occurrence, should be recorded for future reference.

— Maintaining a channel-profile record for the structure, and revising it as significant changes occur, provide an individual record of the tendency toward scour, channel-shifting, degradation or aggradation. A study of these characteristics can help predict when protection of piers and abutments may be required. Being able to anticipate problems and taking adequate positive steps will avoid or minimize the possibility of future serious difficulties.

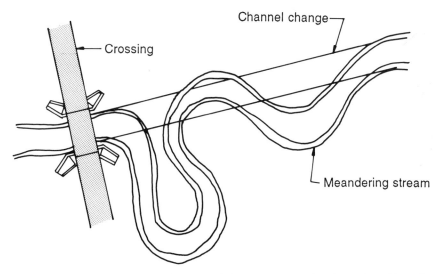

Fig. 5.115 *Channel change:* The channel change steepens the channel profile and increases flow velocity. The entire section may degrade.

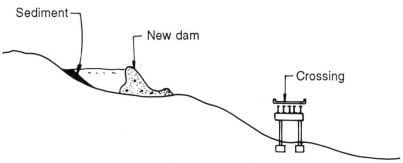

Fig. 5.116 *Sediment deposits:* Sediment previously carried downstream is deposited in the reservoir, which acts as a settling basin. The increased downstream flow may degrade the lower channel.

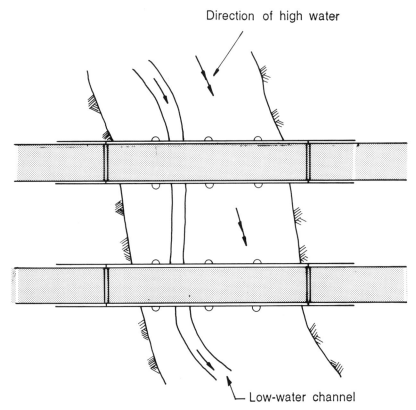

Fig. 5.117 *Pier scour*: Scour around piers is influenced greatly by the shape of pier and skew to flood flow. Note that the direction of flood flow is different from that of normal channel flow.

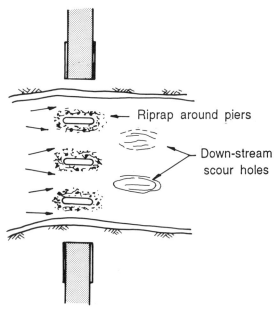

Fig. 5.118 *Loose riprap*: Loose rock riprap piled around piers to prevent local scour at the pier may cause deep scour holes to form downstream.

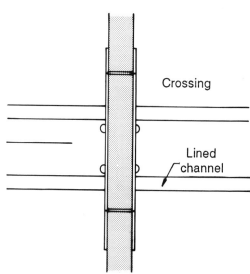

Fig. 5.119 *Lined banks*: Tend to reduce bank scour but such a constriction might increase scour in the bridge opening especially at an adjacent or end pier.

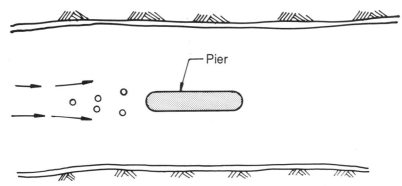

Fig. 5.120   *Scour reduction*: Scour at the pier may be reduced significantly by placing piles upstream in a wedge-shaped pattern as shown (cut-water nosing).

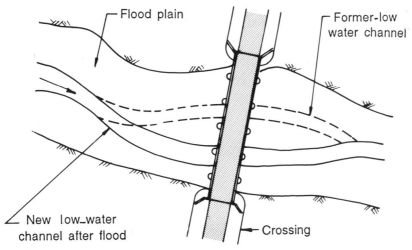

Fig. 5.121   *Channel change*: New low water channel is formed during flood at the river bend.

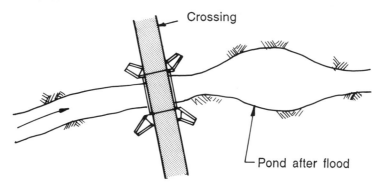

Fig. 5.122   *Horizontal or vertical channel constrictions*: A firm or riprapped bottom or a horizontal constriction can cause a deep scour hole downstream with severe bank erosion resulting in downstream ponding.

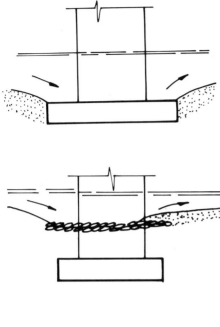

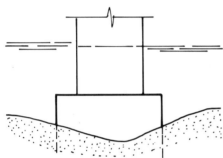

Fig. 5.123 *Scour reduction*: Wide footings or rock aprons tend to reduce scour by deflecting the currents. A deep foundation on the other hand will tend to increase depth of scour.

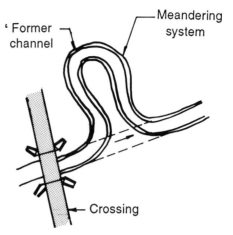

Fig. 5.124 *New channel*: New channel cuts off the neck and changes the river profile. The upstream reach may thus degrade.

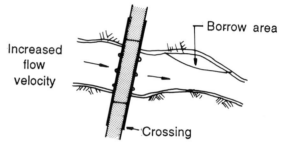

Fig. 5.125 *Gravel removal leading to upstream degradation*: Removing large quantities of gravel from river bottom causes degradation upstream.

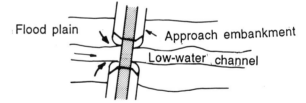

Fig. 5.126 *Scouring during flood*: During flood the waterway constriction may produce general scouring in the vicinity of the bridge and around.

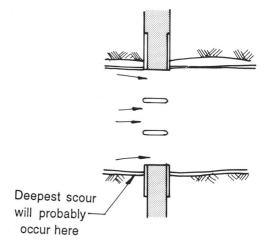

Fig. 5.127 *Scour due to protruding abutments*: Protruding abutments may produce local scour, deepest scour usually takes place at the upstream corner — the severity of scour increases with increased constriction.

Deepest scour will probably occur here

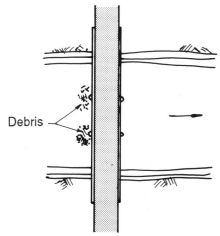

Debris

Fig. 5.128 *Scour due to debris*: Collection of debris around piers, in effect enlarges the size of the pier and causes increased area and depth of scour.

— Existing bank and shore protection plus other existing protective devices should be checked to observe if they are sound and functioning properly. Determine if changes in the channel have caused the present protection to become inadequate and if it may be advisable to place more protection or to revise the existing protection.

— See that the waterway is not obstructed and that it affords free flow of water. Obstructions, such as debris or growth, may contribute to scour and may also be potential fire hazard to the structure. Watch for sand and gravel bars deposited in the channel, which may direct stream flow in such a manner as to cause harmful scour at piers and abutments. Maintenance forces should have a definite program or schedule set up, to provide for removal of debris from around foundations during flood or high water periods to reduce the potential hazard, which might be created otherwise.

— In addition to observing the effect that the waterway and the wave action are having on the bridge and its approaches, observe the surrounding area to see if the bridge and its approaches are causing any problems or potential problems. Items to look for will include possible flooding from inadequate opening at the structure, erosion of banks from improper location or skew of the piers or abutments, etc.

5.9.2.8 Dolphins and Fenders

Dolphins and fenders around the bridge foundations protect the structure against collision by maneuvering vessels. The fender system absorbs the energy of physical contact with the vessel.

1. Dolphin Types

(i) Timber Pile-Cluster Dolphins: This type of dolphin consists of a cluster of timber piles driven into the river bottom with the tops tied together and wrapped tightly with wire rope. It is widely used.

(ii) Steel Tube Dolphins: Steel tube dolphins are composed of one or more steel tubes driven into the river bottom and connected at the top with bracing and fendering systems.

(iii) Caisson Dolphins: Caisson dolphins are sand-filled, sheet-pile cylinders of large diameter. The top is covered by a concrete slab, and fendering is attached to the outside of the sheets.

2. Fender Types

Timber Bents: A series of timber piles with timber walers and braces attached to the pile tops are still used. Steel piles are sometimes used in lieu of timber.

• *What to Look for During Inspection*

(i) Steel: Observe at the splash zone carefully for severe rusting and pitting.

Rusting is much more severe here than at mid tide elevations.

(ii) Concrete: Look for spalling and cracking of concrete, and rusting of reinforcing steel. Be alert for 'hourglass' shaping of piles at the water line.

(iii) Timber: Observe the upper portions lying between the high water and mud line for marine insects and decay. Check the fender pieces exposed to collision forces for signs of wear.

(iv) Structure Damage: Check all dolphins and fenders for cracks, buckled or broken members, or any

other signs of structural failures or damage from marine traffic.

(a) Piling and walers require particular attention since these are the areas most likely to be damaged by impact.

(b) Note any loose or broken cables which would tend to destory the effectiveness of the cluster. Note whether they should be re-wrapped.

(c) Note missing walers, blocks and bolts.

(v) Protective Treatment: Note any protective treatment that needs patching or replacing. This includes breaks in the surface of treated timbers, cracks in protective concrete layers, rust holes or tears in metal shield, and bare areas where epoxy or coal tar preservations had been applied externally.

(vi) Catwalks: Note the condition of the catwalks for fender systems, looking for any deterioration.

5.9.2.9 Underwater Investigations

Underwater investigation is a highly specialized area. This type of bridge-investigation is outside the normal duties of the bridge inspector. Whenever underwater investigations are necessary, they should be conducted by the personnel experienced in this type of inspection. However, the bridge inspector is still responsible for the bridge inspection and for the evaluation of the underwater portions of the bridge.

The importance of such inspection on piled bridges cannot be overstated. Relatively new structures have collapsed due to corrosion of steel piles below the concrete protection. Prestressed and reinforced concrete piles are not immune to failures below water level, while timber piles are known to be vulnerable. Investigation of such corrosion or deterioration is specially important in sea water, which is many times as deleterious as fresh water.

- *What to Look for During Inspection*

(i) Pile Bents: Check piles of all materials below the water line for any signs of deterioration or damage.

(a) *Steel piles*: Steel piles are susceptible to corrosion all over and have been found to be severely perforated any where down the water depth (chloride attack). Where piles are concrete- jacketed in the tidal zone, the diver should carefully check for signs of corrosion for the area just below the concrete jacket all the way to the mud line and below.

(b) *Timber piles*: Timber piles should be observed carefully from the highest water level to the mud line for marine borer and shipworm attacks on the timber and the consequent decay.

(c) *Concrete piles*: Check concrete piles for cracking (pay particular attention to hollow prestressed piles) and spalling due to rust from chloride attack on the reinforcement.

(ii) Dolphins and Fenders: Dolphins and fenders should be examined below the highest water for deterioration and borer attack, and for any damage caused by vessels or large floating objects (impact damage).

(iii) Pier and Abutment Condition: The substructure should be examined for any deterioration of the concrete or of the rubble masonry (loss of protective stone facing) and for any indication of pier movements and abutment movements.

(iv) Scour: The river bottom around the piers and abutments should be checked for development of scour holes, unintended exposure of piles, and the condition and effect of any scour-control-installations. However, it is preferable to also conduct scour investigation from the surface, not only because that is safer but also because more options and judgments can be the basis.

- *The Diver*

Diving confronts a diver with forces and psychological effects which are not encountered in normal environment. Each type of diving equipment and operation gives rise to unique demands for safety precautions. The diver's safety depends upon his knowledge of the factors which constitute safe working conditions, and upon his ability to recognize unsafe conditions. While diving operations will normally be conducted and supervised by specialist personnel, it is incumbent on the bridge inspector to assist in these operations in whatever way possible. In order to assist the diver in his duties the following points should be kept in mind

(a) Define exactly what bridge components the diver is to inspect.

(b) Ensure that he is thoroughly aware of what inspection task he is to accomplish.

(c) Provide him with any reference data, such as previous inspection reports, which may be helpful.

(d) Prepare, as much as practical, the bridge inspection-site beforehand. Often a substantial amount of marine-growth and barnacles can be cleaned from piles and from other relevant areas to be inspected (by working from a boat) in advance!

(e) Ensure that a boat, with a trained crew, is ready, as also the air compressor and the hoses and helmets, duly tested.

5.9.2.10 Maintenance of Submersible Bridges

The submersible bridges need greater maintenance-effort compared to normal bridges, specially after submergence. Normally, every major submersible bridge has its own maintenance manual giving guidelines for information to be collected, checks to be conducted and the locations where the constant vigil is required during and after the submergence. In case there is no manual, the following guidelines may be adopted for the maintenance of submersible bridges :

• The rising and falling flood levels shall be observed and recorded at regular intervals during floods. Any abnormal behaviour in the flood pattern shall be reported. The direction of flow in the river shall be observed during floods, and any obliquity of flow shall be recorded and reported.

• As soon as the water level in the river reaches the specified High Flood Level (HFL), all traffic on the bridge shall be suspended till it recedes to a safer level. The number of traffic interruptions (with durations) in each flood should be recorded and reported.

• During floods, the following observations should be made:

  (i) Record of water gauge (hourly gauge readings),
  (ii) Record of afflux at various levels,
  (iii) Record of velocity at various levels,
  (iv) The traffic census,
  (v) Behaviour of flow pattern in the river during flood, i.e., scour/erosion of banks within say 1/2 to 2 km upstream of the bridge site,
  (vi) Observance of silt content of river water above specified HFL after which the bridge will be submerged, and comparison with the value of silt content in the river water adopted for the design of superstructure,
  (vii) Record of rainfall,
  (viii) Record of silt deposition on the superstructure (within the section also, if it is a box section),
  (ix) Record of maximum and minimum temperatures.

• A constant vigil shall be kept on the foundations during the floods. Soundings may be taken at regular intervals to ascertain the depth of scour around foundations and any abnormal change in the scour pattern recorded, reported and necessary temporary remedial measures taken. After the submergence of the bridge and on receding of the floods, soundings may again be taken to note the scour around foundations, and also any permanent remedial measures (if called for) decided.

• Before and after submergence of the bridge, the Bearings shall be inspected and any damage, if noticed, shall immediately be reported to the higher authorities and necessary remedial measures taken.

• Normally water holes are provided in the superstructure of submersible bridges for the through passage of the water during the submergence. Before floods, these holes should be carefully inspected and cleaned if found choked. After the submergence, the entire superstructure shall be cleaned of the silt deposit and the water holes shall be cleaned free of debris, vegetation, etc., found blocking the water holes in the superstructure. The superstructure shall be carefully examined after the submergence for any signs of distress or cracks and any remedial measures, if necessary, taken.

• The expansion joints should be thoroughly examined (for possibilities of blockage after the submergence of the bridge) and properly maintained.

• Normally, the submersible bridges are provided with collapsible type of railings. These railings should be lowered and fixed again every month after proper lubrication, oiling, etc., so that the possibility of rusting is avoided. The collapsible railing should be lowered as soon as water in the river attains the specified HFL and refixed after the HFL has receded.

5.9.2.11 Maintenance of Movable Bridges and Suspension Spans

• The most common types of movable bridges are the Swing Span, Vertical Lift, Bascule (single or double leaf). Inspection of the trusses, floor-system and other structural elements will require the normal inspection procedures already covered under their respective headings.

• Additional structural elements do exist, however, and must be examined. Counterweights must be examined to see that all elements are sound and secure. Check closely for corrosion of the steel which extends into the concrete. Water may 'pocket' in these locations and penetrate the joint, thus contributing to corrosion. Stains on the concrete around steel embedments should be thoroughly investigated as it may indicate corrosion and loss of cross-sectional area in the steel at the surface of concrete and possibly just beneath the surface. See that the drains in the counterweight pocket are open and functioning properly.

• Counterweight cables plus uphaul and downhaul cables on vertical lift structures must be checked thoroughly for wear and corrosion, and to see that they are adequately lubricated. Check the travel-rollers and guides for the adequate clearance and to see that there is no excessive wear. A significant change in the clearance may indicate pier movement and will require further investigation to determine the cause.

• Check to see that the roller surface is of even grade and that there is adequate clearance at the joints where the movable span meets the fixed span. Also, note differential vertical movement at the joint between the two leaves of the double leafed bascule span under the passage of heavy loads. Shear-locks are subject to heavy wear and pounding under traffic and are many times a troublesome item. Excessive movement should be investigated and reported. Check this joint also for adequate clearance.

• Steel grid decks, both open and closed, are commonly used on movable spans. They should be checked for any evidence of broken welds. These decks become quite smooth on the surface and they should be checked for skid resistance under wet and frosty conditions.

• Examination of the mechanical aspects of the structure must be performed by an inspector qualified in the principles of machinery and familiar with the mechanical functioning and design of the structure being inspected. The machinery should be checked generally for proper lubrication, unusual noise, and looseness in the shafts and bearings.

• Trial openings should be made as necessary to ensure that all operations are functioning properly and that the movable span is properly balanced. No trial opening for inspection is to be made concurrently with an opening for the passage of vessels where the attention of the bridge operator might be divided between the two interests.

• Auxiliary standby power plants are to be started and checked thoroughly in addition to the normal routine periodic operations of the plant. Such routine operations are normally done by the bridge operator on a weekly basis.

• The bridge operator should be consulted as part of the investigation. He is a good source of information on the general handling of the operation and can point out any changes from the normal which may have developed.

• The inspection team, making the inspection, must include a man well qualified in the electrical aspects of the operation. This, of course, may be the same person qualified for the mechanical inspection portion. Many of the mechanical and electrical operations complement each other, and the inspection of these two areas should be a well-planned and coordinated effort.

• Inspection of the electrical system should be thorough and should include such items as the controls, wirings, conduits, motors, and lights. Be watchful for any worn or broken lines which may be hazardous. Check for conditions which may exist that could be hazardous or could be potentially hazardous to the operator or any one using the structure. Safety in this area cannot be over-emphasized.

• During these inspections keep in mind and be watchful for obvious hazards which may involve the safety of the operator and other personnel in performing their normal operational duties and for performing routine maintenance of various items (such as greasing of machinery, and maintaining lights and signals for both channel and highway traffic). Keep in mind that such maintenance jobs may be required in all types of inclement weather which could affect the degree of hazard encountered.

• Submarine cables carrying power and control circuits should be examined in the areas above the water lines at each inspection. The underwater portion should be inspected by divers after unusually high water or at any other time when there is a reason to suspect damage may have occurred.

• Examine traffic gates, barriers, and signal systems for highways and marine traffic, to see that all are functioning properly.

• Examine fenders and dolphins for damage from marine traffic. Inspect all timber sheathing, wales, and piles for decay, for damage from marine borers, and to see that bolts and cables are tight. Observe the overall set-up of the fender system to see if it is reasonably maintained.

Suspension Spans

• Examine the main suspension cables to see that their protective covering or coating is in good condition and is protecting the steel from corrosion. Special attention should be given to the areas adjacent to the cable bands, at the saddles over the towers, and at the anchorages.

• Examine the bands holding the suspenders to the main suspension cable to see that no slippage has occurred and that all bolts appear to be tight.

• Check anchorages carefully for corrosion and to see that there is adequate protection against moisture entering or collecting where it may cause corrosion.

• Inspection of the stiffening trusses, floor system, and cable-bents are to be made in detail as generally covered earlier.

5.9.2.12 Miscellaneous Items

These are described ahead under the following heads

5.9.2.12 (a) — *Light Posts*

5.9.2.12 (b) — *Road Kerbs*

5.9.2.12 (c) — *Drainage (scuppers, etc.)*

5.9.2.12 (d) — *Parapets and Railings*

5.9.2.12 (e) — *Footpaths (walkways) and Facias*

5.9.2.12 (f) — *Utilities*

5.9.2.12 (g) — *Medians*

5.9.2.12 (h) — *Paint*

5.9.2.12 (i) — *Signs*

5.9.2.12 (j) — *Rock Slopes*

5.9.2.12 (k) — *Rock Bolts/Anchors*

5.9.2.12 (a) **Light Posts**

Lighting on bridges will consist of highway lighting, sign lights, traffic control lights, navigation lights and aerial obstruction lights. The last two types of lights are special categories which will be encountered only on bridges over navigable waterways or on bridges having high towers. There will, of course, be many bridges with no lighting at all! The inspection effort will involve only the light pole bases which are tied into the bridge structure proper.

Highway Lighting

Conventional highway lighting on the bridges presents relatively few problems. The most common problems are

(i) *Collisions*: Most (metallic) light standard bases on more modern bridges are supported on parapets, or are located outside the rails. However, on some older urban bridges, light standards may be unprotected and, consequently, they are more vulnerable to damage or destruction from vehicles.

(ii) *Deterioration*: Due to lack of maintenance-painting, the base may have deteriorated to the point where it is no longer functioning as designed.

- *What to Rate*

Rate the light standard base on a 'per span' basis. This includes the base plate and anchor bolts.

- *What to Look For*

Collision: Note any light pole bases that are dented, scraped, cracked, inclined, or otherwise damaged.

Fatigue: Aluminium light standards and castings are most likely to suffer from fatigue. Check for cracking of the base, especially the cast elements.

Corrosion: Check steel pole base for cracks and rusting, and concrete bases for cracking and spalling.

- *Sign Lighting*

Inspect sign lighting for the same defects as conventional lighting, stated above.

- *Sample Rating*
  - Light pole bases and foundations in 'like-new' condition are rated 7.
  - Minor deterioration of bases should be rated 4 or 5.
  - Some minor collision damage or deterioration to light standard bases would be rated from 4 to 6.
  - Serious deterioration or damage of light pole bases and foundations should be rated from 1 to 3.
  - Bases not functioning as designed, which could be hazardous to the public should be rated 1.
  - If there are no light poles on the span, rate it N/A.
  - A lower rating would be given for a problem with traffic control or navigation lighting than for roadway or sign lighting.

5.9.2.12 (b) **Road Kerbs**

Kerbs are provided to protect pedestrains crossing the bridge. Often only a narrow kerb is provided in rural areas.

- *What to Rate*

Rate the physical condition of the kerb. Also rate its ability to function as originally designed.

Typical Types

Kerbs are frequently constructed of the following materials:

— Granite

— Steel

— Concrete

- *What to Look For*

General: Always check kerbs for impact-damage and misalignment. Remember that bridge deck overlays reduce kerb heights and their effectiveness. Sight along the kerbs to check the grade and alignment. Kerb sags or misalignment over piers or abutments may indicate foundation settlement or a problem with the bearings.

Granite: Look for broken or loose sections.

Steel: Check for proper anchorage, proper alignment, and loss of section due to corrosion, specially at drains. Look for metal kerbs protruding into the roadway, which is hazardous.

Concrete: Look for spalling, scaling, cracks, and other forms of concrete deterioration. Check for exposed reinforcing protruding into the roadway.

- *Sample Ratings*

A concrete kerb with minor deterioration is shown in Fig. 5.131. It is rated 5.

Fig. 5.131 Rate 5

Fig. 5.132 Rate 3

An example of a 3-rated kerb is shown in Fig. 5.132.

The concrete kerb in Fig. 5.133 is completely ineffective. The lack of kerb in this bridge will allow roadway salt solutions to run on to the deck below rather than direct it into the scuppers. This is rated 1.

The condition of the kerb in Fig. 5.134 cannot be determined but is known that it is completely ineffective as a traffic delineator or a water diverter. This could represent a hazard to pedestrains and is rated 1 — potentially hazardous.

5.9.2.12 (c) *Drainage (including scuppers and Gratings, etc.)*

Bridge drainage is an important inspection-item since any trapped or ponded water can cause a great deal of damage to a bridge and is also a safety hazard. Therefore, an effective system of drainage that carries the water away as quickly as possible is essential to the proper maintenance of the bridge.

Fig. 5.133 Rate 1

Fig. 5.134 Rate 1

• *Drainage Problems*

Almost all the drainage problems encountered by an inspector are caused by the failure of the drainage system to carry the water away.

(a) Clogging: This is the most apparent problem and occurs at the following junctures.

1. *Inlets* Accumulation of debris compound by design oversights are principal causes for inlet-clogging. The ponds and puddles of water, that form on the bridge deck, pose the problem which constitute a safety hazard and can cause extensive bridge deterioration.

2. *Downspouts* Downspouts and horizontal pipe-runs, which are poorly designed with inadequate slopes and sharp directional changes at the elbows, are conducive to plugging of drains.

3. *Expansion Dams* The gutters under expansion dams fill-up very rapidly, specially where roads are sandy. This could cause the storm-water overflow onto the bridge bearings, end diaphragms, pier caps, bridge seats, etc., resulting in rusting of steel and deterioration of concrete. Of special consequence is deterioration of bearings and the supporting elements.

(b) Corrosion: Drainage water often carries corrosive elements that attack drainage pipes and concrete. Clogging and slow drainage accelerate corrosion.

(c) Entrapped Water: Where a concrete deck is covered with an asphaltic wearing surface, the underlying concrete surface is liable to serious deterioration by entrapped water. On some bridges, drains have been placed under the wearing surface to avoid this situation!

(d) Short Scupper Pipes: Where the discharge pipes are too short, or nonexistent, drain water may splash or blow onto bridge members, causing corrosion, staining and deterioration.

Examples are shown in Figs 5.135 to 5.138.

* *What to Rate*

A rating for the bridge drainage system is determined according to its ability to effectively divert roadway-water off the bridge deck and away from the bridge elements. Scuppers, downspouts, and scupper-piping systems are included in determining a rating for this item.

Continuous or intermittent open steel grating on the bridge is also rated under this item. The grating should be rated based upon its physical condition and its ability to function as designed.

When drainage systems and gratings occur on the same span, code only the lower item.

* *What to Look For*

Inspect grates and grate-supports for loss of section due to corrosion. Always check the profile of the grates across the bridge. Check for cracks and broken welds, debris buildup, and, in general, its ability to support traffic loads.

Inspect the physical condition of the elements comprising the drainage system, such as scupper grates, downspouts, piping, gutters, downspout hoppers, etc. Check to see that these elements are adequately supported with clamps, brackets, supporting concrete, etc.

Check whether the drainage system stands covered-over by a pavement-overlay. Look for debris collecting at scuppers (which indicates clogged condition in system below). Note this in remarks and rate accordingly.

Ascertain if the effluent draining from the bridge is being carried away in such a manner as not to promote deterioration of any bridge element; not contributing to erosion of an embankment and/or draining onto the road-way below. If any or all of these items exist, the rating would be lowered accordingly.

See if drains are located to collect the roadway water. Check any low areas of the deck for proper drainage.

Observe water stains which indicate leaking or clogged drainage system. Watch out for stains due to leaking joints.

Watch for sand accumulation on the deck, at kerbs, and at drains; these indicate drainage system not operating properly.

* *Sample Rating*

A 7 rating should be used for a drainage system in good physical condition that effectively diverts roadway-water away from bridge elements.

Do not downrate gratings merely because they are causing deterioration of the bridge below, as one would

Fig. 5.135 Deck drainage

Fig. 5.136 Open pipe drain

Fig. 5.137   Deck drainage directed away from
pier cap

Fig. 5.138   Roadway drain

do for scuppers. Unlike scuppers, it is not the function of the gratings to protect the lower bridge elements! If any corrosion is accruing below, the item involved should be rated accordingly.

Figure 5.139 shows the half-plugged scupper grate. This scupper has a short underside; it can be seen that the downspout is clean. Therefore, the only problem is that one-half of the existing grate is plugged. Because the scupper can still handle about one-half of its original capacity, it is rated 5.

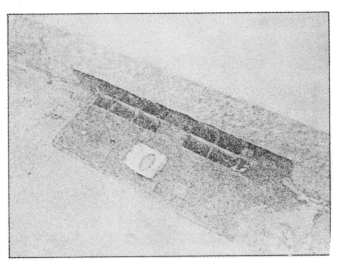

Fig. 5.139   Rate 5

A split downspout is shown in Fig. 5.140. This deficiency is allowing corrosive roadway solutions to be deposited onto the substructure, creating a deterioration problem on these elements; this is rated 3, and is viewed as a serious problem.

A completely plugged scupper is shown in Fig. 5.141. This scupper is useless and is considered potentially hazardous as it will allow water to pond on the bridge deck. This is rated 1.

Gratings in 'like new' condition are shown in Fig. 5.142.

An example of 4-rated grate (minor problem) would be if a couple of bearing bars in the grate were broken.

An example of a 3-rated grate (serious problem) would be if about 10 per cent of the bearing bars had broken at the welds.

A 1-rating should be used for any condition that presents a potential traffic hazard. For example, unstable grating supports.

5.9.2.12 (d) *Parapets and Railings*

Railings and parapets normally do not add structural strength to the bridge. They should have sufficient strength to prevent an out-of-control vehicle from going off the bridge. Some existing bridges may have

Fig. 5.140   Rate 3

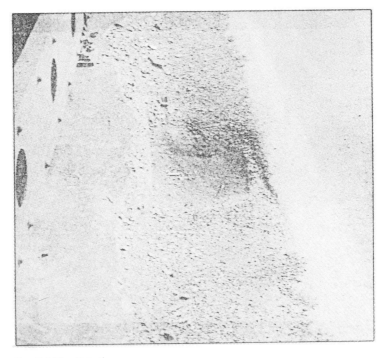

Fig. 5.141   Rate 1

Fig. 5.142   Rate 7.

vehicular guard rails which are inadequate (and therefore unsafe), easily damaged by vehicles, and highly susceptible to deterioration. On the other hand, unyielding guard rails or unprotected parapets pose a hazard to vehicular traffic specially if struck head on.

- *What to Rate*

Rate the railings and/or parapets on the bridge. When both railings and parapets occur on the same span, the element with the lower rating should be coded.

- *Typical Types*
— steel
— aluminium
— concrete
— masonry

- *What to Look For*

Railings and parapets should be inspected for their physical condition and for their ability to function as originally designed.

- Check the vertical and horizontal alignment of all handrails for any indications of settlement in the substructure or any bearing deficiencies.
- Examine all handrail joints to see that they are open and ready to function.
- Special attention should be given to fasteners and anchorages.

- The elements should be examined for deterioration and impact-damage.
- Examine all the handrails to see that they are relatively free of slivers or any projections which would be hazardous to pedestrians.
- Metal elements should be inspected for cracks and section-loss due to corrosion.
- Check for the rust stains on the concrete around the perimeter of steel posts which are set in pockets. Remove grout from around the posts and determine the severity of corrosion, if any.
- On concrete members, look for spalls, cracks, efflorescense, and corrosion of steel reinforcement.
- Check steel and aluminium railings for loose posts or rails. In particular, check the condition of the connections of the posts to the deck, including the condition of the anchor bolts and the deck area around them.
- Note whether barrier railings on the approaches to the bridge extend beyond the edge of the bridge-railing or parapet-end and are anchored to the inside face. This feature reduces the severity of vehicle collision. In situations when parapet-ends are unprotected and no approach-rail exists, a flared, tapered approach-railing should be installed. On two-way bridges, this type of railing should be installed at both ends of the existing railings or parapets.
- Check the horizontal alignment of the railings on skewed bridges for possible lateral movement of the spans.
- Do not consider railing-paint in rating this item. Railing-paint or galvanizing is rated under the "Paint" item.

- *Sample Ratings*

As shown in Figs. 5.143 to 5.148.

Aluminium railing on a concrete parapet is shown in Fig. 5.143. The aluminium is good, the physical condition of the materials is excellent and the anchorage is secure. This parapet with railing is rated 7 — new condition.

Figure 5.144 shows a short section of railing on a small bridge. This railing is in good physical condition and can function as well as it could have when originally designed. It is rated 7. Remember, the railing's ability to meet current standards does not affect the structural rating. If information on railing design is needed, it can be determined through the inventory portion of the system.

The concrete railing in Fig. 5.145 is rated 5. There is minor spalling on the bottom rail, but the railing is still

Fig. 5.143 Rate 7

Fig. 5.144 Rate 7

Fig. 5.145 Rate 5

Fig. 5.146   Rate 5

Fig. 5.147   Rate 3

Fig. 5.148   Rate 1

structurally sound and is functioning as originally designed.

Figure 5.146 shows a railing with a slight impact damage. The connections are still good and the railing can still function, and, therefore, is rated 5.

Figure 5.147 shows a concrete parapet with heavy spalling. Although it can still function, it is rated 3 due to severe deterioration.

Figure 5.148 shows a railing that is partly missing. This is a hazardous situation and is rated 1.

5.9.2.12 (e) *Footpaths (walkways) and Fascias*

Sidewalks and fascias normally do not contribute to the structural strength of the bridge, but may be an integral part of the bridge by virture of their construction. They are provided mainly for public safety. Bridge roadways that are narrower than the approach roadway are potentially dangerous since they restrict vehicular manoeuvreability which can cause accidents. Sidewalks are provided to protect pedestrians while crossing the bridge.

• *What to Rate*

Consideration must be given to the condition of both the sidewalks and the deck-fascia in rating this item. However, only the lower rated of the two elements is coded on the inspection form.

• *Typical Types*

As shown in Fig. 5.149.

• *What to Look For*

Structural sidewalks should be inspected for their ability to span between supports. All sidewalks should be inspected for their walking-surface quality.

When inspecting fascias, look for signs of material deterioration. Beaware that one common function of a fascia is to support railing anchorages. If serious deterioration has occurred and it is no longer capable of supporting railing anchorages as originally intended, it should be rated not higher than 3.

— Fascias can be hazardous if pieces of concrete fall on pedestrians or vehicular traffic the bridge.

— Check concrete sidewalks and fascias for cracks, spalls, and other deterioration applicable to concrete (described earlier).

— Examine the condition of concrete sidewalks at joints, specially at the abutments, for signs of differential movement which could open the joint.

— Check steel sidewalks for corrosion and to see that all connections are secure.

— Check slipperiness of surface during wet or frosty weather to determine corrective action necessary.

— Check sidewalk-drainage for adequate carry–off.

• *Sample Ratings*

Check

— condition of the fascia

— structural strength of structural sidewalks

— walking-surface quality

Only the items in the worst condition are rated on the inspection form.

Safety walks are rated as sidewalks. The safety walk in Fig. 5.150 is free of scaling, spalls, cracking, and other forms of deterioration. It is rated 7 — new condition.

Figure 5.151 shows a 'patched' concrete sidewalk. The only deficiency is minor scaling just off the patched area. This sidewalk is rated 5 — minor deterioration. The patch is considered a permanent repair and does not lower the rating.

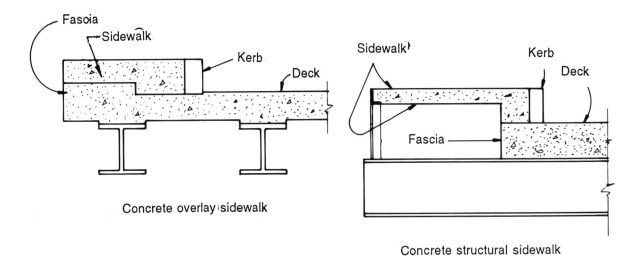

Concrete overlay sidewalk

Concrete structural sidewalk

Fig. 5.149 Footpath (sidewalk) types

Fig. 5.150 Rate 7

The surface of a sidewalk with a large spall and some cracking is shown in Fig. 5.152. This sidewalk is obviously seriously deteriorated. Looking strictly at the walking-quality, this should be rated 3, indicating serious deterioration.

A hole has developed completely through the sidewalk in Fig. 5.153. This is rated 1 — potentially hazardous. If this hole were covered with steel plate, it would still be rated 1 because temporary repairs are ignored in determining ratings.

The concrete fascia shows no signs of deterioration in Fig. 5.154. It is rated 7 — new condition.

Inspection of the fascia in Fig. 5.155 shows spalling near the deck joint and some minor map-cracking on the fascia, extending from the deck joint locations. At this stage of deterioration, the anchorage of the bridge railing is not yet endangered. This is rated 5 indicating a generally minor deterioration problem.

Figure 5.156 shows a fascia with heavy spalling, cracking and efflorescence. Extensive deterioration has obviously occurred, requiring a rating of 3.

A concrete fascia supporting a concrete parapet with pipe railing is shown in Fig. 5.157. The deterioration has continued to the point where the entire parapet railing system could potentially collapse. This is potential hazard and is rated 1.

5.9.2.12 (f) *Utilities*

It is common for commercial and industrial utilities to use a highway right-of-way to provide goods and services to the public. This means that some utilities may be carried by the bridge structures. These may be one or

Fig. 5.151 Rate 5

Fig. 5.152 Rate 3

Fig. 5.153 Rate 1

Fig. 5.154   Rate 7

Fig. 5.155   Rate 5

Fig. 5.156   Rate 3

Fig. 5.157   Rate 1

more of the following: gas, electricity, water, telephone, and sewage, in pipes.

Utility companies provide their own facilities for installation and maintenance. While the gas, light and telephone authorities will usually carryout their own maintenance of their facilities, some of the others are less likely to perform adequate maintenance since they may not be as well staffed. Most utility-lines or pipes are suspended from bridges between the beams or behind the fascias. On older bridges, water pipes and sewer pipes may be found installed along the sides of the bridge or may be suspended under the bridge.

• *What to Rate*

Rate the condition of the utility lines and of their supports on a 'per span' basis, including pipes, ducts, conduits, wires, junction boxes, expansion joints, valves, vents and insulation. Typical utilities are gas, water, electricity, telephone and sewage. Rate the utility and/or support-system that is in the worst condition for each span. Paint is not rated under this item.

• *What to Look For*
— Check utility-lines for leaks, breaks, cracks, rust, and deteriorated coverings. Check for water or sewage leaking onto decks or members and causing a corrosion problem.
— If the abutment and/or pier settlement has occurred, check for utility-line breaks and expansion-joint problems.
— Check that utilities located below the bridge are not reducing the vertical clearance or freeboard.
— Check vents and drains on encasements of pressure pipes. Check junction boxes for moisture, drainage, insulation, and the cover being in place.
— Check for cracked conduits in sidewalks if there are cracks in the sidewalk.
— Check overhead lines for hanging objects.
— Check for seepage leaks where utilities pass through abutments.

— Check for wear or deteriorated shielding and insulation on power cables.

— Determine whether mutually hazardous elements, such as volatile fuels and electricity, are sufficiently isolated from each other. If such utilities are side-by-side or in the same bay, report this condition for either auxiliary encasement or future relocation.

— Check for the presence of shut-off valves on pipelines carrying hazardous pressurized fluids, unless the fluid supply is controlled by automatic devices.

— Determine whether any utility may hinder passage during periods of high water.

— Note whether any utility is located at such a place where there is a possibility that it may get struck and be damaged by traffic, or by debris carried by high water. Check for corrosion, loose connections, lack of rigidity in the supports as well as fatigue-damage due to expansion, contraction and vibration of the utilities. Check for enough padding or bracing to prevent vibrations.

— Determine whether utilities are adequately supported and whether they present a hazard to any traffic which may use the bridge or pass under the bridge.

— Check for utilities interference with bridge maintenance operations.

• *Sample Ratings*

Rate 7 for a new or 'like-new' condition.

Rate 5 for utility-lines and their insulation exhibiting only, minor rust or other deterioration; also for utility supports, minor corrosion without loss of section or rigidity.

Rate 3 for pipes, conduits, ducts, expansion joints, etc., exhibiting major corrosion and loss of insulation; for sagging utility-supports exhibiting severe corrosion and loss of section, lack of rigidity and vibration under traffic loading.

Rate 1 for any sign of leaking or cracked utility lines, exposed wiring (other than lighting) which may pose an electric shock hazard; for utility supports which are severely deteriorated and/or cracked from vibrating utility lines, and indicate the danger of collapse. The condition is most severe if there is a roadway or railroad under the bridge. For this case, the authorities should be notified immediately.

Figure 5.158 illustrates an insulated water main. There is a paper cover that is coming off, but the main is not leaking and the insulation is intact. This is a new condition and is rated 7.

Figure 5.159 illustrates several utilities under the bridge. In cases like this, rate the utility in the worst condition. These are in good condition, except for some minor corrosion, and is rated 6.

Figure 5.160 illustrates a group of utility-ducts whose supports have failed, causing the pipes to break at their joint. Even though the ducts were placed on the bridge for a future utility and are not in use, they should be rated. Since some pieces have already fallen and more may fall in the future, this is rated 1.

Figure 5.161 illustrates cable supports. They are in good condition and are not sagging. They are rated 7.

Figure 5.162 illustrates supports for a pipe. They are holding the pipe rigidly and permitting expansion and contraction of the pipe. There is a minor problem with corrosion. This is a 6-rating.

In Fig. 5.163 the pipe should be supported under every other sidewalk support. The support on the right is broken, and does not support the pipe. This is not functioning and is rated 3. If the pipe were sagging, it would be rated 1.

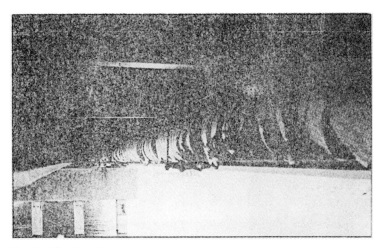

Fig. 5.158   Rate 7

Fig. 5.159 Rate 6

Fig. 5.160 Rate 1

Fig. 5.161   Rate 7

Fig. 5.162   Rate 6

Fig. 5.163 Rate 3

Figure 5.164 shows a broken utility support of a water main. The pipe is sagging and could break, so the support as well as the utility line are both rated 1. The authorities should be notified of a condition such as this immediately, so that the utility authorities can be notified of the problem.

5.9.2.12 (g) *Medians*
A median serves the purpose of physically separating the traffic form opposing directions on the bridge and enhances safety.
- *What to Rate*

Inspect median, kerbs, concrete-backing, stone chips, guide-railing, and median-barriers, when they exist. The lowest rated of these should be coded under this item. Flush-medians, ordinarily described by painted-stripes, are not rated under this item, unless they also have open steel grating, median barrier, guide-railing, or a stone-chip surface. The lowest rated of the elements present is coded on the form.
- *Typical Types*
— raised medians with or without median-railing
— concrete main barriers
— open steel grating medians
— flush-medians with stone-chips
— guide-railing
- *What to Look For*

Inspect the appropriate elements for deterioration, signs of physical distress, proper alignment, proper supports and adequate anchorage.
- *Sample Ratings*

After determining a rating for each of the individual elements that make up the median, the lowest rated of these is entered on the inspection form.

Figure 5.165 shows a typical median. Rating for this median would be determined based on the lowest individual rating of the kerb or concrete backing. (The light standard base is rated under the lighting item, and does not influence the median rating.)

Another median is depicted in Fig. 5.166. The railing has been hit, pulling two posts from their anchorage, and loosening a third. This would be hazardous if hit again and, therefore, the median is rated 1.

5.9.2.12 (h) *Paint*
Since painting is the primary means for protecting steel against corrosion, it is imperative that the conditions of the paint be thoroughly inspected and fully reported upon. Paint protective system is used in metal railing, light pole bases, fascias, kerbs, medians, utilities, structural steel, etc. Continual exposure to weathering and chemical action requires that paint be continually maintained. Spot failures can develop rapidly into large areas of corrosion, which, if permitted to continue, can cause extensive (and ultimately) irreparable damage. The amount of paint maintenance required is influenced by the environment. In damp coastal regions, or in urban industrial areas, annual spot-painting may be necessary, and complete repainting may be required as often as every four years. In dry climates, repainting may not be needed for 10 years, and in some cases the interval may approach 15 years. Other factors affecting the life of the paint are type, quality of its application, structural details, and the amount of exposed areas.
- *What to Rate*

Rate the physical condition of the paint-protective-system on the protected area.

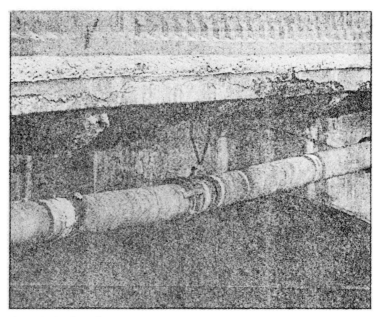

Fig. 5.164   Rate 1

Fig. 5.165

Fig. 5.166

• *What to Look For*

Look for peeling, cracking, chipping, rust-pimples, and scaling. Check to see if the outer coating of paint is worn, and note whether the prime coat or metal surface is exposed. Inspect areas around rivets and bolts specially carefully. Any steel where moisture rests or debris collects is extremely vulnerable to paint-failure.

Look for paint-failure on those surfaces which are most exposed to sunlight. Give particular attention to the ends of beams, inner corners and any other areas that are difficult to paint.

• *Sample Rating*

The rating of paint does not use the same rating definitions that all other items use! Following are the explanations of the rating system to be used for paint. Note that the ratings are tied to both the physical condition of the paint and the amount of corrosion on the member.

— A paint system in good condition should be rated 7.

— A paint system generally in good condition, but requiring touching-up, should be rated 6.

— A 5-rating should be given when the paint generally shows signs of deterioration but no cracks present as yet.

— A 4-rating is used when the paint system is in localized poor condition and minor corrosion is present.

— A 3-rating is used where corrosion is present throughout but no section-loss is evident.

— A 2-rating signifies serious loss of section in a member.

— A 1-rating has paint in very poor condition throughout with significant material loss.

5.9.2.12 (i) *Signs*

Check to see all signs required to show 'restricted weight limit', 'reduced speed limit', or 'impaired vertical clearance' are in their proper place. This inspection is to include 'signs at or on the structure' and any necessary 'advance warning signs'. Check the 'signs' to see that the 'lettering' is clear and legible, and that they are in generally good physical condition.

Any revisions made which will alter the clearances, such as addition of surfacing to the roadway, will necessitate remeasurement of the clearance and correction of the signs and records to reflect the change.

For bridges over navigable channels, check to see that the required navigational signs for water-traffic are in place and in good condition. The inspector must be familiar with the regulations of the concerned authorities and the coast guard to the extent necessary for making these determinations. See that the navigational lights are properly installed in their intended positions and are functioning. See that aerial obstruction lights on high bridges are functioning. Intermediate inspections of the lights must be made at intervals often enough in order to be reasonably sure that they are operating properly.

SIGN POST BASE

Rate the effectiveness of pole-bases/sign-supports, on a 'per span' basis. Also determine two features for the functionality of the sign, namely, its 'clearance above

Fig. 5.167 Rate 7; painted steel railing

Fig. 5.168 Rate 4; painted steel railing

the roadway surface below' and the 'posted load capacity'.

- *What to Look For*
  - (a) *Collision*: Note any sign pole bases that are dented, scraped, cracked, inclined or otherwise damaged.
  - (b) *Fatigue*: Aluminium sign post bases are likely to suffer from fatigue. Check for cracking, especially cast elements.
  - (c) *Corrosion*: Check steel bases for cracks and rusting and concrete bases for cracking and spalling.
  - (d) *Clearance*: Measure minimum clearance between the bottom of superstructure and the top of the roadway below, if applicable.
  - (e) *Load Capacity*: Note any posted load capacity which may be indicated on signs either on the bridge or at its approaches.
- *Sample Ratings*
- sign pole bases and foundations in new or 'like-new' condition are rated 7
- minor deterioration of bases should be rated 4 or 5
- some minor collision damage or deterioration to sign bases would be rated from 4 to 6, depending
- serious deterioration or damage to sign pole bases and foundations should be rated from 1 to 3, as the case may be
- bases not functioning as designed, which could be hazardous to the public, should be rated 1
- if there are no sign poles on the span, rate it NA — not applicable.

5.9.2.12 (j) *Rock-Slopes*

In mountainous areas, where the construction of elevated bridge structures requires cutting of the natural formations, rock slopes may be encountered. If the toe of the slope is sufficiently distanced from a bridge's major structural elements (pier, column, etc.), to allow for a storage area for falling rock, then the inspection of the rock slope is not necessary, since the rock does not endanger the integrity of the structure. When there exists a possibility of rock falling against or onto structural elements, then the rock slopes should be inspected.

- *What to Rate*

The item to rate is the stability of the rock slope itself by observing the amount of deterioration that exists, if applicable.

- *What to Look For*

Check the size, spacing and direction of the visible cracks in the face of the rock; if cracks exist, is water running from them or vegetation growing out of them?

Check the toe of slope to determine if any rock has fallen. Note whether the debris is large blocks or small broken pieces.

- *Sample Ratings*
  - Rate 7, if no visible sign of deterioration is evident.
  - Rate 5, if slope has some cracks with minimum debris evident at the toe.
  - Rate 3, if slope has numerous wide cracks with water seeping through the cracks and significant amount of debris at the toe.

— Rate 1, if the face of the slope indicates wide cracks in all directions and a major fault is evident in the slope, indicating that a rock slide is imminent.

5.9.2.12 (k) *Rock-Bolts/Anchors*

Rock – bolts and anchors are used to stabilize rock formations by providing a continuous structural element (bolt, anchor) between layers of rock which may be discontinuous or deteriorating. Holes are drilled in the rock to a solid base and bolts or anchors are grouted into the holes. The bolts are then tensioned and locked-on. More usually, the bottom portion of an anchor is grout anchored and latter it is prestressed and locked-on. (For more details, *see* Chapter 10.)

- *What to Look For:*
  - Check the condition of all metal elements for signs of deterioration (rust).
  - Check for missing and loose nuts, anchorage lock-on arrangements, anchor-plates, etc.
  - See if the plate beneath the nut is bearing on the rock surface evenly.
  - Check to see if the grout is in place and functioning.
- *Sample Ratings*
  - Rate 7, if no visible sign of deterioration is evident.
  - Rate 5, if some deterioration of the bolt, nut and bearing plate is evident or if the nuts need to be tightened.
  - Rate 3, if significant deterioration exists and if some bolt-nuts are missing.
  - Rate 1, if the bolts have failed and can be moved by hand or by striking with a hammer.

5.9.2.13 *Encroachments*

The number and types of utilities (sewer pipelines, etc.), and other encroachments attached to or enclosed in the bridge, and encroachments in the vicinity, must be on record. If the number, types and installation are not carefully controlled, they may overload the structure or make normal maintenance extremely difficult. Note if any encroachments are obstructing the waterway or are in such a position that they may hinder drift removal during periods of high water. Also, note if the encroachment is located where there is a possibility that it may get hit and be damaged by traffic or by (ice and) debris carried by high water.

See that the encroachments are adequately supported and are not a hazard to any traffic which may use or pass under the structure. Look for wear or deteriorated shielding and insulation on power cables.

Check closely for any adverse effect the encroachments may have on the bridge. Check to see if the vibration or expansion movements are causing

cracking in the support-members or if the paint is being damaged.

Note the bad aesthetic effect which encroachments may have on the bridge. The facet must be considered in permitting the encroachments to remain on a bridge. The general appearance of the vicinity around the structure will be a factor in making this determination.

5.9.2.14 *Aesthetics*

The value of the aesthetic qualities of a bridge varies greatly from area to area. Some agencies attach more worth to this aspect of highway engineering than do others and the importance will also fluctuate from time to time within a given area. Consequently, it is difficult to set any limits in this field and the judgment of the inspector for maintenance requirements must be based on guidelines set up within his own department.

Aesthetic standards set up by each department will have an effect on the degree of maintenance of many items and must be kept in mind during periodic inspections. Paint and concrete conditions are probably the most common items on a bridge where appearance may indicate need for work much earlier than would otherwise be required from a purely protective coating requirement consideration.

5.9.2.15 *Approaches*

Approach pavement condition is to be checked for unevenness, settlement, and roughness. Existence of one or more of these defects may cause vehicles coming onto the bridge to induce undesirable impact stresses in the structure. Cracking or unevenness in a portland cement concrete approach slab may indicate a void under the slab from fill settlement or erosion.

Examine joint between the approach pavement and the abutment backwall.

Also, determine if the joint is adequately sealed to prevent accumulations of non-compressible materials which would restrict the normal movement. (*See* 5.9.2.1 and 5.9.2.2 earlier.)

Conditions of the shoulders, slopes, drainage, and approach guard-rail should be included under this item.

5.9.2.16 *General*

Defects found in various portions of the structure, noted under preceding headings, will require thorough investigation to determine and evaluate their causes. The causes of most defects may be readily evident! However, it may take considerable time and effort to determine the cause of some defects and to fully adjudge their seriousness.

Observe bridges during passage of heavy loads to determine if there is any excessive vibration or deflection. If either is detected, further investigation shall be made until the cause is determined. Careful measurement of line, grade, and length may be required for this evaluation. Seriousness of the condition can then be appraised and corrective action taken as required.

Be watchful for possible fire-hazards. This item will include checking for accumulation of debris such as drift, weeds, and brush. Control houses on movable bridges and storage sheds should be kept free of accumulation which is readily combustible. General good housekeeping should be practised. Storage of inflammable material under or near a bridge should be prohibited.

It is advisable for the engineer making the inspections to conder with the local highway maintenance superintendent or foreman regarding the bridges in his territory. The local maintenance-man sees the bridges at all times of the year under all types of weather. He may point out peculiarities which may not be apparent at the time of the investigation. Stream action during periods of high water, and gap-widths at expansion joints at times of very high and low ambient temperatures, are examples of questionable conditions observed by local maintenance personnel, which may not be seen by the investigating engineer. Some problems or potential problems may be evident at these times but not so apparent at other times. Further investigation may be considered prudent as a result of getting the facts from those more familiar with the conditions.

5.10 EQUIPMENT

In order to accomplish the task required for the field-survey operation, each of the survey crews will require both 'standard' and 'specialized' equipment. It will be the responsibility of the crew leader to check daily to see that the equipment necessary to conduct the various structural inspections is available and in working-order prior to the initiation of the inspection effort.

It is contemplated that each of the bridge inspection crews will have a pick-up truck with flashing arrow panels which will be used to warn motorists of survey activities ahead and to direct vehicles into adjacent travel lanes. In addition, each crew will have either a 'cherry-picker' or a 'snooper', to provide closeup access to the bridges to be inspected. They will need the following tools and equipment.

(i) Standard Tools
Some of the standard tools, which should be available for each inspection crew are
1. 30-meter measuring tape (cloth)
2. Folding 2-meter rule and pocket tape (4m metal)
3. Chipping hammer (107 gr)
4. Scraper (50mm)
5. Keel and chalk
6. Calipers

7. Plumb Bob
8. Straight edge — 1m
9. Ambient temperature thermometer
10. Binoculars (7×35 or 7×50)
11. Camera/flash (35mm) — with zoom lens
12. Safety belts (2)
13. Inspection forms
14. Clip board/hard pencils (2H leads)
15. Screw drivers (large)
16. Heavy duty pliers
17. Protractor
18. Flash light (heavy duty)
19. Pocket knife
20. Wire brushes
21. Ice picks
22. Crack templates
23. Traffic cones (20)
24. Safety vests (4)
25. Steel-toe shoes/rubber soles (4)
26. Hard hats (4)
27. Goggles (2)
28. Leather gloves (4)
29. First-aid kit
30. Red flags (4) — two weighted at free edge
31. Rubber boots (4)
32. Rain suits (4)
33. Walkie-talkie (2)
34. Surveyor's wheel
35. Manilla ropes (2 @ 20 meters)
36. Small (mason's) level
37. 10-meter extension ladder
38. 2-meter folding ladder
39. Tool chest
40. Tool kits for movable equipment repairs
41. Traffic warning signs (for maintenance of traffic)
42. Water receptacle (with cups)
43. Salt tablets
44. Compass
45. Bicycle air pump
46. Area map
47. The 'as-built' drawings of the bridges to be inspected.

(ii) Specialized Equipment
Specialized equipment will be required to conduct some of the field activities needed as part of the inspection operation. The description of some of these items is as follows.

(a) 'Cherry-Picker' (usually restricted to 10.7 m vertical-reach)
The 'cherry-picker' is a movable single-arm retractable truck-mounted inspection platform which will be used to inspect all bridges 10.7 meters in height or less. The inspection is accomplished by setting the vehicle below the bridge and extending the movable platfrom to the element to be inspected. The inspection crew leader and atleast one other team member must be capable of operating and maintaining this equipment. The training of

all users should be undertaken prior to the initiation of the inspection operation.

(b) 'Snooper' (with 'descender' to go down a tall pier)
The 'snooper' is a special truck-mounted platform, attached to hydraulic booms, which allows for the inspection of the underside of the superstructure from the existing bridge deck. This equipment will be used to inspect bridges greater than 10.7 meters in height and those bridges which cannot be inspected from underneath because of the terrain. The inspection crew leader and at least one other member of the team should be trained to operate and service this equipment prior to the initiation of the inspection operation.
(*See* Section 5.12 ahead for some details.)

(c) Ultrasonic Meters
One of the goals of this inspection effort is to determine the degree of deterioration, distress and/or material imperfections that are relevant to the structural integrity of the structure to be inspected. When visual inspection and measurements of defective materials cannot be accomplished by normal manual means, an ultrasonic meter should be used during Activity II work and occasionally during Activity I work, to determine the severity and extent of various types of distress, deterioration and defects in concrete and steel structures. This method measures the time of travel of electronically generated mechanical pulses between defective and sound material and, by virtue of an oscilloscope, registers the imperfection or crack within the material. (For more details, *see* Chapter 15.)

Appropriate instructions and training should be provided for all members of the inspection crew on the use of this meter prior to the initiation of inspection.

(d) Pachometer
Where the concrete structures exhibit serious deteriorations and the as-built drawings are not available, a pachometer will be used to locate the bar reinforcements and measure the sizes of the bars to ± 3 mm (1/8 inch). This information along with measured exterior dimensions of the concrete element may allow for the performance of a structural analysis of individual members and the determination of the load capacity of the structure. (For more details, *see* Chapter 15.)

5.11 CREW TYPES

There have to be two basic types of inspection crews. The first type, which will be larger in number of the two, will inspect bridges less than 10.7 meters height unless they cannot be reached from beneath because of terrain.

The underside of the superstructure of these bridges will be inspected from beneath via the use of ladders and/or 'cherry-pickers' (movable equipment).

The second type, which may consist of a single crew, will inspect bridges greater than 10.7 meters in height and those bridges which cannot be inspected from beneath because of the terrain. The superstructure undersides of these bridges will be inspected from the roadway deck via the use of a 'snooper' vehicle.

Each group will consist of a crew leader and a minimum of three individuals who will provide support to the leader during the inspection process. The crew leader will be an engineer with significant experience in structural design and inspection.

5.12 SOME DETAILS ABOUT BARIN'S* SNOOPER

• *A Specific Solution*
The need to monitor highway and railway bridges in general is becoming an increasing requirement because of the continuous increase of traffic, with particular reference to heavy vehicles. There are available, in the market, inspection machines operating in connection with transit cranes, excavators with electrical, mechanical or hydraulic operation and with manual controls — these are inadequate and unsafe because of their great weight, short extensions of the arms and of the slowness in launching and returning.

A made-to-measure machine must have the specifications as set out below to meet modern needs:
(a) High degree of stability to provide great safety.
(b) Structure suitable for mounting on vehicle of any make (European or North American of similar capacity and performance).
(c) Erection of integral unit on vehicle when machine is factory completed.
(d) Capability of transporting the whole unit in containers, by sea.
(e) Provision of auxiliary motors.
(f) Provision of hydraulic energy.
(g) Provision of levelling equipment for stability.
(h) Launching and return equipment must be provided with hydraulic logic allowing sequential and fast operations.
(i) Crossing wind barriers up to 2.70 m (8.85 feet) high.
(j) The whole unit must be able to move during the inspection exclusively by operation of the individual on board who controls it from the platform at very low, constant speed and without jolts.
(k) The pressure of the stabilizing wheels on tarmac surfaces must be within the permissible limits.

---
* Barin — s. p. a., **via** Ca' Nave, 77, 35013, Cittadella, Padova, Italy.

PLATE 17

Fig. 5.130 Sulphate penetration from sub-soil, ingressed moisture evaporates, leaving behind sulphate patches concrete will disintegrate eventually; Poor maintenance, even trees are snugly growing through the gap between the pier and the median

PLATE 18

Fig. 5.169

Fig. 5.170

Fig. 5.171

Fig. 5.172

PLATE 19

Fig. 5.173

Fig. 5.174

Fig. 5.175

Fig. 5.176

Fig. 5.177

Fig. 5.178

Fig. 5.179

Fig. 5.180

PLATE 20

Fig. 5.181

Fig. 5.182

Fig. 5.183

Fig. 5.184

Fig. 5.186

Fig. 5.185

Fig. 5.187

(l) Having a vehicle of fairly light weight and of dimensions which will not hamper traffic flows during transfers.

(m) Ground weight when the machine is moving must be within legal parameters.

(n) During swing-out, it must maintain the height limit level.

Kindly refer to Figs 5.169 to 5.187 (Plates 18 – 20) and also Figs 5.188–5.196.

• *Performance and Operational Safety*

The design and construction of any inspection machine is determined by the following carefully conceived needs:

— high level of stability

— total weight kept within limits allowing it to be mounted on 2 or 3 axle vehicles

— possibility of mounting the unit on any European or North American vehicle of similar performance

— automatic leveling

— automatic telescopic extension and return with hydraulic logic

— the equipment is moved by slow-speed, no-shock hydraulic motors

— horizontal swing-out (particularly suitable in the railway version) to avoid touching overhead cables

— possibility for inspection of via-ducts with windbreaker barriers up to about 8 ft in height

• *'Explorer' Capability*

The explorer is used to extend the platform and to raise members of the inspection team up to 2.50 m (8.20 feet) above the platform level; its movement is independent.

It can cross the full length of the gangway in order to reach the inner beams and enable the decks to be checked.

• *The Operating Cycle (Figs. 5.174 to 5.187)*

The cycle of operation of the 'Snooper' is described below. (Serial nos below indicate the photographs in Figs. 5.174 to 5.187 respectively.)

1. Road Travel: The equipment is mounted on a standard truck and licensed as part of the category of trucks for specific operations, weighing a total of 18 tons. There is no speed limit, overall dimensions are fully within those laid down by the highway code.

2. Arrival on Site of Vehicle: When the equipment arrives on the site, it is positioned on the emergency lane, taking up a space of 4.00 m (13.12 feet).

3. Prestabilization of Vehicle: Hydraulically controlled extension of the stabilizing beam is followed by movement of the stabilizing and drive wheels until the horizontal position indicated by electrical level is obtained.

4. Automatic Cycle-Equipment Release and Lift: The whole equipment is automatically released and lifted until it reaches the present height.

5. Automatic-Cycle-Launch: Horizontal rotation of the tower and gangway unit by 90°. The opening of the safety handrail at this stage should be noted.

6. Operating Cycle-Launch: Final safety positioning on support jacks.

7. Automatic Cycle-Launch: Tower and platform unit is vertical position.

8. Automatic Cycle-Launch: Completion of cycle with opening out of the platforms from the tower (automatic extension completed in 8 minutes). Operator descends to platform.

9. Manual Extension of Platform

10 and 11. Manual Hydraulic Operation: View from below.

12. Manual Hydraulic Operation: Technicians during 'detail' inspection.

13. Manual Hydraulic Operation: Detail.

14. Automatic Cycle: Horizontal rotation of the fulcrum point by 90°.

SOME CHARACTERISTICS AND RELEVANT TECHNICAL DATA OF SOME OF THE CURRENT MODELS OF BARIN'S SNOOPERS
Models

• ABC 55 (5.5 m platform length, Fig. 5.188)
• ABC 90 (9.0 m platform length, Fig. 5.189)
• ABC 130 (13.0 m platform length, Fig. 5.190)
• ABC 150 (15.0 m platform length, Fig. 5.191)
• ABC 200 (20.0 m platform length, Fig. 5.192)
• DESCENDER (attachment — maxm. descent . . . 30 m, Fig. 5.193)
• ABC 14 (14.0 m platform length, Fig. 5.194)
• ABC 16/S (16.0 m platform length, Fig. 5.195)
• ABC 20 (20.0 m platform length, Fig. 5.196)

ABC 55

Main Characteristics

- Standard 2 or 3 axle Mercedes Benz vehicle or other make with similar characteristics.
- Road overall dimensions and weights of equipped unit within International Codes.
- Stabilization on driving rubber wheels.
- Specific pressure on asphalt not exceeding 11 kg/cm^2.
- Translation on work imparted by the operator on the platform.
- Translation speed up to 8 m/min.
- Launching time of 5 min. only.
- Launch and re-entry in automatic sequence without personnel on board.
- Automatic levelling system in the crosswise horizontal plane and vertical plane while working.
- High degree of stability.
- Intakes on platform for water and compressed air.
- Inspection of slabs with 'Scaffolding ladder'.
- Application of 'Descender' for piers inspection.

Technical Data

Platform length	5.50 m
Platform width	1.10 m
Turret lowering	min. 0.50 m
(Up-Down)	max. 4.00 m
Rotation range underbridge	180°
Max. overcoming of windbreaker barriers	3.10 m
Max. span over sidewalk	1.20 m
Space taken on bridge	2.30 m
Payload on platform	300 kg
Height of device for slabs inspection	2.00 m
Total weight of unit	7.000 kg

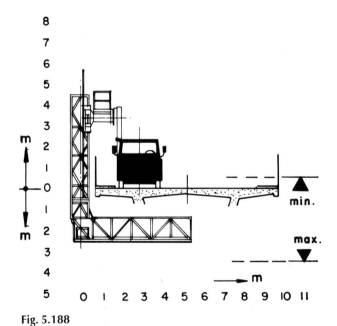

Fig. 5.188

ABC 90

Technical Data

Platform length	9.00 m
Platform width	1.10 m
Turret lowering	min. 1.60 m
(Up-Down)	max. 5.90 m
Rotation range underbridge	180°
Max. overcoming of windbreaker barriers	1.80 m
Max. span over sidewalk	1.20 m
Space taken on bridge	2.50 m
Payload on platform	500 kg
Height of device for slabs inspection	2.00 m
Total weight of unit	14.000 kg

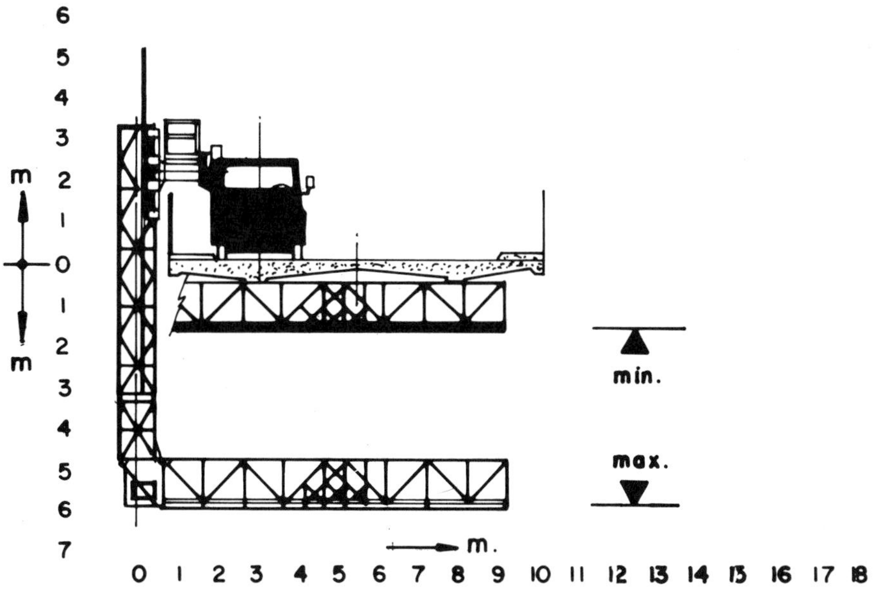

Fig. 5.189

ABC 130

Technical Data

Platform length	13.00 m
Platform width	1.10 m
Turret lowering	min. 1.60 m
(Up-Down)	max. 6.60 m
Rotation range underbridge	180°
Max. overcoming of windbreaker barriers	1.80 m
Max. span over sidewalk	1.60 m
Space taken on bridge	2.50 m
Payload on platform	500 kg
Height of device for slabs inspection	2.00 m
Total weight of unit	18.000 kg

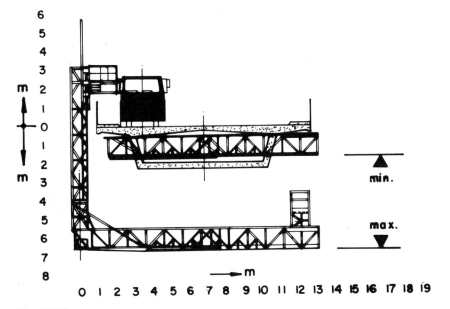

Fig. 5.190

ABC 150

Technical Data

Platform length	15.00 m	Max. overcoming of windbreaker barriers	3.10 m
Platform width	1.10 m	Max. span over sidewalk	1.80 m
Turret lowering	min. 0.40 m	Space taken on bridge	2.50 m
(Up-Down)	max. 7.50 m	Payload on platform	500 kg
Rotation range underbridge	180°	Height of device for slabs inspection	2.00 m
		Total weight of unit	23.000 kg

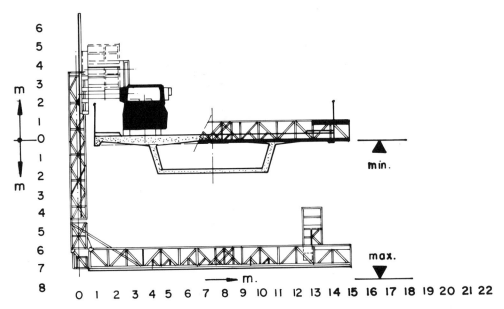

Fig. 5.191

ABC 200

Technical Data

Platform length	20.00 m
Platform width	1.10 m
Turret lowering	min. 2.00 m
(Up-Down)	max. 8.00 m
Rotation range underbridge	180°

Max. overcoming of windbreaker barriers	1.80 m
Max. span over sidewalk	1.50 m
Space taken on bridge	2.50 m
Payload on platform	500 kg
Height of device for slabs inspection	2.00 m
Total weight of unit	24.000 kg

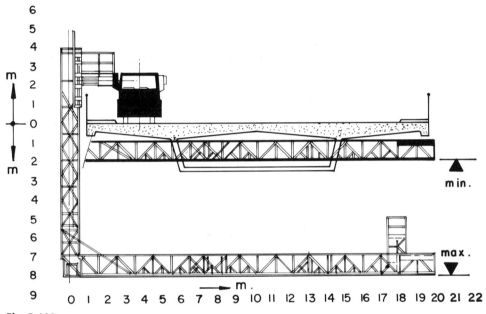

Fig. 5.192

DESCENDER

Technical Data

Attached at tip of ABC platform.

Hand or electric engine operated.

Rectangular or round working cage.

Sliding up/down by means of steel cables.

Safety cable and emergency braking system provided.

Payload on cage	200 kg
Speed of cage ascent/descent	5 m/min
Max. descent from main ABC platform	30.00 m

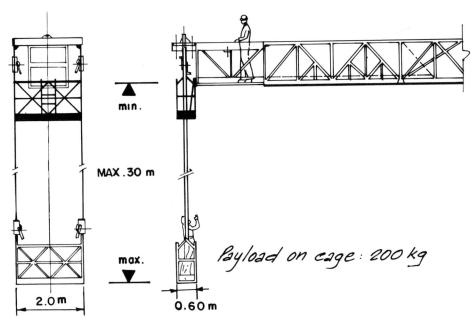

Fig. 5.193

ABC 14

Main Characteristics

- Standard 3 axle Mercedes Benz vehicle or other make with similar characteristics.
- Road overall dimensions and weights of equipped unit within International Codes.
- Stabilization on driving rubber wheels.
- Specific pressure on asphalt not exceeding 11 kg/cm^2.
- Translation on work imparted by the operator on the platform.
- Translation speed up to 8 m/min.
- Launching time of 8 min. only.
- Launch and re-entry in automatic sequence without personnel on board.
- Automatic levelling system in the crosswise horizontal plane and vertical plane while working.
- High degree of stability.
- Intakes on platform for water and compressed air.
- Inspection of slabs with special device.
- Application of 'Spider Descender' for piers inspection.

Technical Data

Platform length	14.00 m
Platform width	1.20 m
Turret lowering	min. 1.50 m
(Up-Down)	max. 7.00 m
Rotation range underbridge	100°
Max. overcoming of windbreaker barriers	2.00 m
Max. span over sidewalk	1.45 m
Space taken on bridge	3.10 m
Payload on platform	800 kg
Height of device for slabs inspection	2.00 m
Total weight of unit	24.000 kg

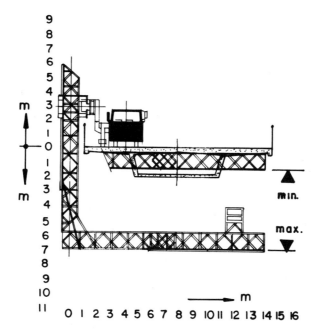

Fig. 5.194

ABC 16/S

Technical Data

Platform length	16.00 m
Platform width	0.90 m
Turret lowering	min. 3.30 m
(Up-Down)	max. 7.30 m
Rotation range underbridge	180°

Max. overcoming of windbreaker barriers	3.00 m
Max. span over sidewalk	3.00 m
Space taken on bridge	3.80 m
Payload on platform	500 kg
Height of device for slabs inspection	2.00 m
Total weight of unit	28.000 kg

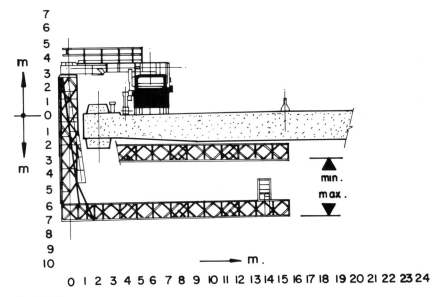

Fig. 5.195

ABC 20

Technical Data

Platform length	20.00 m	Rotation range underbridge	180°
Platform width	1.00 m	Max. overcoming of windbreaker barriers	3.00 m
Turret lowering	min. 4.50 m	Max. span over sidewalk	1.35 m
(Up-Down) L/E Model	max. 10.00 m	Space taken on bridge	4.00 m
Turret lowering	min. 3.00 m	Payload on platform	500 kg
(Up-Down) C/E Model	max. 7.00 m	Height of explorer	3.30 m
		Total weight of unit	24.000 kg

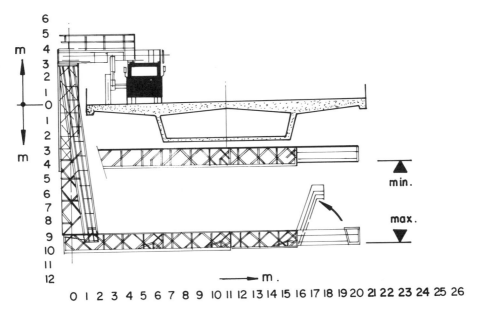

Fig. 5.196

6

BRIDGE STRUCTURE REPAIR

6.1 GENERAL

This chapter essentially deals with rehabilitation by *repairing* the structure. The rehabilitation by 'strengthening' is dealt with separately in Chapters 7, 8 and 9. While the 'strengthening' (design and execution) can involve specialist work, the rehabilitational repair-work usually is a more mundane activity, which may stretch from Routine type of maintenance to preventive type. Nevertheless, as the work concerns a structure, up in the space, each step of each item of work deserves special attention.

6.2 METHODS AND MATERIALS FOR REPAIRING CONCRETE

These are presented in Table 6.1 in a summary format in a workman-like manner, while the individual details are given in subsequent sections.

6.3 REPAIR METHODS

Concrete repairs can be classified either as cosmetic-repairs or rehabilitational-repairs.
The following are commonly used methods for repairs.
  (i) *Dry-Pack Method* for deep and narrow cavities (Mortar-Fill Method),
 (ii) *Preplaced Aggregate Method* (Prepack Method) for restoration of large areas such as walls, foundations and spillways,
(iii) *Partial or Full Depth Concrete Replacement* by casting or patching, using various types of concrete, e.g.
  — ordinary cement concrete, mortar
  — low-slump highly-dense concrete
  — high-alumina cement concrete
  — magnesia-phosphate cement concrete
  — latex-modified concrete
  — epoxy resin mortar or concrete
  — polymer concrete

 (iv) *Shotcrete and Gunite,*
  (v) *Epoxy Mortar Injection,*
 (vi) *Cement Mortar Injection,*
(vii) *Crack Sealing and Filling* by epoxy injection or cementgrout-injection, and
(viii) *Surface Protection by Overlays* of different types of concretes or by various *Sealing Coats.*

6.3.1 Dry-Pack Method

The dry-pack method is used in repairing areas such as narrow slots, cone-bolts, she-bolts and grout insert holes which have a depth equal to or greater than the least surface dimension. Dry-pack is usually a mix of one part of cement to three parts of washed plaster-sand (fine-sand), by weight.

Since colour is important, small amounts of white cement are added to produce matching mortar colour as determined by trials. Enough water is added to produce a mortar which can be moulded into a ball be a slight hand-pressure and will not exude water but will leave the hand dry. The repair holes must be cleaned and dried before filling.

Repairs are done by prewetting the surface and brushing with a stiff mortar or bonding grout. The bonding grout mix is one to one (1:1) mix by volume of cement and fine-sand, mixed to a thick-cream consistency. The dry-pack mix should be promptly compacted into place before the bonding grout dries. The dry-packed mix is compacted with a variety of wooden tools, the exact choice of which should be made by the user to best fit the job requirements. Wooden tools are preferred, as they do not polish the surface layer of the mix and consequently offer better bond between each successive layer of the dry-pack.

6.3.2 Preplaced Aggregate Method (Prepack Method)

This method is used most advantageously on large repair jobs, especially where placement of normal concrete would be difficult. The process consists of removal of deteriorated concrete using pneumatic hand

Table 6.1 Concrete Repair: 'Methods' and 'Materials'

Defects	Repair Methods	Materials
• Live Cracks	— Caulking	Elastomeric sealer
	— Pressure injection with 'flexible' filler	'Flexible' epoxy (resin and hardener mix) filler

	— Jacketing:
	  • Strapping
	  • Overlaying
		
		Steel wire or rod
		Membrane or special mortar

	— Strengthening	Steel plate, post tensioning, stitching, etc
• Dormant Cracks	— Caulking	Cement grout or mortar, Fast-setting mortar.
	— Pressure injection with 'rigid' filler	'Rigid' epoxy (resin and hardener mix) filler
	— Coating	Bituminous coating, tar
	— Overlaying	Asphalt overlay with membrane
	— Grinding and Overlay	Latex modified concrete, highly dense concrete
	— Dry-pack	Dry-pack
	— Shotcrete/Gunite	Mortar (cement), Fast-setting mortar
	— Patching	Cement mortar, Epoxy or Polymer concrete
	— Jacketing	Steel rod
	— Strengthening	Post tensioning, etc.
	— Reconstruction	as needed
• Voids • Hollows and • Honeycombs	— Dry-pack	Dry-pack
	— Patching	Portland cement grout, mortar, cement
	— Resurfacing	Epoxy or Polymer concrete
	— Shotcrete/Gunite	Fast-setting mortar
	— Preplaced aggregate	Coarse aggregate and grout
	— Replacement	as needed
• Scaling Damage	— Overlaying	Portland cement concrete, Latex modified concrete, Asphalt cement, Epoxy or polymer concrete
	— Grinding	– – –
	— Shotcrete/Gunite	Fast-setting mortar, Cement mortar
	— Coating	Bituminous, Linseed oil coat, Silane treatment
	— Replacement	as needed
• Spalling Damage	— Patching	Concrete, Epoxy, Polymer, Latex, Asphalt
	— Shotcrete/Gunite	Cement mortar, Fast-setting mortar
	— Overlay	Latex modified concrete, Asphalt concrete, Concrete
	— Coating	Bituminous, linseed oil, Silane, etc.
	— Replacement	as needed

hammers. Then the cracks are grouted and dowel bars are anchored into the sound concrete to fasten the temperature and shrinkage reinforcement bars (generally 10 mm dia. @ 150 mm centres, in two orthogonal directions). A form is placed around the perimeter of the repairs, then clean coarse-grade aggregate (minimum 16 mm size for thick sections, and minimum 12 mm size for thin sections) is placed and compacted before pumping a especially designed cement sand grout into the aggregate by especial insert fittings placed in the face of the forms.

The coarse aggregate may be of any suitable size, depending on the thickness of the repairs. Normal size varies from 20 to 40 mm down. The grout can consist of 2:2:3 volume ratio of type I portland cement, fly ash and sand, and water (water to cement and fly ash ratio of 0.5 by weight). Fly ash, though preferable, can be replaced by cement.

The 'preplaced-aggregate' concrete offers low drying shrinkage because of its point to point contact between coarse aggregates and high bonding strength, which are essential properties for all concrete repairs. Other features of the preplaced-aggregate method are excellent resistance to freezing and thawing and greater impermeability at later stages in relation to conventional concretes. Preplaced-aggregate method has been used with much success to repair dams, tunnels, retaining walls, bridge piers, and reinforced concrete columns and beams in bridges. It is also used extensively for laying concrete underwater (e.g. seal plug in caissons — the COLCRETE Process).

6.3.3 Partial or Full-Depth Concrete Replacement, and Repairing the Patches and Breakages in Concrete

Various Types of Concrete for Repair
   (i)   Ordinary portland cement concrete or mortar
  (ii)   Low-slump, highly-dense concrete
 (iii)   High-alumina cement concrete
  (iv)   Magnesia-phosphate cement concrete
   (v)   Latex-modified concrete
  (vi)   Epoxy-mortar and epoxy-concrete
 (vii)   Polymer concrete

For fast-setting materials, there are many different products and their properties vary substantially from brand to brand, so care should be exercised in selecting proper products for particular repairs.

(i) Portland Cement Concrete or Mortar
The commonly used patching material is concrete mortar with ASTM — type I, II or III portland cement with or without an admixture. The most widely used admixtures or additives are an accelerator, a retarder, an air-entraining agent or a plasticizer, depending on the requirement and ambient conditions.

The accelerator admixture should not contain more than 1 per cent chloride-ions by weight. Calcium chloride can be used as an accelerator. Superplasticizer may be used to gain workability under low w/c ratio (yielding a slump of even up to 7 inches or more). Higher slump and excessive plasticizer will cause segregation.

Portland cement type III (high early strength cement) has been used widely for patch work, because it is relatively inexpensive, simple to use, and durable. But it has a high shrinkage property, a tendency to crack in patch, and a low rate of strength-gain in cold weather. Caution should be exercised if high early strength cement is specified.

As a patch material, concrete with lower water–cement ratio is preferred. It is suggested to use a water–cement ratio of lower than 0.45 (5 gallons of water per bag of cement, generally one gallon of water is 8.34 pounds at 70° F and one sack of cement weighs 112 pounds) for moderate environments and less than 0.40 in severe environments.

(ii) Low-Slump Highly-Dense Concrete
This concrete could be used for patch work but workability could be a problem in the field. The water – cement ratio should be about 0.35 and the slump of mix should not exceed 1 inch. Cement paste or grout is generally applied before placing the concrete and the repair area is frequently prewetted to improve bond. For low slump concrete, the surface should be completely dried if applying a bonding coat of epoxy-bonding-compound or latex-modified-concrete-grout.

(iii) High-Alumina Cement Concrete
This type of cement has been used in Europe and may be used under certain conditions, and where corrosion is a problem. Fondu Calcium-Aluminate, Lumnite, and others are the available brands in the U.S. market.

(iv) Magnesia-Phosphate Cement Concrete
This premixed cement generally comes with dry magnesia component and a liquid-phosphate package. It appears to perform satisfactorily for patch works. There are few brand names such as Acmaset, Bostik 276, Fast-Crete 100, Horn 240, Set 45, FX 90 and others in the U.S. market.

(v) Latex-Modified Concrete
This is a mixture of concrete with a latex-emulsion-admixture. It has not been used as widely as portland cement mortar and epoxy mortar. However, it is widely used in conjunction with deck overlay systems.

(vi) Epoxy-Mortar and Epoxy-Concrete
Epoxy-mortar is resin plus hardener plus fine-sand filler, and Epoxy-concrete is resin plus hardener plus

fine-sand plus small sized coarse aggregate. Epoxy resin mortar and concrete have been widely used in the last several years. These materials have given satisfactory results for repairing spalls, popouts and partial-depth repairs. But these have a relatively low modulus of elasticity. Also, preparing adequate specifications for particular needs and then meeting these requirements can be a problem because many different types of products are available for a specific purpose. Choosing the type to satisfy the job-conditions and requirements can be a tricky task. Since a wide variety of commercial epoxy patch materials are available (such as Cono/Crete, Colma Dur, FX 700, Epoxy, etc.), it is recommended that a technical representative of the manufacturer should be at the site during the entire operation from mixing to curing in order to provide guidance. Their service will outweigh the additional cost by resulting in satisfactory patches.

Epoxy resin mortar and Epoxy resin concrete are generally used for small shallow patch areas while portland cement mortar and concrete are used for larger areas. The cost of epoxy mortar and epoxy-concrete are substantially higher than portland cement mortar and concrete but their rapid hardening and better bond and strength would make it possible to reduce the required curing time, and thus the bridge can be opened to traffic earlier. This will serve the user's operational cost and reduce the motorist's exposure to the hazard conditions during the construction period.

(vii) Polymer Concrete

Polymer concrete is premixed aggregate with methyl methacrylate (MMA) 'monomer' that is later 'polymerized' in-place. It is fast curing and has high strength and durability. Most prepacked products come in two components, a 'monomer-liquid' and a 'fine-powder'. Coarse aggregate is added to this mix. They should be mixed in the field (just before use) manually or by using a rotating drum mixer. The main advantage of polymer concrete is the minimizing of inconvenience to the public and reducing the recurring maintenance cost. Thus, it is suitable for use in deck repairs where the traffic is heavy and the working hours are limited to off-rush hours. There are a few brands, such as Corrosive 2020 polymer, Crycon, FX 826, Duracryl, Silikal, and others, in the U.S. market.

Concrete Repair (Replacement)

(i) Preparation of Substrate

After adequate precautions for structural stability have been taken, then, all loose and damaged concrete must be carefully removed and sound concrete surface exposed. The concrete substrate (i.e. the concrete surface exposed after removal of all damaged concrete) must be sound. It should be thoroughly cleaned in order to

achieve satisfactory adhesion to new-(i.e. repair-) concrete. Removal of unsound concrete can be accomplished by:

— mechanical methods
— thermal methods
— chemical methods

The choice of a suitable method depends on the situation, specially on the extent and thickness of the concrete to be removed, as well as on the type, location and position of the damage in the structure. The thermal and chemical methods are rarely used and are restricted to special circumstances and hence have not been described here. As for the mechanical methods, the usual ones are:

— chipping method
— milling method
— sand-blasting method
— very high pressure water-jetting method
— grit-blasting method
— steam-blasting method

Cleaning can be done by compressed air jet.

The coarser the demolition process the less expensive it will be. However, dust, noise and vibration may have to be controlled in all mechanical methods.

Cleaning of the exposed sound concrete surface may best be done by clean water-jet. Where the subsequent repair operations involve use of water-sensitive epoxy formulations (prime-coat and epoxy mortar or concrete, for instance) it may be necessary to dry the cleaned exposed concrete.

Removal and Replacement of Concrete: This is necessary when the concrete is found to be delaminated (sounding with hammer can indicate this) or the chloride-ion content is critical, or cracks are found on a chipped surface, or concrete is rust-stained or carbonated down to the reinforcement (and beyond). The removal of damaged concrete is best done with electrically powered or compressed air chipping, ensuring that the reinforcement is not damaged. Flat chisel is normally used to minimize micro crack formation. For a complete removal of a structural element, operations such as sawing, cracking-open the member, thermal-lancing, and blasting may be adopted, depending on convenience and possibility. Special care needs to be taken while removing concrete from prestressed concrete structures. Hydrodemolition (very high pressure water-jet) is the latest method where water is impinged on the concrete in thin jets at a very high pressure. This enables removing of concrete in a more efficient and precise manner, without damaging reinforcement.

The replacement of concrete in larger continuous areas should proceed in the same manner as placing concrete during the construction of the structure.

However, certain features resulting from the combination of old and new concrete should be considered.

Shallow concrete thickness can be restored by mortar but if the deterioration process has reached a level where a shallow surface repair is not enough, a chunky replacement of the damaged concrete is necessary. The technical choice of the repair material depends on volume to be replaced, the depth of the repair, the loading effects to be expected and the conditions of application on site.

Placing concrete in the area to be repaired should be accomplished in such a manner as not to impede concrete flow or entrap air, so as to avoid voids in the concrete. Therefore, the formwork must be sufficiently rigid and tightly fitted to the existing concrete in a manner so as to minimize leakage of cement-paste. The surface of the exposed existing concrete will require adequate preparation, careful cleaning and pre-moistening to saturation. Surface drying is necessary in case a water-sensitive epoxy-formulation is used.

This replacement-concrete should have final properties that match those of the old concrete as closely as possible (strength, modulus of elasticity, creep coefficient, etc.). To avoid temperature and shrinkage cracks, specially in the transition areas, the type of cement, cement-content and the water–cement ratio should be carefully evaluated.

The use of plasticizer (and retarder where necessary) is recommended. Slight recompaction/revibration may be required to improve the bond to the old concrete. However, care should be exercised to avoid disturbing concrete after its initial setting.

For larger concrete volumes, minimizing the temperature difference between old and new concrete may require special procedure (cooling of new concrete and/or warming the old concrete). 'Type' and 'Duration' of curing should be evaluated on a case by case basis.

Removal of Chloride-Contamination in Concrete: There does not exist, within the current state-of-art, any promising method of transforming penetrated chlorides into insoluble compounds so as to render passive the potential for corrosion. Some possible methods of chloride removal are as follows:
— water treatment
— treatment with lime-milk
— electro-osmosis
— mechanical removal of the contaminated concrete layer

The efficiency of the first three methods has not yet been proven, but ongoing research and tests may provide answers in future.

(ii) Schematic Examples of Concrete Rehabilitational Repair
(a) Repairing against Shear Distress by Stitching Rods and Epoxy (Fig. 6.1)
(1) Carefully vacuum drill holes @ 6″ apart and nearly perpendicular to cracks. (2) Insert full length of reinforcing bars (min. 50 cm) in the holes. (3) Fill the holes with epoxy grout under low pressure.

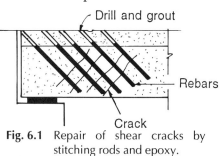

Fig. 6.1 Repair of shear cracks by stitching rods and epoxy.

(b) Repairing against Shear Distress by Jacketing Stirrups (Fig. 6.2)

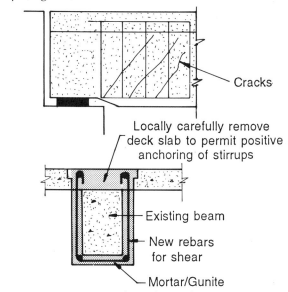

Fig. 6.2 Repair against shear distress by removing concrete as indicated, placing additional stirrups, priming the exposed surface and epoxy mortaring or concreting (or shotcreting/Guniting), temporary supports may be needed.

(c) Repairing against Loss of Section (Fig. 6.3): After removal of the unsound concrete, the beam can be jacketed with a steel form and filled with epoxy concrete.

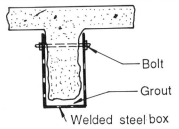

Fig. 6.3 Jacketing by metal sleeve

(d) *Repair-Rehabilitation of Damaged/Spalled Concrete (Fig. 6.4):*

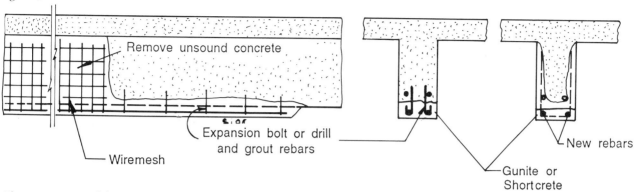

Fig. 6.4   Repair of damaged / spalled concrete. Temporary supports may be needed to relieve dead-load stresses. Needs special stress-check and just-touching support system in case of prestressed concrete.

(e) *Repair by Arresting Crack-Propagation by the Post-Tensioning Principle (Fig. 6.5):* The flexural cracks in reinforced concrete can be arrested and even corrected by the 'post-tensioning' method. It closes the cracks by providing compression force to compensate for tension and adds a residual compression force. This method requires anchorage of the tie-rods (or wires) to the anchoring devices (the guide-bracket-angles) attached to the beam. The rods or wires are then tensioned by tightening the end-nuts or by turning of turnbuckles in the rods against the anchoring devices. However, it may become necessary in certain critical cases to run at least an approximate stress-check to guard against any possible adverse effects.

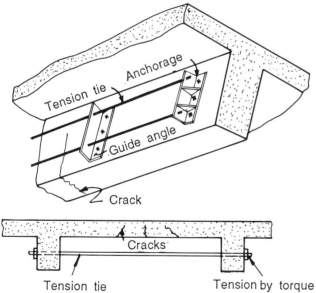

Fig. 6.5   Arresting cracks by lock-on compression by the post-tensioning principle.

(f) *Repair of Cracked Halving-Joints:* In a slab-type deck with halving-joints it is very difficult to physically see the crack-distress in the halving-joint zone other than that at outside faces.

It is not easy to lift the deck off the halving joints. Perhaps one of the appropriate methods would be to drill holes normal to cracks, fill them with a suitable epoxy or epoxy-mortar formulation and then place reinforcement bars (of predetermined sizes and lengths) in them to stitch across the cracks. Such bars can be placed both diagonally (against inclined cracks) as well as vertically (to act as hanger bars). The bars may be placed in the clean holes prior to filling-in the epoxy (so as to save loss of epoxy) but then great care is needed not to entrap any air.

Figure 6.6 shows a distressed halving-joint (badly detailed as well as poorly constructed), which had been kept 'covered'! Figures 6.7 and 6.8 show a temporary repair of halving joints, which was rather quick to execute, cheap and very effective (Plates 21 and 22).

(iii) *Counter-Measures against Corrosion-Deterioration of Concrete Deck-Slabs*

The deterioration of a concrete deck slab (indeed any concrete) in new or replacement construction could substantially be reduced by producing a good quality dense concrete, strict quality control, good field construction practices, adoption of a low water-cement ratio (0.40 to 0.45)and increased concrete cover for reinforcement steel (minimum of 50 mm). The lower water–cement ratio in concrete and adequate cover have significant affects on preventing chloride and carbonation to penetrate the concrete. The use of higher strength concrete will add to the initial cost but it will be cost-effective in terms of life-cycle cost by inhibitting moisture-movement and chemical reactions. The deck slab thickness is recommended to be not less than 25 cm.

The following measures have been used to prevent early corrosion deterioration of concrete deck-slabs subject to de-icing salt attack.

By Using Epoxy Coated Reinforcing Bars: The epoxy coated bars are sometimes used in the top mat of new decks, in the barrier kerbs, parapets, and the top of the bridge seats under the joint, to minimize corrosion damages. The Illinois Department of Transportation assessed the cost of coating the bottom reinforcement mat of deck slabs and found that the cost is less than a one per cent of the cost of a deck slab. Therefore, they decided that it is prudent to epoxy coat all the reinforcement bars in new deck slabs including all bars in parapets, kerbs, side-walks, and medians.

Generally only the bars in the top mat of reinforcement (in both directions) are epoxy coated being nearest to surface. Where the slab thickness is to be increased by placing additional concrete thickness (after adequately scarifying the old concrete, applying the bonding-primer formulation to the cleaned concrete surface and bars, and placing new reinforcement), either shear connectors or special bonding epoxy should be used for composite-action between old and new concrete. Additional bars should be placed at acute corners.

For epoxy coating the reinforcement bars are blast-cleaned to a near white, then heated by inducing coils to the temperature required for application of the epoxy powder (400° F to 450° F). The bars then pass through an electrostatic spray which applies the charged dry powder to the heated bars. This process provides about 7 mills of coating, thus protecting the bars from corrosion-inducing elements — chlorides, moisture, and oxygen. Care should be exercised during shipping and handling of the bars to minimize damages. If any damages occur, they should be repaired before placing the concrete. Where the epoxy layer is punctured (during handling or tieing of bars), severe local corrosion can result, defeating the whole purpose.

According to past research, 'galvanized' reinforcing bars appear to be ineffective in preventing corrosion.

By Overlaying the Deck-Slab Concrete by Latex-Modified Concrete Overlay or Low-Slump Highly-Dense Concrete Overlay: These overlay systems (see ahead for details) have been used for the new and the existing concrete deck slabs to curtail penetration of chlorides into concrete, thus preventing a critical chloride build-up from reaching the reinforcing steel.

6.3.4 Shotcrete and Gunite

Shotcrete (pneumatically applied cement–concrete) and Gunite (pneumatically applied cement–mortar) are suitable for forming the new concrete (i.e. the restoration-concrete) and for strengthening and jacketing of various structural elements.

Pre-treatment of the exposed concrete surface is of prime importance when using shotcrete or gunite. Sand-blasting has proved to be an efficient surface treatment procedure. The exposed sound concrete surface should be sufficiently pre-moistened. No bonding agent is necessary because at the interface a mortar enrichment occurs as a result of aggregate rebound.

Shotcreting in multiple layers requires that the preceding layer achieves a sufficient degree of hardness prior to shooting-on the next layer. Some nominal reinforcement may be required for a thickness larger than 50 mm. This reinforcement should be fixed in position in such a manner that it remains stiff and keeps its position during shotcreting operations (chicken wire mesh is handy).

Curing may be accomplished by an evaporation protection method (e.g. plastic sheet cover), to prevent a rapid drying out. If a freeze-thaw/salt resistant concrete is required, air entrainment admixture may have to be added to the mix. Also, surface protection measures may be necessary in certain cases.

There are two basic gunite (or shotcrete) processes:
— a 'dry-mix' process, where the mixing water is added at the nozzle to which the cement-sand (or cement-sand-small size C.A., as the requirement may be) mixture is brought by compressed air through the delivery hose
— a 'wet-mix' process where all the ingredients, including water, are mixed before entering the delivery hose.

Shotcrete (or gunite), suitable for normal construction requirements, can be produced by either process. However, differences in cost of equipment, maintenance and operational features may make one or the other more attractive for a particular application.

Properly applied shotcrete (or gunite) is a structurally adequate and durable material and is capable of excellent bond with concrete masonry and steel. However, these favourable properties are contingent on proper planing, supervision, skill and continuous attention by the application-crew.

In general, the in-place physical properties of sound shotcretes (or gunites) are better than those of the conventional mortar or concrete having the same composition.

Special variants of shotcretes (or gunites) result from the addition of fibre or synthetic resins. Steel, glass (boron-silicate-glass) and plastics are used for the fibres. The ratio of the fibre to cement will be larger in the initial mixture than in the rebound material. In the case of steel fibres, corrosion protection must be considered, unless the fibres are protected from corrosion. The last layer must not contain steel fibres.

Sprayed dry-mix concrete or mortar has been used for many years for both new construction and repair of old structures. Equipment is available for pneumatic

PLATE 21

Fig. 6.6 Distress exposed at a halving joint

Fig. 6.7 External anchor bars (with plates and nuts) employed to arrest fracture distress at a poorly detailed and overloaded halving joint

PLATE 22

Fig. 6.8 A close-up view of rehabilitation at a halving joint

delivery and application of wet concrete mixes containing coarse aggregate as large as 25 mm. No formwork is necessary in many instances, yet intricate shapes and thin overlays can be successfully constructed, provided good materials are used and proper procedures are followed.

As explained earlier, when cement mortar or concrete is pneumatically sprayed on a surface, the products are known as Gunite or Shotcrete, respectively. The dry process consists essentially of mixing dry sand and cement (and small size coarse aggregate in case of shotcrete) in a mixer, then placing this dry mixture in the delivering equipment, whereafter the mixture is placed under pneumatic pressure. Under pressure, the dry mixture flows through a rubber hose to the nozzle where predetermined amount of water joins-in under high velocity. Compaction is achieved by the force exerted by the impact of the mortar (or concrete in case of shotcrete) on the receiving surface. Gunite (or shotcrete, as the case may be) produces a high quality material with the following desirable characteristics:

1. Low water-cement ratio, resulting in high strength and low permeability.
2. Dense material as a consequence of low water content and high impact velocity.
3. Superior bonding ability, making it specially suitable for repair work of many types of structures.
4. Relatively simple work-area requirements, with only an air compressor and a source of water being necessary. Easy transmittal of concrete (or mortar) through a hose from the gun to the site of application.
5. Shotcrete lends itself to the production of many shapes and thin sections with a minimum or no formwork at all. Wiremesh may be used to attain shape.
6. Resistance to weathering and many types of chemical attack (because of being highly dense). Good abrasion resistance.
7. Good refractory properties with proper aggregates.

Some Details Regarding Shotcreting and Guniting

(i) In general, dry-mix shotcrete, which is pneumatically applied concrete under high pressure, is applied for repair of concrete in bridge work.

(ii) Ordinary portland cement conforming to standard specifications is used in shotcreting.

(iii) Sand for shotcreting shall comply with the requirements stipulated in the standard specification, and shall be graded evenly from fine to coarse. In general, sand should neither be too coarse to increase the rebound not too fine to increase the slump.

(iv)

Water cement ratio for shotcrete should fall within the range of 0.35 to 0.50 by weight and should be wet enough to reduce the rebound.

(v) Test panels, simulating actual field conditions, should be fabricated, and preconstruction testing should be conducted.

(vi) Cement mortar mix for pure guniting should generally be within the range of one part of cement to three parts of sand. For shotcrete, coarse aggregate is included in the mix, and its percentage is normally kept as 20 to 40 per cent of the total aggregate. The mix must be suitably designed.

(vii) It should be ensured from tests on the gunite mortar or the shotcrete, that a strength of about 20 per cent more than that of the parent concrete, is achieved.

(viii) The defective concrete should be cut out till sound concrete is reached. Under no circumstances should the thickness of concrete to be removed, be less then the clear cover to the main reinforcement. No square shoulders should be left at the perimeter of the cut-out portion, and all edges should be tapered. Thereafter all loose and foreign materials should be removed and the surface sand-blasted to make it rough to receive shotcrete (or gunite, as the case may be).

(ix) The exposed reinforcement should be thoroughly cleaned free or rust, scales, etc. by wire brushing, preferably by sand-blasting. Wherever the reinforcements have been corroded, the same should be removed and replaced by fresh reinforcement. Before application of shotcrete or gunite as the case may be, but after thorough saturation of exposed concrete by water, a coat of neat cement slurry should be applied on this surface and the reinforcement.

(x) The additional reinforcement should be preferably welded to the existing reinforcement. In case the existing reinforcement is not weldable and provision of laps becomes necessary, the lapped reinforcing bars should not be tied together. They should be separated by at least twice the diameter of the bar, wherever possible. This will ensure a dense concrete mask around the steel.

(xi) Sufficient clearance should be provided around the reinforcement to permit encasement with sound shotcrete or gunite, as the case may be. Care has to be taken to avoid sand-pockets behind the reinforcement.

(xii) A thickness of 25 to 40 mm of shotcrete can normally be deposited in one operation. If, for some reason, the total thickness is to be built up in successive operations, the previous layer should be allowed to nearly initially-set without becoming hard, before the application of the subsequent layer. It is necessary to apply shotcrete (gunite) on a damp concrete surface. (Gunite thickness is normally restricted to about 50 to 100 mm in total.)

(xiii) It would be desirable to provide welded wire fabrics in the first layer of shotcreting. In case the damage to the concrete member is 'too deep', the specifications for shotcreting as well as requirement of placement of wiremesh may have to be decided as per actual field conditions.

(xiv) The stipulations given in the standard specifications regarding application of shotcrete should be scrupulously followed so as to keep the rebound to a minimum.

(xv) Adequate care has to be taken regarding curing of the shotcrete (or gunite, as the case may be). It would be desirable that green shotcrete (or gunite) is moist-cured for at least ten days.

(xvi) Shotcreting or guniting work should not be done during wind/rain/or hot weather conditions. (Hot weather means the ambient atmospheric temperature at site in shade exceeding about 32 to 35° C.)

6.3.5 Epoxy Mortar Injection for Grouting the Internal Cavities, Hollows and Crevices in Structural Concrete

(a) Epoxy is the resin and hardener (two-component) mix and for its mortar, fine clean sand is added in and mixed well. This 'grout' should be of adequate flowability. (Follow standard manufacturer's specifications for mix proportions and pot-life.)

(b) Procedure-steps for injection are as follows :

1. Locate the hollow spots by sounding with a hammer and mark with highlighter or crayon. Marking an outline of the hollow spot will help in placing drill holes.
2. With epoxy paste, seal the visible spalls, popouts, and cracks within the area of the hollow that would allow epoxy to leak away otherwise.
3. Determine the location of reinforcement steel with a pachometer.
4. Drill a hole near the center of a hollow spot (not on the steel). Keep the hole away from points of possible leakage such as spalls or cracks. Drill holes as perpendicular as possible to make it easier to get a good seal with an injection probe. Check the drill tip often to make sure that it does not become clogged. The hole need not be much deeper than 4 inches and usually less than 2 inches.
5. Inject epoxy resin into hollow spot at 0 to 50 pounds per square inch (psi). Check progress of epoxy with

a hammer. If a hollow spot can not be completely filled by one hole, drill another in the remaining area to be filled and inject it. If epoxy leaks from a previous hole, plug it with a rubber stopper. Keep injection pressure as low as possible. The pressure required will depend on the type of the epoxy used, the temperature, and the thickness of the spot. If the hollow is thin, the pressure will tend to rise faster and the pump speed should be turned back to prevent a pressure build up.

6. Scrape off excess epoxy and place waste in can. Sprinkle the wet epoxy areas with a sand-cement mixture or a dust-sand mixture.

(c) *Epoxy-Mortar and Epoxy-Concrete*: The two-component epoxy (Resin and Hardener) and fine sand (filler) form the 'epoxy-mortar'. Epoxy, Sand and pea-size Coarse Aggregate form the 'Epoxy-Concrete'.

The resin and the harder (i.e. the two components of the epoxy formulation) are mixed together in a specified manner and in a specified ratio and then used as a crack sealant or primer as per exact instructions specified by the manufacturer. Likewise, specific mixing and application instructions have to be followed where 'sand' or 'sand and coarse aggregates' have to be incorporated. The mixture must be used within the duration of its specified 'pot-life' and the portion of the mixture not used within the specified duration of pot-life should not be used (and must be thrown away safely). Although the effectiveness for bond and durability of various standard epoxy-formulations has been well-established, it is essential to ensure about these from the manufacturers who come up with new products ever so often. However, epoxy-formulations are expensive and, where possible, cement–mortar and cement–concrete should be considered seriously.

(d) *Repair of porous/Honey-Combed Concrete*: There are at least two methods.

1. The porous parts of the concrete are replaced by sound, watertight dense concrete.
2. The porous zones are injected and filled with suitable sealing-material.

The former procedure can not be used where there is a continuing inflow of water. In such a case, sealing can be accomplished by the injection method.

6.3.6 Cement-Mortar Injection for Grouting the Internal Cavities, Hollows, and Crevices in Structural Concrete

— This grout usually comprises either of 'neat' cement and water (W/C = 0.4 by weight) or of one part cement with half part fine clean sand and adequate amount of water to obtain the required flowability. 'Neat cement grout' is applied to pressure-fill cracks of about 1.0 to 2.0 mm width. Wider cracks (as in masonry) and fractures can be pressure-filled by 'cement-sand' grout. Usually an expansive chemical compound (for nonshrink effect) is added during mixing of grout. Freedom from chloride and sulphate contamination has to be ensured.

— Procedure for grout-injection is similar to that in case of Epoxy mortar grouting, explained earlier.

6.3.7 Crack Sealing and Filling (Repair of Cracks)

(i) General

(a) In reinforced concrete, cracks wider than about 0.3 to 0.4 mm should be sealed and filled by injection. Before deciding the most appropriate method/material for repairing/sealing a crack, a determination should be attempted on its cause and whether it is active or dormant. Whether the crack is active (i.e. propagates/ breathes), may be determined by periodic observation. A classification of cracks, together with their primary causes, etc., has been covered in Chapter 4 in detail.

(b) Basically, a crack resulting from a rare load-application, and which has ceased to propagate, can be repaired (if it is wider than about 0.3 to 0.4 mm) by pressure-injection with a suitable epoxy-formulation so that the integrity is restored and any adverse influence on the service life of the structure is eliminated or minimized.

(c) In case of cracks which are the result of time dependent effects, such as shrinkage or settlement, the repair should be delayed as much as possible, compatible with the service efficiency of the structure, so that the effect of further deformation is minimized.

(d) Dormant cracks (dead cracks), in excess of about 0.30 to 0.4 mm width, must be cleaned and then filled and sealed, by epoxy-injection for widths up to about 1 mm, and by fine cement grout for wider cracks. (Normal cement grout is easily possible for widths beyond 3 mm.) Live cracks (active cracks) must be periodically monitored for propagation and width-increase, using tell-tales. Where their width exceeds about 0.3 to 0.4 mm, a 'chase' (V-groove) should be made along the crack, the groove and the cracks cleaned by a dry air-jet, and then filled to part of its depth by a flexible filler material (mastic, thermoplastic, etc.) to prevent ingress of moisture and other deleterious materials. After the crack has become dormant, the filler material can be removed and the crack cleaned and filled with a rigid (epoxy) filler. The chase can then be plastered with cement-paste using a non-shrink additive.

(ii) A Flow Chart for Decision-Making on Crack-Repair

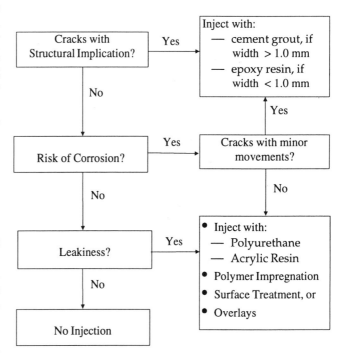

It must be noted that the corrosion-protection effect by covering a reinforcement bar with epoxy resin is not truly dependable. The resin might electrically isolate a corroding bar only where it is still intact (resin-coating some times gets punctured during handling and bar-tying operations). As a consequence, the corrosion propagation-rate may in fact even be higher at the puncture locations in the bar, giving rise to accelerated local corrosion!

With regard to load-induced cracks, it must be noted that injecting them will not strengthen the structure. The cracks will appear again, unless the loads are reduced.

(iii) Materials used for Filling (and hence full-depth sealing) the Cracks

(a) The material used for crack repair must be such as to penetrate easily into the crack and provide durable adhesion between the crack surfaces. The larger the modulus of elasticity of the material, the greater will be the obtainable adhesion strength. The material should be such as not to allow infiltration of water, and should resist all physical and chemical attacks. Currently the following fluid-resins are used for crack-injection (together with hardeners):

— epoxy resin (EP)
— polyurethane resin (PUR)
— acrylic resin
— unsaturated polyester resin (UP)

(b) The formulations of commercially available injection resins vary widely in their properties, and care must be exercised in making proper selection. Important properties of any injection resin are its resistance to moisture penetration and alkaline attack from the cement. Where tensile strength is a requirement, the tensile strength of the resin should approach that of the concrete as closely as possible. Therefore a stiff and highly adhesive resin is desirable. These properties are available in epoxy or unsaturated polyester resins. After hardening of the injection material, the 'stiffness' of the crack will be dependent upon the elasticity of the resin.

(c) The polyurethane or acrylic resin is recommended where moisture resistance is a requirement. Some Epoxy based low-viscous resins will penetrate to the crack-root even when the crack width at the surface is only about 0.2 mm! Comparable results can be obtained from unsaturated polyester and polyurethane resins. Acrylic resins are capable of sealing fine cracks because of their low viscosity. However, in all cases, this requirement can only be obtained with an appropriately long 'reaction-time'. Fast-reactive systems will only close the crack at its surface, which may not be desirable.

Although cement paste is relatively inexpensive; its use is limited to crack widths of approximately 2 mm or more because of its limited viscosity. However, finely ground cements allow injection of cracks with widths down to about 1 mm. Cement glues and mortars are of importance in such applications as injection of voids, hollows, cavities, and honey-combing, and sealing of ducts, etc. For these applications the use of appropriate additives is recommended to reduce viscosity, shrinkage, and the tendency for settlement. Improvement of workability will be obtained if the cement-suspension is formed by using high-speed mixers.

(iv) Crack Injection
(a) Steps
As a rule, the following steps are necessary for injection:
— Thoroughly cleaning the cracks with high pressure clean air;
— Drilling the injection-holes and blowing-clean the holes and cracks;
— Fixing of 'nipples' along the cracks;
— Covering the crack-surface between nipples by a thixotropic liquid sealant;
— Mixing of the injection material;
— Injecting the injection material through the nipples against gravity (unless the crack is horizontal), in a progressive serial order; and
— Re-injection and testing, if required or found necessary.

(b) Injection Equipment
Different Injection equipments are available, depending on whether the materials are premixed or used separately. In the case of 'premixed components' equipment, the resin and hardener are mixed first and subsequently injected into the crack using this equipment. Typical 'premixed components' equipment is a 'hand grease gun', an air pressure tank, a high pressure tank and a hose-pump. With these equipments, rather high pressures can be applied. The pot-life of the mixture is an important parameter in the application by such equipments. Therefore, the length of the crack that can be injected in one go is subject to the volume of material mixed for use and its pot-life.

In the case of 'separate components' equipment, resin and hardener are separately transported to the 'mixing-head' by means of fully automatic dispensing equipment. Therefore, pot-life is only of secondary importance here. Errors in mixing two-component epoxies can have significant effect on the hardening of the resin. Therefore, the used of prepackaged batches (prepared by the manufacturer) is recommended. Generally in the case of the 'separate components' automatic-dosing-devices, errors will not be easily discovered in time to apply any corrective measures!

(c) Injection Process (Figs. 6.9 and 6.10)
A distinction must be made between low-pressure injection (up to approximately 2 MPa) and high pressure injection (up to 30 MPa). The penetration speed of the injection-resin does not increase proportionately with increasing pressure. The viscosity of the resin strongly influences the rate of injection, especially for small crack-widths, and in reaching the crack root.

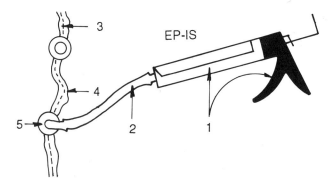

Fig. 6.9 Crack injection

1. Injection gun. 2. Plastic tube. 3. Crack. 4. Thixotropic compound.
5. Flanged injection nipple.

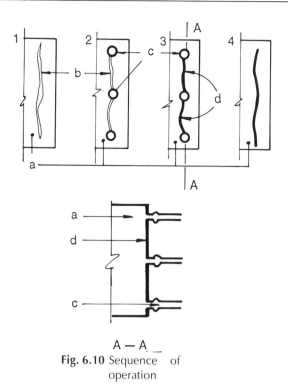

Fig. 6.10 Sequence of operation

1. Untreated crack in face of wall. 2. Crack cleaned and injection nipples bonded on. 3. Crack sealed with thixotropic compound and ready for injection. 4. EP-IS system injected and nipples removed.
(a) Concrete wall. (b) Crack. (c) Injection nipples. (d) Adhesive securing nipples and sealing crack.

(d) Testing

The conventional testing methods are simple drillcore removal and ultrasonic testing.

Coring Technique: The success of an injection operation can be determined by relatively simple procedure of removing cores taken through the crack, and checking the penetration.

Ultrasonic Technique: With ultrasonic pulse velocity measurements, the efficiency of the grouting operation can be evaluated when the propagation of pulse is oriented approximately normal to the crack. It is recommended that during ultrasonic testing, data be gathered not only with regard to the elapsed time for the sound to pass through the member, but also to variations in sound intensity.

The measurement with the existing equipments and methods is, however, not easy to be carried out and the results cannot yet be interpreted reliably. (For details, see Chapter 15.)

(v) Recommendations for Practical Implementation of Crack-Repair

(a) Adequate materials, equipment and experience of the operating personnel, are an essential requirement for the successful injection of cracks. Appropriate test-

ing or certification is required to determine the qualifications of operating personnel.

(b) A system of 'quality control' of the epoxy should be implemented to guarantee a consistent quality for each new application. These are : the determination of infrared composition (IR-Spectrum), the pot-life, the viscosity, the density, and the development of the tensile strength during hardening and of the hardened material. To avoid such expensive routine controls, some resin manufacturers have contracted with independent institutions to provide testing on a statistical sampling basis as a monitoring control. After successful testing, the epoxy batches are provided with a stamp of the testing institute as well as with information regarding durability. Stringent regulations for the use of epoxies in crack-repair, especially where they must resist tensile stresses, is required to ensure the behaviour of structural elements.

(c) No reliable data or appropriate evaluations are currently available for polyurethane and acrylic resins, that might be used as a crack-repair material.

(d) A reduction of adhesive strength of most epoxies will occur when the concrete surface of a crack is wet. There is also the risk of reduction in the quality of epoxy resins when they are used for the repair of structural elements at extreme temperatures. Current experience indicates that epoxy resins can be successfully utilized when the temperature of the structural element is not less than 8° C.

(e) Under relatively hot ambient temperatures, compared to normal temperatures, a considerable reduction in the workability time of the epoxy may result. In such a case, the temperature of the structural element in relation to its influence on the pot-life should be considered and prior testing may be appropriate to establish the timings.

(f) An effective epoxy resin injection can be accomplished even when there is a cyclic crack width variation, as a result of traffic loading, during injection and hardening, provided this variation does not exceed about 0.1 mm. Appropriate traffic limitations, up to maximum of the first three days, depending on temperature of hardening, should be implemented if larger cyclic crack-width variations are anticipated. In case of large crack-width variations resulting from temperature, injection should be coordinated such that the hardening commences at 'maximum crack-width opening' so that the filled crack will be subjected to a compressive stress, at least during temperature variations. Experience indicates that there is no difference in behaviour between an alkaline or carbonized concrete in this context.

(g) The 'deformability' of the usual epoxies, as a rule, is not sufficient to close active-cracks tightly and durably in case these movements cannot be stopped. Under these circumstances the feasibility of either using epoxies that are flexible even after hardening or temporarily filling with mastic or thermoplastic materials should be considered.

(vi) Some Other Methods of Repairing Cracks

(a) *Stitching*: Stitching 'across' the cracks in reinforced concrete members is done either along the cracks or as a series of bands around the member. Reinforcement is placed across the cracks in suitable grooves which are suitably gunited or shotcreted. Steel pins (dogs) are also used to stitch across the cracks.

(b) *Jacketing*: This involves fastening of external material over the concrete members to provide the required performance characteristics and restoring the structural value. The jacketing materials are secured to concrete by means of bolts and adhesives or by bond with existing concrete. Fibre-glass reinforced plastics, ferrocement, and polypropylene can also be used for jacketing.

(c) *For prestressed concrete members*: For PSC members, the simple methods adopted are — sealing and coating to fill out the cracks, grouting of cracks, repairing corrosion locations, and 'vacuum-grouting' using specially formulated epoxies to fill voids. Some of the latest techniques include the use of special material formulations to satisfy the requirements like high-tensile strength, special thermal properties, etc.

Figure 6.11 shows some of the Repair Techniques diagrammatically.

6.4 CONCRETE SURFACE PROTECTION USING OVERLAYS OR SEALING COATS

The protective systems should be effective, durable and reasonably economical. If the system is waterproof and keeps chloride out of the deck or prevents corrosion of reinforcement, it is considered to be effective. If the system provides effective protection for 5 to 10 years under moderate service conditions, it may be considered durable.

The following are the currently recommended alternate overlay protective systems: (i) Low-slump highly-dense concrete, (ii) Latex-modified concrete, (iii) Waterproofing membrane system, (iv) Polymer-Impregnation of Concrete and Polymer Concrete Overlay, and, (v) Protective Sealer. A concrete overlay system requires a minimum of 6 mm scarification over the entire deck slab, and each system requires removal of all delaminated and deteriorated concrete to the scarifying level.

Membranes and other overlay systems, installed on the concrete deck slabs subject to de-icing salts, have shown good performance of retarding active corrosion and sealing off corrosion-attributing moistures and oxygen. However, removal of contaminated concrete will give better results.

(i) Low-Slump Highly-Dense Concrete Overlay

Minimum overlay thickness of low slump concrete should be 45 mm. But, when additional thickness can be achieved without reducing load carrying capacity or adversely affecting the roadway profile, 55 mm thickness is desirable and more cost-effective finally.

The following steps are recommended for the overlay in low slump highly dense concrete:

1. The entire existing concrete deck surface shall be uniformly scarified and prepared to a depth of 6 mm after all the deteriorated concrete is removed to sound concrete level.
2. All exposed surfaces and reinforcing bars should be sandblasted and cleaned before applying grout for bonding new concrete to old concrete. The grout may consist of equal parts by weight of portland cement and sand mixed with sufficient water to form stiff slurry.
3. Concrete shall be placed continuously throughout the deck and patch repair shall be made monolithically with overlay course except for the large full depth repairs.
4. Curing of resurfaced area is very important. Immediately after finishing, the surface shall be covered with a single layer of clean wet burlap and cured for at least 72 hours with temperature above 55 °F.

Either a plasticizer should be added to concrete or it should be air entrained (6 per cent) (Slump: 20 mm with tolerance of plus or minus 6 mm). It is recommended that high early strength portland cement (Type III) should not be used.

Additional care should be taken regarding the following:

(a) All delaminated concrete in the deck should be identified and be removed prior to placing an overlay.
(b) If the half-cell reading is –0.35 volt or more, the concrete should be removed to the top layer of reinforcement steel bars or up to where chloride-carbonation has penetrated.
(c) The protection of concrete in the gutterline and kerb should be made with a high quality concrete sealer. It is preferable to use latex-modified concrete instead of a low slump highly dense concrete because

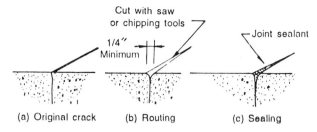

(a) Original crack   (b) Routing   (c) Sealing

Fig. (a) Repair of crack by routing and sealing is a method suitable for cracks that are dormant and not structurally significant. Routing and cleaning before installing the sealant add significantly to life of the repair

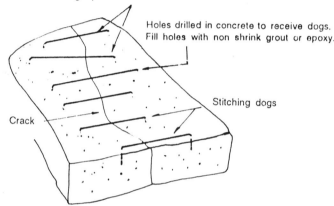

Note variable length, location and orientation of dogs so that tension across crack is distributed in the concrete rather than concentrated on a single plane

Holes drilled in concrete to receive dogs. Fill holes with non shrink grout or epoxy.

Stitching dogs

Crack

Fig. (b) Crack repair by stitching restores tensile strength across major cracks. Where there is a water problem, the crack should be made watertight first to protect the stitching dogs from corrosion

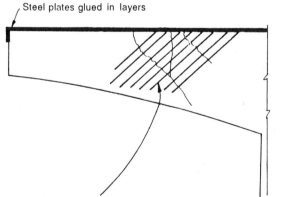

Steel plates glued in layers

Fig. (c) Added reinforcement installed to strengthen repair. Holes are drilled at right angles to the crack, then filled with epoxy and bars are inserted.

Fig. 6.11 Crack repair illustrations

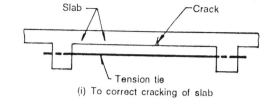

Slab   Crack

Tension tie

(i) To correct cracking of slab

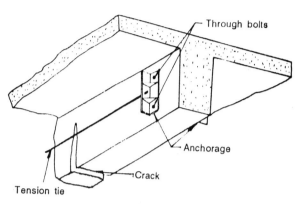

Through bolts

Anchorage

Crack

Tension tie

(ii) To correct cracking of beam

Fig. (d) External prestressing can close cracks and restore structural strength. Careful analysis of the effects of the tensioning force must be made or the crack may migrate to another position.

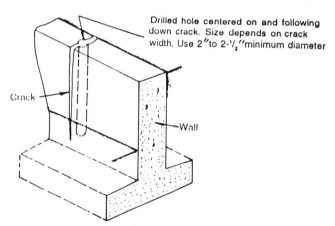

Drilled hole centered on and following down crack. Size depends on crack width. Use 2″ to 2-1/2″ minimum diameter

Crack

Wall

Fig. (e) Drilling and plugging is a repair method well suited to vertical cracks in retaining walls. The repair material becomes a structural key to resist loads and prevents leakage through the crack.

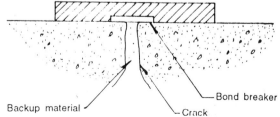

Backup material   Bond breaker

Crack

Fig. (f) Flexible surface sealant can be used over narrow cracks subject to movement, if appearance is not a consideration. Note bond breaker over the crack itself

of the relatively poor skid resistance and thicker depth of overlay of a low slump highly dense concrete.

(ii) Latex Modified Concrete (LMC) Overlay

Latex modified concrete is a mixture of concrete with latex-emulsion-admixture (3.5 gallons per bag of cement or about 15 per cent latex solid by weight of cement). Latex can be used in water and is compatible with portland cement. The particles which are dispersed in water are used in the concrete or mortar mix to improve the physical properties of the concrete, such as bonding, and tensile, compressive and flexural strengths.

An approved and widely used latex-modifier is a styrene butadiene type, developed by the Dow Chemicals Co. in 1957. Recently, several other companies have manufactured styrene butadiene latex modifier.

The preparation of overlay is similar to that of low slump highly dense concrete but with following differences:
(a) High slump (130 mm versus 20 mm).
(b) Before placing the modified concrete, the cleaned surface shall be wetted for at least one hour.
(c) The final surface shall be cured by wet burlap for first 24 hours following by additional minimum 72 hours (prefer 5 days) of air curing. Polyethylene films should be applied over burlap within one hour after placing burlap.
(d) A 35 mm thick latex modified concrete layer will give equivalent chloride resistance of 45 mm thick low slump highly dense concrete. The advantage of a thin overlay is the reduction of dead loads on the bridge.

In preparing LMC overlay plans for existing bridges, the following should be considered:
1. Provision for transition at deck-joints.
2. Necessity to adjust approach slab by LMC overlay or asphalt overlay.
3. Problems of raising existing joint sealers, armoured steel deck joints, and scuppers.
4. Difficulties of providing transitions in intersecting traffic areas.
5. Necessity of additional scarification beyond the 6 mm stated earlier.
6. Repairs of deteriorated concrete deck slabs.

The latex modified concrete or mortar mix is placed over the wet surface and screeded to the proper evaluation. In temperatures above 85° F, placement of concrete should be made in the early morning or at night. In cold weather, the air temperature should be above 45° F for minimum of 8 hours after placement. During the curing period, if temperatures fall below 35° F, the work may be considered unsatisfactory. It is recommended not to place concrete in cold weather unless heating provisions

are provided. Consolidation can be accomplished with a vibrating screed. Spud-vibration should be used at the edges, gutter areas adjacent to joints and bulkheads, and at all depressions. Finishing is then accomplished in the normal manner. Over-finishing of the latex mortar or latex concrete should be avoided.

In no case should existing cracks be simply covered with latex concrete. Such cracks should be cut-out to a minimum depth of 50 mm (12 mm may be used sometimes) and a minimum width of 25 mm, and filled with latex concrete prior to the resurfacing operation. All joints in the surface should be maintained during the placement of latex modified concrete.

Latex modified concrete offers good bond strength to sound substrate of deck, resistance to moisture and salt penetration, and wearing resistance. However, the most common problem is the development of moderate mapcracks (after placement) and bonding failure between latex modified concrete and the existing concrete near the joints. This may be due to improper application of curing method, high temperature, or shrinkage due to high slump.

Defects Possible in Latex Modified Concrete Overlays

(a) *Random cracks*: Most of these cracks appear (shortly after placing) as surface-cracks, and may extend to the full depth after several winters. They may be caused by high air temperatures, high slump and high winds during placement, resulting in differential shrinkage between the overlay and the concrete deck slab.

(b) *Transverse cracks*: These cracks are usually full-depth cracks and occur over the piers in continuous deck slab. They are reflective of the cracks in the deck slab and can cause more damage than random cracks because the moisture could penetrate to the concrete deck slab and lead to spalls, stripping, corrosion, etc.

(c) *Debonding*: Debonding is a hidden separation of overlay from the concrete deck slab below it.

(d) *Stripping*: Stripping is an actual separation and failure between the overlay and the concrete deck slab.

(e) *Spalls*: Failure of overlay and concrete slab, exposing reinforcing steel.

The following are some of the findings from the 1979 Ohio Department of Transportation and FHWA Field Performance Analysis Report:
1. The increase in the number of the random or transverse cracks is rather slow in the first three years but becomes much higher in later years. The percentage of increase may become 100 per cent after 7 years.
2. A continuous bridge deck tends to exhibit a higher degree of surface distress than a simply supported deck. In a continuous deck, the existing tension transverse cracks in the deck over the pier are likely

to be intermittently widened by traffic induced vibration. These lead to more tensile strains in the overlay and form more transverse cracks.

3. Bridges that are open to traffic during latex-overlay-placements generally exhibit a higher degree of surface distress than those which are closed to traffic. Statistically, this increase is insignificant. Therefore it does not warrant closing the deck to traffic and an additional expense in detouring traffic. However, it is beneficial to restrict speed or place the overlay at night or at other times when traffic volume is low, particularly when the original surface conditions require deep scarifying of deck thus reducing its flexural rigidity.

4. A thicker overlay would result in lower chloride content at bond surface.

5. It is estimated that the useful life of a rehabilitated deck with an overlay should be over ten years and of a new construction should be in excess of 15 years.

(iii) Water-Proofing 'Membrane' System (for preventing moisture ingress)

The bituminous wearing surface with 'membrane' has been used widely in Europe and the north eastern United States, and its performance has been acceptable. The advantages of a membrane are its easy installation and its relatively low cost. The disadvantages are as follows:

1. Premature deterioration of bituminous overlay in areas of high volume traffic and inadequate drainage.

2. Blistering, caused by expansion of trapped air and water vapour after placement.

3. Poor bonding at the protection layers in the area near the expansion joint.

4. Local instabilities due to bleeding and bubbling.

5. Replacement of the membrane whenever the surface is removed.

Most of the problems could be corrected by following proper construction procedures. This system could be a cost-effective method for short term performance.

It is suggested that the minimum thickness of asphalt wearing surface over the membrane be a minimum of 65 to 80 mm (Fig. 6.12) and, if placed in two coarses, the lower coarse should be denser or more impermeable than the upper coarse in order to prevent trapping of water in the lower coarse.

Also, placement of a membrane should be made under warm, sunny and moderate wind conditions to provide a drier deck surface which promotes better primer penetration and adhesive bonding, easier handling of membrane and application, and more rapid curing of primers and adhesive, thus preventing blistering.

Most of the 'membranes' are sheets of roof-felt and asphalt-impregnated protection-board. Water proofing membrane retards corrosion activity in the reinforcing steel and significantly extends the service-life of the deck. However, the system requires a total-overlay-replacement in less than 15 years, otherwise the cracks, developed in the overlay, penetrate through the membrane and permit local corrosion activities in the concrete deck slab.

In place of membrane, a dense epoxy-asphalt coarse or non-permeable modified-asphalt-mix (minimum 12 mm) can be used where night-time work is required to minimize traffic interruptions. It replaces the water proofing membrane system. It is easy to install, and capable of supporting traffic shortly after placement. Asphalt-rubber mixtures are used on asphalt pavements by applying a thin layer of hot mixture which contains the rubber made up of 1/4 of the mixture. This method is effective when used to level irregular surfaces. It can be installed at night time without extensive surface preparation, and it permits speedy installation without troublesome blistering or bubbling. It also provides a good base for adhesion of the asphalt wearing surface.

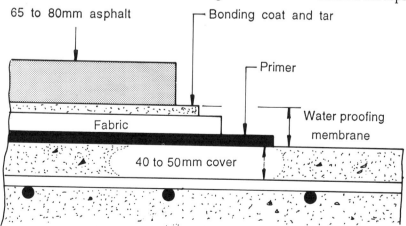

Fig. 6.12 Waterproofing membrane system

(iv) Polymer Impregnation of Concrete and Polymer Concrete Overlay

The impregnation i.e. filling of pores in concrete (of a concrete bridge deck surface) with methyl-methacrylate monomer system, and its in-situ polymerization, is a process to reduce 'chloride' and 'freeze-and-thaw' deterioration of this concrete. (Some details have been described earlier.)

The 'polymer impregnation' of the bridge deck concrete (which is subject to salt-action, but is structurally sound) has demonstrated to be an effective protective method to immobilize chloride-ions in the concrete and to retard or eliminate additional chloride penetration, thus lessening its corrosive action. However, the cost is high, and special safety precautions are required due to flammability and vapour pressure of the 'monomer-mixture'.

However, this concrete is not currently practical due to high initial capital investments and a lengthy process. Further research is required for development of cost-effective practical applications even though it has long service life and reduces maintenance costs.

The following four steps are suggested for the polymer surface- impregnation process:

(a) Prepare the concrete surface (remove contaminations, oil, asphalt, etc.)

(b) Dry the concrete to the depth which would permit the desired monomer penetration, which is normally the top reinforcement level. For this the concrete should be heated until the temperature at the specific depth is 230° F (by using a gas-fired infrared heater and torch) and cooled off to below 100° F before applying monomer.

(c) Impregnate the concrete with the liquid monomer to the desired depth by 'soaking'. Field impregnation is normally performed under greater than atmospheric pressure (by using a pressure chamber) but a heat-blanket-system can be used in place of other drying methods, utilizing enclosures.

(d) Polymerization of the monomer will occur in the concrete, forming a protective barrier, at 158 to 185° by pumping hot water through the pressure-impregnation-chamber or other heating systems.

Polymer Concrete Overlay

A polymer concrete overlay protective system, developed by Brookhaven National Laboratory for FHWA (FHWA-TS-78-225 Polymer Concrete Overlays Interim Users Manual Method B), has been used as an experimental project in several states in the U.S. The overlay consists of an application of monomer resin to the deck-surface, followed by an application of the fine aggregate. The process is repeated until four layers have been placed. The overlay is relatively impermeable and skid resistant. Generally, the resin is sprayed over the deck and fine aggregate is covered over the resin. After polymerization, the excessive aggregate is removed and the process is repeated for other layers. The four layers produce a thickness of about 12 to 15 mm.

The overlay system consists of the following steps for each layer:

• Surface preparation
• Mixing and application of monomer resin
• Fine aggregate application and compaction
• Polymerization of monomer and removal of excess aggregate

The main advantage of this is that it can be placed without the removal of concrete and with minimal interruption to traffic. The disadvantages are: handling problems, high cost, deterioration due to bond-failure between the concrete and overlay and thermally-induced cracks in the overlay.

As a research project, the Oklahoma Department of Transportation placed a 3/4 inch thick polymer concrete overlay on a bridge deck. The polymer concrete overlay was based on 'vinyl ester' which has performed as well as other base materials like polyester styrene, but is less sensitive to water. The promoter, dimethyl aniline was added to resin to improve the chemical reaction and to make polymerization occur at the ambient temperature. The report 'Polymer Concrete Overlay on SH51, Bridge Deck' concludes that the protection of a bridge deck with a polymer concrete overlay appears to be a rather complicated procedure. Besides constraints of weather, temperature, aggregate and resin availability, it is necessary to develop machinery which can accurately and efficiently mix and lay polymer concrete. It appears that polymer concrete *is good for deck repairs and concrete patches but not practical for deck overlays.*

(v) Protective Coating and Sealant

Protective coatings have been used to protect deck slab and top of the substructure and bearing seats. The most common coatings are silicon, polymer, epoxies, linseed oil, and latex modified concrete. After two or three years of exposure, most of the coatings show some distress and no significant evidence of protecting the concrete. Apparently, the protective coatings retard the intrusion of chlorides, moisture and oxygen, and minimize corrosion action for a few years. Linseed oil (or mixtures containing linseed oil) appears to be more effective than others for a deck against freeze-thaw induced scalings. However, it is not very effective against corrosion and water absorption.

For underside-deck protection, epoxy and linseed oil coatings are widely used with 1 to 5 years of expected life. Copolymer acrylic, mixed with a styrene, has also been successfully used with an expected life of 5 to 10 years. In the laboratory-test for the coating's performance on a smooth surface of white concrete, the coatings based on acrylic resin generally perform better than other types. However, there is a wide difference in performance, depending on the type of acrylic-resin used as the major component, and many of the specimens pitted or etches at the completion of the weathering test.

The following are some of the coatings rated as satisfactory:

(a) Methyl-methacrylate (MMA) and ethyl-acrylate copolymer
(b) Polyester-epoxy resin
(c) Styrene-acrylic copolymer
(d) Styrene-acrylic-silicone terpolymer
(e) Alkylalkoxy silane
(f) Urethane

Current research results indicate that epoxy, methyl-methacrylate, and silane show excellent performance in reducing chloride intrusion and water absorption and in providing a barrier to the salt water. The epoxy and methyl-methacrylate require two coats, whereas the silane requires only coat. Generally a five-days air drying period is required before applying the coat. For epoxy coating, the epoxy has low water vapour transmission characteristics, therefore, fresh concrete should be allowed a little longer period so that the internal water can be evaporated before coating. Once the concrete is coated, the epoxy will allow sufficient water vapour transmission to handle vaporization of subsequently absorbed water.

Coating over the linseed oil treated concrete surface appears to be effective except for silane coating. Therefore it is not recommended to use silane over concrete pretreated by linseed oil. Silane and methyl-methacrylate coatings provide added protection against corrosion for embedded reinforcing bars in concrete which has fine cracks.

Besides the use of penetrating sealant, the key considerations reducing chloride intrusion into concrete are lower water-cement ratio of concrete, sufficient cover for the reinforcing bars, lower chloride contents of aggregates and water in the concrete, increased cement-content, use of Blast Furnace slag cement, thorough Compaction, adequate curing and ambient temperature and wind control.

6.5   REPAIRING THE DAMAGE CAUSED BY CORROSION OF STEEL

6.5.1   Corrosion Process

(a) The 'embedment' of steel in concrete normally provides corrosion protection due to alkalinity of the concrete surrounding the steel. Because of its alkalinity, the concrete forms a passive-film on the surface of the embedded steel because of the presence of saturated lime solution in the cement-gel. Moist concrete typically has a pH value in excess of 11. This maintains the passivating film. This film is however de-passivated when the pH level is reduced below a value of approximately 9 (acidic moisture environment) or when a sufficiently high chloride concentration is present.

In case the alkaline passive-film is destroyed, or carbonization (which leads to reduction in pH and increase in acidity in the moisture) has reached the reinforcement and if moisture and oxygen are present, corrosion of the reinforcement will occur. In the absence of moisture (i.e. dry concrete) the corrosion process is inhibited, even if the concrete is carbonated.

(b) When reinforcement corrodes to a certain extent, the surrounding concrete cover tends to crack and split. The cracks are caused by internal bursting pressures developing in the concrete as a result of a net increase in volume by the formation of corrosion products. Spalling of the concrete-cover will then permit the entry of water and other corrosion accelerating agents and the rate of corrosion will accelerate.

NOTE: For more details *see* Chapters 4 and 15.

6.5.2   Protection of Reinforcing Steel

(a) Preparation Prior to Protection

The decision on the necessity to remove the chloride-contaminated concrete where corrosion process has started, will depend upon the amount of chloride-content, availability of moisture and degree of carbonation. This decision requires a case by case evaluation. If the corrosion-protection of the reinforcing steel requires removal of the affected concrete, the reinforcement in the effected zone will have to be exposed completely and fully!

The rusted exposed reinforcing steel is generally sand-blasted or wire brushed, depending on the seriousness. Removal of rust from the remote side of the bars is a difficult operation. A careful check and a repeated treatment of the individual bars is essential.

(b) Restoration of the 'Protection'

A corrosion protection should be applied to the cleaned reinforcement prior to restoration of the concrete cover. If possible, the reinforcing bar should be encapsulated in an alkaline coating. This can be achieved best by embedding the steel in rich and dense cement-mortar, cement-concrete, epoxy-mortar or epoxy-concrete.

The choice of the system depends on the thickness of concrete involved. Concrete or cement-mortar applied pneumatically (Shotcrete or Gunite) is preferable to ordinary concrete or mortar.

Repair with epoxy-mortar (or epoxy-concrete) is advantageous when there exist deteriorating mechanical and/or chemical influences.

(c) Preventive Corrosion Protection

In the case where a thin concrete cover is applied, it may be desirable to seal the surface with an epoxy-resin and solvent containing acrylic resins to prevent ingress of carbonation or corrosive agents.

(d) Cathodic Protection (for more details see Chapters 4 and 15)

The cathodic protection (CP) technique has been used in some countries for many years to product steel pipelines and tanks from corrosion, and has in recent years been applied also for the protection of reinforcing steel in concrete.

Corrosion of steel in concrete proceeds by the formation of an electro-chemical cell action! With the moist concrete acting as the electrolyte, milli-volt currents are set-up between points in steel that lie in stronger and weaker solutions of chloride, and cathodic reaction consumes the dissolved electrons and oxidizes the steel surface.

The presence of chloride-ions produces a local depassivation. By means of an externally applied small direct current (DC), the electrical potential between the steel and concrete is shifted to a non-critical level. Thus the electrons 'impressed' in the electrochemical cell. The potential shift produced by the DC is critical to the cathodic protection. Because of the high resistivity of the electrolyte-concrete, a uniform distribution of the 'protection current' throughout the structure is necessary.

A number of different CP systems have been developed. Two basic categories can be distinguished as follows:
— surface mounted anodes
— embeded anodes

Anode material (metals covered with conductive polymers or activated metals) are used in the form of meshes which are either woven on site to the configuration most suited to the specific requirements of the structure to receive protection or supplied as factory-woven panels e.g. in the form of an expanded-metal band.

The installation procedure of a mounted anode system includes
— assessment of the locations of corrosion
— preparation of the concrete surface at delaminated areas
— Placement of the anode-mesh on the concrete surface to be protected
— establishing an electric connection between reinforcement and DC power source and between anode-mesh and DC power source

— Concrete over-lay is applied using normal pouring, pumping or shotcreting procedures
— the DC power source is activated
— verification tests are conducted to ensure proper operation of the CP system

The primary voltage of the DC power source is dictated by the resistivities in each particular situation. However, in general, it is less than 10 V. The maximum current-density is approximately 20 MA/SqM of the surface of the steel reinforcement.

Prestressing steel, housed in a metal- or plastic-duct, cannot be protected with CP. When applying CP in a prestressed structure, excessive low-prestressing steel potentials against the concrete develop on the steel surface polarized as cathode, causing hydrogen-embrittlement on the high strength steel. *It is accepted that there is still research required to be done before cathodic protection can be safely applied to prestressing steel.*

6.5.3 Protection of Prestressing Steel

In prestressed concrete the repair of the concrete and of the untensioned reinforcement will also require attention. The prestressing force is still active and the stresses transferred to the concrete must be carefully considered, specially when repairing the concrete in the anchorage zones.

Restoring the Corrosion-Protection-System of the 'Tendons' (i.e. restoring the grout)

(a) In the Case of 'Bonded' Tendons: The prestressing steel is protected by the concrete cover and the cement grout in the ducts. Two procedures are possible.

• *Vacuum-procedure*: Where the ducts are not completely filled with cement grout, subsequent grouting is necessary. This can be accomplished by special grouting techniques. The regrouting of a duct requires only one drilled hole for each void. Such holes may already be existing in the form of the drilled-holes used for tendon-inspection or for obtaining samples for chloride content evaluation. Only a diameter adjustment may be required. A comparison between the measured volume of the void and the amount of grout consumed will provide a control measure as to the success of the operation. Where discrepancies occur, further borings will be required. A careful drilling procedure is required to avoid damaging the prestressing steel. Special means have been developed for this purpose, such as, slow drilling speed, special drill head, small impact force, drilling without flushing, sucking away of drilling dust and automatic switch off when the drill-bit reaches the duct. The repair must be accomplished as quickly as possible after opening of the duct to avoid corrosion.

After grouting, a pressure has to be applied to expel the residual air from the voids. There is a risk that for large air cushions, the setting water will be displaced towards the defects and produce paths which will impair the corrosion protection. Therefore, mortar with low setting-characteristics should be used. Special cements are available for this purpose.

In special cases, surplus water in the duct can be evacuated. However, this purpose requires special equipment and knowledge.

• *Grouting of the ducts with special resins:* Where ducts filled with water cannot be drained through drilling or the vacuum process, and drying is not possible, the water can be displaced by use of viscous epoxy resins with a long pot-life and high specific weight.

(b) In the Case of External (unbonded) Tendons: The prestressing steel of external tendons is protected by a tight envelope of plastic pipe or painted steel pipe, and the internal void of the pipe is filled with cement gout or suitable greases. If an inspection indicates deterioration of this protective system, measures must be taken for its restoration. Such measures may be re-painting of steel duct and of protective caps over the anchorages, replacing of plastic pipes, taping over the local pipe damages, filling of voids inside the pipe, etc.

Any material used in the repair procedure must be compatible with existing protection-material and the prestressing steel. Some paints, coating materials and special grouting mortars might contain substances that can produce 'stress-corrosion' and, therefore should not be used.

6.6 REPAIR MATERIALS

1. Portland Cement Concrete or Mortar are most widely used for repair. Additives such as an accelerator, a retarder, an air-entrainer, a plasticiser and a super plasticiser are used to improve setting-time and workability. Superplasticizers (high range water-reducer) are used for good flowability with low water-cement ratio. Type III (high early strength) cement is used where rapid strength gain is required. For pressure grouting, cement-pozzalan-slurry (mix 1:3 by volume) is recommended. Superplasticizer and non-shrink additives may be added to provide fluidity and to offset shrinkage. A whole range is available.

2. Epoxy Resin Mortar is popular and successful for small patch areas. This mortar with a curing agent has a wide range of set-time and strength, depending on manufactured formulations.

3. Polymer Concrete is a two component methyl-methacrylate 'monomer' based mortar. The monomer penetrates the concrete and strengthens the concrete after polymerization.

4. Latex Modified Concrete (LMC) is portland cement mixed with latex emulsion. This is widely used for deck protective systems.

5. Dry Pack refers to a cement mortar with no slump. The mortar mix has a very low water-cement ratio. It forms a ball when you squeeze a small quantity of material in your hand, but leaves very little water.

6. Fast-Setting Cements
(a) ASTM Type III (High early strength) Cement
This has been widely used but it has the disadvantages of high shrinkage and slow strength gain in cold weather.
(b) Chemical-Setting Cement
(i) *High-Alumina Cements (Mono-Calcium Alluminate)*: Widely used in Europe and are used in several states of U.S.A. Concrete made with this type of cement decreases in strength subsequently, and therefore it is not recommended for use for structural concrete unless the strength obtained at 24 hours is about 200% of design strength. It is sulphate resistant and could be considered for use if steel corrosion is a problem.
(ii) *Magnesia-Phosphate Cement*: It is supplied in two packages — dry magnesia and liquid phosphate. It produces a high strength, low permeability, patch repair material. It is suitable for small and large patches.
In general, this fast-setting material is not cost-effective for large repair. However, its use is justified to save curing time where early completion of repair is essential.

7. Bituminous Material
(a) *Bituminous Coatings*: Used for water proofing and protecting concrete for weathering. Coal tars have been used extensively to retard water penetration. The disadvantage of coating is the need for periodic recoating and replacement of the water proofing membrane, if damaged.
(b) *Hot Asphalt Sealer*: Used to seal the cracks in an inexpensive way.
(c) *Guss Asphalt*: This highly impervious asphalt is used for restoring surfaces in thin applications of 1/2" to 3/4".

6.7 THE CATHITE (PHOSPHATE-SOLUTION) METHOD OF REPAIRING CORROSION DAMAGED CONCRETE

The principle on which the Cathite Method of repairing reinforced concrete is based is that of halting the electrolyte-action which is producing corrosion sites at the anode areas.

Normally the reinforcing steel is protected by a stable (passive) oxide-film. The concrete cover-zone acts as a barrier to the penetration of chemicals detrimental to the steel's passive oxide-film. The two main causes of loss of protection are carbonation of the concrete cover to the reinforcement and the concentration of chloride ions at the reinforcement.

The Cathite method of repair involves the removal of the defective concrete, the preparation and treatment of the reinforcement with 'phosphate-solution', followed by a sealer-coat and the reinstatement of the concrete to its original profile using cementitious waterproof mortars and concretes. The mortar and concrete repair-materials are individually designed to match the characteristics of the original concrete.

On completion of the repairs, it is recommended to apply an overall protective coating to suit the environment conditions applicable to the structure.

Thus Cathite Method treats the corrosion sites and halts the corrosive electrolytic action. The coating system minimizes the future deterioration of adjacent areas of concrete.

- *Testing*

The contractor shall thoroughly hammer-test all surfaces of reinforced concrete to locate the areas of delaminated concrete. Additional testing may be required to determine the depth of cover to reinforcement, depth of carbonation and chloride profiles.

- *Cutting*

All defective concrete shall be cut-out and the exposed surface trimmed and cut-back to sound concrete to form a key for the new material. Feather edges shall be eliminated by cutting at right angles to the face of the member by means of sawing or percussive tools. Saw cuts shall be roughened to provide a key. The exposed surfaces shall be closely inspected for incipient cracks or shakes which may indicate loose or shattered materials, which shall be removed. The extent of the cutting is to be agreed with the engineer-in-charge. Pneumatic or electric tools may be used but with care so as not to damage the concrete more than minimum necessary. In the affected areas, concrete shall be removed from behind bars, cutting away sound concrete as necessary so that corroded bars can be properly cleaned, treated or replaced, as required.

- *Anti-Corrosion Treatment*

In the course of cutting of the concrete, the reinforcement shall have seen completely exposed by also cutting behind the bars. They shall be cleaned by grit blasting, wire brushing or other mechanical abrasion after fixing any new reinforcement required. After cleaning, a 'phosphate-solution' shall be applied in sufficient quantity to completely passivate any residual traces of rust on the steel. This shall be followed by a chemical sealer. If, in the application of these chemicals the exposed concrete surface becomes contaminated, shall be cut away.

- *Placing Repair Material*

General: After the cutting has been completed and before the new concrete or mortar is placed, the exposed surfaces shall be washed down with clean water. A bonding coat shall be applied if necessary. All repair materials shall be mixed in accordance with manufacturer's instructions and to the satisfaction of the Engineer-in-Charge. (The surfaces shall be dry in case water-sensitive bonding coat is used.)

Small Repairs: Small repairs are those which are less than about 40 mm in thickness. The defective areas shall be filled with sand and cement mortar in successive coats. The area to be repaired shall be thoroughly wetted. A bonding coat shall be applied consisting of 1 part cement to 1 part sand with styrene butadiene polymer or acrylic polymer included if necessary. While this bonding coat is still damp, a mortar composed of 1 part cement to 1 1/2 parts sand, gauged with water containing 1 part Sika No. 1 to 10 parts water, shall be laid on. A keying coat, consisting of 2 1/2 parts sand : 1 part cement shall be used between coats. The final coat shall consist of 1 part cement : 2 1/2 parts sand, finished with a wood float to minimize crazing and provide a suitable surface to take the surface treatment. The repair mortar shall have minimum compressive strength of * N/mm^2 at 28 days.

Large Repairs: Larger areas shall be reconstructed with concrete of characteristic strength of * N/mm^2. Shuttering shall be constructed and securely fixed to the profiles required (designed to facilitate the placing of the concrete in the particular location). Concrete of correct consistency shall be carefully placed and rammed and vibrated into position using poker vibrators or external vibration, as appropriate, in order to achieve intimate adhesion with the parent-concrete. The concrete mix design shall be based on a water/cement ratio of * with a slump of * mm, and cement content of * kg/m^3. A * mm down aggregate shall be used. Additives of Sika No. 1 (for waterproofing) and a plasticizer or superplasticizer (to improve workability) may be used.

- *Renewing Reinforcement*

Reinforcement will be renewed in places where the original steel bar has corroded to such an extent that its diameter is reduced beyond that acceptable to the Engineer; new steel shall be added as instructed by the Engineer, and adequately anchored/overlapped.

---

* This value to be as required for the mortar / concrete mix strength needed for a particular repair.

- *Curing*

Shuttering shall, if necessary, be sealed to assist in curing and shall remain in position for a minimum of * days. As soon as the shutter is removed, defects shall be hacked back and made good with 1:1 1/2 mortar. A curing agent may be applied or alternatively the surface shall be wet-cured in accordance with standard practice.

- *Surface Treatment*

When all the repairs to a section have been completed, the concrete surface shall be treated with a final treatment coat.

- *Formwork*

Members shall be shuttered to existing concrete profiles. Formwork shall be constructed of plywood and timber or steel, firmly fixed to the structure to avoid distorting through placing and vibrating of concrete.

- *Materials*

(a) Portland Cement shall comply in all respects with BS.12 or equivalent. It will be delivered in original sealed bags and be stored in a proper manner to avoid deterioration.

(b) Fine Aggregate shall comply in all respects with BS.882 or equivalent. The grading shall lie within the medium zone (preferably).

(c) Coarse Aggregate shall comply with the continuous grading limits of BS.882 or equivalent. The aggregate shall be a nominal 10 mm size generally.

(d) Mortar shall be gauged in mixes from 1:1 to 1:3, containing waterproofing Sika No. 1 admixture (used in accordance with the manufacturer's recommendations). It would be anticipated that mortar compressive strengths would probably lie between 20 and 30 N/mm^2.

(e) Concrete shall be volume or weigh batched. The mix shall be designed to give strengths comparable with that of the original concrete and suitable workability to achieve placeability in each location.

6.8 OUTLINES FOR 'REPAIRS' AND 'STRENGTHENING' OF STEEL STRUCTURES (also see Chapters 7 and 15 ahead)

6.8.1 Deck Replacement of Older Steel Bridges

Many of the old truss and arch bridges have either steel plates with a bituminous surfacing or a concrete deck. Due to insufficient water proofing, the steel plates are often corroded.

Bridge decks can be replaced by new concrete decks or by new orthotropic steel decks. Usually, when a reduction in dead load or additional widening (adding a cycle or a pedestrian lane) are necessary, replacement by an orthotropic steel deck is proposed. Bolting is the preferred method in connecting the new deck system to the existing structural members.

Depending on the type of bridge and the load carrying capacity of each structural component, the new concrete deck is placed as a non-composite element, or as a particularly composite element (e.g. in composite action with a stringer and/or cross beams) or as a totally composite element (i.e. in composite action with all main load carrying elements).

The use of light weight concrete is often preferred in such cases where reduction in dead load is an important factor. The use of aluminium decks, however, proved to be an unsuccessful method in Belgium and in the Netherlands. Sometimes to save weight, a type of steel-grid decking is used where the grid can either be left open or filled with concrete.

6.8.2 Strengthening of Structural Members

Strengthening usually involves more conventional techniques, such as installing new diaphragms in existing double compression members (increasing buckling strength), strengthening or replacement of diagonals. Plate girders may be strengthened by external prestressing cables, anchored and fixed on the web in the required profile acting in a similar way as in prestressed concrete.

Strengthening is sometimes concerned with compression failure and has involved the addition of stiffeners to flanges, webs and diaphragms.

6.8.3 Repair of Cracks

Cracks can be due to any one or a combination of the following reasons:
— poor detailing, so that high stress-concentrations are present
— increased traffic loading beyond what was anticipated by the designer
— an unexpected secondary structural action
— inadequate analysis of complex stresses
— a large undetected fabrication flaw

Crack repair methods depend on the root cause of the crack initiation. The structure, and specially those components which influence the overall safety of the structure, should be analyzed. (Also *see* Chapters 4 and 15.)

---

* This value to be as required for the mortar / concrete mix strength needed for a particular repair.

6.8.4 Action to be Taken When a Crack is Detected or Suspected in Welded Steel

(a) Location should be marked distinctly with pointer. Ends of cracks should also be accurately marked to monitor crack propagation.

(b) Length and orientation of cracks should be recorded. Sketch should be prepared indicating locations and details of cracks. If necessary, photographs may be taken for record.

(c) If necessary, the cracks should be examined in detail using non-destructive inspection methods like dye-penetrant ultrasonic techniques.

(d) If a crack is suspected in any location, paint film should be removed and detailed examination carried out using magnifying glass, dye-penetrant inspection or ultrasonic inspection, as necessary.

(e) If more identical details exist in the girder, they should also be inspected in detail.

(f) Cracks should be fully documented on the bridge inspection register and action initiated for early repair.

(g) The cracks and the cracked structural elements should be kept under observation, depending on the severity of cracks, and frequency of inspection should be suitably increased. If the situation warrants, suitable speed restriction may be imposed.

(h) Significance and severity of cracks should be studied on the load carrying capacity of the girder.

(i) Repair plan through a suitable retrofit scheme should be prepared after fully investigating the cause of cracks and then implemented at the earliest.

Repairs can be made by techniques such as drilling holes at the crack-tip (this should only be done in less sensitive locations), cutting out the cracked material and bolting plates in place, cutting out the crack and rewelding with higher class weld (e.g. increasing the size and penetration of a fillet weld), strengthening the connection by introducing stiffening, and by changing the structural action so that loads are supported in a way that prevents high stress range from developing.

6.8.5 Underwater Welding

Arc welding has become an accepted procedure in underwater construction, salvage and repair operations. Underwater welds made on mild steel plates under test -conditions at several United States Navy facilities, have consistently developed over 80 per cent of the tensile strength and 50 per cent of the ductility of similar welds made in air. The reduction in ductility is caused by hardening due to drastic quenching action of the surrounding water. Structural-quality welds have been produced by means of special equipment and procedures that create small, dry atmosphere in which the welding is performed. However, this process is expensive.

Gas-welding under water is not considered to be a feasible procedure.

A word of warning appears appropriate: Although arc-welding and gas-cutting are now common underwater techniques, electric shock is an ever present hazard. This hazard can only be minimised through the careful application of established procedures.

6.8.6 Use of Steel-Arch Superposition Scheme

This can be used to strengthen old truss bridges. The strengthening scheme consists of superimposed arches, hangers and additional floor beams. The concept of combining a truss with an arch is by no means a new system. The idea is that a light arch can carry a significant load if properly supported laterally. In this case, the truss with its cross-beams provides the lateral support while the arch in combination with the hangers and additional floor beams provides the increased load carrying capacity. Additional floor beams and hangers are used for two reasons

— the more uniform the load distribution, the more efficient the arch will be in carrying the load.

— the floor systems of many old truss bridges get deteriorated and are sometimes found underdesigned and unreliable.

The principle of steel-arch superposition scheme is demonstrated in Fig. 6.13 (also *see* in Chapter 7). The thrust of the arch can be resisted by one of the following means:

— the abutments, provided they are adequate and in good condition, or they can readily be repaired or strengthened

— a reinforced lower chord

— superimposed cables or rods

— properly designed and detailed stringers or floor slab

The arch-superimposition scheme can be considered as an overall strengthening measure. The load carrying capacity of the entire structure is upgraded, thus allowing the live load to be increased. There is no need for temporary shoring or jacking for the installation of the superpositioned elements. The increase in dead load can be expected to be in the order of approximately 15 per cent to 20 per cent. The slender arch contributes only modest amounts of additional stiffness to the truss.

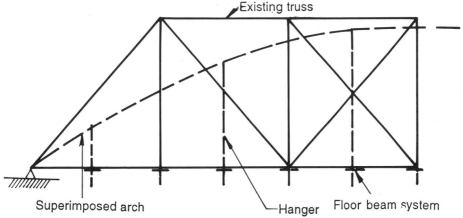

Fig. 6.13 Steel arch superimposition to strengthen old truss deck

6.8.7 Excessive Vibrations

These can be overcome by suitable structural alterna-tions, for which a specialist in dynamic behaviour of structures may have to be consulted.

6.9 OUTLINES FOR REPAIR OF COMPOSITE STRUCTURES

Comparatively very few defects have been reported with well designed and fabricated shear connectors. Problems with concrete decks in composite structures are essentially of the same kind and order of magnitude as those found in regular concrete decks. The same can be said for the main load carrying structural steel components.

Difficulties may be encountered with deck replace-ment or even deck rehabilitation and strengthening op-erations in those composite bridges in which residual relieving stresses have been introduced by sophisticated erection procedures combined with an elaborate casting sequence for the bridge deck. Such cases would be very few and would have to be considered on individual basis as such cases cannot be generalized other than in the commonest of details.

6.10 OUTLINES FOR REPAIR OF TIMBER STRUCTURES

Except for giving treatment to wood and local replace-ment or splicing, there are no special techniques for repairs of timber structures. The distressed members could be either replaced or strengthened with steel plates. (*See* Appendix 1 at the end.)

6.11 OUTLINES FOR 'REPAIR' AND 'STRENGTHENING' OF MASONRY STRUCTURES

Existing masonry bridges are sometimes considered as historical landmarks and need preservation. Strength-ening and widening will, therefore, require maintaining the same appearance. Widening is usually not possible but strengthening can often be done.

Strengthening of Masonry Bridges, ensuring pleas-ant appearance, is a delicate task and needs advice from experts in these fields. The following gives an idea of the general defects and remedial measures for bridges in stone or brick masonry.

(i) Loss of Bond for the 'Crown' Stone: Flat jacks have been successfully used for pushing the stone back to its original position. Generally, low pressure cement grout-ing is done to strengthen the old mortar. The mortar is sometimes replaced by epoxy-mortar also.

(ii) Longitudinal Cracks Along the Direction of Traffic: It is possible to rake mortar joints and refill with epoxy mortar. However, it must be mentioned that the depth of penetration is important as usually it is not possible to suspend traffic. If possible, portions of earth fill could be removed sequentially to ensure that penetration is lim-ited to masonry only. Epoxy injection can be adopted for remedial measures. Generally it is cheaper to grout the cracks with cement than with epoxy.

(iii) *Transverse Cracks*: Injection of epoxy/cement will provide a good bond between stones and brick masonry.

Arches

The 'arch-ring' can be strengthened in two ways — by adding material to the intrados or to the extrados. Add-ing to the intrados causes the least disturbance but is more difficult to complete successfully. Also, it results in a reduction in headroom or clearance, which is often restricted and will, in most cases, be the cause of new damage to the intrados as experienced on many bridges even where the headroom/clearance satisfies legal limits. Extra material may be placed by shuttering and pumping concrete (which is difficult to compact at the crown) or by fixing a mesh to the intrados and shotcreting or guniting. In both cases, any shrinkage of the new concrete will tend to separate the old from the new material, possibly radially. Also, these impervious rings prevent natural drainage between the stones or brickwork of the arch, so that special provisions must be made to deal with water and under severe climatic conditions (such as in mountainous regions) with ice. Sprayed-on concrete will in any case change the appearance of an arch constructed of stone, brick or combination of the two.

A more effective, but at times a more expensive, treatment is to remove the fill and cast the extra required thickness on the extrados of the arch. Usually, a full ring is cast but occasionally only the end quarters are strengthened to act as cantilevers and reduce the effective span of the arch. Usually normal concrete-placing-techniques can be satisfactory. Replacement-backfill may be in normal or lightweight concrete. The latter will reduce dead load on the foundations but may also reduce the factor of safety for stability of the substructure.

Another expedient which is satisfactory where the increase in load carrying capacity is relatively small, especially for small span bridges, is to spread the wheel loads by casting a relieving slab at road level to act as an auxiliary deck.

For cracks in arches, grouting with cement, at pressures of 4 to 6 kg/sq cm is sometime quite effective.

6.12 IMPROVING 'SKID-RESISTANCE' OF DECK SURFACE

A skid-resistant bridge-deck reduces wet-weather skidding accidents and saves lives and property damages. To provide adequate skid-resistance for high speed traffic, two requirements must be satisfied. First, the pavement surface must provide adequate adhesion between the tire and the pavement even in wet-weather conditions. Second, the pavement must have sufficiently rough surface-texture and drainage-potential to prevent a build up of water at the tyre-pavement interface. On a concrete riding surface, the drainage potential can be provided by texturing the concrete surface but the adhesion component depends on the wear and polish-resistance characteristics of the aggregates. A proper 'finishing' procedure for the concrete is required to provide an adequate and durable skid-resistant surface-texture on the bridge deck surface.

6.12.1 On Concrete Decks

The following are some guidelines for a skid resistant surface for a concrete bridge deck:

(a) A 'burlap-drag finish' should not be used as the sole means of providing a surface texture on highway with design speeds of 65 km per hour or greater.

(b) A 'broom-finish' or an 'artificial-grass-finish', while it often offers a good initial surface texture, may not be durable unless a deep texture is produced by heavy bristles and high pressures. When a broom finish or similar finish is specified for use on a project with a design speed of 65 km per hour or greater, the specification requires a deep and durable texture and quality control.

(c) Metal brush finish is recommended as being the most practical and dependable method of providing positive texture in concrete surfaces.

(d) The use of deep surface texture on a bridge deck may warrant increased concrete cover over the top reinforcement steel beyond that normally required for design and construction tolerance. The use of the heavy transverse texture may also increase the concentration of de-icing salts along the kerb line unless proper provisions are made to drain these areas. The last 12 inches of the deck next to the kerb should be left untextured to facilitate drainage.

(e) Special techniques may be required to produce a durable 'groove finish' on latex-modified and low-slump concrete-deck overlays. The use of flexible metal brush on such a dense surface frequently fails to produce a groove depth sufficient for reasonable durability.

(f) Regardless of the finishing, or the texture method used, adequate and durable skid-resistant characteristics cannot be attained unless the fine aggregate has the suitable wear and polish resistance. The crushed aggregates will normally provide a higher skid resistance than the uncrushed natural gravel. Research conducted by the Portland Cement Association indicated that the celestite particles content of fine aggregates should not be less than 25 per cent.

6.12.2 On Timber Decks

A timber-plank bridge deck gives poor resistance against skidding when it is wet or if the surface is bleeding the preservative. The creosote or the heavy oil borne surface-bleeding worsens the skid-resistance. To increase the skid-resistance and to provide a smooth riding surface, asphalt concrete overlays are used extensively on timber decks.

6.12.3 On Grid and Steel-Plate Decks

A bare steel deck or steel grid deck, particularly when wet, has poor skid resistance and exposes traffic to accident hazards. It is suggested that studs be welded on the steel plates or intersection of grids to minimize the hazards. (Also *see* 6.13 below.)

6.13 IMPROVING WEARING SURFACE ON STEEL DECK-SURFACE

It is important to protect deck-plates from corrosion and provide smooth but skid-resistant riding surface. The improvement can be effected by using either the 'bituminous-based thermoplastic' or 'epoxy-based thermosetting' materials. Thermosetting materials can be heated until they become liquid. They set permanently on continued heating and cannot be softened again by reheating. Thermoplastic ones set when they are cooled and can be softened by reheating.

6.13.1 Coal-Tar-Epoxy

Coal-tar-epoxy has been used since 1960 as a seal or bond coat on steel-decks along with inorganic zinc primer or redlead epoxy-primer. The final layer usually consists of a 2" asphalt wearing surface. Coal-tar-epoxy bond-coat appears to work satisfactorily on the Port Mann Bridge of Vancouver, B.C., since its installation in 1964.

6.13.2 Epoxy-Asphalt

Epoxy-asphalt is known as 'Concresive Asphalt' produced by Adhesive Engineering Company in the U.S. It is reported to be performing well on several bridges. It maintains good bond with steel surfaces, adds stiffeness to the steel deck plate resistance against fatigue cracks, has high 'plastic' characteristics, and high resistance to 'wear'. Epoxy-asphalt is a mixture of epoxy resin, asphalt filler and aggregates and the hardener.

The steel deck plate surface is sand blast cleaned to white metal and primed with inorganic zinc primer. After 'spraying' an epoxy tack-coat, a layer of the epoxy asphalt mixture (3/4" to 5/8" thick) will be compacted and 1 1/4" to 2" thick bituminous concrete wearing surface will be placed.

6.13.3 Polyurethane-Tile

This is a relatively new product, manufactured from vulcanized fiberglass-filled elastomeric compound using polyurethane.

6.13.4 Waterproofing-Membrane

This consists of placing bituminous-based fabric-waterproofing-membrane after sand-blasting and primer coating a 3/4" modified non-permeable asphalt layer and 1 1/4" to 2" asphalt wearing surface.

6.14 BEARINGS

Most bearings will not outlast the bridge. Hence, replacement of defective or damaged bearings shall be provided for. However, careful inspection and periodic maintenance of bearings can extend their service-life.

Defects in bearings may result from:
— defective manufacture, and/or use of defective materials
— inadequate design
— inadequate or improper installation (which is not uncommon)
— negligent maintenance

The possible types of defects in bearings have already been indicated in Chapter 5, to which reference should be made.

In principle these may be due to:
— corrosion
— defective seal in pot bearings, defective vulcanization in neoprene bearings
— broken guides
— cracked/broken rollers, plates, pins (lugs) and anchor bolts
— cracks, splits or tears in neoprene material
— accumulation of dirt/debris at the bearing locations
— failure of anchorage system
— movement or creep of parts out of place
— partial contact of bearing plates
— excessive tilt or even shift of its components, etc.

Appropriate corrective action has to be taken after detailed investigation of the defects. Repair or replacement of bearings requires traffic restriction or even temporary suspension of traffic.

Excessive tilts in bearings, like in segmental bearing rollers, should be corrected in time. This can be done by slight temporary lifting up of the structure, shifting the bottom or top plates and lowering back the structure. The cracked or excessively deformed elastomeric bearings also need to be replaced. This also requires lifting of the superstructure. Lifting is normally done with flat-jacks but sometimes cranes may have to be used. In all events of lifting, checking the design of superstructure for stresses induced due to lifting is obligatory in indeterminate structures. These are specialized activities and shall only be undertaken by specialist agencies. Apart from checking the stresses in superstructure (as caused temporarily by the amount of lifting at whichever support, unless the structure is statically determinate), it is equally important to check the stresses caused in the structures below and above the lifting points, due to the lifting-reactions. Many a time, convenient location of lifting-arrangement may not be possible, because the correspondingly induced stresses in the structural elements above and below the lifting points are unacceptable. It is also advisable to connect all the lifting jacks under a deck section, to a common hydraulic-manifold so that in the unlikely event of leakage-failure of any one jack, all others will simultaneously lower their pistons equally, thus preventing any deferential settlement effect.

6.15 EXPANSION JOINTS

Some details have already been described in Chapter 5 earlier, to which reference should be made. Here it is necessary to mention that the operational life of expansion joints, like that of some of the bearings, footpaths and railings, is usually significantly shorter than that of the bridge itself. Their strength and efficiency may be a limiting factor in certain situations. The expansion joints, bearings, railings, parapets, etc., need special attention for repair and replacment or renewal.

The expansion joints are not expected to last throughout the life of the bridge. It is therefore, recommended that joints be replaced on a regular basis without necessarily being left until they have failed. However, there are indications that the neoprene joints presently being installed perform better and give more satisfactory service than the previous-generation joints. The need for further improvement is generally acknowledged.

It is vital to keep the joints watertight in order to prevent the ill-effects of the contaminated drainage. Leaks must not be tolerated.

The joints should be watertight across the full width of the deck, kerbs, footpaths, central verge, etc.

Where watertight joints cannot be provided, or where frequent failure is likely, adequate means of draining the water passing through the joints shall be provided. As far as possible, the water should be kept out of contact with the concrete and steel. This is sometimes difficult to realize. If this measure fails, regular maintenance of the bearings and pedestals will at least prevent chemical damaged to the concrete and steel. Joints may be filled with a flexible sealant of filler. The filler material must be designed to ensure watertightness. Debris may prevent joint-movement if the filler fails and may damage the joint-sides or joint-material,

may spall the sides of joint elements or cause overstress in other bridge elements. Debris also tends to retain moisture and hence contributes due to the deterioration of adjacent bridge components.

Damage to finger-type-joints may manifest in the form of cracks or broken fingers as a result of traffic damage, poor alignment, loose anchorage, etc. The fingers may also cave-in or project up due to unacceptable deformation of the deck or differential settlement of foundations, etc. Cracking and spalling of the pavement or deck in the area adjacent to the joint may cause subsequent failure of the joint by loosening the joint-side-support-material.

Movement of the abutments and piers must also be considered when inspecting joints. Such movements may also either increase or decrease the joint-opening or may even close the joint-opening completely, preventing free heaving of the bridge deck.

All damaged joints should be replaced. The sealant filler shall be replaced periodically. Cracked concrete in the zone of anchoring the joints shall be replaced. Periodic cleaning and removal of debris is a must.

6.16 FOOTPATHS (WALKWAYS) AND CYCLE-TRACKS

In concrete these are constructed in-situ or by using precast elements.

The common distress noted in either case during the service life is in the form of cracks. Some of the precast components may also be found displaced or even missing, at times. (River-side washermen find them quite useful!)

The cracks shall be repaired by injection with either cement grout or epoxy, as appropriate, after detailed investigations described earlier. The broken/missing precast panels shall be replaced with new panels, possibly of better design and strength. It may also be appropriate to provide a mastic topping over the cracked precast panels.

Where major replacement of precast planks is envisaged, it may be desirable to revise the design and provide for in-situ footpaths.

6.17 OTHER MISCELLANEOUS ITEMS— RAILINGS, LIGHT POSTS, ETC.

The repair measures for these will be the same as mentioned earlier for identical items. However, the decision to repair or to replace such items needs to be taken keeping in view the economics and importance of aesthetics and appearance of the bridge as a whole.

6.18 'REPAIR-ADJUSTMENTS' THAT MAY BE REQUIRED FROM HYDRAULIC-CONSIDERATIONS

Bridge structures can get damaged or may show distress or defects due to causes other than those due to traffic loads, environmental factors, ageing, etc. These could be external causes like earthquakes, fire, floods, etc. The most important external cause could be the floods. There are cases where rehabilitation or modification of the bridges becomes necessary on account of changes in the assumed values of hydraulic parameters. The damages can be (i) Due to abnormal floods, (ii) Due to normal floods if the design of the bridge does not cater adequately to the normal design floods, and/or (iii) Due to man-made changes in catchment of watercourse (e.g. the flood-levels may exceed the originally assumed design levels due to back-water effect of a storage constructed on the downstream side, requiring raising of the bridge's superstructure).

The floods can damage both the bridge structure as well as the approaches and the protective-works. The bridge engineer is advised to refer, for example, to IRC 89-1985 'Guidelines for Design and Construction of River Training and Control Works for Road Bridges' (Indian Roads Congress Publication, New Delhi).

If, either due to inadequacy of the originally assumed hydraulic parameters or due to the requirements of traffic as in submersible bridge, the bridge level has to be raised, then this can be done by raising the superstructure by suitable jacking arrangement and building up the substructure in suitable stages by successively resting the superstructure on precast concrete blocks (i.e. pads) which can even then be abedded in the raised height of the piers and abutments. Where, however, the bridge-deck is not to be raised but the bridge has to be protected from floods higher than the design floods instead of designing the bridge as a 'submersible type', then strengthening the same may have to be explored on a case by case basis. At the same time suitable corrective measures may have to be adopted for the decking and approaches, like provision of air-vents between the girders, protection of embankment, strengthening of piers by jacketing, etc.

When the maximum flood velocity, and consequently the calculated scour in the stream, is expected to increase, and as per computations the substructure is found to be unsafe under such conditions, a solution of paving the bed with suitable aprons upstream and downstream can be considered to prevent the scour around the piers and abutments. The piers and abutments may also have to be strengthened by 'jacketing'. If partial submergence of the bridge is unavoidable and if the live load on the bridge substructure is found to develop excessive stresses, then it may be necessary to provide spilling section in the approaches so that the live load is automatically cut off when the water level exceeds a specific limit.

If the damages to the bridge and approaches are of frequent nature, then, after careful investigations, it may be necessary to extend to the length of the bridge to provide adequate waterway under it. If the floods attack mainly one side of the bridge, then additional spans could be provided on the affected side. Sometimes such situation can be handled by adequately designed spurs (or groynes). In some cases, the 'returns' beyond the abutments get damaged and may require replacement by returns with deeper foundations. When a pier gets damaged beyond repair, the span length could be changed either by locating a new pier in between, or if possible, by doubling the span and suitably strengthening the involved structure.

The bridge hydraulics is a highly specialized subject. Hence, the treatment of the damages must be designed and carried out after consulting a specialist. Use of hydraulic model studies for specific problems can be of considerable help in arriving at a proper solution in deserving cases.

ANNEXURE 6.1

"FOSROC"–PRODUCTS

Given here are:
I. Specifications for certain concrete repairs; and
II. Formats of B.O.Q. for various types of repairs; and
III. A partial list of some products for repairs,
as proposed by FOSROC (U.K.)* and FOSROC/FOSAM (Saudi Arabia).**

• The descriptions also give various other useful details, as can be seen. These details are reproduced with courtesy. It is hoped that the engineers dealing with concrete and its repairs will get a bird's eye-view from this list as to various product-formulations from FOSROC that can be relevant to their work—whether for producing good concrete or for repairing the distressed concrete. (Detailed enquiries, including the latest and up todate information regarding new formulations, should be directed to the manufacturers.)

I. FOSROC'S SPECIFICATIONS FOR CERTAIN CONCRETE-REPAIRS AND PROTECTIONS

Problem No. 1: Protection of New or Existing Reinforced Concrete Structures Against Carbonation or Chloride Induced Corrosion

SPECIFICATION

The use of a penetrating, reactive primer and top coat system to minimise ingress of acidic gases, chlorides and water.

1. **Surface Preparation**
    1.1 Prior to application all surfaces must be dry and free from oil, grease, loose particles, decayed matter, moss or algal growth and general curing compounds. All such contamination and laitence must be removed by the use of grit blasting, high pressure water jetting or equivalent mechnical means.

    1.2 Before proceeding to apply the protective coatings, all surfaces which are not to be coated but which may be affected by the application of the coating shall be fully masked and, in particular, flora/fauna shall be protected.

    1.3 Blow holes and areas of pitting shall be made good with RENDEROC FC, a one part modified cementitious material, and allowed to cure in accordance with the manufacturer's recommendations. The application shall be in accordance with the manufacturer's recommendations, particularly with respect to the maximum application thickness of 3 mm.

2. **Coating Materials**
    2.1 The materials are required to provide in-depth protection against carbonation and chloride penetration whilst permitting water vapour transmission from the concrete. The NITOCOTE DEKGUARD system is considered suitable.

    2.2 The Contractor is required to adhere strictly to the manufacturer's recommendations regarding the use, storage, application and safety rules in respect of the approved materials.

    2.3 The exposed concrete surfaces as defined in the documents or as agreed with the Supervising Officer shall be conditioned by the application of NITOCOTE DEKGUARD PRIMER, a penetrating hydrophobic treatment. The primer shall be allowed to dry in accordance with the manufacturer's requirements. The Contractor shall then apply two coats of NITOCOTE DEKGUARD PIGMENTED TOPCOAT in accordance with manufacturer's instructions. The finished coating shall be pinhole free and have a total minimum dry film thickness of 150 microns. The colour and finish is to be as agreed with the Supervising Officer.

    2.4 The coating shall be applied by spray, roller or brush to achieve a finish acceptable to the Supervising Officer. In all operations of storage, mixing and application, the Contractor is to comply with the Health and Safety recommendations of the manufacturer and governing authorities.

    2.5 Where required by the Supervising Officer, trial areas not exposed in the finished work shall first be treated using the selected materials. These trial areas shall be noted on the drawings and shall be carried out using the type of materials, mixing procedures and applications that will be used on the contract and shall be approved by the Supervising Officer before the Contractor commences with the general work.

---

* FOSROC International Ltd., 285—Long Acre, NECHELLS, Birmingham – B 75 JR, England, U.K., Phone: 021-3271911; Fax: 021-3284236; Telex: 051-337308 FOSECO

** P.O. Box 11081, JEDDAH – 21453, Kingdom of Saudi Arabia, Phone: 02-660-2132, 02-665-0187.

2.6 The material employed for the coating shall comply with the following requirements:

Wet film thickness	150 microns per coat
Dry film thickness	75 microns per coat
Reduction in water absorption (ASTM C642)	80% minimum @ 28 days
Carbon Dioxide diffusion resistance (Research Laboratories Taywood Engineering Ltd)	A minimum equivalent to 250 metres of air
Water vapour transmission (Research Laboratories Taywood Engineering Ltd)	Shall be more than $13 \, g/m^2/day$
Reduction in chloride ion penetration	90 % minimum @ 28 days
Freeze/Thaw salt scaling (ASTM C672)	Unaffected by 50 exposure cycles

NOTE: Where test methods are not specified, the procedure for establishing compliance with the above criteria shall be agreed with the Supervising Officer.

Problem No. 2: Carbonation in Reinforced Concrete-Reinforcing Steel Within the Carbonated Zone

SPECIFICATION

The replacement of carbonated concrete by low permeability repair mortars followed by the application of a penetrating, reactive primer and top coat system.

1. **Concrete Preparation**

   1.1 It has been established by testing and investigation work that carbonation of concrete elements has taken place and it is the intention of the work to remove and replace the affected concrete.

   The areas to be repaired are shown on the drawing or are as otherwise indicated by the Supervising Officer. These areas are to be marked out on the concrete elements and agreed with the Supervising Officer before proceeding.

   The areas may be adjusted by the Supervising Officer as work proceeds according to the conditions found.

   Propping shall be provided as noted on the drawings or as agreed with the Supervising Officer.

   1.2 The agreed areas shall be broken out to remove carbonated concrete. Where reinforcement is exposed, breaking out shall continue to expose the full circumference of the steel and to a further depth of 20-30 mm or as directed by the Supervising Officer. Breaking out shall continue along the reinforcement until non-corroded steel is reached

and shall continue 50 mm beyond this point or as directed by the Supervising Officer. Special care shall be exercised to ensure any, reinforcement exposed is not cut or damaged.

1.3 The depth of breakout on the edge of any repair area shall be a minimum of 10 mm and feather edges will not be accepted.

To achieve this the perimeter of the area to be repaired shall first be cut to a depth of 10 mm using a suitable tool.

1.4 After breaking out as specified the exposed surface of concrete shall be tested for carbonation by the use of phenolphthalien. The test shall be carried out on the freshly exposed concrete or at least within 30 minutes of the test surface being exposed. The test shall be carried out on sound, dry and air blown dust free surfaces.

If the concrete substrate still exhibits carbonation, breaking out to remove a further 20 mm shall be carried out and the test repeated. If carbonation is still present the Supervising Officer shall be notified before proceeding further.

1.5 It is essential that no carbonated concrete substrate shall be in contact with, or within 20 mm of, the reinforcing bars. In cases where carbonation has reached within 20 mm of the reinforcement bar face, the concrete shall be broken out to expose the full circumference of the steel and to a further depth of 20-30 mm or as directed by the Supervising Officer.

1.6 Once carbonated concrete has been removed or the extent of breaking out agreed with the Supervising Officer, the exposed surface shall be cleaned free of loose particles, dust and any deleterious matter.

2. **Reinforcement Preparation**

   2.1 All exposed reinforcement shall be cleaned in accordance with the following:

   (a) Where exposed reinforcement is sound and does not show any signs of corrosion other than typical of its original condition, it shall be mechanically cleaned to remove rust and loose millscale.

   (b) Where exposed reinforcement shows signs of corrosion deterioration it shall be cleaned by grit blasting or other approved means to international standards of steel cleanliness such as BS4232 second quality or Swedish Standard SIS 05 5900:1967 Quality SA2½.

   2.2 When corrosion products have been removed and if directed by the Supervising Officer, the diameter of the reinforcement shall be measured. If considered necessary by the Supervising Officer the reinforcement shall be cut out and replaced or additional bars added in accordance with his instructions.

2.3   Reinforcement damaged during the removal of concrete or the preparation process shall, if required by the Supervising Officer, be repaired or replaced.

3.   **Reinforcement Priming**

3.1   Within one hour of preparing and cleaning, the reinforcement shall be coated with a primer to provide active galvanic protection.

3.2   The primer shall be NITOPRIME ZINCRICH, a single pack zinc-rich type primer complying with the requirements of BS4652: 1971.

3.3   The primer should be brush applied on to the cleaned reinforcement ensuring that the full surface area, particularly hidden surfaces of the bars, are fully coated in accordance with the manufacturer's recommendations. It is essential that this protective primer coat is continuous with that of any adjacent repaired area where zinc-rich primers have been used.

4.   **Concrete Bonding Agent**

4.1   Before commencing to apply the concrete repair mortar, the prepared concrete shall be thoroughly soaked with clean water in accordance with the manufacturer's instructions.

4.2   The bonding agent shall be NITOBOND AR for normal conditions or NITOBOND HAR where specific site conditions demand a higher initial 'grab'.

4.3   The bonding agent shall be well worked into the irregularities of the substrate according to the manufacturer's recommendations.

4.4   The repair material shall be applied before the bonding agent dries. The areas coated at one time must be restricted to allow for this and any area where it becomes dry shall be recoated as per the manufacturer's instructions.

NOTE: In situations where the completed repair will be subjected to constant immersion, the bonding agent shall be NITOBOND EP. The application procedure shall be in accordance with Section 5 of Standard Concrete Repair Specification No. 3.

5.   **Repair Materials**

5.1   In areas where the original concrete cover to the reinforcement was equal to or greater than 15 mm or where the extra build up to provide a cover of 15 mm does not affect any other aspect of the structure (e.g. appearance) the repair material shall be RENDEROC TG, a high build polymer modified cementitious shrinkage compensated repair mortar. (Type 1 below.)

5.2   In areas where the cover to reinforcement is less than 15 mm but in no case less than 5 mm, the repair mortar shall be RENDEROC S, a high strength polymer modified cementitious shrinkage compensated repair mortar. (Type 2 below.)

5.3   Where the total thickness of mortar required is greater than that which can be achieved in one layer without sagging or slumping, the material shall be applied in accordance with the manufacturer's recommendations, to achieve the desired profile.

5.4   The materials are required to be compatible with the original concrete substrate. They shall have the properties listed below when tested in accordance with the relevant standards.

5.5   These repair mortars shall not be applied when the ambient temperature is below 5°C or 5°C on a falling thermometer or above 35°C without consulting the manufacturer's technical department.

5.6   The mortar powder is to be of the single pack polymer modified pre-bagged type and only mixes using full bags are to be used. Part bag mixes will not be permitted.

5.7   The Contractor is to ensure that sufficient areas are prepared to provide economical use of material.

5.8   The mixing shall be carried out strictly in accordance with the manufacturer's recommendations using only forced action mixers. The Contractor is to ensure the correct equipment is on site and has been allowed for in the tender.

5.9   The powder should always be added to the water. In no circumstances shall more water be used in the mix than specified by the manufacturer. Remixing and retempering shall not be permitted.

5.10   The material shall be applied by gloved hand or trowel to the prepared and primed surface of the substrate and be well worked in, paying particular attention to packing behind and between the reinforcement.

5.11   Particular care shall be taken in the application of the repair material. Where necessary it should be built up in successive wet on wet layers to the required profile.

If sagging occurs the material must be completely removed and the void filled in two or more successive applications or by the use of formwork in accordance with the manufacturer's recommendations. After applying sufficient mortar to build the surface to the required level or to achieve the required cover to the reinforcement, the surface shall be trowelled smooth to the finished profiles.

Property	Test Method	Renderoc TG Type 1	Renderoc S Type 2
Compressive strength Minimum at 28 days N/mm^2	BS 1881 cubes Demoulded 1 day Cured 20°C in air	30	50

Property	Test Method	Renderoc TG Type 1	Renderoc S Type 2
Flexural strength Minimum at 28 days N/mm^2	BS 6319 prisms Demoulded 1 day Cured 20°C in air	7	8
Elastic modulus N/mm$^2 \times 10^3$	—	7 min 12 max	27 min 33 max
Coefficient of Thermal Expansion $\times 10^{-6}$/°C	—	7.3	7 min 12 max
Water absorption Maximum ml/m^2/sec	ISAT BS 1881: 1970	0.18 10 min 0.06 2 hr	0.01 10 min 0.005 2 hr
Carbon dioxide Permeability Thickness equivalent to 50 m air gap*	Calculated from oxygen diffusion in sweep gas method	8.3 mm max	20 mm max
Chloride ion Diffusion cm^2/sec**	Cell with disc separating limewater and Sodium Chloride-saturated limewater	4.15×10^{-10} max	2×10^{-9} max

NOTES: *Typical concrete 180 mm

        **Typical concrete 50×10^{-9}

6. Curing

6.1 All the areas repaired using cementitious materials specified above shall be fully cured in accordance with good concrete practice.

Details of the method of curing should be submitted to the Supervising Officer for approval.

7. Surface Preparation for Final Protective Coating

7.1 At the discretion of the Supervising Officer, where it is required to provide a uniform texture to the overall exposed concrete surface which has not been repaired, a single pack polymer modified fairing coat of RENDEROC FC shall be applied, strictly in accordance with the manufacturer's instructions.

In preparation for the fairing coat and protective coating all existing surfaces must be free from oil, grease, loose particles, decayed matter, moss or algal growth and general curing compounds that are not deemed compatible with the protective coating. All surface contamination and surface laitence must be removed through the use of grit blasting, high pressure water jetting, or equivalent mechanical means. After this treatment the surface shall be washed down with clean water to remove all dust, thoroughly soaking the substrate.

Blow holes and areas of substantial pitting shall then be filled with RENDEROC FC. The application of this must be in strict accordance with the manufacturer's recommendations, particularly with reference to the maximum application thickness of 3 mm.

8. Coating Materials

8.1 The materials are required to provide an in-depth protection against carbonation whilst permitting water vapour transmission from the concrete elements. The NITOCOTE DEKGUARD system is considered suitable.

8.2 The Contractor is required to adhere strictly to the manufacturer's recommendations regarding the use, storage, application and safety rules in respect of the approved materials.

8.3 The exposed concrete surfaces as defined in the documents or as agreed with the Supervising Officer shall be conditioned by the application of NITOCOTE DEKGUARD PRIMER, a penetrating hydrophobic treatment. The primer shall be allowed to dry in accordance with the manufacturer's requirements. The Contractor shall then apply 2 coats of NITOCOTE DEKGUARD PIGMENTED TOP-COAT in accordance with the manufacturer's instructions. The finished coating shall be pinhole-free and have a total minimum dry film thickness of 150 microns. The colour and finish is to be as agreed with the Supervising Officer.

8.4 The coating shall be applied by spray, roller or brush to achieve a finish acceptable to the Supervising Officer. In all operations of storage, mixing and application the Contractor is to comply with the Health and Safety recommendations of the manufacturer and governing authorities.

8.5 Where required by the Supervising Officer, trial areas not exposed in the finished work shall first be treated using the selected materials. These trials shall be noted on the drawings and shall be carried out using the type of materials, mixing procedures and applications that will be used on the contract and shall be approved by the Supervising Officer before the Contractor commences with the general work.

8.6 The material employed for the coating shall comply with the following requirements:

Wet film thickness	150 microns per coat
Dry film thickness	75 microns per coat
Reduction in water (ASTM C642)	80% minimum @ 28 days absorption
Carbon Dioxide diffusion resistance (Research Laboratories Taywood Engineering Ltd)	A minimum equivalent to 250 metres of air

Water vapour transmission (Research Laboratories Taywood Engineering Ltd)	Shall be greater than 13 g/m^2/day
Reduction in chloride ion penetration	90% minimum @ 28 days
Freeze Thaw/salt scaling (ASTM C672)	Unaffected by 50 exposure cycles

NOTE: Where test methods are not specified, the procedure for establishing compliance with the above criteria shall be agreed with the Supervising Officer.

Problem No. 3: Chloride Induced Corrosion in Reinforced Concrete

Chloride Penetration from External Environments Only

SPECIFICATION

The replacement of chloride contaminated concrete by low permeability repair mortars followed by the application of a penetrating, reactive primer and top coat system.

1. **Concrete Preparation**

   1.1 The areas to be repaired are to be as shown on the drawings or as indicated by the Supervising Officer. The areas are to be marked out in the works and agreed with the Supervising Officer before proceeding.

   The areas may be adjusted by the Supervising Officer as work proceeds according to the conditions found.

   1.2 The agreed areas shall be broken out to remove chloride contaminated concrete. Where this depth corresponds to the depth of concrete cover and thereby exposes reinforcement, breaking out shall continue to expose the full circumference of the steel and to a further depth of 20-30 mm or as directed by the Supervising Officer. Breaking out shall continue along the reinforcement until non-corroded steel is reached and shall continue 50 mm beyond this point or as directed by the Supervising Officer. Special care shall be exercised to ensure that any reinforcement exposed is not cut or damaged.

   1.3 The depth of breakout on the edge of any repair area shall be a minimum of 10 mm and feather edges will not be accepted. To achieve this the perimeter of the area to be repaired shall first be cut to a depth of 10 mm using a suitable tool.

   1.4 This preparation shall be such as to leave a sound exposed concrete substrate free from dust, loose particles and any deleterious matter.

2. **Reinforcement Preparation**

   2.1 All exposed reinforcement shall be cleaned of corrosion products by wet grit blasting or other approved means to achieve a surface finish to comply with international standards of steel cleanliness such as BS4232 second quality or SA2½ of Swedish Standard SIS 05 5900: 1967. Special care shall be taken to clean out properly any pits which have occurred in the steel bars.

   2.2 When the corrosion products have been removed, and if directed by the Supervising Officer, the diameter of the reinforcement shall be measured. If considered necessary by the Supervising Officer the existing reinforcement shall be cut out and replaced or additional bars added in accordance with his instructions.

   2.3 Reinforcement damaged during the removal of concrete or the preparation process shall, if required by the Supervising Officer, be repaired or replaced.

3. **Final Cleaning**

   3.1 Due to the presence of chlorides it is essential that the cleaning process is complete. To ensure this a final wash down of the concrete substrate and reinforcement shall be carried out to ensure the removal of all residual contamination.

4. **Reinforcement Priming**

   4.1 Within one hour of preparing and cleaning, the reinforcement shall be coated with a primer to provide active galvanic protection.

   4.2 The primer shall be NITOPRIME ZINCRICH, a single pack zinc-rich type primer complying with the requirements of BS4652 (1971).

   4.3 The primer should be brush applied on to the cleaned reinforcement ensuring that the full surface area, particularly hidden surfaces of the bars, are fully coated in accordance with the manufacturer's recommendations. It is essential that this protective primer coat is continuous with that of any adjacent repaired area where zinc-rich primers have been used.

5. **Concrete Bonding Agent**

   5.1 The bonding agent shall be NITOBOND EP which shall cure to form a barrier against residual chloride ions inherent within the concrete diffusing back into the areas of repair. The product shall have an open-life to suit Middle East conditions.

   5.2 The bonding agent shall be prepared in accordance with the manufacturer's instructions. It shall be well worked into the irregularities of the substrate according to the manufacturer's recommendations before commencing to apply the repair material.

   5.3 The repair material must be applied before the bonding agent dries. The area coated at one time must be restricted to allow for this and any area where it becomes dry shall be removed before applying another coat and continuing with the repair.

6. **Repair Materials**

6.1   In areas where the original concrete cover to the reinforcement was equal to or greater than 15 mm or where the extra build up to provide a cover of 15 mm does not affect any other aspect of the structure (e.g. appearance) the repair material shall be RENDEROC TG, a high build polymer modified cementitious shrinkage compensated repair mortar. (Type 1 below.)

6.2   In areas where the cover to the reinforcement is less than 15 mm but in no case less than 5 mm, the repair mortar shall be RENDEROC S. (Type 2 below.)

6.3   Where the total thickness of mortar required is greater than that which can be achieved in one layer without sagging or slumping, the material shall be applied in accordance with the manufacturer's recommendations, to achieve the desired profile.

6.4   The materials are required to be compatible with the original concrete substrate. They shall have the properties listed below when tested in accordance with the relevant standards.

Property	Test Method	Renderoc TG Type 1	Renderoc S Type 2
Compressive strength Minimum at 28 days N/mm^2	BS 1881 cubes Demoulded 1 day Cured 20°C in air	30	50
Flexural strength Minimum at 28 days N/mm^2	BS 6319 prisms Demoulded 1 day Cured 20°C in air	7	8
Elastic modulus $N/mm^2 \times 10^3$	—	7 min 12 max	27 min 33 max
Coefficient of Thermal Expansion $\times 10^{-6}/°C$	—	7.3	7 min 12 max
Water absorption Maximum $ml/m^2/sec$	ISAT BS 1881: 1970	0.18 10 min 0.06 2 hr	0.01 10 min 0.005 2 hr
Carbon dioxide Permeability Thickness equivalent to 50 m air gap*	Calculated from oxygen diffusion in sweep gas method	8.3 mm max	20 mm max
Chloride ion Diffusion cm^2/sec **	Cell with disc separating lime-water and Sodium Chloride-saturated limewater	4.15×10^{-10} max	2×10^{-9} max

NOTES: *Typical concrete 180 mm
   **Typical concrete 50×10^{-9}

6.5   These repair mortar shall not be applied when the ambient temperature is below 5° C or 5° C on a falling thermometer or above 35° C without consulting the manufacturer's technical department.

6.6   The mortar powder is to be of the single pack polymer modified pre-bagged type and only mixes using full bags are to be used. Part bag mixes will not be permitted.

6.7   The Contractor is to ensure that sufficient areas are prepared to provide economical use of material.

6.8   The mixing shall be carried out strictly in accordance with the manufacturer's recommendations using only forced action mixers. The Contractor is to ensure the correct equipment is on site and has been allowed for in the tender.

6.9   The powder should always be added to the water. In no circumstances shall more water be used in the mix than specified by the manufacturer. Remixing and retempering shall not be permitted.

6.10   The material shall be applied by gloved hand or trowel to the prepared and primed surface of the substrate and be well compacted in, paying particular attention to packing behind and between the reinforcement.

6.11   Particular care shall be taken in the application of the repair material. Where necessary it shall be built up in successive wet on wet layers to the necessary profile. If sagging occurs the material must be completely removed prior to filling the void in two or more successive applications or by the use of form work in accordance with the manufacturer's recommendations. After applying sufficient mortar to build the surface to the required level or to achieve the required cover to the reinforcement, the surface shall be trowelled smooth to the finished profiles.

7. **Curing**

7.1   All the areas repaired using cementitious materials specified above shall be fully cured in accordance with good concrete practice.

7.2   Details of the methods of curing shall be submitted to the Supervising Officer for approval.

8. **Surface Preparation for Final Protective Coating**

8.1   At the discretion of the Supervising Officer, where it is required to provide a uniform texture to the overall exposed concrete surface area which has not been repaired, a single pack polymer modified fairing coat of RENDEROC FC shall be applied, strictly in accordance with the manufacturer's instructions.

In preparation for the fairing coat and protective coating, all existing surfaces must be free from oil, grease, loose particles, decayed matter, moss or algal growth and general curing compounds. All surface contamination and surface laitence must be removed through the use of grit blasting, high pressure water jetting, or equivalent mechanical means. After this treatment the surface shall be washed down with clean water to remove all dust, thoroughly soaking the substrate.

Blow holes and areas of substantial pitting shall then be filled with RENDEROC FC. The application of this must be in strict accordance with the manufacturer's recommendations, particularly with reference to the maximum application thickness of 3 mm.

9. **Coating Materials**

9.1 The materials are required to provide an in-depth protection against chlorides whilst permitting water vapour transmission from the concrete elements. The NITOCOTE DEKGUARD system is considered suitable.

9.2 The Contractor is required to adhere strictly to the manufacturer's specifications regarding the use, storage, application and safety rules in respect of the approved materials.

9.3 The exposed concrete surface as defined in the documents or as agreed with the Supervising Officer shall be conditioned by the application of NITOCOTE DEKGUARD PRIMER, a penetrating hydrophobic treatment. The primer shall be allowed to dry in accordance with the manufacturer's requirements. The Contractor shall then apply 2 coats of NITOCOTE DEKGUARD PIGMENTED TOPCOAT in accordance with the manufacturer's instructions. The finished coating shall be pinhole-free and have a total minimum dry film thickness of 150 microns. The colour and finish is to be as agreed with the Supervising Officer.

9.4 The coating shall be applied by spray, roller or brush to achieve a finish acceptable to the Supervising Officer. In all operations of storage, mixing and application the Contractor is to comply with the Health and Safety recommendations of the manufacturer and governing authorities.

9.5 Where required by the Supervising Officer, trial areas not exposed in the finished work shall first be treated using the selected materials. These trial areas shall be as noted on the drawings and shall be carried out using the type of materials, mixing procedures and applications as will be used on the contract. These shall be approved by the Supervising Officer before the Contractor commences with the general work.

9.6 The material employed for the coating shall comply with the following requirements:

Wet film thickness	150 microns per coat
Dry film thickness	75 microns per coat
Reduction in water absorption (ASTM C642)	80% minimum @ 28 days

Carbon Dioxide diffusion resistance (Research Laboratories Taywood Engineering Ltd)	A minimum equivalent to 250 metres of air
Water vapour transmission (Research Laboratories Taywood Engineering Ltd)	Shall be greater than $13 \, g/m^2/day$
Reduction in chloride ion penetration	90% minimum @ 28 days
Freeze Thaw/salt scaling (ASTM C672)	Unaffected by 50 exposure cycles

NOTE: Where test methods are not specified, the procedure for establishing compliance with the above criteria shall be agreed with the Supervising Officer.

Problem No. 4: Chloride Induced Corrosion in Reinforced Concrete

Contamination by Chlorides Inherent Within the Concrete

SPECIFICATION

The removal of chloride contaminated concrete from the vicinity of the reinforcing steel and replacement with low permeability repair mortars followed by the application of a penetrating, reactive primer and top coat system.

1. **Concrete Preparation**

1.1 The areas to be repaired are to be as shown on the drawings or as indicated by the Supervising Officer. The areas are to be marked out on site and agreed with the Supervising Officer before proceeding.

The areas may be adjusted by the Supervising Officer as work proceeds according to the conditions found.

1.2 The agreed areas shall be broken out to remove chloride contaminated concrete. Where this depth corresponds to the depth of concrete cover and thereby exposes reinforcement, breaking out shall continue to expose the full circumference of the steel and to a further depth of 20-30 mm or as directed by the Supervising Officer. Breaking out shall continue along the reinforcement until non-corroded steel is reached and shall continue 50 mm beyond this point or as directed by the Supervising Officer. Special care shall be exercised to ensure that any reinforcement exposed is not cut or damaged.

1.3 The depth of breakout on the edge of any repair area shall be a minimum of 10 mm and feather edges will not be accepted. To achieve this, the perimeter of the area to be repaired shall first be cut to a depth of 10 mm using a suitable tool.

1.4 All preparation shall be such as to leave a sound exposed concrete substrate free from dust, loose particles and any deleterious matter.

2. **Reinforcement Preparation**

2.1 All exposed reinforcement shall be cleaned of corrosion products by wet grit blasting or other approved means to achieve a surface finish to comply with international standards of steel cleanliness such as BS4232 second quality or SA2½ of Swedish Standard SIS 05 5900: 1967. Special care shall be taken to clean out properly any pits which have occurred in the steel bars.

2.2 When the corrosion products have been removed, and if directed by the Supervising Officer, the diameter of the reinforcement shall be measured. If considered necessary by the Supervising Officer, the existing reinforcement shall be cut out and replaced or additional bars added in accordance with his instructions.

2.3 Reinforcement damaged during the removal of concrete or the preparation process shall, if required by the Supervising Officer, be repaired or replaced.

3. **Final Cleaning**

3.1 Due to the presence of chlorides it is essential that the cleaning process is complete. To ensure this a final wash down of the concrete substrate and reinforcement shall be carried out to ensure the removal of all residual contamination.

4. **Reinforcement Priming**

4.1 Within one hour of preparing and cleaning, the reinforcement shall be coated with a primer to provide active galvanic protection to the reinforcement.

4.2 The primer shall be NITOPRIME ZINCRICH, a single pack zinc-rich type primer complying with the requirements of BS4652 (1971).

4.3 The primer should be brush applied on to the cleaned reinforcement ensuring that the full surface area, particularly hidden surfaces of the bars, are fully coated in accordance with the manufacturer's recommendations. It is essential that this protective primer coat is continuous with that of any adjacent repaired area where zinc-rich primers have been used.

5. **Concrete Bonding Agent**

5.1 The bonding agent shall be NITOBOND EP which shall cure to form a barrier against residual chloride ions inherent within the concrete diffusing back into the areas of repair. The product shall have an open-life to suit Middle East conditions.

5.2 The bonding agent shall be prepared in accordance with the manufacturer's instructions. It shall be well worked into the irregularities of the substrate according to the manufacturer's recommendations before commencing to apply the repair material.

5.3 The repair material must be applied before the bonding agent dries. The area coated at one time must be restricted to allow for this and any area where it becomes dry shall be removed before applying another coat and continuing with the repair.

6. **Repair Materials**

6.1 In areas where the original concrete cover to the reinforcement was equal to or greater than 15 mm or where the extra build up to provide a cover of 15 mm does not affect any other aspect of the structure (e.g. appearance) the repair material shall be RENDEROC TG, a high build polymer modified cementitious shrinkage compensated repair mortar. (Type 1 below.)

6.2 In areas where the cover to the reinforcement is less than 15 mm but in no case less than 5 mm, the repair mortar shall be RENDEROC S. (Type 2 below.)

6.3 Where the total thickness of mortar required is greater than that which can be achieved in one layer without sagging or slumping, the material shall be applied in accordance with the manufacturer's recommendations, to achieve the desired profile.

6.4 The materials are required to be compatible with the original concrete substrate. They shall have the properties listed below when tested in accordance with the relevant standards.

Property	Test Method	Renderoc TG Type 1	Renderoc S Type 2
Compressive strength Minimum at 28 days N/mm^2	BS 1881 cubes Demoulded 1 day Cured 20°C in air	30	50
Flexural strength Minimum at 28 days N/mm^2	BS 6319 prisms Demoulded 1 day Cured 20°C in air	7	8
Elastic modulus N/mm$^2 \times 10^3$	—	7 min 12 max	27 min 33 max
Coefficient of Thermal Expansion $\times 10^{-6}$/°C	—	7.3	7 min 12 max
Water absorption Maximum ml/m^2/sec	ISAT BS 1881: 1970	0.18 10 min 0.06 2 hr	0.01 10 min 0.005 2 hr
Carbon dioxide Permeability Thickness equivalent to 50 m air gap *	Calculated from oxygen diffusion in sweep gas method	8.3 mm max	20 mm max
Chloride ion Diffusion cm^2/sec**	Cell with disc separating limewater and Sodium Chloride-saturated limewater	4.15×10^{-10} max	2×10^{-9} max

NOTES: *Typical concrete 180 mm
**Typical concrete 50×10^{-9}

6.5 These repair mortar shall not be applied when the ambient temperature is below 5°C or 5°C on a falling thermometer or above 35°C without consulting the manufacturer's technical department.

6.6 The mortar powder is to be of the single pack pre-bagged type and only mixes using full bags are to be used. Part bag mixes will not be permitted.

6.7 The Contractor is to ensure that sufficient areas are prepared to provide economical use of material.

6.8 The mixing shall be carried out strictly in accordance with the manufacturer's recommendations using only forced action mixers. The Contractor is to ensure the correct equipment is on site and has been allowed for in the tender.

6.9 The powder should always be added to the water. In no circumstances shall more water be used in the mix than specified by the manufacturer. Remixing and retempering shall not be permitted.

6.10 The material shall be applied by gloved hand or trowel to the prepared and primed surface of the substrate and be well compacted in, paying particular attention to packing behind and between the reinforcement.

6.11 Particular care shall be taken in the application of the repair material. Where necessary it shall be built up in successive wet on wet layers to the necessary profile. If sagging occurs the material must be completely removed prior to filling the void in two or more successive applications or by the use of form work in accordance with the manufacturer's recommendations. After applying sufficient mortar to build the surface to the required level or to achieve the required cover to the reinforcement, the surface shall be trowelled smooth to the finished profiles.

7. Curing

7.1 All the areas repaired using cementitious materials specified above shall be fully cured in accordance with good concrete practice.

7.2 Details of the methods of curing shall be submitted to the Supervising Officer for approval.

8. Surface Preparation for Final Protective Coating

8.1 At the discretion of the Supervising Officer, where it is required to provide a uniform texture to the overall exposed concrete surface area which has not been repaired, a single pack polymer modified fairing coat of RENDEROC FC shall be applied, strictly in accordance with the manufacturer's instructions.

In preparation for the fairing coat and protective coating, all existing surfaces must be free from oil, grease, loose particles, decayed matter, moss or algal growth and general curing compounds. All surface contamination and surface laitence must be removed through the use of grit blasting, high pressure water jetting, or equivalent mechanical means. After this treatment the surface shall be washed down with clean water to remove all dust, thoroughly soaking the substrate.

Blow holes and areas of substantial pitting shall then be filled with RENDEROC FC. The application of this must be in strict accordance with the manufacturer's recommendations, particularly with reference to the maximum application thickness of 3 mm.

9. Coating Materials

9.1 The materials are required to provide an in-depth protection against chlorides whilst permitting water vapour transmission from the concrete elements. The NITOCOTE DEKGUARD system is considered suitable.

9.2 The Contractor is required to adhere strictly to the manufacturer's specifications regarding the use, storage, application and safety rules in respect of the approved materials.

9.3 The exposed concrete surface as defined in the documents or as agreed with the Supervising Officer shall be conditioned by the application of NITOCOTE DEKGUARD PRIMER, a penetrating hydrophobic treatment. The primer shall be allowed to dry in accordance with the manufacturer's requirements. The Contractor shall then apply 2 coats of NITOCOTE DEKGUARD PIGMENTED TOPCOAT in accordance with the manufacturer's instructions. The finished coating shall be pinhole-free and have a total minimum dry film thickness of 150 microns. The colour and finish is to be as agreed with the Supervising Officer.

9.4 The coating shall be applied by spray, roller or brush to achieve a finish acceptable to the Supervising Officer. In all operations of storage, mixing and application the Contractor is to comply with the Health and Safety recommendations of the manufacturer and governing authorities.

9.5 Where required by the Supervising Officer, trial areas not exposed in the finished work shall first be treated using the selected materials. These trial areas shall be as noted on the drawings and shall be carried out using the type of materials, mixing procedures and applications as will be used on the contract. These shall be approved by the Supervising Officer before the Contractor commences with the general work.

9.6 The material employed for the coating shall comply with the following requirements:-

Wet film thickness	150 microns per coat
Dry film thickness	75 microns per coat
Reduction in water absorption (ASTM C642)	80% minimum @ 28 days
Carbon Dioxide diffusion resistance (Research Laboratories Taywood Engineering Ltd)	A minimum equivalent to 250 metres of air

Water vapour transmission (Research Laboratories Taywood Engineering Ltd)	Shall be greater than 13 g/m^2/day
Reduction in chloride ion penetration	90% minimum @ 28 days
Freeze Thaw/salt scaling (ASTM C672)	Unaffected by 50 exposure cycles

NOTE: Where test methods are not specified, the procedure for establishing compliance with the above criteria shall be agreed with the Supervising Officer.

Problem No. 5: Honeycombed Concrete or Concrete Damaged by Physical Forces Such as Impact or Erosion. (No chlorides or carbonation present)

SPECIFICATION

Concrete replacement using high-strength, free-flowing, cementitious micro-concrete.

1. **Concrete Preparation**

   1.1   The areas of repair shall be marked out on site and agreed with the Supervising Officer.

   1.2   All honeycombed, loose, cracked or friable concrete in these areas shall be removed until sound concrete is reached. Due account shall be taken of any propping or other instructions given by the Supervising Officer regarding sequences of removal and repair.

   1.3   The equipment and methods used to break out the concrete shall be such that no reinforcing steel or other embedded items such as conduits or lifting sockets etc. are loosened or damaged unless so directed by the Supervising Officer.

   1.4   In situations where reinforcement is congested, making removal of concrete by mechanical means difficult, then the use of high pressure water jetting shall be considered and a necessary provision for protecting the rest of the structure shall be allowed for.

   1.5   The prepared void shall be profiled so that entrapment of air during the repair process using fluid microconcrete is avoided.

   1.6   The minimum depth of repair shall be 40 mm throughout. The perimeter of the area to be repaired shall first be cut to a depth of 10 mm using a suitable tool. Feather edges will not be accepted.

   1.7   The prepared concrete surface shall be sound and clean and free of loose particles, dust and debris.

2. **Reinforcement Preparation**

   2.1   All exposed reinforcement shall be cleaned in accordance with the following:

   (a)   Where exposed reinforcement is sound and does not show any signs of corrosion other than typical of its original condition it shall be mechanically cleaned of rust and loose millscale.

   (b)   Where exposed reinforcement shows signs of corrosion deterioration it shall be cleaned by grit blasting or other approved means to international standards of steel cleanliness such as BS4232 second quality or Swedish Standard SIS 05 5900: 1967 quality SA2½.

   2.2   When corrosion products have been removed, and if directed by the Supervising Officer, the diameter of the reinforcement shall be measured. If considered necessary by the Supervising Officer, the reinforcement shall be cut out and replaced or additional bars added in accordance with his instructions.

   2.3   Reinforcement damaged during the removal of concrete or the preparation process shall, if required by the Supervising Officer, be repaired or replaced.

3. **Formwork**

   3.1   Adequate formwork shall be provided in accordance with the relevant codes of practice. This shall be securely fixed to withstand the hydraulic pressures of the fluid microconcrete repair material without distortion or movement during placement.

   The formwork shall be watertight at all joints between panels and between the formwork and the existing concrete surface so as to prevent grout leakage.

   3.2   The formwork shall be constructed from appropriate materials as agreed with the Supervising Officer to achieve the required finish.

   3.3   Formwork surfaces that are to be in contact with the repair microconcrete shall be treated with a suitable mould release agent such as REEBOL. This shall be used in accordance with the manufacturer's recommendations.

   3.4   The entry point of the feed pipe into the form shall be at the lowest point of the void.

   3.5   Sufficient hydrostatic head or pumping pressure shall be maintained to ensure that the void is filled completely and no air remains entrapped.

   3.6   Where necessary, provision shall be made for controllable bleed points to prevent air entrapment and enable the extent of flow of the repair material to be assessed.

4. Pre-Soaking

4.1 The formwork shall be inspected by the Supervising Officer and, if approved, filled with clean water which:

(a) Demonstrates that the formwork is grout-tight and,

(b) Saturates the prepared concrete surfaces

4.2 The formwork shall be completely drained and resealed.

NOTE: In situations where the completed repair will be subjected to constant immersion, the bonding agent shall be NITOBOND EP. The application procedure shall be in accordance with Section 5 of Standard Concrete Repair Specification No. 3 given earlier.

5. Repair material

5.1 The repair material shall:

(a) Be shrinkage compensated in *both* liquid and cured states,

(b) Contain no metallic expansion system,

(c) Be pre-packed and factory quality controlled,

(d) Be a free-flowing cementitious material that has a coefficient of thermal expansion fully compatible with the host concrete and which complies with the following requirements.

Property	Test Method	Conbextra HF
Compressive strength minimum at 28 days N/mm^2	BS 1881 @ w/p ratio of 0.18 @ 20° C	64
Flexural strength minimum at 28 days N/mm^2	BS 4551 @ w/p ratio of 0.18 @ 20° C	10
Coefficient of thermal expansion × 10^{-6}/°C	—	10 min. 12 max.
Thermal conductivity	—	1½ w/m°C
Elastic modulus N/mm^2 × 10^3	BS 1881	29
Anchorage bond stress (ultimate) (requirements for 40 N/mm^2 concrete @ 24 hrs)	CP100 Part 1, 1972, Table 22	Passes
Expansion characteristics plastic hardened	ASTM C 827 CRD 621-82A	Complies Complies
Pressure to restrain plastic expansion	—	appx. 0.004 N/mm^2

CONBEXTRA HF micro-concrete is considered suitable.

Alternatively, the repair material shall:

(a) Be shrinkage compensated in both liquid and cured states,

(b) Contain no metallic expansion system,

(c) Be pre-packed and factory quality controlled,

(d) Have a maximum equivalent sodium oxide content of 3 kg/m^3 (UK Department of Transport Specification for Highway Works 1986, Clause 1704.6),

(e) Be a free-flowing cementitious material that has a coefficient of thermal expansion fully compatible with the host concrete and which complies with the following requirements.

Property	Test Method	Renderoc LA
Flow characteristic of mixed material	UK Department of Transport Standard BD 27/86 para. 4.6(b)	700 mm min. No bleed or segregation
Compressive strength minimum at 28 days N/mm^2	BS 1881 cubes demoulded 1 day cured 20° C in water	55
Elastic modulus N/mm^2 × 10^3	BS 1881 (static)	30 min 37 max
Carbon dioxide permeability thickness equivalent to 50 m air gap*	Calculated from oxygen diffusion in sweep gas method	20 mm max
Chloride ion diffusion cm^2/sec**	Cell with disc separating limewater and sodium chloride saturated limewater	2×10^{-10} max

NOTES: * Typical concrete 180 mm
** Typical concrete 50×10^{-9}

RENDEROC LA micro-concrete is considered suitable.

5.2 The micro-concrete shall be mixed and placed in accordance with the manufacturer's recommendations, particularly with regard to water content, mixing equipment and placing time.

5.3 The placing of the micro-concrete shall be, as far as possible, continuous. The mixing operation shall be timed so that there is minimal interruption in the material flow. If, however, placing is interrupted, the operation shall recommence as soon as possible while the repair material retains its flow characteristics.

6. Formwork Removal

6.1 The formwork shall not be removed until the repair micro-concrete has achieved a compressive strength of at least 10 N/mm^2 or as directed by the Supervising Officer.

7. **Curing**

7.1 Immediately after removal of the formwork the repair area shall be cured in accordance with good concrete practice by the use of high efficiency spray-applied curing membranes such as CONCURE 90 or other approved methods.

7.2 Details of the methods of curing shall be submitted to the Supervising Officer for approval.

Problem No. 6: Chloride or Carbonation-Induced Corrosion of Reinforced Concrete Where Large Volumes are Involved or Where Trowel-Applied Mortars are Considered Less Suitable

SPECIFICATION

Concrete replacement using high-strength, free-flowing, cementitious micro-concrete followed by the application of a penetrating, reactive primer and top coat system.

1. **Concrete Preparation**

1.1 The areas to be repaired are to be as shown on the drawings or as indicated by the Supervising Officer. The areas are to be marked out in the works and agreed with the Supervising Officer before proceeding. The areas may be adjusted by the Supervising Officer as work proceeds according to the conditions.

1.2 The agreed areas shall be broken out to remove all contaminated concrete. Where the depth corresponds to the depth of concrete cover and thereby exposes reinforcement, breaking out shall continue to expose the full circumference of the steel and to a further depth of 20-30 mm or as directed by the Supervising Officer. Breaking out shall continue along the reinforcement until non-corroded steel is reached and shall continue 50 mm beyond this point or as directed by the Supervising Officer. Special care shall be exercised to ensure that any reinforcement exposed is not cut or damaged.

1.3 The prepared void shall be profiled so that the entrapment of air during the repair process using fluid microconcrete is avoided.

1.4 The minimum depth of repair shall be 40 mm throughout. The perimeter of the area to be repaired shall first be cut to a depth of 10 mm using a suitable tool. Feather edges will not be accepted.

1.5 This preparation shall be such as to leave a sound exposed concrete surface free from dust, loose particles and any deleterious matter.

2. **Reinforcement Preparation**

2.1 All exposed reinforcement shall be cleaned of corrosion products by wet grit blasting or other approval means to achieve a surface finish to comply with international standards of steel cleanliness such as BS4232 second quality or SA2½ of Swedish Standard SIS 05 5900: 1967. Special care shall be taken to clean out properly any pits which have occurred in the steel bars.

2.2 When the corrosion products have been removed, and if directed by the Supervising Officer, the diameter of the reinforcement shall be measured. If considered necessary by the Supervising Officer the existing reinforcement shall be cut out and replaced or additional bars added in accordance with his instructions.

2.3 Reinforcement damaged during the removal of concrete or the preparation process shall, if required by the Supervising Officer, be repaired or replaced.

3. **Final Cleaning**

3.1 Due to the presence of chlorides it is essential that the cleaning process is complete. To ensure this a final wash down of the concrete substrate and reinforcement shall be carried out to ensure the removal of all residual contamination.

4. **Reinforcement Priming**

4.1 Within one hour of preparing and cleaning, the reinforcement shall be coated with a primer to provide active galvanic protection to the reinforcement.

4.2 The primer shall be NITOPRIME ZINCRICH, a single pack zinc-rich type primer complying with the requirements of BS4652 (1971).

4.3 The primer should be brush applied on to the cleaned reinforcement ensuring that the full surface area, particularly hidden surfaces of the bars, are fully coated in accordance with the manufacturer's recommendations. It is essential that this protective primer coat is continuous with that of any adjacent repaired area where zinc-rich primers have been used.

5. **Concrete Bonding Agent (where inherent chlorides exist or where the completed repair will be subjected to constant immersion)**

5.1 The bonding agent shall be NITOBOND EP which shall cure to form a barrier against residual chloride ions inherent within the concrete diffusing back into the areas of repair. The product shall have an open-life to suit Middle East conditions.

5.2 The bonding agent shall be prepared in accordance with the manufacturer's instructions. It shall be well worked into the irregularities of the substrate according to the manufacturer's recommendations before commencing to apply the repair material.

5.3 The repair material must be applied before the bonding agent dries. The area coated at one time must be restricted to allow for this and any area where it becomes dry shall be removed before applying another coat and continuing with the repair.

6. **Formwork**

6.1 Adequate formwork shall be provided in accordance with the relevant codes of practice. This shall be securely fixed to withstand the hydraulic pressures of the fluid micro-concrete repair material without distortion or movement during placement.

The formwork shall be watertight at all joints between panels and between the formwork and the existing concrete surface so as to prevent grout leakage.

6.2 The formwork is to be constructed from appropriate materials as agreed with the Supervising Officer to achieve the required finish.

6.3 Formwork surfaces that are to be in contact with the repair micro-concrete shall be treated with a suitable mould release agent such as REEBOL. This shall be used in accordance with the manufacturer's recommendations.

6.4 The entry point of the feed pipe into the form shall be at the lowest point of the void.

6.5 Sufficient hydrostatic head or pumping pressure shall be maintained to ensure that the void is filled completely and no air remains entrapped.

6.6 Where necessary, provision shall be made for controllable bleed points to prevent air entrapment and enable the extent of flow of the repair material to be assessed.

7. **Pre-Soaking (where no inherent chlorides exist)**

7.1 The formwork shall be inspected by the Supervising Officer and, if approved, filled with clean water which:

(a) Demonstrates that the formwork is grout-tight and,

(b) Saturates the prepared concrete surface

7.2 The formwork shall be completely drained and re-sealed.

7.3 When chlorides are inherent in the structure or where the completed repair will be subjected to constant immersion, the pre-soaking shall be replaced by priming with NITOBOND EP to form an impermeable barrier between the host concrete and the replacement micro-concrete. The application of NITOBOND EP shall be in accordance with Section 5 above.

8. **Repair Material**

8.1 The repair material shall:

(a) Be shrinkage compensated in both liquid and cured states,

(b) Contain no metallic expansion system,

(c) Be pre-packed and factory quality controlled,

(d) Be a free-flowing cementitious material that has a coefficient of thermal expansion fully compatible with the host concrete and which complies with the following requirements.

Property	Test Method	Conbextra HF
Compressive strength minimum at 28 days N/mm^2	BS 1881 @ w/p ratio of 0.18 @ 20°C	64
Flexural strength minimum at 28 days N/mm^2	BS 4551 @ w/p ratio of 0.18 @ 20° C	10
Coefficient of thermal expansion $\times 10^{-6}$/°C	—	10 min. 12 max.
Thermal conductivity	—	1½ w/m°C
Elastic modulus N/mm^2 x 10^3	BS 1881	29
Anchorage bond stress (ultimate) (requirements for 40 N/mm^2 concrete @ 24 hrs)	CP100 Part 1, 1972, Table 22	Passes
Expansion characteristics: plastic hardened	ASTM C 827 CRD 621-82A	Complies Complies
Pressure to restrain plastic expansion		appx. 0.004 N/mm^2

CONBEXTRA HF micro-concrete is considered suitable.

Alternatively, the repair material shall:

(a) Be shrinkage compensated in both liquid and cured states,

(b) Contain no metallic expansion system,

(c) Be pre-packed and factory quality controlled,

(d) Have a maximum equivalent sodium oxide content of 3 kg/m^3 (UK Department of Transport Specification for Highway Works 1986. Clause 1704.6),

(e) Be a free-flowing cementitious material that has a coefficient of thermal expansion fully compatible with the host concrete and which complies with the following requirements.

Property	Test Method	Renderoc LA
Flow characteristic of mixed material	UK Department of Transport Standard BD 27/86 para. 4.6(b)	700 mm min. No bleed or segregation

Property	Test Method	Renderoc LA
Compressive strength minimum at 28 days N/mm^2	BS 1881 cubes demoulded 1 day cured 20° C in water	55
Elastic modulus $N/mm^2 \times 10^3$	BS 1881 (static)	30 min 37 max
Carbon dioxide permeability thickness equivalent to 50 m air gap*	Calculated from oxygen diffusion in sweep gas method	20 mm max
Chloride ion diffusion $cm^2/sec**$	Cell with disc separating limewater and sodium chloride saturated limewater	2×10^{-10} max

NOTES: * Typical concrete 180 mm
** Typical concrete 50×10^{-9}

RENDEROC LA micro-concrete is considered suitable

8.2   The micro-concrete shall be mixed and placed in accordance with the manufacturer's recommendations, particularly with regard to water content, mixing equipment and placing time.

8.3   The placing of the micro-concrete shall be, as far as possible, continuous. The mixing operation shall be timed so that there is minimal interruption in the material flow. If, however, placing is interrupted, the operation shall recommence as soon as possible while the repair material retains its flow characteristics.

9.   Formwork Removal

9.1   The formwork shall not be removed until the repair micro-concrete has achieved a compressive strength of at least $10 \, N/mm^2$ or as directed by the Supervising Officer.

10.   Curing

10.1   Immediately after removal of the formwork the repair area shall be cured in accordance with good concrete practice by the use of high efficiency spray-applied curing membranes such as CONCURE 90 or other approved methods.

10.2   Details of the methods of curing shall be submitted to the Supervising Officer for approval.

11.   Surface Preparation for Final Protective Coating

11.1   At the discretion of the Supervising Officer, where it is required to provide a uniform texture to the overall exposed concrete surface, a fairing coat and protective coating shall be applied, strictly in accordance with the manufacturer's instructions.

In preparation for the fairing coat and protective coating, all existing surfaces shall be free from oil, grease, loose particles, decayed matter, moss or algal growth and general curing compounds. All surface contamination and surface laitence must be removed through the use of grit blasting, high pressure water jetting, or equivalent mechanical means. After this treatment the surface shall be washed down with clean water to remove all dust, thoroughly soaking the substrate.

Blow holes and areas of substantial pitting shall then be filled with RENDEROC FC. The application of this must be in strict accordance with the manufacturer's recommendations, particularly with reference to the maximum application thickness of 3 mm.

12.   Coating Materials

12.1   The materials are required to provide in-depth protection against chlorides and carbonation whilst permitting water vapour transmission from the concrete elements. The NITOCOTE DEKGUARD system is considered suitable.

12.2   The Contractor is required to adhere strictly to the manufacturer's specifications regarding the use, storage, application and safety rules in respect of the approved materials.

12.3   The exposed concrete surface as defined in the documents or as agreed with the Supervising Officer shall be conditioned by the application of NITOCOTE DEKGUARD PRIMER, a penetrating hydrophobic treatment. The primer shall be allowed to dry in accordance with the manufacturer's requirements. The Contractor shall then apply 2 coats of NITOCOTE DEKGUARD pigmented TOPCOAT in accordance with the manufacturer's instructions. The finished coating shall be pinhole-free and have a total minimum dry film thickness of 150 microns. The colour and finish is to be as agreed with the Supervising Officer.

12.4   The coating shall be applied by spray, roller or brush to achieve a finish acceptable to the Supervising Officer. In all operations of storage, mixing and application the Contractor is to comply with the Health and Safety recommendations of the manufacturer and governing authorities.

12.5   Where required by the Supervising Officer, trial areas not exposed in the finished work shall first be treated using the selected materials. These trial areas shall be as noted on the drawings and shall be carried out using the type of materials, mixing procedures and applications as will be used on the contract. These shall be approved by the Supervising Officer before the Contractor commences with the general work.

12.6   The material employed for the coating shall comply with the following requirements.

Wet film thickness	150 microns per coat
Dry film thickness	75 microns per coat

Reduction in water absorption (ASTM C642)	80% minimum @ 28 days
Carbon Dioxide diffusion resistance (Research Laboratories Taywood Engineering Ltd)	A minimum equivalent to 250 metres of air
Water vapour transmission (Research Laboratories Taywood Engineering Ltd)	Shall be greater than 13 g/m²/day
Reduction in chloride ion penetration	90% minimum @ 28 days
Freeze Thaw/salt scaling (ASTM C672)	Unaffected by 50 exposure cycles

NOTE: Where test methods are not specified, the procedure for establishing compliance with the above criteria shall be agreed with the Supervising Officer.

Problem No. 7: Non-Active Cracks Within Concrete Elements Caused by Shrinkage or Other Structural Movement

SPECIFICATION

The use of low-viscosity epoxy injection resin to fill and seal non-active cracks and restore structural integrity.

1. **Preparation of Concrete**
   1.1 It has been established by testing and investigation work that cracks have become manifest within concrete elements due to shrinkage and/or structural movement. It has further been established that these cracks are non-active. The intention of this work is to fill and seal these cracks. The extent of the cracks is as indicated on the drawings or as is otherwise indicated by the Supervising Officer. These details are to be marked out on the concrete elements and agreed with the Supervising Officer before proceeding.

   The extent of the work may be adjusted by the Supervising Officer as the project proceeds, according to the conditions found.

   1.2 Grease, oil or other contaminants shall be removed. Algae or other bacterial growth shall be removed by scrubbing with bacticide or detergent and clean water. If necessary, wire brushes shall be used.

   1.3 Loose or spalling areas of concrete, laitence, traces of paint or other coating materials within the marked out scope of the work shall be removed.

   1.4 All cracks shall be thoroughly cleaned out using clean, oil-free compressed air. Both the concrete surface and the cracks shall be allowed to dry thoroughly before continuing.

2. **Fixing Injection Nipples**
   2.1 The injection nipples shall be fixed at intervals along the length of each crack. The distance between each nipple will depend on the width and depth of the crack.

   Spacing shall be close enough to ensure that the resin will penetrate along the crack to the next point of injection. This will normally be between 200 mm and 500 mm.

   Each nipple shall be firmly bonded to the concrete surface by using FOSROC SURFACE SEALANT.

   2.2 FOSROC SURFACE SEALANT shall be supplied in two parts (liquid base and hardener system). The two components shall be thoroughly mixed together for 3 to 4 minutes until a putty-like consistency is achieved.

   2.3 The mixed SURFACE SEALANT shall be applied to the metal base of each surface-fixed nipple. They shall be pressed firmly into place and held for several seconds until secure. The mixed SURFACE SEALANT shall be applied around each embedded nipple, ensuring a complete seal is made. In this way, all the nipples shall be fixed along the length of the crack. In the case of a wall or slab which is cracked all the way through, nipples shall be located on both sides with those at the back placed at midway points between those at the front.

   2.4 The surface of the cracks between the nipples shall be sealed with a band of SURFACE SEALANT 30 to 40 mm wide and 2 to 3 mm thick. Both sides of any cracks which go all the way through a wall or slab shall be sealed in this way.

   2.5 The prepared cracks shall be allowed to cure for 12 to 24 hours. At low ambient temperatures (5°C to 12°C) the curing time will be extended and the applicator shall ensure that the surface sealant has adequately cured prior to continuing.

   2.6 One end of the injection hose shall be attached to the lowest nipple on vertical cracks or to either end of horizontal cracks.

3. **Resin Application**
   3.1 Each crack shall be treated in a single, continuous operation. Sufficient material shall therefore be made ready prior to the commencement of the work.

   3.2 The Contractor is required to ensure that sufficient cracks are prepared to provide economical use of the mixed material.

   3.3 The preparation, mixing and application of the materials shall be undertaken in strict accordance with the manufacturer's recommendations. The Contractor is to ensure that all necessary tools and equipment are on site and have been allowed for in the tender.

3.4   The injection material shall be compatible with the host concrete and shall have the properties listed below when tested in accordance with the relevant standards.

Property	Test Method	
Compressive strength (minimum @ 7 days 40°C)	BS 4551, 2782, 6319	77 N/mm^2
Tensile strength (Minimum @ 7 days)	BS 12	30 N/mm^2
Elastic modulus	ISO/R 527	1600 N/mm^2
Tensile adhesion to damp concrete	—	Failure in concrete
Viscosity @ 40°C	—	300 centipoise max
Specific gravity	—	1080 kg/m^3

NITOKIT epoxy resin supplied in pre-packaged kit form and CONBEXTRA EPLV epoxy resin supplied in pre-weighed bulk form are considered appropriate.

3.5   The material shall not be used when the ambient temperature is below 5°C or is 5°C on a falling thermometer without consulting the manufacturer's technical department.

3.6   The injection resin shall be of a pre-packaged or pre-weighed type and only the use of full units will be allowed. No part packs or on-site batching will be allowed under any circumstances.

3.7   In all operations of storage, mixing and application, the Contractor is to comply with the Health & Safety recommendations of the manufacturer and governing authorities.

4. **Curing**
4.1   The injected system shall be allowed to cure for 24 hours and shall be left undisturbed for this time.

4.2   The nipples and bands of surface sealant shall then be removed and any damaged areas made good to the satisfaction of the Supervising Officer.

Problem No. 8: Cracks Within Concrete Elements Where a Degree of Future Movement is Anticipated

SPECIFICATION

The use of low-viscosity polyurethane injection resin to fill and seal cracks where a degree of future movement is anticipated.

1. **Preparation of Concrete**
1.1   It has been established by testing and investigation work that cracks have become apparent within concrete elements due to structural movement and that a degree of movement may continue. The intention of this work is to fill and seal these cracks. The extent of the cracks is as indicated on the draw- ings or is as otherwise indicated by the Supervising Officer. These details are to be marked out on the concrete elements and agreed with the Supervising Officer before proceeding.

The extent of the work may be adjusted by the Supervising Officer as the project proceeds, according to the conditions found.

1.2   Grease, oil or other contaminants shall be removed. Algae or other bacterial growth shall be removed by scrubbing with bacticide or detergent and clean water. If necessary, wire brushes shall be used.

1.3   Loose or spalling areas of concrete, laitence, traces of paint or other coating materials within the marked out scope of the work shall be removed.

1.4   All cracks shall be thoroughly cleaned out using clean, oil-free compressed air. Both the concrete surface and the cracks shall be allowed to dry thoroughly before continuing.

NOTE: Where water is seeping through the cracks, this shall be stopped by the use of CONBEXTRA WS60, a rapid foaming low-viscosity injection resin before continuing.

2. **Fixing Injection Nipples**
2.1   The injection nipples shall be fixed at intervals along the length of each crack. The distance between each nipple will depend on the width and depth of the crack.

Spacing shall be close enough to ensure that the resin will penetrate along the crack to the next point of injection. This will normally be between 200 mm and 500 mm.

Each nipple shall be firmly bonded to the concrete surface by using FOSROC SURFACE SEALANT.

2.2   FOSROC SURFACE SEALANT shall be supplied in two parts (liquid base and hardener system). The two components shall be thoroughly mixed together for 3 to 4 minutes until a putty-like consistency is achieved.

2.3   The mixed SURFACE SEALANT shall be applied to the metal base of each surface-fixed nipple. They shall be pressed firmly into place and held for several seconds until secure. The mixed SURFACE SEALANT shall be applied around each embedded nipple, ensuring a complete seal is made. In this way, all the nipples shall be fixed along the length of the crack. In the case of a wall or slab which is cracked all the way through, nipples shall be located on both sides with those at the back placed at midway points between those at the front.

2.4 The surface of the cracks between the nipples shall be sealed with a band of SURFACE SEALANT 30 to 40 mm wide and 2 to 3 mm thick. Both sides of any cracks which go all the way through a wall or slab shall be sealed in this way.

2.5 The prepared cracks shall be allowed to cure for 12 to 24 hours. At low ambient temperatures (5°C to 12°C) the curing time will be extended and the applicator shall ensure that the surface sealant has adequately cured prior to continuing.

2.6 One end of the injection hose shall be attached to the lowest nipple on vertical cracks or to either end of horizontal cracks.

3. **Resin Application**

3.1 Each crack shall be treated in a single, continuous operation. Sufficient material shall therefore be made ready prior to the commencement of the work.

3.2 The Contractor is required to ensure that sufficient cracks are prepared to provide economical use of the mixed material.

3.3 The preparation, mixing and application of the materials shall be undertaken in strict accordance with the manufacturer's recommendations. The Contractor is to ensure that all necessary tools and equipment are on site and have been allowed for in the tender.

3.4 The injection material shall be compatible with the host concrete and shall have the properties listed below when tested in accordance with the relevant standards.

Property

Specific gravity of mixed grout	1.0
Viscosity of freshly mixed grout (@ 25°C)	300 centipoise
Reaction time (tropical grade) (@ 25°C)	95 minutes
(@ 35°C)	55 minutes

CONBEXTRA UR63 is considered appropriate.

3.5 The material shall not be used when the ambient temperature is below 5°C or is 5°C on a falling thermometer without consulting the manufacturer's technical department.

3.6 The injection resin shall be of a pre-weighed type and only the use of full units will be allowed. No part packs or on-site batching will be allowed under any circumstances.

3.7 In all operations of storage, mixing and application, the Contractor is to comply with the Health & Safety recommendations of the manufacturer and governing authorities.

4. **Curing**

4.1 The injected system shall be allowed to cure for 24 hours and shall be left undisturbed for this time.

4.2 The nipples and bands of surface sealant shall then be removed and any damaged areas made good to the satisfaction of the Supervising Officer.

Problem No. 9: Surface Protection of Concrete in Contact with Sewage, Associated Chemical and Industrial Waste Seawater and Sebkha* Soils

SPECIFICATION

The use of chemically-resistant surface coatings and joint sealants to protect and prolong the life of concrete structures.

1. **Surface Preparation**

1.1 Prior to commencement, all concrete surfaces to be treated shall be free from oil, grease, loose particles, decayed matter, moss or algal growth. All such contamination and laitance must be removed by the use of grit-blasting, high pressure water-jetting or equivalent mechnical means.

1.2 In the case of new concrete elements or repaired sections, irregularities caused by grout flow between formwork panels shall be removed by an approved method.

As directed by the Supervising Officer, blow holes and areas of substantial pitting shall be filled with NITOMORTAR FC in strict accordance with the manufacturer's recommendations.

1.3 Any residue of curing membranes shall be removed prior to applying the surface coating. Again, this may best be achieved by grit-blasting or high pressure water-jetting. All surfaces shall be prepared to the satisfaction of the Supervising Officer.

2. **Joints**

2.1 All joints in the concrete elements are required to form a watertight seal to prevent the ingress of water-borne contaminants and to offer a high degree of resistance to a wide range of chemicals. NITOSEAL 220 is considered suitable. The joints shall be treated *prior* to the application of the selected surface coating.

2.2 The preparation of the joints shall be exactly in accordance with the sealant manufacturer's recommendations. Particular care shall be paid to the joint shoulders and, where existing joints are being refurbished, to the complete removal of the existing

---

* Salt-layden soil (*Arabic*).

sealant and primer. All preparation work shall be to the satisfaction of the Supervising Officer.

2.3 The joint walls shall be conditioned by the application of a single coat of NITOPRIME 21. Where difficulty is experienced in properly drying the walls, they shall be conditioned by the application of a single coat of NITOPRIME 22. The primer shall be allowed to dry in accordance with the manufacturer's requirements. The Contractor shall then apply NITOSEAL 220 in full accordance with the manufacturer's instructions. The mixing of part packs will not be permitted. The sealant shall be compacted into the joint and surface finished to the satisfaction of the Supervising Officer.

2.4 Any residue of sealant on the adjacent concrete surfaces shall be removed prior to the subsequent application of the protective coating.

2.5 The material employed for the sealing work shall be a pitch extended polyurethane with proven long term resistance (minimum 3 years) to bio-deterioration and shall comply with the following requirements:

MOVEMENT ACCOMMODATION
FACTOR:　　　　　　　　　　20%

SHORE-A HARDNESS
(14 DAYS @ 20°C)　　　　　15-25

SERVICE TEMPERATURE　　　−30°C to +80°C

NOTE: Test methods for establishing compliance with the above criteria shall be agreed with the Supervising Officer.

3. Coating Materials

3.1 The materials are required to form a continuous coating, fully bonded to the concrete substrate, which will offer a high degree of resistance to a wide range of chemicals. Where joints exist in the concrete substrate, the coating materials shall abut the joint sealing material and be fully compatible with it. The following materials are considered suitable:

(a) FOSROC NITOCOTE EP405　　(c) FOSROC NITOCOTE EP430

(b) FOSROC NITOCOTE EP410　　(d) FOSROC NITOCOTE ET550

3.2 (a) NITOCOTE EP405: Exposed concrete surfaces, as defined in the documents or as agreed with the Supervising Officer, where the cured surface film is required to be non-toxic and/or where the substrate is damp or wet and cannot be dried, shall be treated with two coats of NITOCOTE EP405. The coating shall be applied in full accordance with the manufacturer's instructions, shall be pin-hole free and have a finished total minimum dry film thickness of 400 microns. Where cracks exist in the substrate or where directed by the Supervising Officer, glass fibre reinforcement shall be incorporated into the first coat as required. NITOCOTE EP405

shall be applied by brush, roller or airless spray equipment to achieve a finish acceptable to the Supervising Officer.

(b) NITOCOTE EP410: Exposed concrete surfaces, as defined in the documents or as agreed with the Supervising Officer, where the cured surface film is required to exhibit high resistance to a wide range of chemicals and/or will be permanently submerged, shall be primed by the application of NITOPRIME 25, an epoxy bonding aid. The primer shall be applied in a single continuous coat and allowed to cure in accordance with the manufacturer's requirements. The contractor shall then apply a single coat of NITOCOTE EP410.

Where cracks exist in the substrate, or where directed by the Supervising Officer, a second coat shall be applied in conjunction with glass fibre reinforcement. The coating shall be applied in full accordance with the manufacturer's instructions, shall be pin-hole free and have a minimum dry film thickness of 250 microns per coat. NITOCOTE EP410 shall be applied by nylon brush and finally smoothed out using a steel trowel to achieve a finish acceptable of the Supervising Officer.

(c) NITOCOTE EP430: Exposed concrete surface, as defined in the documents or as agreed with the Supervising Officer, where the cured film is required to exhibit high resistance to a wide range of chemicals or will be totally or partially immersed in seawater and, in particular, where the substrate is damp, wet or submerged, shall be treated with two coats of NITOCOTE EP430. The coating shall be applied in full accordance with the manufacturer's instructions, shall be pin-hole free and have a finished total minimum dry film thickness of 400 microns. Where cracks exist in the substrate or where directed by the Supervising Officer, glass fibre reinforcement shall be incorporated into the first coat as required. NITOCOTE EP430 shall be applied by stiff-bristled brush to damp, wet or underwater concrete surfaces. It may also be applied by brush, roller or airless spray equipment to dry concrete surfaces. A finish acceptable to the Supervising Officer shall be achieved.

(d) NITOCOTE ET550: Exposed concrete surfaces, as defined in the documents or as agreed with the Supervising Officer, where the cured film is required to exhibit high resistance to a wide range of chemicals, contaminated ground water and sebkha soils, or will be permanently submerged, shall be treated with one to three coats of NITOCOTE ET550. The coating shall be applied in full accordance with the manufacturer's instructions, shall be pin-hole free and have a minimum dry film thickness of 350 microns per coat. Where cracks exist in the substrate or where directed by the Supervising Officer, glass fibre reinforcement shall be incorporated into the first coat as required. NITOCOTE ET550 shall be applied by brush, roller or airless spray equipment to achieve a finish acceptable to the Supervising Officer.

3.3 In all operations of storage, mixing and application the Contractor is to comply with the Health and Safety recommendations of the manufacturer and governing authorities.

3.4 Where required by the Supervising Officer, trial areas not exposed in the finished work shall first by treated using the selected materials. These trial areas shall be noted on the drawings and shall be carried out using the type of materials, mixing procedures and applications that will be used on the contract and shall be approved by the Supervising Officer before the Contractor commences with the general work.

3.5 The materials employed for coating work shall comply with the following requirements:

	NITOCOTE EP405	NITOCOTE EP410	NITOCOTE EP430	NITOCOTE ET550
Wet film thickness per coat (microns)	200	250	200	530
Dry film thickness per coat (microns)	200	250	200	350
Volume solids	100%	100%	100%	66%
No. of coats	2	1-2	2	1-3
Pot life @ 20°C (minutes)	30-40	90	30-40	240
Touch dry @ 20°C (hours)	6	12	6	4
Recoatable @ 20°C (hours)	6-18	12-30	6-24	24-168

In addition, the coatings shall exhibit excellent resistance to a wide range of industrial chemicals, oil-based products, seawater, contaminated ground water and sebkha soils.

NOTE: Test methods for establishing compliance with the above criteria (as well as for establishing resistance to specific chemicals) shall be agreed with the Supervising Officer.

Problem No. 10: Corrosion in Reinforced Concrete Induced by Chemical Attack

SPECIFICATION

The replacement of chemically contaminated concrete by impermeable, chemically-resistant resin repair mortars.

1. **Concrete Preparation**
   1.1 The areas to be repaired are to be as shown on the drawings or as indicated by the Supervising Officer. The areas are to be marked out in the works and agreed with the Supervising Officer before proceeding.

   The areas may be adjusted by the Supervising officer as work proceeds according to the conditions found.

   1.2 The agreed areas shall be broken out to remove chemically contaminated concrete. Where this depth corresponds to the depth of concrete cover and thereby exposes reinforcement, breaking out shall continue to expose the full circumference of the steel and to a further depth of 20-30 mm or as directed by the Supervising Officer. Breaking out shall continue along the reinforcement until non-corroded steel is reached and shall continue 50 mm beyond this point or as directed by the Supervising Officer. Special care shall be exercised to ensure that any reinforcement exposed is not cut or damaged.

   1.3 The depth of breakout on the edge of any repair area shall be a minimum of 10 mm and feather edges will not be accepted. To achieve this the perimeter of the area to be repaired shall first be cut to a depth of 10 mm using a suitable tool.

   1.4 Where directed by the Supervising Officer, all areas shall be cleaned with a suitable grit or water-blasting technique in order to remove any remaining surface contamination. 1.5 The preparation shall be such as to leave a sound exposed concrete substrate free from *dust, loose* particles and any deleterious matter.

2. **Reinforcement Preparation**
   2.1 All exposed reinforcement shall be cleaned of corrosion products by wet grit blasting or other approval means to achieve a surface finish to comply with international standards of steel cleanliness such as BS4232 second quality or SA2½ of Swedish Standard SIS 05 5900: 1967. Special care shall be taken to clean out properly any pits which have occurred in the steel bars.

   2.2 When the corrosion products have been removed, and if directed by the Supervising Officer, the diameter of the reinforcement shall be measured. If considered necessary by the Supervising Officer the existing reinforcement shall be cut out and replaced or additional bars added in accordance with his instructions.

2.3 Reinforcement damaged during the removal of concrete or the preparation process shall, if required by the Supervising Officer, be repaired or replaced.

3. Final Cleaning

3.1 Due to the presence of chemicals, it is essential that the cleaning process is complete. To ensure this, a final wash down of the concrete substrate and reinforcement shall be carried out to ensure the removal of all residual contamination. The substrate shall be allowed to dry thoroughly before continuing.

4. Reinforcement Priming

4.1 Within one hour of preparing and cleaning, the reinforcement shall be coated with a primer to provide active galvanic protection.

4.2 The primer shall be NITOPRIME ZINCRICH, a single pack zinc-rich type primer complying with the requirements of BS4652 (1971).

4.3 The primer should be brush applied on to the cleaned reinforcement ensuring that the full surface area, particularly hidden surfaces of the bars, are fully coated in accordance with the manufacturer's recommendations. It is essential that this protective primer coat is continuous with that of any adjacent repaired area where zinc-rich primers have been used.

5. Concrete Bonding Agent

5.1 The appropriate bonding agent shall be used, (see Table 1), depending on the final selection of resin mortar, in strict accordance with the manufacturer's requirements.

5.2 The bonding agent shall be prepared in accordance with the manufacturer's instructions. It shall be well worked into the irregularities of the substrate according to the manufacturer's recommendations before commencing to apply the repair mortar.

5.3 The repair material must be applied before the bonding agent dries. The area coated at one time must be restricted to allow for this and any area where it becomes dry shall be removed before applying another coat.

6. Repair Materials

6.1 In vertical or overhead locations the repair material shall be NITOMORTAR HB, a lightweight, high-build epoxy resin non-shrink repair mortar.

6.2 In horizontal locations, or vertical locations where the repaired areas will be subjected to physical abrasion, the repair material shall be NITOMORTAR S, an abrasion resistant epoxy resin non-shrink repair mortar.

6.3 In locations where an impermeable lining or render is required, the repair material shall be NITOMORTAR EL, an abrasion resistant epoxy resin lining mortar.

6.4 Where the total thickness of mortar required is greater than that which can be achieved in one layer without sagging or slumping, the material shall be applied in accordance with the manufacturer's recommendations, to achieve the desired profile.

6.5 The materials are required to be compatible with the original concrete substrate. They shall have the properties listed below when tested in accordance with the relevant standards.

PROPERTY	NITOMORTAR – HB	NITOMORTAR –S	NITOMORTAR –EL
Primer	NITOPRIME 28	NITOPRIME 28 (vertical) NITOPRIME 25 (other)	none
Compressive strength: minimum @ 7 days. N/mm^2	40	70	70
Flexural strength: minimum @ 7 days. N/mm^2	15	20	28
Tensile strength: minimum @ 7 days. N/mm^2	7	10	12
Water absorption: Maximum. %	0.45	0.20	0.10
Compressive modulus: KN/mm^2	4.5	16.2	7.5

6.6 These repair mortar shall not be applied when the ambient temperature is below 5°C or is 5°C on a falling thermometer.

6.7 The resin mortars are to be of the pre-weighed, pre-packed type so that no batching is required on site. Only mixes using full packs are to be used. Part pack mixes will not be permitted.

6.8 The Contractor is to ensure that sufficient areas are prepared to provide economical use of material.

6.9 The mixing shall be carried out strictly in accordance with the manufacturer's recommendations using only suitable mechanical mixers. The Contractor is to ensure the correct equipment is on site and has been allowed for in the tender.

6.10 The liquid components shall always be thoroughly mixed together first and the powder/filler added to the mixed liquids. In no circumstances shall any other components be used in the mix other than those supplied by the manufacturer.

6.11 Particular care shall be taken in the application of the repair material. It shall be applied by trowel or gloved hands to the prepared and primed surface of the substrate and be well compacted in, paying particular attention to packing behind and between exposed reinforcement.

6.12 Where necessary, the repair material shall be built up in successive wet on wet layers to the necessary profile. The maximum thickness recommended by the manufacturer shall not be exceeded. If sagging occurs, the material must be completely removed prior to filling the void in two or more successive applications or by the use of temporary formwork, in full accordance with the manufacturer's recommendations. After applying sufficient mortar to build the surface to the required level or to achieve the required cover to the reinforcement, the surface shall be trowelled smooth to the finished profiles.

7. Surface Preparation for Final Protective Coating

7.1 Where it is required to provide additional protection to the structure in form of a chemically-resistant coating to the overall exposed concrete surface area which has not been repaired. NITOMORTAR FC, an epoxy resin fairing coat, shall be applied, strictly in accordance with the manufacturer's instructions.

In preparation for the fairing coat and protective coating, all existing surfaces must be free from oil, grease, loose particles, decayed matter, moss or algal growth and general curing compounds. All surface contamination and surface laitance must be removed through the use of grit blasting, high pressure water jetting, or equivalent mechanical means. After this treatment the surface shall be washed down with clean water to remove all dust, thoroughly soaking the substrate. The substrate shall be allowed to dry thoroughly before continuing.

Blow holes and areas of substantial pitting shall then be filled with NITOMORTAR FC. The application of this must be in strict accordance with the manufacturer's recommendations, particularly with reference to the maximum application thickness of 3 mm.

Coating Materials

Reference should be made to Standard Concrete Repair Specification No. 9, described earlier in this Annexure.

II. FOSROC'S FORMATS FOR STANDARD 'BILL-OF-QUANTITIES' FOR REPAIR AND PROTECTION OF CONCRETE

General Preamble

All works are to be carried out in accordance with the specification.

Inspection and Diagnosis

Before proceeding with a repair it is important to establish the cause and degree of deterioration or defects in the concrete members and where possible, the extent of these. Where there is any doubt, it is recommended that a diagnostic survey be made by a Specialist in investigation and testing.

Other Specification and Contract Requirements

Due regard must be given to access, environmental considerations and other limitations. Such items include (but are not limited to):

1. Contract conditions
2. Working restrictions
3. Maintaining operations (car parks etc.)
4. Limitation on debris removal
5. Material storage and mixing
6. Protection of user's and architectural features
7. Programming of works
8. Supervision of staff
9. Site testing and sampling
10. Method of measurement
11. Propping
12. Methods of replacing reinforcement
13. Payments
14. Cleaning up

This bill of quantities is designed as a guide and may not include all details relevant to an individual contract. There is a space at the end of each section for other items to be included. If any further information is required please consult FOSROC.

Standard Bill of Quantities

Description of Works and Schedule of Rates for the Repair and Protection of Concrete.

1. Survey

Item	Description	Quantity	Unit	Rate ( )	Total
	Survey (see GENERAL PREAMBLE). May include but not be limited to:				
1.1	*Soundness/hammer test all concrete surfaces*		m²		
1.2	*Cover meter test all concrete surfaces*		m²		

Item	Description	Quantity	Unit	Rate ( )	Total
1.3	*Carry out carbonation depth testing to BRE (UK) information sheet IP6/81 as directed by the Supervising Officer*		each		
1.4	*Carry out chloride depth testing (Quantab, Hach or equivalent) as directed by the Supervising Officer*		each		
1.5	*Other tests* (Specify)				

2. Preparation

Item	Description	Quantity	Unit	Rate ( )	Total
	Preparation/Cleaning				
2.1	*Where required, steam clean/pressure water jet/detergent wash to remove surface contamination as described in the specification or as directed by the Supervising Officer*				
2.1.1	General concrete surfaces		m^2		
2.1.2	Difficult access surfaces		m^2		
2.2	*Grit blast (or equivalent) all concrete surfaces to remove existing coatings, laitance and contaminants to expose any surface defects and prepare the surface for protective coating or overlay*				
2.2.1	General concrete surfaces		m^2		
2.2.2	Difficult access surfaces		m^2		
2.3	*Other preparation* (Specify)				

3. Cementitious Repairs

The cementitious repairs system shall be a tropical formulation, manufactured under approved quality-control conditions for use in the local environment.

Item	Description	Quantity	Unit*	Rate ( )	Total
	Cementitious repairs				
3.1	*Cut back concrete surfaces, grit blast steel to SA 2½, prime and repair with RENDEROC mortar repair system as described in the specification* Up to 15 mm deep 15-40 mm deep 40-80 mm deep 80+ mm deep		 m^2 m^2 m^2 m^2		
3.2	*Cut back concrete surfaces, grit blast steel to SA 2½ prime and repair with RENDEROC LA or CONBEXTRA HF fluid repair system as described in the specification* 40-80 mm deep 80-120 mm deep 120+ mm deep		 m^2 m^2 m^2		

Item	Description	Quantity	Unit*	Rate ( )	Total
3.3	*Other cementitious repairs* (Specify)				

*m^3 may be used instead of m^2 where appropriate.

4. Resin Repairs

The resin repair system shall be a tropical formulation, manufactured under approved quality-control conditions for use in the local environment.

Item	Description	Quantity	Unit*	Rate ( )	Total
	Resin repairs				
4.1	*Cut back concrete surfaces, grit blast steel to SA 2½, prime and repair with NITOMORTAR resin repair system as described in the specification* Up to 15 mm deep 15-40 mm deep 40-80 mm deep 80+ mm deep		 m^2 m^2 m^2 m^2		
4.2	*Other resin repairs* (Specify)				

*m^3 may be used instead of m^2 where appropriate.

5. Crack Injection

The crack injection system shall be a tropical formulation, manufactured under approved quality-control conditions for use in the local environment.

Item	Description	Quantity	Unit	Rate ( )	Total
	Crack injection				
5.1	*Inject and seal cracks with NITOKIT CONBEXTRA EPLV or CONBEXTRA UR63 repair system as described in the specification*				
	Clean surfaces to 50 mm both sides of the crack		lin.m		
	Seal crack surfaces and fix nipples at 200-500 mm centres		lin.m		
	Inject with NITOKIT CONBEXTRA EPLV or CONBEXTRA UR63 resin system		cartridge/ litre		
	Extra over: nipples surfaces sealant cartridges/litres		each lin.m each/litre		
5.2	*Remove nipples and make good*		lin.m each/litre		
5.3	*Other resin repairs* (Specify)				

6. Protection Against Chloride, Sulphate, Carbonation and Atmospheric Acidic Gases

The protective system shall be a tropical formulation, manufactured under approved quality-control conditions for use in the local environment.

Item	Description	Quantity	Unit	Rate ()	Total
	Concrete protection				
6.1	*Preparation*	——— As Item 2 ———			
6.2	*Apply NITOCOTE DEKGUARD Primer and 2 coats of Topcoat as described in the specification*				
	(a) General concrete surfaces: fill blowholes with RENDEROC FC apply Primer apply Topcoat		m² m² m²		
	(b) Difficult access surfaces: fill blowholes with RENDEROC FC apply Primer apply Topcoat		m² m² m²		

7. Protection From Sewage, Chemicals, Contaminated Ground Water, Seawater and Sebkha Soils

The protective system shall be a tropical formulation, manufactured under approved quality-control conditions for use in the local environment.

Item	Description	Quantity	Unit	Rate ( )	Total
	Joint sealant				
7.1	*Preparation*	——— As Item 2 ———			
7.2	*Seal joints with NITOSEAL 220 as described in the specification*				
	Rake out prepare prime and seal to correct dimensions:				
	horizontal vertical		lin.m lin.m		
	Concrete Protection				
7.3	*Apply appropriate NITOCOTE protective coating as described in the specification*				
	(a) General concrete surface: fill blowholes with NITOMORTAR FC		m²		
	apply full coating system		m²		
	(b) Difficult access surfaces:				
	fill blowholes with NITOMORTAR FC		m²		
	apply full coating system		m²		
	(c) Fabric reinforcement of surfaces cracks or other details		m²		

8. Cementitious Overlay for Cathodic Protection of Steel Reinforced Concrete

The cementitious overlay system shall be a tropical formulation, manufactured under approved quality-control conditions for use in the local environment.

Item	Description	Quantity	Unit	Rate ( )	Total
	Cementitious overlay				
8.1	*Preparation*	— As Item 2 —			
8.2	*Thoroughly soak concrete substrate and apply RENDEROC CPO polymer modified cementitious overlay system as described in the specification*				
	10-20 mm deep		m^2		
	20-40 mm deep		m^2		
	40+ mm deep		m^2		

* m^3 may be used instead of m^2 where appropriate

III. A PARTIAL LIST OF FOSROC'S PRODUCTS FOR CONCRETE REPAIR

1. NITOPRIME® ZINCRICH

Active single component epoxy zinc primer
(active steel reinforcement primer)

USES

NITOPRIME ZINCRICH is the recommended anti-corrosion primer for steel reinforcement in the FOSROC CONCRETE REPAIR SYSTEM.

NITOPRIME ZINCRICH actively resists corrosion in the repaired area, and also aids the prevention of corrosion in adjacent areas.

NITOPRIME ZINCRICH benefits from being a single component material and therefore avoids possible errors in mixing.

NITOPRIME ZINCRICH can also be used as a touch-up primer or for coating high voltage transmission towers, overhead gantries, bridges and structural steelwork.

ADVANTAGES

- *Single component system—easy to use, no problems with pot life.*
- *Time-saving-touch dry after 45 minutes and recoatable after 4 hours.*
- *Economical, one pack system ensures no site wastage.*
- *Suitable for use in Middle East conditions.*

DESCRIPTION

NITOPRIME ZINCRICH is supplied as a one component grey-coloured primer coating based on zinc and epoxy resins.

STANDARDS

BS 4652, 1971 specification for metallic zinc rich priming

PROPERTIES

Specific gravity: 2.0.

Volume solids: 30%.

Recommended thickness: 40 µm dry.

Application thickness: 135 µm wet.

Theoretical coverage: 13.3 m^2/1–6.7 m^2/kg.

Practical coverage: Whilst theoretical coverages are quoted for guidance practical coverages may be up to 50% lower depending on nature of substrate and method of application.

Number of coats: 1–2 depending on application.
Pot life: Not applicable.

Drying time:	At 20°C	At 35°C
Touch dry	45 mins	15 mins
Fully dry	4 hours	2 hours
Recoatable	4 hours	2 hours

Application temperature: Minimum 5°C.

INSTRUCTIONS FOR USE

Surface Preparation
Steel—all rust should be removed especially in pits subjected to chloride attack by blast cleaning to SA 2½. Wire brushing should only be used when there is no possibility of blast cleaning.

Mixing
NITOPRIME ZINCRICH should be stirred before use to re-disperse the heavy zinc particles.

Application
NITOPRIME ZINCRICH should be brushed into the exposed reinforcing steel, making sure that the back of the steel is coated also.

A second coat may be applied after 4 hours. The primed surface should not be exposed to the elements for longer than necessary before overcoating or application of the repair mortar. NITOPRIME ZINCRICH will protect steel under clean interior exposure conditions for a period of several months. In non-aggressive exterior conditions a maximum interval of 14 days will be tolerated, but in industrial and/or marine conditions this interval should be reduced to the practical minimum.

Thinning
NITOPRIME ZINCRICH is supplied ready for use. However if necessary a maximum of 5% FOSROC SOLVENT 102 may be added.

Cleaning
Tools should be cleaned with FOSROC SOLVENT 102 prior to curing.

PRECAUTIONS

Health and Safety
Some people are sensitive to epoxy resins so gloves and a barrier cream Kerodex 71, Rozalex No. 9, Debba-wet work should be used when handling these products. If contact with the product occurs it must be removed before it hardens with a resin removing cream such as Kerocleanse 22 and Rozalex 42. Follow this by washing with soap and water. *Do not use solvent.* The use of goggles is recommended, but should accidental eye contamination occur, wash thoroughly with plenty of water and seek medical treatment immediately. Ensure good ventilation.

Fire
NITOPRIME ZINCRICH is flammable. Avoid using near naked flames and do not smoke during use.

Flash point
NITOPRIME ZINCRICH 7°C (Abel Closed Cup) FOSROC SOLVENT 102: 33°C

STORAGE

NITOPRIME ZINCRICH has a shell life of 12 months if stored in a cool place in unopened tins.

PACKAGING

NITOPRIME ZINCRICH is supplied in 2.5 and 12.5 litre containers.

SUPPLY

Contact your local FOSROC office or representative.

ADDITIONAL INFORMATION

NITOPRIME ZINCRICH forms part of the FOSROC CONCRETE RE-PAIR SYSTEM.

The FOSROC CONCRETE REPAIR SYSTEM ensures a successful, economical solution to *your* concrete problem.

NITOPRIME	Active anti corrosion coating for steel
ZINCRICH	reinforcement.
NITOBOND AR	Acrylic emulsion for priming and curing FOSROC cementitious mortars.
RENDEROC TG	Cementitious repair mortar for vertical and overhead repairs.
RENDEROC S	Polymer modified high strength mortar for floor and structural repairs.
RENDEROC FC	Polymer modified re-facing mortar.
NITOCOTE	Impregnating, breathable anti-chloride.
DEKGUARD	and anti-carbonation coating.

2. NITOBOND® EP

Epoxy resin concrete bonding agent
(for bonding old and new concretes)
— concrete bonding agent for all cement based products and barrier coat against inherent chloride and sulphate ions.

USES

For bonding new cementitious materials to existing cementitious surfaces. For use on horizontal surfaces and on vertical surfaces where mortar or concrete can be supported by formwork.

Ideal for extensions and repairs to concrete in factories, loading bays, trucking aisles, bridges, roads, bonded or granolithic floor topping, etc.

ADVANTAGES

- *Can be applied to dry or damp surfaces.*
- *High mechanical strength.*
- *Good positive adhesion.*
- *Special slow set hardener available.*
- *Suitable for use in Middle East Conditions.*

DESCRIPTION

NITOBOND EP is based on solvent-free epoxy resins containing pigments and fine fillers. It is supplied as a two part material in pre-weighed quantities ready for on-site mixing and use.

Coloured components, white base and green hardener, provide visual evidence that adequate mixing is achieved.

PROPERTIES

	Slow set		Strength	
	at 20°C	at 35°C		
Pot life	5.6 hrs.	2½ -3 hrs.	Compressive:	50N/mm^2
Initial		48 hrs.	Flexual	
hardness	48 hrs.	4 days	Tensile:	35N/mm^2
Full cure	7 days			20N/mm^2
Maximum				
Overlay				
Time	24 hrs.	8-10 hrs.	Shear:	25N/mm^2
Below 20°C the rate of cure will be slower Minimum application temperature 5°C			Adhesion to concrete:	In general the bond will always exceed the tensile strength of the concrete.

INSTRUCTIONS FOR USE

Preparation
All surfaces to be treated must be firm, dust-free and clean. All laitence should be removed mechanically.

Where surfaces as contaminated with oil or grease, this should be removed by treatment with FOSROC CHEMICAL DEGREASER followed by water or steam cleaning.

The surface should be washed thoroughly with clean water and all traces of surface water removed.

Mixing
The entire contents of the hardener tin should be poured into the resin container and the two materials thoroughly mixed until a uniform colour is obtained.

To facilitate application at temperatures below 20°C the separate components should be warmed in hot water to a maximum of 25°C before mixing. However the mixed material will need to be used speedily as the pot life will be reduced. (See Properties.)

Alternatively the materials should be stored in a heated building and only removed immediately before use.

Coating
When mixed, NITOBOND EP should be brush or spray applied to the prepared surface. NITOBOND EP is supplied in the Middle East with a special slow curing hardener, to allow for application prior to the fixing of reinforcement. The concrete or screed can then be placed up to 24 hours at 20°C, 8-10 hours at 35°C, after the application of NITOBOND EP slow set.

N.B. Care must be taken never to place concrete onto NITOBOND EP after it has dried.

Cleaning
Tools and equipment should be cleaned with FOSROC SOLVENT 102 immediately after use.

PRECAUTIONS

Health and Safety
Contact with the skin must be avoided. Gloves and barrier creams should be used when handling these products. If contact with the resin occurs, the skin should be washed immediately with soap and water– not solvent. A resin removing cream should preferably be used–e.g. Kerocleanse 22. Eye contamination must be immediately washed with

copious quantities of water and medical treatment sought. Working areas should be well ventilated.

Fire: NITOBOND EP is flammable. No naked flames allowed. Do not smoke during use.

Flash Point: FOSROC SOLVENT 102—33°C.

STORAGE

12 months minimum shelf life if stored below 25°C. If stored in high temperature and high humidity locations the shelf life may be reduced.

PACKAGING AND COVERAGE

1 kg and 5 kg Coverage 2.2–2.6 M^2/kg

These figures are for guidance only. Actual coverage will depend on the texture and porosity of the surface being covered.

FOSROC SOLVENT 102	5 litre cans
FOSROC CHEMICAL DEGREASER	30 litre carboys
FOSROC ACID ETCH	30 litre carboys

ADDITIONAL INFORMATION

NITOBOND EP forms part of the FOSROC CONCRETE REPAIR SYSTEM.

The FOSROC CONCRETE REPAIR SYSTEM ensures a successful, economical solution to your concrete problem.

NITOPRIME	Active anti-corrosion coating for steel
ZINCRICH	reinforcement.
NITOBOND AR	Acrylic emulsion for priming and curing FOSROC cementitious mortars.
RENDEROC TG	Cementitious repair mortar for vertical and overhead repairs.
RENDEROC S	Polymer modified high strength mortar for floor and structural repairs.
RENDEROC FC	Polymer modified re-facing mortar.
NITOCOTE	Impregnating, breathable anti-chloride
DEKGUARD	and anti-carbonation coating.

3. RENDEROC® S

Polymer modified structural repair mortar

USES

RENDEROC S is a high strength, structural repair mortar which can be used to reinstate all types of concrete. Polymer modification has enhanced initial grab and flexural and tensile strengths. The RENDEROC S formulation has similar thermal properties to concrete and can be used to effect large repairs. RENDEROC S forms an integral part of the FOSROC concrete repair system and will be of interest to specialist repair contractors, architects, engineers, local authority engineering departments, military authorities, docks, airport and road engineers.

ADVANTAGES

Simple to Use	— One pack material needs only the addition of water on side.
Low Permeability	— Optimum protection to steel reinforcement from chloride attack.
Superior Bond	— The RENDEROC S formulation plus

NITOBOND AR primer ensure intimate contact with parent concrete.

Non Shrink	— Will not crack even in large volume repair areas.
Locally Produced	— Formulated for use in Middle East conditions.

DESCRIPTION

RENDEROC S is supplied as a grey powder to which water is added on site. It contains carefully blended non shrink cements graded sands and chemical additives which provide a mortar with the superior qualities required for a durable structural repair.

TYPICAL PROPERTIES

Compressive Strength
Tested in accordance with BS 6319 at a water powder ratio of 0.11.

Age (Days)	Typical Compressive Strength in N/mm^2
1	25
3	45
7	50
14	53
28	53

Flexural Strength
9N/mm^2 at 28 days
SETTING TIME at 20°C in accordance with BS 5075.

Initial Set	3 hrs 15 mins
Final Set	4 hrs 30 mins

N B These times will vary with temperature change.

Water Permeability
Surface absorption ml/m^2/S (BS 1881 ISAT)

Time	RENDEROC S	4:1 Sand/Cement
10 mins	0.005	1.11
2 hrs	< 0.001	0.60

CO$_2$ Permeability (Tested at the University of Dortmund)

Thickness equivalent to 50 metre air gap (metres)	3:1 Sand/Cement	
	RENDEROC S	Mortar
	0.05	0.29

Coefficient of Thermal Expansion 9.0/10^{-6}/°C

Adhesion to Concrete
Slant/Shear bond (BS 6319) at 28 days -23 N/mm^2

Fresh Wet Density 2300 kg/m^2

Temperature Limitations
Do not attempt to apply RENDEROC S when ambient temperature is below 5°C and falling.

RENDEROC S has been tested in accordance with the appropriate sections of

BS 1881	BS 5075
BS 4550	ASTM C67
BS 4551	BS 6319

Chemical Resistance
Low permeability substantially slows chemical attack in aggressive environments. For specific data regarding chemical resistance please contact the technical department.

INSTRUCTIONS FOR USE

Before using any repair material FOSROC recommend that the reason for the failure be diagnosed by a suitably qualified authority.

SURFACE PREPARATION

Successful repairs to concrete depend totally on correct surface preparation. Particular care should be taken to ensure all surfaces are completely free of laitence, oil, dust, grease, plaster, paint, corrosion and any other deleterious substances. Laitance should be mechanically removed by high pressure water blasting, grit blasting or a combination of both. Oil and grease deposits should be removed by steam cleaning, detergent scrubbing or the use of a proprietary de-greaser such as FOSROC CHEMICAL DEGREASER (see separate data sheet). Smooth surfaces should be mechanically roughened by scabbling gun to form a good mechanical key. All reinforcing steel must be grit blasted to a profile of SA 2½ prior to the application of the recommended primer.

Edges of the area for repair must be saw cut to 5 mm to avoid feather edging.

Finally the complete area should be blown clean with oil free compressed air.

Priming

All surfaces must be primed prior to repair.

Reinforcing steel must be primer with NITOPRIME ZINCRICH an active zinc/epoxy system which will prevent further corrosion.

All concrete surfaces should be thoroughly soaked but leaving no surface water.

Concrete must be primed with NITOBOND AR NITOBOND AR is an acrylic emulsion which is scrubbed into the surface of the concrete. It will dry in approximately one hour (depending on temperature and substrate porosity) RENDEROC S should be applied when the NITOBOND AR is still tacky but may be applied up to 8 hours later. If the NITOBOND AR is too wet build up may be reduced due to slippage in vertical situations.

MIXING INSTRUCTIONS

Care should be taken to ensure that RENDEROC S is thoroughly mixed prior to use.

Small quantities of RENDEROC S (up to 10 kg) may be hand mixed in a bucket. It is essential that quantities of one bag or more are mixed in a forced action pan mixer or with a slow speed drill (500 rpm) fitted with a spiral paddle (Polyplan No 1707) Free fall mixers are not suitable. For normal applications 2 5-3 0 litres of water should be added.

This will yield approximately 12 litres of RENDEROC S.

APPLICATION

For application to all surfaces FOSROC recommend that RENDEROC S be applied with a wood float to ensure positive contact with the primed substrate. Initial application should be with a tamping or compacting action rather than a floating action. Finishing can be wood float steel trowel or sponge depending which type of finish is required. The wood float will leave a surface suitable for the application of RENDEROC FC. Depending on the nature and position of the repair it may be necessary to build up RENDEROC S in number of layers.

The minimum layer thickness of RENDEROC S is 5 mm and it can be built up to 20-25 mm on vertical surfaces. This can be achieved either by the "wet on wet" technique where each layer is fused to the previous one immediately or by allowing initial set to take place (3-4 hrs) and then applying the next layer onto the previously scored and primed surface.

If sagging occurs at any time due to excessive build-up of RENDEROC S, the material should be completely removed and reapplied in thinner layers.

Where a high build is required in an overhead or vertical situation the associated product RENDEROC HB should be considered.

Curing

RENDEROC S has been designed to self cure but in excessive drying conditions, high temperature or high wind, all repairs should be cured with NITOBOND AR.

Overcoating

RENDEROC FC (see separate data sheet) can be applied to a repair area approximately 3-4 hours after application of RENDEROC S.

Protective or decorative coatings can be applied after 7 days.

CLEANING

RENDEROC S should be removed from tools with clean water immediately after use. Cured material can only be removed mechanically.

PRECAUTIONS

Health and Safety

RENDEROC FC, NITOBOND AR and FOSROC CHEMICAL DEGREASER should not be allowed to come into contact with skin and eyes or be swallowed. Avoid inhalation of solvent vapours from FOSROC CHEMICAL DEGREASER, or dust when mixing RENDEROC FC.

The use of protective clothing and barrier creams, such as Kerodex Antisolvent and Rozalex antipaint is recommended. A dust mask should be used when handling RENDEROC FC powder. Ensure adequate ventilation when handling FOSROC CHEMICAL DEGREASER.

If contact with the skin occurs, wash thoroughly with copious amounts of clean water. If irritation persists, seek medical attention immediately. Do not induce vomiting.

Fire: RENDEROC S is non-flammable.

FOSROC CHEMICAL DEGREASER is flammable.

Flashpoint: 48°C

STORAGE

RENDEROC S has a storage life of 12 months if maintained in dry storage conditions in unopened bags. This may be reduced to 4-6 months in locations of high humidity.

NITOBOND AR has a shelf life of 12 months at 20°C. Protect from frost.

PACKAGING

RENDEROC S is supplied in 25 kg polythene bags with an inner wire tied polythene bag.

NITOBOND AR is supplied in 5 and 25 litre containers.

COVERAGE

One bag of RENDEROC S yields approximately 12 litres of mortar depending on water addition.

This will cover 1.2m^2 at 10 mm thickness.

NITOBOND AR covers 6-8 m^2/ltr depending on the porosity and texture of the substrate.

ADDITIONAL INFORMATION

RENDEROC S and NITOBOND AR form part of the FOSROC CONCRETE REPAIR SYSTEM.

The FOSROC CONCRETE REPAIR SYSTEM ensures a successful, economical solution to *your* concrete problem.

NITOPRIME	Active anti corrosion coating for steel
ZINCRICH	Reinforcement.
NITOBOND AR	Acrylic emulsion for priming and curing FOSROC cementitious mortars.
RENDEROC HB	Cementitious repair mortar for vertical and overhead repairs.
RENDEROC S	Polymer modified high strength mortar for floor and structural repairs.
RENDEROC FC	Polymer modified re-facing mortar.
NITOCOTE	Impregnating breathable anti-chloride
DEKGUARD	and anti-carbonation coating.

4. RENDEROC® TG

High build cementitious repair mortar.
(light weight)

USES

RENDEROC TG (Tropical Grade) has been specially formulated to effect permanent repairs to all types of concrete and masonry. Its lightweight components aid easy application to vertical and overhead repair without the use of formwork. It forms an integral part of the FOSROC concrete repair system and is of particular interest to specialist repair contractors, local authority maintenance division, military agencies, airport, road and port authorities.

ADVANTAGES

Easy to Use	— Needs only the addition of water on site. No errors as in some multi component systems.
Does not Crack	— Large areas can be completed without shrinkage.
No Costly Formwork	— Application can be 30 mm vertical and 20 mm soffit. Pockets up to 100 mm can be filled in one application.
Superior Bond	— Careful formulation and NITOBOND AR primer ensure intimate contact with substrate.
Low Permeability	— Special additives ensure optimum density and water repellancy.
Locally Produced	— Tropical Grade formulated for use in Middle East conditions.

DESCRIPTION

RENDEROC TG is supplied as a powder to which water is simply added on site. It contains special pre-blended non shrink cements, graded sands, fillers and chemical additives which provide a product that has 90% less absorption than conventional mortar and has excellent thermal compatibility with concrete. RENDEROC TG will tolerate wide variations in water content with little effect on ultimate strength. Special fillers allow build of RENDEROC TG on vertical and overhead surfaces. For floor areas subject to high abrasion, the associated products RENDEROC S, NITOFLOR PAVEROC and NITOMORTAR S should be considered.

STANDARDS

Tested in accordance with BS 1881, BS 4550, BS 4551, BS 5075, BS 6319 and ASTM C67.

RENDEROC TG will tolerate wide variations in water content.

Flexural strength: Tested in accordance with BS 4551. 7 N/mm^2 @ 28 days.

Slant Shear Bond strength to concrete: Tested in accordance with BS 6319 Part 4: 1984 23.0 N/mm^2 @ 28 days.

Density: 1750 kg/m^3 at 0.18 water powder ratio.

Water permeability: Surface absorption ml/m^2/second. (BS 1881 I.S.A.T.)

Time	RENDEROC TG	4:1 sand:cement Mortar
10 minutes	0.18	1.11
2 hours	0.06	0.60

Coefficient of thermal expansion: 7.3 × 10^{-6}/C.

Setting time: Dependent on temperature and water content typically 2 to 4 hours.

Temperature limitation: Should not be applied below 5°C.

Chemical resistance: Substantially reduced absorption slows down chemical attack in aggressive environments. For specific data regarding chemical resistance please contact the technical department.

SURFACE PREPARATION

Successful repairs to concrete depend totally on correct surface preparation. Particular care should be taken to ensure all surfaces are completely free of laitence, oil, dust, grease, plaster, paint, corrosion and any other deleterious substances. Laitance should be mechanically removed by high pressure water blasting, grit blasting or a combination of both. Oil and grease deposits should be removed by steam cleaning, detergent scrubbing or the use of a proprietary de-greaser such as FOSROC CHEMICAL DEGREASER (see separate data sheet). Smooth surfaces should be mechanically roughened by scabbling or needle gun to form a good mechanical key. All reinforcing steel must be grit blasted to a profile of SA 2½ removing all corrosion cells, prior to application of NITOPRIME ZINCRICH.

Edges of the areas for repair must be saw cut to 10 mm minimum depth to avoid feather edging, which would lead to desiccation and poor durability.

Finally the complete area should be blown clean with oil free compressed air, or clean water jet.

Mixing
Care should be taken to ensure that RENDEROC TG is thoroughly mixed prior to use.

Small quantities up to 10 kg may be mixed manually in a bucket, not on a spot board. The waterproofing additives within RENDEROC TG induce a degree of water resistance in the unmixed powder and because of this it is advisable to add powder to water. Vigorous agitation with a trowel or spatula will overcome this resistance and a smooth mortar can be produced with two minutes hand mixing. Further powder or water may be added to the mortar to produce the required consistency.

Quantities in excess of 10 kg should be mixed mechanically with a forced action pan mixer. Free fall mortar mixers may not be suitable. A slow speed drill (500 rpm) with a spiral paddle is an acceptable alternative. For normal applications use from 4 to a maximum of 5 litres (7-9 pints) of water per 25 kg bag, dependent on desired consistency This represents from 3-4 volumes of powder to one volume of water.

N.B. For high build situations it is recommended that the water addition should be at the lower of the recommended dosages.

Priming

All surface must be primed prior to repair.

Reinforcing steel must be primed with NITOPRIME ZINCRICH an active zinc/epoxy system which will prevent further corrosion.

All concrete surfaces should be thoroughly soaked but leaving no surface water.

Concrete must be primed with NITOBOND AR. NITOBOND AR is an acrylic emulsion which is scrubbed into the surface of the concrete. It will dry in approximately one hour (depending on temperature and substrate porosity). RENDEROC TG should be applied when the NITOBOND AR is still tacky but may be applied up to 8 hours later. If the NITOBOND AR is too wet, build up may be reduced due to slippage.

Application

RENDEROC TG can be applied as a render at least 30 mm thick on to vertical substrates and up to 20 mm thick onto horizontal soffits without formwork supports. Greater depth can be achieved in one application with the assistance of temporary formwork. High build is achieved by applying wet on wet layers up to 10 mm thick to ensure maximum compaction and substrate adhesion. RENDEROC TG should not be applied in thicknesses less than 10 mm or widths less than 50 mm.

The application thickness achievable is dependent on substrate profile. If sagging occurs due to the applied layer being too thick, the material should be completely removed and reapplied at a reduced thickness. This reduced layer should be surface scored, cured and allowed to take up its initial set prior to priming and application of subsequent layers to achieve the desired profile. When building up RENDEROC TG in successive layers NITOBOND AR should be painted or sprayed over each layer immediately after application and scoring, in order to prevent desiccation. The next layer of RENDEROC should then be applied as soon as the previous layer has reached initial set.

RENDEROC TG should not be applied to substrates which are exposed to moving water or which are permanently wet. Exposure to heavy rainfall prior to final set will result in surface scour.

Application should not be carried out when ambient temperature is below 5°C and falling.

Curing

RENDEROC TG requires curing in accordance with good concrete practice. NITOBOND AR is recommended both as an intercoat and final cure particularly where a surface finish is to be applied.

OVERCOATING

After a suitable curing/drying period has elapsed, 7-14 days dependent on ambient conditions, completed repairs can be overcoated with a suitable protective/decorative finish. NITOCOTE DEKGUARD can be used to provide a uniform finish and protection against further carbonation and chloride attack.

CLEANING

RENDEROC TG should be removed from tools with clean water immediately after use. Cured material can only be removed mechanically.

PRECAUTIONS

Health and Safety

RENDEROC TG NITOBOND AR and FOSROC CHEMICAL DEGREASER should not be allowed to come into contact with skin and eyes or be swallowed. Avoid inhalation of solvent vapours from FOSROC CHEMICAL DEGREASER or dust when mixing RENDEROC TG.

The use of protective clothing and barrier creams, such as Kerodex Antisolvent and Rozalex Antipaint is recommended. A dust mask should be used when handling RENDEROC TG powder. Ensure adequate ventelation when handling FOSROC CHEMICAL DEGREASER.

If contact with the shin occurs, wash thoroughly with copious amounts of clean water. If irritation persists, seek medical attention immediately. Do not induce vomiting.

Fire: RENDEROC TG is nonflammable.

FOSROC CHEMICAL DEGREASER is flammable. Flash point 48°C.

STORAGE

RENDEROC TG has a storage life of 12 months if maintained in dry storage conditions in unopened bags. This may be reduced to 4-6 months in locations of high humidity.

NITOBOND AR has a shelf life of 12 months at 20°C. Protect from frost.

PACKAGING

RENDEROC TG is supplied in 25 kg polythene bags.

NITOBOND AR is supplied in 5 and 25 litre containers.

COVERAGE

One bag of RENDEROC TG yields approximately 16 litres of mortar depending on water addition.

This will cover 1.6 m^2 at 10 mm thickness.

NITOBOND AR covers 6-8 m^2/litre depending on the porosity and texture of the substrate.

---

5. RENDEROC® LA

High performance shrinkage compensated repair concrete
(pre-packaged, free-flowing repair micro-concrete)

USES

For the reinstatement of damaged, defective, honeycombed or deteriorated concrete by partial or total replacement. The material is particularly suitable where steel reinforcing is heavily congested or access is difficult.

ADVANTAGES

Dual expansion	— Unique system compensates for shrinkage in both the plastic and hardened states.
Versatility	— Simple to use product suitable for both small and large voids. Can be placed by pumping or mix and pour techniques.
Self-Compacting	— Eliminates honeycombing in congested sections.
Excellent Bond	— Fluid system ensures full contact with the concrete. Shrinkage compensation reduces bond line stresses.

Reliability — Factory controlled prepackaged product eliminates site batching and ensures consistent physical characteristics.

Durability — High ultimate strength and low permeability ensure long term service.

Low Alkali Content — Minimizes risk of inducing alkali silica reaction.

Priming

Reinforcing steel must be primed immediately after cleaning with NITOPRIME ZINCRICH (see separate data sheet).

Several hours prior to placing the RENDEROC LA, the prepared concrete substrate should be flooded with clean water. Immediately prior to placing, this water should be completely removed and the drainage outlets sealed.

Providing the substrate has been thoroughly soaked, further priming is not normally required. When an epoxy primer is required NITOBOND EP should be used and the presoaking stage omitted. See separate data sheet for details.

Mixing

A mechanically powered mixer must be used. It is essential that the mixing capacity of the machine and the availability of labour is adequate to enable the placing operation to be carried out continuously. The water content is important and should be accurately gauged for each mix. The flow characteristics may be varied to meet particular site requirements by adjustment of the water content.

All water used should be of drinking quality and uncontaminated. Do not exceed 4 litres per 30 kg bag.

The measured water should always be placed into the mixer first followed by the slow addition of RENDEROC LA. Only full packs should be used. Mix continuously for 3-5 minutes, making sure that a smooth even consistency is obtained.

It is recommended that the mixed material be passed from the mixing vessel through a suitable coarse metal screen prior to placing or pumping to confirm the powder is completely dispersed.

Placing

RENDEROC LA should be placed within 25 minutes of mixing to gain the full benefit of the plastic expansion process. Each repair should be poured or pumped in a single, continuous operation. If a break is inevitable, it should not exceed 20 minutes. Poured material should be introduced slowly into the prepared formwork to prevent the entrapment of air. Material should be introduced into the bottom of the void through a suitable flexible feed-pipe or entry port.

If placing by pump, then standard concrete pumping practise should be followed. The pump and pipeline must be primed with a rich cement slurry or mortar, discharging the excess grout to waste. Pumping should be started immediately after priming. Keep the feed hopper full to avoid pumping air into the repair.

Curing

The formwork should be left in place until the compressive strength of the RENDEROC LA has achieved $10 \, \text{N/mm}^2$ or the strength specified by the Supervising Officer. If, in view of the size and geometry of the repair, a large exotherm is anticipated, or if prevailing climatic conditions are such that large temperature gradients could occur, the formwork should be thermally insulated to minimise the risk of thermal cracking. In all cases, the repair should be left to cool to less than 10 deg. C above the ambient temperature before formwork is removed.

Immediately after striking the formwork, all exposed faces should be soaked with clean water and then sprayed with a liquid-applied curing membrane such as CONCURE, or cured in accordance with normal concreting practice.

If a surface coating is to be applied, contact FOSROC for advice on appropriate curing methods.

CLEANING

RENDEROC LA should be removed from tools and equipment immediately after use, with clean water. Pumping equipment should be flushed through with clean water.

PRECAUTIONS

Health and Safety

RENDEROC LA is alkaline and should not come in contact with skin or eyes. Avoid inhalation of dust.

Gloves, goggles and dust masks should be worn. If contact with skin occurs wash with water.

Splashes to eyes should be washed immediately with plenty of clean water and medical advice sought.

Fire

RENDEROC LA is non-flammable.

STORAGE

RENDEROC LA has a shelf life of 12 months if kept in a dry store in sealed bags. If stored in high temperature and high humidity locations the shelf life may be reduced

PACKAGING

RENDEROC LA is supplied in 30 kg sealed plastic bags.

YIELD

The approximate yield per 30 kg bag of RENDEROC LA is 14.5 litres when mixed with 4 litres of water. Allowance should be made for wastage when estimating.

---

6. CONBEXTRA HF

High performance non-shrink cementitious grout and Concrete bulk repair compound
(free-flowing repair micro-concrete)

USES

For free flow grouting over a wide range of application consistencies. Applications include heavy duty support beneath machine base plates, crane rails, stanchion bases and bearing plinths. Also performs as an excellent anchoring grout for masts and holding-down bolts.

For repairs to damaged concrete elements where a highly plasticised replacement material is required. Particularly useful for honeycombed areas, where access is restricted and where vibration of the placed material is difficult or impossible.

ADVANTAGES

Dual Expansion — Unique system compensates for shrinkage in both the plastic and hardened states.

Versatility — Suitable for pumping or pouring over a large range of application consistencies and temperatures.

High Fluidity — Allows high-strength concrete repairs in difficult situations without vibration.

Reliability	— Site batching variations eliminated by the use of factory controlled pre-packaged material.
High Early Strength	— Facilitates rapid installation and early operation of plant.
Durability	— High ultimate strength and low permeability ensure long term service.
Non-Metallic	— Non-metallic expansion system eliminates staining and degradation due to corrosion products.
Hydrogen-Free	— The expansion system does not utilise hydrogen generation.
Chloride-Free	— Composition allows high early strength development without the use of chlorides.
Locally Produced	— Formulated for use in Middle East conditions.

DESCRIPTION

CONBEXTRA HF is supplied as a ready to use dry powder requiring only the addition of water to produce a free flowing non-shrink grout for gap widths of 10-125 mm.

The material is a blend of Portland cements, pre-graded fillers and additives which impart controlled expansion in both the plastic and hardened states whilst minimising water demand. The low water requirement ensure high early strength and long term durability. The filler grading is designed to aid uniform mixing and minimise segregation and bleeding over a wide range of application consistencies.

STANDARDS

CONBEXTRA HF conforms fully to U.S. Corps of Engineers specification for non-shrink grout: CRD-C621-82A which supercedes CRD-C-588-78. Independent test laboratory reports documenting compliance are available on written request.

CONBEXTRA HF is tested using the appropriate sections of the following specifications.

ASTM C-827-78	BS 1881
ASTM C-109-77	BS 4550
ASTM C-191-79	BS 4551
ASTM C-230-67	BS 5075 Part 2

CONBEXTRA HF specified for nuclear plant work is made and tested in accordance with ANSI/ASME N45 Quality Assurance Program Requirements for Nuclear Facilities.

SPECIFICATION CLAUSES

Performance specification: All grouting and/or repair work (specify details and areas of application) must be carried out with a prepackaged cement based product which shall be mixed with water on site to the required consistency. The grout must not bleed or segregate, must be non-metalic and chloride-free. Positive expansion shall occur while the grout is plastic and the grout must also be compensated for shrinkage in the hardened state. The compressive strength of the grout must exceed 30 N/mm^2 at 7 days and 50 N/mm^2 at 28 days, the grout shall fully conform to the requirements of US Army Corps of Engineers specification for non-shrink grout: CRD-C621, and must be stored, handled and placed strictly in accordance with the manufacturer's instructions.

Supplier specification: All grouting and/or repair work (specify details and areas of application) must be carried out using CONBEXTRA HF manufactured by the Fosroc Group of Companies, applied strictly in accordance with the manufacturer's technical data sheet.

TYPICAL PROPERTIES

Compressive strength: Tested in accordance with BS 1881, Part 4, 1970 and curing cubes under restraint. Variation with consistency at 20°C.

Age (days)	Compressive strength N/mm2			
	Trowellable	Plastic	Flowable	Fluid
1	30	24	20	14
7	60	50	44	34
14	70	60	56	46
28	76	66	64	58
180	90	84	82	70

Variation with temperature at flowable consistency.

Age (days)	Compressive strength N/mm2		
	5°C	20°C	35°C
1	4	20	30
7	30	44	46
14	50	56	56
28	64	64	60
180	84	82	70

Flexural strength: Tested in accordance with BS 4551 1980, for flowable consistency at 20°C.

Age (days)	Flexural strength N/mm^2
1	2.5
7	8
14	9.5
28	10
180	11

Coefficient of thermal expansion: 11 × .6 per °C.

Thermal conductivity: 1 ½ w/m °C.

Young's modulus: 29 kN/mm^2.

Freeze-thaw stability: Meets the requirements of BS 5075, Part 2, 1982.

Ultimate anchorage bond stress: Exceeds CP100, Part 1, 1972, Table 22 requirements for 40 N/mm^2 concrete at 24 hours.

Setting time: Variation with temperature, Tested in accordance with BS 4550: Part 3, 1978 for flowable.

Temperature °C	Initial set (hours)	Final set (hours)
5	9	11
20	5½	7½
35	3	4½

Fresh wet density: The fresh wet density will vary between 2100 kg/m^3 dependent on consistency.

Expansion Characteristics: Positive expansion when measured to ASTM C827 overcomes plastic settlement in the unset material. Longer term expansion in the hardened state is designed to comply with the requirements of CRD 621-82A to compensate for drying shrinkage.

Time for expansion:

	Start ..	Finish
Plastic state	15 mins	Initial set
Hardened state	Initial set	3 days

Temperature above 20°C may slightly reduce these times.

Pressure to restrain plastic expansion:

Approximately 0.004 N/mm^2.

NOTE: The Fosroc technical department should be consulated if more detailed information on the test methods and physical properties are required.

INSTRUCTIONS FOR USE

Preparation

Underplate grouting: Reference should be made to the design table (Figure 2) when determining the grout consistency and the fluid head necessary to achieve adequate flow, based on the gap width and flow distance required. Use the lowest practical fluidity for maximum strength development and to enable the grout to flow through the gap as a continuous front.

The unrestrained surface area of the grout must be kept to a minimum. Generally the gap width between the perimeter formwork and the plate edge should not exceed 150 mm on the pouring side and 50 mm to the opposite side.

The formwork should be constructed to be leakproof as CONBEXTRA HF is a free flowing grout. This can be achieved by using foam rubber strip or mastic sealant beneath the constructed formwork and between joints.

In some cases it is practical to use a sacrificial sami-dry sand and cement formwork. The formwork should include outlets for the pre-soaking water.

Figure 2		Max. flow	distance	in mm
Grout consistency	Gap width mm	540 mm head	100 mm head	250 mm head
Trowellable	40	100	250	600
	50	150	400	1100
Plastic	10	100	200	450
	20	200	500	250
	30	350	900	2600
	40	600	1500	3000
	50	900	2400	3000+
Flowable	10	180	360	1200
	20	400	950	2600
	30	700	1500	300
	40	1100	2200	3000+
	50	1400	3000	3000
Fluid	10	350	900	2500
	20	750	1900	3000
	30	1250	3000	3000+
	40	2000	3000+	3000+

N.B. This table is based on the following factors.

— Temperature 20°C.
— Minimum gap width 10 mm
— Water saturated substrate
— Minimum unrestricted flow width 300 mm

Foundation surface: This must be free from oil, grease or any loosely adherent material. If the concrete surface is detective or has laitence, it must be cut back to a sound base. Bolt holes or fixing pockets must be blown clean of any dirt or debris.

Base plate: It is essential that this is clean and free from oil, grease or scale. Air pressure relief holes should be provided to allow venting of any isolated high spots. FOSROC CHEMICAL DEGREASER is recommended for degreasing.

Levelling shims: If these are to be removed after the grout has hardened, they should be treated with a thin layer of grease.

Pre-soaking: Several hours prior to grouting, the area of cleaned foundation should be flooded with fresh water. Immediately prior to grouting any free water should be removed, particular care should be taken to blow out all bolt holes and pockets.

Mixing

Select the water content required from the table below.

Consistency * on mixing	Approx. water at 20°C (litres/25 kg bag)
Trowellable	3.5
Plastic	4.2
Flowable	4.5
Fluid	4.8

* Trowellable is the consistency at which the grout will just pour and can be trowelled to a 20°C slope without slump. The other consistencies are as defined in CRD C621-82A.

This table is intended for guidance only, and if strict compliance with CRD-C621-82 is required, control testing in accordance with the specification is recommended.

For best results a mechanically powered grout mixer must be used, for quantities up to 50 kg a slow speed drill fitted with a high shear paddle is suitable. Larger quantities will require a high shear vane mixer.

Do not use a colloidal impeller mixer.

It is essential that machine mixing capacity and labour availability is adequate to enable the grouting operation to be carried out continuously. This may require the use of a holding tank with provision for gentle agitation to maintain fluidity.

The selected water content should be accurately measured into the mixer. Slowly add the total contents of the CONBEXTRA HF pack, mix continuously for 5 minutes, making sure that a smooth, even consistency is obtained. Pass. the mixed grout through a 5 mm sieve to remove any lumps prior to placing.

Placing

Place the grout within 45 minutes of mixing to gain the full benefit of the expansion process.

CONBEXTRA HF may be placed in thickness up to 125 mm in a single pour when used as a underplate grout. For thicker sections it is necessary to fill out CONBEXTRA HF with well graded silt free aggregates to minimise temperature rise. (See specific applications.)

Continuous grout flow is essential. Sufficient grout must be available prior to starting and the time taken to pour a batch must be regulated to the time taken to prepare the next one. Pouring should be from one side of the void to eliminate the entrapment of air or surplus pre-soaking water. The grout head must be maintained at all times so that a continuous grout front is achieved.

Where large volumes have to be placed CONBEXTRA HF may be pumped. A heavy duty diaphragm pump is recommended for this purpose. Screw feed and piston pumps may also the suitable.

Curing

On completion of the grouting operation, exposed areas which are not to be cut back should be thoroughly cured by means of water application. CONCURE * curing membrane or wet hessian:

Cleaning

CONBEXTRA HF should be removed from tools and equipment immediately after use with clean water. Cured material can be removed mechanically, or with FOSROC ACID ETCH.

SPECIFIC APPLICATIONS

Grouting under base plates: Use the least fluid mix consistent with figure 2. This is usually flowable or fluid consistency.

Grouting anchor bolts: Use plastic or trowellable consistency if anchor bolts are to be grouted as a separate operation.

Grouting large volumes: For sections thicker than 125 mm, it is necessary to fill out the CONBEXTRA HF with graded 10 mm silt free aggregate to minimise temperature rise. (See mix design below.)

Concrete repair: All loose, damaged or spalling material should be removed. Where necessary, reinforcing steel should be properly cleaned and treated with NITOPRIME ZINCRICH. The concrete substrate should be pre-soaked's described above or should be primed using NITOBOND EP. The formwork should be constructed to be leakproof and provision should be made for suitable access at the highest point of the repair area. Where necessary, provision should be made for air-release holes.

For sections deeper than 125 mm it may be necessary to fill out the CONBEXTRA HF with graded 8-12 mm silt free aggregate should be tested to ASTM C-289-81). If any doubt exists. Fosroc Technical Department should be consulted.

The quantity of aggregate required will vary depending on the nature of the repair area. Generally, for sections 125-250 mm deep, where high fluidity is required, the following mix design should be considered:

25.0 kg CONBEXTRA HF

4.3 kg clean water

10.0 kg 8-12 mm aggregate

NOTE: Water damaged may vary depending on the conditions of the aggregate.
For sections deeper than 250 mm, the mix design should be modified. However, the quantity of aggregate should never exceed 1 part aggregate to 1 part CONBEXTRA HF (by dry weight). Please contact Fosroc Technical Department for details.
For such mixes, CONBEXTRA HF and aggregate should be dry-blended in a conventional concrete mixer before adding water, the repair material may be poured or pumped into the prepared formwork.

Low temperature working: In cold conditions down to 2°C, warm water (30-40°C) is recommended to accelerate strength development.

At ambient temperatures below 10°C the grout consistency should not exceed flowable and the formwork should be maintained in place for at least 36 hours. Normal precautions for winter working with cementitious materials should then be adopted.

High temperature working: At ambient temperature above 35°C, the grout should be stored in the shade and cool water used for mixing.

PACKAGING

CONBEXTRA HF is supplied in 25 kg plastic bags.

YIELD

This depends on the required consistency. The approximate yield per 25 kg bag at each consistency is as follow.

Consistency	Trowellable	Plastic	Flowable	Fluid
Yield (litres)	12½	13	13¼	13½

PRECAUTIONS

Health and Safety
CONBEXTRA HF is non-toxic, but it is alkaline in nature. Gloves should be worn. Splashes of grout to the skin or eyes should be washed off with clean water. In the event of prolonged irritation, seek medical advice.

Fire: CONBEXTRA HF in non-flammable.

STORAGE

CONBEXTRA HF has a shelf life of 12 months kept in a dry store in sealed bags. If stored in high temperature and high humidity locations the shelf life may be reduced.

7. NITOFLOR PATCHROC®

Fast setting concrete patching compound.

USES

NITOFLOR PATCHROC has been specially designed to provide a very fast set repair in concrete floors and pavements and is particularly useful where traffic must be kept flowing. NITOFLOR PATCHROC will accept pedestrian and vehicular traffic in 1 hour.

NITOFLOR PARCHROC will be of particular interest to industrial maintenance engineers, transport managers, airport authorities or anywhere that potholes in concrete cause a hazard to pedestrian and wheeled traffic.

ADVANTAGES

Fast setting	— Minimum disruption to traffic.
Easy to use	— Needs only the addition of water.
Easy to lay	— Virtually self compacts.
Non shrink	— Does not crack.
High adhesion	— Easy to use NITOBOND AR primer ensures no debonding.
Durable	— Tough surface-resistant to wear and weather.
Locally made	— Specially formulated for use in Middle East conditions.

DESCRIPTION

NITOFLOR PATCHROC is supplied as a preblended dry powder to which water is added to produce an easily trowelled mortar. NITOFLOR PATCHROC is a blend of inorganic cements, modifying additives to control the rate of set and strength gain plus selected hard wearing aggregates. NITOFLOR PATCHROC does not contain calcium chloride.

The blend of cements eliminates shrinkage with the consequent reduction of stress at the bond line. Long lasting adhesion is ensured by the use of NITOBOND AR primer.

NITOFLOR PATCHROC has been designed for the emergency patching of small areas of concrete the associated product NITOFLOR PAVEROC should be used.

PROPERTIES

	10°C	20°C
Working Life	20 mins	10 mins
Setting Time	40 mins	20 mins
Traffic Time Vehicular Pedestrian	2 hours 2 hours	1 hour 1 hour
Compressive Strength (N/mm^2) at 20°C		

at 1 hour	1 day	3 days	28 days
20	60	65	70

INSTRUCTIONS FOR USE

Before using any repair material FOSROC recommend that the reason for the failure be diagnosed by a suitably qualified authority.

Full instructions can be found in the FOSROC publication 'APPLICATION GUIDELINES FOR THE USE OF FOSROC CONCRETE REPAIR SYSTEMS.'

Surface Preparation

Successful repair to concrete depend totally on correct surface preparation. Particular care should be taken to *ensure all surfaces are completely free of laitance, oil, dust, grease, plaster, paint, corrosion and any other* deleterious substances. Laitance should be mechanically removed by high pressure water blasting, grit blasting or a combination of both. Oil and grease deposits should be removed by steam cleaning, detergent scrubbing or the use of a proprietary product such as FOSROC CHEMICAL DEGREASER (see separate data sheet). Smooth surfaces should be mechanically *roughened by scabbling or needle gun to form a good key.* All reinforcing steel must be grit blasted to a profile of SA2½ prior to the application of the recommended primer.

Edges of the areas for repair must be saw cut to 12 mm to avoid feather edging

Finally the complete area should be blown clean with oil free compressed air

NITOFLOR PATCHROC may be used from a minimum of 12 mm thick to a maximum of 50 mm thick. If the depth of the pothole is greater than 50 mm individual layers should be scratch keyed and allowed to set for a minimum of 1 hour before priming and placing the next layer.

Priming

All surfaces must be primed prior to repair. Before priming the substrate should be thoroughly soaked with clean water and any excess removed from the surface preferably by oil free compressed air. Concrete must be primed with NITOBOND AR, NITOBOND AR is an acrylic emulsion which is scrubbed into the surface of the concrete. It will dry in approximately one hour (depending on temperature and substrate porosity) NITOFLOR PATCHROC should be applied to the primed surface as soon as possible after mixing it ideally with the primer having been allowed to become tacky.

Mixing

NITOFLOR PATCHROC must be mixed in a forced action mixer or alternatively with a slow speed drill (500 rpm) fitted with a spiral paddle (Polyplan No: 1707). Free fall mixers must not be used

The mixing ratio of NITOFLOR PATCHROC is 25 kg of powder to 2 5/2 75 litres of water.

The recommended amount of water should be put into the mixer and the contents of the NITOFLOR PATCHROC bag added. The mixing times should be 5 minutes to allow the special plasticisers to disperse. Do not mix for longer than 5 minutes, or delay in placig the mixed material otherwise the available working life will be reduced in general it is recommended that only 1 bag be mixed at any one time to ensure that NITOFLOR PATCHROC is placed within its working life.

APPLICATION

Place the mixed material evenly over the primed surface and tamp in place with a wood float to ensure full compaction. Strike off the surface to the correct level and finish with a steel trowel to fully close the surface.

If a textured finish is required this can be done with a roller or sweeping brush.

CURING

NITOFLOR PATCHROC requires curing in accordance with good concrete practice

As soon as trowelling has been completed apply either NITOBOND AR or CONCURE 90 curing membranes.

In winter protect the surface of the NITOFLOR PATCHROC from frost by the use of insulated covers.

Protect the surface from heavy rainfall to prevent surface scour.

CLEANING

NITOFLOR PATCHROC should be removed from tools with clean water immediately after use. Cured material can only be removed mechanically.

PRECAUTIONS

Health and Safety

NITOFLOR PATCHROC, NITOBOND AR and FOSROC CHEMICAL DEGREASER should not be allowed to come into contact with skin and eyes or be swallowed. Avoid inhalation of solvent vapours from FOSROC CHEMICAL DEGREASER, or dust when mixing NITOFLOR PATCHROC.

The use of protective clothing and barrier creams, such as Kerodex Antisolvent and Rozalex antipaint is recommended. A dust mask should be used when handling NITOFLOR PATCHROC powder. Ensure adequate ventilation when handling FOSROC CHEMICAL DEGREASER.

If contact with the skin occurs, wash thoroughly with copious amounts of clean water. If irritation persists, seek medical attention immediately. Do not induce vomiting.

Fire: NITOFLOR PATCHROC and NITOBOND AR are non flammable. FOSROC CHEMICAL DEGREASER is flammable, Flash point 48°C.

STORAGE

NITOFLOR PATCHROC has a shelf life of 12 months in unopened bags kept in a dry store. In high humidity locations the shelf life may be reduced to 4-6 months.

NITOBOND AR has a shelf life of 12 months in unopened containers but must be protected from frost.

PACKAGING

NITOFLOR PATCHROC is packed in 25 kg bags.

NITOBOND AR is packed in 25 litre drums.

CONCURE 90 is packed in 210 litre drums.

FOSROC CHEMICAL DEGREASER is packed in 30 litre carboys.

COVERAGE

25 kg of NITOFLOR PATCHROC yields approximately 12 litres of mortar which will cover $1m^2$ at 12 mm thickness.

NITOBOND AR covers 6-8 m^2/litre depending on porosity and texture of substrate.

8. NITOMORTAR® HB

High build epoxy mortar
(structural epoxy mortar)

USES

Fast and permanent repairs to spalled concrete structures particularly in vertical or overhead situations. Emergency repairs of concrete column, roof and beam faces to effect a profiled finish and provide protection to reinforcing steel.

ADVANTAGES

Versatile	— May be used for high build layers in vertical and overhead situations.
Speed	— Early development of strength minimises maintenance disruption.
Quality	— Pre-weighed quality controlled materials ensure consistency and reduce risk of site errors.
Chemically resistant	— Unaffected by a wide range of acids, alkalis and industrial chemicals.
Strong and permanent	— Typically twice as strong as good concrete. Resistant to abrasion and impact.
Waterproof	— Will cure under damp conditions. Cured surface is impermeable to water.
Appearance	— Natural colour sympathetic to aesthetic requirements.

DESCRIPTION

NITOMORTAR HB is based on a high quality solvent-free epoxy resin system. The filler is specifically designed to give excellent "hanging" properties for vertical or overhead work. NITOMORTAR HB, when applied to a prepared and primer surface, cures to give an impermeable render unaffected by most forms of chemical attack.

The product is supplied as a three-pack material in pre-weighed quantities ready for on-side mixing and use.

PROPERTIES

Pot Life: 45 minutes at 20°C or 20 minutes at 35°C.

Initial Hardness: 8 hours at 35°C.

Full Cure: 4 days at 35°C.

Below 20°C the curing time will be increased.

Minimum application temperature: 5°C.

Chemical Resistance: Performance of NITOMORTAR HB blocks continually immersed at 20°C

Citric Acid 10%	Excellent
Tartaric acid 10%	Excellent
Acetic acid 5%	Satisfactory
Nitric acid 10%	Good
Hydrochloric acid 25%	Excellent
Sulphuric acid 10%	Very good
Sodium hydroxide 50%	Excellent
Diesel fuel/petrol	Excellent
Sugar solutions	Very good
Lactic acid 10%	Very good
Hydrocarbons	Very good
Phosphoric acid 10%	Very good

Mechanical Characteristics:

	Test method	NITOMORTAR HB N/mm^2	Typical concrete N/mm^2
Compressive strength	BS 6319: Part 2: 1983	45	20
Flexural strength	BS 6319: Part 3: 1983	20	7
Tensile strength	BS 6319: Part 7: 1983	7.5	3.5
Water absorption	No test method	0.45%	5.0%

INSTRUCTIONS FOR USE

Preparation
All grease, oil, chemical contamination, dust, laitance and loose concrete must be removed by scabbling or light bush hammering to provide a sound substrate. Alternatively the surface should be etched with FOSROC ACID ETCH. After the reaction has ceased, the surface should be thoroughly washed with clean water and allowed to dry.

Priming
NITOPRIME 28 should be used to prime surfaces before all applications of NITOMORTAR HB.

Mix the primer in the proportions supplied, adding the entire contents of the hardener tin to the base tin. Once mixed this should be brushed well into the prepared concrete. After about 30 minutes in 20°C, or 20 minutes at 35°C. If the concrete has absorbed the resin a second primer coat should be applied.

NITOMORTAR HB should be placed onto the primed surface between 30 minutes and 4 hours at 20°C or between 20 minutes and 2 hours at 35°C, after application of the primer. The optimum time is the point at which the primer has started to gel but still has surface 'tack'. The usable life of the mixed primer is approximately 60 minutes at 20°C or 40 minutes at 35°C.

Mixing
The pre-weighed quantities of the NITOMORTAR HB components are mixed to produce a trowellable mortar. For minimal wastage, two pack sizes are available to allow full utilisation of the material within pot life of 45 minutes.

Standard Pack: Mixed by hand. Contents of the base and hardener containers should be emptied into the plastic bucket and mixed thoroughly. The aggregate should then be added and the three components blended together, ensuring that the aggregate is thoroughly wetted out with resin.

Industrial Pack: Mechanical mixing is necessary. The base and hardener components should be thoroughly mixed in the base container and then emptied into a paddle mixer. FOSROC technical department can advise on suitable types. Add the aggregate slowly with the mixer running and continue for 2 to 3 minutes until all the components are thoroughly blended.

Trowelling
The mixed material should be applied to the primed surface with a wood float pressingly firmly into place to ensure positive adhesion. The surface may be closed using a steel trowel. Application thicknesses up to 50 mm per layer may be used. When larger areas are being rendered-chequer board application is recommended. For small repairs maximum thickness may be increased to 75 mm. If subsequent layers need to be applied they should follow between 8 and 24 hours at 25°C, or between 4 to 8 hours at 35°C, after the previous application.

The surface should be left roughened to provide a mechanical key for subsequent layers. Priming should be carried out between layers.

Cleaning

Uncured Primer and NITOMORTAR HB can be removed with FOSROC SOLVENT 102. Tools and equipment should be cleaned with FOSROC SOLVENT 102 immediately after use.

PRECAUTIONS

Health and Safety

Some people are sensitive to epoxy resins and skin contact should be avoided. Gloves and barrier cream should be used when handling these products. Suitable barrier creams include Kerodex 71, Rozalex 9 and Debba-Wet Work. If contact with the resin occurs, it must be removed before it hardens. This is best done with a resin removing cream such as Kerocleanse 22, and Rozalex 42 followed by washing with soap and water. Solvent should not be used for cleaning hands. Contact with eyes must be avoided and accidental contamination should be washed thoroughly with plenty of water and medical treatment sought immediately. The use of goggles is recommended.

Fire

FOSROC SOLVENT 102 and NITOPRIME 28 are flammable. Ensure adequate ventilation. Do not use near a naked flame or smoke during use.

Flash Points

FOSROC SOLVENT 102	33°C.
NITOPRIME 28	27°C.

STORAGE

12 months shelf life at 25°C. Store in a dry place. If stored in high temperature and high humidity locations the shelf life may be reduced.

PACKAGING AND COVERAGE

	Size	Coverage
NITOPRIME 28		
HANDY Pack	1 kg	5m^2 approx.
INDUSTRIAL Pack	5 kg	25m^2 approx.
NITOMORTAR HB	4 kg	4 litres approx
	8 k.g	8 litres approx..
FOSROC SOLVENT 102	5 litre cans	

FOSROC ACID ETCH 30 litre carboys

9. NITOKIT®

Epoxy resin crack injection system
(in kit-form)

USES

For injecting cracks in concrete and masonry wherever there is a need to consolidate a structural element and to exclude water from contact with reinforcement.

The low viscosity characteristic of the two component epoxy resin ensures that the finest of cracks (down to 0.2 mm) can be filled and an excellent bond, even to damp concrete, can be achieved.

Ideal for small scale repairs on site, for in situ elements or precast sections prior to installation.

Caution must be exercised with cracks which are live or filled with dust, water or salts.

ADVANTAGES

Complete kit ready for use.

Convenient disposable cartridges.

Low viscosity penetrating resin.

Excellent bond to concrete, brick and masonry.

Injection accessory pack available.

Enables smooth finish which blends with surrounding concrete or masonry.

DESCRIPTION

NITOKIT Epoxy Resin Crack Injection System:

NITOKIT is a system designed for effecting small scale repairs in construction elements and is delivered on site in complete kit form ready for use.

The Basic Kit contains:

10 Cartridges containing 0.175 litre of NITOKIT Epoxy Resin
10 Tubes each containing 0.070 litre of NITOKIT Hardener
20 Injection nipples with locating pins
20 Nipple caps
10 Connecting hoses
Instructions for use

An Accessory Pack is also available to complete the system. This contains:

1 Injection gun
10 Pairs disposable gloves
2 Pairs goggles
1 Litre plastic mixing bucket
1 × 5.5 litre pack NITOKIT Surface Sealant

Normally at least one Accessory Pack is required for each job, but this may be used to service approximately 10 Basic Kits.

NITOKIT Surface Sealant

This is a polyester resin compound specially designed for use with NITOKIT. It combines the dual function of sealing the surface of the crack to prevent leakage of the resin and bonding on the injection points.

It is supplied as a liquid resin together with a powder hardener which are mixed together to give a stiff putty-like consistency.

NITOKIT Surface Sealant has the added advantage that it can be rubbed down with suitable hand or power tools to give a smooth finish which readily blends with concrete or masonry around it.

Two versions of the product are available according to temperature at the time of use.

The usable life of each of these products is as follows:

Rapid	15-35 minutes at 5°C-25°C.
Standard	15-35 minutes at 25°C-40°C.

PROPERTIES

The characteristics of the injection resin when mixed are given in the following table.

		Rapid	Std.
Pot Life	10°C	70	110
(minutes)	20°C	25	50
	30°C	10	25

Viscosity	10°C	300	300
(in mPa S)	20°C	110	110
	30°C	70	70
Cure Time	10°C	24	36
(hours)	20°C	6	16
	30°C	4	6

Typical properties when fully cured (7 days at 23°C) are as follows:

Compressive strength

(ISO/R 604) $80 \, N/mm^2$

Tensile strength

(ISO/R 527) $26 \, N/mm^2$

Elastic Modulus

(ISO/R 527) $3200 \, N/mm^2$

Tensile adhesion to damp concrete–Excellent (failure in concrete)

INSTRUCTIONS FOR USE

Surface Preparation

NITOKIT Surface Sealant has to retain the injection system under pressure. Care must be taken to provide a bond surface which is clean dry and sound. Wear goggles provided in pack.

1. Remove heavy deposits of grease and dirt by scrubbing with detergent solution and washing with plenty of clean water to ensure complete removal of the detergent. Dirt alone may be removed with wire brushes or similar mechanical means. Both surface and cracks must be allowed to dry thoroughly.

2. Remove deteriorated concrete, laitance and paint. The best treatment is shot-blasting with sand steel-shot or a proprietary abrasive. Where shot-blasting is impracticable, use rotary wire brushes or a sander.

3. Blow the cracks and treated surfaces with a clean dry (filtered) compressed air-blast to ensure complete removal of all dust and loose particles.

Mixing the Surface Sealant

Mix only the quantity of sealant that can be applied within the usable life. Pour a small quantity of the resin into the mixing bucket provided and slowly add the powdered catalyst from the polythene bag and stir until a smooth putty-like consistency is obtained. Mix further quantities as required and retain some material for making good after removing the injection nipples.

Application of Surface Sealant

Immediately after mixing, apply the compound to the surface cracks. Overlap the cracks on both sides by 15-20 mm. In general, air to form a band of sealant 30-40 mm wide and 2-3 mm thick. Use the sealant to bond on the injection nipples supplied using the locating pins to ensure correct siting. Depending on the width and depth of the cracks, the distance between nipples should be 200 to 500 mm.

Fit nipples to both sides of a wall that is cracked all the way through: those at the back should be midway between those in front.

Take care not to block the nipple holes with NITOKIT Surface Sealant.

Seal the surfaces of the cracks using the NITOKIT Surface Sealant.

Seal both sides of any crack going all the way through a wall.

Application of the injection system may be commended as soon as the NITOKIT Surface Sealant has cured (at least 1 hour).

NOTE: At low ambient temperatures the curing time must be extended in order to allow sufficient strength to build up in the sealant.

Application of NITOKIT Epoxy Resin Injection System

1. Pierce the end of the hardener tube with a sharp pointed tool. Cut the end from one of the plastic nozzles provided and fix over the end of the hardener tube.

2. Pierce the seal on the threaded end of the resin cartridge and insert the hardener complete with plastic nozzle.

3. Completely and slowly empty the tube of hardener into the resin cartridge and remove tube and nozzle. If hardener tube is emptied too quickly overspill will occur.

4. Close the cartridge with a plastic cap supplied with the pack and mix the contents by slowly inverting the cartridge 30 times. Do not shake vigorously otherwise air will be included.

5. Use up the resin/hardener mix as soon as possible within the pot life shown.

6. Insert cartridge in the injection gun. Make sure there are sealing rings at both ends of the hose and attach one end firstly to the nipple and secondly to the cartridge.

7. Commence injecting at the lowest nipple, this is particularly important when treating vertical cracks.

8. Inject resin/hardener mix into the lowest nipple until it begins to flow out of the adjacent nipple.

   Disconnect the hose and close off the first nipple with a cap.

9. Then inject mix through the adjacent second nipple until it flows out of the third.

10. Repeat this process until the entire length of the crack has been injected.

11. Allow the injected system to cure.

12. Where the cracks have penetrated completely through the wall section, check out the sequence above on the nipples on the reverse side.

Making Good

Remove the nipples. Make good any holes or voids with more sealing compound and allow to cure. The NITOKIT Surface Sealant used to seal the cracks can then be ground off with an angle drive grinder of softened with a blow lamp and peeled off:

Cleaning

Use FOSROC SOLVENT 102.

PRECAUTIONS

Health and Safety

Some people are sensitive to epoxy resins so gloves and a barrier cream KERODEX 71, Rozalex 9, Debba-Wet Work or similar, should be used when handling these products. If contact with the resin occurs, it must be removed, before it hardens with a resin removing cream such as Kerocleanse 22 and Rozalex 42. Follow by washing with soap and water. DO NOT USE SOLVENT Goggles must be used but should accidental eye contamination occur, wash thoroughly with plenty of water and seek medical treatment immediately. Avoid inhalation of the powder and ensure good ventilation.

Fire

NITOKIT Surface Sealant is flammable. Do not use near naked flames or smoke during use.

Flash Point

NITOKIT Surface Sealant 29°C
FOSROC SOLVENT 102 33°C

STORAGE

12 months shelf life if stored at 15-35°C.

PACKAGING

NITOKIT   Basic Kit
          Accessory Kit
          NITOKIT Surface Sealant Pack

10. CONBEXTRA EPLV

Low viscosity structural epoxy crack injection grout
Tropical Grade
(in bulk-form)

USES

For injection into non-moving cracks in concrete or masonry where reinstatement of structural strength is required.

ADVANTAGES

Low Viscosity	— Penetrates down to crack widths of 0.25 mm.
Good Adhesion	— Adhesion fully to dry concrete.
Tough	— Withstands high hydrostatic pressures.
Impermeable	— The cured resin is impermeable to water.
Chemical Resistant	— The cured resin withstands attack by a wide range of industrial chemicals, chlorides, sulphates, etc..
Chloride Resistant	— The cured resin is unaffected by the presence of chloride ions.
Non-Shrink	— Full stability during and after cure.
High Strength	— Exceptional compressive, flexural and tensile strengths ensure durabilityand long-term service.
Minimum Creep	— Material designed for low creep characteristics under sustained loading.
Reliability	— Premeasured, factory-controlled material ensures consistent mechanical properites. No site batching is necessary.
Local Manufacture	— Tropical grade formulated for Saudi Arablan weather conditions and is rapidly available from local stocks.

DESCRIPTION

CONBEXTRA EPLV is a solvent-free epoxy resin crack injection material designed for free-flow or pressure injection into gap widths between 0.25 and 10 mm. The product is a tropical formulation, manufactured under approved quality-control conditions for use in the Saudi Arabian environment.

CONBEXTRA EPLV is supplied as an all liquid, two component pack containing premeasured quantities of epoxy base and epoxy hardener. The two components are supplied in the correct mix proportions designed for easy whole-pack mixing on site. Packs should never be split and no other materials should ever be added. The mixed injection resin is transparent and slightly yellow in colour.

SPECIFICATION

Performance specification: All injection grouting (specify details and areas of application) shall be carried out using a pre-packaged, two component epoxy resin grout. The grout shall be mixed on site using the entire contents of each base and hardener tin, in the proportions

supplied by the Manufacturer. The compressive strength of the cured injection grout shall be not less than 75 N/mm^2 after 7 days. The resin system shall comply with ASTM C881-78, Type III. Grade 1, Class B or C. The grout shall be stored, handled and placed strictly in accordance with the Manufacturer's instructions.

Supplier specification: All injection grouting (specify details and areas of application) shall be carried out using CONBEXTRA EPLV (tropical grade), manufactured by FOSAM CO LIMITED in the Kingdom of Saudi Arabia and applied strictly in accordance with their technical data sheet.

PROPERTIES

Pot Life
The time for which complete packs will remain fluid once the base and hardener are mixed together will vary with temperature. Typical values are shown in Table 1.

TABLE 1: Pot life in minutes at different temperatures.

20°C	70 minutes
40°C	20 minutes *

*NOTE: This time can be increased by storing the material at cool temperatures before mixing.

Viscosity:
The viscosity of freshly mixed CONBEXTRA EPLV will vary with temperature. Typical values are shown in Table 2.

TABLE 2: Viscosity of freshly mixed material at different temperatures.

20°C	4 poise
40°C	3 poise

Exotherm:
The temperature rise developed in the mixed grout is a function of the volume to surface area ratio, the ambient temperature and the mass and thermal conductivity of the surrounding material. The freshly mixed resins should always be used at once.

Application temperatures: CONBEXTRA EPLV is a tropical formulation manufactured by FOSROC/FOSAM in the Kingdom of Saudi Arabia specifically for use in local conditions.

Compressive strength gain: Tested in accordance with BS 4551, BS 6319 where applicable. The results are shown in Table 3:

TABLE 3: Compressive strength gain of CONBEXTRA EPLV (tropical grade) @ 40°C., expressed in N/mm^2.

1 day	55 N/mm^2
3 days	68 N/mm^2
7 days	77 N/mm^2

Tensile and Flexural Strengths:

Ultimate tensile strength to BS 12:	30 N/mm^2
Ultimate flexural strength to BS 6319:	55 N/mm^2
Ultimate Young's modulus	16 kN/mm^2
Specific gravity	1080 kg/m^3

INSTRUCTIONS FOR USE

Preparation
All non-moving cracks to be treated must be identified and clearly marked out prior to the commencement of the work. A full inspection must be made to ensure that all surrounding concrete is sound and that all cracks to be injected are not caused by corrosion of steel

reinforcing bars. Where any such corrosion exists it should be treated with a suitable FOSROC method of repair.

All loose particles and any other surface contamination should be removed from the cracks and at each side of the cracks to reveal a band of clean concrete approximately 50 mm wide.

Injection nipples should be bonded at intervals of 200-500 mm depending on the crack width. They must be correctly positioned above the cracks using location pins and bonded in place using SURFACE SEALER, a high-strength, epoxy resin surface sealant. Special care must be taken not to block the nipples holes with the sealant.

SURFACE SEALER should be used to form a band, 40 mm wide and 3 mm thick, above the cracks between each of the fixed injection nipples. It must be allowed to cure for not less than 6 hours before continuing.

Mixing

The entire contents of the 'hardener' tin should be emptied into the 'base' tin and the two components thoroughly mixed for 3 to 5 minutes. Part packs must not be used. Once mixed, CONBEXTRA EPLV should be used immediately.

Injection

The mixed grout should be inserted into standard resin injection equipment capable of injection pressure of at least 0.4 N/mm^2 (4 Bar). The injection gun must be securely fixed by means of an access hose to the first injection nipple and the mixed resin applied in this way.

Injection should always commence from the lowest nipple in a vertical crack or from either end of a horizontal crack. CONBEXTRA EPLV should be injected until it flows from the adjacent nipple. The hose should be disconnected and the injection nipple sealed off. Injection should continue in this manner, progressing from one nipple to another, until the entire length of the crack has been filled.

Making good

The nipples and the surface sealant may be removed after 2 or 3 days. Any remaining small holes or voids may be filled with SURFACE SEALER. Where a final protective coating such as NITOCOTE DEKGUARD is to be applied, the finished installation should be allowed to cure for not less than 5 days before application of the coating.

Cleaning

All tools and application equipment should be cleaned with FOSROC SOLVENT 102 immediately after use.

PRECAUTIONS

Health and Safety

Some people are sensitive to resins. Gloves and a barrier cream such as Exaderm or similar, should be used when handling CONBEXTRA EPLV. If contact with the skin occurs, the resin must be removed before it hardens with a resin removing cream such as Kerocleanse 22 or Rozalex 42. Follow by washing with soap and water. Do not use solvent. The use of goggles is recommended but should accidental eye contamination occur, wash thoroughly with plenty of clean water and seek medical treatment immediately.

Fire

FOSROC SOLVENT 102 is flammable with a flashpoint of 33°C.

STORAGE

6 months shelf life at 35°C.

PACKAGING & YIELD

CONBEXTRA EPLV	1.5 litre packs
SURFACE SEALER	5.0 kg packs (Theoretical coverage $3.12 \text{ kg/m}^2/\text{mm}$)
FOSROC SOLVENT 102	5.0 litres tins

11. CONBEXTRA® UR 63

Solvent free two-part polyurethane resin for sealing cracks
(semi-flexible polyurethane)

USES

Can be injected into cracks in concrete or masonry in dry or damp conditions to form an elastic seal. Caution must be exercised with cracks which are live or filled with dust, water or salts.

Use with CONBEXTRA WS 60 to seal cracks in wet conditions in basements, subways and tunnels.

ADVANTAGES

Low Viscosity	— Penetrates fine cracks and cavities.
Good Adhesion	— Adheres strongly to dry or moist concrete.
Flexible	— Strong but flexible to withstand differential structural movement.
Tough	— Withstands high hydrostatic pressures.
Impermeable	— On curing it forms a hard mass impermeable to water.

Suitable for use in Middle East conditions.

DESCRIPTION

CONBEXTRA UR 63 is a two-part liquid polyurethane. When mixed in the proportion supplied they react to form a tough, slightly flexible resin. CONBEXTRA UR 63 has a good adhesion to concrete and masonry and when injected into cracks it allows some movement without loss of bond.

TYPICAL PROPERTIES

Specific gravity of mixed grout	1.0
Viscosity of mix at 20°C	3.0 poise

Cure Properties (Tropical Grade)

	25°C	35°C
Temperature		
Pot Life	35 mins	20 mins
Reaction Time	95 mins	55 mins

INSTRUCTIONS FOR USE

Surface Preparation

Care must be taken to provide a clean and sound surface for bonding and injecting.

1. Remove heavy deposits of grease and dirt by scrubbing with FOSROC CHEMICAL DEGREASER and washing with plenty of clean water to ensure complete removal of the detergent. Dirt alone may be removed with wire brushes or similar mechanical means.

2. Remove deteriorated concrete, laitence and paint. The best treatment is grit-blasting with sand, steel-shot or a proprietary abrasive. Where grit-blasting is impracticable, use rotary wire brushes.

3. Blow the cracks and treated surfaces with oil free compressed air to ensure complete removal of all dust and loose particles. Ensure that wet surfaces are blown dry.

In the presence of running water, the flow must be stopped by injecting CONBEXTRA WS 60 which produces a rapid setting foam (see the Product Data Sheet). When the flow of water is stopped the cracks are reinjected with CONBEXTRA UR 63.

The surface of the crack must be sealed and injection nipples bonded in place before CONBEXTRA UR63 is injected using FOSROC SURFACE SEALANT.

Mixing the Surface Sealant

Add the hardener component to the base component and mix thoroughly until a uniform colour is achieved.

Application of Surface Sealant

Immediately after mixing, apply the SURFACE SEALANT to the surface cracks. Overlap the cracks on both sides by 15-20 mm. In general aim to form a band of sealant 30-40 mm wide and 2-3 mm thick. Use the sealant to bond on injection nipples using locating pins to ensure correct siting. Depending on the width and depth of the cracks the distance between nipples should be 200 to 500 mm.

Fit nipples to both sides of a wall that is cracked all the way through: those at the back should be midway between those in front.

Take care not to block the nipples access holes with the SURFACE SEALANT.

MIXING AND APPLICATION OF CONBEXTRA UR 63

Application of the injection system may be commenced as soon as the SURFACE SEALANT has cured (at least 8 hours at 35 deg. C). This will be extended at lower ambient temperatures. Mix the entire contents of the base and hardener components together until the liquid becomes clear.

The product is applied by standard resin injection equipments. The injection pressure should be at least 0.4 N/mm^2 (4 Bar).

MAKING GOOD

Remove the nipples. Make good any holes or voids with SURFACE SEALER and allow to cure. The remaining SURFACE SEALER used to seal the cracks can then be ground off or softened with a blowlamp and peeled off. Do not allow to burn.

CLEANING

Use FOSROC SOLVENT 102.

PRECAUTIONS

Health and Safety

Some people are sensitive to resins. Gloves and a barrier cream such as Exaderm or similar should be used when handling all resins. If contact with the skin occurs, the resin must be removed before it hardens with a resin removing cream such as Kerocleanse 22 or Rozalex 42. Follow by washing with soap and water. Do not use solvent. The use of goggles is recommended but, should accidental eye contamination occur, wash thoroughly with plenty of clean water and seek medical treatment immediately. Ensure good ventilation and do not smoke during use.

Fire

FOSROC SOLVENT 102 and CONBEXTRA UR 63 are flammable.

Flash point

FOSROC SOLVENT 102

STORAGE

Six months shelf life if stored at 20°C

PACKAGING

Usually available in 5 litre packs.

12. CONBEXTRA® WS 60

Rapid foaming and setting resin for stopping flow of water

(water-stopping polyurethane)

USES

STOPS the flow of water when injected into cracks in concrete in the presence of moving water.

CONBEXTRA WS 60 with CONBEXTRA UR 63 provides an effective system for crack sealing in wet conditions.

ADVANTAGES

Seals against water	Reacts and produces foam resistant to water.
Rapid acting	Rapid reaction to produce water stop in presence of flowing water.
Reinjectable	Can be reinjected with CONBEXTRA UR 63 to produce a permanent seal.

DESCRIPTION

CONBEXTRA WS 60 is a two part liquid polyurethane. When mixed in the proportions supplied it reacts rapidly with water to form a foam barrier.

TYPICAL PROPERTIES

Specific Gravity	1.18
Density of foam	0.025
Viscosity at 20°C	2 poise
Pot life, in absence of water,	at 20°C approx. 8hours. at 30°C approx. 4 hours.
Reaction time with water –	5 to 20 seconds dependent upon temperature.

INSTRUCTIONS FOR USE

If water flow permits then the surfaces should be free from oil, grease and other contaminants.

CONBEXTRA WS 60 can be applied using either injection plugs or adhesion packers (nipples) bonded to the surface with NITOKIT SURFACE SEALANT.

INSTRUCTIONS FOR USE OF NITOKIT SURFACE SEALANT

Surface Preparation

NITOKIT Surface Sealant has to retain the injection system under pressure. Care must be taken to provide a clean and sound surface for bonding.

1. Remove heavy deposits of grease and dirt by scrubbing with detergent solution and washing with plenty of clean water to ensure complete removal of the detergent. Dirt alone may be removed with wire brushes or similar mechanical means.

2. Remove deteriorated concrete, laitance and paint. The best treatment is grit-blasting with sand, steel-shot or a proprietary abrasive. Where grit-blasting is impracticable, use rotary wire brushes.

3. Blow the cracks and treated surfaces with oil free compressed air to ensure complete removal of all dust and loose particles.

MIXING THE SURFACE SEALANT

Mix only the quantity of sealant that can be applied within the usable life. Pour a small quantity of the resin into the mixing bucket provided and slowly add the powdered catalyst from the polythene bag and stir until a smooth putty-like consistency is obtained. Mix further quantities as required and retain some material for making good after removing the injection nipples.

APPLICATION OF SURFACE SEALANT

Immediately after mixing, apply the compound to the surface cracks. Overlap the cracks on both sides by 15-20 mm. In general, aim to form a band of sealant 30-40 mm wide and 2-3 mm thick. Use the sealant to bond on injection nipples using locating pins to ensure correct siting. Depending on the width and depth of the cracks the distance between nipples should be 200 to 500 mm.

Take care not to block the nipple holes with NITOKIT Surface Sealant.

Seal the surfaces of the cracks, using the NITOKIT Surface Sealant.

Application of the injection system may be commenced as soon as the NITOKIT Surface Sealant has cured (at least 2 hours).

NOTE: At low ambient temperatures the curing time must be extended in order to allow sufficient strength to build up in the sealant.

MIXING AND APPLICATION OF CONBEXTRA WS 60

Thoroughly mix the accelerator with the base resin. Take care to exclude moisture as much as possible and place in an enclosed container after mixing. There will be a skin on the surface but the liquid underneath will be satisfactory for use.

CONBEXTRA WS 60 should be used with standard injection equipment having closed containers.

When flowing water has been stopped, reinject with CONBEXTRA UR 63 to give permanent seal. See the Data Sheet.

MAKING GOOD

Remove the nipples. Make good any holes or voids with more sealing compound and allow to cure. The NITOKIT Surface Sealant used to seal cracks can then be ground off, or softened with a blow lamp and peeled off. Do not allow to burn.

CLEANING

Use FOSROC SOLVENT 102.

PRECAUTIONS

Health and Safety
Some people are sensitive to resins so gloves and a barrier cream, Kerodex 71, Rozalex 9, Debba-Wet or similar, should be used when handling CONBEXTRA WS 60. If contact with the resin occurs, it must be removed, before it hardens, with a resin removing cream such as Kerocleanse 22, and Rozalex 42. Follow by washing with soap and water. Do NOT use solvent. The use of goggles is recommended but should accidental eye contamination occur, wash thoroughly with plenty of water and seek medical treatment immediately. Ensure good ventilation and do not smoke during use.

Fire
FOSROC SOLVENT 102 and NITOKIT RESIN are flammable.

Flash point
FOSROC SOLVENT 102	33°C
NITOKIT RESIN	29°C

STORAGE

Six months shelf life if stored at 20°C.

PACKAGING

Usually available in 5 litre packs.

13. NITOCOTE® EP410

Epoxy resin tank and surface lining material
(chemically resistant epoxy coating)

USES

NITOCOTE EP410 is a hygienic and highly chemically resistants coating for brick and concrete walls, concrete and metal tanks, sluices and ducts. Suitable for applications in process plants and sewage works.

ADVANTAGES

Excellent chemical resistance.
Excellent adhesion.
Equally effective on concrete or metal substrates.
Hygienic smooth finish.
Tough abrasion resistant film.
Excellent immersion properties.
High build for maximum protection.
Long maintenance-free life.
Suitable for use in Middle East conditions.

DESCRIPTION

NITOCOTE EP410 is a two-component, solvent free, high build epoxy formulated to provide a thixotropic coating suitable for application to vertical surfaces. The film is opaque and pigments are added to ensure that complete mixing is achieved, indicated by a uniform green colour.

Certain exposure conditions may cause some discolouration. This does not affect product performance.

PROPERTIES

Specific gravity:	1.45
Volume solids:	100%

Recommended thickness per coat:
   Dry film thickness (dft) 250 μm.
   Wet film thickness 250 μm

Theoretical coverage: $4m^2$/litre ($2.75m^2$/kg) for a dft of 250 μm.

Practical coverage: Theoretical coverages are quoted for guidance. Practical coverages may be lower, depending on substrate and application method.

Number of coats: 1 (2 with glass fibre reinforcement where special circumstances dictate).

	at 20°C	at 20°C
Pot life	1 ½ hours	35 mins
Drying times:		
Touch dry:	12 hours	6 hours
Fully cured:	7 days	6 days
Recoatable:	12-30 hours	6-18 hours

Minimum application temperature: 7°C
Chemical resistance: The fully cured coat is resistant to:

Ammonium hydroxide 30%	Slight attack[*]
Caustic soda 50%	No attack
Citric acid 50%	No attack
Detergents	No attack
Fatty acids (higher)	Slight attack[*]
Hydrochloric acid 30%	No attack
Lactic acid 10%	No attack
Nitric acid 10%	No attack
Oil, mineral	No attack

Petrol	No attack
Sodium hypochlorite 10%	Slight attack*
Sulphuric acid 20%	No attack
Water	No attack

* indicates slight attack under continuous immersion, but NITOCOTE EP410 will withstand lengthy exposure to chemical splash, spray and fumes. "Attack" refers to any etching or swelling observed but ignores discolouration.

For details in respect of other chemicals, or in conditions where temperatures exceed 30°C, FOSROC technical department should be consulted.

INSTRUCTIONS FOR USE

Preparation
All surface should be clean, dry and free from dust.

Treat oil or grease contamination with FOSROC CHEMICAL DEGREASER followed by water or steam cleaning.

Concrete surfaces: Should be grit blasted or wire brushed.

Steel surfaces: Should be cleaned back to bright steel, sand or grit blast to SA2½. Small areas can be cleaned with a wire brush.

Priming
Concrete surfaces: All prepared concrete surfaces should be primed using NITOPRIME 25. This is a two pack epoxy resin primer supplied in pre-weighed quantities ready for mixing. Complete packs should be mixed and brushed in a thin continuous film over the concrete surface. The primer should be allowed to cure overnight, but not more than 24 hours at 20°C, 16 hours at 35°C, before applying the NITOCOTE EP410. The usable life after mixing is 60-80 minutes at 20°C, 30-40 minutes at 35°C.

Metal surfaces: All metal surfaces should be primed immediately after preparation using NITOPRIME 28. NITOCOTE EP410 should be applied within 8 hours at 35°C.

Mixing
The contents of the base can should be stirred thoroughly to disperse any settlement. The entire contents of the hardener can should be poured into the base container and the two materials mixed thoroughly until a uniform consistency and colour is obtained.

It is recommended that mechanical mixing is employed, using a stirrer on a heavy duty slow speed electric drill.

Application
Apply NITOCOTE EP410 to the primed surface immediately after mixing. Use a nylon brush and finally smooth out using a steel trowel. A continuous coating of uniform thickness should be obtained.

Use of glass fibre reinforcement
NITOCOTE EP410 may be used in conjuction with glass fibre fabric to increase coating thickness or where it is necessary to bridge fine cracks in the substrate. The fabric should be laid directly onto the first coat whilst it is still wet and should be pressed in and smoothed out with a stiff nylon brush or split washer roller. Second and subsequent coats may then be applied as necessary, allowing 16 hours but not more than 24 hours at 20°C between each coat. Suitable grade of fabric is 110g/m^2 open weave glass cloth.

Repairing and overcoating
Areas which have been previously coated with NITOCOTE EP410 and subsequently damaged can be readily overcoated. The existing coated surface must be well abraded, using sand blasting or wet applied abrasive paper, to ensure a good bond. Overcoating may then be carried out as for new work.

NOTE: Although NITOCOTE EP410 may be applied at temperatures down to 7°C, the curing time increases significantly in the lower ranges. For cold weather working, it is recommended

that materials are stored in a heated building and only removed immediately before use. Accelerated heating methods must not be used.

Cleaning
Tools and equipment should be cleaned with FOSROC SOLVENT 102 immediately after use.

PRECAUTIONS

Health and Safety
Some people are sensitive to epoxy resins so gloves and a barrier cream, Kerodex 71, Rozalex 9, Debba-Wet Work or similar, should be used when handling these products. If contact with the resin occurs, it must be removed, before it hardens, with a resin removing cream such as Kerocleanse 22, and Rozalex 42. Follow by washing with soap and water. Do not use solvent.

The use of goggles is recommended but should accidental eye contamination occur, wash thoroughly with plenty of water and seek medical treatment immediately. Ensure good ventilation and do not smoke during use.

Fire
NITOCOTE EP 410, NITOPRIME 25, NITOPRIME 28 and FOSROC SOLVENT 102 are flammable. Ensure adequate ventilation when using primers and solvents. Do not smoke during use and do not use in the vicinity of a naked flame.

Flash Points

NITOCOTE EP 410	65°C
NITOPRIME 25	39°C
NITOPRIME 28	27°C
FOSROC SOLVENT 102	32°C

STORAGE

Shelf life 12 months if stored below 35°C.

PACKAGING AND COVERAGE

		Approx Coverage
NITOCOTE EP 410	5.6 kg packs	(10-15m^2)
NITOPRIME 25	1 kg packs	(4.5-5.8m^2)
	5 kg packs	(22.29m^2)
NITOPRIME 28	1 kg packs	(3.5-5m^2)
	5 kg packs	(1.7-25m^2)
FOSROC SOLVENT 102	4 litres cans	

14. NITOCOTE® EP430

Epoxy resin coating for underwater applications
(for damp, wet or underwater concrete)

USES

As a corrosion resistant underwater coating for concrete or metal surfaces. Suitable for applications to marine structures, tunnels, docks, harbours, aqueducts, dams, drilling rigs, sewage works.

ADVANTAGES

Will bond and cure underwater.

Equally effective on dry, damp or wet surfaces.

Abrasion and chemical resistant.

High build, solvent-free.

Colour blends with concrete.

Smooth, easy to clean finish.

Eliminates flame drying or rain protection before coating.

DESCRIPTION

NITOCOTE EP430 is an underwater coating based on solvent-free epoxy resins containing pigments and fine fillers.

It is supplied as a two-pack material in pre-weighed quantities ready for on site mixing and use.

The material can be applied directly to damp, wet or underwater prepared concrete or metal surfaces and cures to provide a smooth hygienic and corrosion resistant surface.

Colour: Light grey.

PROPERTIES

Specific gravity: 1.67 (approx.).

Volume solids: 100%

Recommended thickness per coat:
Dry film thickness (dft) 200μm.

Wet film thickness 200μm.

Theoretical coverage:
$5m^2$/litre ($3m^2$/kg) for a dft of 200μm.

Practical coverage: Theoretical coverages are quoted for guidance. Practical coverages may be lower, depending on substrate and application method.

Number of coats: 2.

Pot life:		at 20°C	at 35°C
		30-40 mins	15-20 mins

Drying time:		at 20°C	at 35°C
	Touch dry	6 hours	3 hours
	Fully dry	7 days	7 days
	Recoatable	6-24 hours	3-12 hours

Application temperature: Minimum 5°C.

Resistance of film:
The fully cured coat is resistant to:
Distilled water
Chlorinated water
Brine 20%
Marsh water
Sewage water
Kerosene
Sea water
Sodium hydroxide 10%
Petrol †
Gas oil †
_____
† May cause surface discolouration.

INSTRUCTIONS FOR USE

Preparation
All surfaces: Must be clean and free from dust or loose material.

Treat oil or grease contamination with FOSROC CHEMICAL DEGREASER followed by water or steam cleaning.

Concrete surfaces: Should be sand blasted, water jet blasted or wire brushed.

Steel surfaces: Should be grit blasted to SA2½ or wire brushed to remove all mill scale and rust.

Underwater surfaces: Should be cleaned free of slime, algae or other contamination.

NOTE: Although NITOCOTE EP430 may be applied at temperatures down to 5°C, the curing time increases significantly in the lower ranges. For cold weather working. It is recommended that materials are stored in a heated building and only removed immediately before use. Accelerated heating methods must not be used.

Mixing
All application: The contents of the base can should be stirred thoroughly to disperse any settlement.

The entire contents of the hardener can should be poured into the base container and the two materials mixed thoroughly until a uniform consistency is obtained.

It is recommended that mechanical mixing is employed, using a stirrer on a heavy duty slow speed electric drill.

Underwater applications: Where the material is to be applied underwater, mixing must be carried out thoroughly above water. The mixed material can then be taken underwater for use.

Application
All applications: Apply material with a brush. The first coat must be firmly applied and be well scrubbed into the surface, ensuring a continuous coating of uniform thickness.

The second coat will cover more readily than the first and should be applied 6-24 hours later.

Where an appreciable time lapse occurs between coats, e.g. work carried out in a tidal zone or in sewage penstocks at off-peak times, it is essential that the previous coat is cleaned free of any contamination and lightly abraded before applying the next coat. Care should be taken to ensure that a continuous coating is obtained.

Underwater applications: Adopt general application procedures but when coating large areas underwater, use pressure fed brush equipment for maximum ease and efficiency.

Use of glass fibre reinforcement
NITOCOTE EP430 may be used in conjunction with glass fibre mat to increase coating thickness or where it is necessary to bridge fine cracks in the substrate. The mat should be laid directly onto the first coat whilst it is still wet and should be pressed in and smoothed out with a stiff nylon brush or split washer roller.

Second and subsequent coats may then be applied as necessary, allowing 6-24 hours at 20°C between each coat.

Repairing and overcoating
Areas which have been previously coated with NITOCOTE EP430 and which have been damaged can be readily overcoated. The existing coating surface must be well abraded, using a stiff wire brush or medium/coarse wet applied abrasive paper, to ensure good bond is achieved. Overcoating may then be carried out as for new work.

Cleaning
Tools and equipment should be cleaned with FOSROC SOLVENT 102 immediately after use.

PRECAUTIONS

Health and Safety
Some people are sensitive to epoxy resins so gloves and a barrier cream, Kerodex 71, Rozalex 9, Debba-Wet Work or similar, should be used when handling these products. If contact with the resin occurs, it must be removed, before it hardens, with a resin removing cream such as Kerocleanse 22, and Rozalex 42. Follow by washing with soap and water. Do not use solvent. The use of goggles is recommended but should accidental eye contamination occur, wash thoroughly with

plenty of water and seek medical treatment immediately. Ensure, good ventilation and do not smoke during use.

Fire

NITOCOTE EP430 is non-flammable.

Flash point
FOSROC SOLVENT 102 32°C.

STORAGE

12 months shelf life if stored below 25°C.

PACKAGING

NITOCOTE EP430: 2.5 kg packs (1.5 litres).

FOSROC SOLVENT 102: 5 litre cans.

ANNEXURE 6.2

"SIKA" PRODUCTS

Given here are details of some of the products manufactured by SIKA* (SIKA AG, Tuffenwies 18, CH8048 — Zurich, Switzerland; Phone: 01-436-4040, Telex: 822-254 Sik ch, Fax: 01-432-3362), representing:

    A. Concrete ADMIXTURES (Plasticizers, Retarders, Super-Plasticisers and Water-Reducers, Accelerators),
    B. Shotcrete ADMIXTURES,
    C. Curing-Compounds, Mould-Release agents and Surface-Retarders,
    D. Mortar ADMIXTURES,
    E. Cementicious Mortars,
    F. Polymer-Modified Cement-Mortars,
    G. Epoxy-adhesives and Mortars,
    H. Protective Coatings and Impregnations for Concrete,
    I. Floor-Treatment and Toppings,
    J. Concrete and Steel Protectors,
    K. Steel Protection,
    L. Joint-sealing compounds and Water-bars,
    M. Synthetic Water proofing Membranes, and
    N. Equipment (for applying some of these products).

• The descriptions also give various useful details regarding their use, features, form, consumption, etc. These details are reproduced with courtesy. It is hoped that the engineers dealing with concrete and it repairs will get a bird's eye-view from this list as to various product - formulations from SIKA that can be relevant to their work-whether for producing good concrete or for repairing the distressed concrete. (Detailed enquiries, including the latest and up-to-date information regarding new formulations, should be directed to the manufacturers.)

A. CONCRETE ADMIXTURES

1. Sikament® 163

DESCRIPTION: High-range water-reducing polymer type dispersion admixture. ASTM C-494 Type F.

USE: Highly effective water-reducing agent and super-plasticizer for high quality concrete in hot climates. Promotes accelerated hardening with high early and ultimate strengths.

Scope: Flow concrete in slabs and foundations, walls, columns, slender components. As substantial water-reducer for high early strength concrete in prestressed concrete, precast concrete, bridges and cantilever structures.

----
* and SIKA (Riyadh, Saudi Arabia)

Application: Make use of substantial water-reduction. Variable dosage facilitates easy adaptation to site conditions.

SPECIAL FEATURES: Compatible with all types of Portland cement, including S. R. C. Accidental overdosing will cause extension of initial setting time, however no excessive air entrainment. No Chlorides.

FORM/DENSITY: Brown liquid; 1.2 kg/l.

CONSUMPTION: Between 0.6–2.5% by weight of cement, depending on scope of work.

STORAGE: Store in frost free conditions. Shelf life at least 12 months in unopened original containers.

PACKAGING: 230 kg drums.

2. Sika® Antifreeze 1%

DESCRIPTION: Cold weather concreting admixture.

USE: Concrete admixture for the production of high quality concrete at moderately low temperatures.

Scope: For concreting work in cold conditions, e.g. slight daytime frost, expected overnight frost, expected cold periods, when under normal conditions work has to stop.

Application: Observe relevant codes of practice

SPECIAL FEATURES: Water-reduction without loss of workability. Increased frost resistance. Improved strengths. Chloride free.

FORM/DENSITY: Yellow liquid; 1.25 kg/l.

CONSUMPTION: 1% by weight of cement.

STORAGE: Store above –13° C. Shelf life unlimited if stored in unopened original container.

PACKAGING: 200 kg drums.

Sika Antifreeze 1 % is also available in powder form and added to the dry mix.

DOSAGE: 1 % by weight of cement.

STORAGE: Store in dry conditions. Shelf life unlimited in unopened original containers.

PACKAGING: 0.5 kg or 25 kg bags.

3. Sikament® -NN
(Naphthalene based)

DESCRIPTION: Superplasticizer and high-performance water-reducing admixture. ASTM C-494 Type F.

USE: Increases workability to produce flowing concrete. When used as a water reducing agent, high early strength and high final strength concrete is obtained. Water reduction up to 30%.

Scope: High quality concrete, prestressed concrete, concrete secondary products, high early strength concrete. Slender components with densely packed reinforcement.

Application: Make use of water reducing action for pump-crete and high early strength concrete. Self-levelling concrete, no water reduction.

SPECIAL FEATURES: Self-levelling concrete, use aggregates with suitable grading and sufficient fines content. If necessary can help save cement. No Chloride content. Compatible with all types of Portland cement, incl. S. R. C.

FORM/DENSITY: Brown liquid; 1.2 kg/l.

CONSUMPTION: Between 0.8–30% by weight of cement, depending on jobsite requirements.

STORAGE: Store in frost free condition. Shelf life at least one year in unopened original containers.

PACKAGING: 230 kg drums.

4. Sikament® FF

(Melamine based)

DESCRIPTION: Highly effective water-reducing agent and superplasticizer for promoting accelerated hardening and free flowing concrete. ASTM C-494-81 Type F.

USE: As superplasticizer for flow concrete in floor slabs, foundations, slender components with dense reinforcement, walls and columns etc. As water-reducer for the manufacture of pre-cast concrete elements, bridges and cantilever structures, prestressed and precast concrete.

Scope: High quality concrete, prestressed concrete, high early strength concrete.

Application: As a superplasticizer substantial improvement without increased water or risk of segregation. As a water-reducer up to 20% water reduction and 40% increase in 28 day strengths. High early strength after 8 hrs.

SPECIAL FEATURES: Improved surface finish and density. Increased watertightness and frost resistance.

FORM/DENSITY: Brown liquid; 1.24 kg/l.

CONSUMPTION: 0.6–3% by weight of cement, depending on type of effect sought.

STORAGE: Store in frost free conditions. Shelf life at least 12 months in unopened original containers.

PACKAGING: 250 kg drums.

5. Sikament® 520

DESCRIPTION: Highrange water-reducing and set retarding admixture. ASTM C-494 Type G.

USE: Highly effective superplasticizer with set retarding effect to produce free-flowing concrete in hot climates. As a substantial water-reducer to produce high early and ultimate strength concrete.

Scope: Wherever high quality concrete is demanded under difficult placing and climatic conditions.

Application: Make use of water-reduction of up to 20% for high early strength concrete.

SPECIAL FEATURES: Compatible with all types of Portland cement, including S. R. C. Accidental overdosing will increase set retarding effect. Keep concrete moist during this period. Can be added to the mixing water or separately to the freshly mixed concrete. No Chlorides.

FORM/DENSITY: Dark brown liquid; 1.18 kg/l.

CONSUMPTION: Between 0.8-2.5% by weight of cement depending on ambient temperature, quality of cement and aggregates and scope of work.

STORAGE: Store in frost free conditions. Shelf life at least 1 year in unopened original containers.

PACKAGE: 230 kg drums.

6. Plastiment® BV-40

DESCRIPTION: Water reducing, normal setting admixture. ASTM C-494 Type A.

USE: Improves workability, density and strengths. Reduces shrinkage and creep. Water reduction 10-15%.

Scope: Building and civil engineering construction, prestressed concrete, precast concrete and wherever high quality concrete is required.

Application: Make use of water reducing action.

SPECIAL FEATURES: No Chloride content, no noticeable air entrainment, no strength loss with overdosage, but set retardation (keep concrete moist).

FORM/DENSITY: Brown liquid; 1.2 kg/l.

CONSUMPTION: 0.2–0.5% by weight of cement.

STORAGE: Store in frost free conditions. Shelf life at least 1 year in unopened original containers.

PACKAGING: 200 kg drums.

7. Plastiment® A 750

DESCRIPTION: Water-reducing admixture ASTM C-494, Type A.

USE: Concrete admixture for the production of normal setting, high quality concrete.

Scope: For high quality structural concrete under difficult placing conditions. Where high concreting performance is required and high-quality exposed concrete surfaces are of importance.

Applications: Variable dosage allows good adaptation to jobsite conditions.

SPECIAL FEATURES: Improved surface finish. Increased strength at equal workability. Normal setting time at recommended dosage. Overdosage causes extension of initial setting time and additional air entrainment. Chloride free. Compatible with all types of Portland Cement, including S. R. C.

FORM/DENSITY: Brown liquid; 1.1 kg/l.

CONSUMPTION: 0.3–1.0% by weight of cement, depending on desired results.

STORAGE: Store in frost free condition. Shelf life at least 2 years if stored in unopened original containers.

PACKAGING: 200 kg drums.

8. Plastiment® AR 340

DESCRIPTION: Water-reducing and retarding admixture. ASTM C-494 Type A+D.

USE: As powerful concrete plasticizer and retarder for high quality concrete.

Scope: For fair-faced concrete, ready mix concrete, high concrete temperatures, difficult placing conditions etc.

Application: Variable dosage allows good control of setting time, slump and workability (water-reduction).

SPECIAL FEATURES: Longlasting control of slump loss at high concrete temperatures, increased setting time, accelerated hardening, water-reduction without loss of work-ability.

Compatible with all types of Portland Cement, including S. R. C. Chloride free.

FORM/DENSITY: Brown liquid; 1.1 kg/l.

CONSUMPTION: Between 0.2–0.8% by weight of cement, depending on desired results, ambient temperature, quality of cement and aggregates.

STORAGE: Store in frost free conditions. Shelf life at least 2 years in unopened original containers.

PACKAGING: 200 kg drums.

9. Plastiment® -R

DESCRIPTION: Water-reducing and retarding admixture. ASTM C494-81 Type D.

USE: Highly effective set retarding admixture with plasticizing effect. As a substantial water-reducer without loss of workability.

Scope: Where high quality concrete is required under difficult conditions, e.g. high temperatures, ready-mix concrete, prestressed concrete.

Application: Variable dosage facilitates easy adaptation to different requirements. Make use of water-reduction.

SPECIAL FEATURES: Longlasting control of slump loss. Reduced shrinkage and creep. Improved surface finish. Compatible with all types of Portland Cement, including S. R. C. Chloride free. Does not entrain excessive amounts of air.

FORM/DENSITY: Brown liquid; 1.1 kg/l.

CONSUMPTION: Between 0.2–0.5% by weight of cement, depending on effect sought, ambient temperature, quality of cement and aggregates.

STORAGE: Store in frost free conditions. Shelf life at least 2 years in unopened original containers.

PACKAGING: 200 kg drums.

10. Plastiment® -VZ

DESCRIPTION: Wear reducing, set retarding admixture. ASTM C-494 Type D.

USE: Improves workability, density, strengths and retards setting time. Reduces shrinkage and creep. Water reduction 5–7% (with use to Sika AER 12–15%)

Scope: For hot weather concrete, ready mixed concrete, mass concrete, prestressed concrete, pump-crete.

Application: Make use of water reducing action.

SPECIAL FEATURES: No Chloride content. Reduces exothermic temperature. Setting time can be controlled by consumption.

FORM/DENSITY: Almost clear liquid; 1.18 kg/l.

CONSUMPTION: below 20 °C 0.2%, 20–29 °C 0.25%, 29–38 °C 0.35%, over 38 °C 0.5% by weight of cement.

STORAGE: Store in frost free conditions. Shelf life at least 1 year in unopened original containers.

PACKAGING: 200 kg drums.

11. Plastocrete® -N

DESCRIPTION: Waterproofing admixture with water reducing action and limited air entrainment.
ASTM C-494 Type A.

USE: Increase watertightness, strengths and resistance to frost and deicing salts.

Scope: For all types of building and civil engineering construction, such as basements, reservoirs, retaining walls, canals, funnels, bridges, swimming pools, dams, etc.

Application: Make use of water reducing action. Mixing time and other rules of concreting practice can remain unchanged.

SPECIAL FEATURES: For waterproof concrete observe relevant rules of mix design and concrete practice.

FORM/DENSITY: Brown liquid; 1.1 kg/l.

CONSUMPTION: 0.5% by weight of cement.

STORAGE: Store in frost free conditions. Shelf life in undamaged original containers at least 1 year.

PACKAGING: 200 kg drums.

Plastocrete-N is also available in powder form and added to the dry mix.

DOSAGE: 0.5% by weight of cement.

STORAGE: If stored in dry conditions, shelf life at least one year.

PACKAGING: 25 kg bags.

12. Sikacrete® PP1

DESCRIPTION: New-generation admixture based on silica-fume technology.

USE: To increase the density, durability and compressive strength of concrete.

Scope: Wherever durable, dense high strength concrete is required.

Application: To be added to the dry mix, with a recommended mixing time of 90 seconds.

SPECIAL FEATURES: Contains latently active silicone dioxide which increases the portion of set cement in the concrete during hardening. Increases workability over a longer period of time. Greatly increases durability. Reduces permeability of water and gases (improved resistance to carbonation). Greatly reduces chloride infiltration. Very high ultimate strengths. Chloride free. Can be used together with other Sika admixtures.

FORM/DENSITY: Grey powder; 0.7 kg/l (bulk).

CONSUMPTION: Recommended dosage 5–10% by weight of cement.

STORAGE: Store in dry conditions. Shelf life in undamaged original containers at least one year.

PACKAGING: 15 kg bags.

Sikacrete PP1 is also suitable for use with shotcrete and especially in the field of mechanically applied concrete repair mortars.

13. Sika® Retarder

DESCRIPTION: Set retarding admixture. ASTM C-494-81 Type B.

USE: Concrete plasticizer with a highly efficient set retarding effect that at the same time promotes accelerated hardening after setting.

Scope: Retarder/Plasticizer in structural and mass concrete where controlled extension of setting time is required, e.g. large volume pours, long hauls, avoidance of cold joints, difficult placing conditions, re-vibrated concrete, high temperatures.

Application: Variable dosage allows accurate control of setting times.

SPECIAL FEATURES: Controlled extension of setting times. Improved workability without increased water content. Improved adhesion to reinforcement. Increased strengths and reduced shrinkage and creep. Chloride free. Compatible will all types of Portland Cement, including S. R. C. Do not use with antifreeze admixtures.

FORM/DENSITY: Yellow liquid; 1.20 kg/l.

CONSUMPTION: 0.2–2.0% by weight of cement, depending on desired results, quality of cement and aggregates and ambient temperature.

STORAGE: Store in frost free conditions. Shelf life at least 2 years in unopened original containers.

PACKAGING: 200 kg drums.

Sika Retarder is also available in powder form and added to the dry mix.

DOSAGE: 0.2–2.0% by weight of cement.

STORAGE: Store in dry conditions. Shelf life at least 2 years in unopened original containers.

PACKAGING: 25 kg bags.

14. Sika® -AER

DESCRIPTION: Air entraining admixture based on synthetic tensides. ASTM C-260-81, BS 5075.

USE: Increases resistance to frost and deicing salts and improves workability. Water reduction 6–8%.

Scope: For mass concrete and concrete pavements. Is used for controlled air-entrainment with other Sika admixtures.

Application: Mixing time and other rules of concreting practice can remain unchanged.

SPECIAL FEATURES: Sika AER can be premixed with other Sika admixtures. Control of entrained air advisable.

FORM/DENSITY: Transparent liquid; 1.0 kg/l.

CONSUMPTION: 0.03–0.15% by weight of cement.

STORAGE: Store in frost free condition. Shelf life at least 1 year in unopened original containers.

PACKAGING: 200 kg drums.

B. SHOTCRETE ADMIXTURES

1. Sika® Shot-3

DESCRIPTION: Ready for use gunite.

USE: For repair and strengthening of deteriorated concrete structures. Repair of fire damage. Construction of water retaining structures (swimming pools, tanks, etc.), tunnel support systems, repairs, relinings. Facing to diaphragm walls, rock sealing and support. Air and fire controls in mines. Roof construction (domes, shell and barrel vaults), etc.

Application: Dry mix process only.

SPECIAL FEATURES: Ready for use, no concrete mixer needed, just feed it into gunite machine. No chloride content, constant quality, high strengths, excellent bonding properties, reduced rebound. Improved impermeability, good chemical resistance, less dusting, short setting time.

FORM/DENSITY: Grey powder, apparent density—1.8 kg/l.

CONSUMPTION: Average approx. 20–24 kg/m^2 per cm thickness.

STORAGE: If stored in unopened original container in dry condition shelf life is at least 6 months.

PACKAGING: 30 kg bags.
Also available: Sika Shot-3-SR sulphate resistant type.

2. Sigunit® Powder

DESCRIPTION: Set-accelerating and waterproofing shotcrete admixture in powder form.

USE: Reduces setting time and time to develop sufficient strength. Improves watertightness.

Scope: For shotcrete linings for instant support. Linings on damp or wet substrate, waterproofing and overhead application. For pits and slopes, flow channels as well as for repairing concrete constructions.

Application: To be added to dry mix.

SPECIAL FEATURES: Final strength usually lower than without Sigunit. No chloride content, does not attack steel. High initial strength. Reduces rebound. Excellent adhesion to reinforcing steel, caustic.

FORM/DENSITY: Light-grey powder, 0.8 kg/l.

CONSUMPTION: 2–4% by weight of cement, depending on type of cement, site conditions and desired results.

STORAGE: Store in dry conditions. Shelf life at least 2 years in unopened original containers.

PACKAGING: 25 kg bags.

3. Sigunit® -L

DESCRIPTION: Set-accelerating and waterproofing shotcrete and gunite admixtures in liquid form.

USE: Reduces setting time and time to develop sufficient strength and reduces rebound considerably. Improves watertightness.

Scope: For shotcrete and gunite linings in tunnels, caverns, for swimming pools, concrete repair work, slopes, pits, etc. For instant support. Linings on damp or wet substrate, waterproofing and overhead application.

Application: To be added to the gauging water, using a dosage pump and injected into the spray nozzle.

SPECIAL FEATURES: High compressive and flexural strenghts. No shrinkage cracks. Less harmful dust for nozzle man. Excellent adhesion to substrates. No chloride content. Alkaline, does not attack steel. To be used with special dosage pumps.

Sigunit-L (N6)
Sigunit-L-61   for dryspray system
Sigunit-L-62
Sigunit-L-20   for wetspray system

FORM/DENSITY: Light-yellow liquid, 1.23–1.56 kg/l.

CONSUMPTION: 4–7% by weight of cement, depending on type of cement, site conditions and desired results.

STORAGE: No shelf life in unopened original container. Store in frost free condition.

PACKAGING: 200 kg and 250 kg drums

4. Sikacem® -Gunit 133/143

DESCRIPTION: One-component normal setting (133) or accelerated (143) cementitious, polymer modified silica-fume based gunite mortar.

USE: As high-performance repair mortar for dry-spray application, for new construction, repair and maintenance of e.g. powerstations and penstocks, tunnels and galeries, bridges, industrial and residential buildings, etc.

Application: Dry spray method only. Recommended equipment: Aliva 240 or 246.

SPECIAL FEATURES: Ready for use, one-component, superior workability. Can be trowelled after application. Very high density, excellent carbonation barrier. Excellent adhesion to substrate. Low modulus of elasticity. Up to 50 mm coat thickness in one application, low rebound. Extremely economical.

FORM/DENSITY: grey powder; 1.7 kg/l (dry bulk), - 2.2 kg/l (when mixed).

CONSUMPTION: Approx. 22–24 kg/m^2 per 1 cm thickness.

STORAGE: Store in dry conditions. Shelf life at least 6 months in unopened original containers.

PACKAGING: 25 kg plastic bags.

5. Sikacem® -Gunit 103/113

DESCRIPTION: One-component normal setting (103) or accelerated (113) silica-fume based gunite mortar.

USE: As ready for use gunite for dry-spray application, where high strength gunite is required, levelling gunite on large areas, reprofiling of concrete.

Application: Dry spray method only. Recommended equipment: Aliva 240 or 246.

SPECIAL FEATURES: Ready for use, one-component, excellent workability. Can be trowelled. High density. Good adhesion on substrate. Sulphate resistant. Up to 50 mm coat thickness in one application. Low rebound.

FORM/DENSITY: grey powder; 1.7 kg/l (dry bulk), 2.2 kg/l (when mixed).

CONSUMPTION: Approx. 22–24 kg/m^2 per 1 cm thickness.

STORAGE: Store in dry conditions. Shelf life at least 6 months in unopened original containers.

PACKAGING: 25 kg plastic bags.

C. CURING-COMPOUNDS, MOULD-RELEASE AGENTS and SURFACE-RETARDERS

1. Sika® Form-Oil Concentrate

DESCRIPTION: Form release oil concentrate.

USE: Form release emulsion for all types of formwork, especially wood. Facilitates removal and cleaning of shutters.

Application: It is recommended that wooden shutterings to be re-used several times be sealed with Sika Formoil Concentrate diluted 1:3 with water to 'varnish' the surface. Dilute the oil up to 1:4 with water for subsequent applications. Diluted Sika Formoil Concentrate is applied onto the forms approx. 3 hours before concrete is being poured.

Coverage: One litre covers between 30–180 m^2 of shutterings, depending on condition and material (wood, plastic, steel).

SPECIAL FEATURES: Stir well before use. Protect fresh coat from rain and frost. For fair faced concrete prefer Separol to Sika Formoil Concentrate. Do not water the applied coating.

FORM/DENSITY: Yellowish liquid; 0.9 kg/l undiluted.

STORAGE: Protect from frost. Shelf life at least 1 year, when stored in unopened original container.

PACKAGING: 180 kg drums.

2. Antisol® -E/Antisol® /Antisol® White

DESCRIPTION: Curing compounds based on emulgated paraffin (Antisol E - ASTM C309-81, Type 1, Class A). synthetic resin and solvent (Antisol-ASTM C309-81, Type 1, D, Class B), and white pigmented (Antisol White-ASTM C309-81, Type 2, Class B).

USE: Antisol range forms a membrane on the concrete surface to reduce premature water evaporation of fresh concrete. Without disturbance to the normal setting action, the concrete is then allowed to cure and achieve maximum properties. Replaces other curing measures, such as water-curing, hessian, etc. Membrane will gradually disappear.

Scope: Highways, run and taxiways, aprons and hard-standings, roof decks, retaining walls, prestressed beams, etc.

Application: To be applied by hand-operated or power-driven spray-gun as soon as concrete surface water has evaporated (between 1/2 to 2 hours), depending on temperature.

SPECIAL FEATURES: Stir well before use. Protect from rain for at least 2–3 hours. May impair adhesion of subsequent coatings, screeds or renderings.

FORM/DENSITY: White or reddish liquids, 0.95–1.15 kg/l.

CONSUMPTION: 0.15-0.20 kg/m^2, depending on wind, humidity and temperature.

STORAGE: Emulsion types store frost free, solvent containing types store away from open flames. Shelf life in unopened original containers at least 6 months.

PACKAGING: Antisol-E: 200 kg drums.

Antisol and Antisol White: 170 kg drums.

3. Separol®

DESCRIPTION: Form release oil, ready for use.

USE: Form release oil for steel, plastic and wooden formwork. Facilitates removal and cleaning of shutterings. Number of lifts of shutters is substantially increased.

Application: Apply by brush, roller or spray gun.

Coverage: Plastic and steel approx. 50 m^2/kg. wood approx. 25 m^2/kg.

SPECIAL FEATURES: Combined chemical and physical action. Ensures fair faced concrete finish. Protects fresh coat from rain. Reduces concrete contamination.

FORM/DENSITY: Brown liquid; 0.8 kg/l.

STORAGE: At least 2 years in well sealed original containers. Protect against sunlight and frost.

PACKAGING: 25 kg and 170 kg drums.

4. Rugasol® -1-Paste/Rugasol® -2-Liquid

DESCRIPTION: Concrete surface retarders for exposed aggregate finish. (Rugasol-1-Paste–brush type, Rugasol-2-Liquid–spray type).

USE: To retard the setting of the superficial cement coat next to the treated surface as to allow easy removal of the cement paste after stripping the forms, thus providing exposed aggregate finish.

Scope: For exposed aggregate concrete e.g. precast panels, non skid treatment of concrete pavements, construction joints (instead of chipping), etc.

Application: Rugasol-1-Paste by brush, roller or spray equipment on formwork. Rugasol-2-Liquid by brush, roller or spray equipment to formwork or cast concrete. Remove forms as early as possible, then wash away cement paste, using brush if needed (8-24 hours after casting concrete).

SPECIAL FEATURES: Neat finish of exposed aggregate concrete. Easy roughening of construction joints. Easy to apply. More economical than chipping. Protect freshly treated formwork of concrete from rain.

FORM/DENSITY: Rugasol-1 red paste, Rugasol-2 yellow/greenish liquid; Rugasol 1–1.1 kg/l., Rugasol 2–1.1 kg/l.

CONSUMPTION: Rugasol 1 and 2 approx. 0.1–0.2 kg/m^2.

STORAGE: Store in cool conditions. Shelf life in unopened original containers at least 1 year.

PACKAGING: Rugasol 1:10 and 25 kg pails, Rugasol 2:25 kg pails.

D. MORTAR-ADMIXTURES

1. Sika-1® /Sikalite®

DESCRIPTION: Normal-setting, waterproofing admixture in liquid form. NWC No. 830252.

USE: For waterproof internal and external cement renderings, screeds, masonry mortar and concrete.

Scope: Waterproof structures such as tunnels, basements, culverts, canals, watertanks, manholes, swimming-pools, etc.

Application: Dilute Sika 1 with gauging water. Normal dilution ratio is 1:10 (1 part Sika-1, 10 parts water). When using very wet sand dilution ratio must be adjusted.

Apply Sika-1 mortar to sound substrate with adequate mechnical key, by means of a trowel or, according to Manufacturer's instructions. Leakages must first be sealed with Sika-4a. Sika-1 can also be added as a waterproofing admixture for concrete.

SPECIAL FEATURES: Stir well before use.

FORM/DENSITY: Yellow Liquid; 1.05 kg/l.

CONSUMPTION: Approx. 0.3 kg/m² per cm mortar thickness.

STORAGE: Unlimited shelf life when stored in well sealed containers. Store in frost free condition.
PACKAGING: 30 kg, 100 kg and 200 kg drums.

- Sikalite is the equivalent product to Sika-1, but supplied in powder form.
  Dosage: 2% by weight of cement, use 1 × 1 kg bag of Sikalite for each bag of cement.

STORAGE: Unlimited shelf life when stored in unopened original packing in dry conditions.

PACKAGING: 20 or 100 × 1 kg bags.

2. Intraplast® -Z/Intracrete-EH

DESCRIPTION: Expanding grout admixtures which introduce micro bubbles into the mix and produce wet volume expansion and increased fluidity without segregation.

USE: To increase fluidity and cohesion in cement grouts (addition of sand is possible).
Scope: To grout prestressed cable ducts; rock and soil anchoring, bearing plates, pre-placed aggregates, tunnel linings, etc.

Application: Add directly to the cement/water mix. followed by sand if required. For mechanical batching observe 4 minutes mixing time.

SPECIAL FEATURES: Provides increased frost resistance, cohesion and fluidity; volume expansion in the wet state. Intraplast-Z produces hydrogen, whereas Intracrete-EH develops nitrogen, both of which are not harmful to prestressed cables. Do not use cements containing fly ash.

FORM/DENSITY: Intraplast-Z grey powder, Intracrete EH white powder; Intraplast-Z—0.7 kg/l, Intracrete-EH—1.2 kg/l.

CONSUMPTION: 2% by weight of cement for both products.

STORAGE: Store in dry conditions. Shelf life in unopened original containers at least 6 months.

PACKAGING: 20 x 1 kg or 160 x 1 kg bags.

3. Sika® -2

DESCRIPTION: Extra quick-setting waterproofing admixture.

USE: To stop strong localized leakages in concrete, brickwork and rock. Setting time at 18°C approx. 15-20 seconds.
Application: By hand or trowel, Mix 1 part of undiluted Sika-2 with 2 parts of fresh Portland Cement.

SPECIAL FEATURES: Must not be mixed with other Sika products. Alkaline. Does not attack steel. Wear rubber gloves!

FORM/DENSITY: Red liquid; 1.25 kg/l.

CONSUMPTION: –40% by weight of cement, according to job requirements.

STORAGE: Shelf life at least 2 years when stored in well sealed containers. Does not freeze above approx. 10° C.

PACKAGING: 30 kg and 250 kg drums.

4. Sikanol® -M

DESCRIPTION: Normal setting, liquid ready for use mortar plasticizer.

USE: As additive to produce high-quality mortars.
Scope: For brickwork and block-work mortar as well as cement/sand screeds, rendering and plaster.

Application: Add to the gauging water prior to its addition to the cement/sand mix.

Dosage: 0.1–0.2% by weight of cement (100-200 gr. per 100 kg cement).

SPECIAL FEATURES: Improves workability. Produces higher strengths and good adhesion. Reduces surface tension in mortar, thus allowing mixing water to become more effective. Sand: Cement ratio can be increased without loss of cohesion.

FORM/DENSITY: Brown liquid; 1.1 kg/l.

STORAGE: Shelf life in unopened original container, frost free, approx. 3 years.

PACKAGING: 200 kg drums.

5. Sika® -4a

DESCRIPTION: Quick-setting waterproofing admixture.

USE: To stop localized leakages in concrete, brickwork and rock, and to fix bolts, anchors, etc. Setting time at 18°C approx. 15-45 seconds.

Mild leakages: Seal with 0.5 cm thick Sika-4a cement coat.

Severe leakages: Drain off by means of plastic hoses to be embedded in Sika-4a cement paste.

Application: Use only fresh, lumpfree Portland Cement. Apply by hand or trowel.

SPECIAL FEATURES: Caustic. Does not attack steel. Use only wrought-iron or plastic valves on Sika-4a containers. Wear rubber gloves!

FORM/DENSITY: Clear liquid; 1.13 kg/l.

CONSUMPTION: Dilute 1:1 to 1:4 in gauging water, depending on site conditions (approx. 40% of cement weight).

STORAGE: Unlimited shelf life when stored in well sealed containers. Does not freeze above approx. 20°C

PACKAGING: 30 kg and 250 kg drums.

6. Sika® Latex

DESCRIPTION: Water resistant bonding agent for mortar on synthetic rubber base. NWC NO. 830253, BS 6319.

USE: Increases adhesion, flexural and tensile strengths, resilience and chemical resistance. Resists permanent exposure to water.

Scope: For renderings, screeds, bonding mortars in concrete construction joints, patching mortars, cement paints, tile adhesives, etc.

Application: Follow manufacturer's instructions.

FORM/DENSITY: White liquid; 1.05 kg/l.

CONSUMPTION: 1:1 to 1:4 in gauging water, depending on application.

STORAGE: At least 1 year shelf life when stored in well sealed container. Store in frost free condition.

PACKAGING: 30 kg and 200 kg drums.

7. Sikacem® 810

DESCRIPTION: Waterproof, reactive, synthetic polymer dispersion as bonding agent and mortar improver.

USE: Improves bond, flexural, tensile and compressive strengths; greater abrasion resistance; increases workability and stability of green mortar. Increases impermeability to water and resistance to chemical attack etc.

Scope: For bonding slurries, rendering mixes, repair and patching mortars, bedding mortars, cement screeds and toppings subject to heavy wear.

Application: Follow manufacturer's instructions.

FORM/DENSITY: Light grey liquid of low viscosity when stirred; 1.1 kg/l.

CONSUMPTION: 1:1 to 1:4 in gauging water, depending on application.

STORAGE: At least 6 months shelf life in unopened original containers. Store in frost free conditions.

PACKAGING: 10 kg and 20 kg pails, 100 kg drums.

8. SikaTop® 77

DESCRIPTION: Oil and water proof synthetic adhesive emulsion for mortar, renderings and grout. NWC NO. 830252, BS 6319.

USE: Improves elasticity, chemical and abrasion resistance and reduces shrinkage. Greatly increased mechanical strengths. Excellent oil and water resistance.

Scope: For renderings, floor toppings, repair and patching mortars, construction joints, abrasion resistant mortars as well as injection grout for cracks with low static pre-tensions.

Application: Follows manufacturer's instructions.

FORM/DENSITY: White liquid; 1.03 kg/l.

CONSUMPTION: Dilute 1:1 to 1:4 in gauging water, depending on application.

STORAGE: At least 1 years shelf life in unopened original containers. Store in frost free condition.

PACKAGING: 30 kg and 200 kg drums.

E. CEMENTITIOUS-MORTARS

1. Sika® Grout 210/214

DESCRIPTION: Ready for use, 1-component, cement bound, non-shrink expanding, self-levelling and grouting mortars. (Type 210 for temperatures of 0°-15° C, Type 214 for temperatures above 10° C).

USE: For grouting fittings, anchor bolts, machine beds, rail beds, columns, bridge beams, bearings and cavities.

Application: Substrates must be clean, sound, free of laitance, dust, oil, grease, remains of previous coatings, rust or scale.

Mixing Ratio: Water: Sika Grout –1:6-1:8 by weight, depending on site conditions.

SPECIAL FEATURES: Sika Grout is non corrosive, non inflammable, ready for use and easy to apply. Excellent flowability. Controllable consistency and high strengths. Non toxic. Impact and vibration resistant.

FORM/DENSITY: Concrete grey powder; 2.28 kg/l mixed.

STORAGE: At least 6 months shelf life in unopened original container when stored in dry conditions.

PACKAGING: 25 kg bags.

2. Sika Refit®

DESCRIPTION: Ready for use, one component mortar for concrete surface repair.

USE: To carry out repairs to concrete surfaces e.g,. pre-cast concrete elements, fair faced concrete, columns and beams etc.

Application: Apply Sika Refit by trowel to the properly prepared substrates. For deep patches 2 layers may be required. Sika Refit may be filled with sand 0-3 mm at a max. ratio of 2 parts sand to 1 part Sika Refit for the first, thicker layer.

Mixing Ratio: Sika Refit: Water 3:1

SPECIAL FEATURES: Easy and economical to use, practically no wastage. Good adhesion, high strength, little tendency to crack if applied according to manufacturer's instructions.

FORM/DENSITY: Light or dark grey powder; 1.25 kg/l bulk density, 1.75 kg/l mixed.

STORAGE: Shelf life at least 6 months when stored in unopened original container in dry conditions.

PACKAGING: 10 kg pails, 10 × 1 kg bags per carton.

3. Sika® -4a Fixing Mortar

DESCRIPTION: Quick setting, ready mixed waterproof mortar.

USE: For the rapid fixing and installation of steel frames, rag bolts, studs, light load anchors, etc.

Application: Concrete surfaces must be clean and saturated. Mix only enough material which can be used within its setting time of 2-4 minutes.

SPECIAL FEATURES: Economical and easy to use mortar. Facilitates rapid installation work.

FORM/DENSITY: Grey powder; 1.1 kg/l (bulk density).

STORAGE: Store in dry conditions. Shelf life in unopened original containers at least 6 months.

PACKAGING: 5 and 10 kg pails.

4. Sika® Grout Aid

DESCRIPTION: Sika Grout Aid is an expansive admixture for site mixed high strength non-shrink grout.

USE: High strength non shrink grout to grout-in bridge bearings, anchors, machine beds, steel columns, inverted construction method, etc.

Application: Substrate must be sound, clean and free from oil, grease or other contaminants and if possible saturated surface dry. After mixing the components, pour mortar immediately into properly prepared cavity. Make sure to leave opening so that displaced air can escape.

Consumption: 91 kg Sika Grout Aid per m³ of mortar.

SPECIAL FEATURES: Economical. Excellent flow and filling properties. No bleeding. Good dimensional stability. High strengths.

FORM/DENSITY: Beige powder; 1.21 kg/l (apparent density).

STORAGE: Store in dry conditions. Shelf life in unopened original container approx. 1 year.

PACKAGING: 20 kg bags.

F. POLYMER-MODIFIED CEMENT MORTARS

1. Sika Top® Seal 107

DESCRIPTION: Two component, flexible, cement based, polymer modified protective and waterproofing slurry.

USE: Waterproofing slurry with good adhesion, abrasion/erosion and excellent deicing salt resistance. On concrete, mortar, steel, brickwork.

Scope: Rigid waterproofing membrane for watertanks, basements, in and outside. Protective coating on terraces and balconies, small flat roofs, park decks, bridges, retaining walls, etc. Protective coating against influences of deicing salts on concrete structures and for restoration of such structures.

Application: On clean sound substrates, oil, grease, rust and laitance free. Absorbent substrates have to be wetted thoroughly prior to the application of SikaTop-Seal 107. Lump-free mixing of predosed components, preferably using low speed mechanical mixer. Apply either by brush or trowel. Can also be sprayed. If required, SikaTop-Seal 107 may be reinforced with Sika glass fibre cloth Type 107.

Mixing Ratio: A:B = 1:4 (by weight), brush application. A:B = 1:4.5 (by weight), trowel application.

SPECIAL FEATURES: Easy application. Excellent adhesion. Impermeable. Good resistance to deicing salts. Good abrasion and erosion resistance. Non corrosive. Min. application temperature + 8 °C. Apply a minimum of 2 coats. Not suitable for 'cosmetic' repairs. Potlife at 20 °C approx. 30-40 min.

CONSUMPTION: Protection against deicing salt: 2.0 kg/m² per coat. Waterproofing membrane: 1.5 kg/m² per coat up to 1 m water-head, 2.0 kg/m² per coat above 1 m water-head.

FORM/DENSITY: Comp. A (white liquid), Comp. B (grey powder); A + B: 2.0 kg/l.

STORAGE: Comp. A store in frost free condition, Comp. B store in dry condition. Shelf life at least 6 months if stored in unopened original container.

PACKAGING: 25 kg sets.

2. SikaTop® Armatec 110 EpoCem

DESCRIPTION: Three-component, cement-based epoxy-modified anticorrosion coating and bonding agent.

USE: As an anticorrosion coating for reinforcement steel in repairs to reinforced concrete and for the preventive protection of reinforcement steel in thin reinforced concrete sections.

As bonding agent for use on concrete, mortar, steel in concrete repair when using SikaTop patching and repair mortars or other site batched polymer modified mortars. For bonding old to new concrete.

Application: On clean, sound concrete, mortar, stone, free from grease, laitance or loose material. Steel surfaces must be clean and free from all traces of grease, oil, rust or mill scale (sandblast or wirebrush). Mix mechanically for 3 minutes using low speed mixer (250 rpm) to entrain as little air as possible. Leave for 5 –10 minutes until mixture attains brushable, slow dripping consistency. Apply by brush, roller or spray gun.

Mixing Ratio: A:B:C: = 1:2.4:13.7 by weight.

SPECIAL FEATURES: Easy application, excellent adhesion to steel and concrete. Contains corrosion inhibitors. Effective water and chloride penetration barrier. Excellent bonding coat for subsequent application of repair mortars. Solvent-free, non flammable.

Long potlife (90-120 min. at 5 °C – 30 °C). Long open time. Min. application temperature + 5 °C

CONSUMPTION: As an anticorrosion coating: 2 coats at 0.5-1.0 mm each (approx. 3–4 kg/m^2 depending on method of application). As a bonding coat for repair mortar or concrete: 1 coat of not less than 0.5 mm thickness (not less than 1.2 kg/m^2).

FORM/DENSITY: Component A (white liquid), Component B (yellowish liquid), Component C (light grey powder); A + B + C (grey); 2.0 kg/l.

STORAGE: Shelf life in unopened original containers at least 6 months if stored in the frost free and dry conditions.

3. SikaTop® 121 Adhesive Mortar

DESCRIPTION: Two-component, thixotropic, polymer modified cement mortar.

USE: Versatile adhesive and filling mortar for the protection and repair of concrete and metal surfaces.

Scope: Adhesive for wall and floor tiles, insulation panels, etc. Patching mortar for concrete (irregularities, spallings, honeycombings, etc.). Bonding slurry for following repair mortars.

Application: Porous substrates must be clean and saturated with water. Lump free mixing of predosed sets by means of mechanical low speed mixer.

Mixing Ratio: As adhesive mortar A:B = 1:4.5, as spatula mortar A:B = 1:5 (use only 90% of comp. A) by weight.

SPECIAL FEATURES: Easy to apply, good adhesion, waterproof, no shrinkage cracks, non toxic, high strengths.

FORM/DENSITY: Comp. A (white liquid), Comp. B (grey powder); A + B, 2.0 kg/l.

STORAGE: Comp. A store in frost free condition. Comp. B store in dry condition. Shelf life in unopened original containers at least 6 months.

PACKAGING: Predosed sets of 25 kg.

4. SikaTop® 122 Repair Mortar
(Fibre Type)

DESCRIPTION: Two-component, thixotropic, polymer modified cement mortar with coarser aggregates.

USE: Versatile high-strength repair and levelling mortar for concrete. Contains polyamide fibres.

Scope: Filling cavities, repairing damaged concrete, mortar and stone. For all types of structures, particularly hydraulic structures, tunnels, bridges, etc.

Application: Substrate must be sound and clean and saturated with water. Lumpfree mixing of predosed sets by means of mechanical low speed mixer. Apply by trowel.

Mixing Ratio: A:B = 1:6 by weight.

SPECIAL FEATURES: Good adhesion, high mechanical strengths, short setting time, good water and oil resistance, high abrasion resistance, non-toxic.

FORM/DENSITY: Comp. A (white liquid), Comp. B (grey powder); A + B: 2.15 kg/l.

STORAGE: Comp. A store in frost free condition. Comp. B store in dry condition. Shelf life in unopened original containers at least 6 months.

PACKAGING: Predosed sets of 35 kg.

5. SikaTop® 122-HB
Lightweight, High Build Repair Mortar

DESCRIPTION: Two-component, thixotropic, high build, polymer modified lightweight repair mortar.

USE: Repair mortar with excellent bond strength, greatly reduced water, carbondioxide and chloride permeability. Specially formulated for high build applications overhead and vertical.

Scope: For fast repairs especially overhead and vertical of any kind of concrete or mortar surface.

Application: Substrate must be sound, clean and free from oil, grease or other contaminations and saturated to SSD condition. Mechanically mix with forward action mixer or drill and paddle. Do not use normal concrete mixer. Apply by trowel or hand.

Mixing Ratio: A:B = 1:5.5 (by weight).

SPECIAL FEATURES: Excellent adhesion, fast and easy application at layer thickness of approx. up to 40 mm. Very high sag resistance. Good resistance to chlorides, carbon dioxide, deicing salt, etc. Pre-bagged. Non toxic (WRC/NWC approved No. 830255), lightweight, good yield. Use only with bonding bridge.

FORM/DENSITY: Comp. A (white liquid), Comp. B (grey powder); A + B, 1.4 kg/l.

STORAGE: Comp. A store in frost free conditions. Comp. B store in dry conditions. Shelf life in unopened original containers at least 6 months.

PACKAGING: Predosed sets of 10 kg and 32 kg.

6. Sika® 101

DESCRIPTION: One-component cement based waterproofing slurry and moisture seal.

USE: Waterproofing slurry for the protection of concrete structures against infiltration of water under low pressure.

Scope: Basements, culverts, manholes, watertanks, etc.

Application: 2 coats by brush or mortar spraying machine onto properly prepared substrate.

SPECIAL FEATURES: Impermeable, good adhesion to sound concrete, non-toxic, chloride free.

FORM/DENSITY: Grey powder; 1.0 kg/l (bulk density).

CONSUMPTION: Approx. 1.5 kg/m^2 and mm thickness.

STORAGE: Store in dry conditions. Self life in unopened original containers at least 6 months.

PACKAGING: 25 kg bags.

G. EPOXY ADHESIVE AND MORTARS

1. Sikadur® 31/Sikadur® 31 SBA*

DESCRIPTION: Thixotropic, solvent-free, 2-component adhesive on epoxy resin base.

USE: As thin layer adhesive for structural connections, or as mortar for patching and thin protective coating on a wide range of common construction materials, e.g. cement bound, metal, ceramics, polyester, epoxy.

Application: After thorough mixing of the two predosed components A (resin) and B (hardener) with low speed mixer, apply on properly prepared substrate. Brush well into substrate where damp.

Coverage: 1.7-3.4 kg/m^2 (1-2 mm thickness)

Mixing Ratio: 2:1 by weight and volume for L.P. type
　　　　　　3:1 by weight and volume for Normal and Rapid types.

SPECIAL FEATURES: Normal, Rapid and Long Potlife types available. Potlife Normal type at 20° C–40 min.; 30° C–20 min.; Long potlife type at 20° C—90 min.; 35° C—55 min Adhesion on dry and damp but not wet substrates. High resistance to aging and a wide range of aggressive chemicals. Good abrasion resistance.

Minimum application temperature: + 5° C (Normal)
　　　　　　　　　　　　　　+ 20° C (L.P.)

Max. coating thickness: 3 cm.

COLOUR/DENSITY: Comp. A: White, Comp. B: Black, Mix: Concrete grey; 1.7 kg/l.

STORAGE: At least 1 year shelf life when stored in unopened container below 25° C.

PACKAGING: Normal and Rapid: 2 kg and 5 kg sets, L.P. 5 kg sets.

*Sikadur 31 SBA is available in various grades, specially formulated as structural adhesive for segmental bridge construction. Conforms to FIP specifications.

---

2. Sikadur® 32

DESCRIPTION: Solvent-free, 2-component bonding agent on epoxy resin base.

USE: Viscous liquid for binding fresh concrete to old, hardened concrete or mortar.

Application: After thorough mixing of the two predosed components, A (resin), B (hardener), apply by brush, roller or spray equipment to roughened substrate. Place new concrete while Sikadur 32 is still tacky.

Coverage: 0.3-0.8 kg/m^2 (0.2-0.6 mm thickness).

Mixing Ratio: A:B = 2:1 by weight.

SPECIAL FEATURES: Normal and Long Potlife types available. Potlife Normal type at 20° C–25 min; 30° C–15 min. Potlife L.P. type at 20° C–30 min.; 30° C–40 min. Can be applied on dry and damp. but not wet substrates

Minimum application temperature: + 5° C (Normal)
　　　　　　　　　　　　　　+ 20° C (L.P.)

COLOUR/DENSITY: Comp. A: Grey; Comp. B: Yellowish; Mix: Light Grey; - 1.40 kg/l.

STORAGE: At least 1 year shelf life when stored in unopened container below 25° C.

PACKAGING: 1 kg and 5 kg sets.

---

3. Sikadur® 52

DESCRIPTION: Low viscosity, solvent-free, 2-component injection liquid on epoxy resin base.

USE: Pure synthetic resin with low viscosity for filling up of or injection into cracks and cavities with a maximum width of approx. 5 mm.

Application: Mix predosed Comps. A (resin) + B (hardener) and pour or inject into cracks or cavities, using appropriate equipment and method (ask for technical advice).

Mixing Ratio: A:B = 2:1 by weight and volume.

SPECIAL FEATURES: Normal and Long Potlife types available. Potlife Normal type at 20° C–20 min.; 30° C–10 min. L.P. type at 20° C–60 min.; 30° C–30 min. No shrinkage. Excellent adhesion on dry and damp surface. High mechanical strengths and good chemical resistance.

COLOUR/DENSITY: Yellowish; 1.1 kg/l.

STORAGE: At least 1 year shelf life when stored in unopened container below 25° C.

PACKAGING: 2 kg sets.

---

4. Sikadur® 41

DESCRIPTION: Thixotropic, solvent-free, 3-component mortar on epoxy resin base.

USE: Trowel grade mortar with medium sized aggregate filling for repairing, bonding, patching and coating, also in vertical applications. Adheres to cement-bound materials, masonry, wood, ceramics, polyester, epoxy.

Priming: On damp substrates use A + B component of Sikadur 41.

Application: Mix predosed Comp. A (resin) + B (hardener), add Comp. C (aggregates), stir again and supply by hand or trowel to properly prepared substrate.

Mixing Ratio: A:B:C = 3:1:4 by weight Normal and Rapid types
A:B:C = 2:1:3 by weight L.P. type.

SPECIAL FEATURES: Normal, Rapid and Long Potlife types available. Potlife Normal type 20° C–60 min,; 30° C–20 min,; Rapid type 20° C–60 min.; 30° C–30 min., L.P. type 20 °C - 120 min.; 30 °C -60 min. Hardens at low temperature and high humidity. Excellent compressive and flexural strengths. High abrasion, impact and chemical resistance. Good adhesion on dry and damp but not wet substrate.

Min. application temperature: + 5° C (Normal)
+ 20° C (L.P.)

COLOUR/DENSITY: Comp. A: White, Comp. B: Black, Comp. C: Sand; Mix Concrete Grey 1.9 kg/l.

STORAGE: At least 1 year shelf life when stored in unopened container below 25° C.

PACKAGING: Normal and Rapid types: 4 kg and 10 kg sets, L.P. type 10 kg sets. Industrial packing on request.

5. Sikadur® 42

DESCRIPTION: Castable, self-levelling, solvent-free, 3-component mortar on epoxy resin base.

USE: Castable mortar containing aggregates of special qualities and grading for the grouting of bearing plates, machine bases, fixing bolts and anchors, crane rail tracks, and as self-levelling repair, patching and coating mortar. Adheres to cement bound materials, masonry, metals, wood, ceramics, polyester, epoxy.

Application: Mix predosed Comp. A (resin) and Comp. B (hardener), add Comp. C (aggregates), mix again and pour into prepared cavities or on to substrate.

Mixing Ratio: A:B:C = 2:1:12 by weight.

SPECIAL FEATURES: Normal and Long Potlife types available. Potlife Normal type at 20° C–40 min.; 30° C–20 min.; L.P. type at 20° C–120 min.; 30° C–60 min. Excellent flowability. No shrinkage after hardening. High strengths. Suitable for dry and damp, but not wet substrates.

Max. thickness of layer at 30° C: 4 cm.

Min. application temperature: + 5° C (Normal)
+ 20° C (L.P.)

COLOUR DENSITY: Comp. A + B: Transparent; Comp. C: Sand grey; Mix: Concrete grey; 2.0 kg/l.

STORAGE: At least 1 year shelf life when stored in unopened container below 25° C.

PACKAGING: Normal type 2 and 10 kg sets, L.P. type 10 kg sets.

6. Sikadur® 43

DESCRIPTION: Solvent-free, 3-component mortar on epoxy resin base.

USE: Trowel grade, non thixotropic mortar with high degree of aggregate filling for economical heavy-duty repairing, patching and cavity-filling. For cement bound materials, masonry, etc.

Priming: For good adhesion especially on dry substrates, use Sikadur 43 A + B Comp. as primer. On wet substrates use Sikadur 31 as primer. Brush primer well into substrate.

Application: Mix predosed Comp. A (resin) + B (hardener), add Comp. C (aggregates) and mix again. Apply by hand or trowel to properly prepared substrate.

Mixing Ratio: A:B:C = 2:1:30 by weight.

SPECIAL FEATURES: Normal and Long Potlife types available, Potlife Normal type at 20° C–45 min.; 30° C–20 min.; L.P. type at 20° C–60 min; 30° C–45 min. Economical and easy to apply. Curing not affected by high humidity. High mechanical strength. Abrasion and skid resistant. No shrinkage.

Min. application temperature: + 5° C (Normal)
+ 20° C (L.P.)

COLOUR/DENSITY: Comp. A + B: Yellowish, Comp. C: Sand grey, Mix: Sand coloured; 2.0 kg/l.

STORAGE: At least 1 year shelf life when stored in unopened container below 25° C.

PACKAGING: 10 kg sets. Industrial packing on request.

7. Sika® Pronto 11
Self-levelling mortar

DESCRIPTION: Sika Pronto 11 is a 2-component, solvent-free, modified methacrylate mortar system.

USE: For 'rapid' repairs of factory floors, car-parks, roadways, loading ramps, bridge-decks, airport runways and aprons, etc. wherever short down-time is essential.

Application: Mix mechanically with mortar mixer until uniform blend is obtained. Pour Comp. A into mixer first, add Comp. B and mix for approx. 3 min.

For repairs greater than 25 mm depth, add coarse (approx. 10–12 mm) oven dried, clean, well graded aggregate at a rate of approx. 50% of the standard unit size. If trowel grade mortar is required, Sika Pronto 11 can be extended with approx. 5 litres of oven dried quartz sand. Place Sika Pronto 11 to properly prepared substrate, forcing material into substrate, working towards center. No primer is required. Follow detailed application instructions.

Coverage: 2.10 kg/m² per mm thickness.

Mixing Ratio: A:B = 1:8 by weight.

SPECIAL FEATURES: High early and final strength. Very fast curing (short potlife, approx. 10 min. at 20° C). Good adhesion and chemical resistance. Easy to mix. No primer required. Low odor. Min. application thickness approx. 5 mm, min. substrate temperature +10° C. Open to traffic in 1-3 hrs. On dry substrate only.

COLOUR/DENSITY: Concrete grey; 2.10 kg/l.

STORAGE: At least 1 year in unopened original container, stored in dry and cool conditions.

PACKAGING: 16.85 kg sets.

8. Tecasyn® -S

DESCRIPTION: 2-component jointing mortar for acid-resistant tiles, based on epoxy emulsion.

USE: As an acid and water-resistant tile grout where joints are exposed to permanent damp, frequent washing with high-pressure water, corrosion by diluted acids and lyes, e.g. in dairies, breweries and bottling plants for soft drinks, laboratories and food processing plants.

Application: Pour all of component B into component A, mix using low speed electric drill. Apply by filter knife, scraper and metal float into property prepared tile joints. Clean off surplus material with a scouring pad or sponge. Potlife at 20 °C – 3 hrs.

Coverage: Depending on joint width and depth. Allow at least 10% for wastage. Min. joint width 5 mm.

SPECIAL FEATURES: High mechanical strength. Excellent bonding to all tiles. Good chemical resistance. Rapid hardening. Easy to use, excess material can be cleaned off with cold or lukewarm water before material has cured.

Mixing Ratio: A:B = 4:1 by weight.

COLOUR/DENSITY: Grey; 1.6 kg/l.

STORAGE: At least one year shelf life if stored in unopened original containers protected from heat and frost.

PACKAGING: 15 kg sets.

9. Sikadur® 53

DESCRIPTION: Solvent-free, filled, low-viscosity 2-component injection and pouring mortar on epoxy resin base.

USE: As 'water-chaser' for injection and pouring into damp, wet or submerged cracks and for the repair of concrete, steel or wooden structures within formwork. Min. crack width 0.5 mm.

Application: Mix predosed comp. A + B thoroughly. Allow waiting time of 15 min. before injecting or pouring Sikadur 53 into cracks of formwork under water. In underwater applications, best results are obtained if Sikadur 53 is placed through hose into formwork.

Mixing Ratio: A:B = 8:1 by weight.

SPECIAL FEATURES: Excellent adhesion on property prepared wet and immersed substrates. Curing is not affected by humidity. No shrinkage. High density ensures complete water displacement. Good resistance to a wide range of aggressive chemicals. Potlife at 20 °C:20 kg approx. 30 mm.

COLOUR/DENSITY: Mix (A + B), green; 2.0 kg/l.

STORAGE: Shelf life in unopened original containers stored below 25° C at least 6 months.

PACKAGING: 20 kg sets.

10. Sikadur® Injection Gel

DESCRIPTION: 2-component, solvent-free high-modulus moisture insensitive, high-strength structural epoxy-resin paste adhesive. ASTM C881, Type 1, Grade 3, Class B + C.

USE: For structural crack repairs not exceeding 6 mm width, mechanical grouting of bearing pads, machine and 'robotic' base plates, bolts, dowels, pins. Waterproofing tunnels, basements, cable vaults. Re-anchoring of veneer masonry. Woodtruss repairs. Preventive maintenance, e.g. crack grouting on new or existing structures to seal off reinforcing steel from corrosive elements.

Application: To hand mix, pre-mix both components. Proportion components A (resin) and B (hardener) and mix with low speed drill and apply by trowel or suitable injection equipment, depending on application. Sikadur injection Gel is also suitable for application with automatic pressure-injection equipment.

Coverage: 1.5 kg/m² per mm thickness.

Mixing Ratio: A:B = 1:1 by volume.

SPECIAL FEATURES: Non abrasive texture permits application with automated pressure-injection equipment. High-modulus, high-strength adhesive, ideal for vertical and overhead applications. Fast setting. Excellent lubricity for deep penetration. Good adhesion to most structural materials. Min. substrate temperature +5° C. Potlife at 20° C approx. 30 min.

COLOUR/DENSITY: Grey; 1.4 kg/l.

STORAGE: At least 2 years shelf life if stored in unopened original containers below 25° C.

PACKAGING: 5 kg sets, 4-gal, units, 13-fl. oz. disposable coaxial cartridges (16 per case).

H. PROTECTIVE COATINGS AND IMPREGNATIONS FOR CONCRETE

1. Sikagard® 62

DESCRIPTION: Slightly thixotropic, solvent-free, 2-component high-build coating on epoxy resin base.

USE: As protective coating, mainly for concrete structures against weathering and mild to medium range chemical attack in sewage treatment works, tanks and silos, dairies and dyeworks, laundries, garage maintenance bays, etc.

Application: Pour component B into component A and stir well, using lowspeed electric drill until uniform consistency is obtained. Apply by brush or roller onto the property prepared substrate.

Coverage: 0.3 – 1.0 kg per coat, depending on substrate and required coat thickness.

Consumption: For 150 µ dryfilm thickness approx. 0.225 kg/m²

Mixing Ratio: A:B = 3:1 by weight
2:1 by volume

SPECIAL FEATURES: Colour stable (except white) high-build coating. Hard but resilient, excellent adhesion to concrete;

abrasion resistant; one-coat application possible; high mechanical strength. Attention to dew point. Max. coat thickness: vertical 0.2 mm, horizontal 1 mm.

COLOUR/DENSITY: Available in a wide range of colours; 1.3 kg/l.

STORAGE: At least 12 months shelf life if stored in unopened original containers below 25° C.

PACKAGING: 5 kg and 10 kg sets.

2. Sikagard® 63

DESCRIPTION: Slightly thixotropic, solvent-free, 2-component high-build coating with excellent chemical resistance on epoxy resin base.

USE: As protective coating, mainly on concrete, where good protection against chemical attack is required, e.g. in sewage treatment plants, petrochemical plants, dyeworks, tanks and silos etc.

Application: Pour component B into component A and stir well, using lowspeed electric drill until uniform consistency is obtained. Apply by brush or roller onto the property prepared substrate.

Coverage: 0.3–1.0 kg per coat, depending on substrate and required coat thickness.

Consumption: For 150 μ dryfilm thickness approx. 0.230 kg/m².

Mixing Ratio: A:B = 3:1 by weight
2:1 by volume

SPECIAL FEATURES: Exceptional chemical resistance; no shrinkage; high chemical strength; abrasion resistant; good adhesion; easily cleaned; attention to dew point. Minimum dryfilm-thickness for heavy duty corrosion protection is 0.6 mm in 1–3 coats.

COLOUR/DENSITY: Available in olive and red; 1.35 kg/l.

STORAGE: At least 12 months shelf life if stored in unopened original containers below 25° C.

PACKAGING: 5 kg and 10 kg sets.

3. Sikagard® 70
Impregnation for cement based substrates

DESCRIPTION: One-component, solvent-containing, Siloxane based, hydrophobic impregnation with good penetrative properties.

USE: As an impregnation for mineral substrates, e.g. concrete, mortar, brickwork, asbestos cement, etc. against driving rain and dampness. It is mainly used to protect facades and fair faced concrete or cementitious mortar surfaces.

Application: Apply 1–2 coats by brush, roller, or spray equipment to the properly prepared dry substrate. Sikagard 70 is ready for use.

Coverage: 0.2 kg/m² and coat.

SPECIAL FEATURES: Deep penetration into the substrate. Cures tackfree and is highly alkali resistant. Good chemical adhesion to silicates in the substrate. No vapor barrier but excellent impregnation against driving rain. Good protection against deicing salts and freeze/thaw cycles. May be overpainted with emulsion paints. Reduces efflorescence and capillary water absorption.

COLOUR/DENSITY: Colourless; 0.8 kg/l.

STORAGE: At least one year shelf life if stored in unopened original containers in cool conditions.

PACKAGING: 20 kg pails and 170 kg drums.

4. Sikagard® 75 EpoCem
Superfine Epoxy-Cement Sealing Mortar

DESCRIPTION: 3-component epoxy-modified, cementitious flowable mortar.

USE: As a thin-film sealer coat for vertical and horizontal surfaces on concrete, mortar, stone. For making good spalled, pitted and honey-combed concrete etc.

Application: Pour Comp. A into Comp. B container and shake vigorously for 30 seconds. Pour binder mixture into suitable mixing vessel and C Comp. Mix thoroughly for 3 min. using electric stirrer. Apply Sikagard 75 EpoCem by trowel or spatula as appropriate. To achieve fine surface finish, work over with moistered rubber sponge or distemper brush.

Coverage: 2.5–3.5 kg/m² per 1–1.5 mm applied thickness.

SPECIAL FEATURES: Ideal surface preparation for subsequent application of other Sika Epoxy coatings. Reduced waiting time for follow-up epoxy coating applications. Waterproof. Water-vapor permeable. For internal and external use. Easy to apply. Min. temperature of substrate +8° C, max. +25° C.

Mixing Ratio: A:B:C = 1:2.73: 15–17.7 by weight.

COLOUR/DENSITY: Dark grey; 2.0 kg/l. (mix).

STORAGE: At least 6 months shelf life if stored in unopened original containers in dry and frost free conditions.

PACKAGING: 23 kg sets.

5. Sikagard® 64

DESCRIPTION: Thixotropic, 2-component high-build coating on coal-tar/expoxy resin base.

USE: As economical high-quality protective coating mainly on concrete against weathering and chemical attack, in sewage treatment plants, effluent pits, foundations in aggressive soils, channels, culverts, tanks and pipes.

Application: Pour component B into component A and stir well, using lowspeed electric drill until uniform consistency is obtained. Apply by brush or roller or airless spray gun onto the properly prepared substrate.

Coverage: 0.5–1.0 kg per coat, depending on substrate and required coat thickness.

Consumption: For 150 μ dryfilm thickness approx. 0.230 kg/m².

Mixing Ratio: A:B = 1:3 by weight
1:2.6 by volume

SPECIAL FEATURES: Economical. Hard but resilient, high film thickness. Excellent adhesion to concrete. Good chemical resistance. Attention to dew point. Not suitable for contact with drinking water.

COLOUR/DENSITY: Available in black and red/brown; 1.35 kg/l.

STORAGE: At least 12 months shelf life if stored in unopened original containers below 25° C.

PACKAGING: 10 kg sets.

6. Sikagard® 67

DESCRIPTION; Slightly thixotropic, solvent-free, 2-component high-build coating on epoxy resin base.

USE: As all purpose protective emulsion coating for concrete, rendering, stone, asbestos cement, wood, e.g. cantilevers, galleries, retaining walls, road tunnel walls, balconies, workshops, etc.

Application: Stir component A well, then pour into component B and mix well, using lowspeed electric drill until uniform consistency is obtained. Apply by brush, roller or spray gun onto the propely prepared substrate. High absorbent substrates have to be wetted first.

Coverage: 0.15–0.25 kg per coat, depending on substrate.

Consumption: For 100 µ dryfilm thickness approx. 0.3 kg/m². Apply at least 2 coats.

Mixing Ratio: A:B = 1:1 by weight
                       3:4 by volume

SPECIAL FEATURES: Solvent-free, no fire hazard. Good adhesion to damp substrates. Weather proof and mildew resistant, odourless, easy to apply, thixotropic, no runs on vertical surfaces. Protect from rain for up to 18 hrs. after applications, depending on temperature.

COLOUR/DENSITY: Available in a wide range of colours; 1.3 kg/l.

STORAGE: At least 12 months shelf life if stored frost free and below 25° C in unopened original containers.

PACKAGING: 5 kg and 10 kg sets.

7. Sikagard® 660 Glaze

DESCRIPTION: 1-component water-dilutable protective coating on synthetic resin base.

USE: As surface treatment for aesthetic purpose on concrete and mortar, to uniform different greys on fair faced concrete structures and to protect concrete structures against efflorescence and discoloration.

Application: By brush, roller or airless spray gun onto properly prepared substrates. Porous substrates to be impregnated with Sikagard 65. First coat can be diluted with water 10-20%.

Coverage: 0.2–0.3 kg/m² for 2 coats, depending on substrate.

SPECIAL FEATURES: Excellent weather resistance and colour stability. Easy to apply. No change in the character of fair faced concrete mat finish. Carbonation barrier.

COLOUR-DENSITY: Light concrete grey, with colour paste to match colour on site; 1.4 kg/l.

STORAGE: At least 6 months shelf life if stored frost free and below 25° C in unopened original container.

PACKAGING: 20 kg pails.

8. Sikagard® 550 Elastic Top

DESCRIPTION: One-component, plasto-elastic coating with crack bridging properties based on ethylene-copolymer dispersions.

USE: As a protective/decorative coating on a wide range of concrete structures subject to cracks. Also applicable on cement renderings, brick work, asbestos cement, etc.

Application: Stir Sikagard 550 Elastic Primer and Sikagard 550 Elastic Top thoroughly. On dense substrates Sikagard 550 Elastic Primer should be diluted with up to 30% Sika Thinner 12. Apply Sikagard 550 Elastic Primer by brush, roller or airless spray onto properly prepared (dry, solid, free from loose and friable particles) substrate. Apply Sikagard 550 Elastic Top either by brush, roller or airless spray. 1 × Sikagard 550 Elastic Primer, 2-3 × Sikagard 550 Elastic Top. Min. application temperature of substrate +5°C.

Coverage: Material consumption per coat Sikagard 550 Elastic Primer 0.2–0.25 kg/m², Sikagard 550 Elastic Top 0.3–0.4 kg/m².

Consumption: For 150 µ dryfilm thickness approx. 0.4 kg/m²

SPECIAL FEATURES: Sikagard 550 Elastic Top has crack bridging properties and provides coats of varying texture. With a layer thickness of 0.3 mm, crack movements of at least 0.3 mm width can be absorbed. Sikagard 550 Elastic Top is resistant to the usual aggressive agents in the atmosphere. As Sikagard 550 Elastic Primer contains solvents, provide adequate ventilation in confined areas and keep away from fire. Not to be applied on permanently wet surfaces.

COLOUR/DENSITY: Available in various colours; Sikagard 550 Elastic Primer —0.96 kg/l. Sikagard —550 Elastic Top 1.30 kg/l.

COLOUR/DENSITY: Sikagard 550 Elastic Primer 20 kg pails, Sikagard 550 Elastic Top 25 kg pails.

9. Igol® A

DESCRIPTION: One-component, ready for use Phenol-free bitumen coating. Conforms to BS 3416 Type 1.

USE: For the protection of concrete and steel structures, e.g. drinking water reservoirs, food-stuff silos, watertanks, pipes, etc.

Application: Apply by brush, roller or spray at least 2 coats. When spraying, first coat should be diluted with 5% Sika Thinner13.

Coating systems: Cementitious substrates 2 coats, metal substrates 2–3 coats of Igol A.

Consumption: Cementitious substrates 0.3–0.5 kg/m² for 2 coats, metal substrates 0.2–0.4 kg/m² for 2–3 coats.

SPECIAL FEATURES: Phenol free, suitable for contact with drinking water, ready for use, no mixing. Resistant to chlorinated water. Excellent adhesion to properly prepared substrates. Do not use in exposed applications. Flammable, contains solvents. Provide ventilation if applied in confined spaces.

COLOUR/DENSITY: Black; 0.9 kg/l.

STORAGE: Shelf life at least 12 months in unopened original containers, when stored in dry condition below 35° C.

PACKAGING: 5 kg pails, 25 and 180 kg drums.

10. Igol® T

DESCRIPTION: One-component, ready for use coal tar coating.

USE: As protective coating both externally and internally in substructures e.g. foundations, basements, canals, reservoirs, sewage works, etc.

Application: Apply at least 2 coats by brush, roller or spray onto dry substrates only. When spraying, dilute first coat with 5% Sika Thinner 12.

Coating Systems: Cementitious substrates 2 coats, metal substrates 3 coats of Igol T.

Consumption: Cementitious substrates 0.3–0.5 kg/m² for 2 coats, metal substrates 0.3–0.4 kg/m² for 3 coats.

SPECIAL FEATURES: Good resistance to diluted acids and alkalis, mildly aggressive soils, salts, sewage and waste water. Ready for use. Good adhesion. Not suitable for contact with food stuff and drinking water. Contains solvents, ensure good ventilation if applied in confined spaces.

COLOUR/DENSITY: Black; 1.1 kg/l.

STORAGE: Shelf life at least 12 months in unopened original containers, when stored in dry condition below 35° C.

PACKAGING: 25 and 200 kg drums.

11. Igolflex®

DESCRIPTION: Flexible, solvent-free, high-build coating for waterproofing on bitumen synthetic rubber base.

USE: Ready for use protective, and waterproofing compound for foundations and underground concrete structures. Externally installed vapourbarrier for unventilated structures which will receive internal synthetic coatings.

Application: By trowel or special spray pump on properly prepared substrate in 1 or 2 coats; protect from rain for 2-3 hours. For porefree waterproof membrane always apply in 2 coats.

Coverage: 2-3.5 kg/m² depending on site conditions and job requirement.

SPECIAL FEATURES: Easy application. Thixotropic, odorless, nonflammable. Good weather resistance and aging

properties. Can bridge hair and shrinkage cracks. Permanently plasto-elastic. Good adhesion, also on damp surface. Temperature resistance 30° C to +80° C. Resistant to water, seawater, acid, alkaline and sulphate waters.

COLOUR/DENSITY: Black paste; 1.0 kg/l.

STORAGE: 1 year shelf life if stored in original unopened container. Protect from frost.

PACKAGING: 25 kg and 180 kg drums.

12. Igasol® Liquid

DESCRIPTION: Solvent-free bituminous emulsion containing rubber.

USE: Protective and water resistant bituminous coating for concrete foundations, retaining walls, columns, rendering, plaster, wood, etc. Separation coating for concreting work.

Application: By brush or spray equipment.

Coverage: 1—coat applications 0.2—0.3 kg/m²
       2—coat applications 0.3—0.5 kg/m².

SPECIAL FEATURES: Ready for use, solvent-free, odorless. Can be applied to damp surfaces. Stir well before use. Dry content approx. 60%. Non toxic.

COLOUR/DENSITY: Black liquid; 1.0 kg/l.

STORAGE: At least 9 months shelf life if stored in well sealed original container. Protect from frost.

PACKAGING: 25 kg and 180 kg drums.

13. Conservado® 5

DESCRIPTION: Water-repellent, one-component impregnation on silicone resin base. Conforms to A. F. Specs. SS-W-00110a.

USE: As a water-repellent, colourless surface treatment on cement-bound and other porous mineral material exposed to rainfall (facings, etc.). Reduces capillary water absorption without preventing passage of vapour. Extends life and preserves appearance of surface by improving frost resistance and minimizing efflorescence, penetration of waterborn dirt and corrosive substances.

Application: By brush, roller or spray-equipment.

Coverage: On renderings 2 coats at 0.2-0.35 kg/m². On concrete 2 coats at 0.2-0.3 kg/m².

SPECIAL FEATURES: Minimum application temperature +5° C. Apply only to dry surfaces. Resists weathering for many years. Not suitable against water under pressure.

COLOUR/DENSITY: Clear; 0.8 kg/l.

STORAGE: 1 year shelf life when stored in well sealed container.

PACKAGING: 20 kg and 150 kg drums.

14. Conservado® HP

DESCRIPTION: 1-component, coloured protective coating on liquid Hypalon base.

USE: As protective coating on concrete and rendering (thin shelled concrete structures, prestressed concrete units, foundation walls, balcony parapets, masonry, asbestos cement, roofs and linings.

As corrosion protective coating to metal in industrial plants, gutters, metal roofs, etc.

Application: By brush and roller. From second coat on also by high pressure gun. 2-3 coats are normally applied.

Consumption: On concrete 0.2-0.25 kg/m² and coat, on metal and new asbestos cement 0.1-0.15 kg/m² and coat.

SPECIAL FEATURES: Ready for use. Forms tough, elastic, weather resistant coat. Waterproof, oil and good chemical resistance. Not suitable for contact with drinking water.

COLOUR/DENSITY: Available in various colours: 1.1 kg/l.

STORAGE: At least 12 months shelf life if stored in unopened original containers, below 25° C.

PACKAGING: 10 kg pails.

15. Icosit® Aqualastic
Elastic, crack bridging coating

DESCRIPTION: One-component, solvent-free, emulsion-type crack- bridging, flexible coating system on styrene acrylic dispersion base.

USE: As a seamless, impervious coating for roofs, particularly for repair and refurbishment. Suitable for localized repair and sealing works, also for internal sealing of gutters. May be applied to the following substrates: bituminous felt, concrete, asbestos cement, aluminium, zinc, steel, copper, PVC, polyester etc.

Application: By brush, trowel, squeegee or airless spray onto the properly prepared substrate in one coat. Reinforcement suitable for large cracks, or large crazed areas available.

Coverage: -2.160 kg/m² for a dryfilm thickness of 1 mm incl. 20% wastage.

SPECIAL FEATURES: Icosit Aqualastic forms an elastic, crack bridging, impervious membrane , suitable for occasional light foot traffic. Good adhesion to many substrates. Low water vapor diffusion resistance. Resistant to urban, rural, industrial and marine environment. No primer required. Min. dry film thickness 1 mm. Min. application temperature +5° C.

COLOUR/DENSITY: Available in 4 standard colours; 1.34 kg/l.

STORAGE: At least 6 months shelf life if stored in unopened original containers in cool conditions.

PACKAGING: 17.5 litre pails.

I. FLOOR TREATMENTS AND TOPPINGS

1. Sikafloor® 81/82 EpoCem
Self-Levelling epoxy-cement floor toppings

DESCRIPTION: 3-component, solvent-free, cementitious, epoxy- modified self-levelling floor topping.

USE: Self-levelling toppings of 1.5–3 mm thickness (Sika-floor 81) and 4–7 mm thickness (Sikafloor 82) to level or patch concrete floors, both unfinished and after grinding or planing. As floor topping on damp, non suspended substrates, where aesthetic appearance is not a prime consideration. As levelling layer beneath Epoxy coatings. Extended with quartz sand, used as patching and repair mortar for surfaces to be coated with epoxy resins. Designed for use on mineral based substrates.

Application: Substrate must be clean, sound, free from loose material, laitance, oil and grease, and surfaces dry (no standing water). Shake Comp. A briefly, pour into Comp. B container and shake vigorously for approx. 30 seconds. (Use mixed Comp. A + B as primer.) Pour mixture (A + B) into suitable container, add Comp. C and mix thoroughly for 3 min. with electric stirrer. Pour the mix onto still wet primer coat and level using flooring trowel. Use spiked roller to ensure uniform thickness and remove entrained air. If non slip finish is required, broadcast with Sikadur 501 quartz sand.

Coverage: Primer coat (Comp. A + B): 0.25-0.30 kg/m² depending on substrate. Topping mix Sikafloor 81 (Comp. A + B + C): 4 kg/m² for a thickness of 2 mm, Sikafloor 82 (Comp. A + B + C): 9 kg/m² for a thickness of 4 mm. Min. application temperature +8° C.

SPECIAL FEATURES: Solvent free, easy application. Good flowability. Waterproof. Can be overcoated with Epoxy floor coatings within 24 hours. Permeable to water vapor. Excellent adhesion and mechanical strengths. Excellent resistance to water and oil. Non corrosive.

Mixing Ratio: A:B:C = 1:2.73:14–16 by weight for Sika-floor 81
1:2.73:20—22.7 by weight for Sikafloor 82

COLOUR/DENSITY: Grey; 2.0 kg/l, Sikafloor 81.
2.1 kg/l, Sikafloor 82.

STORAGE: At least 6 months shelf life if stored in unopened original containers in dry and frost-free conditions.

PACKAGING: Sikafloor 81:21 kg sets.
Sikafloor 82:28 kg sets
EpoCem Module (A+B Comp.) 4 kg.
Industrial packing on request.

2. Sikafloor® 93
Non-slip levelling floor topping

DESCRIPTION: Solvent-free, 3-component, pigmented, self-levelling mortar screed on epoxy resin base.

USE: For dense, non-slip, abrasion resistant, self-levelling floor coatings of 3–4 mm thickness in workshops and factories, in wet areas of the beverage industries, meat processing and bottling plants, dairies, warehouses, loading bays, ramps and hangers, etc. On mineral based substrates.

Priming: Not required.

Application: Mix Comp. A + B (A = resin, B = hardener), add Comp. C (aggregate) and mix again, using suitable equipment. Apply base coat in even layer using notched trowel (6-8 mm notches). Remove entrapped air with spiked roller. Depending on desired finish, broadcast uniform layer of sand into level base layer (different grades available). As soon as curing rate allows foot traffic, remove excess sand and seal floor surface with one coat of Sikafloor 93 (A + B Comp. only).

Coverage: Base coat 4 kg/m² for 2 mm thickness, non slip surface finish 3-4 kg/m² for 1-2 mm thickness (sand). Sealer coat 0.5 kg/m² (A + B Comp. only).

SPECIAL FEATURES: No priming required. Easy application system, designed for one-day application. Good chemical resistance, impermeable to liquids. Non slip, non skid even when contaminated with oil, grease or water. Excellent abrasion resistance and high mechanical strengths. Apply to mineral based substrate only. Substrates must be ventilated or protected by vapor barrier. Min. age of concrete/mortar screeds, 4 weeks. Min. application temperature +8° C.

Mixing Ratio: A:B:C = 10:3:16 by weight.

COLOUR/DENSITY: Available in 5 standard colours; 2.0 kg/l.

STORAGE: At least 6 months shelf life if stored in unopened original containers in cool conditions.

PACKAGING: A + B Comp. in 220 and 180 kg drums or 30 kg and 25 kg pails respectively. C Comp. in 25 kg bags.

3. Sikafloor® 2415/2430

DESCRIPTION: 2-component, solvent containing coatings on epoxy resin base.

USE: For colourless and coloured floor improvement, as impregnator and sealer, for heavily exposed floor and wall surfaces of concrete and cement screeds, e.g. factories and warehouses, car parks, aircraft maintenance hangars, car repair shops and food industries.

Application: Never exchange B component of sikafloor 2415 and 2430. Mix A and B components with electric slow speed mixer. Before applying sikafloor 2415/2430 by brush onto dry properly prepared substrate, let mixed coating mature for appox. 15 minutes.

Coating system: Depending on site conditions and intended results, different coating systems may be applied. priming with sikafloor 150 may be necessary.

Minimum application temperature of substrate: + 10° C.

Waiting time between coats: 6-12 hours, depending on temperature, never exceed 48 hrs.

Potlife: ~6 hrs. at 20° C.

Mixing Ratio: Sikafloor 2415 A:B = 80:20, Sikafloor 2430 A:B = 70:30 by weight.

Coverage: Sikafloor 2415/2430 – 0.10 – 0.20 kg/m² per coat, depending on site conditions.

SPECIAL FEATURES: High abrasion, good chemical and temperature resistance. Non toxic when dried. Provide adequate ventilation if applied in confined area.

COLOUR/DENSITY: Sikafloor 2415 colourless; 0.92 kg/l., 2430 various grey colours: 1.16 kg/l.

STORAGE: Shelf life in unopened original container approx. 1 year.

PACKAGING: 2415 in 10 and 20 kg sets, 2430 in 10 and 25 kg sets.

4. Sikafloor® 2630

DESCRIPTION: Water-soluble, solvent-free, 2-component coating on epoxy resin base.

USE: Sealer and coating for concrete, cement screed and mastic asphalt (indoors only) for normal and medium mechanical wear, e.g. in food stuff industries.

Application: Stir each component thoroughly, mix together, preferably using electric slow speed mixer and apply onto dry, properly prepared substrate by brush, roller or airless spray. First coat should be applied by brush.

Coating System: Normally 1–2 coats Sikafloor 2630. Priming with Sikafloor 150 may be necessary. Minimum application temperature of substrate: + 10° C. Waiting time between coats at 20° C approx. 15 hrs., never exceed 48 hrs.

Potlife: At +20° C max. 2 hrs., never to be exceeded. End of potlife not recognizable by change in viscosity.

Mixing Ratio: A:B = 70:30 by weight.

Coverage: 0.250 kg/m² per coat, depending on surface conditions.

SPECIAL FEATURES: Cures by evaporation of water and reaction of the two components. Dries to a silky gloss finish that reduces skidding. Good abrasion, weather, water-resistance.

Temporary resistant to mineral fuels, lubricants and numerous industrial chemicals. Non toxic when cured. Provide adequate ventilation if applied in confined areas to allow water to evaporate. Protect from rain for approx. 24 hours.

COLOUR/DENSITY: Available in a number of RAL colours: 1.3 kg/l.

STORAGE: Shelf life in unopened original containers approx. 12 months.

PACKAGING: 5 and 14 kg sets.

5. Purigo® 5S
Surface Hardener and Dust Proofer

DESCRIPTION: Sodium silicate based surface binder and 'dust proofer'.

USE: As a surface binder and dust proofer on old or new concrete floors and granolithic paving.

Application: New concrete must be at least 3 weeks old. Substrate should be dry, sound, clean and free from oil, grease dust, etc. Apply 2 coats undiluted either by plastic watering can or by brush or squeegee. Do not allow to form puddles.

Coverage: One litre covers approx. 4-6 m² per coat.

SPECIAL FEATURES: Purigo 5S reacts chemically with free lime present in the cement to produce a 'case hardened' surface. Improves resistance to attack from oil and grease. Reduces dusting and moisture penetration. Ready for use. Easy application.

COLOUR: Transparent liquid.

STORAGE: No shelf life if stored in unopened original containers.

PACKAGING: 200 kg drums.

6. Kemox® A

DESCRIPTION: Metallic hardening compound for concrete floors.

USE: As hard-wearing, tough and durable industrial floor surfaces that are static dispersing and spark-resistant.

Application: Kemox A is applied to the freshly placed concrete or screed according to manufacturer's instructions.

SPECIAL FEATURES: *Forms monolithic abrasion resistant floors of high impact resistance. Long wearing, easy to clean. Extremely economical industrial floor where extreme wear resistance is required.*

Consumption: *Depending on intended wear between 1.4 kg/m² (light duty) to 5.5 kg/m² (extra heavy duty).*

FORM: Dry mix metallic grey aggregate.
STORAGE: No shelf life if stored in dry conditions.

PACKAGING: 25 kg multi-wall polyethylene-lined paper bags
25 kg net weight steel pails.

7. Sikafloor® 6516/6530

DESCRIPTION: 1-component, ready-for-use, solvent containing, reaction curing polyurethane sealers.

USE: For improving, impregnating and sealing (Sikafloor 6516 transparent, Sikafloor 6530 coloured) heavily worn concrete floors and screeds in factories, warehouses, car parks, aircraft hangers, garages, etc.

Application: Stir thoroughly, using low speed electrical drill and apply by roller, brush, or spray onto properly prepared substrate according to instructions.

Coating Systems: For floors with normal wear 1 coat Sikafloor 6516 or 6530 thinned 1:1 with Thinner R and one coat of Sikafloor 6516 or 6530 unthinned. For floors with heavy wear 1 coat Sikafloor 6516 thinned 1:1 with Thinner R and 2 coats of unthinned Sikafloor 6516 or 6530.

Potlife: At 20° C and 40–60% relative air humidity approx. 12 hours. Higher temperatures or higher air humidity reduce potlife.
Coverage: Sikafloor 6516–0.1–0.2 kg/m² and coat.
Sikafloor 6530–0.15–0.2 kg/m² and coat.
depending on substrate conditions.

SPECIAL FEATURES: Treated floors become highly resistant and completely dustfree. Good abrasion resistance, easy to clean. Resistant to water, mineral fuels, many vegetable and synthetic oils and grease as well as numerous industrial chemicals. Not to be used on vacuum concrete.

COLOUR/DENSITY: Sikafloor 6516 yellowish/transparent; ~1.0 kg/l.
Sikafloor 6530 available in 4 standard colours; ~1.4 kg/l.

STORAGE: Shelf life in unopened original containers stored in cool, dry conditions at least 12 months.

PACKAGING: Sikafloor 6516 in 10 kg and 25 kg pails.
Sikafloor 6530 in 12.5 kg and 25 kg pails.

8. Sikafloor® 7530
Antiskid Floor Coating

DESCRIPTION: Sophisticated, low-solvent content, 2-component epoxy resin based floor coating with good skid resistance.

USE: Skid resistant floor coating for concrete subjected to normal and medium heavy wear in industry, especially wet areas, e.g. washrooms, textile plants, laundries, food processing industries, power stations, garages, warehouses, etc.

Priming: For normal exposure dilute first coat of Sikafloor 7530 with Thinner 15. For heavy exposure apply 1 coat of Sikafloor 94 Primer.

Application: Stir comp. A (resin) well, before mixing comp. A and B (hardener) intensively using electric stirrer and apply onto properly prepared substrate observing primer application instructions, using brush or roller. Top coat is applied by short piled lambskin roller and finished off with a textured foam rubber roller. To improve skid-resistance in wet or greasy environment, 10% quartz sand may be added to the top coat. Blinding of top coat with quartz sand would further improve skid resistance.

Coverage: 0.9 kg/m² for an average dryfilm thickness of 0.35 mm (actual thickness of peaks ~0.7 mm).

Mixing Ratio: A:B = 90:10 by weight

SPECIAL FEATURES: Tough, hard, skid and abrasion resistant coating. Good resistance to mechanical and chemical exposure. Easy to clean. Min. application temperature +10° C.

COLOUR/DENSITY: Available in 3 standard colours; 1.87 kg/l.

STORAGE: At least one year shelf life if stored in unopened original containers in cool conditions.

PACKAGING: 12.5 kg sets.

9. Sikafloor® 91

DESCRIPTION: Solvent-free, heavy duty, 3-component floor mortar on epoxy resin base.

USE: Trowel grade mortar with high degree of aggregate filling. For heaving duty industrial floor coverings, e.g. workshops, warehouses, loading ramps, garages. On cement bound (dry and damp) substrates, metal etc.

Priming: Use Sikafloor 94 Primer.

Application: Mix Comp. A (resin) + B (hardener), add Comp. C (aggregates), stir again and spread on still tacky substrate.

Coverage: According to coat thickness (4–10 mm)–2 kg/m² and mm thickness; Primer consumption–0.2–0.3 kg/m². Minimum application temperature: 5° C.

SPECIAL FEATURES: High mechanical strengths. Good abrasion and skid resistance. High resistance against a wide range of chemicals. Waterproof.

Potlife at 20° C–45 min.; 30° C–20 min.

Mixing ratio: A:B:C = 2:1:21 to 2:1:30 by weight.

COLOUR/DENSITY: Various colours available; 2.2 kg/l.

STORAGE: At least 1 year shelf life when stored in unopened container below 25° C.

PACKAGING: A + B Comp. in 10 and 20 kg pails or 180 kg and 200 kg drums, C Component in 50 kg bags.

10. Sikafloor® 92

DESCRIPTION: Solvent-free, 3-component, self-levelling floor topping on epoxy resin base.

USE: Self-levelling mortar for seamless floor toppings in industries with medium degree of exposure, power plants, exhibition halls, shops, pharmaceutical industries, laboratories, clean rooms, etc. On cement bound (dry and damp) substrates.

Priming: Use Sikafloor 94 Primer.

Application: Mix Comp. A + B (A = resin, B = hardener), add Comp. C (aggregate), stir again and spread with bedding comb on even and still tacky surface and detrain entrapped air with spiked roller.

Coverage: 4 kg/m² at 2.3 mm coat thickness.

Primer consumption 0.3–0.5 kg/m².

Min. application temperature: 5° C.

SPECIAL FEATURES: High mechanical strengths. Good abrasion resistance. Smooth, dense surface for high hygene requirements. Good decontamination properties. High chemical resistance. Waterproof, Potlife at 20° C –50 min.

Mixing Ratio: A:B:C = 100:27:150 by weight.

COLOUR/DENSITY: 6 colours available; 1.67 kg/l.

STORAGE: At least 6 months shelf life when stored in unopened container below 25° C.

PACKAGING: 30 kg sets, industrial packing on request.

11. Sikafloor® 95

DESCRIPTION: Solvent-free, 2-component, self-levelling floor coating on polyurethane-modified epoxy resin base.

USE: Crack-bridging, abrasion, skid and chemically resistant flooring material for bottling plants, dairies, pharmaceutical plants, laundries, etc., where medium wear is expected.

Priming: Use Sikafloor 94 Primer.

Application: Mix Comp. A + B and apply onto properly prepared, primed substrate using tooth trowel and spiked roller.

Coverage: Sikafloor 94 Primer 0.2–0.5 kg/m². Sikafloor 95-1.7 kg/m² for 1.5 mm layer thickness.

SPECIAL FEATURES: Excellent abrasion and skid resistance (also wet), good elasticity, no shrinkage, easy, fast application. Waterproof, seamless, and crack-bridging. Application temperatures of substrate: min. 10° C, max. 30° C. Potlife at 20° C–30 min., 30° C–15 min.

COLOUR/DENSITY: 5 colours available; 1.15 kg/l.

STORAGE: Shelf life in unopened original container approx. 6 months.

PACKAGING: 8 kg and 30 kg sets.

12. Sikafloor® 96/97 Antistatic

DESCRIPTION: Antistatic coloured coating system on epoxy resin base for non-conductive substrates (concrete, mortar, etc.). Meets DIN 51953, IOS 1853, BS 2044-2050, ANSI/UL 779.

USE: As coloured, antistatic flooring system for e.g. electronic laboratories, manufacturing and assembly plants for electric and electronic components and appliances, ammunition manufacturing plants and depots, solvent bottling plants, hospitals, computer rooms and wherever a dust and static-free environment is essential.

Application: Stir each component thoroughly, mix together using electrical slow speed mixer and first apply Sikafloor 96 AS conductive coat onto propely prepared substrate (primed with Sikafloor 94), by brush or short haired roller. Install necessary earth-connection. Apply high-build topping Sikafloor 97 AS after approx. 12 hrs. Min. substrate temperature +10° C. Pay attention to dew point.

Potlife: Sikafloor 94 Primer   20 min. at 20° C
*   Sikafloor 96 AS   8 hrs. at 20° C*
*   Sikafloor 97 AS   40 min. at 20° C*

Mixing Ratios: Sikafloor 94 A:B = 2:1 by weight
*   Sikafloor 96 AS A:B = 3:1 by weight*
*   Sikafloor 97 AS A:B = 4:1 by weight*

Coverage: Sikafloor 94 Primer   0.2–0.3 kg/m²
*   Sikafloor 96 AS   0.1 kg/m²*
*   Sikafloor 97 AS   2.0–2.5 kg/m² for*
a 1.5–2 mm coat thickness.

SPECIAL FEATURES: Produces electrically conductive floor coatings to international standards. Good chemical and abrasion resistance. Easy to clean. Attractive cost/benefit ratio.

COLOUR/DENSITY: Sikafloor 96 AS black; 1.0 kg/l.
Sikafloor 97 AS grey **and** brown/red; 1.4 kg/l.

STORAGE: Shelf life in unopened original containers stored below 25° C at least 6 months.

PACKAGING: Sikafloor 96 AS and 97 AS in 10 kg sets.

13. Sikafloor® 94 Primer

DESCRIPTION: Low viscosity, solvent-free, 2-component epoxy primer.

USE: As a primer, penetrating sealer and bonding agent on concrete, mortar, stone, timber, etc. As a bonding agent for epoxy mortar screeds and self-levelling floor toppings, as well as for impregnating and sealing cement based floors in warehouses, garages, boiler rooms, etc.

Application: Mix Comps. A (resin) and B (hardener) until uniform mix is obtained. Pour mix onto the substrate, then brush or roll out until coverage is uniform (not too thick). Application by airless spray gun is also possible.

Coverage: ~ 0.2-0.3 kg/m² depending on substrates.

SPECIAL FEATURES: Excellent penetration, high mechanical strengths. Easy batching by weight and volume. Min. application temperature +5° C.

Mixing Ratio: A:B = 2:1 by weight and volume.

COLOUR/DENSITY: Yellowish transparent: ~ 1.05 kg/l.

STORAGE: At least one year shelf life if stored in unopened original container in cool conditions.

PACKAGING: 10 kg sets. Industrial packing on request.

14. Chapdur®

DESCRIPTION: Non metallic floor hardener on mineral base.

USE: For superficial reinforcing of floor slabs and screeds, where good abrasion resistance is required, e.g. warehouses, quays, engineering workshops, parking lots, service stations, garages, etc.

Application: Chapdur is applied to freshly placed floor slabs and screeds according to manufacturer's instructions.

SPECIAL FEATURES: Easy to apply, forms monolithic structure with floor slabs or screeds. Very good abrasion resistance. Moh's hardness: 7-8 (steel core). Wear resistance according to Taber, weightloss 1% after 500 revolutions, 1.4% after 1000 revolutions. (Test report of CEPTP LAB.642.6.785.)

CONSUMPTION: 3-4 kg per m² according to job requirements.

COLOUR/DENSITY: Grey or red powder; apparent density ~1.4 kg/l.

STORAGE: No shelf life if stored in undamaged original container.

PACKAGING: 50 kg bags.

J. CONCRETE AND STEEL PROTECTORS

1. Inertol® Poxitar/Poxitar F

DESCRIPTION: Highly resistant 2-component protective coating on epoxy/coal tar pitch base with mineral fillers. Poxitar F is suitable for application on wet substrates and underwater. Conforms to BS 5393; KG3B and KG3D

USE: As internal and external protective coating on steel, concrete, above and below ground, as well as submerged, e.g. sewage works, hydraulic steel strcutures, ports, locks, weirs, chemical industry etc.

Application: Stir A component thoroughly, add B component and mix to uniform consistency, preferably using electric stirrer. Apply by brush, roller, high pressure spray gun or airless spray onto dry (Inertol Poxitar F) and properly prepared substrate. Use of Primer and number of coats depend on application. Min. application temperature of substrate +10° C.

Mixing Ratio: Inertol Poxitar: A:B = 88:12, Inertol Poxitar F A:B = 85:15 by weight.

Potlife: Inertol Poxitar: 6 hrs. at 20° C. Inertol Poxitar F: 90 min. at 20° C.

Coverage: Material consumption incl. 20% loss for medium dry film thickness of 150 microns: Inertol Poxitar: 0.42 kg/m², Inertol Poxitar F: 0.37 kg/m².

SPECIAL FEATURES: Cures to a hard, robust weather and abrasion resistant coating. Resistant to water, seawater barnacles, diluted acids and lyes, neutral salts, mineral oils. Not permanently resistant to aromatic hydrocarbons, tar oil. Inertol Poxitar F can be applied on wet surfaces of under water.

COLOUR/DENSITY: Black and red; Inertol Poxitar–1.7 kg/l, Poxitar F–1.8 kg/l.

STORAGE: Shelf life in unopened original containers approx. 6 months.

PACKAGING: 15 and 30 kg sets.

2. Inertol® 49W/49W Thick

DESCRIPTION: Inertol 49 W and 49 W Thick are 1-component, solvent-containing, phenol-free coatings on bitumen base. Inertol 49W complies to BS 3416 Type II, Inertol 49W Thick is approved by the National Water Council of Great Britain.

USE: For the protection of steel and concrete in contact with drinking water.

Application: Stir product thoroughly. At low temperatures or for dense substrates up to 3% Thinner B should be added. Apply by brush, high pressure cup gun or airless spray onto dry and properly prepared substrate.

Coating System: Steel: 3–4 × Inertol 49 W Thick. For heavy exposure: 1 × Friazinc R, 3 × Inertol 49 W Thick.

Concrete: without special exposure, 3–4 × Inertol 49 W; heavy exposure, 2–3 × Inertol 49 W Thick.

Coverage: Material consumption with 20% loss for medium dry film thickness of 80 microns: Inertol 49 W Thick –0.21 kg/m². Inertol 49 W Thick Red –0.24 kg/m². 60 microns: Inertol 49 W 0.14 kg/m².

SPECIAL FEATURES: Inertol 49 W Thick is also available in red colour for easy identification of intermediate coats. Resistant to contact with water, chlorinated water, diluted acids and lyes as well as neutral salts. Not resistant to petrol, aromatic hydrocarbons and fats. Taste and odorless after drying. During application in confined areas provide adequate ventilation.

COLOUR/DENSITY: Black, Inertol 49 Thick also red. Inertol 49 W–0.9 kg/l. Inertol 49 W Thick–1.2 kg/l. Inertol 49 W Thick Red –1.4 kg/l.

STORAGE: Shelf life practically unlimited if stored in unopened original container.

PACKAGING: 25 kg pails.

3. Inertol® 100 G

DESCRIPTION: 2-component, high-build reaction coating on epoxy hydrocarbon resin combination base with mineral fillers.

USE: As highly resistant anticorrosion coating for pipes, fittings and other cast iron and steel parts. For the protection of submerged concrete surfaces in sewage and water treatment plants and marine structures etc.

Application: Thoroughly stir base component, add hardener and mix with electrical slow speed stirrer. Apply to properly prepared substrate by suitable brush or roller. Can also be sprayed with airless or cup gun equipment.

Potlife: 4 hrs at 20° C.

Coverage: Material consumption incl. 20% loss for medium dry film thickness of 150 microns ~0.440 kg/m².

SPECIAL FEATURES: Glossy, tough-hard abrasion and weather resistant finish. Economical coating system suitable also for contact with drinking water (complies to German Federal Health authorities regulations for drinking water). Resistant to water, seawater, diluted lyes and acids, neutral salts. Not resistant for longer exposure to solvents and tar oils. Permanent dry and wet heat resistance approx. +60° C.

COLOUR/DENSITY: Available in many RAL colours; ~1.8 kg/l.

STORAGE: Shelf life at least 6 months when stored in dry and cool conditions. (12.5 and 30 kg sets.)

4. Icosit® K24 Thick

DESCRIPTION: Icosit K24 Thick is a heavy duty, 2-component coating on epoxy resin base with special fillers. Conforms to BS 5493:KU 1B/KF 1B and MIL-C-45568.

USE: Protective coating for concrete, asbestos cement, steel and light metal etc., in and outdoors, especially heavy corrosion protection for steel structures, machines, tankers, pipe lines, sewage treatment plants, etc.

Application: Stir base component, add hardener and mix well, preferably by mechanical mixer. Apply by brush or spray equipment onto property prepared substrates.

Coating System: Use of primer and number of coats depend on application and substrates. Min. application temperature of substrate +10° C.

Mixing Ratio: *A:B = 1:1 by weight.*

Potlife: *At 20° C approx. 6 hours.*

Coverage: Material consumption with 20% loss for medium dry film thickness of 80 microns: 0.27 kg/m².

SPECIAL FEATURES: Cured coating is hard but not brittle, abrasion resistant, as well as highly resistant to heavy mechanical and chemical attack. Contains solvent. Provide adequate ventilation during application in confined areas. Non toxic when cured.

COLOUR/DENSITY: Icosity K24 Thick is available in a wide range of RAL colours; 1.6 kg/l.

STORAGE: Shelf life approx. 1 year, in unopened original container.

PACKAGING: 8 kg and 30 kg sets.

K. STEEL PROTECTION

1. Icosit® 6630 M

DESCRIPTION: Icosit 6630 M is a 1-component high-build coating based on a synthetic resin combination with mica fillers. Conforms to BS 5393: HU2B/HF2D resp. HU2D/HF2F.

USE: As a weather resistant, robust, thick coating on steel and galvanized steel. Specially developed for the corrosion protection of lattice masts and transmission towers, outdoor switch gear and similar structures, where brush application is necessary. Suitable as refresher coat on old 1-pack coatings.

Application: Apply ready for use material by round or distemper brush onto properly prepared substrate. Min. application temperature 5 °C. Avoid dew point.

Coating System: Depending on substrates, site conditions and desired results, various coating systems can be applied.

Coverage: Material consumption with 20% loss of medium dry film thickness of 80 microns ~0.250 kg/m².

SPECIAL FEATURES: Highly thixotropic. High film thickness in one coat possible (80–120 microns dry-film). If properly applied, no stripe coat for edges, welding seams, screws, etc. necessary. Excellent resistance to weather and aggressive atmosphere. Resistant to rain and dew after a few hours. Temporary resistance to sea water, sodium chloride, diluted acids and alkalis and caustic soda. Not resistant to permanent exposure to alcohol, fatty oils, fuels, mineral oils, etc. Not suitable for permanent submerged exposure of liquids. Temperature resistance up to +60° C, short term +80° C.

COLOUR/DENSITY: DB 601 green, RAL 7033 cement grey; 1.6 kg/l.

STORAGE: Shelf life at least one year when stored in unopened, original containers in dry conditions.

PACKAGING: 12.5 kg pails.

2. Friazinc® R

DESCRIPTION: 2-component, highly pigmented, zinc rich primer based on epoxy resin.

USE: As protective coating or as primer for following top coats. Especially suitable for objects subjected to mechanical wear, e.g. weirs, interior of pressure pipe lines, gates, penstocks, etc.

Application: Stir base component well, add hardener and mix thoroughly if possible mechanically. Apply by brush (soft), spray cup gun or pressure vessel on properly prepared surface, in cool weather add max. 5% Thinner K.

Coverage: Material consumption incl. 20% loss for medium dry film thickness of 60 microns ~0.350 kg/m², 80 microns (spray application) ~0.470 kg/m². Min. substrate temperature +10° C.

Mixing Ratio: A:B = 95.5: 4.5 by weight.

SPECIAL FEATURES: Fast curing, resistant to weathering abrasion and barnacles. Physiologically harmless when fully cured. Heat resistance (dry) up to 150° C, (wet) up to 50° C.

Potlife: At 20° C ~8 hrs.

COLOUR/DENSITY: Grey and green; 3.0 kg/l.

STORAGE: Shelf life in well sealed original containers approx. 1 year.

PACKAGING: 5 kg, 12.5 kg and 25 kg sets.

3. Inertol® 88

DESCRIPTION: Inertol 88 is a unique 1-component solvent containing coloured protective high-build coating based on a combination of oils, bitumen and synthetic resins.

USE: As protective coating for steel and galvanized surfaces, e.g. roofs, steel structures in chemically aggressive industrial atmosphere, e.g. coking plants, blast furnace plants, boiler houses, for steel chimneys, etc. Also suitable for structures occasionally exposed to condensed water.

Application: Stir ready for use material thoroughly and apply by brush onto dry, properly prepared substrate. For large areas it is recommended to apply top coat by spray.

Coating System: Depending on substrates, site conditions and intended results, various coating systems can be applied. For spray application below +20° C, add up to max. 5% Thinner B. No primer required.

Coverage: Material consumption with 20% loss for medium dry film thickness of 70 microns ~0.230 kg/m².

SPECIAL FEATURES: Good adhesion. Resistant to weather, industrial atmosphere, occasional condensed water. Temperature resistance: dry heat up to +180° C, damp heat up to +60° C. In humid atmosphere with high concentrations of SO_2 or NH_3 use oxide red shade as top coat. During application in confined areas, provide adequate ventilation. Min. application temperature +5° C.

COLOUR/DENSITY: Available in 5 colours; 1.6 kg/l.

STORAGE: Shelf life in unopened original container at least 12 months.
PACKAGING: 17.5 and 25 kg pails.

4. Icosit® EG1

DESCRIPTION: Icosit EG1 is a 2-component, solvent containing epoxy combination with micaceous iron oxide pigments. Conforms to BS 5493: System SK 2.

USE: As corrosion protection with high chemical and mechanical resistance on steel and galvanized surfaces e.g. in bridge and tank construction for industry and harbour facilities in aggressive atmosphere, in water, seawater and sewage. Particularly suited as transportable mechanically resistant priming or top coat after workshop application.

Application: Thoroughly stir base component, add hardener and mix, preferably mechanically. Apply on properly prepared substrates, by brush (brush in one direction only to avoid brush marks, or use roller) or airless spray. Min. application temperature +5° C.

Potlife: At 20° C ~8 hrs.

Coating System: Depending on substrates, site condition and desired results, various priming and coating systems can be applied.

Coverage: Material consumption with 20% loss for medium dry film thickness of 80 microns: 0.28 kg/m².

SPECIAL FEATURES: Cured coating is tough, plastic and hard but not brittle, resistant to shock, impact and abrasion. Non toxic. Resistant to weathering, water, sewage, seawater, flue gases, deicing salts, acid and alkali vapors, oils, grease and short term exposure to fuels and solvents.

COLOUR/DENSITY: Icosit EG 1 light and dark grey; 1.6 kg/l.

STORAGE: Shelf life at least 1 year in unopened original containers.

PACKAGING: 12.5 kg and 30 kg pails.

5. Icosit® 5530/EG

DESCRIPTION: Icosit High Build 5530 is a solvent containing, one component high build coating based on PVC-acrylic resin combination binder and mica filler. Conforms to BS 5493, HU2B/HF2D, respectively HU2D/HF2F and SH7.

USE: As corrosion production coating on steel structures or galvanized surfaces in rural, urban, industrial and marine atmosphere. Also suitable as maintenance coating on old one-component corrosion protection coatings.

Application: Icosit High Build 5530/EG are supplied ready for use. Apply by airless spray or brush onto properly prepared substrates.

Coating System: Depending on substrates, site condition and desired results, various priming and coating systems can be applied.

Coverage: Material consumption with 20% loss for medium dry film thickness of 80 microns: Icosit High Build 5530 ~0.30 kg/m², EG ~0.35 kg/m².

SPECIAL FEATURES: Highly thixotropic. High film thickness in one coat possible. Good resistance to acidic and alkaline industrial atmosphere, short term exposure to seawater, various diluted acids and alkalis, fuels, oils, not for constant immersion. Temperature resistance up to 60° C.

COLOUR/DENSITY: Available in a wide range of RAL colours. Icosit High Build Primer ~1.5 kg/l, Icosit High Build 5530 ~1.3 kg/l., EG ~1.3 kg/l.

STORAGE: Shelf life at least 1 year in unopened original containers.

PACKAGING: 12.5 kg and 30 kg pails.

L. JOINT-SEALING COMPOUNDS AND WATER-BARS

1. Sikaflex® -1A

DESCRIPTION: Elastic, one-component, gun-applied sealant on polyurethane basis. Conforms to BS 4254 (67), UK Agreement Board 83/1106, US-FS-TT-S-00230 C Type II, Class A, JIS-A-5754, 5758, Label du SNJF (FRANCE) CEBT NORM 85401 (FRANCE), Canadian Board 19-GP-16a Type II.

USE: For expansion joints and pointing in building and civil engineering constructions. For all types of pointing, bedding, gap sealing in building and industry (masonry, tiling, joinery, metalwork, roofing, glazing, water supply and drainage). Movement capacity 20% of average joint width.

Priming: Use Sika Primer 1 on cleaned joint sides. No priming is required on glass, ceramics, anodized aluminium, stainless steel, epoxy and polyester. Adjust joint depth with suitable back-up material.

Application: Cartridges and sausages by hand or air pressure gun.

SPECIAL FEATURES: Curing time depends on temperature and humidity. Service temperature –30° C to +70° C. Excellent adhesion, durability and weather resistance. Easy application. Suitable for contact with drinking water. Not suitable for surfaces directly exposed to heavy traffic.

COLOUR/DENSITY: White, lightgrey, concrete grey, brown; 1.2 kg/l.

STORAGE: At least 9 months shelf life when stored in unopened containers in dry, cool conditions.

PACKAGING: 310 ml cartridges (12 per carton)
600 ml Sika Unipacs (20 per carton)
310 ml Sika Unipacs (12 per carton)

2. Sikaflex® -11FC

DESCRIPTION: Elastic, fast curing, one-component, non sagging adhesive sealant on polyurethane basis.

USE: Flexible bedding and fixing of light-weight building components. Adheres to most of the usual building materials. Air and watertight sealing of connections in building installations, silos, containers, ships, cars, etc. Movement capacity 10% of average joint width.

Priming: Select primer according to type of substrate.

Application: By hand or air pressure gun.

SPECIAL FEATURES: Easy to apply, fast curing. Excellent adhesion, stability and weather resistance. Not suitable for expansion joints. Slight discolouring of white Sikaflex-11FC possible. Service temperature – 40 °C to +80 °C. Suitable for contact with drinking water.

COLOUR/DENSITY: White, light grey, black; 1.2 kg/l.

STORAGE: At least 6 months shelf life when stored in unopened original containers in dry and cool conditions.

PACKAGING: 310 ml cartridges (12 per carton).

3. Sikaflex * -12SL

DESCRIPTION: Elastic, one-component, self-levelling sealant on polyurethane basis. Conforms to US-FS-TT. S-0230C, Type 1 Class A.

USE: For horizontal movement and construction joints subject to light traffic e.g. concrete pavements, access roads, yards and parking areas, around machinery and equipment etc.

Movement capacity 20% of average joint width.

Priming: Use Sika Primer 1 on clean joint sides of all cement bound substrates, bricks, tiles, epoxy mortar etc. Adjust joint depth with suitable back-up material.

Application: Straight from the sachet or using pail gun.

SPECIAL FEATURES: Curing speed depends on temperature and humidity. Service temperature –30° C to +70° C. Excellent adhesion to most substrates. Good resistance to weather and aging. Suitable for contact with drinking water. Easy application.

COLOUR/DENSITY: Grey; 1.2 kg/l.

STORAGE: At least 6 months shelf life when stored in unopened container in dry, cool conditions.

PACKAGING: Aluminium-lined 500 ml sachets (20 per carton).

4. Sikaflex® -14CA

DESCRIPTION: Plasto-elastic, 2-component, pourable sealant on polyurethane base. Cold applied, self-levelling.

USE: To seal construction and expansion joints in civil engineering, e.g. concrete floor joints in underground car-parks, construction joints between bituminous and cement-bound substrates, construction joints between bituminous pavings and concrete parapets, construction joints in mastic asphalt pavements. Movement capacity 15% of average joint width.

Priming: Use Sika Primer 14 for cement and bituminous bound substrates. Sika Wash Primer for metals. Use bond-breaker or suitable back-up material.

Application: Mix both pre-dosed components for 3 minutes using low-speed mixer (400-600 rpm) and pour sealant into properly prepared joint, using plastic watering can.

SPECIAL FEATURES: No heating, pourable, easy to use. Good weather and aging resistance. Good adhesion to cement and bitumen bound substrates. Limited stress on joint faces (plasto-elastic behaviour). Service temperatures –35° C to +70° C. For traffic joints, sealant is to be left approx. 3 mm below surface.

COLOUR/DENSITY: Grey and black; 1.07 kg/l.

STORAGE: At least 8 months shelf life in unopened original containers stored in cool, dry conditions.

PACKAGING: 10 kg set.

5. Sikaflex® -15 LM

DESCRIPTION: Permanently elastic, one-component polyurethane based, gun applied joint sealant. Conforms to DIN 18540, BS 5889:1980, BS4254:1967, US-FS-TT-S-00230C Type 2, Class A, US-FS- TT-S-00227E Type 2, Class A.

USE: Low modulus joint sealant for construction and expansion joints with high movement in building and civil engineering construction, e.g. joints on exterior walls of precast concrete panels, lightweight building materials, renderings, curtain walls etc. Movement capacity 30% of average joints width.

Priming: Use Sika Primer 1 on cleaned joint sides. No priming required on glass, ceramics, anodized aluminium, stainless steel, epoxy, polyester and clean sound concrete. Adjust joint depth with suitable back-up material.

Application: Cartridges and Unipacs by hand or air pressure gun.

SPECIAL FEATURES: Curing time depends on temperature and humidity. Service temperature –10° C to +75° C. Excellent adhesion, durability and weather resistance. Easy application. Not suitable for joints subjected to traffic.

COLOUR/DENSITY: White, beige, light grey, concrete grey, brown, black; 1.31 kg/l.

STORAGE: At least 9 months shelf life when stored in unopened original containers in dry, cool condition.

PACKAGING: 310 ml cartridges (12 per carton)
310 ml Sika Unipacs (12 per carton)
600 ml Sika Unipacs (20 per carton)

6. Sikaflex® T68 (W)

DESCRIPTION: Elastic, two-component sealant on polyurethane tar base. Conforms to US-FS-SS-S-200 D.

USE: For horizontal expansion joints in roads, runways, truck and bus terminals, gas stations, bridges, etc. Movement capacity 25% (**concrete**) of average joint width.

Priming: Use Sika Primer T68 on cleaned joint sides. Adjust joint depth with suitable back up material.

Application: Prior to mixing, thoroughly stir comp. A, then add all of comp. B and mix for approx. 3 min. with low speed mixer (no air entraining). Pour with suitable can.

SPECIAL FEATURES: Fast hardening at normal temperatures thus allowing to take traffic load in short time. Good adhesion, oil, flame and weather resistance. Excellent tear and abrasion resistance due to high module. Service temperature –25° C to +50° C (temporary +70° C)

COLOUR/DENSITY: Black; 1.5 kg/l.

STORAGE: At least 9 months shelf life when stored in unopened original container in dry and cool conditions.

PACKAGING: 1, 8 kg and 10 kg sets.

7. Sikaflex® T68NS

DESCRIPTION: Elastic, two-component, non-sagging, gun applied sealant on polyurethane tar base.

USE: For vertical and horizontal expansion joints and pointing in civil engineering, e.g. retaining walls, tunnels, sewers, sewage treatment plants, basements, etc. Movement capacity 25% of average joint width.

Priming: Use Sika Primer T68 on cleaned joint sides.

Application: Mixing procedure same as T68 (W). Apply by hand or air-pressure gun.

SPECIAL FEATURES: Resistant to fuels. Good adhesion and weather resistance. Not suitable for exposed joints in facings. Not suitable for drinking water. Not suitable for surfaces subject to traffic. Service temperature –30° C to +50° C (temporary +70° C).

COLOUR/DENSITY: Black; 1.45 kg/l.

STORAGE: 6 months shelf life when stored in unopened original container in dry and cool conditions.

PACKAGING: 3.5 kg sets.

8. Sikacryl-GP®

DESCRIPTION: One component multi purpose sealant and gap filler on acryline emulsion base.

USE: To fill internal gaps and joints with low movement, e.g. around doors and windows, fixtures and fittings. As a joint filler between different building materials. Movement capacity 10–15% of average joint width.

Priming: Sika Primer 80 for all substrates.

Application: By hand or air-pressure gun.

SPECIAL FEATURES: Ready for use. Excellent adhesion to most building material. Odourless, may be overpainted. Do not use on soft plastic, bare metals, bitumen or tar. Not suitable for underwater joints.

COLOUR/DENSITY: white; 1.6 kg/l.

STORAGE: Shelf life at least 9 months when stored in un-opened original container in frost free conditions but below 25° C.

PACKAGING: 310 ml cartridges (12 per carton)

9. Sikalastic®

DESCRIPTION: Elastic 2 component, gun-applied sealant on polysulphide base. Conforms to DIN 18540, US-FS-TT-S-00227E Type 2, Class A.

USE: For vertical and horizontal expansion joints in many types of building and civil engineering constructions, e.g. retaining walls, underpasses and tunnels, bridges, precast concrete elements, high- and low-rise buildings, etc. Movement capacity 30% of average joint width.

Priming: Use Sika Primer 80 for most building materials. Consult Primer chart. Adjust joint depth with suitable back-up material.

Application: Use a low-speed stirrer to mix predosed components A + B. Use hand- or air-pressure gun to apply sealant, without entrapping air. Tool off to concave finish.

SPECIAL FEATURES: Excellent adhesion to most building materials. High movement capacity. For vertical and overhead joints. Good chemical resistance. Service temperature –30° C to +90° C.

COLOUR/DENSITY: Grey; 1.6 kg/l.

STORAGE: At least 8 months shelf life when stored in un-opened original container in cool conditions.

PACKAGING: 3.5 kg sets (6 per carton).

10. Sikasil® -E

DESCRIPTION: Elastic one-component, non sagging, gun-applied joint sealant on silicone rubber base. Acetoxy type.

USE: For glazing (single or double glass windows) on metal or hard PVC; insulating glass production, for glass partition walls, aquariums, curtain walls, glass-block walls, light prefabs, heating and cooling installations, sanitary installations, car an shipbuilding etc. Movement capacity 20% of average joint width.

Priming: No priming required on glass, building materials with glazed surfaces (ceramic, clinker slabs, tiles, etc.), stove enamelled and powder coated metals, anodized aluminium.

Sika Adhesive Cleaner 1 on light alloys, chrome steel, epoxy, PU, rigid PVC, polyester, etc. Sika Wash Primer on stainless steel and galvanized iron, non anodized aluminium.

Application: Cartridges, by hand- or air-pressure gun.

SPECIAL FEATURES: Fast curing, service temperature – 40° C to +150° C. Excellent adhesion, durability, weather and UV-resistance. Dirt repellent. Easy application. Cannot be painted. Adhesion check prior to application on plastic or plastic coated metals.

COLOUR/DENSITY: Transparent, white; 1.05 kg/l.

STORAGE: At least 9 months shelf life when stored in un-opened containers in dry, cool conditions.

PACKAGING: 310 ml cartridges (12 per carton).

11. Igas® Black

DESCRIPTION: Plasto-elastic, one-component, trowel-grade sealant on bitumen/rubber basis.

USE: For vertical and horizontal joints in civil engineering and building construction, with good adhesion on cement-bound materials, masonry, metal. Joint width 10–30 mm. For hydraulic structures, water-retaining structures, box culverts, irrigation channels, marine structures, retaining walls, tunnels, basements, sewage treatment plants. Movement capacity 15% of average joint width.

Priming: Igas-Duro Primer.

Application: Heat indirectly to 60° C – 80° C (never exceed 140° C) and apply by trowel.

SPECIAL FEATURES: Service temperature range –30° C to +70° C (dry), 40° C (wet). Good weather resistance in all climatic conditions. Suitable for permanent immersion in water and sea water. Resists high water pressure when adequately supported and reinforced. Suitable for contact with drinking water. Not resistant to oils and fuels. Also available as Igas Black Hot Climate.

COLOUR/DENSITY: Black; 1.40 kg/l.

STORAGE: At least 1 year shelf life when stored in well sealed container.

PACKAGING: 30 kg and 115 kg drums.

Igas® Black is also available in preformed profiles (many different sizes) as Igas® Profile Black.

12. Igas® -K

DESCRIPTION: Plasto-elastic, one-component, fuel resistant, hot- poured sealant on tar/synthetic resin basis. Conforms to BS2499- 73 Type B1, ASTM D 1854-85, ASTM D 3581-85, ASTM D 3569-85.

USE: For horizontal expansion and dummy joints in concrete pavements, in particular for airport runways, hardstandings, aprons and hangars, re-fueling areas, stores for fuel oils, maintenance bays and garages, roads etc.

Min. joint size 10 mm. Movement capacity 15% of average joint size.

Priming: No priming required.

Application: Heat in suitable mastic melter at 120°C–140°C (never exceed 150° C) and apply with pouring machine or spout can. Adjust depth with suitable backup material.

SPECIAL FEATURES: Sets by cooling after application. Service temperature range –20° C to +70° C (dry), 40° C (wet).

Resistant to water, sea water, sewage. Fuel and oil resistant; suitable for surfaces subject to fuel spillage. Not suitable for drinking water.

COLOUR/DENSITY: Black; 1.4 kg/l.

STORAGE: Unlimited shelf life when stored in well sealed container.

PACKAGING: 30 kg and 200 kg drums.

13. Sikadur® 51

DESCRIPTION: 2-component trowel-applied jointing compound on epoxy resin base.

USE: For sealing joints with little or no movement, particularly for heavy duty concrete floors and for repairing cracks.

Movement capacity 5% of average gap width.

Priming: Generally not required. If surface slightly damp, use Sikafloor 94 Primer.

Application: Add all of Comp. B to Comp. A and mix well, using electric low speed drill. Apply by trowel, spatula or mastic gun to properly prepared joint.

SPECIAL FEATURES: Easy to apply; non-sag on vertical joints, fast setting, excellent adhesion to most materials. Good chemical resistance and mechanical strengths. Remains flexible. Not suitable for expansion joints. Do not expose to water and chemicals for at least 10 days.

COLOUR/DENSITY: Grey; 1.55 kg/l.

STORAGE: At least 12 months shelf life when stored in unopened original containers below 25° C.

PACKAGING: 3 kg sets.

14. Sikadur® -Combiflex

DESCRIPTION: High-performance joint sealing system consisting of Sika Norm Hypalon elastomeric sealing strip and Sikadur 31 thixotropic epoxy adhesive.

USE: For sealing of joints and cracks in building and civil engineering constructions, where excessive movements and/or irregular shape prevent the use of conventional sealants. Good adhesion on most of the common construction materials, also on damp substrates. For tunnels, pipelines, basements, water- retaining structures, silos, roofs, etc.

Application: After thorough surface preparation, apply epoxy adhesive by brush or spatula, working first coat well into substrate. Activate sealing strip with Colma-Cleaner and embed in epoxy adhesive. Weld sealing strips by hot-air. Minimum application temperature 5° C.

SPECIAL FEATURES: Curing of adhesive within 1-3 days, depending on type of adhesive and temperature. Service temperature range −20° C to +70° C. Resistant to weathering, water and wide range of chemicals.

COLOUR/DENSITY: Sealing strip: Thickness 1.0, 1.5, 2.0 mm in 20 m rolls of various widths.

Grey; SN HYP 10 ~1.5 kg/m^2, SN HYP 15 ~2.3 kg/m^2, SN HYP 20 ~3 kg/m^2.
Adhesive: Grey; 1.65 kg/l.

STORAGE: Adhesive 12 months shelf life in well sealed container below 25° C, Hypalon strip 9 months when stored in dry, cool conditions.

PACKAGING: Different packagings, ready for use sets.

15. Igas® -R

DESCRIPTION: Plasto-elastic, one-component, hot poured sealant on bitumen rubber basis. Conforms to Swiss Specs. SNV 671 625a, ASTM D 1190-74.

USE: For horizontal expansion and dummy joints in concrete pavements, e.g. roads, parking areas, canal linings, runways, taxiways, multi storey carparks, bituminous pavings, and for pointing of concrete tiles for roofs and terraces. Minimum joint size 10 mm. Movement capacity 15% of average joint width.

Priming: Use Igas Primer 10.

Application: Heat in suitable mastic melter to 175° C–185° C and apply with pouring machine or spout can. Adjust joint depth with suitable backup material.

SPECIAL FEATURES: Sets by cooling after application. Service temperature range −20° C to +70° C (dry), 40° C (wet). Good weather-resistance in all climatic conditions. Resistant to water, sea water and sewage. Not suitable for contact with coaltar. Not suitable for pavements subject to severe fuel spillage.

COLOUR/DENSITY: Black; 1.2 kg/l.

STORAGE: At least 2 years shelf life when stored in well sealed container.

PACKAGING: 25 kg pails.

16. Igas® Duro Primer

USE: For Igas Black and Igas Profile.

STORAGE: 12 months shelf life when stored in unopened original containers below 25° C.

PACKAGING: 1 kg and 5 kg tins, 25 kg drums.

17. Igas® Primer 10

USE: For Igas R.

STORAGE: 12 months shelf life when stored in unopened original containers below 25° C.

PACKAGING: 5 kg, 10 kg, 25 kg pails.

18. Sika® Primer 1

USE: For Sikaflex 1A, 11FC, 12SL, 15LM on cement bound substrates, bricks, tiles, epoxy mortar etc.

STORAGE: 6 months shelf life when stored in unopened original containers below 25° C.

PACKAGING: 0.2 kg tins (6 tins per carton), 1 kg tins, 20 kg and 180 kg drums.

19. Sika® Wash Primer

USE: For Sikaflex 1A, 11FC, 12SL, 15LM, T68 on iron, steel, aluminium, zinc and non ferrous metals.

STORAGE: 6 months shelf life when stored in unopened original containers below 25° C.

PACKAGING: 0.2 kg tins (6 tins per carton), 20 kg pails.

20. Sika® Primer 21

USE: For Sikaflex 1A, 11FC, 12SL, 15LM on rigid PVC.

STORAGE: 6 months shelf life when stored in unopened original containers below 25° C.

PACKAGING: 0.2 kg tin (6 tins per carton), 20 kg pails.

21. Sika® Waterbars

DESCRIPTION: Flexible PVC waterbar sections for the sealing of construction, contraction, expansion and subsidence joints subject to low or high water pressure, in insituconcrete members of hydraulic and basement structures. With ribbed flanges for anchoring and sealing. Plain-web sections, or sections with special form flexibility to accommodate large expansion and shear movements.
Conforms to BS 2571 and BS 2782, US Corps of Engineers Specs, CRD-C572-74.

SPECIAL FEATURES: Site jointing by means of heat welding. Special welding equipment available on request. Metal clips for easy fixing to formwork available. Prefabricated sections (L-T-pieces, corner and cross pieces) can be supplied on order. High quality, high-strengths PVC.

TECHNICAL DATA: Tensile strength: min. 14 N/mm^2. Elongation at rupture: min. 300%. Service temperature range: –35° C to +50° C. Density: ~1.3 kg/l. Non-aging. Resistant to aggressive water, diluted acids and alkalis. Only B-type waterbars are suitable for permanent contact with bituminous materials.

More than 40 different standard profiles available.

Normal types supplied in yellow colour, oil and bitumen-resistant types in green colour.

M. SYNTHETIC WATERPROOFING MEMBRANES

Flexible High-Polymer membranes for all aspects of water-proofing

Roofing:	Sikaplàn PVC 12D/15D
	Sikaplan PVC 12G/15G (reinforced)
	Sikanorm Hypalon 12A/15A
	Sikanorm Hypalon 10/20
	Sikanorm CSM 13G/15G (reinforced)
Civil Engineering:	*Tunnelwaterproofing:*
	Sikaplan 15V/20V Tunnel
	Sikaplan 9.6V/14.6V Tunnel
	(with Signal layer)
	Underground structures:
	Sikaplan 12/15/20
	For contact with bitumen:
	Sikaplan 15B/20B
	Pit and Pond liners:
	Sikaplan 10T/12T

All Sika high polymer membranes are at least double layerd, calendered membranes of outstanding quality.

Sika supplies all necessary auxiliary material for all types of applications.

Please ask for the relevant detailed documentation.

Sikaplan and Sikanorm membranes are used worldwide. Extensive reference list available.

N. EQUIPMENT

Aliva machines for guniting and shotcreting (dry and wetspray), site mixed or pre-bagged mortars (Sigunit, Sika-Shot 3, Sikacrete Gunit 103, Sikacem Gunit 133, SikaTop-Epocem 110, etc.)

Portable mixers of varying capacity for SikaTop ready for use polymer modified cement mortars, Sika Grout, SikaTop-EpoCem, Sikadur epoxy mortars and adhesives.

Injection guns and ancillary materials for sikadur expoxy resins.

Cartridge and sausage guns for Sikaflex sealants.

Heating and pouring equipment for Igas sealants.

Welding equipment for PVC waterbars.

Leister Heatwelders for Sikaplan PVC and Sikanorm Hypalon membranes.

7

BRIDGE-STRUCTURE 'STRENGTHENING'

7.1 DESIGN OF STRENGTHENING

7.1.1

(a) With due consideration given to 'which loads' act on 'which sections' and on 'what span-arrangements', the strengthening of structural members can be attempted by some of the following means:
— 'replacing' poor quality or defective material by better quality material
— 'attaching' additional load-bearing material
— 're-distribution of the loading actions' through 'imposed deformation' on the structural system
(b) The new load-bearing material will usually be
— high quality concrete
— reinforcing steel bars (longitudinals, laterals, stirrups, etc.)
— thin steel plates and straps (externally bonded by epoxy)
— various combinations of these

7.1.2 The main problem in strengthening is to achieve 'compatibility' and a 'continuity' in the structural behaviour between the original material structure and the new material/repaired structure.

7.1.3 It has to be clearly understood that the strengthening effect (e.g., increased section properties) can participate only for live loads and subsequently imposed load actions (e.g. removal of temporarily applied load, reversed prop reactions, etc.) and possibly the dead loads applied subsequently.

7.2 DESIGN

(a) The strengthening of structure should be 'designed' and implemented in accordance with appropriate codes. If special codes for strengthening exist,

they will of course be of assistance to the designers and contractors. However, this is seldom the case, and many problems in connection with strengthening are not dealt with in the regular codes.

(b) Typical problems of this kind are the 'transfer of shear stresses between the old concrete and the new concrete applied for strengthening', and the 'post-tensioning of the existing structure' which in some respects is different from the post-tensioning of fresh structure, etc.

7.3 INTERACTION BETWEEN OLD AND NEW CONCRETES

The 'joint' (i.e., the interface) between the old concrete and the new concrete must be capable of transferring horizontal stresses without relative movement of a magnitude that will affect the static behaviour of the structure significantly.

Furthermore, the 'joint' must be durable for the environment in question, i.e. the composite structural action must not change with time.

When using large concrete volumes, the possibility of additional stresses as a result of hydration-heat has to be taken into account. Temperature differences can be minimized by special measures, e.g. pre-heating of the old structural element and/or cooling of fresh concrete.

Differences in creep and shrinkage characteristics between the old and the new structural elements will require careful evaluation. Cracks may be develop as a result of 'constraint' forces. Therefore, it becomes necessary to correctly 'detail' and 'anchor' the reinforcements. To implement the strengthening measures, it will be necessary to employ suitable mortars or concretes with low creep and shrinkage properties as well as minimal development of hydration-heat. At the same time, an effort should be made to match, as closely as possible, the strength and modulus of elasticity of the new material with that of the old material.

Vibrations due to traffic during hardening of the new concrete can have a negative influence on its strength and its bond with the old concrete. Speed-limits should therefore be introduced during the hardening phase (the traffic should be stopped if necessary).

7.4 STRENGTHENING WITH 'REINFORCEMENT'

The strengthening of the reinforcement subject to tensile force, be it due to flexure or shear on the section, can be achieved by:

(a) Replacing of reinforcement (severely damaged say by corrosion) and mechanically tying-in additional reinforcement in the old cross-section and/or placing it in an additional concrete layer but mechanically anchoring it to the old cross-section (stirrups to be well anchored into the compression flange).

(b) Epoxy-bonded steel plates.

(a) Strengthening with Reinforcing Bars: In the simplest case, a strengthening of the concrete tension zone is possible by the addition of reinforcing steel. Reinforcement should be added after unloading i.e. after relieving the existing stresses* and after the concrete cover has been removed or after adequate recesses have been cut in the existing concrete so as to accommodate the added reinforcement. Afterwards the concrete must be restored. This reinforcing steel should be effectively anchored into the old concrete at its ends and should be also tied-in at intervals. The anchoring can be done by providing sufficient anchorage-length into the old concrete, or by steel plates and bolts with anchoring disc.

Any severely damage reinforcing bars must be replaced. After duly supporting the structure, the effected bars can be removed and the new reinforcing bars joined in to the old ones by lapped splices, welding, or 'coupling' devices. Transverse reinforcement is needed to insure a ductile behaviour of the component.

Staggering of lapped-splices is recommended if possible.

Lapped-splices in a structural element can produce problems of congestion, interference with the proper compaction of concrete, etc. These difficulties may be overcome by the use of welded splices and increased section areas of steel since the welding can lower the strength of the high-yield strength bars unless special electrodes and controls are adopted.

The provisions of CEB/FIP Model Code or equivalent regarding the weldability and the method of welding of steel should be observed.

(b) Strengthening by means of Epoxy-Bonded Steel Plates: This has been covered in detail in Chapt. 9 ahead to which reference may be made.

7.5 POSSIBLE EFFECT OF 'LOSS OF COVER' AND 'LOSS OF BOND' IN REINFORCED CONCRETE

Loss of concrete cover on reinforcing bars does not substantially reduce the strength but the loss of bond between concrete and reinforcing bars can increase the tensile stress in the region of bond-loss because stress may not redistribute. The concrete may develope large stress concentrations at the inside re-entrant corners, culminating in spalling. This can reduce the strength of the member eventually.

When large spalls and cracks occur, any further deterioration would reduce the capacity and cause loss of composite action. The large spalls can result in 'catenary effects' (hanging wires between two

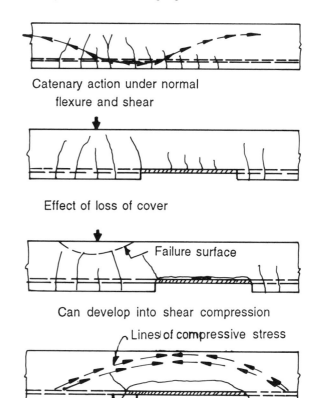

Catenary action under normal flexure and shear

Effect of loss of cover

Failure surface

Can develop into shear compression

Lines of compressive stress

Tensile stress Exposed rebar

Tied arch action

Fig. 7.1 Behaviour of cracked R.C. beam — effect of loss of cover and bond.

_____

* While a reinforced concrete structure can be supproted even at close intervals in a designed sequence, it is dangerous to do so in a p.s.c. case where extreme care is necessary since stresses can get aggravated.

supports). This catenary effect makes the spalled concrete beam act like a tied-arch. Therefore, the end anchorage of beam would influence the cracks in the beam and the carrying capacity. In the limit, its failure can be sudden shear-compression failure. In repairing these spalls, it is necessary to use bonding compound in the areas to be patched to provide good bond with the existing concrete.

There is reserve strength in most old structures since most of them were designed conservatively by using lower allowable stresses and higher safety factors. If designed by load-factor-method, we get a more realistic capacity of these structures, especially in longer spans where dead load uses up a big portion of the allowable stress. Before strengthening the existing structures their capacity should be checked by the load-factor-method to assess whether the structure could sustain the desired live loads. If the investigation shows over-stresses, then strengthening of the structure should be considered by adding steel plates in the tension flanges and on webs, and if feasible, even post tensioning the beam externally.

7.6 STRENGTHENING STRATEGY IN CASE OF PRESTRESSED CONCRETE

In selecting a strengthening method, the following factors should be considered:

1. Load Carrying Capacity Based on Remaining Section: It is desirable to upgrade a damaged girder to the service load capacity. However, if the bridge is to be widened or raised at a later date, the girder may be repaired temporarily to less than the service load capacity. The simple repair will cost less and provide a minimum level of service for the public for a short period until widening, etc. is taken up.

In many cases, a damaged prestressed girder still has an ultimate load capacity that may be more than that in the other types of structures under similar distress, therefore the selection of repair method has to rely heavily on the engineering judgment and not on calculations alone! It is also suggested that in certain extreme cases a fatigue stress-check be made for the remaining strands (i.e. range of stress-variation as live load comes and goes) although it is seldom critical in prestressed concrete.

2. Severity of Damage: All loose concrete should be removed, the fractured area should be identified and sealed, and all compression cracks should be carefully examined for deciding the repair strategy.

3. Availability of materials, equipment and personnel, should be considered for particular repair schemes.

4. Cost comparison between alternative repairs should be made.

5. Public Inconvenience and Safety during Repair should be considered: Since maintenance of the traffic is emerging as one of the most costly items for bridge repairs, special care should be exercised to select practical and cost-effective traffic control schemes. This might affect the choice of repair schemes.

6. Aesthetics: A pleasant looking bridge repair is desirable.

Having stated these, it is then necessary to inspect the damage and see whether it is
    (i) Minor damage
    (ii) Moderate damage, or
    (iii) Severe damage

(i) Minor Damage

Minor damage requires superficial patching up by the use of epoxy grout or shotcrete methods. Where necessary, all damaged and delaminated concrete should be removed by hand tools to avoid the possibility of further damage.

The exposed surface should be cleaned and applied with epoxy bonding compound primer before patching. All cracks should be sealed by the epoxy pressure injection which should be made from the bottom, rising toward the top, along the cracks, against gravity.

(ii) Moderate Damage

Moderate damages with extensive spalls are generally patched with epoxy grout or concrete mortar applicable to minor repairs. It is recommended that welded wire fabrics be attached to drilled dowels placed at about 2 feet spacing or to other reinforcement steel in the patched area. Surface preparation is similar to that in the concrete repairs case.

If the strands or reinforcement steel are exposed, care should be exercised so as not to damage the steel during cleaning. The exposed strands should be coated with epoxy bonding compound or slurry cement grout before patching.

In order to restore the structure to its original capacity, pre- loading may be necessary during the repair operations. Pre-loading is an application of temporary vertical loads to restore the equivalent of partial or full 'prestress-effect' in order to reduce the tension stress under live loads. Preload can be applied by the use of loaded trucks or vertical hydraulic jacking. (More details ahead.)

(iii) Severe Damage

Repair of a severely damaged girder requires structural analysis and design-check based on the conditions of the damage and the best engineering assumptions and judgment. It is important to have a complete review of the calculations taking account of the damage, because it would help selection of a cost-effective repair method for the bridge to safely carry the load.

PLATE 23

Fig. 7.7 Strengthening the deck by introducing a midspan diaphragm (to better absorb and distribute the damage from the fast-moving vehicles on the road below the deck)—inner girder had been heavily damaged and was repaired as can be seen. Strengthening was done in addition

Fig. 7.8 Another view of strengthening arrangement referred to in Fig. 7.7 (Extensive epoxy repair to the damaged inner girder can be seen)

The preloading method should be seriously considered in making concrete repairs in order to restore the equivalent of partial or full 'prestress effect' as possible. The repair procedure for the use of patching concrete, epoxy pressure injection, shotcrete, and additional welded fabric with drilled anchors are similar to other methods of concrete repairs, described earlier.

7.7 RESTORATION OF STRENGTH OF PRESTRESSED CONCRETE GIRDERS

In order to restore strength to the damaged girders, the damaged strands can be effectively spliced or strengthened by external forces. The following are some of the splicing and strengthening methods. *See* Figs 7.2–7.8 (Figs 7.7 and 7.8, Plate 23).

Method 1 The use of two post tensioning rods (one on either side of a web) and Jacking (concrete) corbels located outside of the damaged areas. Construction procedure is as follows:

(a) Apply the calculated preload, then repair the damaged concrete;

(b) After repaired concrete has gained strength, remove preload;

(c) Construct jacking corbels and finally post tension the rods as per strengthening design.

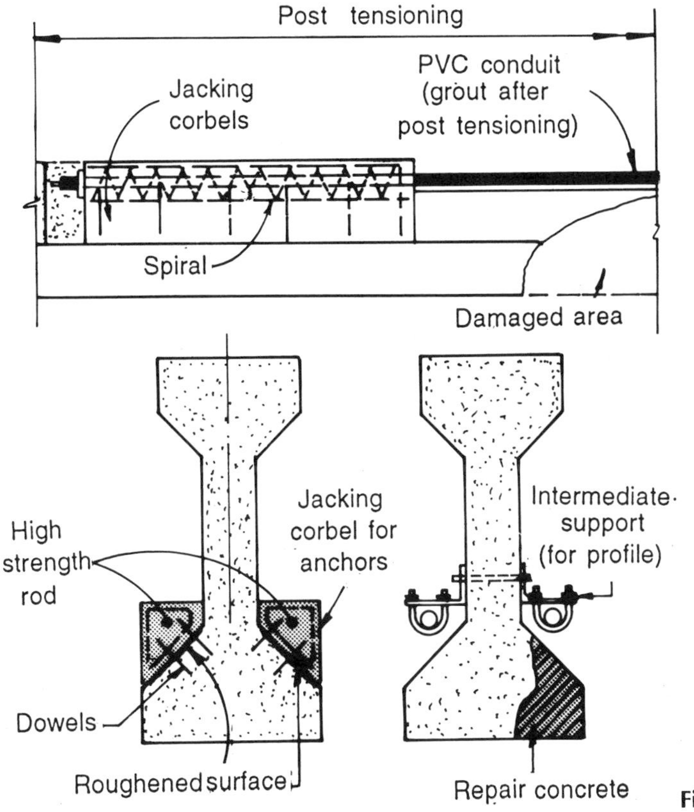

Method 2 *Adding External Reinforced Concrete*: The use of conventional reinforcement steel with concrete corbels throughout the girders.

(a) After applying the preload, the damaged concrete should be repaired and the concrete corbels constructed with required reinforcement steel.

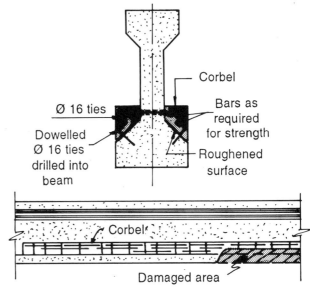

Fig. 7.3 Restoration by external reinforced concrete corbels

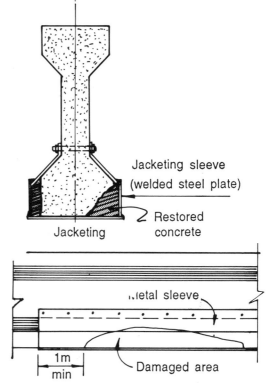

Fig. 7.4 Restoration by installing steel metal sleeve (Jacket) and restoring concrete within

Fig. 7.2 Restoration by post-tensioning

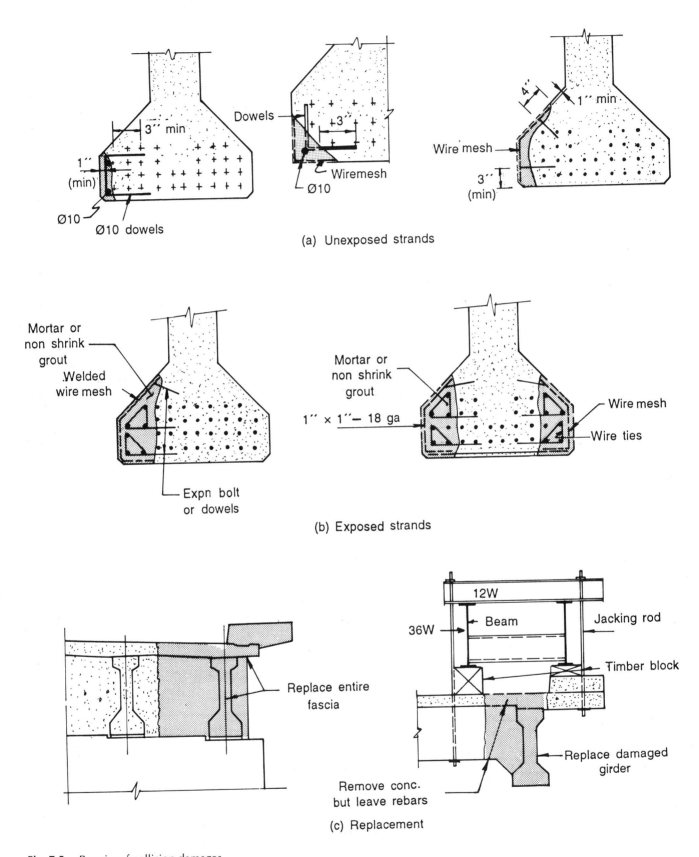

(a) Unexposed strands

(b) Exposed strands

(c) Replacement

Fig. 7.5 Repairs of collision damages

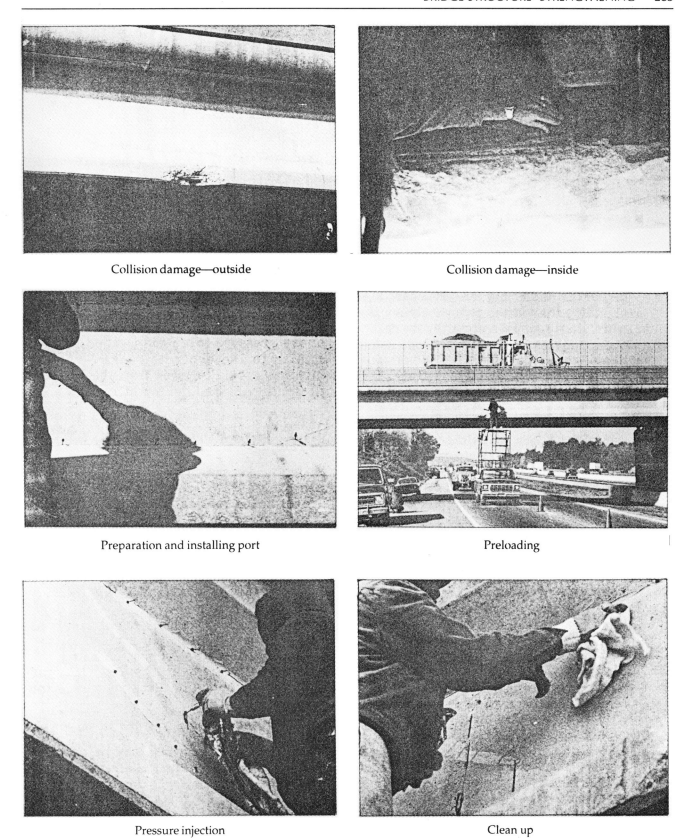

Collision damage—outside

Collision damage—inside

Preparation and installing port

Preloading

Pressure injection

Clean up

Fig. 7.6   Repair and strengthening by epoxy-pressure-injection and epoxy-concrete

(b) Upon concrete attaining the required strength, the preload is removed.

Method 3 Addition of a Metal Sleeve Jacket: This method does not restore the prestress loss except that a partial prestress may be gained by the preloading. The preload should be applied prior to the repair of the damaged concrete and removed upon completion of repairs. Then a metal sleeve jacket should be installed around and beyond the damaged areas (1 m min.). The gap between the metal sleeve and girder is filled with epoxy grout by pressure injection.

7.8 GIRDER REPLACEMENT

This is a high-skill operation. The removal of a damaged girder could be done from below or above the structure. In general, the removal consists of cutting through the existing slab and diaphragms and lifting the damaged girders. However, the beam could be removed from the underside of the deck by saw cutting the slab along the girder without damaging the existing reinforcement, installing a temporary support, then sliding down the damaged girder from below. The method has the advantage of minimum disruption to the deck reinforcement but careful supervision is required during construction. Extreme care is needed in case the slab is prestressed too (redifusion of prestress on reduced section).

It is recommended that a sufficient length of existing reinforcement bars be left in place for splicing with new reinforcement. Otherwise, a drilled dowel or coupler should be used for that purpose. Also, protective shielding and false work are required during removal operation for the safety of the public and workers. Generally, the protection details are designed by the contractor and are subject to the approval of the agency supervising the work.

The exposed concrete surface should be cleaned and coated with epoxy bonding compound before pouring new (slab and diaphragms) concrete.

In designing 'replacement girder', new girder should be designed to conform with the original design method and to provide strength equal to that of the original girder by using the same strand arrangement and type (preferable).

It is important that the camber of the new girder be matched with that in the old girder. Excess new-girder-camber can result in an inadequate slab thickness. Girder-camber must be controlled by prestress, curing time or dimensional changes. It is suggested to consider differential shrinkage stresses in designing new composite deck slab on the existing girders.

It is preferable to pour the new slab and diaphragms simultaneously in order to avoid overloading of existing girders in the structure. Extra bracing of the girder at the time of slab pouring is required.

Method of construction should be selected to minimize inconvenience and danger to the public. Grouts and concretes having high early strength should be considered. Plans must carefully spell out the materials and procedures in order to insure that this result is attained.

When pouring new concrete, vibration from live loads from the existing structure should be minimized until a specified minimum concrete strength is obtained.

7.9 ADDING SUPPLEMENTARY MEMBERS

(a) *Girder:* An inadequate structural system can be strengthened by placing additional stringers or floor beams 'between' the existing members. Installation of a new girder can be installed from below the structure by jacking. Any gap between the top flange of the beam and the underside of the deck slab can be filled by

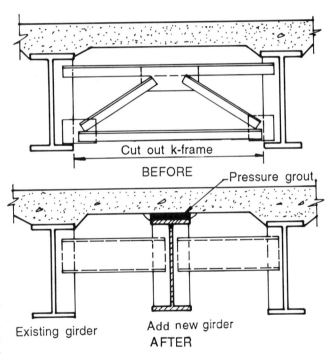

Fig. 7.9 Additional girder

drilling holes through the deck slab and pressure grouting or by pressure grouting from below.

(b) *Transverse Support for Girder:* Under certain favourable conditions where the vertical clearance and the geometric requirements allow, the load carrying capacity can be increased by shortening the effective span length of the bridge. Installation of auxiliary piers or a transverse floor beam system with a main girder

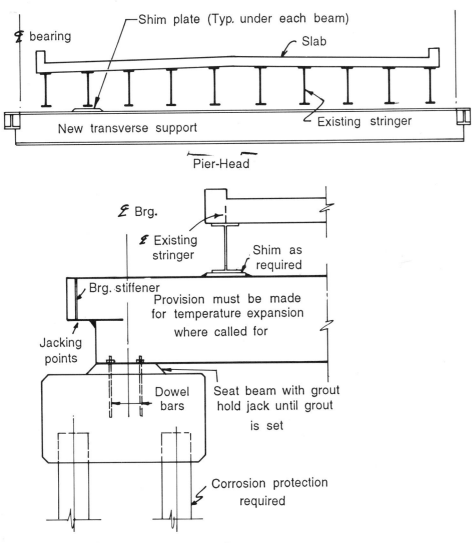

Fig. 7.10   New transverse beam

would convert a simple span bridge in to a continuous span bridge and reduce the effective span length. This would introduce new and different stresses in the beams and a thorough analysis of the new system would have to be made to strengthen the deficient member to carry the revised loading before placing a new transverse support. Special attention should be given to the member 'over' the new support.

This principle can also be used for increasing the capacity of a truss by installation of a pier at the first or second interior panel points, subject to stress check.

(c) *Truss*: Deck type truss spans can be strengthened by adding a center-truss. The center-truss should not be too stiff but it should have equal deflection under the load as the old outer trusses. Otherwise excessive stresses will be induced in the center-truss before other trusses could deflect enough to duly stress their members. If

that 'equal' deflection is not possible, heavy bracing will be required.

(d) *Compression Members*: Reinforcing the compression members requires a thoughtful study and analysis. The chord members of most old bridges are symmetrical in section generally, and, they most likely are eccentrically loaded. This problem may be solved by the addition of steel plates at the proper location. A relatively small piece of steel plate placed at the critical point often will increase the capacity of the member substantially.

An additional steel plate could be added to the cover plate, usually between the existing lines of the rivets. This plate should be balanced by placing additional steel on the lower flanges. Also, a full length side plate can be added to either the web plate of the section or between the vertical legs of the upper and lower angles. This provides an adequate way of transmitting stress to the connection or the adjacent members. If the existing cover plate is so much wider in proportion to its thickness that it has a limited resistance to buckling, then the condition could be corrected by adding a cover plate connected by bolts along the center line and through the angles.

Often, reinforcing a compression member will reduce its slenderness ratio and result in an increased allowable stress.

The most difficult problem in strengthening a compression member is introducing dead load stress to the new added material. Otherwise, the full value of the allowable stress can not be achieved in new material. If the maximum allowable stress is obtained in new material, the older member will be overstressed unless residual dead load stress is removed. So while the old material reaches a maximum stress, the new material will develop the stress only due to live loads.

There are a few methods to introduce the dead load stress into the new material. The one commonly used in heavy members is to calculate the shortening required to produce the desired stress under dead load, and then drill the holes in the old and new members in such a way that the new members will be shortened by the drifting method (i.e. cooling the old heated members) before they are bolted together.

(e) *Superimposing Arches, Hangers, etc.*: Scale model tests of truss rehabilitation utilize the concept of 'superimposing arches, hangers and additional floor beams' to the existing trusses. It is assumed that the 'truss' would provide 'lateral support' while the light-weight-arch will provide additional load carrying capacity. The additional floor beams and hangers may be necessary to strengthen the deteriorated and/or under-capacity floor beams and to provide a balanced distribution of the live loads. The model appears to be satisfactory in regard to easy construction, minimum disruption of traffic, light-weight, and localized repairs. The design can be based on the tied-arch or normal arch concept, depending on the existing bridge condition to resist the thrust forces. This retrofit would increase the capacity of the bridge

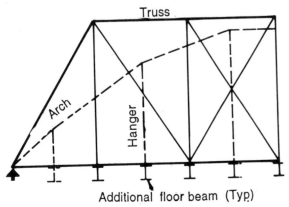

Fig. 7.11 Arch–hanger–floor beam system

but it would not improve the geometry of the bridge if it was narrow.

7.10 STRENGTHENING WITH SUPPLEMENTARY PRESTRESS

Given below are introductory details regarding strengthening by External Prestress. However, for more details, *see* Chapter 8.

7.10.1 In certain specific cases, strengthening by supplementary prestressing can be highly effective. Both reinforced concrete and prestressed concrete structures can be strengthened by this method. The influence of the supplementary prestress on serviceability and ultimate limit states can be varied within wide ultimate limits by selecting different methods of introducing the prestressing force and using different alignments of the tendon.

7.10.2 Choice of System for Supplementary Prestressing

For supplementary prestressing, so far only post-tensioned prestressing has been used. While prestress can be applied for strengthening a structure, it may be noted that slight deviations (for instance in resulting fibre stresses) may become necessary vis-a-vis those normally allowed by the Codes of Practice for design. Sound Engineering Judgement has to be exercised here.

7.10.3 As for prestressing means, both unbonded as well as bonded tendons can be used. If short prestressing elements are required, a post-tensioning system with minimal slipping in the anchorage (anchor-set) should be chosen. Short prestressing elements can be sensitive to deviations due to construction tolerances (eccentricity, inclination and tolerance of the anchorage elements, the prestressing jack, etc.).

At tendon profile deviation points (where saddles are employed to achieve radius and reduce friction), small radii of curvature in the tendon profile should be avoided.

7.10.4 The strengthening by means of post-tensioning can normally be 'designed' as an ordinary prestressed member design. When calculating prestress losses, however, it should be noted that the effect of creep and shrinkage may generally be less than in normal design, due to the age of the old concrete. The stress in an unbonded tendon in the ultimate state will be only slightly larger than that after prestress losses, owing to lack of bond with adjacent concrete.

7.10.5 Protection Against Corrosion and Fire

The post-tensioning tendons should be protected against possible corrosion and fire to the same extent as in a newly built structure. The requirements for concrete cover for bonded tendons are much the same as for the regular prestressed concrete case in the area. (The protection can be in high strength concrete or mortar, possibly applied pneumatically.)

7.10.6 Anchorages and Deflectors

Since the post-tensioned tendons are not really embedded in the structure in the conventional manner, special attention must be given as to how the prestressing force is introduced. The space requirement of the anchorage arrangement and the prestressing tendons has to be taken in to consideration. When strengthening an existing structure, it is not generally possible to provide reinforcement against concrete 'spalling' or 'bursting' behind the anchorages in the same manner as for a regular prestressed concrete structure. Spalling can be prevented by means of transverse prestressing. This prestressing has the further function of creating contact pressure between new and original concrete, such that necessary shear stresses can be transferred through the joint. To insure full interaction between the tendons and the rest of the structure, the same method can be used along the entire beam but the required shear stress is often so small that it can be dealt with by means of non-tensioned reinforcement. Another method could be to locate the anchorages in compressive zone and design the anchor plates for suitably reduced bearing stress.

7.10.7 There are several methods available for 'anchoring' ('attaching') the supplementary prestressing tendons.

(i) 'Anchoring at Deck-Ends' (involving relocating the abutment back-walls (Fig. 7.12)): The advantage of this system is that the introduction of 'concentrated local forces' into the existing structure is avoided. But it has the disadvantage that all tendons have to run from one abutment to the other.

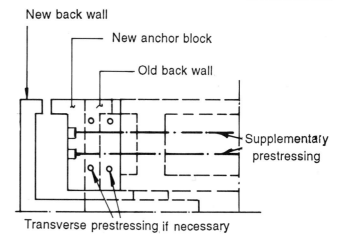

Fig. 7.12   Anchoring of supplementary prestressing elements at the end of the girder

(ii) 'Providing Concrete-Jacket Local Anchor Blocks' at intermediate sections, either in concrete or in steel, fixed to the webs of the girders (Fig. 7.13): This method provides a good distribution for the force in the supplementary tendons, but creates high stresses locally where the prestressing force is introduced. Because of very short transverse dowels, the fixation of the tendon anchor blocks (brackets) can be problematic.

(iii) 'Anchoring at Existing Diaphragms' (Fig. 7.14): Existing diaphragms require extensive and careful coring so that the tendons can pass through them and could be anchored at the backsides of the selected diaphragms. If the latter do not have sufficient capacity to transmit the applied prestressing force, it may be necessary to provide a structural steel frame to transfer the longitudinal prestressing force (*see* Fig. 7.15).

(iv) Using Deflectors or deviation-saddles (Fig. 7.16): Where a polygonal profile is needed for the tendons, saddles or deflectors have to be provided to achieve the desired profile. These devices can be either in concrete or in steel. They are attached to the existing webs and/or flanges by short prestressing bolts or other type of anchors. These short bolts or dowels are very sensitive to anchorage seating movements. A large radius of tendon curvature should be used.

Vertical or inclined tendons can be used to increase shear resistance. One arrangement is shown in Fig. 7.17.

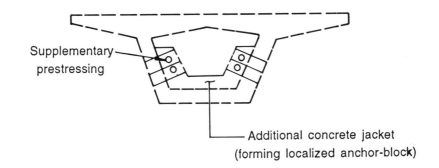

Supplementary prestressing

Additional concrete jacket
(forming localized anchor-block)

SECTION

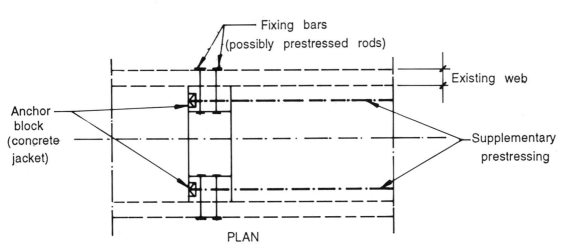

Fixing bars
(possibly prestressed rods)

Anchor
block
(concrete
jacket)

Existing web

Supplementary
prestressing

PLAN

Fig. 7.13 'Anchoring' of supplementary prestressing tendons by use of concrete jacket (forming localised anchor-block)

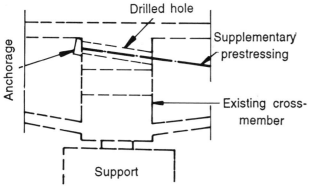

Drilled hole

Anchorage

Supplementary
prestressing

Existing cross-
member

Support

Fig. 7.14 'Anchoring' of supplementary prestressing tendons at existing diaphragms (sectional elevation)

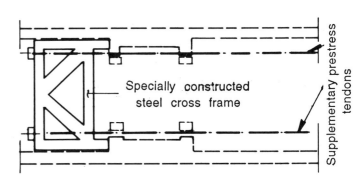

Specially constructed
steel cross frame

Supplementary prestress
tendons

Fig. 7.15 'Anchoring' with auxiliary steel frames

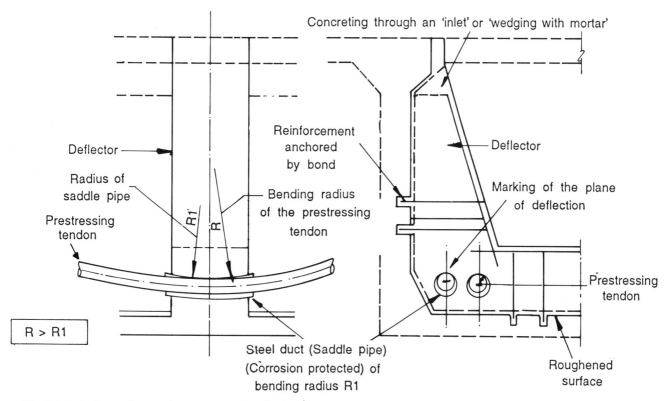

Fig. 7.16   Deflector for supplementary prestressing tendons

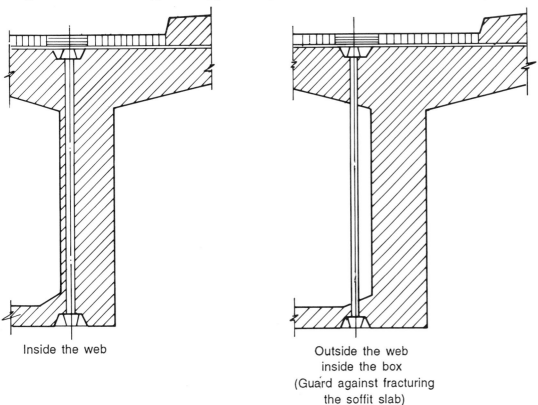

Inside the web

Outside the web
inside the box
(Guard against fracturing
the soffit slab)

Fig. 7.17   Supplementary prestressing for shear strength

NOTE: *See* Chapter 8 for more details about Strengthening of Concrete Bridges by External Prestressing.

7.11 STRENGTHENING BY INCREASING THE SECTION

In this method of strengthening, the effected element has first to be 'de-loaded'—keeping an adequately detailed supporting system ready in advance such that the latter supports the element when it needs this support–mechanism.

In case of a reinforced concrete element the supporting system can be safely brought up to bear the element at as many points as convenient, since this will only reduce the moments (and shears) and relieve the reinforcement that is otherwise taking these load effects; thereafter the affected concrete can be removed, additional reinforcements placed / tied in, and fresh concrete, as required, poured.

But in case of prestressed concrete, the supporting system should only be brought to merely 'lightly touch' the soffit of the prestressed element without actually taking any reaction until the prestress is de-stressed first. Only as the prestress is gradually reduced, does the element gradually require (and find) the supports. Instead, if the supports are made to positively bear the prestressed structure, it is likely to crack and explode since the prestress will continue to act while the gravity load is being reduced by the supporting effect. (All this, of course, can be done only in those of the prestressed structures where prestress can be de-stressed, which is seldom the case.) After the initially prestressed structure has been de-stressed and supported 'gradually' in the above manner, the effected concrete can be removed, additional reinforcement placed as required, fresh concrete-as required-poured and then the new prestress applied as per sequence of the 'design of strengthening'. As the prestressing proceeds, the temporary supports are correspondingly and gradually relieved of the gravity load in the process.

Where feasible, precast concrete panels may be used for increasing the section in the process described above. Such panels will, of necessity, have to have very rough surface on their interface with the existing concrete and a suitable epoxy (resin) mortar is used here to improve bond and restore interface shear-transfer. Additionally, male and female shear keys in the precast panel and the original concrete, respectively, will help. (Sandblasting can be used for cleaning and roughening the mating concrete surfaces.)

7.12 STRENGTHENING BY 'IMPOSING DEFORMATION'

In certain situations (e.g., prestressing tendon snapped; reinforcement destroyed; considerable concrete section lost on account of a broad-side-on collision damage from a high-speed truck mounted with a protruding crane-arm, etc.), it may be easier to strengthen a damaged deck by externally imposing on it temporarily a precalculated load (or deformation) at a pre- determined location while the section is in its damaged conditions, then restoring the concrete and reinforcement - but not the lost prestress, and finally removing the imposed load (or deformation). The imposed load, when applied, acts on the 'damaged' i.e. reduced section properties, and so also do the permanent loads. Stresses due to them are locked-in. The imposed load is removed after the section properties have been restored. This 'removal' of the imposed load tantamounts to applying its moment, etc., in the reverse direction but on the restored section properties, and stresses caused by this are automatically superimposed on the earlier arrested stresses. The net effect, on clever balance, can be improved fibre stresses, which can then withstand the subsequent Live load effect on the restored section properties happily. In this way the strengthening may be achieved without new prestressing.

An unfortunate side-effect of the growing sophistication of the modern bridge is the increasing complexity of repairing the damage. In a post-tensioned cast-in-place 2-span continuous prestressed concrete box girder deck structure, the box girder soffit was hit by an unknown vehicle of excessive height, probably by the outstanding telescoping jib of a heavy mobile crane, travelling at a high speed on the expressway below.

The soffit of the concrete box girder was destroyed over a width of about 0.80 m for the full width of the bridge deck at approximately the 0.4th of the left span from abutment. The leading web was damaged, exposing the lowest of its three stressing cables. The longditudal reinforcing bars in the soffit slab of one inner cell were severed. After a detailed inspection of the damage and after the necessary structural calculations were carried out, it was decided that the bridge could be satisfactorily repaired and restored to its original design capacity. The rehabilitation procedure, in principle, comprised of temporarily loading the damaged deck by a precalculated load, repairing the damages, and removing this load only after the repairconcrete had attained the desired strength. The reversed stresses locked in on account of removal of this load after restoring the section properties countered the distress to the extent required.

By means of imposed deformation, overstressed sections of a structure can be partially relieved. With this the load carrying capacity of the whole structure is improved. A self-equilibrated stress condition can be induced in the structure by relative displacement (raising and/or lowering) of some of the supports or by the introduction of new intermediate supports.

In certain instances the strengthening can be achieved by additionally prestressing for a short length across the damage. Such prestressing tendons have to be duly anchored in to concrete-jackets or steel-brackets made permanent parts of the structure and the effect of such prestress at other sections has to be investigated first to ensure no adverse effects are created. (In continuous structures, such prestress will lead to parasitic prestress moments and shears (which can be quickly calculated by the flexibility method for instance).

It is important to note that relieving some sections of the structure will increase action-effects (bending moment, shear, torsion, etc.) in other sections. A strengthening of these sections may be required. Another important factor is time: relative settlement of the supports, shrinkage and creep of the old structure and the new supporting elements will influence the distribution of the action-effects in the structure.

7.13 PRESTRESSING THE STRUCTURAL STEEL BEAMS

Prestressing here is the application of a predetermined straight or eccentric force to a steel member so that an external loading will be counter-balanced.

There are three general ways of prestressing steel beams: the first is using end anchored high strength wire; the second is to stress cover plates; and the third is to cast a composite concrete slab.

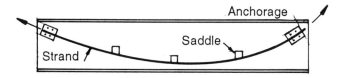

Fig. 7.18 Prestressed beam with draped prestressing tendon

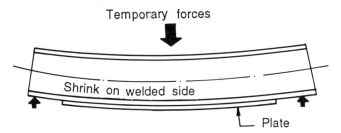

Fig. 7.19 Prestressing by temporarily deflecting a beam and attaching cover plates quickly

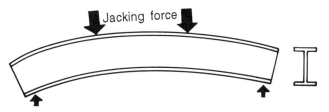

Jacking forces are applied to the cambered beam

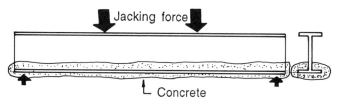

Concrete is placed while jacking forces are maintained

Fig. 7.20 Predeflected beam (preflexed) precompressing the concrete composited with steel tension flange

This technique can be used were shallow depths are required. The beam is temporarily deflected in the direction of the loads. While it is deflected, the bottom flange in encased with concrete. The jacking force is released when the concrete gains the required strength to obtain precompressed concrete on the tension flange. The beam is then erected in the field, then the top flange is encased monolithically with deck slab concrete (sometimes even the web too).

Since the second and third methods are not suitable for bridge rehabilitation, we will discuss only the first method (prestressing by high strength steel). The prestressing method works in much the same way as in a concrete structure which is post-tensioned.

One of the common methods of prestressing is to place draped tendons below the cetroid of the beam or even below the beam with hold downs, and anchor the tendons at each end of the beam. Other is simple straight post-tension tendons attached to the bolted anchor brackets which is located a few inches above the bottom flange. The moments developed by the variable eccentricity prestressing force will cancel out some of the moments produced by the external loads. Prestressing induces stresses in the steel member which are similar but opposite in character to those produced by the dead load and live loads.

These procedures can be developed to carry or atleast reduce the stresses, thereby increasing the live load capacity. If it is designed to carry live load only, then the tension cable should be drawnup snug under the dead load. The live loads then are carried jointly by the strand and the beam. If the dead load countering is also required, the strand is stressed to reverse the beam weight effect also.

The stress distribution in a member at a section due to prestressing at the section consists of an axial stress and a bending stress caused by the eccentricity, and it is shown by the following equation:

At a particular section, the fibre stress is

$$f = H/A \pm He/Z$$

where: H = 'Horizontal' component of the prestressing force at the section,

e = Eccentricity of the force w.r.t. c.g. of section.

z = Section modulus of the member at the section to the fibre under consideration.

In a simply supported girder, the vertical components of the tension cable cause upward forces on the girder 'at the point of contact and 'downward forces at the points of anchorage'. The upward forces should be designed to counteract the downward dead loads. Direct or axial stress should be 'prevented from being transferred into the girder' by providing a horizontal compression member to which the draped post-tesioning cable is 'anchored'. The compression member is independent and free of the girder and is in no way connected to it during the post-tesioning, so that it can 'deform without transferring the axial force of the cable into the girder'.

The draped cable produces moments and shears (and bending and shear stresses) that are similar but opposite in direction to those produced by the loads. The axial stress is eliminated by the use of a free compression member. The reduction or elimination of the dead load moments permits increased load carrying capacity of the structure without increasing the section modulus. In addition, it is possible to counteract the existing and new dead loads by a beam that is free of axial stress under the applied loads.

The post-tensioning cable may be held-down by pins, brackets, or other devices structurally connected to the beam. Supports for the independent compression member are made such that it can deform freely under the axial force without transferring it into the existing

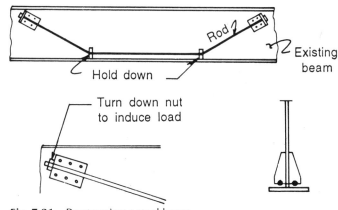

Fig. 7.21 Prestressing a steel beam

beam, and at the same time the beam could provide lateral support to prevent buckling of the compression member.

The King Truss type prestressing procedure is similar to the previous method. This procedure works by tensioning the strands that are connected below the bottom flange with one or more posts. Threaded end-connections are provided so that proper tension can be induced into the system.

Additional capacity can be obtained by changing the configuration of the truss and by adjusting the tension in the bottom chord. The installation should be monitored by controlling the number of turns of the nuts at the anchors and by measuring the deflection induced in the existing member.

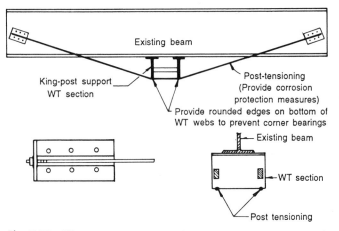

Fig. 7.22 King-truss type prestressing

The advantages of prestressing over the reinforcing method (of increasing section modulus) of a steel member are

- It may be more economical
- Normal traffic may be maintained and if detour is needed, the period will be short
- Jacking of the beams to stress-free the members (for the purpose of connection of new flange plates) is eliminated
- In many cases, such as with existing riveted plate girders, increasing the section modulus may not be feasible while the technique described offers a feasible (and perhaps economical) solution

Disadvantages of such prestressing are

- Relaxation of the steel tendon can occur
- Tendons need to be protected against corrosion
- Without relieving the axial compression force, the beam will act as a beam-column and the deflection may cause a significant change in stress distribution
- Possible cracking of the concrete deck

8

STRENGTHENING OF CONCRETE BRIDGES BY EXTERNAL PRESTRESSING

8.1 GENERAL

The potential of this simple concept for strengthening existing bridges is quite obvious — axial as well as upward acting load-balancing forces can be applied to bridge girders with only the minimum of interference of the tendons with the existing structure. Usually drilling holes through diaphragms, placing 'deviator' blocks, and adding some form of 'force transfer buttress' for the end-anchorages are the only concrete works required when adding external tendons to existing bridge structures. The work can normally be carried out under traffic, which is a prerequisite in many situations. Strengthening by external post-tensioning adds virtually no extra weight to the structure, an advantage particularly in cases where the foundations are already fully utilized, nor does it change the appearance of the bridge. The application is by no means limited to concrete bridges. Any material with reasonable compression characteristics can be strengthened by external tendons, including steel, timber and masonry. (As explained under 'Prestressing the Steel Beams' in Chapter 7, where axial compression may need to be eliminated, special independent compression members can be used during prestressing so as to incorporate only vertical load effects due to this prestress.)

8.2 APPLICATION FOR STRENGTHENING AND REPAIRS

Typically, additional force and strength need to be added to an existing beam or girder. The external prestressing provides additional uplift and pre-compression, therefore, providing additional strength to the structural member. Access has to be provided for the

'anchoring' and 'deviation' points. The deviations are used if additional force is required in a direction different from the tendon axis. Normally, structural steel brackets and deviation blocks are provided and attached to the structure with bolts (see Fig. 8.2) or saddles are attached to the bottom of the member (see Fig. 8.3). As an option, a hole is core-drilled through the member and a heavy steel pipe is grouted in, providing a deviation point support for tendons on both sides of the member. If sufficient clearance is available, the low points can be placed below the existing member soffit for additional drape and uplift. The fire and corrosion protection can be achieved by many different means, the most effective being an encasement in concrete. A known example of this type of strengthening is the rehabilitation of the Pier 39 Garage in San Francisco. The tendon arrangement can be similar to a cable-stay if uplift is the major objective and if geometry and usage of the structure permit such a configuration.

8.3 BASIC ELEMENTS OF EXTERNAL POST-TENSIONING

8.3.1 The Tendons

Tendons can be made-up of high-strength bars, wires or strands. While bars are normally only used for short (preferably straight) tendons, wires and strands can be applied almost universally. Of the two, however, strands are more commonly used. In the following, therefore, the discussion in general will be limited to external tendons made up of 7-wire prestressing strands. The corrosion protection of the strands is usually provided by an outer tube made of either steel or High-Density Polyethylene (HDPE) and an alkaline environment in the form of cement grout injected into the

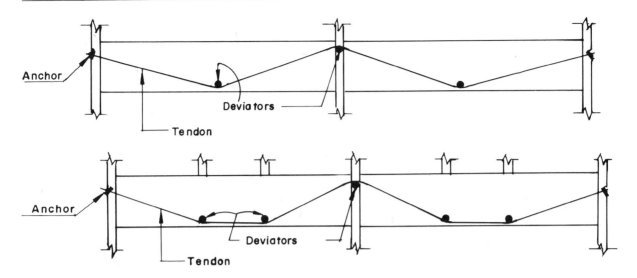

Fig. 8.1   Basic arrangements of external tendons

Fig. 8.2   Structural steel bracket bolted to the side of a
beam at a low point; The PT is encased in
concrete for protection

Fig. 8.3   Structural steel saddle is attached to the bottom
of a beam, serving as a low point deflector or
deviator

tube so that all strands are completely embedded in it
(Fig. 8.4). Sometimes wax or grease based corrosion inhibitors are injected instead of cement grout.

A further improvement of the corrosion-protection
is achieved by using bundles of greased and PE-
sheathed strands ('monostrands') instead of bare
strands and injecting cement grout around them (Fig.
8.4).

This system offers the possibility to monitor the tendon throughout its life, to re-stress or even replace it

during normal bridge operation since strand-by-strand
replacement is possible. It also allows some savings in
the quantities of strands since the friction losses are
considerably smaller than for tendons with bare strands.

The external tendons consist of the following main
elements (Fig. 8.5)
— a bundle of prestressing strands (either bare or indi-
    vidually greased and plastic-sheathed) as the tensile
    member;
— a plastic or steel tubing for strand bundle;

Type 1:   Bare Multistrand System

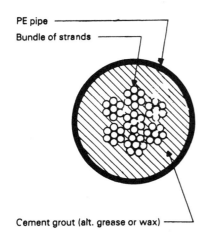

PE pipe

Bundle of strands

Cement grout (alt. grease or wax)

Type 2:   Monostrand System

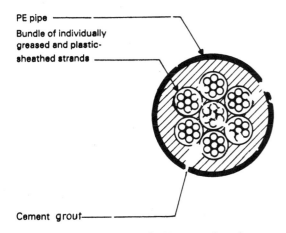

PE pipe

Bundle of individually
greased and plastic-
sheathed strands

Cement grout

Fig. 8.4   Cross-sections of VSL external tendons
Type 1 and Type 2

— end and intermediate anchorages, and couplers;
— a grouting compound.

In the anchorage zones, specially suited corrosion preventive compounds are used.

The main criteria for selecting the tendon-type (Fig. 8.4) are

— environmental conditions and tendon exposure: as for internal bonded prestressing, for the external tendons it seems logical to select the degree of corrosion protection according to the environmental conditions and the exposure of the tendons. It is recommended to use Type 1 and Type 2 Tendons, as shown, for different environments;

— need for tendon force adjustment during lifetime of the structure: in this case Type 2 is recommended;

— tendon friction during stressing operation: the friction with tendon Type 2 is much smaller than with Type 1. In the case of long tendons running over several spans with sizable angular changes, Type 2 offers technical and economical advantages.

Table 8.1 summarizes the selection criteria. It should be mentioned that other factors may influence the decision (such as price, availability of materials, local practice, etc.)

For External Tendons generally cold-drawn 7-wire prestressing strands of ϕ 13 mm (0.5'') and 15 mm (0.6''), normally of low relaxation quality, are used. The geometrical and mechanical properties are given in Table 8.2

In case of Type 2 Tendons (comprising of a bundle of individually greased and plastic-sheathed monostrands), the sheathing around a monostrand is of polyethylene (or alternatively, polypropylene) and has a minimum thickness of 1 mm for straight cables and 1.5 to 2 mm for curved cables.

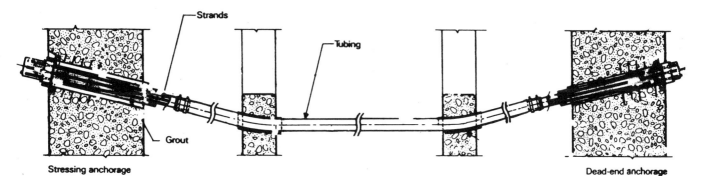

Strands

Tubing

Grout

Stressing anchorage

Dead-end anchorage

Fig. 8.5   Composition of VSL external tendon (schematic)

Table 8.1 Main Technical Criteria for Selection of Tendon 'Type'

	External Tendon	
	Type 1	Type 2
• *Environmental Conditions*		
— always dry or always underwater	•	
— moist conditions	•	
— humid, wet/dry alternately	•	
— aggressive environment		•
• *Need for Adjusting Tendon Force during Service-life*		
— no	•	
— yes		•
• Tendon Friction		
— shorter tendon and small total angular deviation	•	
— longer tendon and larger total angular deviation		•

Table 8.2 Strand Types

	13 mm (0.5")		15 mm (0.6")	
Strand type →	(A) Euronorm 138-79 Super	(B) ASTM A 416-85 Grade 270	(C) Euronorm 138-79 Super	(D) ASTM A 416-85 Grade 270
Parameter ↓				
Nominal Diameter (mm)	12.9	12.7	15.7	15.2
Nominal Steel Area (mm^2)	100	98.7	150	140
Nominal Mass per m (kg)	0.785	0.775	1.18	1.10
Yield Strength (N/mm^2)	1.580*	1.670**	1.500*	1.670**
Ultimate Strength (N/mm^2)	1.860	1.860	1.770	1.860
Min. Breaking Load (kN)	186.0	183.7	265.0	260.7

*0.1% Proof Load Method

**0.1% Extension Method

Table 8.3 gives the nominal breaking loads for the VSL External Tendons according to the four strand types as detailed in Table 8.2. The characteristics of the strand may, however, slightly deviate from these values, depending on the manfacturer and applicable standard.

Table 8.3 Characteristic Breaking Loads

φ 13 mm (0.5") Strand				φ 15 mm (0.6") Strand			
Cable Type	Max Number of Strands	Breaking load (kN)		Cable Type	Max Number of Strands	Breaking load (kN)	
		Strand Type A	Strand Type B			Strand Type C	Strand Type D
5.3	3	558	551	6.3	3	795	782
5.4	4	744	735	6.4	4	1,060	1.043
5.6	6	1,116	1,102	6,6	6	1,590	1,564
5.7	7	1,302	1,286	6.7	7	1,855	1,825
5.12	12	2,232	2,204	6.12	12	3,180	3,128
5.19	19	3,534	3,490	6.19	19	5,035	4,953
5.22	22	4,092	4,041	6.22	22	5,830	5,735
5.31	31	5,766	5,695	6.31	31	8,215	8,082
5.37	37	6,882	6,797	6.37	37	9,805	9,646
5.43	43	7,998	7,899	6.43	43	11,395	11,210
5.55	55	10,230	10,104	6.55	55	14,575	14,339

8.3.2 Plastic (PE) Tubing Around the Bundle of Strands

The strand bundle (consisting of either 'bare' or 'greased and plastic coated' strands) is usually encased in a plastic tube. Alternatively, steel tubes may be used. In certain areas, such as at deviation saddles or where parts of the tendon are embedded in concrete, regular corrugated steel duct as normally used for post-tensioning cables may be chosen. The latter, however, is only applicable when the tendon does not need to be replaceable and tendons of Type 1 are used.

In general, the plastic material is polyethylene and meets the requirements of appropriate standards such as DIN 8074 and 8075, ASTM D 1248 and 3035, or equivalent. Alternatively, polypropylene may be used. The ratio of internal diameter to wall-thickness is approximately 16:1. In general carbon black is added as ultraviolet stabilizer. This material is chemically inert against practically any foreseeable agent (see e.g. DIN 16934). It has shown excellent durability behaviour in structural applications.

In the case of steel tubes, a higher internal diameter to wall thickness ratio can be used (approx. 30:1 to 50:1). The dimensions used are primarily dictated by the availability of standardized tubes. The outer surface of the tubing is normally provided with a paint, giving additional corrosion protection.

The plastic or steel tubing represents the prime barrier against corrosive attack. It is connected to the anchorages and the saddles, thus providing an effective and continuous envelope around the prestressing steel.

8.3.3 Anchorages

The type of anchorage depends on such requirements as adjustability, replaceability, load-monitoring, installation procedure, access, environmental conditions and static considerations. Fig. 8.5 to 8.11 show examples of various types of basic anchorage and a replaceable anchorage in VSL system, for instance.

Equally important is the proper force-transfer from the anchorage to the 'existing' structure. The high concentrated forces must be safely introduced into the existing structure, avoiding, where possible, eccentricities that could cause local bending and shears. This usually involves strengthening of webs and end- diaphragms by attaching new concrete elements to them by prestressing bars or strand tendons. Due regard must also be given to 'sufficient clearances' behind the stressing anchorage to accommodate the jack and cable extension during the prestressing operation. Where this space cannot be provided, it is also possible to use a 'centre stressing anchorage', with dead anchorage at both ends of the tendon (Fig. 8.12). Because of the labour-intensive nature of the work required at the anchorages, the most economical external post-tensioning arrangement is the one involving as few anchorages as possible. The exposed anchorages are properly coated for corrosion protection.

8.3.4 Couplers

These are needed only for stage-stressing, and are generally not needed unless the total length of cable is long enough to require using these so as to reduce friction effects. These are shown in Figs. 8.13 and 8.14 for the VSL system, for instance.

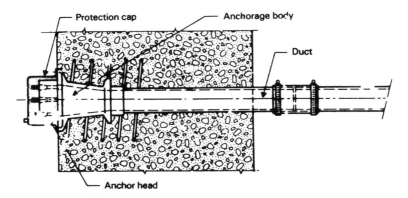

Composition: Anchor head, wedges, cast iron anchorage body; spiral reinforcement, protection cap with grout/air vent.

Features: Simple and economical anchorage where no replaceability, adjustability, load monitoring is required. Preferable for bare strand tendons (Type 1). Suitable as stressing or dead-end anchorage.

Fig. 8.6   Anchorage Type EC (VSL system)

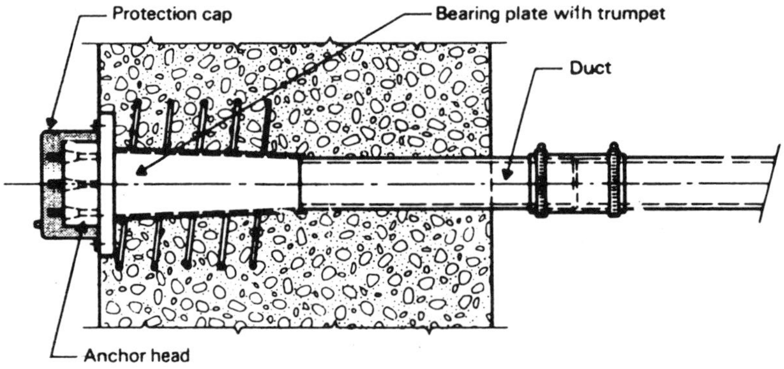

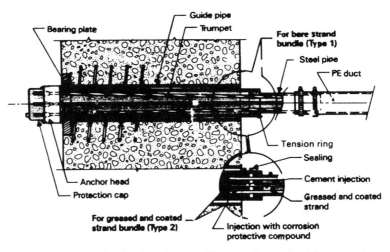

Composition: Anchor head, wedges, steel bearing plate, trumpet, spiral reinforcement, protection cap with grout/air vent.

Features: Simple and economical anchorage where no replaceability, adjustability, load monitoring is required. Preferable for bare strand tendons (Type 1). Suitable as stressing or dead-end anchorage.

Fig. 8.7 Anchorage Type E (VSL system)

Composition: Anchor head, wedges, steel bearing plate with guide pipe and spiral reinforcement, trumpet, protection cap with grout/air vent.

Features: Replaceable dead-end anchorage; stressing anchorage without adjustability or load monitoring.

Fig. 8.8 Anchorage Type A (VSL system)

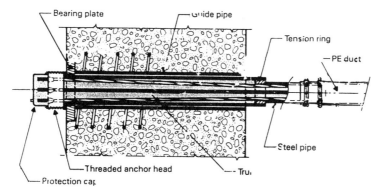

Composition: Threaded anchor head, wedges, steel bearing plate with guide pipe and spiral reinforcement, trumpet, protection cap with grout/air vent.

Features: Stressing or dead-end anchorage, for load monitoring or smaller adjustments. Adjustments by shimming. Replaceable.

Fig. 8.9   Anchorage Type A_m (VSL system)

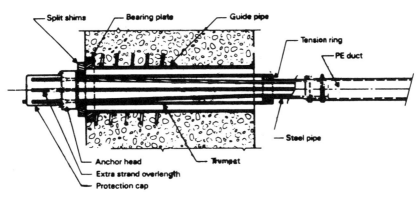

Composition: Anchor head, wedges, split shims, steel bearing plate with guide pipe and spiral reinforcement, trumpet, protection cap with grout/air vent.

Features: Large sized guide pipe enabling push-through for trumpet/anchor head assembly. Replaceable stressing or dead end anchorage. Detensionable if extra strand overlength is provided.

Fig. 8.10   Anchorage Type A_S (VSL system)

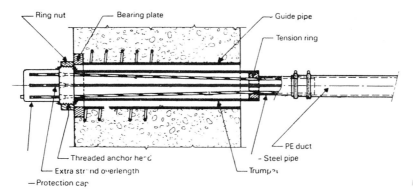

Composition: Threaded anchor head with ring nut, wedges steel bearing plate with large guide pipe and spiral reinforcement, trumpet, protection cap with grout/air vent.

Features: Fully adjustable, detensionable and replaceable stressing anchorage.

Fig. 8.11   Anchorage Type A_R (VSL system)

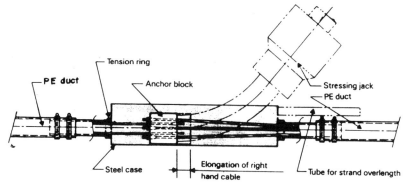

Composition: Anchor block with wedges, retainer plate on passive side to secure wedges, steel case. Tube for strand overlength if detensionability or adjustability required.

Features: For tendons with insufficient access for stressing at end anchorages (e.g. strengthening of structure) or for circular tendons.

Fig. 8.12   Centre-stressing anchorage Type Z (VSL system)

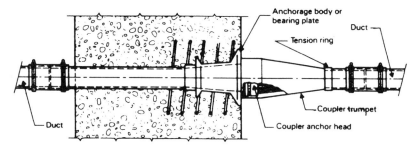

Composition: Coupler anchor head, wedges on active side, compression fittings on passive side, cast iron anchorage body or steel bearing plate with trumpet and spiral reinforcement, coupler trumpet.

Features: Coupling of a new tendon to an already placed and stressed tendon.

Fig. 8.13   Coupler anchorage Type K (VSL system)

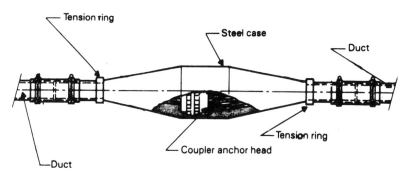

Composition: Coupler anchor head, wedges on active side, compression fittings on passive side, steel case.

Features: Splicing of a new tendon to a previously placed, but unstressed tendon.

Fig. 8.14   Movable coupler anchorage Type V (VSL system)

8.3.5 Deviation Saddles

(i) Saddles serve the purpose of supporting the tendon at locations of angle changes, i.e. at the high and low points of draped tendons, and to transfer the resulting deviation forces to the structure. They typically consist of 'prebent' steel tubes embedded in the concrete or attached to a steel structure by stiffening plates. Over piers and abutments, the saddles can normally be integrated into existing diaphragms with relatively minor alterations. The low point deviation saddles, however, usually require new concrete or structural steel cross beams with or without web stiffeners. (Fig. 8.15). The detailing of the saddles is critical to the overall performance of the external tendon. In particular the connection between the free tendon length and the saddle, and the jointing of the protective sheathing must be designed carefully to avoid sharp kinks in the prestressing steel and grout-leakages, respectively. In order to avoid excessive contact pressure between strands and sheathing which could damage the PE tube, a minimum radius of the saddle must be observed. Figure 8.16 shows three typical saddle arrangements. If no PE tube is placed in the saddle (Fig. 8.16 (i) and (ii), the minimum radius will depend on other factors, e.g. bending stresses in the strands, lateral pressure on the strands or on the concrete, and friction between the strands and the steel tube. If greased and sheathed strands ('monostrands') are used, the strands are only stressed slightly to take out any slack prior to grouting around them. The full prestress is applied after this surrounding grout has hardened. In this way the contact pressures between strands, and between strands and outer sheath are small, thus avoiding excessive friction losses and potential damage to the monostrand sheathing, and allowing strand replacement, should this ever be required.

(ii) Figures 8.17 and 8.18 show the actual details at two deviation saddles. Figure 8.19 shows the detail at a block deviator.

(iii) If tendon replacement is a design requirement, the saddle arrangement must be chosen accordingly, e.g. double sheathing : Fig. 8.16 (iii).

(iv) Minimum radii: Limits must be respected because otherwise either the prestressing steel or the protective sheathing could suffer. Although some tests exist indicating reasonable values, which may be used for preliminary designs, more research work is required in this respect. It is therefore advisable to verify the feasibility of a particular practical solution by tests, if possible. Otherwise a judicious resolute workman-like choice will be called for.

Also *see* Figs 8.20–8.22, plates 24, 25.

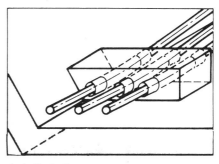

(a) R.C. Block deviator

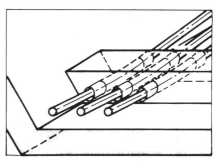

(b) R.C. Cross Beam deviator

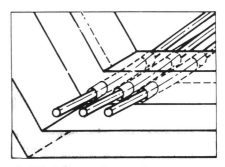

(c) R.C. Cross Beam & Web Stiffener type of deviator

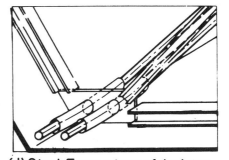

(d) Steel Frame type of deviator

Fig. 8.15 Examples of various types of deviators

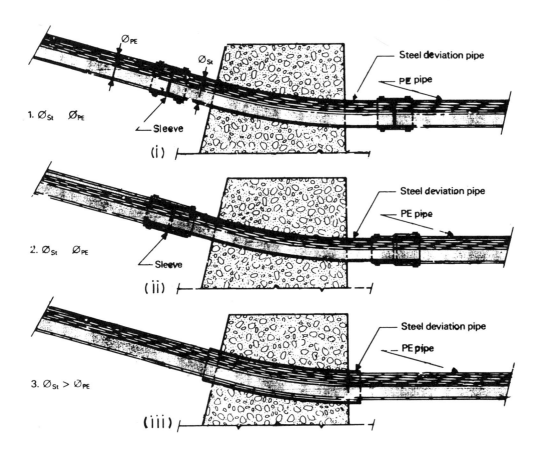

ϕ_{st} = dia. of steel tube

ϕ_{PE} = dia of PE tube

Fig. 8.16 Various saddle arrangements

8.3.6 Additional Tendon Supports

In order to avoid excessive wobble, additional supports must be provided at regular intervals. These are most often steel brackets or clamps; sturdily attached to the webs of the superstructure. Where this is not possible, temporary supports must be provided until the tendons have been stressed. Possible problems (that could arise when no permanent intermediate supports are provided at all, e.g. dynamic resonance of the tendons or second-order effects at large bridge girder deflections) should be carefully investigated.

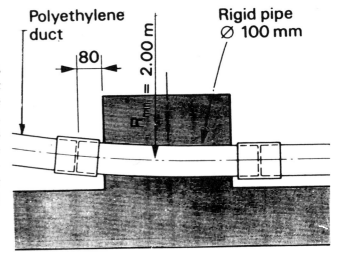

Fig. 8.17 Detail at a R.C. deviation saddle

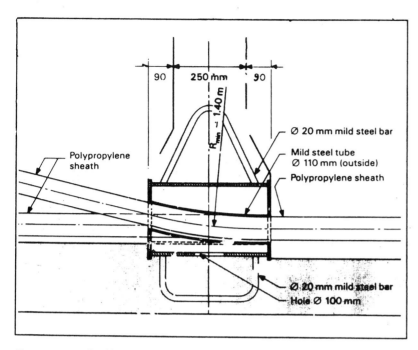

Fig. 8.18 Embedded steel pipe saddle detail

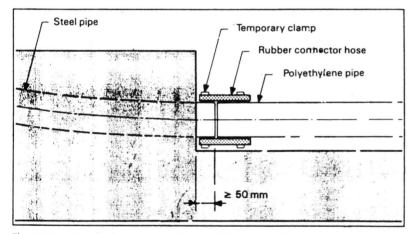

Fig. 8.19 Detail of pipe connection at R.C. deviator block

8.3.7 Installation

Most often the protective sheathing sections are first placed, coupled, and fixed to 'saddles' and 'intermediate supports'. Then the strands are pushed-in hydraulically, one-by-one, and fixed to the anchorages. Sufficient space must be available behind one of the two anchorages to position the pushing and cutting equipment. For very long tendons, a second pushing machine can be placed some distance along the tendon to overcome the friction resistance (the PE sheath is then temporarily left open for a length of approximately 2 m and later closed by coupling sleeves).

Instead of pushing individual strands, the entire strand-bundle can be pulled into the preinstalled PE sheath. This method is used when the tendon length or the total angular deviation exceeds the capacity of the pushing machine(s), and when the monostrands system is used. Sufficient space to allow the cable to be drawn-in without sharp bends must then be provided behind the anchorage. However, if the available space permits, then there is the possibility to pre-fabricate the complete tendon, including anchorages and sheathing, and to push or pull it through oversize steel tubes in the anchorage blocks and saddles. This is normally only feasible for relatively short and light tendons.

Hence, basically there are two different methods used for the installation of External Tendons:

(a) Installation of the empty tube in the final position followed by insertion of the strands.

(b) Installation of completely prefabricated tendons.

(a) Fabrication in the Final Position: Besides the fixing of bearing plates (anchorages) and deviation points, it is necessary to provide (temporary) intermediate tendon supports along the length of the tendon prior to the placing of the tubes.

The tube (steel or PE) is prepared in suitable sections (lengths) and placed in its previously fixed supports. The tube sections are connected by welding or by using couplers. At the ends the tube is tightly connected to the anchorages.

When the tube is securely fixed in the final position, it is ready to receive the strands. The strands are inserted by pulling the prepared strand-bundle (as one unit or in groups) through the tube by a winch.

(b) Prefabrication: The method of complete tendon prefabrication is usually applied to short, light tendons where easy access on site allows the placing of the entire prefabricated tendon.

Prefabrication may be carried out either in a factory or in a prefabrication area at the site, depending on the means of transport, the time between manufacture and installation, and the availability of adequate space on site. The standard lengths of tube are connected to achieve the required total length. PE tubes are connected by welding or with couplers.

The strands are inserted by pushing the tube over the prepared strand-bundle or pushing individual strands through the tube.

The bearing plates or anchorage bodies and the supports at the deviation points are fixed to the structure. The prefabricated tendon is then placed into its final position either manually or by mechanical means using hoists or winches. Intermediate temporary supports along the straight lengths are provided to keep the tendon in its correct position.

8.3.8 Stressing

The external tendons are stressed with the appropriate multistrand jack. All the strands are stressed simultaneously but individually locked-off. The stressing operation normally follows the procedures established by the specifications, by local codes of practice or by the FIP recommendations.

Type 1 Tendons (bare strands) are stressed at a steady rate in one or several increments until the required stressing force is reached. Grouting is carried out 'after' completion of the stressing operation.

Type 2 Tendons (greased and plastic-coated monostrands) are stressed in two stages. In the first stage, an initial force is applied which removes the slack in the tendon. Then the tendon is grouted. After the grout has attained the required strength, the stressing operation is then continued. In second stage, the stressing force is raised in uniform fashion to its final value.

Depending upon the anchorage type chosen, the tendon force can be checked, adjusted or released, using the same multistrand stressing jack.

8.3.9 Grouting

The anchorages incorporate a grout connection which can be used as inlet or as outlet. Further, grout connections are also provided at the deviation points.

Grouting commences at the lower end of the tendon and proceeds at a steady rate until grout of the same consistency is ejected at the deviation points and finally at the other end of the tendon. For long tendons, the grout is injected at subsequent inlets along the tendon. When using greased and plasticsheathed strands inside a steel or PE tube (tendon Type 2), the tendon is grouted after initial tensioning by injecting cement grout into the tubing only. The anchorage zones are filled with a non-hardening corrosion preventive compound.

8.3.10 Corrosion Protection Systems

It is known the prestressing steel needs careful protection against the various possible types of corrosion attack. For internal, bonded prestressing, this protection is provided by the alkaline environment of the cement grout and the surrounding concrete. Experience has shown, however, that there are several aspects to which attention must be paid, in both design and construction, to make the protection really effective.

A corrosion protecting strategy is summarized in Table 8.1. It is in line with more recent recommendations in various national standards. In addition to the given design measures, adequate materials and good workmanship are needed. From comparison with the practice of the past and experience gained with existing structures, it has been recognized for some time that improvements are necessary with regard to concrete quality, detailing and the amount of reinforcement.

As for internal tendons, it seems advisable to apply a corrosion protection strategy which is based primarily on environmental conditions and also safety considerations (e.g. with regard to fire, strand-failure, etc.). But more protective means are necessary for external cables. Many different solutions have been adopted in the past

PLATE 24

Fig. 8.20 Showing construction of R.C. block deviators for strengthening a box-deck by external prestressing (view from inside cell)

PLATE 25

Fig. 8.21 Showing construction of R.C. cross beam deviators (inside a cell) for strengthening a box-deck by external prestressing

Fig. 8.22 VSL's new external post-tensioning system utilizes individually greased and sheathed monostrands for increased corrosion protection and reduced tendon friction. The deviators for profiling the external prestressing cables are also clearly seen, as also the suspenders

(a) Zinc Coating: Its corrosion resistance depends upon the type of galvanization and its applied thickness. Zinc coated prestressing steel has been used in France on several occasions. There is doubt, however, as to whether zinc coating provides a permanent corrosion protection. It seems to be durable only under very favourable environmental conditions. Coatings have been damaged during handling and installation. Another problem arose when zinc accumulated in the stressing anchorage inside the wedges.

(b) Polymer Coating: This technology, in which polymers are bonded to the steel by fusion, has been developed in the United States primarily for the protection of reinforcing steel. Polymer coated strands have also been available for some time, and a number of applications are reported. It remains to be seen whether this system will prove to be a viable solution for prestressing steel. Problems could occur due to the fact that only the outer strand surface is protected, the king wire and inner surfaces of the six surrounding wires in a 7-wire strand having no coating! At the anchorages, the coating is locally interrupted by the 'indentations' of the wedge teeth. It is also possible that, as with zinc coated strands, problems may occur in the anchorages. Special care must be taken to prevent damage to the coating during handling and installation.

(c) Protective Sheathing: The protective sheathing represents an envelope around the prestressing steel. Suitable materials are steel or plastic tubes (polypropylene [PP] or polyethylene [PE]). In order to achieve an effective protection system, proper solutions are required for 'coupling' these tubes with each other, with the anchorages, and with saddles.

Injection of the remaining voids inside the sheathing with cement grout has proven to be economical and reliable. In the case of restressable anchorages, cement grout must be replaced at least locally by grease or similar soft plastic material. Besides being rather expensive, these products are difficult to inject (e.g. preheating up to 100°C required) and special measures are needed to prevent leakages.

In this category, individually greased and plastic-sheathed monostrands offer many advantages. They are manufactured under factory conditions. The prestressing steel is therefore effectively protected against corrosion during transportation, storage on site and installation, provided that proper care is taken not to damage the sheathing. Monostrands can be used either individually or in bundles as multistrand tendons. In the latter configuration they are usually placed inside a plastic or steel tube. The remaining voids are filed with cement grout.

8.4 TWO TYPICAL SITUATIONS WHERE STRENGTHENING BY EXTERNAL PRESTRESSING CAN BE ELEGANTLY ACHIEVED

(a) The Flexural and/or Shear Strength of the Bridge Superstructure Insufficient: Various reasons may be responsible for the need to strengthen the structure, such as badly corroded or already fractured prestressing tendons, general deterioration of the structure, upgrading of the bridge to higher traffic loads, addition of permanent loads due to deck widening or additional surfacing layers, or errors in the original design. Straight external tendons, placed near the centroidal axis, are often the most economical solution in this case, because the construction is relatively simple due to the absence of costly deviation saddles. Straight continuous tendons in excess of 400 m length have successfully been installed in such situations.

In bridge girders with constant depth, this configuration only provides a uniform compression stress while in haunched girders opposing bending moments will also be produced. For the flexural capacity in the Ultimate Limit State (ULS), the tendons can be treated as a resisting force with the corresponding internal lever arm about the centroid of the compression block. If, in a girder with constant depth, the weakness is mainly in the spans, straight tendons are preferably placed near the bottom of the cross section where they are more effective in ULS. However, the stress increase in the prestressing steel at ultimate is relatively small in unbonded tendons so that normally designs are based on the effective prestress rather then the yield strength. Obviously, straight tendons do not markedly increase the shear capacity of the structure.

Generally more effective is the arrangement of external tendons with a drape following approximately the moment diagram, thus partly balancing the gravity loads by deviation forces. Drapped tendons also significantly enhance the shear strength of the structure so that additional vertical prestressing or add-on concrete to strengthen the webs is normally not required. This configuration is also more effective in SLS (Service Load Stage) since it reduces the bending stresses at span and pier sections. Although drapped tendons, by their nature, have greater friction losses than straight ones, they can still be arranged in quite long continuous units if the monostrand system is used because monostrand possesses a very low friction coefficient. Thus the number of anchorages can be reduced.

(b) The Superstructure Exhibits Excessive Mid-Span Deflections and in Certain Cases is Accompanied by Large Cracks: The global ULS capacity of the structure may be

sufficient, but to reach it, large plastic rotations have to take place, possibly exceeding the available rotation capacity at the plastic hinges. This situation can be encountered usually in bridges that were built by the cantilever method and camber control was inefficient so that, in time, the deflections magnify because of reduction in E-modulus of concrete. There are many possible reasons for such deficiencies, most of which can be attributed to errors in the original design and lax construction control on the camber. Until recently, the effects of temperature gradients through the deck were ignored by most bridge design codes. Also the long-term creep-related 'redistribution of moments' from the pier sections to the spans was often underestimated or even ignored. Both these effects are 'imposed deformations', leading to increased positive-moments in the spans up to the point when cracking reduces the stiffness of the girder as well as the rate of further moment increase! In segmental bridges it is often the opening of construction joints, rather than well distributed cracks, that release the induced tensile stresses in the concrete, resulting in high steel stresses at the opened joints, so that eventually local yielding of the bonded prestressed and non-prestressed steels (tension failure) may occur. The mid-span section of the girder then behaves almost like a plastic hinge, resulting in a deflected shape similar to that of a series of dipping cantilevers.

Apart from the mainly aesthetic problem of such deflections, the wide cracks pose a serious threat to the structure by allowing accelerated corrosion to take place. Equally serious is the increased danger of a fatigue-failure as a consequence of the much higher 'variation' in the live-load-induced-stress in the prestressing steel which has then a lower margin left (after cracking of the section). The philosophy for rehabilitation must therefore primarily aim to reduce the steel stress due to global effects, the traffic-induced 'stress-range' and the crack-width. To that effect the application of external post-tensioning is no different from the situation described earlier. By their unbonded nature, unbonded tendons experience very small stress fluctuations under variable loads (lack of strain-transfer), and are, therefore, also ideally suited to remedy the fatigue problem. However, one should not expect that the opposing forces from the draped external tendons would reverse the in-span deflection by more than a fraction, nor will the wide cracks or the opened joints completely close after applying the additional prestress. This is because a cracked section is substantially stiffer for unloading due to opposing moments, than for continued loading, and because the long-term creep induced part of the deflection cannot be reversed by significant degree!

Another common cause for excessive mid-span deflection is the over-estimation of the concrete E-modulus in the design, and the construction of the free cantilevers with insufficient pre- chamber to compensate for long-term deflections. This situation can arise without the presence of any alarming cracks, so that the excessive deflection is then mainly an aesthetic problem! In this case too, the deflection itself cannot be completely reversed by external post-tensioning. The appearance of the bridge, and the riding comfort for vehicles, can be improved by means of a variable-thickness surfacing layer, the additional weight of which can be resisted by the external prestressing tendons!

In segmental balanced cantilever bridges with mid-span hinge joints, excessive long term or temperature-related deflections can lead to 'blocking' of the joints. This in turn can seriously endanger the durability of the structure since the blocked hinges may impose large local stresses onto the concrete, causing bursting cracks and spalling. Possible retrofit strategies in such situations would be to add some straight external tendons near the top of the section, causing a constant positive moment and, thus, regaining part of the deflection, or to convert the bridge into a continuous girder by casting a concrete closure pour and providing external continuity tendons in the spans. The latter of course will depend on the expansion joint layout of the bridge the residual stress configuration at different sections as well as the tolerance of differential settlement of foundations.

8.5 ACTUAL STRENGTHENING EXAMPLES* FROM PRACTICE — USING EXTERNAL PRESTRESSING

• *Practical Example 1:* The Los Chorros Viaducts *(Venezuela)*

(a) Background: The two parallel 320 (300) m long Los Chorros viaducts at Caracas, Venezuela, with 40/60/120/ 60/40 (20) m spans were built between 1969 and 1971. The central portion was constructed by the cantilever method, with 60 m haunched cantilevers (Fig. 8.23). A bridge survey carried out by the authority responsible for the bridge in 1972 showed unusually large mid-span deflections in the 120 m main span. Surveys in 1973, 1974 and 1978 showed an alarming rate of increase of this deflection which prompted the authority to investigate possible lines of action. By 1982 the midspan deflection measured on one of the viaducts was 410 mm (Fig. 8.23). The deflected shape exhibited a sharp kink at mid-span, where cracks of 1 to 1.5 mm were clearly visible, indicating that the flexural stiffness

---

* Ref: VSL Publications, with grateful acknowledgement.

had decreased significantly in the vicinity of the closure pour.

(b) Diagnosis: The basis of planning the rehabilitation scheme carried out by VSL International at Berne was a thorough physical inspection of the bridge, followed by the study of the available documents and a retrospective analysis. These investigations lead to the conclusion that the most likely causes for the excessive deformation were

— higher than anticipated superimposed dead loads
— underestimation of the redistribution of bending moments from piers to span due to creep, shrinkage and steel relaxation
— no consideration of temperature differential through the deck
— insufficient pre-chamber built-in during construction

(c) Rehabilitation Concept: The design philosophy for the rehabilitation consequently had the following objectives

— to improve the riding comfort and aesthetic appearance of the bridge with a minimum of extra weight
— to close the cracks by epoxy injection and additional prestress
— to restore sufficient safety against a fatigue-failure
— to provide a sufficient safety factor in ULS

After the preliminary study of a number of alternative schemes, the following concept was chosen:

— replacement of the heavy concrete parapets by light-weight guide-rails with improved alignment

— provision of variable-thickness surfacing to 'bridge' the sharp kink at mid-span to a 5800 m radius
— provision of external prestressing tendons to carry the additional surfacing weight and to achieve uncracked behaviour for permanent and local (fatigue) live loads. This would also restore the original stiffness of the mid-span region, thus arresting the continued deformation, and increase the ULS safety factor to an acceptable level.

(d) Design Consideration: An arrangement of twelve draped tendons VSL 12 × 0.6", four along each web of the twin cell box girder, was chosen (Fig. 8.23 to 8.29). For economic reasons the tendons were continuous, without intermediate anchorages. The tendons were deviated over the piers by steel tubes set into holes drilled through the diaphragms (Fig. 8.25 to 8.26) and by two steel deviation frames in the main span (8.24). Since there was not enough space to stress the tendons at the expansion joints (grids B — *see* Fig. 8.23), the tendons where anchored in the 60 m side span by means of reinforced concrete buttresses stressed to the webs. Standard VSL anchorages of the type shown earlier where used. In order to achieve a smooth transfer of the prestressing forces, the anchorages where staggered over a length of 6 m (Fig. 8.28). Together with the top and bottom slab, two new transverse R.C. struts resist the transverse forces arising from the eccentricity of the anchorages with respect to the webs (Fig. 8.27). Since the force transfer to the webs is by shear-friction, the existing concrete surfaces where carefully roughened. Short

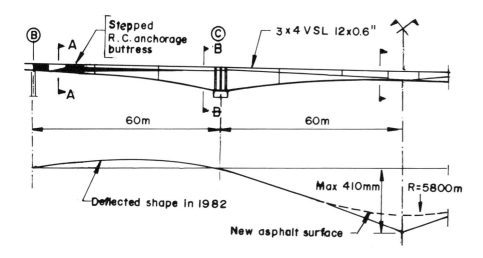

Longitudinal section showing external tendons
Deflection curve and new asphalt surface to bridge mid-span kink

Fig. 8.23   Los chorros viaduct

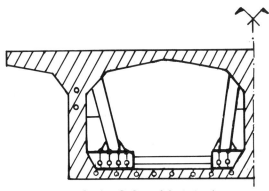

Fig. 8.24

Section C–C: steel deviation frame

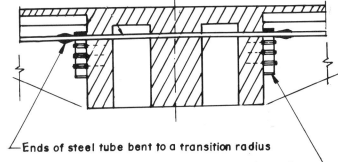

Steel tube (straight within existing cross beams above pier)

Ends of steel tube bent to a transition radius

R.C. bracket to support bent tube ends

(Longitudinal view of bridge)

Fig. 8.25   Saddle arrangement above pier

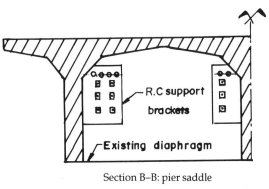

R.C support brackets

Existing diaphragm

Section B–B: pier saddle

Fig. 8.26

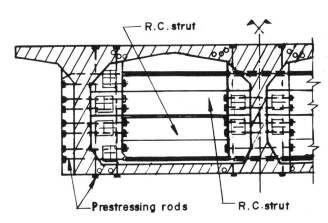

R.C. strut

Prestressing rods

R.C. strut

Section A–A: R.C. anchorage buttress

Fig. 8.27

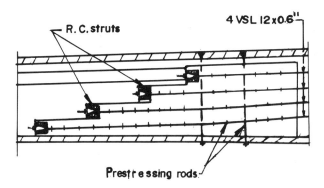

R.C. struts

4 VSL 12 x 0.6"

Prestressing rods

Detail of stepped R.C. buttress for anchorages

Fig. 8.28

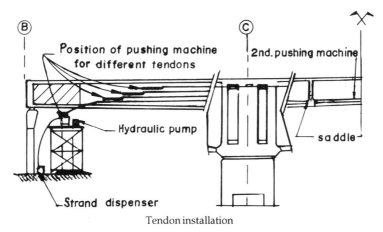

Tendon installation

Fig. 8.29

VSL-prestressing rods through the walls provide the required 'normal' force.

(e) Execution of the Work: After the installation of the PE tubes and the steel tubes at the saddles, the strands where pushed in one by one. In order to overcome the frictional resistance, a second pushing-machine was installed near mid-span (Fig. 8.29). The PE tube was left open at that location and later closed by coupling sleeves. In this way the 220 m long tendons were placed without any problems. The tendons were stressed simultaneously from both ends and then cement grout was injected for corrosion protection. Finally, the new parapet guide rails where installed and the surfacing

was replaced as stated earlier. The rehabilitation work was carried out between January and May 1988.

• *Practical Example 2: Motorway Viaduct, Höll (Switzerland)*

This 404 m long 2-lane viaduct originally consisted of a series of simply supported spans of 31 to 45 m length and double-T cross section. Amongst other deficiencies, the bridge suffered from cracks near the supports and deterioration of bearings and expansion joints. It was decided to convert the bridge to a continuous girder by constructing prestressed concrete jackets over the piers, including web thickening, additional diaphragms and a bottom compression slab (*see* Fig. 8.30). At the same time the deck was to be 'widened' which required a deck slab

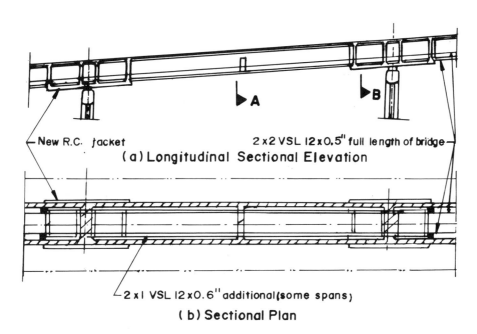

Fig. 8.30   Höll bridge

'thickening' throughout. In order to resist the additional load and to generally upgrade the capacity, four straight external tendons VSL 12 × 0.6" (monostrands system) where installed near the bottom of the cross section, running along the inside faces of the webs, continuous over the entire length. Two additional 52 m long straight mono-strand tendons, VSL 12 × 0.6", were installed in each of four spans, also near the bottom (Fig. 8.30 and 8.31). The resulting negative moment in the support regions is compensated by the additional bonded pre-stressing tendons near the top slab, integrated into the R.C. jackets. The long tendons are anchored in the strengthened end-diaphragms at the abutments and new access chambers behind the abutment diaphragms are constructed with sufficient clearance for the stress-ing-jacks and strandends (Fig. 8.32). The short tendons

are anchored in the new diaphragms integrated into the R.C. jackets over the piers. All tendons are pulled into the preinstalled PE sheaths. A cable draw-in duct is provided at the abutment to facilitate drawing-in the long tendons (Fig. 8.32).

• **Practical Example 3:** *Motorway Bridge Across the Rhone at Massongex (Switzerland)*
This 230 m long box girder bridge, with spans 3 × 27–38 –72–38 m (Fig. 8.33) exhibited 113 mm mid-span deflection in the 72 m haunched main span, accompanied by a number of cracks up to 0.5 mm wide. The deflection had increased linearly with time and it was feared that this trend may continue unless strengthening was implemented. It was found that the effects of differential temperature had not been considered in the design. Furthermore, it was expected

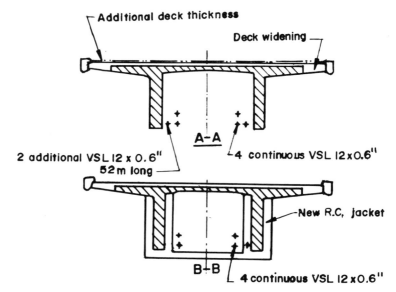

Fig. 8.31   Höll Bridge — cross sections

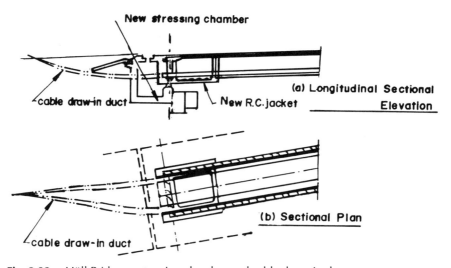

Fig. 8.32   Höll Bridge — stressing chamber and cable draw–in duct

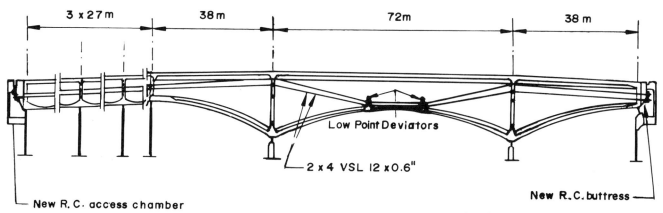

Fig. 8.33   Massongex Bridge — longitudinal section showing external prestressing (unequal scales)

that the amount of pre-compression provided by the existing tendons was appreciably lower than that assumed in the design. A total of eight external tendons, VSL 12 × 0.6", were installed (Fig. 8.33). The tendons are draped with deviator tubes placed in the existing pier-diaphragms and two low-point deviator R.C. cross beams in the main span (Figs 8.34 and 8.35). The anchorages are located in prestressed buttresses added to the abutment diaphragms and new access chambers are constructed, similar as for the Höll bridge described above. The tendons are of the mono-strand type, with the PE sheath installed before hand and attached with a single prestressed strand strung along the profile, thus avoiding any further intermediate supports. The strand-bundles where pulled-in from behind one abutment. After taking the slack out of the strand by

slightly stressing them, cement grout was injected in to the PE tubes. The full prestressed was applied after hardening of the grout. The expected upward deflection due to the external tendons was only 14 mm, i.e. roughly 10 per cent of the deflection could be reversed. Similarly the added prestress was not expected to fully close the cracks so that they would need to be epoxy-injected. However, further deflection and crack growth would be arrested by the added prestress.

• **Practical Example 4:** *Roquemaure Bridge near Avignon (France)*

Owner. . . . . Autoroutes du Sud de la France, Vedene

Engineer. . . . . Etudes Ouvrages d'Art (Bouygues), St. Quentin en Yvelines

Additional
Post-   { VSL France s.a.r.l.,
Tensioning. . . . . . .   { Boulogne-Billancourt

Execution. 1975–1976

This bridge is part of Motorway A9 Orange-Narbonne in Southern France, which it carries across the river Rhone near Avignon. The structure is 420 m. long, and has spans of 50 – 4 × 80 – 50 m. The 21.60 m wide superstructure consists of a double-T section with a depth between 5.40 m at piers and 1.80 m at mid-spans.

It was built in 1971 to 1974 by the free cantilevering method, with cast-in-place segments up to 6.12 m in length. In 1975 a surveillance campaign revealed the presence of major cracks (as wide as 8 to 10 mm) at the mid span sections. After examining the damage and after checking the design, it was found that no temperature gradient had been considered and the cover of the prestressing tendons was insufficient. The structure, therefore, had to be repaired in the shortest possible period. A complete interruption of the highway being unacceptable, the owner had to allow for the repair work to be done under light traffic. Therefore, the consultant proposed to apply the external longitudinal prestressing after the cracks had been grouted with

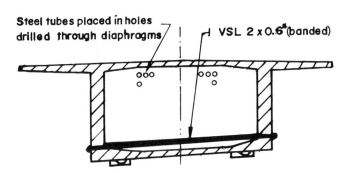

Fig. 8.34   Massongex Bridge — section at pier

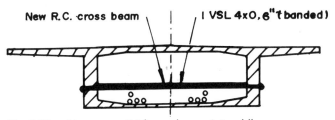

Fig. 8.35   Massongex Bridge — low point saddle

resin. In addition the following measures had to be taken:

— construction at either end of the bridge of prestressed concrete cross beam incorporating the anchorages of the new tendons and transmitting the external prestress forces to the superstructure of the bridge;

— construction of a working chamber behind each cross beam from which the strands would be fed into the ducts and where the cables could be stressed;

— installation of hangers beneath the bridge deck for carrying the cable ducts.

The longitudinal prestressing force required amounted to 54,000 KN after losses. VSL proposed to use 8 tendons each of 55 strands of 0.5" (Unit 5–55, ultimate force 9, 169 KN) running from one end of the bridge to the other without any coupler (cable length 438 m); four spare ducts where also installed in case additional prestressing should be needed, Fig. 8.36. Placing of the tendons was the most demanding part of the job. Using preassembled strand-bundles was excluded from the beginning because of the limited space available in the working chamber and because of the length and weight of the tendons. Thus only push-through method (using individually sheathed mono-strands, external Tendon-Type 2, Fig. 8.4) was applicable. Tests made by VSL inabled the best way of operating to be found. They showed that two intermediate pushing posts where required. According to the access possibility, pushing sections of 135, 160, and 135 m were selected. Two pushing machines had to be placed at the first intermediate post in order to obtain the required pushing force.

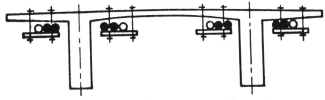

Fig. 8.36 Cross-section of Roquemaure Bridge with added external tendons

When a certain number of strands had been introduced, the pushing force available was no longer sufficient for overcoming the friction in the steel tubes, and an auxiliary strand running between the first two posts was used, to which the strands to be installed where coupled. The first machine pushed the auxiliary strand while the second machine pulled it. After pulling, the auxiliary strand was pushed back to the first post and the operation then repeated. When the pushing operation was finished, the openings in the tubes were closed by previously mounted coupling sleeves.

Before the cables where stressed, each individual strand was tensioned to 1 N/mm² by means of a monojack to bring all strands to the same length and take up the slack.

Stressing had to be done on two cables simultaneously and in both ends for symmetry reasons. The time available for stressing 4 (of the 8) tendons was fixed at six hours and therefore great mobility of the equipment was required. Five VSL jacks ZPE-1000 (one as spare) and corresponding accessories, as well as five pumps where engaged. The jacks, each weighing 2.5 tonnes, were mounted on specially constructed hydraulic carriages. Stressing was done in steps of 5 N/mm². The cable extension amounted to 3, 150 mm. The cable of the end cross beams (16 No. EE5-12 i.e. 12/0.5", each side) were stressed in groups of two at the same time as the longitudinal tendons.

In view of the quantity of material to be injected and of the length of the cables, the use of a special grout mix with retarded hardening, and consisting clinker and resin, was required by the client. Grouting was executed in sections, which made movable equipment necessary. The grout mix was injected over a distance of 180 m before the installation had to be moved. Two cables were grouted per day, requiring 12 m³ of grouting material.

• *Practical Example 5:* *Ruhr Bridge, Essen-Werden, Germany*

Owner. . . . . City of Essen
Engineer (Repair). . . Prof Dr G. Ivanyi, Essen.
Contractor (Repair). . Polensky & Zollner AG, Bochum.
Additional
Post-   { SUSPA Spannbeton Gmbh,
Tensioning. . . . . { Langenfeld
Execution. 1985-1986

This two-span post-tensioned concrete bridge (spans 66.40-47.00 m) has a multi cell box super structure with a deck width over the intermediate pier of 34.41 m. This width increases on both sides towards the abutments.

In the bottom slab and in the webs of the larger span, numerous cracks due to bending had developed, making rehabilitation measures necessary specially in view of the corrosion protection of the post-tensioning cables in the cracked area. Grouting the cracks (which were up to 0.4 mm wide) was disregarded since it was established that the cracks originated from temperature gradients; thus new cracks would have appeared again near the grouted cracks. Furthermore, it was established that the stress variation in the post-tensioning steel considerably exceeded the allowed value.

Thus a static strengthening was required, not only the corrosion protection measures. In view of the limited space available inside the box cells, which did not allow for adding reinforcement, strengthening the larger span by means of post-tensioning cables offered the best solution. Straight unbonded tendons were selected, the number of which had to be the smallest possible.

A total of 24 tendons VSL type 5-16 (16/0.5" strands, breaking force 2,833 KN each), with threaded anchor heads, and an average length of 75 m, were required. At the abutment side these were anchored in anchor blocks added to the web prolongations, while buttresses were provided behind the pier diaphragm, which itself also had to be post-tensioned to take the additional forces.

The monostrands were placed in PE ducts which were provided with two movable joints to absorb temperature movements. As strand deviations could not be avoided and inaccuracies in the boring had to be expected and in view of the large elongations, the likelihood of damage on the strand coating due to transverse pressures caused by strand-deviations was evaluated in tests. The coating remained safe in these tests.

Borings had to be carried out with high accuracy. Boring distances were 10 to 12 m (in pier diaphragm). Oblique boring (in plan view) was also required through a span diaphragm and subsequently through a web.

The strand-bundles were prepared in the workshop. The diaphragm cables were placed by means of a movable crane. The longitudinal tendons were stressed in the anchorages behind the pier diaphragm. Elongations measured 440 mm on average. Post-tensioning work lasted for seven days. All cracks closed after stressing.

The rehabilitation work took five months.

• *Practical Example 6: Bridge over Wangauer Ache near Mondsee (Austria)*

Owner . . . . Republic of Austria, Federal Road Administration, Vienna.

Engineer (Repair). . . Kirsch-Muchitsch, Linz

Contractor (Repair). . .Hofman U. Maculan, Salzburg

Additional

Post-Tensioning. . . . Sonderbau Gesmbh, Vienna

Execution. 1987–1988

This twin bridge is part of the highway—Vienna-Salzburg. It was built in 1962–1964. In recent years improvements have been made several times, but thorough inspection revealed a large number of deficiencies, making a general rehabilitation necessary. In particular, a lack of longitudinal prestressing force was detected, which had become obvious in opened construction joints. Thus rehabilitation had to include also the installation of additional tendons. Since a closure of the motorway was not acceptable, one structure was strengthened first, followed by the other.

The twin bridge has span of 25–6 × 28–2 × 41.25–3 × 28–25 m, i.e. a total length of 384.50 m. Each superstructure has a double-T cross section, 13.05 m wide and 2.20 m deep. Since the existing longitudinal post-tensioning was bonded, and no spare ducts were available, the additional post-tensioning had to be placed on the outside of the webs. However, the total length of 386 m was considered to be a problem.

Four VSL tendons EE 5–12 (i.e. 12/0.5" strands) per web were selected as the additional post-tensioning. Polyethylene tubes of 90 mm diameter and 4 mm wall thickness were chosen as sheathings. At the ends of the superstructures, end diaphragms each post-tensioned by 3 VSL tendons EP 5-7 (i.e. 7/0.5" strands), were provided.

The ducts were fixed to the webs at the quarter points of the spans by means of clamps. In between, additional cable supports were provided in order to avoid wobble. These supports were hung from the deck slab. In order to prevent the strands from abrading the polyethylene duct at clamping points, steel tubes were placed inside the polyethylene ducts at those points. These steel tubes also act as stiffeners at the joints of the polyethylene ducts.

The strands were installed by the VSL Push-Through Technique. Two pushing machines were placed one behind the other and driven by a hydraulic pump of corresponding power. The strands had to be cut to length by hand before they could be pushed through. However, only the first 3 to 4 strands could be fully pushed through without squeezing. Therefore, strand installation was completed by hand from a joint opened in the duct about 250 m from the push-through machines.

Before stressing, all joints were checked for tightness. Stressing was done from both ends. A friction coefficient of only 5 per cent was observed. Grouting was provided for corrosion protection, not bond, as is the case in external prestressing. It was performed from the lowest point in the middle of the cable, towards both ends. Vent hoses were provided at distances of 40 to 50 m. Two grouting pumps were required.

• *Practical Example 7: PC Beams Strengthening, Minnesota (U.S.A.)*

A precast concrete structure consisted of cantilever beams supporting roof planks. When the roof planks were installed, flexural cracks occurred in the beams over the supports. The cause become quite obvious when it was 'discovered' that the top and the bottom prestressing steels had been reversed without being detected during the precasting operation!

Prestressing agencies were contacted to propose an economical solution. External tendons offered just that. Between one and six strands of 0.5" (13 mm) diameter were installed per beam to satisfy the ultimate capacity and to close the cracks. The tendons were straight at the top of the beams, just below the roof planks with no deviation points needed. The attachments consisted of galvanized structural steel brackets, designed by VSL. The fire and corrosion protection of the strands was to be provided by the owner.

• *Practical Example 8: Stadium Beams, Georgia (U.S.A.)*

Due to some design changes, 36 out of the 52 cantilever beams, supporting precast beams in the seating area, required strengthening to meet all serviceability and ultimate capacity requirements.

Various agencies were asked to propose a solution, preferably without any architectural impact. It was suggested by some to use one 1 1/4" (32 mm) bridge cable on each side of a beam, stressed to 75 per cent of capacity, resulting in a total force of 288 kips (1280 KN). It was important to stay above the future ceiling-line with all components of the cables. Deviation saddles and anchoring brackets would have been required for this option.

VSL proposed a single grouted tendon passing through the center of the column above and into center of the beam (*see* Fig. 8.37). A hole was core-drilled through the column 48" × 48" and into the beam 48" wide × 34" deep.

A constant angle was maintained, using a guide-bracket at the top of the beam during the drilling operation. A tendon with 10 strands of 1/2" diameter (i.e. 10/0.5") was inserted into the hole. The anchorage consisted of the usual compression fittings, a bearing-plate and a retainer plate. To increase the bond, an epoxy bonding agent was applied to the hole in the beam and a wave spacer was used similar to that in a

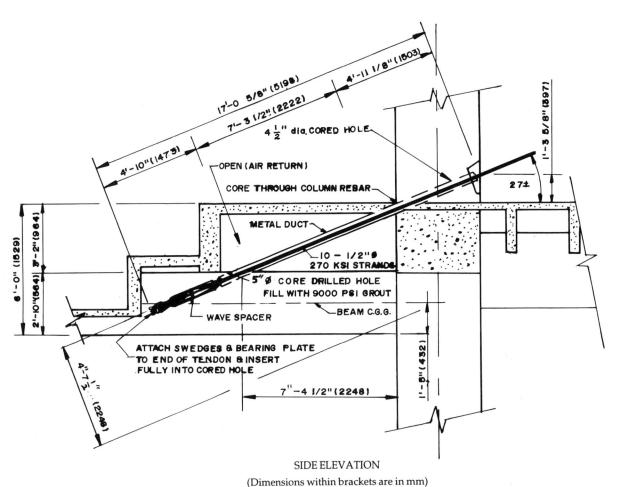

SIDE ELEVATION

(Dimensions within brackets are in mm)

A single tendon per beam provided the necessary strengthening.

Fig. 8.37

rock-anchor. The first stage grouting was achieved by pouring a 9000 psi (62 Mpa) strength grout into the hole in the beam after the sheathed tendon was inserted. The tendon was then stressed and 'shimmed' from the block-out in the column above. 'Shimming' was required to compensate for the wedge-seat slip loss as the strand elongation was minimal because of being a short free length tendon. A heavy metal conduit provided encasement for grouting and fire and corrosion protection for the tendon as it passes through the air-conditioning-return-vent. After the elongations were verified, the tendons were grouted through the anchorage (like in regular rock-anchors). An independent laboratory conducted load tests of several of the beams to verify that all design criteria were met.

The solution of using a single straight tendon per beam was more economical with minimal effects on the beams and the structural members around them.

• *Practical Example 9: Girder Repair, Wisconsin (U.S.A.)*

Several girders were unable to carry the required loading due to design deficiency. A design review revealed that additional prestressing force was necessary to provide the required capacity.

Again, an external post-tensioning concept was used. Some girders required 10 strands of 1/2" (13 mm) diameter (5 per side), and one required 28 strands of 1/2" (13 mm) diameter (14 per side). The tendon-path had two low point deviation blocks in one span to provide uplift against the concentrated loads from above. The girder 'width' was increased from 24" to 48" and the 'depth' was increased from 60" to 72", actually encasing the post-tensioning.

VSL designed the post-tensioning system, the anchorage brackets and deflectors at the high and low points. The prestress consisted of monostrand external (unbonded) tendons. This approach maximized the depth available for the increased drape and allowed for the tendons to be stressed individually after the concrete encasement reached the required strength.

The deficiency of these girders was of such a magnitude that, without the effective use of the post tensioning, the girders would not have been able to support the structural steel tower above. The solution used resulted in an economical option and provided the structural capacity intended for the structure.

9

STRENGTHENING OF CONCRETE STRUCTURES BY EXTERNALLY BONDED STEEL PLATES

9.1 INTRODUCTION

There is a growing need to strengthen existing concrete structures. This applies both to buildings and bridges and also to other structures.

Two fundamental types of strengthening/repair are possible with epoxy resin adhesives, namely

(a) Increasing the section depth of a structural element by adding a new layer of concrete on top of an existing cross-section and bonding the two with epoxy resin adhesive.

(b) Enlarging the total reinforcement cross-section by bonding on additional steel plates for flexural strength and even for shear strength.

Its range of application can be defined roughly as follows

- Refurbishing of structures or parts of them by rectifying construction shortcomings that impair their safety, e.g., as a result of faulty dimensioning, reinforcing bar corrosion, overloading, etc.

- Strengthening of a structural element by increasing its load-bearing capacity

- Altering a load-supporting structure's system, e.g., changing spans by shifting or removing a support, conversion of continuous beams to single-span beams, and vice-versa, etc.

The first attempts to strengthen structures with bonded reinforcements were carried out in France, in 1964–65. Initial practical applications date back to 1966–67 in France and South Africa. The first bridges were reinforced this way in Japan and Russia at the beginning of the seventies. In Switzerland the method has been applied increasingly to both buildings and bridges for nearly 18 years.

9.2 INFORMATION FROM SOME OF THE TESTS CONDUCTED AT EMPA
(The Swiss Federal Materials Testing Institute)

In 1973 the Federal Materials Testing Institute started conducting tests with bonded reinforcements, the objective being to accumulate a body of dimensioning data for concrete structures strengthened with external reinforcements. Three types of tests were involved: *preliminary* testing, *main* testing and *long-term* testing.

The preliminary testing was concerned with the width and thickness of the reinforcing plates, the anchoring length and the problem of the plate butt joints. The shear reinforcing plates were carefully arranged in the area of high transverse shear stress. Following sand blasting, the plates were primed with two coats of a solvent-type epoxy resin primer.

The bonded steel plates along the soffit generally cannot be carried past the supports for obvious physical reasons. Therefore tests have also been done accordingly so that the critical section was between the support and the end of the bonding tension plate.

The load/deflection diagrams are virtually identical. Rupture occurred in all cases as a result of concrete strain in the compression zone, approximately at mid span, with simultaneous flow in the external reinforcement. In the area of constant bending moment, the steel plate sprang away from the adhesive when the load was removed; outside of this area, failure occurred in the concrete.

Even though the first applications of bonded reinforcement go back more than ten years and no failures of such reinforced structures have yet been reported, the method is still relatively new. So it is necessary to investigate the long-term behaviour as well in order to apply the method with complete confidence. For this reason, the third part of the testing programme was concerned with the behaviour under permanent loading. Variables are the load level, corrosion protection and weathering. The testing programme, which began in November 1977, is scheduled to cover about 15 years.

Brief Summary of these Test Results

Reproduced below, in the authors' own words (courtesy), is the summary of the tests reported in Reference 2 indicated at the end of this chapter.

"The object of the investigations was to develop the basic body of data required for applying the strengthening method of bonding external steel reinforcing elements to existing reinforced concrete structures. Instead of being restricted to the investigations required to establish dimensioning data, the work also took technical questions relating to design and execution into account as far as possible in order to ensure the practical applicability of the method of externally bonded reinforcement.

"The investigation work commenced with bending tests on reinforced concrete beams of rectangular cross-section (150 mm × 250 mm) and a span of 2.0 m which had been strengthened with an externally bonded steel plate. The purpose was to establish the most favourable anchorage zone design. Variable test parameters were the geometric dimensions of the steel plates; they were selected so as to keep the cross-sectional areas of the steel constant for all test beams. The anchorage zones of the steel plates were located outside the supports. Here again the anchorage lengths were varied in relation to plate width to provide the same size bonding area on all beams.

"Accurate and extensive strain measurements on the steel plates in the anchorage zone made it possible to determine the magnitude and distribution of the bond stresses r_a between concrete and steel. It was found that the actual bond stresses are far lower than the peaks expected on the basis of theory. Conversely, the force transmission or anchorage length l_a required turned out to be much longer than theoretical considerations indicate. Finally, the test results showed clearly that, for a given cross-sectional area, thin, wide plates behave more favourably than thick, narrow ones.

"Another static bending test was then conducted on a T-beam, with 83 per cent of the total flexural reinforcement bonded externally, to investigate cracking and problems related to the anchorage of shear plates.

"In this test, the first visible cracks occurred in the web at an applied load of $F = 78$ kN. Following alteration of the load in steps, including reductions to determine the spring-back resilience, the steel reinforcing plate reached its yield point at $F = 517$ kN. Thereafter the load was increased on the beam until classical bending

failure occurred at $F = 630$ kN with destruction of the concrete compression zone. At the same instant one end of the steel plate broke away from the concrete. The deflection at midspan had reached a remarkable 230 mm, or $l/26$ of the beam's span l.

"It can be concluded from the results of the extensive measurements that the T-beam behaved elastically until the stress level in the steel reached = 400 N/mm². But because no irregularities or local disturbances occurred when the external load was raised above this level, the load-bearing system as a whole can be regarded as fully effective up to the point at which the concrete compression zone failed. This result was only achievable, however, because design measures, particularly with regard to anchorage of the shear plates, caused the load-bearing system to function on the truss principle.

"Moreover, the type of cracking that occurred, and the width of the cracks, indicated clearly that an excellent bond existed between the externally bonded steel plate and the concrete.

"Finally, a comparison of the reading obtained from this beam with those from conventionally reinforced T-beams demonstrated that it is possible to obtain structural components by bonding steel plates to concrete elements that do not differ significantly from conventional reinforced concrete components in terms of behaviour under statically applied loading.

"Fatigue tests were then carried out on beams of rectangular cross-section with tensile reinforcement consisting of externally bonded steel plates. The purpose was to obtain an insight into the failure mechanism under this type of loading and to probe the general fatigue loading limits. Of the eight beams subjected to fatigue testing, three stood up to 10^7 load cycles at steel stress levels of 140/100 N/mm², 160/80 N/mm² and 200/40 N/mm² without failing. In two other beams, for which the stress limits were 240/20 N/mm² and 280/20 N/mm², fatigue failures occurred in the steel after 1.26×10^6 and 2.38×10^5 load cycles respectively. Only in the case of three beams, for which the stress limits were set at 220/20 N/mm², 230/10 N/mm² and 260/20 N/mm², did a failure occur in the primer. On several of these specimens it was observed that the failure or the breaking away of the steel plate began in the vicinity of a flexural crack in the concrete and propagated towards the end of the plate as the load cycle count increased. The load cycle counts upon failure of these beams were 2.22×10^6, 8.79×10^5 and 1.79×10^6, respectively.

"A further fatigue test was conducted on a T-beam with externally bonded reinforcement for both tension and shear. Four fatigue phases were applied with the calculated steel stress limits of 240/120 N/mm², 300/150 N/mm², 360/180 N/mm² and 400/200

N/mm^2. Each of these stress levels was to be applied to the specimens for 2×10^6 load cycles. Near the end of the 2nd fatigue phase, however, a fatigue failure occurred in one of the internal reinforcing bars close to a support. Each of the supports was then shifted 550 mm towards the centre of the beam, which made it possible to complete all four fatigue phases. No further damage was noted in the beams or the bonds. A final static failure test was, therefore, conducted to determine whether any externally undetectable damage had occurred in the course of the fatigue tests. The results of these static failure tests showed, however, that the beam had survived the four fatigue phases without any damage outside of the aforementioned failure of an internal reinforcing bar; it was still capable of sustaining the full plastic moment. Ultimately the concrete compression zone failed, after the tensile plate had been deformed far past its yield point.

"A report is also given on initial results from long-term tests on reinforced concrete beams with externally bonded reinforcement. The parameters varied in these tests are the magnitude of the load, the type of weathering, and the type of corrosion protection. Observation of the test beams over a longer period — about 15 years — is planned. The results of the first year of testing show that, except for a few minor rust spots on the steel plates without corrosion protection, no significant damage has yet occurred in comparison with the original condition. To supplement these results, mention is also made of the outcome of a ten-year creep test on a reinforced concrete beam strengthened with an externally bonded plate. These results are similar in every respect to the performance one would have expected from a conventionally reinforced concrete beam.

"Further investigations were carried out to determine how butt joints behave under mechanical loading. Uniaxial tensile tests were conducted on concrete prisms with steel plates bonded on in butt joint formation with a cover plate across. Parameters varied were the cover plate thickness t and width b and cover plate overlap length l_u. Reinforcing plates and cover plates of identical thickness and width were used for a given test.

"The strain readings from these tests reveal that part of the load is transmitted by the concrete, and not by the cover plate over the joint. The shorter the overlap length l_u, the higher the proportion of load carried by the concrete.

"In addition, the results from uniaxial tensile tests carried out on purely metallic covered butt joints likewise showed that the overlap length l_u, the plate width b and thickness t are the main determinants of a joint's load-bearing capacity, which increases with rising shape factor $\lambda = \sqrt{l_u\, b/t}$.

"Besides these small-scale tests, tests were also conducted on three reinforced concrete beams with a length of 3.70 m of rectangular cross-section (0.15×0.25 m) in order to obtain detailed information on butt joint behaviour. The beams were reinforced on the tension side with an externally bonded steel plate with a butt joint at midspan. Two of these beams were subjected to a static load test to failure, the third to a multiphase fatigue test with constant load amplitude. In both the static tests and the fatigue test the cause of failure was that the cover plate broke away from the reinforcing plate. Extensive strain readings on the reinforcing plates and cover plates prior to failure provided information on the strain pattern in the joint zone, however. It was used to calculate the bond stress distribution between cover plate and reinforcing plate on the one hand and between reinforcing plate and concrete on the other. As had been expected, pronounced stress peaks occurred above all at the ends of the cover plate and reinforcing plates . . ."

9.3 SOME INTERESTING EXAMPLES OF REPAIR/STRENGTHENING WORKS ACTUALLY CARRIED OUT IN PRACTICE

Examples of Strengthening Building Floor-Slabs

Füsslistrasse Telephone Exchange Building, Zurich
The load-bearing thickness of the first floor above ground was increased by pouring a 12 cm thick layer of lightweight, mesh- reinforced concrete on top of the existing 11 cm thick reinforced concrete slab. The shear-transmitting bond between the two layers is achieved with an all-over layer of araldite adhesive.

The reinforcement cross-section of the second floor above ground was increased by bonding additional steel plates to the underside of the floor slab and the joists.

The effectiveness of the methods selected, i.e., the load bearing behaviour of the strengthened floors, was monitored at regular intervals. The checking programme, which covered a two-year period, was also carried out by EMPA.

Shopping Centre at Spreitenbach, Switzerland
In this case, about 2300 m^2 of the floor above the underground garage had to be reinforced to increase its admissible loading. A total of 216 steel plates were bonded to the underside of the floor slab. Their dimensions are: thickness 6 mm, width 160 and 200 mm, length 6600 to 7900 mm, max. weight per plate 75 kg.

Example of Strengthening Reinforced Concrete Frames

ALBA Aluminium Smelter, Bahrein

The prefabricated concrete frames serve as supports for the melting ladles; in addition, they transmit the materials handling loads to the foundation soil. Since these handling loads proved substantially greater than assumed in the calculation, cracks developed in the frame crossbeams. Thus it became necessary to increase the total reinforcement cross-section.

In all, 700 concrete frames had a single steel plate measuring 280/8-2850 mm bonded to the underside of the crossbeam. Afterwards, the cracks were injected with an epoxy resin.

It is interesting to note the temperature levels to which the adhesive is continuously subjected in the melting shop. They vary between +25°C and +70°C. The repair work was done in 1974.

Examples of Repair/Strengthening in Bridges

Some Bridges in France and Japan

As early as the end of 1960s, as inclined bridge on the French A6 motorway over the 'Chemin Departmental 126' was strengthened by bonding-on additional tensile and shear reinforcement plates. Both Bresson (4) and Vidal reported on this work in detail.

Bridges over watercourses are exposed to particularly severe weathering. JP Sevene reports on the repair and reinforcement of a bridge of French 'Route Nationale 186' crossing the Saint Denis Canal. The bridge had been built at the beginning of the century (8).

At the beginning of the 1970s, the roadbuilding authorities of Tokyo undertook a project involving the repair and reinforcement of elevated motorways in the entire metropolitan area. Today, nearly 10 km of a total of about 100 km of elevated motorway has been strengthened with bonded reinforcements.

Quinton Bridges in U.K.

This project involved four bridges on the M5 motorway at the Quinton interchange, west of Birmingham. The bridges are all designed alike; the superstructure, with spans of 16.5-27-16.5 m, consists of a voided slab 90 to 105 cm thick. During a routine inspection, cracks were discovered on the underside of the end and central sections. When the static calculations were reviewed, it was found that the tensile reinforcement was dimensioned too weakly at certain points. Two possible reinforcement methods were weighed, namely the installation of 'prestressing elements' or 'external reinforcement with bonded-on steel plates'. A comparison of the feasibility of the two methods argued clearly in favour of the bonded reinforcements, in spite of the fact that the technology was new.

Tests were carried out at the Transport and Road Research Laboratory to determine the effect of traffic loads during execution of the reinforcement work (5). However, no clear conclusions were drawn.

The work was done without interrupting traffic. The end sections were strengthened with steel plates about 6 mm thick. A double layer of 6 mm thick plates was employed down the middle of the central span; along the sides of this span it was necessary to employ up to three layers of 12 mm thick steel plates. The plates, measuring about 300 cm long and 25 cm wide, were fastened additionally to the soffits with screw plugs spaced at intervals of 90 or 45 cm.

The pretreatment of the underside of the concrete proved very time-consuming because of the unevenness and shoulders present.

In spite of the difficulties and the fact that this was the first application of bonded reinforcements, the reinforced bridge slabs are behaving as expected.

Swanley Bridges in U.K.

These bridges are part of the M25–M20 motorway intersection. They are three-section structures with a continuous slab and inclined piers. Shortly after they were opened, cracks were discovered on the underside of one of the end sections. A review of the design showed that the reinforcement cross-section was inadequate in one end section and above the adjacent pier. The missing reinforcement was made up with bonded steel plates.

The steel plates used were 6 mm thick, 250 mm wide and 3 or 6 m long. On the end section of slab, three layers of plates were applied. Each strip of reinforcement was 12 m long, and 15 strips were distributed over the entire width of the bridge.

As many as four steel plates were bonded to one another above the pier on top of the slab. Here again the strip length was 12 m, and 21 strips were distributed next to each other over the entire width. The greatest thickness of the laminated plate was 30 mm.

All together, 449 plates were applied within 20 days, including pretreatment of concrete and plates. Temperature and humidity were measured continuously during the job. After the adhesive had cured, the bridges were subjected to a load test; in addition, the interaction of the new reinforcement with the superstructure was checked under dynamic loading (7).

Gizenen Bridge, Muotta Valley, Switzerland

This bridge, in central Switzerland, built in 1911, had to be strengthened to withstand planned future loading. Again in this case, it was decided that bonded reinforcements would to be most suitable and quickest way to perform the restoration.

The first step was to pick out damaged parts of the bridge slab and repair them with epoxy resin mortar. The joint in the roadway slab at mid-span of the bridge deck was structurally closed and a new cross beam poured in at the same time.

The bonding work was started immediately after the bonding surfaces of the concrete and the steel plates had been sandblasted; 15 mm thick steel plates were bonded to the main girders, 10 mm thick plates to the transverse girders. The plates were 200 and 150 mm wide respectively. The plates running longitudinally are butted together under the suspension posts; the joint is covered with a butt strap of the same cross-section.

The entire bonded surface totals about 22 m^2. The Federal Material's Testing Institute ran loading tests on the strengthened bridge. As a final step, damaged parts of the concrete arches and suspension posts were also to be repaired with epoxy resin mortar.

9.4 SOME COMMENTS ON EXECUTION OF BONDING WORK

Reference 8, listed at the end of this chapter, gives detailed instructions on matters relating to execution detail, as issued by the 'Institut Technique du Batiment et des Travaux Publics', Paris. Given below is a short summary presented in Reference 1.

Steel

Basically, any grade of structural steel is suitable for bonded reinforcing plates. For thick plates it is advisable to use higher quality steels, e.g., grade 52. Particularly where joints and girders are being reinforced, weldable qualities should be used (end anchors, spot welding on shear plates, etc.). For obvious reasons, the steel plates should be kept as thin as possible. For a given cross-section, this means they must be wider. Where thin plates are used, the force is transmitted into the concrete over shorter path than is the case with thicker plates.

Plate gauges below 3 mm are not suitable, because (among other things) sandblasting can deform them.

In France, R L'Hermite (8) prescribes plates from 3 mm to a maximum of 4 mm. Such plates adapt well to unevenness in the concrete surface without creating perceptible stresses perpendicular to the joint surface after the adhesive has cured. If thicker cross-sections are required, several plates are laminated together (4).

On some of the jobs described above (carried out in Switzerland and England), steel plates between 6 and 16 mm thick were employed. In some cases, 6 mm and 12 mm thick plates were bonded together in several layers. But thicker and therefore stiffer plates cannot be used unless the concrete surface is sufficiently smooth.

Concrete

One of the factors involved in the dimensioning of a bonded joint is the concrete's resistance to shear. This is directly proportional to its tensile strength. It is not possible to state exactly when there is no longer any point in reinforcing a concrete cross-section. But a useful rule of thumb is a minimum tensile strength of about 20 kg/cm^2.

Pretreatment (Fig. 9.1 Plate 26)

It is advisable to sandblast the steel surfaces to be bonded or to roughen them by shotblasting. A measure of the roughness required is Swedish Standard SA 3, Roughness No. 7.

According to the French recommendations, the surface should be protected with a solvent-type primer with an epoxy resin base immediately after the blasting operation. When the primer has cured, the plates should be wrapped in polyethylene film for storage at the job site.

On the other hand, many plate reinforcement jobs have been carried out without priming the pretreated steel surfaces. The adhesive is then applied immediately after blasting. This procedure is probably acceptable in buildings, but where the structural elements being reinforced are exposed directly to weathering—as in the case of bridges, structures near the ocean, etc. — a primer should be used as additional protection for the steel. This is true even if an extended period (hours or days) elapses between blasting and application of the adhesive.

The reason this precaution has to be taken is that the concrete cross-sections requiring to be reinforced usually have large enough cracks in the tension zone that can serve as the starting point for corrosive attack.

The primer selected should offer optimum adhesion, corrosion protection and thermal resistance. Solvent-type products with epoxy resin or EP/PUR bases are suitable.

Pretreatment of the concrete surfaces can be carried out by sandblasting, shotblasting, grinding, or roughening with a pneumatic needle-gun or granulating-hammer. The grain structure of the concrete must become visible.

Shoulders should be dressed or ground off. Cavities and gravel pockets should be picked out and filled with an epoxy resin mortar. Broken edges should be treated the same way. The surface to be bonded should be as even as possible; deviations of + 5 mm are admissible over a length of 4 m. If thicker plates are used, however, closer tolerances should be maintained.

Pretreatment of the concrete also includes injection of the cracks with low-viscosity epoxy resin systems.

PLATE 26

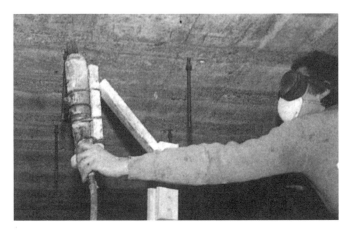

Fig. 9.1 Preparation of surface to be glued by needle-hammer or sandblasting

Fig. 9.2 Mixing the glue components

Fig. 9.3 Installing the steel sheet reinforcement

Fig. 9.4 Application of the adhesive

Fig. 9.5 Applying pressure with steel props

PLATE 27

Fig. 9.6 Strengthening a bridge to accommodate an increased live load—carried out without stopping traffic flow

Fig. 9.7 Strengthening of beams to accept increased sheer forces

Fig. 9.8 Strengthening of steel structures—increasing the stiffness of a galvanized I-beam

Bonding Procedures (Figs. 9.2 to 9.8, Plates 26 and 27)

The concrete surfaces should be dust-free, clean and dry, and the steel surfaces should be free of film rust. Whether the plates have been primed or not, they should be degreased with a solvent immediately prior to application of the adhesive.

The bonding work, i.e., mixing of the adhesive, should not be started until it is certain that the successive worksteps can be carried out without any interruption. To avoid mixing errors on the job site, one should use only adhesive in the proper working packs.

The adhesive is applied to the steel plate, and then the plate is pressed on to the concrete in such a way that the adhesive contacts the concrete all over and surplus adhesive is pressed out at the edges. Pressing is done with timbers or steel joist's and the usual ceiling braces. To mark sure contact is perfect between steel and concrete in the case of tension plates, it may be advisable to drive in wooden wedges from both sides between the timber (or steel joint) and the steel plate. It is also possible to press the steel plates on with plugs anchored in the concrete, high-tensile bolts or prestressed bars. This type of installation is appropriate, for instance, in the case of very high ceilings, the underside of bridges, and vertical surfaces. The adhesive joint is generally between 1 and 3 mm thick.

The procedure recommended in France is somewhat different. First a solventless primer is applied to the concrete. As soon as it has started to dry, a layer of adhesive is applied to both the concrete and steel surfaces and then the plate is pressed on.

Adhesive

See details ahead.

Maturity of Bond through Adhesive

Assuming adequate cure temperatures, bonded reinforcements can be subjected to full loading after 7 days. But it should be kept in mind that the new reinforcement participates increasingly in the transmission of loads as the adhesive gains in strength.

Bonding work should be done only by contractors with suitably trained personnel. Detailed knowledge of the necessary hygienic precautions is also very important.

Testing the Adhesive

It is also advisable to test the curing and strength properties of the adhesive while the bonding work is being carried out. In its ruling dated 1 October 1979 approving the use of bonded reinforcement, the Institute for Construction Technology, Berlin, prescribed the following tests

- Bending tensile strength, using prisms $4 \times 4 \times 16$ cm (three-point loading)
- Compressive strength, using cubes $4 \times 4 \times 4$ cm
- Shear strength, using 3 bonded steel plates $5 \times 100 \times 100$ mm, with the middle plate projecting 10 mm.

Three specimens per test; cure for 7 days at 23°C.

Work Following the Bonding Procedure

After the adhesive has cured, the reinforcement plates should be lightly tapped with a hammer to determine whether any cavities exist. Larger cavities can be filled by injection.

Exposed surfaces of the steel plate reinforcement should be protected against corrosion, e.g., with two coats of a solvent-type epoxy resin primer followed by a lightfast PUR cover coat (paint).

9.5 BUTT JOINTS IN THE TENSION-CARRYING BONDING PLATES

For practical reasons, butt joints between plates cannot be avoided. At the ends of the plates the forces are transmitted to the concrete over very short distances. L'Hermite considers an anchoring length of only 20 cm sufficient for a plate 3 mm thick and an adhesive layer of 1 mm (8). But he advises doubling this length for safety reasons.

Covering of butt joints by cover plates results in a change in rigidity; this should be kept to a minimum. Plate joints should be avoided at locations of high deformation. Cover plates should have the same thickness as the reinforcement plates. In the case of butt joints, it is important to note that a substantial part of the tensile force is transmitted via the adjacent concrete cross-section.

9.6 RECOMMENDED RELEVANT DETAILS OF BONDING ADHESIVE, ANTI-CORROSIVE PRIMER, AND ANTI-WEATHERING PAINT

The Epoxy Resin Adhesive

For structural applications—which include bonded plate reinforcement—the adhesive selected must survive careful testing in terms of their mechanical properties, adherence to wet concrete and behaviour in damp surroundings and at high temperatures to make sure they are suitable for structural use.

In addition, the reactivity of the adhesive must be adapted to the temperature range on in situ curing. Reactivity, minimum curing temperature and cure-time must be mutually matched in such a way that the adhesive is sufficiently crosslinked when loads are first applied. As for application of the adhesive, its pot life and the contact time are the main variables. The 'contact time' is the time period available from application of the adhesive to the steel plate (immediately after the components are mixed) until the plate is attached to the concrete.

The product employed for the EMPA tests was XB 3074 A/B, a medium-reactivity adhesive with an araldite base. It is recommended for application in the +15°C to +30°C temperature range. This adhesive has performed well both in earlier tests carried out at the EMPA on bonding of new to old concrete and in actual practice. The resin component of adhesive XB 3074A is based on a bisphenol-A epoxy resin, the hardener component XB 3074B on a modified polyamine hardener.

Table 9.1 indicates some of the properties of the adhesive.

Table 9.1 Properties of Epoxy Resin Adhesive XB 3074 A/B

Viscosity at 25°C	mPa s	6.5×10^4
Pot life at 20°C	min.	60
Contact time at 20°C	min.	45
Compressive strength	kg/cm^2	800
Bending tensile strength	kg/cm^2	250
Modulus of elasticity	kg/cm^2	62 000

Anti-Corrosive Primer

To protect the steel plate against corrosive attack, a primer is applied on both the bonding side and the opposite side. This commercially available product (Nuvokat) is a solvent-type epoxy isocyanate-based primer. It can be applied by either spraying or brushing. The mixing ratio of resin to hardener is 4 : 1 (parts by mass). The manufacturer of the primer states the other specification details as follows.

Manufacturer:	Dr W. Maeder AG. Ch-8956 Killwangen
Primer type:	Nuvokat primer, zinc chromate, yellow
Base:	Two component epoxy system with isocyanate hardener
Resin:	Nuvokat-Primer Yellow, 630.1.4.0005 (1977)
Hardener:	Nuvokat 857.0.0.0001/2 (1977)
Thinner:	990.0.0.0152 (approx. 10% thinner for brushing).

Pot life:	about 4 hours at 20°C.
Tensile strength at room temp.:	layer thickness 25 to 40 μm: β: = 155 kg/cm^2 layer thickness 60 to 80 μm: β: = 152 kg/cm^2

Two primer coats are recommended each at least 30 μ*m* thick. The second coat can be applied after the first has dried for 24 hours. A final drying period of 5 days should be observed. Prior to priming, the steel plates should be sandblasted, cleaned and degreased with trichloroethane. After final drying, the cured primer should be roughened lightly with sand paper to provide a good bond between primer and adhesive.

Anti-Weathering Paint

Since some of the strengthened reinforced concrete beams may be weathered out-of-doors, it is necessary to protect the air-side of the steel plate against corrosion with a weatherproof paint on top of the primer. The commercial enamel employed has following specifications, as supplied by the manufacturer:

Manufacturer:	Dr W. Maeder AG. Ch-8956 Killwangen
Paint type:	Nuvovern LW enamel, glossy, chromium oxide, green 563.8.7.0001 (1977)
Base:	Two component aliphatic polyurethane enamel
Hardener:	857.0.0.0002
Mixing ratio:	Resin: hardener : : 2 : 1 (mass parts)
Thinning:	25% to 50%, depending on desired spraying or brushing viscosity
Pot life:	about 12 to 16 hours at 20°C.

This paint should also be applied in two coats of about 30μm layer thickness each, after the surface to be painted has been thoroughly cleaned and degreased with trichloroethane.

9.7 SOME IMPORTANT CONSIDERATIONS THAT CAN HAVE FAR-REACHING EFFECTS ON THE STRENGTH OF REPAIR

Effect of 'Surface Conditions' on Bond

If the high demands placed on adhesives in structural applications are to be met, it is essential that precise directions be issued on working procedures and adhesive use. Such directions are extremely important wherever load-bearing elements and components have to be bonded together with adhesive. Failure of the bond would have very serious consequences in such cases. For these reasons, the surface conditions of the elements being bonded and climatic conditions during the bonding phase are of utmost importance.

Besides the need for high strength values — in fact, far above the respective figures for the concrete — excellent adhesive properties, high resistance to chemical attack and volume stability (little shrinkage or swelling), the construction adhesives have to be unaffected by temperature and moisture variations both during the cure phase and with regard to final properties.

Humidity can even make it impossible to carry out the bonding procedure. The trouble is that the chilling resulting from the evaporation of solvents can cause water to condensate and be adsorbed by the surface of the adhesive. This water film, which is invisible to the naked eye and hardly detectable by ordinary means, interferes with the contact between the adhesive and the bonding surface. The constructions adhesives now offered on the market for structural applications contain no solvents.

Humidity can affect the adhesive's bonding properties adversely in yet another way. If parts are moved from cold surroundings to warm ones immediately before the adhesive is applied, a film of condensed water can form on the surface of the parts and prevent proper bonding. Hence the parts being joined should never be colder than their surroundings.

Many resin/hardener systems are vulnerable to high humidity levels. Where thin bonding layers of such moisture-sensitive adhesives are applied, a tacky film can form on the adhesive surface and severely impair adhesion.

Structural adhesives should exhibit mechanical behaviour closely resembling that of cement mortar, but without its drawbacks (such as shrinkage).

Some epoxy resin adhesives are claimed to cure even under water. But little is yet known about the use of such adhesives for transmitting forces.

Since the effectiveness of a bond between parts depends largely on the quality of the boundary layers of the surfaces being joined, they must be prepared with painstaking care. In the case of concrete structural elements, the surface of the concrete usually consists of a layer enriched with hardened laitance. Because this layer has a high water/cement ratio and therefore lower strength than the underlying concrete, it must be removed prior to bonding. Various mechanical methods are used in practice to accomplish this, as indicated earlier.

These include sandblasting, grinding, hosing down with water at very high pressure, and roughening and loosening of the hardened laitance with wire brushes or pneumatic bush-hammers. Experience has shown, however, that sandblasting and granulating with the bush-hammer are the most suitable methods, because they remove the lower-strength components without damaging the harder components such as aggregate particles excessively. Another advantage of these methods is that, together with the laitance, they also remove any contamination from the surface of the concrete—such as that left by form oils. Thus another possible source of defective bonding is eliminated automatically.

Apart from the condition of the surfaces, material qualities are also of utmost importance. The strength of the boundary layer should be checked with pull-off tests involving small standardized steel plates bonded to the concrete surface.

Up to the present day, no process seems to exist for determining the moisture content of structural concrete on a nondestructive basis with sufficient accuracy. It would be useful to know the moisture content not only at the surface, but also down to a certain depth within the structural element. However, accurate investigation of the adhesion of construction adhesives to concrete would require an estimation of the extent to which application of the adhesive alters the moisture condition of the concrete, because the layer of adhesive can act as a vapour barrier.

Another point to bear in mind is that condensate water can form on the concrete surface when the adhesive mix is applied, even if the concrete has been declared dry. This can happen above all when warm adhesive (the curing of the adhesive is an exothermic process) is applied to cold elements.

Nevertheless, bonding work on job sites has often be carried out under unfavourable climatic conditions, and varying temperature conditions, and hence the race for better and more amenable adhesives that can perform under adverse conditions.

Influence of Adhesive Thickness on the Strength of Structural Bonds

It was recognized quite early that the strength of the bond depends to a large extent on the adhesive thickness. This thickness can be taken as the average distance between the two surfaces being bonded. The tensile shear strength of the adhesive is initially proportional to the square root of this thickness. However, the tensile shear strength reaches a maximum and then starts diminishing again as the adhesive thickness is increased further. The reason is that, for the same displacement, the displacement angle is smaller in thicker bond layers than thin ones, so that lower shear stresses are to be expected in thicker bond layers. Nonetheless, tests have shown that bonds with thicker layers — despite lower shear stresses — are not able to achieve tensile shear strengths any higher than those with thin layers. The reasons put forward for this are

- As the bond layer becomes thicker, the shear strength of the layer declines as a result of reduced hindrance to transverse contraction
- The bending moment increases with increasing bond layer thickness because of greater bond eccentricity. This creates normal stresses, particularly at the end of the overlapped joint, which act on the bond surfaces at right angles.
- In the case of thick bond layers, there is a greater probability that defects will occur. Such non-homogeneities produce local stress concentrations.
- Shrinkage and heat create tangential and normal stresses as a function of layer thickness.

The same conclusions have also been drawn on the basis of tests.

The strength of the bond decreases with increasing bond layer thickness. Thinner layers prove stronger and have greater resistance than thick ones in tests on overlapped joints. In all of the tensile shear tests the bond layer thicknesses varied between 0.1 and 0.6 mm. In cases where steel was bonded to steel, the bond layers varied in thickness from 0.2 to 0.6 mm.

Thick bond layers behave more softly with regard to creep and therefore permit greater overall deformation. Thin layers behave more favourably when exposed to heat than thick ones.

The fundamental physical behaviour is no different in concrete-concrete or steel-concrete bonds than in steel-steel bonds. But the average bond layer thicknesses are somewhat greater, because the adhesive has to equalize the unevenness of the concrete surfaces as well. Layer thicknesses between 1 and 2 mm are usual. In special cases, such as the bonding of prefabricated concrete elements in bridge structures bond layer thicknesses as large as 20 mm have also already been employed successfully using Epoxy Mortars. In such construction, shear keys are incorporated and the precast bonded segments are prestressed across the joints.

9.8 BACKGROUND INFORMATION ON THE ADHESIVES

Among plastics generally the epoxy resins are classified as thermosets. They harden as the result of a chemical reaction between the original components (resin and hardener). Once the cure is complete, the reaction products can no longer be melted down.

The epoxies cure by a polyaddition process. Since no by-products are formed, the cure takes place virtually without shrinkage.

Epoxy resin systems always consist of the actual epoxy resin and a reaction partner responsible for curing (hardener, crosslinking agent). Nearly all of the systems used in the construction industry are of the cold-curing, liquid, two-component type.

The most commonly used epoxy resins are condensation products of two petrochemicals, epichlorohydrin and bisphenol-A. The most common hardeners are either polyamines or polyamides, which guarantee adequate crosslinking (hardening) even at outdoor temperatures.

In addition to resin and hardener, ready-to-use adhesives may also contain fillers, pigments and additives. Adhesives are offered in two components (resin and hardener), and are delivered to jobsites in working packs. Field experience has shown that special formulations have to be developed for each field of application.

The highly reactive epoxy resin adhesive system LMB 1474 (resin component) and LMB 1475 (hardener component) used for the initial tests, i.e., preliminary and minor testing, was replaced by a less reactive product during the course of the initial testing work for practical reasons. Details of the highly reactive adhesive will therefore be omitted.

Its successors, adhesive I and adhesive II, are based on a binder system of medium reactivity. These adhesives are generally applied within a temperature range of +15°C and +30°C. System XB 3074 A/B (adhesive II) is a slightly improved version of system LMB 1824/LMB 1815 (adhesive I). Both adhesives are based on the same binder system (resin/hardener). The only difference is that the XB formulation has a slightly higher binder content, which improves the adhesive's ability to flow under pressure. Table 9.2 lists the composition and attributes of the two adhesive systems. The very small difference in binder content has no effect on the strengths. For this reason, the manufacturer, Ciba-Geigy Ltd., Basle, has determined the other properties with adhesive XB 3074 A/B only.

When the epoxy resin adhesive is utilized, two characteristic time periods are important to the user: 'pot life' and 'open time'.

The *pot life* is the period of time within which the epoxy resin formulation must be applied at a given ambient temperature. After this time has elapsed, the cure process is already so far advanced that the mixture can no longer be allowed to be applied and it starts hardening in the mixing pot.

The *open time* starts when the adhesive has been applied to the parts being bonded. They have to be

Table 9.2 Composition and Properties of Adhesive Systems I and II

Characteristics	System I				System II		
	Resin	Hardener	Adhesive		Resin	Hardener	Adhesive
	LMB 1824	LMB 1815			XB 3074A	XB 3074B	
Composition							
Binder (mass%)	59.2	35.5			61.24	35.5	
Fillers/thixotropy agents (mass%)	40.8	64.5			38.76	64.5	
Mixing ratio (part by mass)	100	60			100	62	
Properties of the Components							
Density at 25° C (kg/m^2)	1540	1640			1540	1640	
Viscosity at 25° C (mPas)	7×10^4	6×10^5			4.5×10^4	6×10^5	
Properties of the Adhesives							
Viscosity at 25° C (m Pas)			7.7×10^4				6.5×10^4
Binder content (mass%)			50.4				51.4
Fillers/thixotropy agents (mass%)			49.6				48.6

joined together within this time. If the open time is exceeded, the adhesive strength is sharply reduced. High humidity together with low curing temperatures shortens the open time. Under these conditions, the amino groups of the hardener react at the adhesive/air interface with the moisture and the CO_2. These groups then no longer effectively react with the epoxy of the resin component, which can impair adhesion.

In the case of steel/concrete bonds, the adhesive should be applied to the steel surface whenever possible. The boundary layer of the adhesive is then destroyed by a rough surface of the concrete when the plate is set in place. If the adhesive has to be applied to the concrete surface for practical reasons, the surface of the adhesive should be roughened with the toothed spatula just before the plate is pressed on.

9.9 SOME CAUTIONS

(a) Although the bonded plate reinforcing process for strengthening structures made of reinforced concrete has been employed successfully for over 10 years in a number of countries under various conditions (the method can also be employed in combination with prestressed concrete elements and structural steel), the static limit of the process is established by the load-bearing capacity of the concrete compression zone and the technical limit by the shear strength of the adhesive at high temperatures.

(b) In principle, the method outlined is elegant for strengthening an existing concrete structure that is otherwise healthy.

(c) The method is still relatively young and more information is awaited for ascertaining the long-term behaviour of structures repaired by this method.

(d) To date there is limited experience in this methodology.

(e) The technology is considerably sophisticated and requires extreme care in

- Preparing the 'bonding surfaces'
- Controlling the humidity across these surfaces
- Preparing the bonding agent (with its limited pot life and effective setting time, i.e., open time)
- Preparing dependable 'anchoring' arrangements for shear-plates and tension-plates
- Ensuring the structural safety if the tension plate has to stop short of supports, as it may not be possible to cross the supports
- Ensuring force-transfer at butt-joints of bonded plates

(f) The tests reported in References 1 and 2 are on very simple laboratory cases, although some examples of repairs to actual bridges also have been reported.

(g) However, most important is to remember that such repairs are possible only where the structural strength of an otherwise 'healthy concrete' structure has to be enhanced for increased loading, etc., but is meaningless where the concrete is already sick owing to inadequate quality-control and construction-practice that went into its very making. In other words, this repair/strengthening technique can be employed where the concrete is structurally healthy and not already disintegrated or plagued by profusion of cracks (e.g., those caused in its pre-hardening stage due to excessive plastic shrinkage).

REFERENCES

1. HUGENSCHMIDT, F., 'Strengthening of Existing Concrete Structures with Bonded Reinforcements', Sept. 1981, EMPA (Switzerland).
2. LANDER, M. and C. WEDER, 'Concrete Structures with Bonded External Reinforcement', Dubendorf 1981, EMPA (Switzerland).
3. BRESSON, J., 'Nouvelles Recherches et Applications Concernant l'Utilisation des Collages dans les Structures', Beton Plaque, Annales de l'Institute Technique du Batiment et des Travaux Publics, Supplement No. 278, February 1971, Concrete and Reinforcement Series No. 122.
4. BRESSON, J., 'Renforcement Par Collage d'Armatures du Passage Inferieur du CD 126 sous l'Autoroute du SUD' Annales de l'Institut Technique du Batiment et des Travaux Publics, Supplement No. 297, Sept. 1972, Concrete and Reinforcement Series No. 122.
5. IRWIN, C. A. K., 'The Strengthening of Concrete Beams by Bonded Steel Plates', Transport and Road Research Laboratory, Department of Environment, Supplementary Report 160 UC, Crowthorne, Berkshire, 1975 (UK).
6. MACDONALD, M. D., 'The Flexural Behaviour of Concrete Beams with Bonded External Reinforcement', Transport and Road Research Laboratory, Department of Environment, Supplementary Report 415, Crowthorne, Berkshire, 1978 (UK).
7. SOMMERARD, T., 'Swanley's Steel-Plate Patch-Up', New Civil Engineer 1977, No. 247, 16 June 1977, pp 18/19.
8. L'HERMITE, R., and R. DEVARS DU MAYNE, 'Le Collage Structural et le Renforcement par Resines des Structures de la Construction', Annales de l'Institut Technique de Batiment et des Travaux Publics, No. 349, April 1977, General Construction Series No. 62 (13 papers are summarized under the title 'Le Renforcement du Beton par Resines' and 'Les Applications du Renforcement du Beton').
9. External Reinforcement of Concrete Structures', CIBA-GIEGY Aspects, June 1981.

Addresses of Various Important Related Agencies

ASTM   American Society for Testing and Materials, 1916 Race St., Philadelphia, Pa. 19103, USA.

ACI   American Concrete Institute, Box 19150, Redford Station, Detroit, Michigan 48219, USA.

BSI   British Standards Institution, 2 Park Street, London W1A 2BS, GB.

DIN   Deutsches Institut fur Normung e.v., Burggrafenstrasse 4-10, D-1000 Berlin 30, BRD.

NEN   Nederlands Normalisatie-instituut, Kalfjeslaan 2, Postbus 5059, 2600 GB Delft, Nederland.

DAS   Deutscher Ausschuss fur Stahlbeton, Dusseldorf, BRD.

CSA   Canadian Standards Association, Rexdale, Ontario, Canada.

CIRIA   Construction Industry Research and Information Association, 6 Storey's Gate, London SW 1P 3AU, GB.

FIP   Federation International de la Precontrainte, Wexham Springs, Slough SL3GPL, GB.

10
REHABILITATION OF BRIDGE FOUNDATIONS

10.1 INTRODUCTION

Various practical cases that can be actually encountered in the field are discussed in this chapter with a view to rehabilitating the distressed bridge-foundations from a practicing-professional's stand-point, without the mile-wide and inch-deep academic monologues. Practical aspects are discussed in extensive yet clear-cut detail in step-by-step 'tool-kit' manner, subduing theory to practice, and honing with workman-like details that only extensive practicing-experience can unfold.

At the outset it must be clearly understood that it is wrong to generalize a method for repairs and/or strengthening of foundations. Each case has to be analyzed individually and may require special investigations. Most repair-works for foundations fall in the categories of 'protection' and 'strengthening'. Some examples are:
— scour and erosion protection
— repair of foundations built on soft ground subjected to erosion
— strengthening of foundations affected by overloading of existing piles due to settlement in the soft ground
— remedying the effects of the horizontal movements of abutments built on soft ground and/or retaining soft embankment
— strengthening of foundations as required by widening of a channel or a road
— extending existing foundations

10.2 INDICATIVE SUMMARY OF VARIOUS POSSIBLE COUNTER-MEASURES AGAINST DISTRESS AND FOR REHABILITATION OF BRIDGE SUB-STRUCTURE

The following is a summary of possible counter-measures

~through modification of bridge
- Sheet pile protection
- Underpinning or Jacketing
- Pier Protection
- Driving deeper pile and grouting undermined area
- Raising the bridge
- Replacing a solid parapet with an open parapet and railing for overtopping of stream during flood

~through replacement of bridge
- Large waterway openings with longer spans and fewer piers
- Deeper pile foundation
- Raising the bridge
- Relocating the bridge

~through roadway and channel
- Constructing overflow sections and channels
- Constructing relief-culverts
- Realignment of the channel
- Removal of debris from the channel

~through armour against instability and scour

— *using flexible revetment*
- Riprap or broken concrete
- Rock and wire mattress
- Gabions
- Precast concrete blocks
- Planted vegetation
- Car bodies

— *using rigid revetment*
- Concrete pavement
- Sacked concrete (bagged concrete)
- Precast concrete slope paving
- Asphalt concrete paving

- Grouted riprap
- Concrete filled fabric mat

~through flow-control structures

- Spurs (rock or earth embankment) and sheet piles
- Retards (timber piles, sheet piles, rail, or tetrahedron-field)
- Guide Bund (rock or earth embankments, and sheet piling)
- 'Drop' structure—check dam, cutoff wall

Let us commence with various preventive measures for minimizing flood damages and then go into the rehabilitational details of protection and strengthening.

10.3 PREVENTIVE MEASURES TO MINIMIZE FLOOD DAMAGES

There are several methods by which the flood damages to foundations can be minimized. Providing deep foundations (e.g. caissons and sheet pile protected piles/footings) in the initial design itself and constructing spur-dykes are obvious solutions. But sheet pile protected footings and piles may be unsuitable if scour depth is significant. Piles, when exposed to significant depth, may be unsuitable from buckling considerations particularly if the floods bring floating debris (e.g. floating trees, logs and rolling boulders). Obviously, in designing bridges over waterways, the estimate of depth of scour is required because scour will occur unless the stream bed is inerodible rock. The following three design factors should be considered for preventive-design

— the flood frequencies and magnitudes
— the flow pattern for each flood with geometry of given design
— the resulting scour (If a scour prediction can not be made explicitly, the foundation should be designed more conservatively to accommodate unexpected scour depth.)

The following preventive measures are suggested to minimize damages

(i) Abutment-Scour Problems can be Reduced by providing

- Riprap protection
- Sheet piling curtain
- Spur dyke
- Deep foundation

(ii) Pier-Scour Problems can be minimized by

- Alignment with stream-flow direction
- Placing foundation below estimated scour depth

- Providing riprap at expected scour depth, where possible
- Providing slope for scour-holes so as not to disturb natural flow of water
- Using pile foundations or caissons to provide stability
- Sheet pile protection

(iii) Erosion and silting Prevention

Stream banks should be checked frequently for erosion. In order to minimize erosion, channel widening may be considered. This will permit discharge of the same volume of water with a lower velocity, so long as abutment design can accept it.

Accumulation of silt would reduce the freeboard between the water surface in a stream and the soffit of the bridge deck. Streams which are susceptible to such silting should be watched for the presence of deposits this material should be removed from the stream-bed before serious damage occurs.

Gravel 'bars' often form in a stream and divert the water from its normal channel and cause erosion of the banks and scour of foundations. The removal of gravel bars would reduce the erosion and scour slightly.

10.4 REPAIR OF UNDERWATER BRIDGE SUB-STRUCTURES

It is futile to list all the possible causes of deficiency of underwater structures, as they are too numerous. The same also applies to the combinations of conditions requiring repairs. Given the wide range of materials and repair techniques, the choice of the most appropriate technique cannot be generalized. Some of the repair-works for the foundations are indicated below for information and guidance, but as already mentioned each case has to be decided on its own merit

(a) Erosion Problems

Stone rip-rap is placed on a mattress at or beneath bed level. The weight of the mattress preferably shall not be less than 150 kg per sqm. The slope of the protective rip rap should be between 1 in 3 and 1 and 3.5. Heavier stones should be used for rip-rap in case steeper slope is necessary.

(i) Underpinning: with replaced aggregate and pressure grouting, cast-in-place concrete or concrete filled fiber bags.

(ii) Placing Riprap: (Fig. 10.2) Placed around abutments and piers to guard against progressive erosion—dumped riprap is widely and effectively used to fill scour holes and protect bank slopes on meandering streams and bends. It is proven to be effective when adequate quantities and proper sizes of riprap are used.

Fig. 10.1 Underpinning

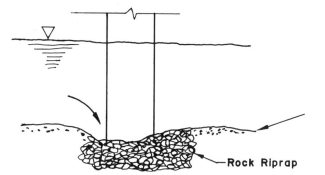

Fig. 10.2 Dumped riprap

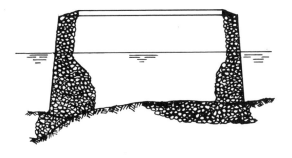

Fig. 10.3 Concrete jacketing

(iii) Jacketing the Foundation: Excavate to depth and then jacket with concrete (Fig. 10.3).

(b) Protection against Scour

Scour of the substructure support is one of the more frequent factors that may cause or lead to structural failure or foundation distress. Scour is the removal of stream bed, backfill, slopes, or other supporting material, by stream tidal action, dredging, propeller back-wash, etc. The degree of damage depends on such factors as the character of the stream-bed, the intensity of discharge, the volume of water, obliquity of stream-flow, and the shape of the structure.

(i) Void Filling, Dumped Riprap, and Giving 'Cut' and 'Ease' Waters in the Encasement (Fig. 10.4).

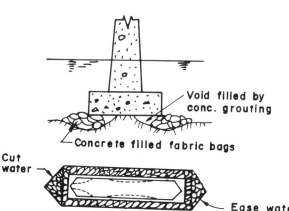

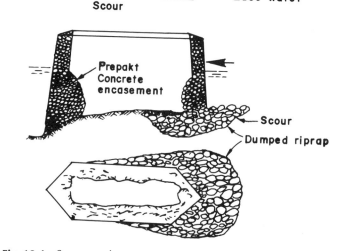

Fig. 10.4 Scour repair

(ii) Sheet-Pile Cofferdam Enclosure and Cutoff-Wall (Fig. 10.5).

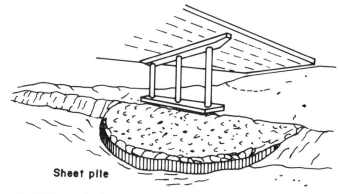

Fig. 10.5 Cofferdam

(iii) Placing Guide Bunds (special spurs): This is for partial diversion of the stream to reduce the scour depth within the bridge-openings and protect abutments (Fig. 10.6).

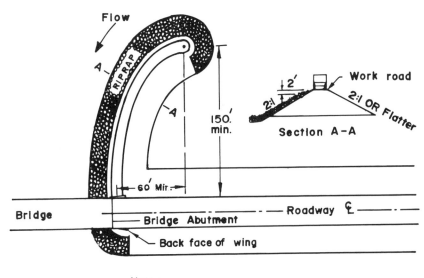

Fig. 10.6   Guide bund

Guide Bunds are used at bridge embankments to increase the effective waterway and improve flow conditions through the openings. *Guide Bunds* can reduce the scour depth at bridge sites. Therefore, the use of *Guide Bunds* provides protection to the existing bridge abutments and piers against scour.

If the cause of scour can be identified, such as change in the alignment of the stream, an inadequate waterway, or the presence of debris, the repairs made to scoured areas are much more likely to be successful. Determining the most effective solution to a scour problem is difficult and may require model studies at times.

Spur dykes, jetties, deflectors and other devices may be constructed to direct water away from a fill, bridge pier, or abutment. Caution is needed because only correctly designed and constructed training works are helpful in controlling scour and erosion. Repair damage caused by channel-scour may be as simple as the replacement of displaced material, or it may involve redesign of the foundation, construction of training works, or sheet piling, or other modifications of the structure or channel.

At sites were soil erosion has occurred because of stream or tidal action, it is common practice to place rock or riprap material in the void or to protect the replaced soil with riprap, bagged concrete, riprap, or grouted or wire-enclosed boulders. Piers and abutments may be protected or repaired by placing sheet piling to keep material in place or to prevent further scour. Sheet piling should be driven to a depth where non-erodible

soil conditions or rock exist. The overhead clearance required for driving the sheet piles here may be a major disadvantage in using sheet piling. If supporting material has been removed from under a large area of the footing, consideration should be given to redesigning the foundation, including filling the void with concrete. In some cases the footing may be extended by using sheet piling as 'forms' for the extension and as stay-in-place protection against further scour. If scour has exposed supporting piles, it may be necessary, particularly if they are short, to drive supplemental piles that are part of the extended footing.

To restrict scour around piers, it is also common to utilize what is known as 'garlanding technique'. In this very heavy concrete blocks or stones of designed weight are placed around the pier foundations below the bed level after necessary excavation. The size of the garland and the weight should be properly designed.

It is advisable to consult experts before deciding undertaking the solution of a serious repair problem.

(c) Foundation on soft substrata subject to erosion, can be protected by reinforced concrete curtain walls enclosing the footing or piles.

(d) Increasing the bearing capacity of the soil by injecting cement or chemical grout taking care that grout pressure does not exceed the overburden pressure.

(e) Rock or ground anchors are often used for abutments were protective slope has to be removed for (say) widening of navigable channel, road, etc. Normally the protective sheet-pile-wall is driven first in the case of ground anchors. Design and execution of rock/ground anchors requires great care and should take into account all factors likely to affect the bearing capacity and durability of the anchoring system. Prestressed anchors are commonly used (*see* details ahead).

(f) Adding to the Existing Foundations

This would be necessary while widening an existing bridge.

(g) Liquefaction of Foundation Soil

Some foundation failures during the earthquakes could be the result of excessive soil movements specially due to soil liquefaction. There are two approaches which may mitigate this types of failure

(i) Eliminate or improve soil conditions that tend to be responsible for seismic liquefaction.

(ii) Increase the ability of the structure to withstand large relative displacements similar to those caused by liquefaction or large movements.

Some methods are available for stabilizing the soil at the site of the structure. Each method should be individually designed, making use of established principles of soil mechanics to ensure that the design is effective and that construction procedures will not damage the existing bridge. Possible methods for soil stabilization include

— lowering of ground water table
— consolidation of soil by vibrofloatation or sand-compaction
— placement of permeable over-burden
— soil grouting or chemical injection

At a site subjected to excessive liquefaction, methods to improve the structure may be ineffective unless coupled with methods to stablilize the site itself.

(h) Underwater Work

While dealing with underwater work, it will be relevant to refer to underwater inspection* also. Inspection of the underwater portions of the structure can be difficult because of the harsher environment, poor visibility, deposition of marine organisms, etc.. To do an effective underwater inspection, it is necessary to deploy properly trained and equipped supervisory personnel. The quality of inspection under water has to be no less than the quality of inspection above water. Clearing the marine growth from the underwater portions of a bridge is almost always necessary. Visual inspection is a primary work of detecting underwater problems. In turbid waters too the inspector should examine to detect flaws, damages and deterioration. In some cases ultrasonic thickness gauges, computerized tomography or TV monitors may be required. After the initial identification of a trouble spot i.e. the damage, for the purpose of detailed examination and carrying out repairs it

may be necessary either to exposed the member by means of a cofferdam and dewatering or by providing a small air-lock chamber.

Normally inspection of underwater components of a bridge is carried out by divers who are usually not engineers. It would be useful to train some of the engineers in diving techniques so that they can interpret the observations in more scientific way. Underwater photographic techniques are also available wherein damages are detected by divers who can then take photos of effected areas. Similarly, underwater cameras (mounted on diver's head-gear) can be used to scan the various components of submerged portion of the structure continuously and signals can be read on a TV monitor kept on the bridge deck. A new technique, using acoustic microscopy measurements, has been developed for underwater study. In this, measurement of small electrical potential differences caused by corrosion current in sea water is combined with acoustic inspection to determine the crack width and depth. Computerized tomography is yet another recent method to locate voids and steel reinforcement in underwater concrete. A gamma ray source is collimated to form a flat fan of rays which are attenuated as they pass through the approach to a set of detectors. The source detector apparatus is rotated to obtain a series of projections through the cross-sections.

Placing of concrete underwater can be carried out with the help of conventional underwater bucket or by tremie concreting, although under certain conditions placing of prepacked concrete or bagged concrete or pumped concrete may be more suitable. In all such underwater repairs the surfaces of the pile or wall or pier have to be cleaned of the dirt and other foreign material and after removing the cracked and unsound concrete the surface is to be prepared for receiving new concrete. Suitable priming coat by materials like moisture-compatible epoxy resin is helpful to ensure proper bonding.

It is often useful to provide temporary cofferdam which can be fixed to the pile and the water within can be pumped out. Joint of the jacket at its ends has to be properly detailed and treated with epoxy. Grouting with quick-setting cement or epoxy can also be carried out where necessary.

Methods used by divers to perform under-water sealing and repairing of cracks by epoxy injection are similar to methods used above water except that the epoxy surface-sealer takes several days to harden sufficiently to withstand injection pressure. For underwater use, epoxies *must be 'water-insensitive'*. Before the application of epoxy surface-sealer, crack-cleaning is necessary. If oil or other contaminants are present in the cracks, and the epoxy is used for restoring the strength

---

*See details described in Chapter 5.

of the cracked concrete instead of simply blocking the free entry of water in the crack, bonding will be improved by mixing detergents or special chemicals with a water jet to clean the crack interiors. After all cracks are prepared and sealed and the nipples positioned, the low viscosity epoxy adhesive is injected under pressure into the crack-network. A surface mounted positive-displacement pump is used to dispense the two components of the adhesive to the submerged injection sites where the adhesive is mixed in the injection-head as it is pressure-pumped into concrete. Water temperature must be above 4 degrees centigrade. Most of the adhesives 'cure' to full strength in about 7 days. Cracks up to 2 mm width must be sealed with straight epoxy-resin-hardener, without a filler. For wider cracks, the addition of a filler is generally required. Nowadays, for underwater repair works, devices called 'Habitat', are used. 'Habitat' is a multicell metal unit open at the bottom with waterproof joints. This is installed around the member to be repaired. With compressed air the Habitat chamber is kept dry so that the divers can undertake repairs (*see* some details ahead).

Various methods are currently used to prevent the corrosion of steel piles in sea water, including application of protective coatings, cathodic protection, encasement of the steel in concrete, or a combination of these procedures (*see* some details ahead).

10.5 STRENGTHENING OF FOUNDATIONS

1. By Reinforcing
The existing foundation can be strengthened by widening and encasing existing footing with reinforced concrete and/or adding new piles around the perimeter of the footing. The new footing is then tied into the existing foundation. *See* Fig. 10.7.

Reinforcing can also be done by increasing the capacity (adding extra piles for instance), *see* Fig. 10.8.

2. By Drilling Mini Piles behind Existing Wall
In this case the new piles relieve an unstable wall from lateral soil thrust. It can be constructed to prevent or stop landslides.

3. By Drilling Mini Piles to Underpin an Existing Foundation
The mini piles are rotary-drilled through the structure to be underpinned. The load between the structure and the piles is transferred through concrete friction. It will increase the load bearing capacity of existing foundation and of new pile foundations.

4. Underpinning
Underpinning is used for constructing new footings under existing foundations. It can also be used for strengthening the existing foundation which otherwise can no longer safely support the existing loads. This condition causes soil settlements. Generally, it is used

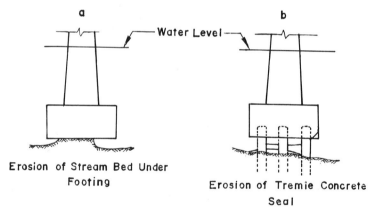

Erosion of Stream Bed Under Footing

Erosion of Tremie Concrete Seal

Typical Distress Condition At Pier Foundation

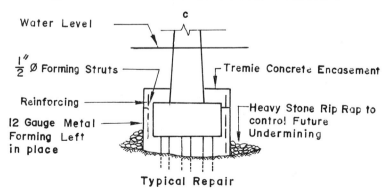

Typical Repair

Fig. 10.7 Reinforcing

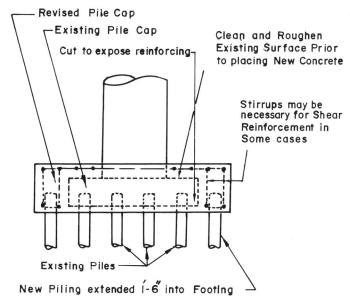

Revised Pile Cap
Existing Pile Cap
Cut to expose reinforcing
Clean and Roughen Existing Surface Prior to placing New Concrete
Stirrups may be necessary for Shear Reinforcement in Some cases
Existing Piles
New Piling extended 1-6" into Footing

Fig. 10.8 Increasing capacity

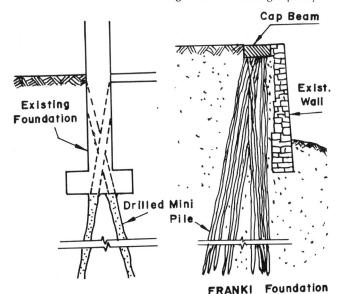

Existing Foundation

Drilled Mini Pile

Cap Beam

Exist. Wall

FRANKI Foundation

Fig. 10.9 Drilling mini piles

for constructing new foundations in small trench excavations underneath the existing footing. New piles are constructed near the existing footing and then loads are transferred by a 'needle beam'. (*See* Figs 10.10 and 10.11)

In the 'trench-pedestal' method, minimum sized sections of the trench are excavated in stages underneath the existing footing and new pedestals are constructed in small sections in stages by placing reinforcement and concrete in the stage excavations. Sometimes it is necessary to jack up the existing footing during excavation and this construction to restore proper levels and provide support. In all cases, a new pedestal footing should be built, section by section, to provide some support for the existing footing.

Sheeting and bracing for the excavation should be provided as necessary to prevent horizontal movement of the surrounding ground. Also, it may be required to provide for dewaternig and avoid disturbance of bearing materials.

In order not to damage adjacent footing areas due to potential cave-in, a trench should be excavated by using

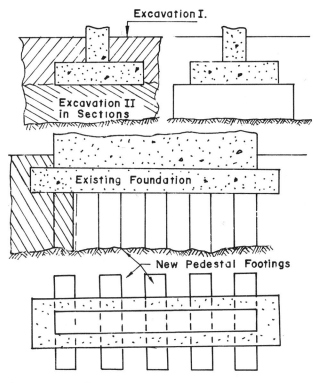

Excavation I.

Excavation II in Sections

Existing Foundation

New Pedestal Footings

Fig. 10.10 Underpinning

the slurry-wall-method or some other protective system. The slurry-wall-method is applicable where the ground water table is high and soil is unstable for a narrow and deep trench. In slurry-wall construction, as each bucket of soil is removed, bentonite slurry is poured into the excavated area to maintain hydraulic pressure on each side of the excavation. Finally in the slurry-filled excavation, the prefabricated reinforcement is placed and tremie concrete is pumped in (bentonite will be recovered).

Needle Beam Method : (Fig. 10.11)

This method requires the construction of a new foundation or driving piles adjacent to the existing footing. A horizontal supporting 'needle' beam is then installed underneath the existing foundation. The needle beams are supported on piles, cribs, grillage, posts or new footings.

The final contact with structure is made by wedging steel bearing plates or dry pack concrete. A careful examination of the structure and settlement should be made before and during underpinning in both the design phase and the actual construction. Maximum reduction of the existing load should be made before excavation. Also, stability and settlement of the structure and pressures in the 'braced-excavation' should be evaluated. For underpinning piles, the possible side frictional losses should be considered in estimating their capacities.

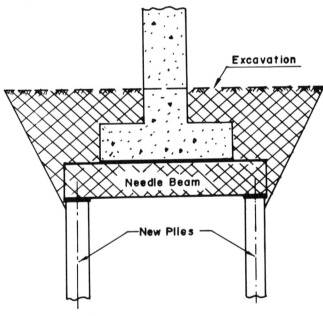

Fig. 10.11   Needle beam

10.6 DISTRESS AND REHABILITATION OF ABUTMENT WALLS AND OTHER RETAINING WALLS

(a) Due to and Against Overturning Effect: If the vertical loads are not enough to overcome the overturning moment from horizontal earth pressure etc., wall would have rotational movement. Proper drainage of the back-fill is required to reduce the lateral pressure.

Repairing this includes drilling new weep holes in the wall, placing a slurry wall, installing a tie-back and soil-anchor system. (For details *see* ahead.)

(b) Due to and Against Sliding Effect: If the vertical loads do not develop sufficient friction force to resist movement between the base and foundation soil, lateral movement will take place.

Repairing this defect includes installing a pile system with soldier beams and a sheet pile or tie back system.

(c) Due to and Against Settlement Effect: If the shearing resistance of foundation material is too low, rupture of the soil occurs. To improve this condition, cement grouting and chemical grouting are recommended.

(d) Due to and Against Slope-Failure Effect: Slope-failure of the embankment can occur when the soil lacks sufficient cohesion and/or internal friction and the foundation is not placed at a sufficient depth below ground. To improve this condition, 'tie- back' and 'soil anchor' systems have been used.

(e) Against Cracking of Abutment Wall and Retaining Wall: When structurally inadequate, the walls may crack in tension or totally fail.

Remedies for this defect are strengthening the wall by installing sheet piles behind the walls or building new wall with 'tie-back' system in front of the existing walls. The latter can be directly tied-back in concert with horizontals and vertical soldier beams.

10.7 STRENGTHENING AND PROTECTING BY VARIOUS TYPES OF WALLS

Various types of 'walls' that can be adopted for this work are :
— R.C. cantilever walls (suitable for heights up to about 8 m)
— R. C. counterfort walls (suitable for heights above 8 m)
— buttress and gabion walls (suitable for heights up to about 8 m)
— reinforced earth walls (suitable for heights above 8 m)

— tied-back (i.e. ground-anchored) rigid walls or flexible sheet piles
— etc.

The reinforced concrete 'cantilever' and 'counterfort' type walls are commonplace and therefore we shall discuss the details of only some of the other types here.

• However, it must be implicitly understood that, in the event of a scour emergency, the undermined portions should first be attended to as described below. The various protective or strengthening walls can then be (designed and) built subsequently, as this will need some time.

Grouting against Scour in an Emergency

Obstruction to the flow of the river by the bridge piers and abutments, and water in the stream during high velocity flow and flood periods, are factors causing undermining and serious scour at bridge foundations if the foundations are not laid at proper depths and do not have adequate protection.

If serious scour (which has not been provided for) occurs, then the undermined area should be grouted with aggregate and pressure injected cement or cast-in-place concrete (e.g. 'prepaked concrete', etc., as discussed earlier in Chapter 6). Large size boulders/rocks/mass-concrete-cubic-blocks can also be dumped around the foundation after rehabilitating the undermined portions.

Empty cement bags of burlap, filled with mass concrete, can also be used by divers to plug the undermined zones. These bags can alternatively be filled with dry mixed sand and cement (e.g. 1:1 by volume, depending on the situation) and placed underwater to fill the undermined portions.

(A) Buttress Wall (in gabions or mass concrete)

A wall or an abutment can be stabilized by constructing 'gabions' or a reinforced or mass concrete 'block buttress' at the toe of the failure slope or distressing wall. The strength of the buttress varies, depending on weight of the wall and the failure planes of soil. The rock filled gabion has advantages over a concrete buttress:

1. It is less expansive than a concrete structure.
2. A gabion wall does not require any drainage system. A concrete buttress requires weep holes to drain the water from behind the wall.

(B) Gabion Wall (Fig. 10.13)

Gabion walls have successfully been used for embankment protection and as retaining walls. Subject to various stresses, they are able to settle, twist and conform to channel and foundation shifts and changes! At the same time, gabions must have sufficient strength to contain the weight of the rocks with which they are filled and the weight of the additional filled gabions that may be

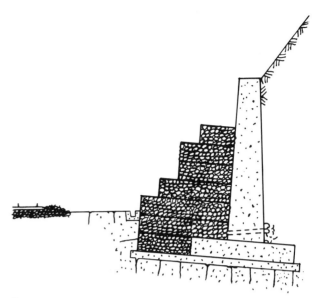

Fig. 10.12 Buttressed wall

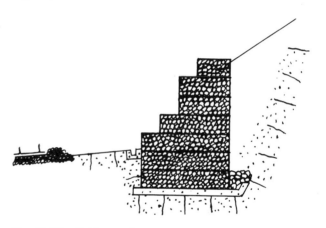

Fig. 10.13 Gabion wall

placed upon them. They should also be able to resist the impact of high velocity currents and soil pressures. A gabion basically comprises of a wire-mesh-net filled with rocks or boulders, forming a three dimensional crate. Rock or boulder size is 4" to 8" and sometimes larger if filled at place.

A gabion wall is considered to be a mass gravity structure. The (galvanized or polyvinyl-chloride PVC coated) wire mesh, on its own, is ignored as any reinforcement; it only gives the shape and form to the crate.

When water quality is in doubt (pH more than 12 or less than 8) or where high concentration of organic acids is suspected, PVC coated wire mesh gabion should be used. The active soil pressure behind the wall can be calculated using the Coulomb Wedge theory and the mass of the wall is designed to balance the force exerted by the soil, When designing a vertical-faced-wall, it should be battered at an angle of about 1:6 to keep the resultant force toward the back of the wall. The coefficient of friction between the base of a gabion wall and a cohesionless soil can be tan ϕ (ϕ = angle of internal friction for the soil), and the angle of wall friction may be taken as 0.9 ϕ. Where the retained material is mostly sand or silt, a filter cloth or granular filter is recommended to prevent any leaching of the soil. Determine the unit weight of the gabion by assuming the porosity to be 0.3. The specific gravity of the common rock fill material ranges between 2.2 (sandstone) to 3.0 (basalt). Along all exposed gabion faces the outer layer of stones should be hand placed to ensure proper alignment and a neat compact appearance.

A gabion wire-mesh shall be of a single-unit construction. The meshes for front, base, back and lid, shall be woven into a single unit. The meshes for ends and diaphragm (s) shall be factory connected to the base mesh.

The galvanized wire mesh net shall be fabricated to be non- revealing type, which is defined as the ability to resist pulling-apart at any of the wire-twists or connections forming the mesh when a single wire in the section of the mesh is cut and the section of mesh then subjected to the load. The gabion length shall be 1½, 2, 3, or 4 times its width. The width shall not be less than 36 inches. However, all gabions, furnished by the manufacturer, shall be uniform width. Where the gabion length exceeds 1½ times its width, the gabions shall be divided into equal cells by diaphragm (s) of the same wire mesh and gauge as the gabion body.

The assembled gabion units shall be placed in their proper location. All adjoining gabions shall be placed tightly along the perimeter of the gabion contact surfaces to obtain a monolithic structure.

Gabions shall be filled by earth-handling-equipment, such as backhoe, gradall, crane and/or by hand. Care shall be taken when placing fill-material to ensure that the coating on the wire mesh of the gabions is not broken or damaged. Gabions shall be filled in layers, about a foot at a time.

Two connecting wires shall be placed between each layer in all cells along all exposed faces of the gabion structure. All connecting wires shall be looped around 2 mesh-openings and the wire terminals shall be securely twisted to prevent their loosening.

The cells in any row shall be filled in stages so that local deformation will be avoided. At no time shall any cell be filled to a depth exceeding 1 foot more than in the adjoining cell. Along all exposed gabion faces, the outer layer of the stone shall be carefully placed and packed by hand to ensure proper alignment and a neat, compact, square appearance.

The last layer of stone shall be leveled with the top of the gabion to allow proper closing of the lid and provide an even surface for the next course. Gabions shall be packed full, without excessive bulging.

(C) Precast Concrete Modular Type Retaining Wall (Doublewal) (Fig. 10.14)
'Doublewal' is a recently developed gravity retaining wall system. This system consists of precast, interlocking, reinforced concrete modules of two face panels, held rigidly by connecting beams. Once in place, without using fasteners, the units are backfilled with filler material. Then it functions as an economical gravity-type retaining wall. The main advantages of this system are: rapid construction; comparatively low cost; and reduced quantities of excavation and special back-fill.

The foundation for the wall is graded level and compacted to the specifications, depending on the soil conditions. A concrete toe-footing and a rear concrete -leveling-pad are then cast-in-place according to the profile, line and grade. Then the first course of concrete modular units is set with a crane. All modular units above the first course interlock with lower courses.

Vertical joints are staggered with each successive course. Front and rear bearing pads are used between horizontal courses, filter fabric is placed behind the vertical joints, and horizontal joint filler is installed at the front face. The joints at corners or angle points are closed.

The interior of each successive course of precast modular units is filled with granular fill material. Units 4 feet or less in height are filled on one layer and then thoroughly consolidated with vibratory tamping devices. The units which are more than 4 feet in height are filled in two equal layers and thoroughly consolidated after each layer is placed.

Backfill around the outside of the wall is placed and compacted in layers according to the usual specifications. When erecting a battered wall, placement of backfill behind the wall should closely follow erection of successive courses of units. At no time should the difference in elevation between the backfill and the top of the last erected course exceed six feet.

The Doublewal structure is designed to resist the overturning moment created by the soil pressure behind the wall based on Coulomb Equation assuming partial wall friction (2/3 ϕ).

The factor of safety for overturning should be atleast 1.5. The Doublewal system functions somewhat differently than conventional retaining wall in that as the wall starts to rotate, the vertical loads are translated into heel and toe (mainly toe) to achieve a safety factor of 1.0 for overturning; and frictions are developed between the soil inside the wall and the wall itself (as the wall tilts).

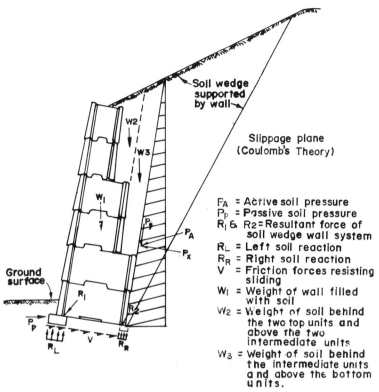

Fig. 10.14 Doublewal

A full scale field test of twenty-two feet high Doublewal confirmed the validity of the design assumption based on pressure calculated from Coulomb Equation with wall friction.

The resistance against sliding is achieved by the soil friction at the base. When the horizontal force exceeds the allowable soil to soil friction, the excessive horizontal forces is transmitted to the toe and heel. If there is excessive force at the toe and heel, consideration should be given to design a footing for the wall units.

The 'Doublewal Corporation' provides complete construction plans and detailed designs, including external stability considerations for overturning moments and sliding. However, it is the engineer's responsibility to study and analyze other items, such as allowable bearing pressure, foundation settlements, backfill characteristics, and performance requirements. The engineer should analyze and review the adequacy of the soil, footing and foundation requirements.

The 'Doublewal' System is patented by 'Doublewal Corporation', 59–East Main St. Plainville, Conn. 06062, U. S. A.

(D) Fabric Reinforced Embankment (geo-grid concept)

This is based on the concept of the earth mass getting reinforced by synthetic fabric layers placed between the fill layers. The fabric is covered with about 1 foot of fill

and the outer edge of the fabric is folded back around the berm and lapped over with the next layer, the fold forms part of the vertical surface of the wall. Then the backfill is leveled with a dozer, and compacted. The same procedure of unrolling a new fabric-strip is repeated. At the 'face' of the wall the fabric keeps the soil from getting pushed out, the fabric deriving its tensile strength through friction with the enveloping soil layer. The wall-face should be shotcreted after the completion of the embankment.

(E) Ground-Anchored Bulkheads (tied-back sheetpiles and walls)

1. General: Anchored bulkheads are formed of 'flexible' sheeting restrained by tieback and by penetration of the sheeting below the dredge line. Since active pressures are 'redistributed' on the sheeting (or wall) by its deflection, moving away from the position of maximum moment, the computed maximum moment can be reduced to allow for flexibility of the sheeting wall. To limit differential water pressures, it is recommended to provide weep holes or a drainage system at a level above mean water level.

— for anchorage, a tie-back may be carried to a buried 'deadman', anchorage, to a 'pile-anchorage', to a parallel 'wall'–anchorage, or it may be a drilled and grouted anchor. If a 'deadman' must be positioned close to wall, the

anchorage resistance is decreased and an additional passive reaction is required for stability at the wall-base. It is necessary to protect tie-rods by wrapping, painting, galvanizing, or encasement, to resist corrosion. Where backfill could have a major settlement, to avoid overburden loading, The tie rod should be enclosed in a rigid tube or should be supported;
— the following precautions should be taken during construction:
(i) Removal of soft material or placement of fill in the passive zone should be made before driving the sheet piles.
(ii) Deposit backfill by working away from the wall rather than toward it to avoid trapping soft material near sheating.
(iii) Before anchorage is tied-in, the sheeting is acting as a cantilever-wall, safety under such condition during construction stages should be checked and ensured in advance.
(iv) When granular backfill is scarce, a sand-dyke may be placed to form a 'plug' across the potential failure surface of the active-wedge. When such a dike rests on firm foundation soil, the lateral pressure on the bulkhead will be only the active pressure of the dyke material.

2. *The Tieback or Ground-Anchor System*: The tieback system as a permanent wall is a rather recent development. Sometimes it is the most effective way to provide horizontal resistance and to stabilize and control soilslide. It has been used to stabilize existing retaining walls,

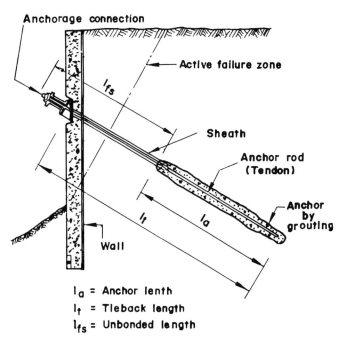

Fig. 10.15 Tieback or ground anchor

to resist uplift as a tie-down, and as a permanent wall near a stub-abutment when widening the road in front. Tiebacks are used in cohesionless soil and rock but should not be used in fill-material or soft-cohesive-soils for obvious reasons;
— permanent tiebacks can eliminate the need for foundation piles, large footings and improved backfills. However, problems can come from corrosion of the high-strength tie-rod (tendon) and lond term creep at the anchor zone. As a protective measure, grease-tube is generally provided in the unbonded length of the tendon against possible corrosion;
— tiebacks are suitable only in particular soil conditions and the anchor system should demonstrate its reliability in that soil by proper inspection and testing during and after installation. The anchor of a working tie-back should be tested by a minimum pull of 1.5 times the design pull value;
— depending on the field conditions and availability or equipment and materials, one can decide upon selection of tendon-system, systems of corrosion-protection and grouting, and method of drilling for the tieback (soil or rock anchor).

3. *Some Pertinent Details*
(i) *General*: A tieback system is a structural system which uses an anchor in the ground to secure a tendon which applies a force to a structure.

Vertical or near vertical tiebacks are called tiedowns. Tiebacks are also referred to as ground-anchors.

Tiebacks are used for 'temporary' or 'permanent' applications. A temporary tieback is used during the construction of a project; its service life is usually less than two years. A permanent tieback is required for the life of a permanent structure.

The tendon is made up of prestressing steel with sheathing, and anchorage. The anchorage transmits the tensile force in the prestressing steel to the ground. Cement grout, or polyester resin, or mechanical anchors are used to anchor the steel in the ground. The anchorage is made up of an anchor-head on nut, and a bearing plate.

The anchor-head or nut is attached to the prestressing steel, and transfers the tieback force to a bearing plate which evenly distributes the force to the structure. Anchor-heads can be 'restressable' or 'nonrestressable'. A restressable anchor head is one where the tieback force can be measured and increased anytime during the life of the structure. The load can not be adjusted when a nonrestressable anchor-head is used. A 'coupling' can be used to transmit the anchor force from one length of prestressing steel to another.

The 'anchor-length' is the design length of the tie-back through which tieback force is 'transmitted to the ground'. The 'tendon bond length' is the length of the tendon which is bonded to the anchor grout. Normally the 'tendon bond length' is equal to the 'anchor length'. The 'unbonded length' of the tendon is the length which is free to elongate elastically. The 'jacking length' is the portion of the tieback which is required for testing and stressing of the tieback. The 'unbonded testing length' is the sum of the 'unbonded length' and the 'jacking length'. A sheath or bond breaker is installed over the unbonded length to prevent the prestressing steel from bonding to surrounding grout. The anchor diameter is the design diameter of the anchor.

Anchor-grout is used to transmit the tieback force to the ground. The anchor-grout is also called the 'primary grout'. Secondary grout is injected into the drill hole after stressing to provide corrosion protection for unsheated tendons.

Tiebacks carry various loads during their lifetimes. The 'design-load' is the maximum anticipated working load that is applicable to the tieback. The 'test-load' is the maximum load applied during testing. The 'lock-of load' or 'transfer-load' is the load transferred to the tieback upon completion of stressing. The 'alignment load' is a nominal load maintained on a tieback during testing to keep the testing equipment in position. The 'lift- load' is the load required to lift the anchor-head or nut from the bearing plate: The 'residual load' at any time is the load carried by the tieback at that time. The 'load transfer rate' is the tieback capacity per unit length of anchor.

The basic types of tieback are — pressure-injected, low-pressure-grouted, straight-shafted, single-under-reamed, multi-underreamed, and postgrouted.

Pressure-injected tiebacks are used in sandy or gravelly soils. Grout pressures in excess of 150 psi (1034 kPa) are used to achieve high load transfer rates.

Low-pressure-grouted, straight-shafted tiebacks are installed in rock, cohesive soils, and sandy or gravelly soils. They can be made using a variety of drilling and grouting techniques. The grout pressure is less than 150 psi (1034 kPa).

Single-underreamed tiebacks are installed using large encased drill holes in cohesive soils. Sand-cement grout or concrete is used in grouting the tieback and the grout or concrete is not placed under pressure.

Multi-underreamed tiebacks are used in stiff cohesive soils and weak rocks. The spacing of the underreams is selected in order to induce strength against

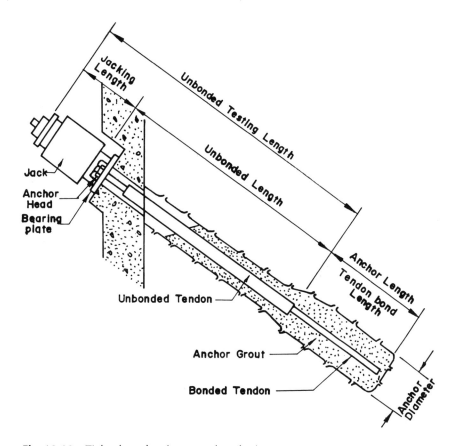

Fig. 10.16 Tieback anchor (or ground anchor)

shear failure along the cylinder determined by the tips of the underreams.

Post-grouted tiebacks are primarily used in cohesive soils. In granular soils and rocks, post-grouting is used to increase the rate of load transfer.

A tieback is a structural element (a tendon) which uses a grouted anchor at its end in ground to secure itself in order to apply its force to a structure. Figure 10.17 shows the components of a tieback.

The 'anchor-length' is the length of the tieback which is bonded to the soil, and it is where the tieback force is transmitted to ground. Each tieback has an unbonded length between the 'anchor' and the 'structure'. There the tieback tendon is not bonded to the soil, and is free to elongate elastically. Force is applied to the tieback by post-tensioning it.

Tiebacks are relatively new construction elements. They have been developed, in a large part, by speciality contractors who design and build temporary excavation support systems, etc. Each contractor has evolved his own method of performing the work, and many of the techniques are proprietary.

Parmanent tiebacks have been used to support structures in Europe since the mid-1960's; and since the early 1970's in the United States. They can be protected from corrosive attack and they are tested to evaluate both their short term performance and their long term load holding capacity.

(ii) Design of Ground Anchors: An initial evaluation must be made to determine if tiebacks can be used at a particular site or whether or not they will be able to develop the necessary capacity without excessive movements or loss of load. The capacity of soil tiebacks are estimated using empirical relationships developed for the particular tie-back type. Generally all rock materials can be considered as 'suitable ground' in which to found anchors.

Permanent tiebacks are routinely installed in noncohesive soils with a standard penetration resistance greater than 10 blows per foot. They should not be anchored in fill. Past experience including testing and monitoring of actual installations, indicates that the permanent tiebacks installed in sandy soil will have satisfactory long-term performance.

Permanent tiebacks are not routinely installed in soft to medium type cohesive soils because their long term load capacity is questionable.

These soils can often be avoided by installing the tieback at a steeper angle and to a depth where better soil or rock can be found. Soil strength, Atterberg limits,

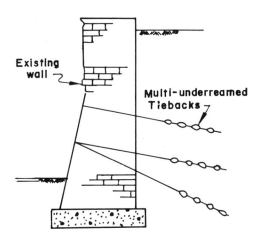

Fig. 10.17

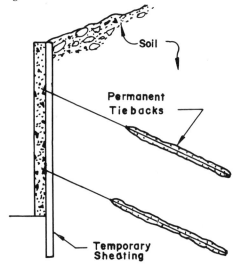

Permanent tieback wall.
Fig. 10.18

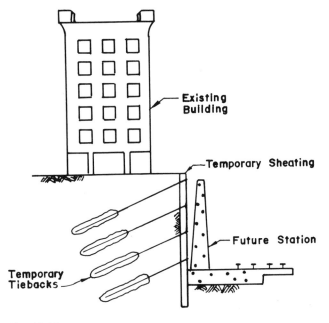

Fig. 10.19

natural water content, and experience in similar soil, will provide the best indication of the long term performance of a permanent tieback installed in a cohesive soil.

Tiebacks installed in soils with a high organic content, in normally consolidated clays, and in cohesive soils with an unconfined compressive strength less than 1.0 ton/ft^2 (96 kPa) and remoulded strengths less than 0.5 ton/ft^2 (48 kPa) may be creep-susceptible. Tiebacks installed in soils that exceed these strengths and have a consistency index (I_c) greater than 0.8, have not experienced significant loss of load or movement with time. The consistency index I_c is given by the relationship:

$$I_c = \frac{W_L - W}{W_L - W_p}$$

where W_L = Liquid limit of the soil
W = Natural water content
W_p = Plastic limit of the soil

In order to establish load holding characteristics and to establish confidence in long-term performance, a tieback test-program is recommended if permanent tiebacks are to be anchored in a cohesive soil or in a sandy soil with a standard penetration resistance less than ten blows per foot. (Tiebacks need to be tested in any soil or rock, in any case.)

The most economical tieback installation will be obtained if the design specifications permit the contractor to select the tieback type, the construction method and

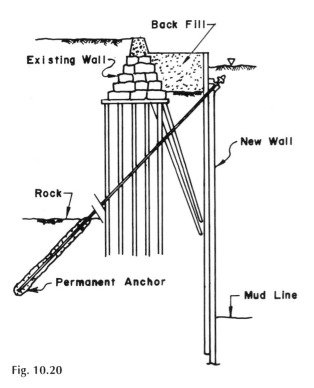

Fig. 10.20

the tieback capacity. The designer should specify the minimum unbonded length, the minimum total tieback length, and the loading diagram for determining the tieback loads. In lieu of specifying a load diagram, a unit tieback capacity for each tieback level could be specified. Finally, each working tieback should be tested to verify that the anchor will carry the design load.

(iii) Corrosion Protection of Ground-Anchors: Permanent tiebacks have been installed routinely since the mid-1960's in Europe and since the early 1970's in the United States. They are performing well in a variety of environments. Most tiebacks use cement grout for protection over their anchor length. Portier and Herbst reported that there is no evidence of a corrosion failure where the tieback tendon was encased in grout. Corrosion failures have occurred along the unbonded length of unprotected tendons, with most of them located within 6.65 feet (2 m) of the anchor head. A significant number of the tieback-corrosion-failures have occured in tendons fabricated using quenched and tempered prestressing steels. These steels do not meet ASTM specifications for prestressing wires, strands, or bars, and may not be used.

Most permanent tiebacks can be protected by portland cement grout along the 'anchor length', and a grease-filled-tube or heat shrinkage-sleeve over the 'unbonded length'. Grout protected tiebacks should be electrically insulated from the structure they support, and the tendon should have a minimum of 0.5 inch (12.5 mm) of grout protection around it.

If the soil surrounding the anchor length has a pH less than 4.5, or a resistivity less than 2000 ohm-cm, or if chlorides are present, then a 'local corrosion' system could develop on the tendon. Figure 10.22 shows a local corrosion system. When the aggressive environments are encountered, then the tendon should be completely encapsulated in a plastic or steel tube. Figure 10.23 shows an encapsulated tieback. The encapsulation will interrupt any long-line and stray current corrosion system, and prevent the local corrosion system from developing.

Figures 10.21 and 10.23 show two ways to provide corrosion protection for the anchorage and the tendon below the bearing plate. Care must be taken to insure that this area is well protected since most known corrosion failures have occurred near the anchor head. The corrosion protection under the anchorage should be designed to accommodate small movements.

The American Water Works Association (AWWA) describes how the pH, resistivity, and chloride content can be measured in the field. The soluble sulphate content of the soil is determined in the laboratory. If the soluble sulphate content exceeds 2000 mg/kg, then the

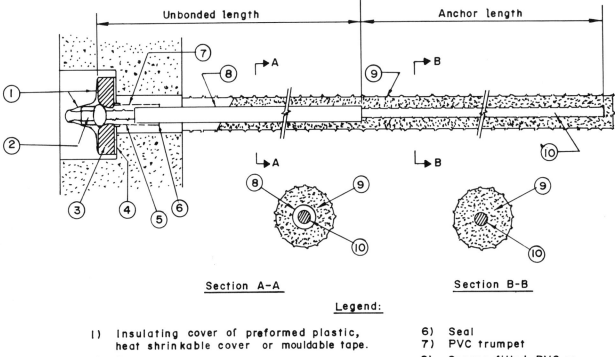

Legend:

1) Insulating cover of preformed plastic, heat shrinkable cover or mouldable tape.
2) Nut.
3) Bearing plate.
4) Bearing plate insulation.
5) Anticorrosion grease.

6) Seal
7) PVC trumpet
8) Grease filled PVC or polyethylene sheath
9) Anchor grout.
10) Tendon.

Fig. 10.21 Insulated simple corrosion protected tieback

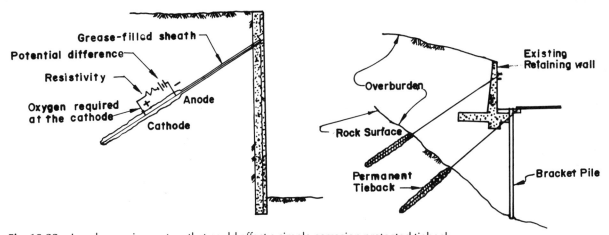

Fig. 10.22 Local corrosion system that could affect a simple corrosion protected tieback

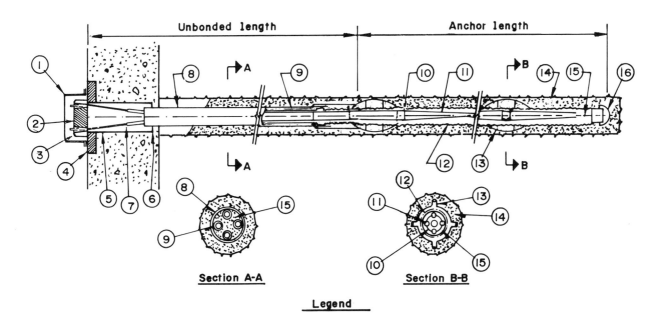

Unbonded length   Anchor length

Section A-A   Section B-B

Legend

1. Anchorage cover.
2. Anchor head and wedges.
3. Anticorrosion grease or grout.
4. Bearing plate
5. Trumpet
6. Seal.
7. Anticorrosion grease or grout.
8. PVC or polyethylene tube.
9. Individually greased & sheathed strands
10. Spacer
11. Strand tendon.
12. Corrugated polyethylene or PVC.
13. Centralizer.
14. Anchor grout.
15. Grout or polyester resin.
16. End cap

Fig. 10.23   Encapsulated strand tieback

ASTM Type V cement should be used. When the pH of the soil is less than 4.5 or when nearby buried concrete structures are suffering from chemical attack, ordinary portland cement alone should not be used for the anchor-grout.

(iv) Specification for Ground-Anchors: The specification should establish a quality level without eliminating suitable proprietary tieback systems or methods. The designer may require the prequalification of the tieback contractor. The prequalification can be based on experience, or a list of 'acceptable contractors' be included in the specifications.

— an alternative type of prequalification warrants evaluation. This method would require the submission and approval of the tieback system details, and those of its corrosion protection, prior to bid. The submission must be detailed enough to enable the designer to determine if his design is satisfied. This form of prequalification would also allow the contractor to know if his proprietary techniques could be acceptable, and the owner would be able to take advantage of cost savings, if any. Preparation and review of the submittal would not require a great deal of time, and this contracting procedure would encourage alternate tieback types, and continued tieback development.

(v) Testing of Ground-Anchors: Every tieback should be tested to verify that it will carry the design load without significant movement. Tiebacks are one of the few structural systems where every member can and should normally be tested before placing them into service. Three types of tests are recommended; performance, proof, and creep tests.

— a hydraulic jack and pump are used to apply the pull load on the tieback. The entire tieback tendon should be simultaneously loaded during testing. Movement of the tieback is measured with a dial gauge or a vernier scale supported on a reference which should be independent of the tieback structure. Movement cannot be accurately monitored by measuring the jack-ram travel;

— the first few tiebacks and a selected percentage of the remaining tiebacks should be performance-tested. The performance- test is used to establish the load-deformation behavior for the tiebacks at a particular site. It is also used to separate and identify the causes of tieback movement, and to check that the unbonded length has been established. The movement patterns developed during the performance-test are used to interpret the results of a simpler proof-test;

— Each production (i.e. working) tieback which is not performance-tested, should be proof-tested. A proof-test is a simple test which is used to measure the total movement of the tieback. Proof-testing is done by measuring the load applied to the tieback and its movement during incremental loading;

— creep-tests are performed on tiebacks installed on cohesive soils. They are normally made on the initial two performance-tested tiebacks. During a creep-test, each increment of load is held and the elongations are recorded and plotted;

— tieback tests are used to identify the load-deformation behavior of each tieback, and provide data that will enable the engineer to make a decision as to their adequacy. The total movement curve is helpful in quickly identifying any unusual behavior. However, the primary purpose of the test is to verify that the tieback will carry the load without excessive movement. The tieback behavior during the load-hold or the creep-test provides the best indication of the load carrying ability of the tieback.

(vi) Conclusion Regarding Tiebacks (ground-anchors): Permanent tiebacks are effective tools to support a variety of different structures. They can be installed in rock and sandy soils without much concern about their long-term performance. Permanent tiebacks can also be made in stiff cohesive soils. However, a careful testing program, and a proven tieback system, should be used in cohesive soils. The tieback tendons can easily be protected from corrosion. Each tieback should be tested to verify that it will carry the design load for the service life of the structure.

(F) Internally Braced Flexible Walls
To retain a foundation or to hold a trench-excavation, frequently laterally braced sheet-piles are used. This restrains lateral movement of the soil and causes loads on the braces which exceed those expected from active earth pressure. Braces may be either long raking-braces or short horizontal cross-braces between trench walls.

Design Factor	Comments
1. Water Loads	Often greater than earth load on impervious walls. Recommend piezometers during construction to monitor water levels. Should consider possible lower water pressure as a result of seepage through or under wall. Dewatering can be used to reduce water loads. Seepage under wall reduces passive resistance.
2. Stability	Consider possible instability in any berm or exposed slope. Sliding potential beneath the wall or behind tiebacks should be evaluated. Deep seated bearing failure under weight of supported soil to be checked in weak soils. Stability should consider weight of surcharge or the weight of other facilities in close proximity to excavation.
3. Piping	Loss of ground caused by high ground water table in silty and fine sandy soils. Difficulties occur due to flow beneath wall, through bad joints in walls, or through unsealed sheetpile 'handling' holes. Dewatering may be required.
4. Movements	Movements can be minimized through use of stiff wall supported by preloaded tieback or braced system.
5. Dewatering-Recharge	Dewatering reduces load on the wall system and minimizes possible loss of ground due to piping. May cause settlements and will then need to recharge outside of support system.
6. Surcharge	Construction materials usually stored near wall systems. Allowance should always be made for surcharge.
7. Prestressing of Tiebacks or Struts	Necessary to remove slack from system and minimize soil movements.

(. . . . continued on next page)

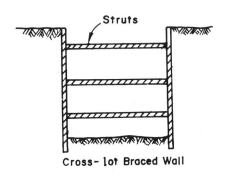

Cross-lot Braced Wall

Fig. 10.24 Braced wall

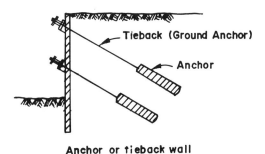

Anchor or tieback wall

8. Construction Sequence	The amount of movement is dependent on the depth of excavation. The amount of load on the tie-backs is dependent on the amount of wall movement which occurs before they are installed. Movements of walls should be checked at every major construction stage. Upper struts should be installed early (Important).
9. Temperature	Struts are subject to load fluctuation due to temperature loads; may be important for long struts.
10. Frost Penetration	In very cold climate, frost penetration can cause significant loading on the wall system; Design of upper portion of the system should be conservative. Anchors may have to be heated. Freezing temperature can also cause blockage of flow and thus unexpected buildup of static water pressure.
11 Earthquakes	Seismic loads may be induced during earthquake.
12. Factors of Safety	Suggested Minimum Design Factor of Safety for Overall Stability

Item	Under Permanent Loading	Under Temporary Loading
• Earth Berms	2.0	1.5
• Cut Slopes	1.5	1.3
• Bottom heave above foundation level	1.5	1.5
• General stability	1.5	1.3
• Bottom heave at foundation level	2.0	1.5

NOTE: These values are suggested guidelines only. Design safety factor depends on project requirements.

(G) Cantilever Sheet Piles

An 'un-tied-back' (un-anchored) cantilever sheet pile derives support from the passive resistance below the dredge line to support the active pressure from the retained soil above dredge line.

This type of un-tied-back wall is suitable only for heights up to 5 m or so and can be used only in granular soils or stiff clays. For details, refer to the *U.S. Steel-Sheet-Piling Design Manual*.

(H) Cofferdams

Cofferdams can be built by interlooking sheet piles in a single-wall system or in a double-wall system,

depending on the sturdiness requirements. The area enclosed or screened by the cofferdam can (temporarily) remains shielded from the body and current of water so dammed, thus permitting any work (e.g. excavation and laying foundation, or supporting a construction-support-system, etc.) within the shield area.

Double-wall (i.e. cellular) cofferdam consists of a line of circular cells connected by intersecting smaller arch or parallel semi-circular wails connected by straight diaphragms, or a succession of clover leaf cells. The stability depends on the ratio of width to height, presence of an inboard berm, and type and drainage of cell-fill material. Ordinarily, a cell-wall may tilt by 0.002 to 0.03 radians. The active and passive pressures act on exterior face of the sheeting. A cell must be stable against sliding on its base, shear failure between sheeting and cell-fill, shear failure on centerline of cell, and it must be resist bursting pressure through interlocking tension. These factors are influenced by the type of foundation.

Clean coarse-grained free-draining soils are preferred for cell-fill materials. They may be placed hydraulically or dumped thorough water without needing compaction or specific drainage.

(I) Reinforced Earth Structure (Fig. 10.25)

'Reinforced Earth' method was developed in 1967 by the French Engineer, Henry Vidal, and has been used to construct retaining walls and bridge abutments since.

'Reinforcement Earth' is the proprietary name for the 'composite-mass' formed by the combination of a granular soil and linear metallic reinforcement strips; and as a block, it behaves somewhat like a monolith. In a 'Reinforced Earth' wall, the metal reinforcing strips are placed horizontally within the backfill and attached to the vertical precast concrete facing panels in the front. The basic concept of 'Reinforced Earth' lies in developing sufficient friction between thin reinforcement strips and the enveloping earth backfill, without causing internal shear-failure. Design of the wall is based on the distribution of stresses in granular soil and the angle of internal friction. This friction can be developed without any slippage if the granular soil has a high enough (at least 15 to 25 degrees) internal friction angle.

'Reinforced Earth' system has three main components: (i) The facing precast R.C. panels; (ii) Reinforcing strips of galvanized steel, and (iii) A granular backfill.

The panels are precast with light reinforcing bars and about 4000 psi concrete. The galvanized reinforcing strips have transverse ribs, and they are cut into the required lengths (generally 70 per cent of the wall-height, and about 8.5 feet minimum). The galvanized steel strips are bolted to protruding tiestrips cast into the back side of the facing panels (eye and toggle system).

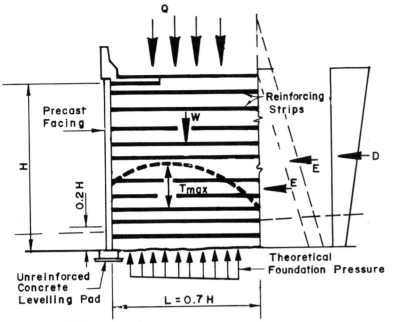

Fig. 10.25 Reinforced earth

W = Weight of Reinforced earth.
T   Measured stress distribution in reinforcing strips.
Q   Surcharge load.
D   Seismic load
E   Earth pressure.

'Reinforced Earth' construction is a repetitive process. It begins with placing a lift of facing panels on an unreinforced concrete leveling pad. The first row of panels is 'braced' for support. The backfill consists of sand and gravel or crushed stone, and is spread and compacted. The contents of backfill material are very important factors for wall's performance. Soil backfill is usually granular material with not more than 15 per cent of fines (by weight) passing a No. 200 mesh sieve. The backfill material should not contain materials which promote corrosion. A higher percentage of fines and moisture will reduce the strip's grip (pullout) capacity and could result in the wall deformation by decreasing the internal stability. Fine grained soils are not generally used as backfill for 'Reinforced Earth' walls. If such soil is used, special quality-control should be exercised during construction for variable fines-content in the soil and its sensitivity to moisture. The coarse grained micaceous soil will face reduction of strength with increased moisture contents. A good drainage system is required behind the wall to prevent saturation, loss of bond and development of hydraulic pressure. Generally the backfill length is about 80 per cent of the wall height.

Next, the first layer of galvanized steel strips is laid perpendicular to the facing panels. They are bolted to the connections on the back of the panels. Another layer of backfill, about 2 feet deep, is spread and made composite, keying the first strips into position and initiating the frictional bonding of earth and steel strips. Following rows of panels are installed into place and the entire procedure is repeated until the desired structural height is achieved.

The 'reinforced earth' wall is used for reconstruction at landslides, as embankment wall, and as bridge abutment, etc. It is an economical construction form, with an attractive wall face.

For further information contact the Reinforced Earth Company, Rosslyn Center, 1700 North Moore St. Arlington, Va. 22209-1960, U. S. A. Reference may also be made to the author's book: 'Concrete Bridge Practice — Analysis, Design and Economics', Tata McGraw-Hill Publishing Co., New Delhi, India.

'Retained-earth' is another proprietary system similar to 'reinforced earth', except that it uses galvanized welded wire mesh in place of galvanized linear steel strips. For further information on 'retained-earth', contact the VSL Corp., 101-Albright Way, Los Gatos, CA 95030, U. S. A.

10.8 REHABILITATION OF PILES

10.8.1 Pile-Protection

This can be achieved by:
— coating: bitumen painting
— barrier: Cladding, Jacketing, Plasticshell
— sheathing
— cathodic-protection

(a) Timber Piles
Deterioration occurs in the splashzone by 'marine borer' attack and at the top of piles by fungus. The most common protection for timber piles, is creosote treatment to retard deterioration. In areas where 'limnoria' attack is expected, the piles are treated with either chromated

copper arsenate or ammoniacal copper arsenate, followed by creosote treatment. For 'toredo' attack areas, creosote or coal-tar treatment is adequate with proper field treatment at the cut area.

(b) Concrete Piles

Deterioration occurs due to corrosion of reinforcing steel, freeze and thaw cycles, and wet and dry cycles in the splash zone. To repair concrete piles, it is suggested to remove unsound concrete before placing new concrete with a minimum of 80 to 100 mm cover. (For details *see* ahead.)

(c) Steel Piles

Corrosion of steel piles can be prevented by coatings, claddings, concrete jacketing, sheathings, and cathodic-protection.

The coatings and claddings provide a shield from moisture and oxygen, thus preventing formation of differential aeration cells for rust. The cathodic-protection provides a sacrificial metal (anode) to reverse the flow current and the corrosion will occur at this anode, instead of, in the steel. This system is effective in the submerged areas but ineffective at the splash zone and in areas exposed to air.

Coal-tar epoxy (U.S. Steel Structure Painting Code: SSPC No. 16) is frequently used as protective coating for steel piles in the splash zone. Generally epoxy is a two-component system and requires two coatings but if it is shop-painted, only one coat may be used. Hand applied (epoxy type) cladding materials are used in the splash zones and underwater. Other coating systems include metallic zinc, aluminium, and phenolic mastic.

Concrete-jacketing is susceptible to crack on impact, therefore, this may not be effecive where excessive movement of the structure is expected by action of wave, impact, or debris. Jacketing of a pile is normally done from a couple of feet below the mean water line to several feet above the high water line (or to the pile-top.) It is recommended that the concrete cover be at least 100 mm, and the concrete be rich aluminuos cement and non-reactive selected aggregate mix, air-entrained or plasticized.

The most widely used 'sheathing' system is the metal sheathing known as "Monel 400", a copper-nickel alloy manufactured by the U.S. International Nickel Company. This alloy shows excellent resistance for corrosion in a marine environment. However, it is susceptible to damages by impact, scratching, cutting or wear out. Stainless steel can be used as a sheathing materials, but it is expensive.

Another type of sheating material is an epoxy polymaid mastic, which is used in the splashzone by trowelling this two component system. Commercially available products in the U. S. are 'Sikagard 694' from Sika Corp., 'A-788 SplashZone Compound' from Koppers Co., and others. Flexible-sheet plastic-material, such as vinyl and glass reinforced epoxy, can be placed by clamps (also the premoulded split-shapes of this plastic can be clamped in place), and sealed with an epoxy polyamide-adhesive.

(d) Cathodic Protective System for Piles

Cathodic protection systems can be classified into 'impressed' systems (which require direct electric current) and 'sacrificial' systems (which use a sacrifical anode with an electric current). Refer to Chapter 4 and 15 for details.

The combination of cathodic protection and coating systems are sometimes recommended because cathodic protection systems will take care of the damaged coated areas underwater and the coating will reduce the operational cost of the system. However, care must be exercised in selecting coating systems which are compatible with the cathodic protective system, to avoid blistering and loss of adhesion. Coal-tar-epoxy and epoxy offer excellent protection in conjunction with the cathodic protection systems. Zinc-based coating is not recommended for use with impressed cathodic protection systems. (Reference to be made to specialist literature in this field.)

10.8.2 Pile Repair

The repair of piles requires field-inspection and information to determine the condition of the piles, interpretation of the obtained data, the course of remedial-action and evaluation of the repair-technique.

Considerable damages can occur where piles are exposed to alternate wetting and drying cycles at splash and tidal zones and where steel piles are butt-welded.

The thickness of steel piles should be measured from mean low water line to several feet below mud line or to where the original pile thickness is intact. Since the loss of thickness is not directly proportional to loss of capacity, the remaining capacity should be computed based on the new moments of inertia and the section areas of the existing piles. The location of damaged areas will influence the capacity and stability of the piles. Therefore, a careful analysis has to include: (i) accurately determining the reduced cross sectional areas sectional areas and moments of inertia, and (ii) the length of each corroded section. If the reduced cross-sections are adequate to carry the loads to saticfy the minimum service level, it is preferred to apply a protective coating system to prevent further corrosion instead of replacing or strengthening weak members.

- *Pile repairs can be carried out by*
  1. Jacketing
  2. Underpinning (driving new piles)
  3. Splicing for strengthening
  1. Pile-Jacketing (see Fig. 10.26)

This system has been used extensively for concrete and steel piles to protect against corrosion and damage due to impact at, above, or below water lines. It is also used in timber piles to protect from marine borer attack by jacketing the pile with concrete or by wrapping with metal sheating.

Before jacketing, the deteriorated concrete and steel should be removed and the surface cleaned by means of sandblasting, water-jet-blasting, or hard-wire brushing. Removed steel should be replaced and lapped with that in good condition and wire fabric should then be installed around the exposed damaged of parts of piles. Then grout mortar is to be filled after placing a jacketing-form around.

Commercially available products include the 'Symon Jacket' with Z-bead system of the Zymon crop., U. S. A., which is composed of a fiberglass sleeve made of fiberglass polyester (FRP); the 'Fabriform' system of the Intrusion-Prepakt, Inc. (U. S. A.), which is essentially nylon-fabric forms holding concrete mortar while permitting the release of excess mixing water; the FX-70 system of Fox Industries (U. S. A.); and others.

2. Pile-Underpinning

This underpinning requires an opening through the deck to drive new piles into the foundation soil. The new piles will be incorporated into the structure as explained earlier in this chapter. However, if the superstructure is poor in condition, total superstructure replacement would eliminate this underpinning.

3. Pile-Splicing

This requires removal of the locally damaged portions of an individual pile and then splicing on a new section to remaining, portions of the old pile. This strengthening of piles does not 'protect' them. Therefore, a protection system should be installed after strengthening the piles. (Details given in the previous section.)

10.9 APPROACH EMBANKMENT SETTLEMENT—PROBLEMS AND CURE

Approach embankment settlement can be a major foundation problem in highway construction. Unlike stability problems, the results are seldom catastrophic, but the cost of perpetual maintenance of continuing settlement can be immense. The difficulty in preventing these problems is not so much the lack of technical expertise as it is the lack of communication between personnel involved in the roadway design and those involved in the structure design! It has to be understood without bias in order to do a 'practising professional' job.

- *The Problem*

The design of the 'roadway-embankment' can utilize a wide range of soil materials and permit substantial amounts of settlement without affecting the performance of the highway. Roadway designers necessarily permit such materials to reduce project costs by utilizing cheap locally available soils. However, structures are necessarily designed for little or no settlement to maintain specified highway-clearances and to ensure integrity of the structural members. The 'approach-embankment' represents a 'transition' between the roadway embankment and the structure.

In most agencies the responsibility for 'approach-embankment' design is not defined, which results in roadway criteria being used for it. This is wrong; the 'approach-embankment' requires special materials and placement-criteria to prevent internal consolidation and to moderate external consolidation.

- *The Solution*

Step 1: Estimate the settlement of the approach-embankment caused by consolidation of the subsoil. Many and varied procedures exist for computation of embankment settlement for cohesiveness and cohesive soils (refer to any standard practical handbook on foundation and soils, for details of computations). If the settlement is serious enough, then proceed to reduce/eliminate it as explained in *Step 2* below.

Step 2: Reduce settlement within the approach embankment, at least shrink the time for settlement. A well constructed soil embankment, using quality control with regard to material and compaction, will not consolidate by a significant amount. Standard specifications and construction drawings should be prepared for the approach-embankment area (normally designated as 50 feet behind the wing wall). The structural designer should have the responsibility for selecting the appropriate approach-embankment cross-section, depending on the structures's foundation-type.

Special attention must be given to the interface area between the structure, and the approach-embankment as this is where the infamous 'bump' at the end of the bridge occurs. The reasons for the bump are twofold; poor compaction of embankment material near the structure and migration of fine soil into drainage material. Poor densificaton is caused by restricted access of standard compaction equipment. (Proper densification can be achieved by optimizing the soil gradation in this area to permit maximum density with a minimum effort.)

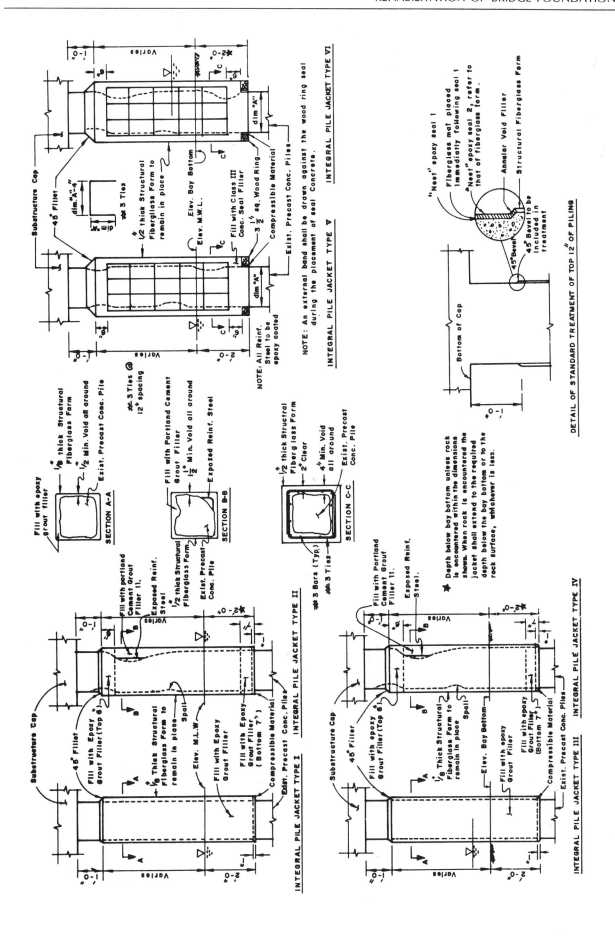

Fig. 10.26(a) Pile-jacketing
Courtesy: D. O. T. Florida, U. S. A.

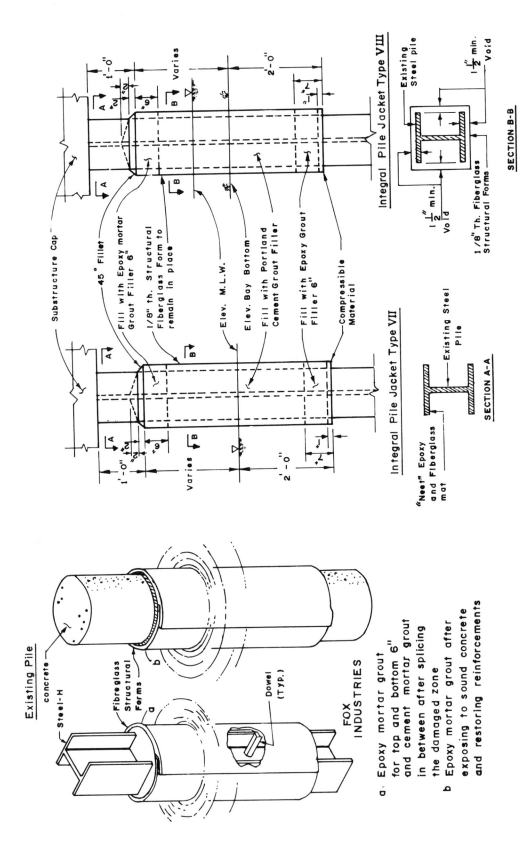

Fig. 10.26(b) Pile-jacketing
Courtesy: D. O. T. Florida, U. S. A.

Reducing Settlement Time

Often a major design consideration, when faced with a settlement problem, is the time it will take for the settlement to occur. Low permeability clays and silt-clays can take a long time to consolidate (water squeeze-out). The settlement time is generally what will get the chief engineer 'excited', since this affects construction schedules, and increases project costs due to inflation, etc.

The two most common methods used to accelerate the settlement to reduce the settlement-time are

1. Surcharge treatment
2. Sand or Wick Drains Treatment of subsoil

1. Surcharge Treatment: The embankment is built up 1 to 10 feet above the final grade evaluation and allowed to remain for a predetermined waiting period (typically 3–12 months). The length of waiting can be estimated using consolidation test data. The actual settlement occurring during embankment construction is then monitored with geometrical instrumentation. When the settlement with surcharge equals the settlement originally estimated for the embankment, the surcharge may be removed as illustrated in Fig. 10.27.

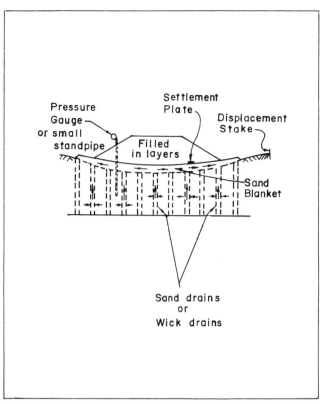

Fig. 10.28

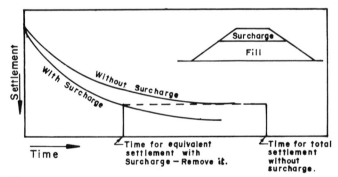

Fig. 10.27

Note that the stability of a surcharged embankment must be checked to insure that an adequate safety factor exists to permit placement of the surcharge load, even though only for less than a year.

2. Sand or Wick Drain Treatment of Subsoil: Some highly plastic clays of extremely low permeability can take many years for settlement to be completed. Surcharging alone may not be effective for reducing settlement-time sufficiently. In such cases 'sand' or 'wick' drains can be used to accelerate the settlement. Surcharging should always be considered first, since sand and wick drains are much more expensive. The reason such drains accelerate the settlement is that they shorten the 'drainage path' the water must travel to escape from the impervious soil, as illustrated in Fig. 10.28.

The 'settlement time' is proportional to the square of the 'length of the drainage path'; thus if the drainage path length can be cut by half, the time is reduced by factor of four! The drains and sand-blanket must have high permeability to allow water to squeeze out of the subsoil (due to the fill pressure) to go up the drains and out through the blanket. 'sand' drains are typically installed by jetting or augering 12 to 18 inches diameter holes to the bottom of the compressible soil and backfilling with a high-permeability graded sand. Typical sand drains spacing is 8 to 15 feet, center to center.

'Wick' drains are small pre-fabricated drains, consisting of a plastic core, which are wrapped by a piece of filter fabric. Wick drains are approximately 4 inches wide and about 1/4 inch thick. Two examples are the 'Alidrain' and 'Mebra Drain' (both proprietary, U. S. A.) Wick drains are installed by pushing or vibrating a mandrel into the ground with a wick drain inside. When the bottom of the compressible soil is reached, the mandrel is withdrawn and the wick drain left in the ground. To minimize smear of the clays, it is recommended that the cross-sectional area of the mandrel be limited to a maximum of about 12 square inches. Since the wick drains have a much smaller surface area than the sand drains, more are required. (About twice as many wick drains will be required versus sand drains.)

Use of wick drains is quickly gaining acceptance. Wick drains are now used almost exclusively in Japan.

Wick drain projects will typically be 20–30 per cent less costly than if sand drains were used. This is primarily due to much faster speed of installation of wick drains versus sand drains.

• *Practical Aspects of Embankment Settlement*

Few engineers realize the influence of embankment placement on the subsoils. The total weight of an embankment has an impact on the type of foundation-treatment that may selected. For instance, a relatively low-height embankment of ten feet may be effectively surcharged because the surcharge weight could be 30 to 40 per cent of the proposed basic embankment. However, when the embankment exceeds 50 feet, the influence of a 5 or 10 feet surcharge trapezoid of soil on top of this heavy 50 feet mass is small and probably not cost-effective. Conversely, as the embankment height (and therefore, its weight) increases, the use of a spread footing abutment becomes more attractive. A 30′ high and 50′ long approach embankment weighs about 15000 tons compared to the insignificant weight of the abutment load which in most cases will not exceed 1000 tons! Besides weight, the width of an embankment has an effect on total settlement. Wider embankments cause a pressure increase deeper into the subsoil. As might be expected, wider embankments will cause more settlement and take longer for consolidation to occur. *See* Fig. 10.29, Plate 28.

10.10 APPROACH-EMBANKMENT STABILITY (OR INSTABILITY), AND HORIZONTAL MOVEMENT AND TILTING OF ABUTMENT (OR RETAINING WALL) AS A RESULT OF VERTICAL SETTLEMENT OF THE RETAINED EMBANKMENT

10.10.1 General

'Approach-embankment' stability must be assumed prior to consideration of other foundation related items. Embankment foundation problems involve the support of the embankment by natural soil. Problems with embankments and structures occasionally occur which could be prevented by initial recognition of the problem and appropriate design. Stability problems most often occur where the embankment is built over soft weak soils such as low strength clays, silts, or peats. Once the soil profile, soil strengths, and depth of water table have been determined by field explorations and field and laboratory testing, the stability of the 'approach-embankment' can be analyzed and necessary measures taken.

10.10.2 There are **three major types of stability problems** that should be considered in the design of approach-embankments over weak foundation soils.

I) **Circular Arc Failure:**

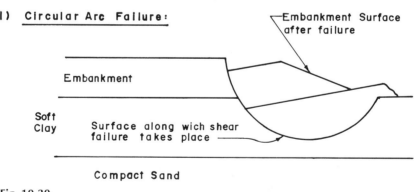

Fig. 10.30

2) **Sliding Block Failure:**

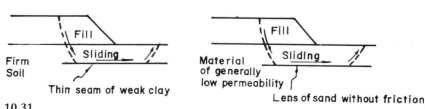

Fig. 10.31

PLATE 28

Fig. 10.29 Failure (localised extensive settlement) of subgrade in the 'approach embankment' of the bridge. The extent of damage can be seen even from the serpentine deformation exhibited by the lane-marking white lines

3) Lateral Squeeze of Foundation Soil:

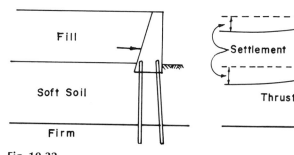

Fig. 10.32

These are: (i) Circular arc failure, (ii) Sliding block failure, (iii) Lateral squeeze of foundation soil.

10.10.3 The following are recommendations on how to recognize, analyze and solve each of these three problems.

NOTE: The stability problems outlined above are 'external' stability problems. "Internal" embankment stability problems can be taken care of by adequate selection of embankment materials and placement requirements. Internal stability may be 'ordered' in project specifications by specifying minimum gradation and compaction requirements, etc.

  (a) Understanding the Effects of Water on Slope-Stability
Next to gravity, water is the most important factor in slope stability.

  (i) Effect of Water on Frictional Soil: In cohesionless soil, water does not significantly effect the angle of internal friction (ϕ). The effect of water on cohesionless soil below the water table is to decrease the inter-granular (effective) pressure between soil grains (due to buoyancy), and this decreases the frictional shearing resistance and the bearing capacity.

  (ii) Effect of Water on Clay: An increase in absorbed moisture is a major factor in the decrease in strength of cohesive soils. Water is absorbed by clay materials and the high water content decreases cohesion of all clayey soils.

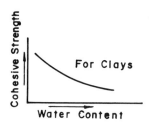

Fig. 10.33

• *Fills on Clay*
Excess pore pressures are created when fills are placed on clay or silt. As the pore pressure dissipates, consoli-

dation occurs and the clay or silt strength increases. This is the reason the factor of safety increases with time in the case of clay surcharged with fill.

• *Cuts in Clay*
As a cut is made in clay, the 'effective' stress is relieved (reduced). This will allow the clay to expand and absorb water, which will lead to an increase in pore pressure and decrease in the clay strength with time. This is the reason the factor of safety of a clay 'cut' slope decreases with time. Cut-slopes in clays should be designed using effective strength parameters and the effective stress which will exist 'after' the cut is made.

  (iii) Effect of Water on Shales, Claystones, Siltstones, etc. (slaking mechanism): Sudden moisture increases in dry soil can produce a pore pressure increase in trapped pore air and local soil expansion, leading to strength decrease. The "slaking" or sudden disintegration of hard shales, claystones and slitstones, results from this mechanism. If placed as rock-fill, water percolating through such fill causes these materials to disintegrate to a clay soil, which often leads to settlement and/or shear failure of the fill.

  (b) Understanding the effect of Fill Settlement on Deformation of Abutment (and wall) and How to Estimate amount of Horizontal movement of Abutment (and wall)
The amount of horizontal movement the abutment may undergo towards the fill can also be estimated in design. The following table contains case-history-information for nine structures where measurements of abutment movements were made (in the U. S.).

  This data provides a basis for estimating horizontal abutment movement for similar problems, provided a reasonable estimate of the post-construction fill-settlement is made, using data from consolidation tests on high quality undisturbed Shelby-tube samples. Note that the data for the structures listed in the table ahead shows horizontal abutment movement to range from 6 to 33 per cent of vertical fill-settlement, with the average being 21 per cent.

Foundation on	Fill-Settlement (in.)	Abutment-Settlement (in.)	Abutment-Tilting (in.)	Ratio of Abutment-Titling to Fill-Settlement
Steel H-Piles	16	Unknown	3	0.19
Steel H-Piles	30	0	3	0.10
Soil Bridge	24	24	4	0.17
Cast-in-place Piles	12	3.5	2.5	0.19
Soil Bridge	12	12	3	0.25
Steel H-Piles	48	0	2	0.06
Steel H-Piles	30	0	10	0.33
Steel H-Piles	5	0.4	0.5 to 1.5	0.1 to 0.3
Timber Piles	36	36	12	0.33

Therefore, the horizontal abutment movement that may occur, can reasonably be estimated as 25 per cent of the vertical fill-settlement, i.e.,

Horizontal Abutment Movement

$$= 0.25 \times \text{(Fill-Settlement)}$$

(c) Possible Design Solutions To Prevent Abutment-Tilting

The best way to handle the abutment-tilting problem is to prevent it by getting the fill settlement out before the abutment piles are driven or its foundation is made. If the construction time-schedule or other factors do not permit the settlement to be accomplished before the piles can be driven, then the problem resulting from abutment-tilting can be provided for by the following design provisions:

1. Use sliding plate expansion shoes large enough to accomodate the anticipated horizontal movement.

2. Make provisions to fill in the expansion joint (over the abutment) by inserting either metal plate fillers or larger neoprene joint fillers.

3. Design piles for downdrag forces due to settlement.

4. Use steel H-piles for the abutment piling since steel H-piles are capable of taking large tensile stresses before failing.

5. Use backward battered piles at the abutment and particularly for the wingwalls.

Movement should also be monitored so that predicted movement can be compared to actual, and corrective measures tried.

10.11 STABILIZING THE SLOPES OF CUTS AND FILLS *(Method of stabilizing slopes of cuts and fills is shown in Fig. 10.34.)*

10.12 STABILIZING A SOIL-MASS

Soil stabilization has been used for increasing its strength, decreasing its permeability, reducing com-pressibility and increasing resistance to erosion. This can be generally achieved by drilling piles through weak strata of soils, by replacing unsuitable soils with selected materials and by treating or injecting additives into soil to improve its properties.

(A) Drainage Control Method

Since soil strength decreases with the increase of water in a soil, proper surface and subsurface drainage is needed. Some of the methods adopted are as follows

1. Place a drainage gutter at the top and behind the walls and place catch basin adjacent to the structure.

2. Surface treatment by 'seeding' and 'paving'. Place drainage facilities to collect and redirect surface water to prevent water penetration into unstable mass.

3. Seal the joints and cracks to avoid seepage of water from behind the wall.

4. Drill horizontal or sloping drain-holes to drain out water trapped in soil.

5. Place drainage pipes or trenches to intercept ground water.

6. Place vertical sand drain in wells (and pump out the collected water) to lower the water table and increase the bearing capacity by expulsion of water and consolidation of the soil.

(B) Seepage Control Method

1. A sheet-pile cut off-wall is suited specially for stratified soils with high horizontal and low vertical permeability or pervious hydraulic fill-materials. Interlocking steel sheet piling is used for deeper cut offs. Steel sheeting must be carefully driven to maintain adequate interlocks between sheets. Steel H-pile soldier-beams are used to minimize deviation of sheeting in driving.

2. Grouting is applicable where the depth or the character of the foundation materials make a sheet-pile-wall or cutoff-trench impractical. Utilized extensively in major hydraulic structures, it may be used as a supplement below cut off-sheeting or trenches. A complete grouted cut off is often difficult and costly to attain, requiring the pattern of holes to be staggered in rows

Scheme	Applicable Methods	Comments
1. Changing Geometry. EXCAVATION 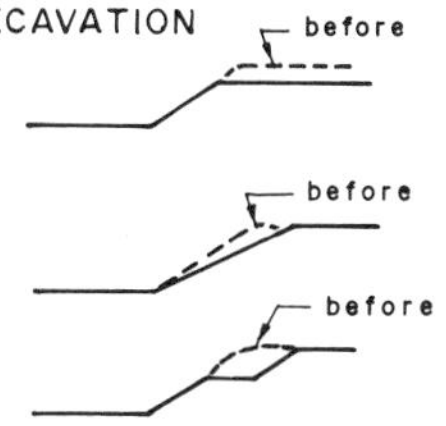	1. Reduce slope height by excavation at top of slope. 2. Flatten the slope angle. 3. Excavate a bench in upper part of slope.	1. Area has to be accessible to construction equipment. Disposal site needed for excavated soil. Drainage sometimes incorporated in this method.
2. Earth Berm Fill: 	1. Compacted earth or rock berm placed at and beyond the toe. Drainage may be provided behind berm.	1. Sufficient width and thickness of berm required so failure will not occur below or through berm.
3. Retaining Structures: 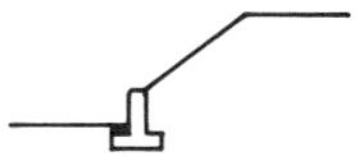	1. Retaining wall crib or cantilever type.	1. Usually expensive. Cantilever walls might have to be tied back.
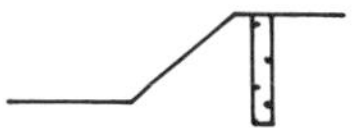	2. Drilled, cast-in-place vertical piles, founded well below bottom of slide plane. Generally 18 to 36 inches in diameter and 4- to 8-foot spacing. Larger diameter piles at closer spacing may be required in some cases to mitigate failures of cuts in highly fissured clays.	2. Spacing should be such that soil can arch between piles. Grade beam can be used to tie piles together. Very large diameter (6 feet $\pm$) piles have been used for deep slides.
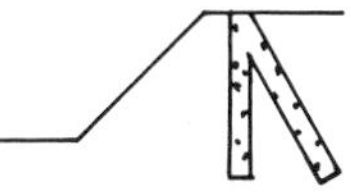	3. Drilled, cast-in-place vertical piles tied back with battered piles or a deadman. Piles founded well below slide plane. Generally, 12 to 30 inches in diameter and 4- to 8-foot spacing.	3. Space close enough so soil will arch between piles. Piles can be tied together with grade beam.
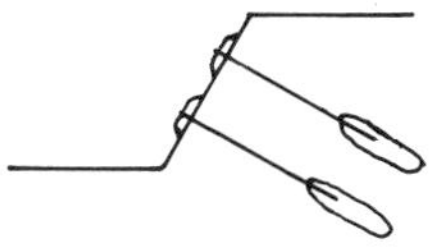	4. Earth and rock anchors and rock bolts.	4. Can be used for high slopes, and in very restricted areas. Conservative design should be used, especially for permanent support. Use may be essential for slopes in rocks where joints dip toward excavation, and such joints daylight in the slope.
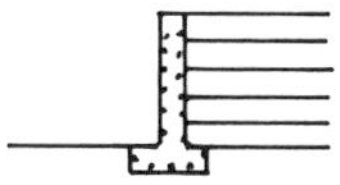	5. Reinforced earth.	5. Usually expensive.

Fig. 10.34 Methods of stabilizing the slopes of cuts and fills

with a carefully planned injection sequence and pressure control.

3. A 'slurry seal' is suited for construction of an impervious cut off below ground water or for stabilizing the trench excavation when concrete foundations are placed with tremie method. In constructing slurry wall, a vertical-sided trench is excavated below ground water level and a slurry (with specific gravity generally between 1.2 and 1.8) is pumped into the trench. Slurry may be formed from the mixture of powdered bentonite with fine grained material removal from the excavation, and water.

For a permanent cutoff-trench-wall, well-graded backfill-material is dropped through the slurry in the trench to form a dense mixture which is essentially incompressible after the trench has been backfilled. The foundation-wall is formed by placing concrete by tremie in the trench (bottom up), displacing the slurry upward.

(C) Increasing Shear-Strength of the Soil

The following methods are used to increase the shear-strength of soil

1. **Injection** and grouting is commonly used for soils of moderate permeability. The common grouting materials are cement-grouting, bitumen-grouting, clay-grouting and chemical-injection. (See details ahead.)

2. **Compaction** is a common procedure used for embankments to squeeze out water, to remove and replace portions of unstable mass, and rearrange the soil particles by densification.

3. **Soil and rock anchors** are used for steep-slopes of weathered rock, deteriorated retaining walls, tieback bulkhead systems, and tie-down anchors to resist uplift force, as discussed earlier.

4. **Freezing** can be used for temporary soil stabilization during construction. In this process in situ pore water is converted into ice by liquid nitrogen grout-refrigeration process. The ice binds the soil particles together to increase the strength and stiffness of the soil mass, and makes it impermeable. This process is slow and costly.

5. **Drainage by the electro-osmosis** method is applied for clays, silts and fine soils by passing direct electric current through saturated soil in an attempt to stabilize the slopes of foundation excavations.

6. **Riprap protection** will prevent erosion and undermining of the footing.

Cement Grouting*: Grouting will increase strength, decrease compressibility and permeability, and seal large voids. This technique has been used for a variety of structures to change the character of the foundation soil to improve the deteriorated condition. The selection of the grouting method depends on the type of foundation materials, durability of grout, desirable improvement of foundation properties and cost limitation. Studies should first be made to determine the overall stratification from the test-borings, to locate cracks, cavities and loose strata, to determine seepage problem, to evaluate the performance of grouting using alternative methods and materials.

Cement grouting is effective if the voids are large than the size of the material in the grout mortar; it is also effective in coarse-gravel and fissured-rocks. Cement grout is pumped into soil or rock, under pressure, to solidify and form a strong and durable founation-monolith. Disadvantage of cement grouting is the overflow of grout into surrounding areas. It is difficult to check whether the intended void is completely filled or not.

The design of the grout-mix is fundamental to the success of a cement grouting program. A wide variety of properties can be given to cement based grout, depending on the additives and the amount of water included in the mix. Fluid grout can be made to flow long distances in rock cracks and crevices. In contrast, paste-like grouts, with controlled set-times, can be used to carefully limit the grout flow range; special chemical additives have been developed to set the grout very quickly, even in flowing water conditions.

In addition the water–cement ratio, additives such as sand, flyash, bentonite, accelerators, retarders, water-reducing agents, etc., must be considered. The flow properties of the grout under pressure must be balanced to the injection conditions. For example, water loss in grout under pressure may undesirably stop flow.

Placement of cement mortar grout should be carried out according to the plans. Splitspacing–drilling methods and vertical staging are often employed.

Compaction cement grouting techniques have been used to stabilize faulty supporting soils to provide subsurface densification and improve inadequate soils. Compaction-cement-grouting requires the injection of a stiff granular mortar grout into subsurface soil under high pressure to displace, densify and improve the adjacent soil.

The key element in any grouting system is the effective control and placement of the grout. In compaction cement grouting, this control is achieved by the use of a granular grout mix of variable flowability. The use of more fluid mixes, under pressure required for effective compaction, may result in loss of control in the grout location, splitting or fracturing the ground and migration of the grout away from the intended repair area. Injection of grout mortar, with slumps of 100 to 200 mm. (although normally considered a stiff mix when

---

*See 'Note' to Fig. 10.35.

pumping concrete), has been shown to easily disperse through consolidated soils, and run 1 to 3 meters away from the grout pipe tip even under low pressures in some soils.

The selection of the mortar-mix proportions is extremely important to achieve pumpable mix that still exhibits high internal shearing resistance to flow and can avoid sand blockages in the lines under high pumping pressures.

Compaction cement grouting techniques have been used for the following:

- To stabilize poor backfill materials behind retaining walls and under footings
- To fill large voids under foundations
- To treat and displace peat areas underlying structures
- To eliminate settlements in weak strata
- To replace and to densify tunnel crown soils during soft ground tunneling operations (to reduce surface settlement and eliminate the need for conventional underpinning)

Also, this technique could be applied to improve slope-stability in conjunction with the reinforcing methods, described earlier.

The success of compaction cement grouting techniques depends on the correct understanding of the cause of foundation-distress or settlements, and proper evaluation of the subsurface conditions, as well as competent application of the grouting.

In addition, a monitoring program should be carried out during grouting, to observe any potential surface-heave and associated movements, and to verify the effectiveness of the grouting operation by observation of slight surface heaves.

Chemical-Grouting (Fig, 10.35): Chemical grouting has been used not only to rehabilitate a structure but as a construction method of cut-and-cover work and tunneling when conventional underground construction methods prove too costly or technically impractical. The injection of a chemical fluid into granular soils, fissured rock or concrete, which later gels, binds the soil particles together. Chemical grouting has been used to solve soil-stabilization problems on bridges. Chemical grouting can provide structural support to existing building, highways, railway structures, utilities, and excavations, and also can control groundwater. Generally it is used for ground-water-proofing and consolidation.

Chemical grouting is usually classified as follows

$$c = \text{consolidation}$$
$$w = \text{waterproofing}$$

1. *Aqueous systems*
   (a) Silicate derivatives (c, w)
   (b) Lingnossulphite derivatives (w)

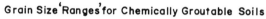

Grain Size 'Ranges' for Chemically Groutable Soils

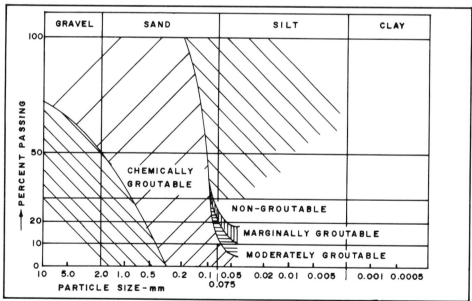

Note: One rule of thumb suggests that soils with less than 12% fines (passing a No. 200 sieve) are easily groutable. Soils with 12 to 20% fines may be only moderately groutable and soils with 20% to 25% fines are only marginally groutable The accompanying figure shows typical ranges of chemically groutable soils.

Fig. 10.35 Chemically groutable soils (Hayward Baker Co., U. S. A.)

(c) Acrylamide derivatives (*w*)

(d) Phenoplasts (*c, w*)

2. *Non-aqueous systems*

Synthetic polyester, polyvinyl, polyepoxy (c, w)

3. *Colloidal solutions in water*

(a) Inorganic-bentonite (*w*)

(b) Organic-aliginate (*w*)

4. *Emulsions*

(a) Bituminous emulsions (*w*)

(b) Polyester, epoxy, vinyl (*w*)

• *Joosten's method of chemical grouting:* Two fluids, sodium silicate and calcium chloride, are injected successively by actually introducing a solution of silicic acid followed by calcium chloride. Reaction between these two solutions is almost instantaneous and calcium silicate is precipitated in soil voids.

• *Single-stage sodium silicate chemical grouting:* Premixed sodium silicate with a setting agent such as sodium aluminate in water is injected into boreholes through pipe. (Also available are a mixture of acrylamide and methylenebisacrylamide, epoxy and polyester resin grout.)

The success of chemical grouting techniques depends on the correct evaluation of site problems, as well as competent application of grouting. In addition, a concurrent monitoring program should be carried out during grouting, to observe any potential surface-heaves and to verify the strength and location of the grout underground. Where the soil has variable permeability, it may be necessary to change the type of grout during the grouting process by using a more penetrating type to compensate for zones of low permeability in the soil. Different types of grout should be used when one particular type of grout becomes ineffective and possibly causes rupture and heave. A more fluidy grout (with greater penetration) may be used as the operation progresses. It is important to select the proper grout for different types of soil. Fluidy grouts are good only in fine soils.

The U. S. FHWA Research Report '*Chemical Grouts For Soils*' (FHWA-RD-77-51) concludes the following concerning grout suitability

1. There is a series of grouts available that are suitable for a wide range of soils

(a) Binghamian-type materials, with cement or clay in suspension, are applicable to coarser soils.

(b) Somewhat viscous colloidal solutions (silica or lignochrome gels, organic or inorganic colloids) are suitable for soils of average particle size.

(c) Pure non-colloidal solutions (organic monomers in aqueous solution) are suitable for very fine soils.

2. Grout users must know how to place the grout where it is needed.

3. When using a series of grouts at one site, one must know when to change grout. The right moment comes when more viscous grout is no longer effective, penetrating the ground only through force and rupturing it. One must monitor the heave benchmark precisely to define the moment to change grout. Grouting not only concerns the selection of a grout, but involves the technology used for injection. Both the grout and the injection-technology must be adapted to the terrain. Large sites must often be carefully evaluated for the purpose of such matching.

The controls used at a given worksite should insure the matching of the grout and the injection-technology to the site

(a) Concerning the grout, it should be verified that the specification given by the manufacturer and by the operator are the same.

(b) Concerning the application, injection pressures, flow-rates, ground movements, potential resurgences, etc., it should be checked (to ensure) that the grout will penetrate the ground without ruptures.

10.13 A FIELD-MAN'S QUICK ESTIMATE OF SOIL BEARING CAPACITY FROM CORRECTED SPT N-VALUE

See Fig. 10.36 for details.

10.14 COUNTER-MEASURES AGAINST BED-DEGRADATION

(a) General: A channel-bed may tend to degrade (i.e. erode and slowly lower its grade, in time) due to various causes imposed upstream and downstream of the bridge. This action can slowly destabilize the approaches and embankments and, in course of time, even effect the bridge foundations. Dumped riprap (properly keyed), concrete-filled (i.e. grouted) riprap, concrete-filled bags and fabric mats, Gabion-protection and concrete-slope-protection systems can be adopted as effective measures for small and medium streams. (Larger streams may need model-testing of various protective systems, some of which can be surprisingly effective even in lesser streams.) These systems may comprise of one or more of the following flow-control-structures:

(a) Spurs and dykes

(b) Retards

(c) Guide Bunds (or Special Spurs)

(d) Check Dams

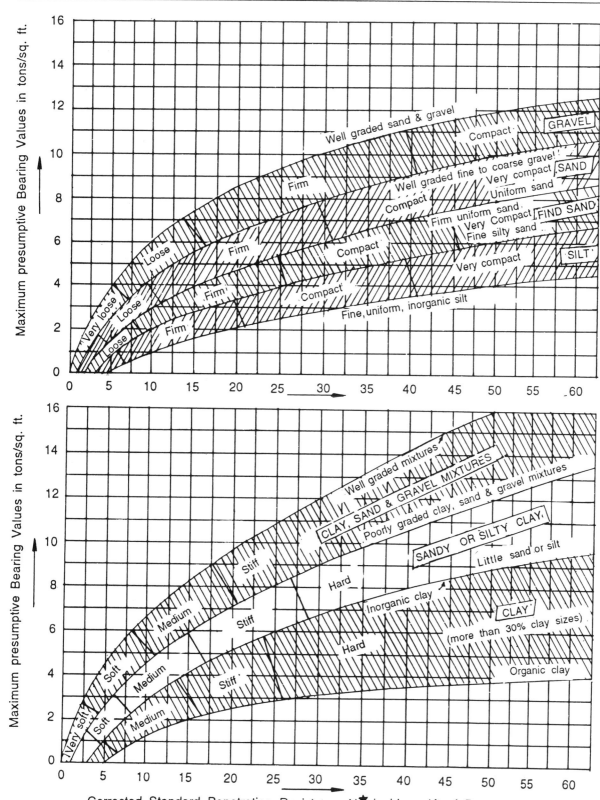

Corrected Standard Penetration Resistance N★ **in blows / ft. of Penetration**

★ Number of blows of standard 140-lb. (pin-guided) weight falling 30 in. per blow required to drive a split-barrel sample spoon (with a 2-in. outside diameter) 12 inches into subsoil.

Fig. 10.36 Bearing capacity versus corrected* standard penetration resistance 'N'

* Corrected for effects of overburden and sand/clay effect

The important property of any flow-control-structure is its permeability. While an impermeable structure would deflect and redirect the stream, permeable structures would essentially reduce the velocity of flow of the stream. The permeable structure provides opening through which water may pass through gaps between its piles, wire mesh, etc.

(b) Spurs and Dykes (see Fig. 10.37): A spur is a linear structure projected from the bank into a channel to alter flow directions, induce deposition, or reduce flow velocities along the bank. Spur-structure types include the following: timber piles, sheet piles, steel piles, and earth or rock embankments.

For bank protection and control of flow at the bends, spurs are more cost-effective than riprap-revetments. The permeable spurs generally provide adequate performance and are effective when placed at right angles to the banks or are inclined toward downstream. Impermeable spurs can cause flow disturbances and bank erosion. Spurs are commonly used in shallow channels.

While conventional impermeable riprap spurs are expensive and may cause lateral stream corrosion, permeable spurs offer flexibility and easy maintenance, and are less costly.

(c) Retards (see Fig. 10.37): A Retard is normally used at the toe of the bank to check velocity and to induce deposition. This changes the trend of errosion, replaces the loss of material and helps maintain the flow-alignment. The types include timberjacks, steel jacks, fence, tetrahedrons, piles and embankments.

Retards provide better flow-lines along the bank-line, and can withstand the direct impingement of the flow on the bank. Retards have been effective for small to medium width channels (flow velocity not exceeding about 2 m/s) and for channels with sand-beds.

The required permeability of a retard proportional to the radius of curvature of the bend so that sharper the bend, greater should be the impermeability of the retard to control lateral erosions.

A retard is best suited in humid regions where natural vegetation would help the bank-slope-stabilization, but is susceptible to damage by ice and debris in cold regions.

(d) Guide Bunds (see Fig. 10.37): The functions of the Guide Bunds is to prevent erosion at abutments from flood flow concentration at upstream face of embankments, and to redirect flood water parallel to the desired channel alignment so that the full bridge water-way-opening is utilized. The guide bund does not eliminate scour, but reduces the scour-damages, and moves the scour-holes away from the abutment.

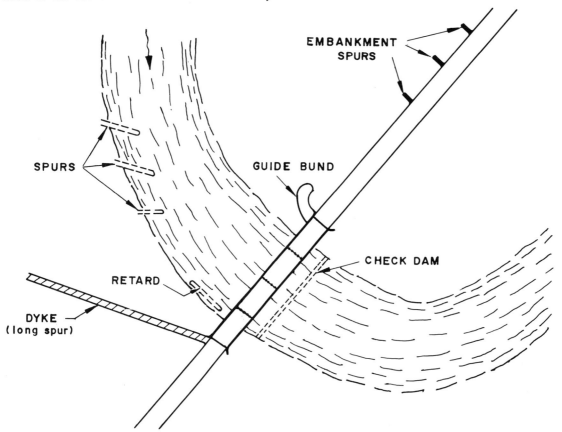

Fig. 10.37 Placement of flow control structure

Most commonly used guide bund is a revetted earth or rock embankment, with quarter of an ellipse shape in plan. It is recommended also for the control of meanders migrating towards bridge or embankment.

(e) Check Dams (see Fig. 10.37): A check dam has been a successful method for protecting bridge foundations by maintaining original stream-bed-grade. When it is constructed just downstream of bridge, it may cause downstream erosion. Therefore, care should be exercised in designing check dam to minimize the downstream degradation. It can be built of sheet pile, rock riprap, gabions, concrete or concrete-filled fabric mats, etc.

As for countering the effect of bed-degradation on bridge foundations, the stabilizing measures are:

— underpinning them
— jacketing them
— armouring by laying riprap, concrete-filled sacks or mats, very large concrete cubes (1 to 2 m side dimension), or concrete tetrahedrons, on the bed around the foundations

10.15 COUNTER-MEASURES AGAINST BED-AGGRADATION

Certain channel beds tend to aggrade (i.e. raise their level) in time due to various causes imposed upstream and downstream of the bridge. This action can raise the water level, consequently reducing the vertical clearance below soffit of bridge deck, and flood the approaches.

Counter-measures against aggradation include the following

(a) Channelization of the bed
(b) Bridge modification (raising or replacement)
(c) Continuous dredging maintenance

(a) Channelization: This appears to be the best solution to remedy aggradation problems, but extensive channelization (including dredging and clearing poor bed materials) has not been a complete success because it does not alter the natural instability of the river system.

(b) Bridge Modification: Increasing span length or raising superstructure can be very costly.

(c) Dredging Maintenance on Routine Basis: Routine maintenance of the stream bed is most cost effective where aggradation is caused by a temporary source. As an alternate to continuous maintenance, controlled sand and gravel mining can be done from debris-basin upstream of the bridge, but caution should be exercised and detailed analysis should be made not to alter the natural balance of the stream system.

11

STRUCTURE DEFICIENCIES, REMEDIES AND PREVENTIONS—A SUMMARY

11.1 SUMMARY OF VARIOUS POSSIBLE STRUCTURE DEFICIENCIES, REMEDIES AND PREVENTIONS

Presented here as a ready-reckoner for quick reference (summarized in a tabular format) are the various possible deficiencies in various components of a bridge, their possible remedies, and the preventive actions against them.

This tabular presentation is purely a re-statement of various details already covered so far, but when presented in this summary form, reference becomes far more convenient.

Note: In Table 11.2 at the end of this chapter are indicated the probable Weightages of various causes of damage in various elements of a regular concrete bridge in order to assist in 'feeling' as to where can the bridge be more susceptible to damage.

Table 11.1   Structure Deficiencies, Remedies and Preventions

Structure Subsystem	Structure/ Component	Deficiency	Remedy	Prevention	Remarks
Wearing Surface	—Asphalt	Map cracking (alligator cracking)	Remove to deck and replace	Early detection and repair	
		Edge cracks, reflection cracks, lane joint crack, corrugations	Clean thoroughly, fill with emulsion or liquid asphalt and seal and treat after curing with liquid asphalt and dry sand	Early detection and repair	Observe for cause and whether they result from deck slab or super-structure distress
		Shrinkage cracks and slippage cracks	The same as above, followed by a surface treatment of the entire surface	Early detection and repair	Same as above
		Distortion	Determine limits, apply tack coat, spread dense graded asphalt concrete and roll. Then sand seal.	Same as above	Same as above
		Disintegration	Potholes same as alligator cracks. For ravelling, clean and fog seal: apply surface treatment after curing	Same as above	Same as above

(Table contd.)

Structure Subsystem	Structure/ Component	Deficiency	Remedy	Prevention	Remarks
	—Concrete	Scaling	For light scaling seal with concrete curing membrane. For medium, heavy and severe scaling 'sound' the area to locate extent, remove unsound material with hammer and chisel, and clean thoroughly to a depth guaranteeing good bond. Prime with grout and fill with mortar, sealing the joints with thin grout	Chlorinated rubber - curing compounds or waterproofing membrane covered with asphalt concrete overlay (not if steel corrosion is detected)	Prior to all preventive actions make sure the steel reinforcement is not corroded.
		Spalling	Survey area for extent and check with voltmeter for corrosion. If area is small (and no corrosion) repair as in scaling. If area is large, scarify to 6.25 mm chipping out unsound material and clean with sand and air blast. Coat area with bonding grout and place new concrete (with water reducing admixtures and accelerators) firmly into cleaned area and cure	Same as above	Same as above
		Cracks	Seal with epoxy concrete compounds. Same as above Severe cracking should be sealed with a good waterproofing membrane and overlayed with bituminous wearing course if cracking is not associated with corrosion. Otherwise apply remedial action of large areas of spalling above.	Same as above	Same as above

(Table contd.)

Structure Subsystem	Structure/ Component	Deficiency	Remedy	Prevention	Remarks
Deck Slabs	—Concrete Deck slabs	Scaling	Thin epoxy mortar patch. Proprietary fast setting cements can also be used	Air-entraining concrete, good construction deck drainage and deck flushing	
		Cracking	If dormant crack— inject low viscosity epoxy. If live crack; first check reasons. If no hazard, epoxy can be injected. Flexible fillers and mastics can be used. Enough concrete may be removed for a full depth patch, depending on size and extent	Same as above. If due to approach pavement thrust, develop relief joints	
		Spalling	For early spalling patch with polymer modified or fast setting concrete, as well as 'controlled set portland cement'. For advanced spalling use bituminous concrete wearing surface over a waterproofing membrane, a P.C. concrete mortar or latex modified concrete. For very advanced spalling replace deck. If combined with corrosion of reinforcement, use cathodic protection	Same as above, plus more impermeable concrete, protective waterproofing, good design, cathodic protection and polymer-impregnated concrete	Do not cover with protective waterproofing, if reinforcement is corroded
	—Orthotropic ribbed steel decksheet	Failure of the welds	Air arc gouging and rewelding	Frequent and close weld inspection	
		Corrosion at the interface	Remove wearing course, provide waterproofing membrane and resurface	Provide waterproofing during construction and restore on a 2-year basis	
	—Steel Grid Decksheet	Failure of welds	Correct failure by rewelding. For extensive failure, the replacement of sections may be required	Frequent and close weld inspection	

(Table contd.)

Structure Subsystem	Structure/ Component	Deficiency	Remedy	Prevention	Remarks
Expansion Joints	—Finger joint	Debris and moisture accumulation	Clean by flushing	Correct drainage and resultant accumulation	
		Loosening of beams and supporting angles	Remove loose or faulty bolts or rivets and rebolt. Correct any breakage of deck concrete. Also rebuild device, if broken and re-align if mis-aligned.	Adequate design. Frequent and close inspection.	
		Jamming or locking of joint	Reverse shift or trip expansion fingers. If due to pavement thrust, provide relief joints	Same as above	
	—Armoured joint	Leakage	Remove debris from opening and install preformed compression joint	Reseal regularly	
		Wedging of debris	Remove and clean. If crushing of deck edges remove and replace damaged concrete. During repair provide increased joint opening	Clean regularly	
		Loosening of supports	Cleaning joints, repositioning super-structure and sealing joints	Same as above	
	—Sliding plate joint	Permanent deformation and cracking of plate	Remove exp. device and unsound concrete and repair fully by repositioning	For T-on-an-angle section, support plate on non-shrinkage concrete and steel supports.	
		Closing of joint and damage to steel anchorage	Trim sliding plate, if no hazardous substructure movement	Periodic flushing	
		Differential movement of plates	Arrest movement		
	—Elastomeric joint	Water leakage	Replace neoprene material/assembly	Close supervision during construction	
		Wear of neoprene material	Same as above	Same as above	
		Popping up of plugs	Replace plugs	Same as above	
		Failure of sealant	Place new sealant	Same as above	
		Loosening of support	Remove broken or loose bolts and re-establish device	Same as above	

(Table contd.)

Structure Subsystem	Structure/ Component	Deficiency	Remedy	Prevention	Remarks
	—Compression seal joint	Breaking of bond between compression seal and joint face	Remove and instal a wider seal	Proper design and installation	
		Wear of the neoprene material	Replace neoprene material	Same as above	
		Forcing of neoprene above road surface	Same as above plus redesign new element	Same as above	
	—Gland sealant joint	Accumulation of debris	Clean and instal new sealant. Repair joint edges, if required	Same as above	Examine whether repositioning of the superstructure has occurred and restore or prevent
Primary Members	—Steel members	Rusting of element	If cross-section area reduced repair by adding web stiffener or flange cover	Frequent inspection and painting	
		Cracking	Immediate repair. Repair to be designed according to position and nature of crack.	Use details and techniques that prevent high fatigue stresses	May need to stop traffic or drill hole to delay advance of crack
		Impact damage	Depending on extent, straighten, strengthen or replace member	Protective guardrail and reflective painting (as convenient or possible)	
	—Concrete members	Deterioration (as in deckslab concrete)	Same as in deckslabs; may require formwork and supports		
		Deterioration of the bottom slab or beams	Very difficult particularly in tension area. Require relief of dead load and repair as required	Good drainage to prevent water from reaching bottom slab	
		Impact damage of stringers or girders	Especially in T-beam construction. If no reinforcement is broken can be repaired in place. Otherwise, the jacking or falsework and the replacement of the beam cannot be avoided.		
		Uncontrolled discharge from drains	If only light scaling, sandblast and seal with epoxy. Deep scaling requires patching to restore cover of the reinforcement	Extend deck drains below beam level	

(Table contd.)

Structure Subsystem	Structure/ Component	Deficiency	Remedy	Prevention	Remarks
Secondary Members	—Steel members	Corrosion of members	If cross-sectional area reduced repair as per primary member.	Frequent inspection and painting	
		Cracking of diaphragms and bracing	Same as primary members, only may not require the closing of the bridge.	Use details and techniques that prevent high fatigue stresses.	
		Collision damage of exterior struts	Straightening, strengthening or beam replacement	Protective guardrail and reflective painting (as convenient or possible)	
	—Concrete members	Deterioration due to spalling of concrete and corrosion of the reinforcement	Repairs as in spalling repairs for decks. May require formwork and supports	Good design and maintenance of drainage system	
		Impact damage of exterior diaphragms and struts	Same as primary members only jacking and/or falsework may not be required	Protective guardrail and reflective painting (as convenient or possible)	
Bearings	—Steel plates, rollers, rockers, fixed and sliding bearings, pot bearings, etc.	Build up of debris	Clean up to prevent build up	Frequent inspection and maintenance	
		Loss of protection system (paint or galvanizing)	Clean up and replace protection system	Same as above	
		Corrosion and steel delamination	Clean and repair if minor, if severe to prohibit function replace bearing	Same as above	
		Sheared and/or heavily corroded anchor bolts, retainer plates, etc.	Replace sheared and/or heavily corroded parts	Same as above	
		Disintegration and/or deterioration of the material adjacent to the bearing	Remove defective material and replace with new. This may require properly designed support system	Frequent inspection and maintenance	
		Movements of bearing	Unless danger of immediate collapse, no action—in-depth investigation	Same as above	
		All of the above plus stress cracks in structural steel	If no danger of immediate collapse, no action—in-depth investigation	Same as above	

(Table contd.)

Structure Subsystem	Structure/ Component	Deficiency	Remedy	Prevention	Remarks
		Bearing immobilized due to corrosion or debris	Clean bearing and area around it. Repair and replace protective system (paint) Lubricate, if applicable	Same as above	
	—Pot bearing	*In addition* to its previously stated deficiencies it may have —non-uniform compression —leaking piston seal —failure between teflon and substrate —failure between stainless steel and sole plate —cuts or deterioration of teflon —deterioration of stainless steel plate	In-depth investigation for any of these additional deficiencies separately and in coordination with each other	Good design and construction; and frequent inspection and maintenance	
	—Elastomeric bearings	Build-up of debris	Clean up to prevent build up	Frequent inspection and maintenance	
		Disintegration and/or deterioration of material under or near the bearing	Remove deflective material and replace. This procedure may require a properly designed temporary support and jacking-up arrangement of deck	Frequent inspection and maintenance	
		Excess shear deformation, non-uniform compression, ozone cracking, failure of bond between elastomeric element and plates, and rounding of the edges of the elastomeric element	If no immediate hazard, no action. In-depth investigation	Same as above	
	—Pin and hanger bearings	Loss of protection system (paint)	Clean, and repair or replace deteriorated protection system.	Same as above	
		Corrosion and delamination	Repair minor deterioration after thoroughly cleaning and priming	Same as above	
		Cracking or pin failure	Replace pin using an appropriately designed support system	Same as above	

(Table contd.)

Structure Subsystem	Structure/ Component	Deficiency	Remedy	Prevention	Remarks
Foundation: Piles and Caissons	—Exposed concrete piles and caissons	Scaling or spalling	Deteriorated concrete removed, reinforcement cleaned and concrete restored, using non-shrink concrete mix. Also protect exposed part with collar protection	Design to withstand erosive forces	
	—Exposed steel piles and caissons	Corrosion of metal parts	If little, clean, paint and water proof. If loss of section, reinforce, splice or provide concrete splice above and below corroded part	Provide protective collar in the wet-dry cycle sections	
Foundation: Spread-Footings	—Concrete footing	Deterioration of the concrete	Chip away deteriorated section and air blast clean. Install formwork and/or anchors as required. Restore footing to its dimensions using low slump concrete. Dam water away from repair, if applicable. Do not prime with shot-crete	Prevent exposure of footing with good foundation protection	
		Footing cracks	If structural design & details ok, then, Open V-shape channel along crack, flush and blow out clean. Fill crack with cement-and-fine-sand-grout. Then clean and fill Vee with grout, using a bonding compound or a neat cement paste for bonding	Frequent inspection of the footings	
Abutments and Piers	—Concrete abutments and piers	Settlement or movement	Cause shall be determined. If stabilized, re-level the bridge—May be serious enough to close bridge and investigate		

(Table contd.)

Structure Subsystem	Structure/ Component	Deficiency	Remedy	Prevention	Remarks
		Vertical cracking	Determine cause. If pile supported, may need to drive additional piles and tie them to structure. If spread footing, underpin. Both of the above may additionally require backfill, jacking of the superstructure resetting of bearings, adjusting exp. mechanisms		
		Surface deterioration	Same as in other concrete components. If corrosion of reinforcement present, provide cathodic protection		Cathodic protection for substructure has already been used and can functions well for many years (2 to 10) but can be cumbersome
		Deterioration at the waterline	Same as above, except where water is present (coast areas), temporary removal of water is required	Use protective non-corrosive fender construction	
		Spalling under pressure	Determine cause. Relieve pressure. Subsequently repair spalling as in other concrete components.	Proper design and construction	
		Backwall undermining	Replace the joint, if joint generated then backfill and compact behind joint	Design good joint drainage	
	—Stone masonry abutments and piers	Mortar deterioration	Cut out areas of mortar deterioration, replace joint material. Where bonding may be difficult an encasement of reinforced concrete may be used	Frequent inspection and maintenance	
	—Sheet piling abutments	Damage or corrosion at the wet-dry cycle zone	Remove and replace damaged piling. Remove and replace also deteriorated backing material	Coat piling with waterproofing material to prevent corrosion	

(Table contd.)

Structure Subsystem	Structure/ Component	Deficiency	Remedy	Prevention	Remarks
	—Mechanically stabilized earth walls (e.g. Tied-back)	Construction inadequacy	Good construction practice	Good construction practice	
		Erosion of backfill	Remove, replace and recompact eroded backfill	Verify reason for erosion and remedy	
	—Steel frame piers	Corrosion	Remove rust and protect by painting. If section is lost reinforce with additional planting	Protect with paint or other waterproofing agent	
	—Pile bents	Impact damage	If concrete, remove damaged concrete, clean reinforcement, apply bonding compound and replace with low shrinkage concrete. May require support or falsework	Protect with guardrail or paint with reflective paint	
		Corrosion of steel members	As per corrosion of steel frame piers	Protect with paint or other waterproofing agent	
Pier-Caps and Bridge Seats	—Concrete pier-caps and bridge-seats	Concrete deterioration and corrosion	Construct a temporary bent (or suitably arrange) to transfer the load and repair concrete and steel deterioration as required in previous components	Reduce leakage of joints and of the drainage system. Inspect and flush drainage system frequently	
	—Steel pier caps	Corrosion	Remove rust by sand-blasting or chipping hammer. If significant loss of section—plate, reinforce or replace, as required	Proper painting schedules and frequent inspection	
Footpaths (sidewalks) and Fascias	—Concrete footpaths (sidewalks) and fascias	Loss of structural ability	Same as in concrete primary members	Good design and construction	
		Material deterioration	Same as in concrete decks	Good design and construction	
		Signs of differential movement at joints	Do nothing, if serious, do in-depth investigation		
		Impact damage	Restore as per deck repairs	Use reflective paint on vertical face	
	—Steel Sidewalks and Fascias	Corrosion	Clean, and restore the protective system (paint)	Frequent inspection and painting	

(Table contd.)

Structure Subsystem	Structure/ Component	Deficiency	Remedy	Prevention	Remarks
		Loose connections	Tighten or replace rivets and bolts		
		Impact damage	Restore as per deck repairs	Use reflective paint on vertical face	
Railings and Parapets	—Concrete or steel parapets and railings	Same as in sidewalks and fascias	Same as in sidewalks and fascias	Same as in sidewalks and fascias	
Drainage System and Utilities	—Bridge drainage	Clogging	Clean inlets, downspouts and dams as required	Well designed and constructed drainage system	
		Corrosion	Replace corroded element. Reconstruct, to avert potential future clogging	Frequent inspection and painting	
		Short Scupper Pipe	Replace or extend as required		
	—Utilities	All types of verified malfunction	Inform the appropriate utility company (for power, for gas, the appropriate Municipality for water and sewage, etc.)	Good and timely inspection and maintenance	
Streambed and Banks	—Streambed and Banks	Erosion	Remedy by regrading, compacting and protection	Appropriate stream protection	
		Deposition	Removal beyond active Wadi bed	Appropriate stream protection	
Culverts	—Cast-in-place or precast concrete culverts	Cracking	Same as in footing cracks	Early detection and repair	
		Undermining of Inlet/Outlet	Replace lost material, underpinning of culvert may be required	Frequent inspection and early repair	
		Floatation or displacement	Complete replacement of structure.	Good foundation construction	
		Joint defects (infiltration/ exfiltration)	Seal cracks or joints as required	Good joint drainage designed and constructed	
		Surface deterioration	Sealing or spalling corrected in accordance with similar work. Formwork will also be required	Good design and maintenance practice	

Table 11.2 Weightage of Probable Causes of Damage in Different Bridge Elements

Damage Appearance in Bridge-Element	Weightage of Probable Cause of Damage			
	Struct. Defic. (%)	Corrosion (%)	Other (%)	Total (%)
—Surface of Bridge (Cracks in surfacing)	1	—	1	2
—Expansion Joints (Loose expansion joints, other distresses)	—	—	1	1
—Parapets and Railing (Cracks in concrete)	1	2	—	3
Sidewalk/Median (Cracks in concrete)	1	1	—	2
—Wing Wall (Cracks in concrete; any movements to be monitored)	3	1	—	4
—Abutment (Cracks in concrete)	2	1	—	3
—Bearings (Displacement, distress, Wrong position)	1	—	1	2
—Edge Beam (Parapet Beam, etc.) (Cracks in concrete)	—	2	—	2
—Deck Slab (Cracks in concrete)	5	2	1	8
—Primary Members of Deck (Shear and flexural cracks, cracks in slab soffit, cracks in cantilevered part, etc.)	50	8	—	58
—Pier Caps (Shear and flexural cracks, cracks in concrete)	3	2	—	5
Piers (Cracks and spalling, impact)	1	8	1	10
TOTAL	68	27	5	100

12
BRIDGE MONITORING

12.1 GENERAL

After the rehabilitation–strengthening of the structure is completed, it is essential that it be kept under observation and its conditions monitored regularly so that any distresses are located promptly and corrective measures taken well in time. It is essential that the form of monitoring is specified and inspections carried out according to a calender which should be prescribed. The various methods of monitoring the bridge structures are briefly indicated below.

12.2 METHODS OF MONITORING

During the distressed stage of the bridge and after the distressed bridge has been rehabilitated or strengthened, it is necessary to carefully monitor its behavior for a certain period of time to ascertain its performance and the efficacy of the measures adopted. The monitoring would involve carrying out certain laboratory and field tests as well as condition-surveys and measurements to detect strains, movements, changes in reactions and deformations.

The usual methods adopted for monitoring the behaviour of a structure are:

(a) Observing deflection by periodically taking levels. Deformations can also be monitored by water-levels in tubes connected to tank filled with water.

(b) Visual observations for cracks, deflections, overall integrity, profile, functioning of bearings and hinges, corrosion-stains, etc. Particular note must be made of the cracking pattern, the crack widths and lengths, and whether cracks are due to plastic-settlement or plastic-shrinkage of concrete when the concrete was still plastic, foundation settlements, structural deficiency, reactive aggregates, corrosion, sulphate attack, etc. Signs of reactive aggregates, delaminations, spalling, hollow or dead sound when tapped with hammer, honeycombing and expansion of concrete should also be observed. Frequency and levels of inspections have to be specified, depending on individual case circumstances.

(c) The change in the width of the cracks with the passage of time and live load need to be observed through 'tell-tales' to know whether the cracks are static or live, or even increasing.

(d) Plumb-bobs can be used to measure deviation from plumb. Special tilt-meters or inclinometers also could be used for detecting tilts and inclinations.

(e) Opening of Joints, particularly, near the hinges, Expansion Joints, etc., needs to be observed.

(f) Redistribution of support-reactions may also be measured in some cases to ascertain impending behaviour.

12.3 INSPECTIONS

The first and the foremost requirement is to carry out principal inspections at more frequent intervals than those for normal structures, say immediately after distress is noticed and on completion of the remedial measures and, during the service of the structure at short intervals (bimonthly, half-yearly, yearly depending on the case. These need to be repeated often after carrying out some of the investigative tests, particularly when any signs arousing suspicion are discovered. Use of mobile inspection units to have an access to each and every part of the bridge is a must for the principal inspections. The techniques and underwater inspection described earlier should also be adopted. (Details have been given in Chapter 5.)

12.4 CORROSION MONITORING

The use of permanent electrodes for accurate measurement of the corrosion potential of steel in concrete is sometimes necessary. Use of current-density or rebar-probes and corrosion-rates-monitoring-probes can be made to meet the particular requirements. Careful selection of permanent monitoring equipment is required. The locations should be kept to minimum and should be in the areas of most-active-corrosion-rates. Relatively thin steel wires are embedded in the structure near the reinforcement with permanent electrical connections to the tell-tales so that electrical resistance can be measured. Corrosion of tell-tales would cause an increase in the electrical resistance. Certain devices can be permanently embedded in the concrete for facility of subsequent measurements of the extent and the rate of corrosion in future years. However, the evaluation of such instruments is yet to be perfected. A new probe has been recently developed for evaluation and control of steel corrosion in marine concrete structures to provide information about the corrosion conditions for both the embedded and the exposed steel. The probe gives information on the passivity of the embedded steel, electrical resistivity and the oxygen available as well as the corrosion rates. (*See* details in Chapter 15.)

12.5 STRAIN MEASUREMENTS

The measurement of strains at critical sections or joints is another method of monitoring the behavior of critical structural elements. Strain gauges are fixed at predetermined points (Sometimes dial-gauge using Demec gauges are also used. However, the experience shows that these gauges do not work efficiently in outdoor atmosphere.)

12.6 USE OF LASERS

Application of lasers in structural monitoring is finding increasing use. In its simplest form the system consists of 'threading' a laser-beam though series of apertures in the plates fixed along the length of the beam, say along the soffit of a girder or soffits of series of adjoining girders—Fig. 12.1. Similarly, a laser-beam can also be directed vertically along the piers and abutments. This

beam, after passing through a series of apertures in place fixed along its path, reaches the light-sensitive receivers at the farthest end. The failure of the beam in reaching the receiver requires further investigation because it could be due to some structural deformation of the members supporting the plates or due to some other reason which may uncover an impending distress.

A system of a series of such laser-beams can be provided in a structure, and an arrangement made to sound an alarm in case of blockage of light of any laser beam. Further refinement of the system could be made by attaching detectors to the structures along the path of the laser-beam, whereby any movement of the structure at the location of each detector would be continuously tracked relative to the structure at each detector-location, measured, recorded and analyzed with the help of a programmed computer (controlling timing and the operational sequence of the various detectors). Readings to the accuracy of even 0.1 mm are possible and continuous and constant 24 hours-a-day monitoring of a structure for its integrity and soundness is possible.

12.7 VIBRATION CHARACTERISTICS

Measurements of vibration characteristics of a structure can also be adopted in some cases for monitoring the continued structural integrity. Surface-mounted vibrating-wire-gauges can help in monitoring the frequencies.

12.8 INSTRUMENTATION

Instrumentation can be provided for proper monitoring of (particularly long span) bridges to study their behaviour during their service. The measurements may include strength, settlements, temperature effects, deflections, movement at hinges, etc.

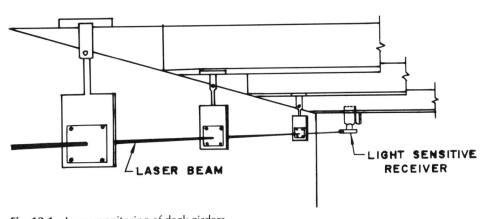

Fig. 12.1 Laser monitoring of deck girders

12.9 MONITORING SOIL-BEHAVIOUR AND FOUNDATION DISTRESS

12.9.1 General

Mointoring the behaviour of the soil and the structure, by using instruments, is necessary for the following purposes

- To diagnose the specific nature of distress and provide economical remedies
- To evaluate adequacy of new construction methods and construction control
- To provide safety
- To verify satisfactory performance and design adequacy

When a bridge or a wall shows evidence of soil distress, remedial action needs to be taken to prevent further damage. These measures will depend on the magnitude and the location of any movement of the structure as well as other evidences of structural or soil failure. Without proper and accurate information, most engineers will use a conservative solution. When proper instruments are installed, the engineer can collect necessary data to select an economical remedial measure. Also, the instruments can possibly be used to monitor the effectiveness of any repairs.

The following variables may be measured by using appropriate instruments:

Variable	Instrument
— Deformation	<ul><li>Survey</li><li>Vertical pipe–guage</li><li>Settlement platform</li><li>Portable gauge</li><li>Heavy gauge</li></ul>
— Horizontal Movement	<ul><li>Survey</li><li>Inclinometer</li><li>Crack gauge</li></ul>
— Tilt	<ul><li>Tiltmeter</li></ul>
— Groundwater and Pore Pressure	<ul><li>Observation-well Piezometer</li></ul>
— Load and Strain	<ul><li>Calibrated jack</li><li>Load-cell</li><li>Strain-gauge</li></ul>
— Earth Pressure	<ul><li>Earth pressure cell</li></ul>

12.9.2 Geotechnical Instrumentation

(a) Monitoring Deformation
The movement of a soil can cause vertical settlement, horizontal deformation and tilt. Normal survey methods can be used to monitor the magnitudes of horizontal and vertical deformations of the structure and of the ground surface. Other portable equipments and gauges, which can be used to monitor deformations, are graduated-scale, survey-tape, wire, calipers, dial-gauge and crack-gauge.

(b) Monitoring Settlements
The most widely used instrument to measure settlement is a vertical-pipe-settlement-gauge, commonly known as settlement- platform.

(c) Monitoring Horizontal Movement (Fig. 12.2)
To measure horizontal movement of a soil mass, an inclinometer is used. A simple 'Poor-man's' type inclinometer is inexpensive and consists of thin wall PVC pipe containing a series of rigid rods of various lengths. After inserting this instrument in a vertical hole, as any horizontal movement occurs, a rod will move until it can not move anymore so that the series of rods will form the curvature or bend to show horizontal movement! (Schematic in Fig. 12.2.)

(d) Monitoring Tilt
A tiltmeter is used to monitor the tilt of bridge substructure and walls.

(e) Monitoring Groundwater Level and Pore Pressure
The ground water level and the pore pressure indicate the flow of water through soil. They effect the strength, pressure, stability and settlement of the soil. Monitoring the deformation and pore-pressure will give valuable information for planning and designing counter-measures against foundation-distress, rock-slope failure, embankment-slide, etc. The pore pressure can be monitored by an inexpensive 'observation-well', which is simply driven in the place where monitoring is required. Open stand pipe and closed hydraulic type piezometer are two types of simple and inexpensive piezometers. (*See* Fig. 12.4)

(f) Monitoring Earth Pressure
To monitor the magnitude and distribution of earth pressure on walls and bridge abutments a pressure-gauge can be used.

(g) Monitoring Structural Integrity through Measuring Loads and Strain
To establish the integrity of the structural system, it is necessary to take load-cell and/or strain-gauge measurements for instance, for piles, tie-back anchors, rock-anchors, etc. Pile load tests are generally performed by using a calibrated jack along with load-cells and strain-gauges.

Load-cell measurements are needed for tie-backs and anchor-rods to determine whether the required anchorage has been attained? These measurements are

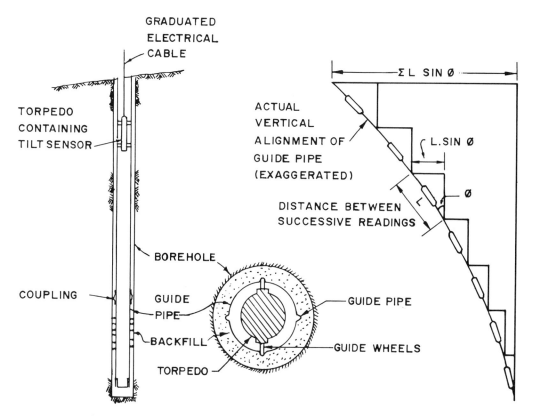

Fig. 12.2 Inclinometer

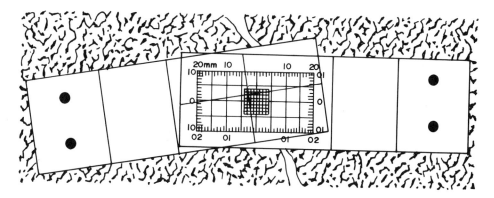

Fig. 12.3 Crack gauge (Avongard calibrated crack monitor)

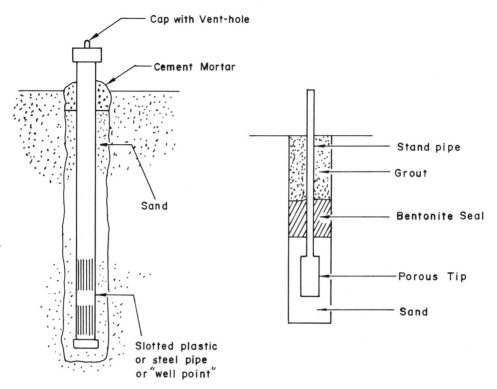

Fig. 12.4 Well and piezometer

also used to monitor any excess or loss of tension force in the rod (due to anchorage slippage, improper installation, and soil movement). Strain-gauges will show any reactive deformation of bridge piers, retaining walls, piles and bridge members.

(h) Implementation
The implementation of geotechnical instrumentation is usually undertaken by geotechnical engineers but the necessity of such instrumentation should be kept in mind during the project development. The following steps are suggested for any program of geotechnical instrumentation

1. Define the condition and problems
2. Define the purpose of the instrumentation
3. Select the variable to be monitored
4. Make prediction of behaviour
5. Select instruments and install them
6. Observe and record data
7. Analyze and evaluate the data
8. Develop countermeasures

12.10 TRAINING

Monitoring of distressed bridges as well as rehabilitated bridges requires certain amount of skill and specialization. The engineers maintaining and inspecting such bridges, therefore, will need to be trained for such jobs.

12.11 DATA BANK

Monitoring also requires setting up a data bank as a reference frame.

13
LOAD CAPACITY EVALUATION OF EXISTING BRIDGES

13.1 INTRODUCTION

Often times it become incumbent to estimate the load carrying capacity of an existing bridge, particularly if it has developed material-deterioration and loss of section-strength, be these due to poor condition-survey (inspection) and consequent lack of maintenance (repair/strengthening), accidental damages, overload, under-design, poor quality-construction or poor-detailing in the first place.

The methodology for rating the live load capacity of an existing bridge has to allow for combining the probability theory, statistical data and engineering judgement into a rational decision making tool. In particular, the procedure has to allow the engineer to use specific information from actual detailed inspection of the structure in a manner so as to permit improving his judgement. This has been proposed in the 'AASHTO: Guide Specifications for Strength Evaluation of Existing Steel and Concrete Bridges 1989'. In the following, these Guide Specifications of AASHTO have been adopted with grateful acknowledgement. The abbreviation GSSE is used when referring to these AASHTO Specifications.

These guidelines have been developed by the AASHTO for evaluation of almost all types of bridges in the United States. Current American bridge evaluation practices were considered in developing the methodology. Although the basic concepts can be universally applied, it would be prudent to consider the effects of local practices on the load limit values obtained by the methodology before applying these guidelines to bridges outside the United States.

This methodology is intended for evaluating almost all existing bridges. Steel spans include simple and continuous girder bridges and trusses and floor systems. Concrete spans recognized include slab, girder, T-beam and box-beam bridges with short to medium span lengths. Prestressed beams, although of recent vintage, are also included here.

NOTE: The above referred Guidelines, GSSE, are intended by AASHTO to produce Rating Factors (R.F.) for routine evaluation and 'posting' considerations. For the purpose of finding as to whether a particular special Live Load Truck can be allowed on a bridge (with some constraints if necessary, e.g. reduced speed, travel-path, etc.), the Load-Factors described in GSSE may be suitably altered (using engineering judgement). For such cases, the exact actual truck details, viz., size, weight and configuration of axles and wheels, should be used as the 'prevailing Truck Load' and then the Rating Factor calculated. Only if this is found acceptable, can the special permit for such a truck be issued. (Procedure for evaluating the Rating Factor is described ahead.)

13.2 CAUTION

The procedure for Rating the Live Load Capacity of an existing bridge requires knowledge of both the actual physical condition of the bridge as well as of the actually applied loads (dead load and the prevailing Live Load). A safe level of rating a bridge pre-supposes that the 'nominal' section-strengths should be estimated based on a detailed inspection/investigation of the actual physical condition of the bridge, taking due account of any repair/strengthening 'already carried out' and 'whether-will-be-carried-out-on-a-continuing-basis'.

Further, knowledge of actual traffic loads and 'degree of control' on truck-weights, their axle-loads and configuration, including signs of over-weight vehicle-combinations, and use of more accurate methods of structural-analysis for apportionment of Dead and Live Load Effects, will render a more realistic rating of the bridge. These points are of paramount importance and cannot be over-emphasized. The Load-Factors (σ) and the Resistance-Factors (ϕ) that must be applied, should rationally recognize the corresponding uncertainties in making these judgements on section-strength, accuracy of analysis and load effect apportionment and the loads 'themselves'.

13.3 THE RATING EQUATION

13.3.1 The evaluation of a structure is based on the simple principle that the available capacity of a structure to carry loads must exceed the capacity required to support the applied loadings! To perform an evaluation, therefore, it is necessary to know something about the available capacity, the applied loading and the response of the structure to that loading. Knowledge and information with respect to each of the these items is never complete; and therefore, evaluation can never be done precisely!

To compensate for this lack of knowledge and information, engineers have used safety factors to insure that failure does not occur!

In recent years, the load factor method has been introduced in design and rating to provide more uniform safety. The method implicitly recognizes that dead load effects may require lower safety margins than comparable live (truck) load effects due to the latter's higher relative uncertainty. This probabilistic approach to safety is logically extended in the load factors and resistance factors used here.

The rating check is done by comparing the factored load-effects (both dead and live) with the factored section-resistance at all critical sections. The comparison indicates the rating factor, which determines the suitability of the given bridge for the prevailing loads. If the bridge rating is not acceptable, several options for a more detailed analysis are given. Each of these options are associated with an increasing level of effort and may be done if the rating engineer warrants their use. An initial screening level, however, is provided for routine investigation.

13.3.2 The evaluation is carried out, in principle, by comparing the 'factored Dead and Live Load Effects' with 'factored structural strength', at critical sections, in bending, shear, etc.. The load factors on Live load and Dead Load are so selected as to account for uncertainties in them and the inaccuracies in the method of 'structurally analyzing their effects'.

The load factor on dead load includes the possible variations in section dimensions, details and densities. The load factor on live load accounts for possible uncertainties in the expected maximum vehicle load effect, impact and transverse distribution of live load, during the time-period between inspections. The factor applied on structural strength, called Resistance Factor ϕ, also called 'strength reduction factor' or 'the fraction to be applied for evaluating the dependable value of the section-strength', accounts for uncertainties in the strength-

prediction-theories, material-properties, and deterioration influences over time periods between inspections. Furthermore, the 'Load-Factors' and the 'Resistance Factors' are so selected as to provide an overall safety margin which would lead to an adequate level of safety, considering the aforementioned uncertainties and deficiencies. If the Rating Factor then works out to an acceptable value, the bridge can be posted as safe for the prevailing live-load.

13.3.3 Safety and Economics

Each of the steps in the evaluation process may be performed in any one of several ways. The GSSE-proposed guidelines are general enough to accommodate the practices of different engineers and/or agencies. The load factors and resistance factors presented in the guidelines were developed on the principle that the accuracy of an evaluation was dependent, in part, on the methods used to perform the evaluation. For economic reasons, it is desirable to keep the evaluation effort to a minimum. If the capacity of a bridge can be shown to be sufficient by making some reliable approximations, there is no need to resort to an extensive evaluation procedure. On the other hand, if the sufficiency of the bridge cannot be reliably established using an approximated method, then the engineer may wish to resort to a more sophisticated approach in order to demonstrate the sufficiency of the bridge. Therefore, the evaluation process outlined in the guidelines is a cyclic process in which one or several of the steps may be repeated in what may then seen a cascading process within acceptable bounds of assumptions for analysis.

The various options provided in the guidelines along with corresponding load/resistance factors have been developed so as to maintain an adequate level of safety based on calibration with existing performance 'experiences'. The evaluation procedures presented here therefore provide a balance between safety and economics without getting lost in analysis alone!

The lowest Rating Factor produced by the AASTHO Guidelines (GSSE) will by higher for a well-maintained and non-deteriorated bridge having a well enforced traffic. It will fall to a lower (and possibly unacceptable) value if the bridge is heavily deteriorated or has non-redundant components subjected to heavier and higher density truck traffic. A gradation of rating between these two extremes will be obtained, depending on the condition of the bridge, type and volume of traffic, the quality of inspection and the regularity of maintenance. Thus, a deficient bridge may be made to rate higher if certain preventive measures such as load control, inspection, repair, etc. are undertaken. A variety of options may exist and the engineer could choose one of them depending on the economics of the situation and the amount of effort the authority is willing to expend.

13.3.4 The Equation

The basic structural engineering equation states that the resistance of a structure must equal (or exceed) the demand placed on it by loads. Stated mathematically,

$$R \geq \sum_k Q_k$$

where R = resistance
Q_k = effect of load k

The solution of this simple equation encompasses the whole art and science of structural engineering including the disciplines of strength of materials, structural analysis, and load determination! This equation applies to design as well as to evaluation.

In bridge structural evaluation, the objective is to determine that the 'withstandable load effect' is not less than that due to 'prevailing loads'. The ratio of the 'effects' of these two is defined as the Rating Factor, R. F. Hence, if R. F. is ≥ 1.0, the bridge can 'withstand' the 'prevailing Live Load'. (The 'effects' could be those in bending or shear, etc., as explained ahead.)

Any rational approach to the analytical solution of the basic structural engineering equation requires that the modes of failure be identified and also establish the corresponding section resistances. The location, type, and extent of the critical failure modes must be determined. The checking equation must be solved for each of these potential failure checking modes.

Since neither section-resistance nor the load-effect can be established with certainty, safety factors must be introduced that give adequate assurance that the 'limit-states' are not exceeded. This may be done by stating the equation in a factored load and factored resistance 'format'.

Separate load factors and resistance factors, that will account for each of the major sources of uncertainty, may be introduced in the equation. The basic Rating Equation used in the stated guidelines is simply a special form of the basic structural engineering equation with load factors and resistance factors introduced to account for uncertainties that apply to the bridge evaluation problem. It is written as follows:

$$\phi \cdot R_u = \sigma_D \cdot D + \sigma_L \cdot (R.F.) \cdot L \cdot (1+I) \cdot C_f$$

$$\text{or} \quad R.F. = \frac{\phi \cdot R_u - \sigma_D \cdot D}{\sigma_L \cdot L \cdot (1+I) \cdot C_f} \qquad (Eqn\ 13.1)$$

where ϕ = Resistance Factor or Strength reduction factor or the factor to be applied on the nomi-

nal ultimate strength R_u of the section so as to obtain a 'dependable' value of the ultimate strength of the section (see 13.8 and Table 13.3 ahead)

R_u = Nominal 'ultimate' strength of the section under consideration, in bending, shear, etc., depending on which potential mode of failure is being considered (see 13.7 ahead)

σ_D = Load Factor on Dead Load; may have different values for different dead loads (see 13.5 and Table 13.1 ahead)

σ_L = Load factor on live load (see 13.5 and Table 13.1 ahead)

C_f = Correction factor; which equals 1.0 if transverse distribution of Live loads is done as per AASTHO Bridge Design Specifications; different values for different types of materials, if structural analysis technique is refined e.g. Finite Element Method, etc. (see 13.6 and Table 13.2 ahead)

D and L = Dead load and Live Load 'effects' (e.g. bending moment, shear force, etc.) at the section under consideration, in working load state, analyzed by regular structural analysis; see 13.9 ahead. (If apportionment or transverse distribution of live load effect among the deck longitudinals is done by the Distribution Factors as per AASTHO Bridge Design Specifications, which is quick but approximate, then C_f is taken as 1.0, but if structural analysis is done more accurately (e.g. by Finite Element Method, etc., so that the transverse distribution and material-property considered is automatically more accurate than by AASTHO method of simple 'Distribution Factors') then C_f values shall be taken from Table 13.2 which takes into account the 'material' and the 'bias' in the AASTHO Distribution Factors.)

I = Impact Fraction (see 13.4 (c) ahead)

R. F. = Rating Factor, i.e. the ratio of the 'Withstandable Load effect' to that of the 'Prevailing Load'. The smallest of the R. F. values (in bending, shear, etc.) from among the various critical sections in a span shall represent the R. F. of that span. If this value equals or exceeds 1.00, then the span can withstand the prevailing live load, otherwise the bridge is deficient for that Live load and corrective/preventive actions are needed.

13.3.5 Important Note

The Rating Factor (R. F.) should be calculated at each critical section (of the structural element under consideration) for each potential failure mode, e.g. bending, shear, etc. (Generally the R. F. is calculated only for the deck-girders and deck-slab; but many have to be done even for some substructure elements, though very rarely.) The lowest R. F. value is then assumed as the Rating Factor for the entire span.

Normally only Dead load and Live Load Effects are considered in the R. F. evaluation, no other load effects.

13.4 FIXING THE PREVAILING LOADS

(a) Dead Loads

The dead load shall be estimated from data available from the inspection at the time of analysis. The load factor on dead load accounts for normal variations of material densities and dimensions. Nominal dimensions and densities shall be used for calculating dead load effects. For overlays either cores shall be used to establish the true thickness or an additional allowance of 20 per cent should be placed on the nominal overlay thickness indicated at the time of initial analysis. The recommended unit weights of materials to be used in computing the dead load are as follows

Material	Unit wt (kgs per Cu-m.)
— Asphalt surfacing	2350
— Concrete, plain or reinforced (normal)	2400
— Prestressed concrete	2500
— Steel	7850
— Cast iron	7210
— Timber (treated or untreated)	800
— Earth (compacted): sand, gravel, or ballast	1950

(b) Live Loads

Normally the live load to be assumed on the deck for calculating the Prevailing Live Load effects should be the 'legal' Live Load prescribed for the bridge. Such a Live Load should nearly match 'in effect' the Live Load for which the bridge had been designed, even if its truck-weight, axle-loads and their configuration, may be slightly different. The spacings between axles and axle-weights for this prevailing vehicle should be selected from actual 'Truck Weight Surveys'. This so selected Live Load, called the Prevailing Live Load, may actually well be lighter or heavier than the Design Live Load in a given country, depending on the level and pace of its infrastructural development and the strictness of control exercised on the truck weights. Hence, it is tremendously important to select the Prevailing Live Load Truck (its weight, axle-loads wheels per axle and

axle-spacing) from an actual Truck Weight Survey, even if it turns out to be heavier than the Design Live Load. Then alone can it be uncovered as to whether the bridges being rated are dependable for the actually prevailing Live Load!

In computing the Live Load effects, one vehicle shall be considered in one lane, loading as many lanes simultaneously as would give the most critical effect at a particular section. It is unnecessary to place more than one vehicle in a lane since the load factors, given ahead, have already been modeled by the AASHTO (GSSE) Specifications for this possibility. (Where maximum load-effect at a particular section is produced by simultaneously loading a number of lanes, the reduction factors on Live Load effect shall be operated as is generally specified in the applicable design specifications.)

The actually 'probable' maximum footpath loading should be used in calculations since this will actually vary from location to location.

(c) Impact

An impact allowance shall be added to the Static Live Load effect used for rating the bridge, as indicated in equation 13.1 earlier. Compared to the usual impact allowances prescribed in the applicable design specifications, the actual impact effect may be lesser under enforced speed restrictions. For smooth approach and deck-surface conditions, the impact fraction, I, may be taken as 0.10; for a rough surface with bumps, a value of 0.20 may be more appropriate; and, under extremely adverse conditions, (e.g. high speed, span less than about 12 m, pavement heavily distressed) a value of 0.3 would be more realistic.

13.5 FIXING THE VALUES OF 'LOAD FACTORS' FOR DEAD AND LIVE LOADS

GSSE recommends the Load Factors as shown in Table 13.1 ahead. These are intended to represent conditions existing at the time the AASHTO specification was written based on field data obtained from a variety of locations using weigh-in-motion and other data gathering methods. The load factor on Live Load accounts for the likelihood of extreme loads side-by-side following in the same lane and the possibility of overloaded vehicles. Since one aim of these specifications is to protect the investment in the bridge structure, the load factors for Live Load do recognize the presence of overweight trucks on many highways! An option to reflect effective overload enforcement is contained here by a reduced live load factor. The presence of illegal loads has been noted, and if such vehicles are present in large numbers at the site, the higher load factors may lead to poor i.e.

unacceptable ratings and enforcement efforts should be instituted.

Table 13.1   Load Factors (σ)

Loading	Load Factor σ
Dead Load	$\sigma_D = 1.2$

Allow an additional allowance of 20% on overlay thickness if nominal thickness is used. No such allowance is needed when actual measurements are made for thickness.

Live Load (Category-wise)	
1.	Low volume roadways (ADTT* less than 1000), reasonable enforcement and apparent control of overloads	$\sigma_L = 1.30$
2.	Heavy volume roadways (ADTT* greater than 1000), reasonable enforcement and apparent control of overloads.	$\sigma_L = 1.45$
3.	Low volume roadways (ADTT* less than 1000), significant sources of overloads without effective enforcement.	$\sigma_L = 1.65$
4.	High volume roadways (ADTT* greater than 1000), significant sources of overloads without effective enforcement.	$\sigma_L = 1.80$

NOTE: *If unavailable from traffic data traffic data, Estimates for ADTT may be made from ADT as follows: urban areas, ADTT = 5% of ADT; rural areas, ADTT=25% of ADT. In the absence of accurate data on overloads, a site may be assessed as reasonably enforced if fewer than 5% of the Trucks exceed the local legal gross weight limits.
 • ADT = Average Daily Traffic
 • ADTT= Average Daily Truck-Traffic

13.6   CORRECTION FACTOR (C*f*)

The 'fraction' of vehicle load 'effect' apportioned to a particular longitudinal girder of the deck, owing to Transverse Live Load distribution in the deck, may be estimated in accordance with current AASHTO Design specifications (the simple method of Distribution Factors). In such a case, the value of Correction Factor C_f in the rating equation, mentioned earlier (Eqn 13.1), shall be taken as 1.00. However, if the apportioning of the Live Load Effect is done by adopting a more accurate technique of Structural Analysis (e.g. Finite Element method, plain grid method, etc.), then this shall be reflected by adopting different values for the Correction Factor C_f, as indicated in Table 13.2, 'to adjust for the expected bias in AASHTO Distribution Factors for different material types'.

Table 13.2

Distribution of Load Effects Estimated by	Values of Correction Factor C*f* in case of		
	Steel	Prestressed Conc.	Reinforced Conc.
AASHTO Distribution Method	1.00	1.00	1.00
Tabulated Analysis with simplifying assumptions	1.10	1.05	0.95
Sophisticated Analysis (finite elements, grillage analogy, orthotropic plate, etc.)	1.07	1.03	0.90
Field Measurements	1.03	1.01	0.90

*One example would be the use of the Distribution Factor Charts in the Ontario Highway Bridge Code (Canada).

13.7   EVALUATING THE NOMINAL ULTIMATE SECTION-STRENGTH (R_u) IN BENDING, SHEAR, ETC., OF A CRITICAL SECTION

The nominal ultimate section-strength, R_u (in bending, shear, etc.), of a critical section may be estimated in accordance with the relevant clauses in the applicable Design Specification i.e. Code (as for instance the current AASHTO Design Specifications, etc.). In these calculations, the actual physical section-condition (such as loss of concrete or steel section, etc.) should be taken into account.

(a) Concrete: The strength of sound concrete shall be assumed to be equal to either the values taken from the plans and specifications or the average of construction test values. When these values are not available, the ultimate crushing strength of sound reinforced concrete may be assumed to be 3000 psi. A reduced ultimate strength shall be assumed (no less than 2000 psi, however) for unsound or deteriorated concrete unless evidence to the contrary is gained by field testing.

(b) Reinforcing Steel: The area of tension steel to be used in computing the ultimate flexural strength of reinforced concrete members shall not exceed that available in the section and as shown in the plans. The steel yield stresses to be used for various types of reinforcing steel are given below:

Reinforcing Steel	Yield Stress F_y (psi)
Unknown Steel	33,000
Structural Grade	36,000
Intermediate Grade (Grade 40)	40,000
Hard Grade (Grade 50)	50,000
Grade 60	60,000

(c) Structural Steel: Nominal unit stresses depend on the type of steel used in the structural member. When tests are performed to assess yield stress, the mean values shall be reduced by 10% to produce nominal values for strength calculations. Nominal values shall be nominal strength computed without any resistance factor applied.

13.8 FIXING THE VALUE OF RESISTANCE FACTOR, φ

The Resistance Factor, φ, also called the Capacity-Reduction factor or the factor which when multiplied on the 'nominal ultimate strength' (R_u) of a critical section (in bending, shear, etc.) gives 'dependable value' of that strength, has to take into account both the uncertainities in estimating the section-strength as well as any bias or conservativeness introduced in such estimates. Since further changes may occur to the section during the inspection interval, the section-strength estimate will also depend on the quality of inspection and maintenance. Even the level and detail of inspection is important since it may reveal the actual section details to be used in the section- strength calculations.

AASHTO (GSSE) has proposed certain Resistance Factors, φ, for various combinations of conditions encountered. These are shown in Table 13.3. These φ values may be modified slightly after taking due note of the Condition-Survey-Rating scale vis-a-vis that used by FHWA/AASHTO.

Table 13.3 Resistance Factor, φ

Superstructure Condition	Redundancy		Inspection		Maintenance		Steel and Pre-stressed Concrete	Reinforced Concrete
	Yes	No	Careful	Estimated	Vigorous	Intermittent		
Good or Fair	X		X		X		0.95	0.95
	X		X			X	0.95	0.85
	X			X	X		0.95	0.95
	X			X		X	0.90	0.85
		X	X		X		0.85	0.80
		X	X			X	0.75	0.70
		X		X	X		0.85	0.80
		X		X		X	0.75	0.70
Deteriorated	X		X		X		0.95	0.90
	X		X			X	0.85	0.80
	X			X	X		0.90	0.85
	X			X		X	0.80	0.75
		X	X		X		0.80	0.80
		X	X			X	0.70	0.70
		X		X	X		0.75	0.75
		X		X		X	0.65	0.65
Heavily Deteriorated	X		X		X		0.85	0.80
	X		X			X	0.75	0.70
	X			X	X		0.80	0.75
	X			X		X	0.70	0.65
		X	X		X		0.70	0.70
		X	X			X	0.60	0.60
		X		X	X		0.65	0.65
		X		X		X	0.55	0.55

NOTE: For ratings using data obtained from plans only, the capacity reduction factor (i.e. the Resistance Factor, φ) should be based on judgement of the engineer, supplemented by any additional information obtained.

13.9   ESTIMATING THE APPLIED DEAD LOAD AND LIVE LOAD EFFECTS (e.g. BENDING MOMENTS, SHEAR FORCES, ETC. DUE TO THEM) AT A CRITICAL SECTION, UNDER SERVICE-LOAD CONDITION

The calculations for estimating the applied Dead Load and Live Load effects (e.g. bending moments, shear forces, etc.) at a (critical) section under service-load condition shall be done by any of the usual methods of Structural Analysis. The transverse distribution of loads among the deck longitudinal may be done either by the simplified AASHTO method using the AASHTO Distribution Factors or the entire analysis may be done by a more accurate method e.g. Finite Element Method or Plane Grid Method, etc. The Load Factors on Dead Load and Live Load Effects shall be taken as explained in 13.5 and Table 13.1, and, the Correction Factor, C_f, shall be taken as explained in 13.6 and Table 13.2.

13.10   CHECK FOR SERVICEABILITY OF THE STRUCTURE

Even if the Rating factor (R. F.) is acceptable, the exercise is not complete unless the structure is also found to be serviceable under the Prevailing (Dead And Live) Loads. The structure is 'serviceable' if the computed maximum flexural crack-width in concrete at the critical section is acceptable and, where critical, the maximum deflection and fatigue-effect in reinforcement (i.e. whether the range of variation in its tensile stress is acceptable) also need to be checked.

It has to be remembered clearly that the guidelines described here for ensuring the Rating Factor of the Bridge are primarily concerned with the structural safety of the bridge in its 'ultimate-load state' as it addresses itself essentially to load-capacity evaluation. This means that the 'normally allowable stresses under service-load state' may well be exceeded under service-load-state. This is permissible (as in any Load-Factor approach) so long as the consequences (such as the flexural crack-width in concrete, deflection, and fatigue effect in reinforcement) in service-load condition are acceptable. Hence, the determination of load-capacity (Rating Factor) by the aforementioned Load-factor approach should be followed by the serviceability checks under service-load condition.

Following limits for crack-widths (Table 13.4) may be considered as a guide.

Table 13.4   Allowance Crack Widths

Environmental Class	Max. Crack Width, Reinforced Concrete (mm)	Max. Crack Width, Prestressed Concrete (mm)
Aggressive	0.25 + 10%	0.10 + 10%
Moderate (temperate)	0.35 + 10%	0.15 + 10%
Passive (inert)	0.45 + 10%	0.20 + 10%

The limits of crack–widths of observed shear cracks (and bending cracks in the web away from the main flexural reinforcement) in girders may be allowed to exceed the limits given in the Table 13.4 by another 10 per cent if the risk of corrosion is negligible.

However, all dormant cracks wider than about 0.30 mm in reinforced concrete and wider than about 0.20 mm. in prestressed concrete must be cleaned and sealed and filled with appropriate viscosity Epoxy resin formulation or cement grout, depending on the crack-width. Live cracks shall be initially filled with flexible filler and subsequently by a rigid filler. (*See* details in Chapter 6.)

The crack-width check may be done by the usual empirical formulae given in the relevant design specifications (Codes of Practice), as for instance the B. S. specifications.

The fatigue effect in the tension reinforcement, under the service-load bending moment due to prevailing Live Load (particularly under repetitive heavy Trucks), may be guarded against by the simple check given, for instance, in clause '8.16.8.3 of the AASHTO Standard Specifications for Highway Bridges — 1989'. (Fatigue is caused when the 'range of variation' in stress, due to repetitive presence and absence of Live-Load, is considerable. Hence, this is more significant in (short-span) slabs than in (long-span) girders, because in the former the majority of the moment is that due to Live Load which keeps pulsating while in the latter the permanent moment (that due to Dead Load) is considerable, leaving lesser margin for the Live Load stress and hence smaller range of variation in the total tensile stress in the reinforcement.)

13.11   NUMERICAL EXAMPLE OF AN ACTUAL BRIDGE

As an example, the Rating Factor (R. F.) calculation for the superstructure of a simply supported 15 m span, 2-lanes wide, 4-girdered, beam-and-slab cast-in-situ R. C. bridge-deck, are indicated below:

(i) The section-details at the critical sections (i.e. the actually existing reinforcement, concrete-strength, dependable section-dimensions and actual span) adopted in the calculations are based on the details in the as-built drawings and a detailed condition-survey of the actual bridge.

(ii) The maximum shear force and the maximum bending moment due to Dead loads and Live load (latter taken as the prevailing 5-axle 46 T truck, per lane) are calculated by the FRAME program at the critical sections in the inner as well as the outer girders. (The critical sections are the one-eighth span section from support for shear and close to mid span section for bending.) Hence, the computation of shears and moments due to Dead and Live Loads has been done more accurately than by the AASHTO method and the transverse distribution of Live Load has been achieved more accurately than by the simple AASHTO approach of 'Distribution Factor on Wheel-Line Effect'. Hence, the value of the Correction Factor (C_f) has been taken as 0.9 (Table 13.2)

(iii) In accordance with the guidelines discussed in this chapter
— Impact Fraction = 0.2 (span being more than 12 m and wearing coarse being in poor condition)
— Load Factors (Table 13.1)
• on structure Dead Load . . . 1.2
• on Overlays . . . $1.20 \times 1.2 = 1.44$
• on Live Load . . . 1.65 (because of significant sources of over-load, truck weight enforcement not being effective, and roadway being of low-volume traffic)

(iv) The section-strengths, R_u, at the critical sections (i.e. ultimate Moment of Resistance at the critical section for Bending and ultimate Shear Capacity at the critical section for shear) are calculated as indicated in the text earlier.

(v) The Resistance Factor (ϕ) is appropriately selected from Table 13.3.

(vi) *Rating Factor (R. F.)*: The Rating Factor (R. F.) is now calculated, first in Bending mode and then in Shear mode, using Eqn. 13.1, as follows.

(a) *in Bending Mode*
$$R. F. = \frac{\phi \cdot R_u - \sigma_D \cdot D}{\sigma_L \cdot L \cdot (1 + I) \cdot C_f}$$
. . . considering the bending effects ~
$$R.F. = \frac{3131 - 875}{834 \times 1.2 \times 0.9} = \underline{2.5 > 1.0\ OK}$$
where $\phi \cdot R_u$ = 3131 KNm
$\sigma_D \cdot D$ = 875 KNm
$\sigma_L \cdot L$ = 834 KNm

$(1 + I)$ = 1.2
C_f = 0.9

(b) *in Shear Mode*
$$R. F. = \frac{\phi \cdot R_u - \sigma_D \cdot D}{\sigma_L \cdot L \cdot (1 + I) \cdot C_f}$$
. . . considering Shear effects ~
$$R.F. = \frac{491 - 189}{275 \times 1.2 \times 0.9} = \underline{1.02, > 1.0\ OK}$$
where $\phi \cdot R_u$ = 491 KN
$\delta_D \cdot D$ = 189 KN
$\delta_L \cdot L$ = 275 KN
$(1 + I)$ = 1.2
C_f = 0.9

From the above two Rating Factors it is obvious that the superstructure will be more critical in Shear than in Bending mode of failure since the smaller of the R. F.s is the one in case of shear. However, even this is greater than 1.00, indicating that the withstandable Dead plus Live Load effects is more than that due to the Prevailing Dead plus Live Loads. Hence, the load carrying capacity is satisfactory from ultimate load i.e. ultimate-limit-state consideration. However, in the next step we should check the serviceability criteria of crack width, deflection and fatigue-effect on reinforcement before the bridge can be rated as 'satisfactory' (not requiring any cautious or posting of any special-care warnings).

NOTE: Incidentally, it may be noted in the present case that if majority of the actually plying Live Load trucks were heavier than the one assumed as the prevailing Live Load, then the minimum R. F. value here is likely to fall below the critical value of 1.00. In such case, the Live Load Capacity of the bridge would not be satisfactory and would have to be 'posted' as such. This would immediately call for relevant cautions (e.g. Truck weight reduction, speed reduction, reducing the number of lanes that can be loaded simultaneously, imposing rumble strips and bumps some distance away from the abutments, etc.) and possibly repair and strengthening actions.

(vii) *Serviceability Checks*: Maximum flexural crack width at the critical bending section, under Service load condition, works out to 0.39 mm. which is nearly OK and hence acceptable in the limit.

Maximum deflection and the fatigue consideration, as described in the text earlier, were calculated, and found acceptable, hence OK.

Hence, the load capacity of the bridge in question is satisfactory vis-a-vis the prevailing Live load assumed in the calculation.

14
THE 'BAILEY' BRIDGE

The Bailey Bridge is a pre-engineered ready-to-assemble standard-panel-system Bridge Deck, particularly useful in emergencies

14.1 THE BAILEY SYSTEM

This bridging system takes its name after its inventor, the late engineer Bailey from the old British Army Corps of Engineers. Bailey bridges are built on site from a pre-engineered system of ready-to-assemble components. Utilising standardized prefabricated components, Bailey bridges can be built to match a wide range of vehicular bridging applications. Due to their excellent versatility and overall value, thousands of Bailey bridges have been installed throughout the world.

14.2 FEATURES

- Adaptable — to match each application
- Fast — from planning to opening for service — only a few weeks
- Cost — is offset by the advantages of 'quick-to-construct' and 're-use' factors
- Easy — to handle, transport, assemble, install, reuse (disassemble and reassemble elsewhere)

14.3 STANDARDIZED COMPONENTS

The heart of the Bailey system is the ingenious set of precision-made interchangeable components from which all Bailey structures are assembled.

All components are manufactured in such a manner that they are fully interchangeable with other components of the same type, with all site connections made by means of pins, clamps or bolts. No site welding required.

14.4 MATERIALS AND FINISH

Structural steel is used throughout. Most load-bearing components as supplied by the U. S. Co. 'Bailey Bridges Inc. use low alloy high tensile ASTM A242 steel with a yield stress of 50,000 psi. Corrosion protection is achieved with a two-coat rust inhibitive paint system. Final colour is lusterless light grey. Mabey & Johnson (U.K.) supply all their structural components conforming to British Standard 4360:1986 to grades 55 C, 50A and 43A. All their structural components are galvanized in accordance with British Standard 5493. All fastenings (bolts, nuts, pins, etc.) should be electroplated in accordance with British Standard 1706 or equivalent.

14.5 CARRIAGEWAY WIDTH

Standard one, two and three lane width carriageways can be built. Even extra-wide single lane widths can be built. Mabey and Johnson (U.K.) can, for instance arrange:

- *Two-Lane Bridging:* 7.35 m between kerbs, 8.00 m between side girders;
- *Single Lane Bridging:* 4.20 m between kerbs, 4.73 between side girders. (All structural components, including the decking, are fully interchangeable between the 2-lane bridging and the single lane (extra wide) bridging.)

14.6 VARIABLE CONFIGURATIONS

Bailey bridge components can be assembled in different configurations to efficiently accommodate a wide range of span and capacity requirements. Panels, the primary Bailey component, are pinned together at the job-site to make girders of almost any length. Various girder strengths are achieved by assembling either a single row of panels, or two or three rows side-by-side. Panels may also be stacked in double storey height to further increase strength. For greatest strength, spans over 40 ft may be chord-reinforced.

14.7 SPANS

Usable spans range from 15 ft to 200 ft clear, and more! For highway use, typical Bailey bridge clear spans are from 50 ft. to 190 ft. Multiple span bridges of any length are possible by incorporating intermediate supporting piers. Even half bays can be incorporated in the spans.

14.8 EASE OF ASSEMBLY AND INSTALLATION

Most Bailey bridges are assembled and installed in a matter of days by a small crew. Common hand tools are utilized. All connections are pinned, bolted or clamped. No welding or heavy equipment is necessary. Dissembly is similarly easy, and the components can be stored in minimal space until reused.

Bailey bridges are often installed by the 'cantilever-launching' method, in which the assembled bridge, together with a front 'launching nose', is pushed across the gap, without falsework or heavy equipment. The 'cantilever' method allows bridges to be quickly erected over 'flooding rivers' or 'deep gorges'. Additionally, some Bailey bridges may be hoisted into place by conventional crane system.

14.9 STANDARD BRIDGES FOR ARMIES FOR TOP-QUICK BATTLE-FRONT USE

The U. S. M2 Bailey bridging, supplied by the Bailey Bridges Inc. (U. S. A.), is genuine U. S. Army equipment, manufactured in the U. S. A. to the highest military specifications. It is the 'standard-panel-bridge' of the U. S. Army. Bailey Bridge decks are used by the armies of many countries around the world.

14.10 SAFETY FACTOR

Certified evidence can be provided by the Suppliers from full-scale testing of the bridge system in its various constructions to the effect that the bracing systems for the main girders will adequately restrain the compression members against buckling at a minimum of 1.7 times the published recommended working loads.

14.11 GENERAL

(i) All costs for bridging generally include an extra quantity of spares (e.g. nuts, bolts and pins, etc.), about 7.5 per cent.

(ii) Construction equipment includes the following
   — launching nose for specific spans
   — rollers
   — 60 tonne jacks
   — hand tools
   — reconditioned, nonreturnable, containers of stores

(iii) The suppliers normally provide the necessary assembling and dissembling training in their work-shops, and, in addition, also provide the services of their Erection Engineer at the client's job-site (generally free of cost) for a few weeks to transfer the assembling — dissembling technology and procedures.

(iv) The suppliers can also redesign their standard panel assemblage for the client's special loading, if required.

14.12 COST

The costs cannot be generalized as they will depend on many variables, e.g. the pressure on the supplier's order-book, the availability of ready-stores with the supplier, the urgency of the order, the magnitude of the requirement of the client, cash-flow constraints (if any), etc. Purely as a rough guide, it might be of interest to know that the ex-UK-port costs, per square meter of deck-plan-area, 'quoted' in September 1991, for a 2-lane 'Mabey' Universal Bridge Deck, 7.3 m wide (with steel deck roadway and special transoms), designed to Saudi Highway Bridge Loadings, worked out to approximately:

— £ 490 for 31.5 m span (1-off) in DSR2H++Construction (90T gross weight)
— £ 540 for 40.5 m span (1-off) in TSR2H++Construction (128T gross weight)
— £ 650 for 49.5 m span (1-off) in DDR2H++Construction (186T gross weight)
— £ 725 for 58.5 m span (1-off) in TDR2H++Construction (245T gross weight)
— £ 873 for 72.0 m span (1-off) in QDR4H++Construction (368T gross weight)

In addition to these costs, the 'quotation' for the appropriate Launching and Erection Equipment (to launch and construct the bridge decks at site) worked out to approximately 30% of the above costs (ex-UK-port), the respective structural steel tonnages being about 18T, 25T, 38T, 54T and 75T. However, these quotations appear to be rather high.

NOTE: The above deck cross-section-designations are as follows

DSR2H++	—	Double-Single-Reinforced-Two-Heavy++
TSR2H++	—	Triple-Single-Reinforced-Two-Heavy++
DDR2H++	—	Double-Double-Reinforced-Two-Heavy++
TDR2H++	—	Triple-Double-Reinforced-Two-Heavy++
QDR4H++	—	Quadruple-Double-Reinforced-Four-Heavy++

The sign ++ refers to the Heavier two end panels (for higher Shear). A diagrammatic explanation of these designations is explained ahead.

Fig. 14.1

14.13 ADVANTAGE

Since the panels can be re-arranged and integrated to form decks of shorter or longer spans, the client may like to buy stores for say only one or two of 'long span' decks (say 2 off, each) and thereby be ready for putting up the decks of a relatively large number of intermediate spans (possible by appropriate permutation and combination of the panels). The client should then ask the supplier to provide him the necessary 'Key Arrangement Drawings' showing which panels to combine together to form the various intermediate spans.

14.14 SUPPLIERS: 'BAILEY BRIDGES, INC., U. S. A.', AND 'MABEY AND JOHNSON LTD., U. K.'

These agencies have been continuously supplying pre-fabricated-panel bridging for many years to civilian and military customers. They maintain an extensive inventory of all Bailey equipment, which allows orders to be met rapidly. Full technical support, including on-site assistance, is available from them. Their addresses are

(i) Bailey Bridges, Inc., P. O. Box 1186, San Luis Obispo, California 93406, U. S. A.

$$\begin{bmatrix} \text{Ph: (805)543--8083} \\ \text{Fax: (805)543--8983} \end{bmatrix}$$

(ii) Mabey and Johnson Ltd., Floral Mile, Twyford, Reading, RG10- 9SQ, England, U. K.

$$\begin{bmatrix} \text{Ph: (0734)403921} \\ \text{Fax: (0734)403941} \end{bmatrix}$$

14.15 'MABEY'-UNIVERSAL

The 'panel' is the basic unit from which main girders are formed. Measuring 14'9" (4.5m) by 7'9" (2.36m), the Mabey Universal panel is claimed to be longer and stronger than any other of this type. The panel size has been deliberately designed to permit many of the spans, normally encountered, to be in single panel-height construction. (*See* Fig. 14.2.)

The effective bending moment capacity of the truss is increased by introducing 'reinforcement' in the form of identical panel-chords. Bolted top and bottom to a single panel, the bending moment capacity is doubled. The introduction of further panel lines increases both the bending and shear capacity.

14.16 BRACING SYSTEM

The Universal Bridge is braced in single storey by the stiff frame formed by the panel trusses bolted to the transoms and stabilized by rakers. (*See* Fig. 14.3.)

On multiple panel trusses, a bracing frame is bolted to the top chords of the panels and covers 80 per cent of

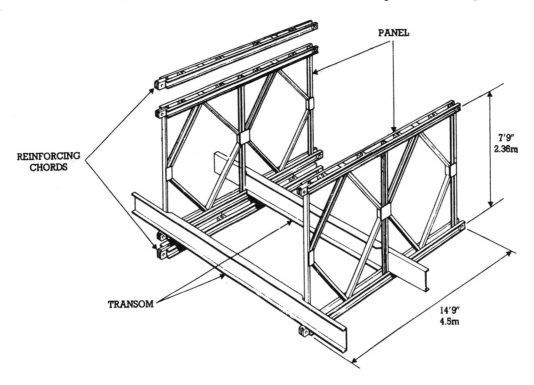

PANEL

7'9"
2.36m

REINFORCING CHORDS

TRANSOM

14'9"
4.5m

Fig. 14.2

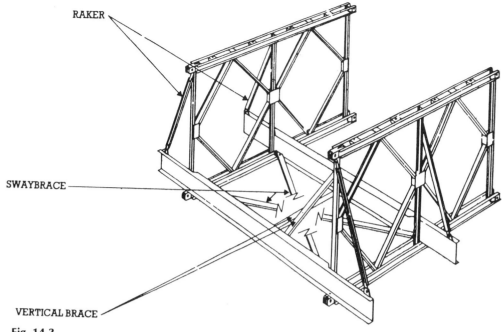

RAKER

SWAYBRACE

VERTICAL BRACE

Fig. 14.3

the bay length. Full scale tests have shown that this frame may be omitted for single-storey bridges, but must always be used on double-storey constructions. Lateral stability is provided by the use of a fixed-length swaybrace.

In addition to the positive bolted connection of the transom to the vertical element of the panel, a vertical bracing system stabilizes transoms at their mid-point.

14.17 DECK-SURFACE SYSTEM

The bridge is complete when the deck-surface units have been added. This is usually done after the bridge has been installed on its abutments (and piers). (*See* Fig. 14.5.)

The geometry of the bridging is such that all transoms are equally spaced so that deck surface units and swaybrace units are all of the same length.

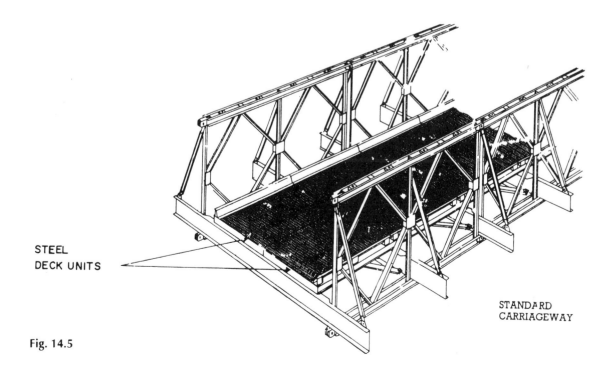

STEEL
DECK UNITS

STANDARD
CARRIAGEWAY

Fig. 14.5

14.18 HIGHWAY DECKING SURFACE

The deck surface units are designed to be bolted together transversely by means of three bolts along each side. Combining standard deck surface units in multiples and introducing longer transoms, give the four highway widths shown, catering for single lane H15 loading up to three lanes of HS20, HS25, MS250 and HA. (*See* Fig. 14.7.)

Each deck surface unit, either the 1606 or 803, is attached to the transoms or floor-beams by four clamp screws. These may be completely tightened up from a location on the deck and the need for temporary access to the underside of the bridge done away with.

The deck units are of an orthotropic construction with the patterned steel plate stiffened by longitudinal joist efficiently distributing wheel loads.

Running Surface

The deck surface units can be travelled upon directly, but, should a flexible wearing surface be required, a layer of asphalt can be laid after the bridge has been built.

For increased anti-skid properties, an epoxy coating can be applied.

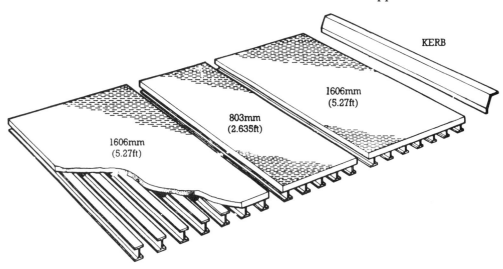

HIGHWAY DECK

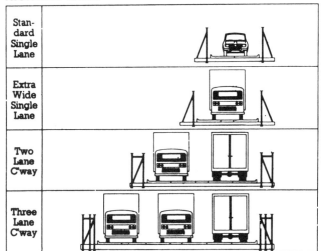

Fig. 14.7

PLATE 29

Fig. 14.4

PLATE 30

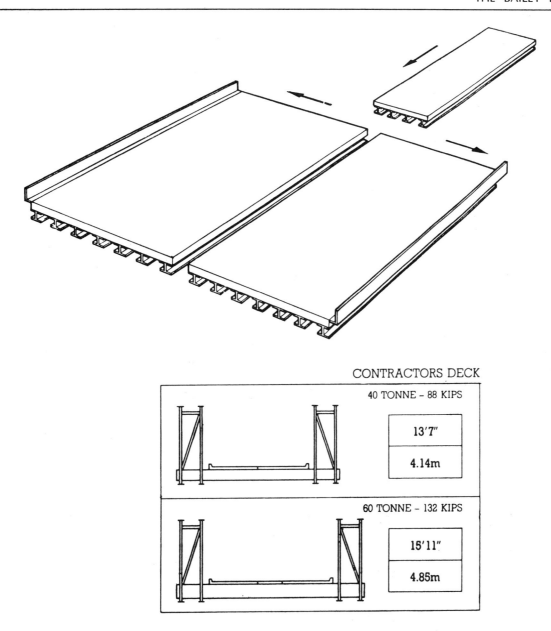

CONTRACTORS DECK

40 TONNE – 88 KIPS

| 13'7" |
| 4.14m |

60 TONNE – 132 KIPS

| 15'11" |
| 4.85m |

Fig. 14.8 For very heavy special haul trucks

14.19 CONTRACTOR'S VEHICLE DECKING

As well as providing solutions for highway bridging, the Mabey Universal seems particularly suitable for off-highway applications. These could be very remote logging-roads or haul-routes for quarries. Special rugged decking systems have been developed to carry the very heavy axle loads common to the construction field.

Deck systems have been produced to carry 40 tonne and 60 tonne axles, respectively. The systems have been rationalized into two roadwidths 13'7" (4.14 m) for 40 tonne axles and 15'11" (4.85 m) for 60 tonne axles.

14.20 GIRDER-CONSTRUCTIONS—
CROSS-SECTION DESIGNATION

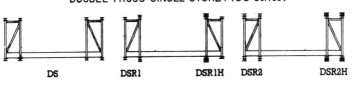

SINGLE TRUSS   SINGLE STOREY (S S Series)

SS   SSR   SSRH

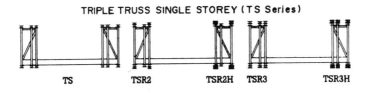

DOUBLE TRUSS SINGLE STOREY (DS Series)

DS   DSR1   DSR1H   DSR2   DSR2H

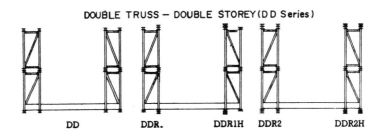

TRIPLE TRUSS SINGLE STOREY (TS Series)

TS   TSR2   TSR2H   TSR3   TSR3H

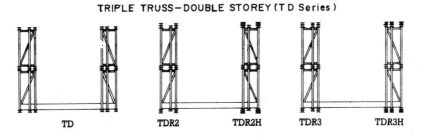

DOUBLE TRUSS — DOUBLE STOREY (D D Series)

DD   DDR.   DDR1H   DDR2   DDR2H

TRIPLE TRUSS—DOUBLE STOREY (T D Series)

TD   TDR2   TDR2H   TDR3   TDR3H

NOTE :—
    R = Reinforcing Chord
    H = Heavy Reinforcing Chord

Fig. 14.9

14.21 BRIDGE DECK DIMENSIONS: MABEY SYSTEM

See Fig. 14.10 and Table 14.1 for explanation.

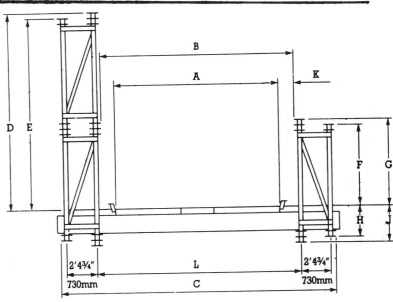

Bridge Deck Dimension "MABEY" System (Fig. 14.10 and Table 14.1)

Fig. 14.10 Bridge deck dimensions (Mabey system)

Table 14.1 Bridge Deck Dimensions

Roadway Width		A	B	C	D	E	F	G	H	J	K	L
Standard Single Lane	Imperial Units	10'7"	12'1¾"	19'2⅜"	14'6½"	14'1½"	5'11½"	6'4½"	2'2⅜"	2'7⅜"	9⅜"	12'10¾"
	Metric Units	3226	3703	5850	4431	4304	1817	1944	670	797	239	3932
Extra Wide Single Lane	Imperial Units	13'2⅝"	15'6"	22'5⅝"	14'6¼"	14'1¼"	5'11¼"	6'4¼"	2'2⅝"	2'7⅝"	13⅝"	16'3"
	Metric Units	4029	4724	6850	4426	4299	1812	1939	675	802	347	4953
Two Lane Carriageway	Imperial Units	23'9"	26'3"	32'6½"	13'10⅜"	13'5⅜"	5'3½"	5'8½"	2'10½"	3'3½"	15"	27'0"
	Metric Units	7241	8001	9920	4226	4099	1612	1739	875	1002	380	8230
Three Lane Carriageway	Imperial Units	34'3½"	37'3⅞"	43'7⅜"	13'7"	13'2"	5'0⅛"	5'5⅛"	3'1¾"	3'6¾"	18"	38'0⅞"
	Metric Units	10453	11375	13294	4141	4014	1527	1654	960	1087	460	11604
13'7" 40 Tonne Single Axle 4.14m	Imperial Units	13'6¾"	15'6"	21'9½"	13'8⅜"	13'3⅜"	5'1½"	5'6½"	3'0½"	3'5½"	11¾"	16'3"
	Metric Units	4135	4724	6642	4174	4047	1560	1687	927	1054	295	4953
15'11" 60 Tonne Single Axle 4.85m	Imperial Units	15'11"	17'10¼"	24'1¾"	13'7"	13'2"	5'0"	5'5"	3'1¾"	3'6¾"	11¾"	18'7¼"
	Metric Units	4850	5440	7359	4139	4012	1525	1652	962	1089	295	5670

Dimensions are subject to Manufacturing Tolerances.

14.22 PROPERTIES AND WEIGHTS OF MABEY-UNITS

(See Table 14.2 below for explanation.)

Table 14.2

| | Properties | | | | | | Weights Per Bay | | | | | |

Properties

Metric Units			Kips Units				Table A*			Highway			
Moment	Shear (t)		Moment	Shear (K)			t	K		10'9"	13'6"	24'	34'6"
tm	Std	Heavy	K. ft.	Std.	Heavy					3.28 m	4.12 m	7.32 m	10.5 m
402	71	101	2912	156	223	SS	1.180	2.602					
805	71	101	5825	156	223	SSR	1.967	4.337					
898	71	101	6497	156	223	SSRH	2.084	4.595		**Table B***			
805	128	183	5825	282	403	DS	2.379	5.245	K	5.450	6.954	13.898	22.667
1207	95	137	8732	211	302	DSR1	3.166	6.981	t	2.472	3.154	6.303	10.280
1300	93	132	9410	204	292	DSR1H	3.282	7.236					
1610	128	183	11650	282	403	DSR2	3.953	8.716					
1796	128	183	12997	282	403	DSR2H	4.185	9.228		**Table C***			
1207	192	274	8738	422	604	TS	3.553	7.834	K	3.975	4.926	8.661	12.923
2013	160	229	14563	353	504	TSR2	5.127	11.305	t	1.803	2.234	3.928	5.861
2198	156	223	15908	345	493	TSR2H	5.359	11.816					
2415	192	274	17476	422	604	TSR3	5.913	13.038					
2694	192	274	19493	422	604	TSR3H	6.263	13.810		**Contractor's**			
1610	237	256	11650	524	564	DD	5.052	11.140		13'7"		15'11"	
2415	178	192	17476	393	422	DDR1	5.828	12.850		4.14 m		4.85 m	
2601	172	185	18820	379	408	DDR1H	5.945	13.108					
3220	237	256	23301	524	564	DDR2	6.605	14.564		**Table D***			
3592	237	256	25991	524	564	DDR2H	6.838	15.078	K	9.534		12.603	
2415	356	384	17476	786	846	TD	7.450	16.427	t	4.324		5.716	
4026	296	320	29127	654	706	TDR2	9.002	19.850					
4397	290	313	31816	641	690	TDR2H	9.235	20.363		**Table E***			
4831	356	384	34952	786	846	TDR3	9.779	21.562	K	6.182		7.611	
5388	356	384	38986	786	846	TDR3H	10.128	22.332	t	2.804		3.452	

Design Weights (per bay)		
	Decked*	Undecked*
Highway	A + B	A + B − C
Contractor's	A + D	A + D − E

Conversion Table
K = kip = 1000 lbs
t = tonne = 2205 lbs = 1000 kg = Short Ton

- Design Notes
  1. Bay weights are based upon latest available production drawings for standard loadings. Add 2½ per cent for galvanising and finishings.
  2. The figures tabulated above are consistent with a minimum factor of safety of 1.7 for both bending and shear.
  3. The figures tabulated for shear take account of the maldistribution of load between panel lines owing to the differential stiffnesses within the truss line, e.g. DSR1, DDR1H.
  4. All double storey bridges must have end posts at abutments.

- Heavy End Panel (MU 110)
  Although panels can have their bending resistance increased by the addition of reinforcements, it is not possible to supplement shear resistance in a similar fashion. As the highest shear forces occur at the end of the span it is more economic to fabricate end panels with an increased shear capacity to be situated at the ends of the bridge rather than introduce another complete line of panels to the truss.

  High shear panels (MU 111) may be used as alternatives, where fatigue strength does not require the use of MU 110.

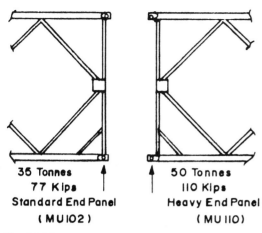

36 Tonnes
77 Kips
Standard End Panel
(MU102)

50 Tonnes
110 Kips
Heavy End Panel
(MU 110)

Fig. 14.11

14.23 'MABEY' CONSTRUCTION TABLES
(*See* Tables 14.3 to 14.8.)

Notes 1. *'Highway Loading'*: Construction tables for various highway loadings are based on the vehicle and lane loadings as specified in the relevant tables. 'Constructions' are chosen for a minimum fatigue life of 1,00,000 cycles of loading. Where a longer life is required the advise of Mabey Engineers should be sought. A maximum span to depth ratio of 25:1 has been assumed. Full account has been taken of impact factors and eccentricity.

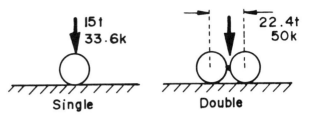

Not Less Than 1.2m (4ft.)

15t
33.6k

22.4t
50k

Single   Double

Fig. 14.12   Maximum axle loads

2. These tables are based on the use of MU 102 and MU 110 End-Panels.

3. *'Single Vehicle Loadings' Tables (Fig. 14.12)*: These are based upon a Uniformly Distributed Design Load of 5 tonnes/m (3.4 kips/ft) and account for an impact factor of 25 per cent and a maximum eccentricity of 10 per cent of the road-width from the centreline. Note that they are intended for guidance only. Refer to Mabey Engineers with specific vehicle details and required frequency of use for a more accurate design check.

4. Carriageway-width for Standard-Single-Lane, extra-wide-single-lane, two-lanes, etc., together with various other pertinent dimensions (whether a single-storey or a double-storey arrangement, depending on the span-length and loading), are given in Fig. 14.10 and Table 14.1 earlier.

Table 14.3

Standard Single Lane

The column headings for the Single Vehicle Loading are given as Kips / Tonnes.

Bays	Feet	Metres	HS20	HS25	MS250	HA	45/20	56/25	67/30	78/35	90/40	100/45	112/50	124/55	135/60	145/65	157/70	168/75	180/80
4	59.1	18.00	SS	SS	SS	SS	SS	SS	SS	SS	SS	SS	SS	SS	SS	SS	SS	SS	SS
4.5	66.4	20.25	SS	SS	SS	SS	SS	SS	SS	SS	SS	SS	SS	SS	SS	SS	SS	SS	SS
5	73.8	22.50	SS	SS	SSR+	SSR	SS	SS	SS	SS	SS	SS	SS	SS	SS	SSR	SSR	SSR	SSR
5.5	81.2	24.75	SS	SSR	SSR+	SSR+	SS	SS	SS	SS	SS	SS	SSR	SSR	SSR	SSR+	SSR+	SSR+	SSR+
6	88.6	27.00	SS	SSR	SSR+	SSR+	SS	SS	SS	SS	SS	SSR	SSR	SSR	SSR	SSR+	SSR+	SSR+	SSR+
6.5	96.0	29.25	SSR	SSR	SSR+	SSR+	SS	SS	SS	SS	SSR	SSR	SSR	SSR	SSR+	SSR+	SSR+	SSR+	SSR+
7	103.3	31.50	SSR	SSR	SSR+	SSR+	SS	SS	SS	SSR	SSR	SSR	SSR	SSR+	SSR+	SSR+	SSR+	SSR+	SSR++
7.5	110.7	33.75	SSR	SSR	SSR+	SSRH+	SS	SS	SSR	SSR	SSR	SSR	SSR	SSR+	SSR+	SSR+	SSR+	SSRH++	SSRH++
8	118.1	36.00	SSR	SSR	SSR+	DSR1+	SS	SSR	SSR	SSR	SSR	SSR	SSR+	SSRH+	SSRH+	SSRH+	SSRH++	DSR1	DSR1+
8.5	125.5	38.25	SSR	SSR	SSR+	DSR1+	SSR	SSR	SSR	SSR	SSR	SSR	SSR+	SSRH+	DSR1	DSR1	DSR1+	DSR1+	DSR1+
9	132.9	40.50	SSR	SSR	SSR+	DSR1+	SSR	SSR	SSR	SSR	SSR	SSR+	SSRH+	DSR1	DSR1	DSR1	DSR1+	DSR1+	DSR1+
9.5	140.3	42.75	SSR	SSR+	SSRH+	DSR1H+	SSR	SSR	SSR	SSR	SSR	SSRH+	SSRH+	DSR1	DSR1	DSR1	DSR1+	DSR1+	DSR1H++
10	147.6	45.00	SSR	SSR+	SSRH+	DSR2	SSRH	DSR1	SSR	SSR	SSRH	SSRH+	DSR1	DSR1	DSR1	DSR1+	DSR1H+	DSR1H	DSR2
10.5	155.0	47.25	SSR	SSRH+	DSR1	DSR2	SSRH	DSR1	SSR	SSRH	SSRH	DSR1	DSR1	DSR1	DSR1	DSR1H+	DSR2	DSR2	DSR2
11	162.4	49.50	SSR	DSR1	DSR1	DSR2	DSR2	DSR2	SSRH	SSRH	DSR1	DSR1	DSR1H	DSR1H+	DSR1H+	DSR2	DSR2	DSR2	DSR2H
11.5	169.8	51.75	SSRH	DSR1	DSR1	DSR2H	DSR2	DSR2	DSR1	DSR1	DSR1	DSR1	DSR1H	DSR2	DSR2	DSR2	DSR2	DSR2	DSR2H
12	177.2	54.00	DSR1	DSR1	DSR1	TSR2	DSR2	DSR2	DSR1	DSR1	DSR1	DSR1H	DSR2	DSR2	DSR2	DSR2	DSR2H	DSR2H	DSR2H
12.5	184.5	56.25	DSR1	DSR1	DSR1H+		DD	DD	DSR1	DSR1	DSR1	DSR1H	DSR2	DSR2	DSR2	DSR2H	DSR2H	TSR2	TSR2
13	191.9	58.50	DSR1	DSR1H	DSR2	TSR2	DD	DDR1	DSR1	DSR1	DSR1H	DSR2	DSR2	DSR2	DSR2H	DSR2H	TSR2	TSR2H	TSR2H
13.5	199.3	60.75	DSR1H	DSR2	DSR2	TSR2H	DDR1	DDR1	DSR2	DSR1H	DSR2	DSR2	DSR2	DSR2	DSR2H	TSR2	TSR2H	TSR2H	DDR1
14	206.7	63.00	DSR2	DSR2H	DSR2H	DDR1	DDR1	DDR1	DSR2	DSR2	DSR2	DSR2	DSR2H	DDR1	DDR1	DDR1	DDR1	DDR1	DDR1
14.5	214.1	65.25	DSR2	DDR1	DSR2H	DDR1H	DDR1	DDR1	DSR2	DSR2	DSR2	DSR2H	DDR1	DDR1	DDR1	DDR1	DDR1	DDR1	DDR1H
15	221.5	67.50	DSR2	DDR1	DDR1	DDR2	DDR1	DDR1	DDR1	DSR2	DSR2	DSR2H	DDR1	DDR1	DDR1	DDR1	DDR1H	DDR1H	DDR2
15.5	228.8	69.75	DDR1	DDR1	DDR1	DDR2	DDR1	DDR1	DDR1	DDR1	DDR1	DDR1	DDR1	DDR1	DDR1	DDR1H	DDR2	DDR2	DDR2
16	236.2	72.00	DDR1	DDR1	DDR1	DDR2	DDR1	DDR1	DDR1	DDR1	DDR1H	DDR1	DDR1	DDR1H	DDR1H	DDR2	DDR2	DDR2	DDR2
16.5	243.6	74.25	DDR1	DDR1	DDR1H	DDR2	DDR1	DDR1	DDR1	DDR1	DDR1	DDR1	DDR1H	DDR1H	DDR2	DDR2	DDR2	DDR2	DDR2
17	251.0	76.50	DDR1	DDR1H	DDR2	DDR2H	DDR1	DDR1	DDR1	DDR1	DDR1	DDR1H	DDR1H	DDR2	DDR2	DDR2	DDR2H	DDR2H	DDR2H
17.5	258.4	78.75	DDR1H	DDR2	DDR2	DDR2H	DDR1	DDR1	DDR1	DDR1	DDR1H	DDR1H	DDR2	DDR2	DDR2	DDR2	DDR2H	DDR2H	DDR2H
18	265.7	81.00	DDR1H	DDR2	DDR2	DDR2H	DDR1	DDR1	DDR1	DDR1	DDR1H	DDR2	DDR2	DDR2	DDR2	DDR2H	DDR2H	DDR2H	TDR2H

Highway Loadings: HS20, HS25, MS250, HA. Single Vehicle Loading (Kips / Tonnes): 45/20, 56/25, 67/30, 78/35, 90/40, 100/45, 112/50, 124/55, 135/60, 145/65, 157/70, 168/75, 180/80.

+ Denotes heavy end panels in end bays only.
++ Denotes heavy end panels in two end bays only.

Table 14.4

Extra Wide Single Lane

Bays	Span (Feet)	Span (Metres)	Highway Loadings HS20	HS25	MS250	HA	Single Vehicle Loading 45 kips / 20 t	56 / 25	67 / 30	78 / 35	90 / 40	100 / 45	112 / 50	124 / 55	135 / 60	145 / 65	157 / 70	168 / 75	180 / 80	Bays
4	59.1	18.00	SS	SS	SS+	SS+	SS	SS	SS	SS	SS	SS	SS	SS	SS	SS	SS	SS	SS	4
4.5	66.4	20.25	SS	SS	SSR+	SSR+	SS	SS	SS	SS	SS	SS	SS	SS	SS	SS	SS	SS	SS	4.5
5	73.8	22.50	SS	SSR+	SSR+	SSR+	SS	SS	SS	SS	SS	SS	SS	SS	SSR	SSR	SSR	SSR	SSR+	5
5.5	81.2	24.75	SS	SSR+	SSR+	SSR+	SS	SS	SS	SS	SS	SS	SSR	SSR	SSR	SSR+	SSR+	SSR+	SSR+	5.5
6	88.6	27.00	SSR	SSR+	SSR+	SSR+	SS	SS	SS	SS	SS	SSR	SSR	SSR	SSR+	SSR+	SSR+	SSR+	SSR+	6
6.5	96.0	29.25	SSR	SSR+	SSR+	SSRH++	SS	SS	SS	SS	SS	SSR	SSR	SSR+	SSR+	SSR+	SSR+	SSR+	SSR+	6.5
7	103.3	31.50	SSR	SSR+	SSR++	DSR1+	SS	SS	SSR	SSR	SS	SSR	SSR	SSR+	SSR+	SSR+	SSR+	SSR+	SSR++	7
7.5	110.7	33.75	SSR	SSR+	SSR++	DSR1+	SS	SSR	SSR	SSR	SS	SSR	SSR	SSR+	SSR+	SSR+	SSR+	SSR+	SSRH++	7.5
8	118.1	36.00	SSR	SSR+	SSRH+	DSR1+	SSR	SSR	SSR	SSR	SS	SSR	SSR+	SSR+	SSR+	SSR+	SSRH++	SSRH++	DSR1+	8
8.5	125.5	38.25	SSR	SSR+	DSR1	DSR2	SSR	SSR	SSR	SSR	SS	SSR+	SSR+	SSRH+	SSRH++	SSRH++	SSRH++	DSR1+	DSR1+	8.5
9	132.9	40.50	SSR	DSR1	DSR2	DSR2+	SSR	SSR	SSR	SSR	SS	SSR+	SSRH+	DSR1	DSR1	DSR1+	DSR1+	DSR1+	DSR1+	9
9.5	140.3	42.75	SSR	SSRH+	DSR1	DSR2+	SSR	SSR	SSR	SSR	SSR	SSRH+	DSR1	DSR1	DSR1	DSR1+	DSR1+	DSR1H+	DSR1H++	9.5
10	147.6	45.00	SSR	DSR1	DSR1	DSR2H+	SSR	SSR	SSR	SSR	SSRH+	DSR1	DSR1	DSR1	DSR1+	DSR1H+	DSR1H+	DSR2	DSR2	10
10.5	155.0	47.25	SSRH+	DSR1	DSR1+	DSR2H+	SSR	SSR	SSRH	SSRH	DSR1	DSR1	DSR1	DSR1	DSR1H+	DSR2	DSR2	DSR2	DSR2H	10.5
11	162.4	49.50	DSR1	DSR1	DSR1H+	TSR2	SSR	SSR	SSRH	DSR1	DSR1	DSR1	DSR1H+	DSR1H+	DSR2	DSR2	DSR2	DSR2H	DSR2H	11
11.5	169.8	51.75	DSR1	DSR1	DSR2	TSR2	SSRH	SSRH	DSR1	DSR1	DSR1	DSR1	DSR2	DSR2	DSR2	DSR2H	DSR2H	DSR2H	DSR2H	11.5
12	177.2	54.00	DSR1	DSR1H+	DSR2	TSR2H	SSRH	SSRH	DSR1	DSR1	DSR1	DSR1H	DSR2	DSR2	DSR2	DSR2H	DSR2H	DSR2H	TSR2	12
12.5	184.5	56.25	DSR1	DSR2	DSR2	DDR1	SSRH	DSR1	DSR1	DSR1H	DSR1H	DSR2	DSR2	DSR2	DSR2H	DSR2H	TSR2	TSR2	TSR2H	12.5
13	191.9	58.50	DSR1H	DSR2	DSR2	DDR1	SSRH	DSR1	DSR1	DSR1	DSR2	DSR2	DSR2	DSR2H	DSR2H	TSR2	TSR2H	TSR2H	TSR2H	13
13.5	199.3	60.75	DSR2	DSR2H	DSR2H	DDR1H	DSR1	DSR2	DSR1H	DSR2	DSR2	DSR2H	DSR2H	DSR2H	TSR2	TSR2H	TSR2H	DDR1	DDR1	13.5
14	206.7	63.00	DSR2	DSR2H	DDR1	TSR3H	DSR2	DSR2	DSR2	DSR2	DSR2	DSR2H	DSR2H	DDR1	DDR1	DDR1	DDR1	DDR1H	DDR1H	14
14.5	214.1	65.25	DSR2H	DDR1	DDR1	DDR2	DSR2	DSR2	DSR2	DSR2	DSR2H	DSR2H	DSR2H	DDR1H	DDR1	DDR1	DDR1H	DDR2	DDR2	14.5
15	221.5	67.50	DSR2H	DDR1	DDR1	DDR2	DSR2	DSR2	DSR2	DSR2H	DSR2H	DDR1	DDR1	DDR1	DDR1	DDR1H	DDR1H	DDR2	DDR2	15
15.5	228.8	69.75	DDR1	DDR1	DDR1H	DDR2H	DD	DDR1	DDR1	DDR1	DDR1	DDR1	DDR1	DDR1	DDR1H	DDR2	DDR2	DDR2	DDR2	15.5
16	236.2	72.00	DDR1	DDR1H	DDR2	DDR2H	DDR1	DDR1	DDR1	DDR1	DDR1	DDR1	DDR1H	DDR1H	DDR2	DDR2	DDR2	DDR2	DDR2	16
16.5	243.6	74.25	DDR1	DDR2	TDR2	TDR2	DDR1	DDR1	DDR1	DDR1	DDR1	DDR1H	DDR1H	DDR2	DDR2	DDR2	DDR2	DDR2H	DDR2H	16.5
17	251.0	76.50	DDR1H	DDR2	DDR2	TDR2H	DDR1	DDR1	DDR1	DDR1H	DDR1H	DDR1H	DDR2	DDR2	DDR2	DDR2H	DDR2H	DDR2H	DDR2H	17
17.5	258.4	78.75	DDR2	DDR2	DDR2H	TDR3	DDR1	DDR1	DDR1	DDR1H	DDR1H	DDR2	DDR2	DDR2	DDR2H	DDR2H	DDR2H	TDR2H	TDR2	17.5
18	265.7	81.00	DDR2	DDR2H	DDR2H	TDR3	DDR1	DDR1	DDR1H	DDR2	DDR2	DDR2	DDR2	DDR2	DDR2H	DDR2H	DDR2H	TDR2H	TDR2H	18

+ Denotes heavy end panels in end bays only.
++ Denotes heavy end panels in two end bays only.

Table 14.5

Two Lane Carriageway

Span			Highway Loadings				Single Vehicle Loading													Bays
Bays	Feet	Metres	HS20	HS25	MS250	HA	45	56	67	78	90	100	112	124	135	145	157	168	180	(Kips)
							20	25	30	35	40	45	50	55	60	65	70	75	80	(Tonnes)
4	59.1	18.00	SSR+	SSR++	DS+	SSR+	SS	SS	SS	SS	SS	SS	SS	SS	SS	SS	SS	SS	SS	4
4.5	66.4	20.25	SSR+	SSR++	DS+	SSR+	SS	SS	SS	SS	SS	SS	SS	SS	SS	SS	SSR	SSR+	SSR+	4.5
5	73.8	22.50	SSR+	SSR++	DSR1++	SSR++	SS	SS	SS	SS	SS	SSR	SSR	SSR	SSR+	SSR+	SSR+	SSR+	SSR+	5
5.5	81.2	24.75	SSR++	DSR1++	DSR1++	DSR1+	SS	SS	SS	SS	SSR	SSR	SSR	SSR+	SSR+	SSR+	SSR+	SSR+	SSR+	5.5
6	88.6	27.00	SSR++	DSR1++	DSR1++	DSR1++	SS	SS	SSR	SSR	SSR	SSR	SSR+	SSR+	SSR+	SSR+	SSR+	SSR+	SSR+	6
6.5	96.0	29.25	SSR++	DSR1++	DSR1++	DSR2+	SS	SSR	SSR	SSR	SSR	SSR+	SSR+	SSR+	SSRH+	SSR+	SSR+	SSR++	SSR++	6.5
7	103.3	31.50	SSRH++	DSR1++	DSRH++	DSR2+	SSR	SSR	SSR	SSR	SSR+	SSR+	SSR+	SSRH+	SSRH++	SSRH++	SSRH++	SSRH++	DSR1+	7
7.5	110.7	33.75	DSR1+	DSR1++	DSR2++	DSR2H++	SSR	SSR	SSR	SSR+	SSR+	SSR+	SSRH+	SSRH++	DSR1+	DSR1+	DSR1+	DSR1+	DSR1+	7.5
8	118.1	36.00	DSR1+	DSR1H++	DSR2++	DSR2++	SSR	SSR	SSR	SSR+	SSRH+	SSRH+	DSR1	DSR1+	DSR1+	DSR1+	DSR1+	DSR1+	DSR1+	8
8.5	125.5	38.25	DSR1+	DSR2+	DSR2++	DSR2H++	SSR	SSR	SSR+	SSRH+	SSRH+	DSR1	DSR1	DSR1+	DSR1+	DSR1+	DSR1H++	DSR1H++	DSR1H++	8.5
9	132.9	40.50	DSR1++	DSR2+	DSR2++	TSR2+	SSR	SSR	SSR+	SSRH+	SSRH+	DSR1	DSR1	DSR1+	TSR2	DSR2	DSR2	DSR2	DSR2	9
9.5	140.3	42.75	DSR1H++	DSR2+	TSR2+	DDR1+	SSR	SSR	SSRH+	DSR1	DSR1	DSR1	DSR1+	DSR1+	DSR1H+	DSR2	DSR2	DSR2	DSR2+	9.5
10	147.6	45.00	DSR2	DSR2+	TSR2+	DDR1+	SSR	SSRH	DSR1	DSR1	DSR1	DSR1+	DSR1H+	DSR1H+	DSR2	DSR2	DSR2	DSR2	DSR2+	10
10.5	155.0	47.25	DSR2	DSR2+	TSR2+	DDR1+	SSRH	DSR1	DSR1	DSR1	DSR1	DSR1H+	DSR2	DSR2	DSR2	DSR2	DSR2	DSR2	DSR2H+	10.5
11	162.4	49.50	DSR2H+	DSR2H+	DSR2H++	TSR3H+	DSR1	DSR1	DSR1	DSR1H++	DSR1H+	DSR2	DSR2	DSR2H	DSR2	DSR2H+	DSR2H+	TSR2	TSR2	11
11.5	169.8	51.75	DSR2H+	TSR2+	TSR2H+	DDR2	DSR1	DSR1	DSR1	DSR2	DSR2	DSR2	DSR2	DSR2H	DSR2H	DSR2H+	TSR2	TSR2	TSR2H	11.5
12	177.2	54.00	TSR2	TSR2H+	TSR2H+	DDR2	DSR1	DSR1	DSR1H	DSR2	DSR2	DSR2	DSR2H	DSR2H	TSR2	TSR2	TSR2H	TSR2H	DDR1	12
12.5	184.5	56.25	TSR2H+	DDR1	DDR1+	DDR2H	DSR1	DSR2	DSR2	DSR2	DSR2H	DDR2	TSR2	TSR2	TSR2H	TSR2H	DDR1	DDR1	DDR1	12.5
13	191.9	58.50	TDR2	DDR1H+	TSR3H	DDR2H	DDR1H	DSR2	DSR2	DSR2H	DSR2H	DDR2	TSR2	TSR2H	TSR2H	DDR1	DDR1	DDR1H	DDR1H	13
13.5	199.3	60.75	DDR1H	DDR2	TSR3H+	TDR2	DDR1H	DSR2H	DSR2H	DSR2H	TSR2	DDR2	TSR2H	DDR1	DDR1	DDR1H	DDR1H	TSR3H	TSR3H	13.5
14	206.7	63.00	DDR2	DDR2	DDR2	TDR2H	DDR2	DSR2H	DDR1	DDR1	DDR1	DDR2	DDR1	DDR1H	DDR1H	TSR3H	TSR3H	TSR3H	DDR2	14
14.5	214.1	65.25	DDR2	DDR2H	DDR2	TDR2H	DDR2	DDR1	DDR1	DDR1	DDR1	DDR2H	DDR1H	DDR1H	DDR2	DDR2	DDR2	DDR2	DDR2	14.5
15	221.5	67.50	DDR2	TDR2	DDR2H	TDR3	DDR2	DDR1	DDR1	DDR1	DDR1	DDR2H	TSR3H	DDR2	DDR2	DDR2	DDR2	DDR2	DDR2H	15
15.5	228.8	69.75	DDR2H	TDR2	TDR2H	TDR3H	DDR1	DDR1	DDR1	DDR1H	DDR1H	DDR2	DDR2	DDR2	DDR2	DDR2	DDR2H	DDR2H	DDR2H	15.5
16	236.2	72.00	TDR2	TDR2H	TDR2H	TDR3H	DDR1	DDR1	DDR1H	DDR2	DDR2	DDR2H	DDR2	DDR2	DDR2	DDR2H	DDR2H	TDR2	TDR2	16
16.5	243.6	74.25	TDR2H	TDR3	TDR3	—	DDR1H	DDR1H	DDR2	DDR2	DDR2	DDR2H	DDR2	DDR2H	DDR2H	DDR2H	TDR2H	TDR2H	TDR2H	16.5
17	251.0	76.50	TDR3	TDR3	TDR3H	—	DDR2	DDR2	DDR2	DDR2	DDR2H	DDR2H	DDR2H	DDR2H	TDR2H	TDR2H	TDR2H	TDR2H	TDR3	17
17.5	258.4	78.75	TDR3	TDR3H	—	—	DDR2	DDR2	DDR2	DDR2H	DDR2H	DDR2H	DDR2H	TDR2	TDR2H	TDR2H	TDR3	TDR3	TDR3	17.5
18	265.7	81.00	TDR3H	—	—	—	DDR2	DDR2	DDR2	DDR2H	DDR2H	DDR2H	DDR2H	TDR2H	TDR2H	TDR3	TDR3	TDR3	TDR3H	18

+ Denotes heavy end panels in end bays only
++ Denotes heavy end panels in two end bays only

Table 14.6

Three Lane Carriageway

Bays	Span Feet	Span Metres	Highway Loadings HS20	HS25	MS250	HA	45 (20)	56 (25)	67 (30)	78 (35)	90 (40)	100 (45)	112 (50)	124 (55)	135 (60)	145 (65)	157 (70)	168 (75)	180 (80)	Bays
												Single Vehicle Loading (Kips / Tonnes)								
4	59.1	18.00	DS+	DS+	TS	DSR1+	SS	SS	SS	SS	SS	SS	SS	SS	SS	SS	SS	SS	SS	4
4.5	66.4	20.25	DS+	DSR1++	TS	TS	SS	SS	SS	SS	SS	SS	SS	SSR+	SSR+	SSR+	SSR+	SSR+	SSR+	4.5
5	73.8	22.50	DSR1++	TS	TS+	DSR2+	SS	SS	SSR	SSR	SSR	SSR+	SSR+	SSR+	SSR+	SSR+	SSR+	SSR+	SSR+	5
5.5	81.2	24.75	DSR1++	TS	TS	DSR2++	SS	SS	SSR	SSR+	SSR+	SSR+	SSR+	SSR+	SSR+	SSR+	SSR++	SSR+	SSR+	5.5
6	88.6	27.00	DSR1++	DSR2++	DSR2++	DSR2H++	SSR	SSR	SSR	SSR+	SSR+	SSR+	SSR+	SSR+	SSR+	SSR+	SSR+	SSR++	DS	6
6.5	96.0	29.25	DSR1++	DSR2++	DSR2++	TSR2++	SSR	SSR	SSR+	SSR+	SSR+	SSR+	SSR+	SSR+	SSR++	SSRH++	DSR1+	DSR1+	DSR1+	6.5
7	103.3	31.50	DSR1++	DSR2++	DSR2++	TSR2H++	SSR	SSR	SSR+	SSR+	SSR+	SSR++	SSR++	SSRH++	DSR1+	DSR1+	DSR1+	DSR1+	DSR1+	7
7.5	110.7	33.75	DSR2+	DSR2++	DSR2H++	TSR3++	SSR	SSR+	SSR+	SSR+	SSRH+	SSRH+	DSR1+	DSR1+	DSR1+	DSR1+	DSR1+	DSR1+	DSR1H++	7.5
8	118.1	36.00	DSR2+	DSR2H+	TSR2++	TSR3H++	SSR+	SSR+	SSRH+	SSRH+	DSR1+	DSR1+	DSR1+	DSR1+	DSR1+	DSR1H++	DSR1H++	DSR1H++	DSR1H++	8
8.5	125.5	38.25	DSR2+	DSR2+	TSR2+	DDR2+	SSRH+	SSRH+	DSR1	DSR1+	DSR1+	DSR1+	DSR1H+	DSR1H++	DSR2	DSR2+	DSR2+	DSR2+	DSR2+	8.5
9	132.9	40.50	DSR2+	TSR2	TSR2H++	DDR2+	SSRH+	SSRH+	DSR1	DSR1	DSR1	DSR1+	DSR1H+	TSR2H	DSR2	DSR2+	DSR2+	DSR2+	DSR2+	9
9.5	140.3	42.75	DSR2H+	TSR2++	TSR3+	TDR2	DSR1	DSR1	DSR1+	DSR1+	DSR1H+	DSR1H++	DSR2	DSR2	DSR2+	DSR2+	DSR2+	DSR2H+	DSR2H+	9.5
10	147.6	45.00	TSR2+	TSR2H++	TSR3+	TDR2	DSR1	DSR1	DSR1+	DSR1H+	DSR2	DSR2	DSR2	DSR2+	DSR2H+	DSR2H+	DSR2H+	DSR2H+	TSR2	10
10.5	155.0	47.25	TSR2H+	TSR3H+	TSR3H+	TDR2+	DSR1H+	DSR1H+	DSR1H+	DSR2	DSR2	DSR2H+	DSR2H+	DSR2H+	TSR2	TSR2H	TSR2H+	TSR2H+	TSR2H+	10.5
11	162.4	49.50	TSR3H+	DDR2	DDR2	TDR2H+	DSR2	DSR2	DSR2	DSR2H	DSR2H	TSR2	TSR2	TSR2	TSR2H	TSR2H+	DDR1	DDR1	DDR1	11
11.5	169.8	51.75	TSR3H+	DDR2	DDR2	TDR3	DSR2	DSR2	DSR2H	DSR2H	TSR2	TSR2	TSR2H	TSR2H	DDR1	DDR1	DDR1H+	DDR1H+	DDR1H+	11.5
12	177.2	54.00	DDR2	DDR2H+	DDR2H+	TDR3	DSR2	DSR2	DSR2H	DSR2H	TSR2	TSR2H	TSR2H	TSR2H	DDR1	DDR1	DDR1H+	DDR1H+	DDR1H+	12
12.5	184.5	56.25	DDR2	DDR2H+	TDR2	TDR3H	DSR2H	DSR2H	TSR2	TSR2	TSR2H	TSR2H	DDR1	DDR1	DDR1H	DDR1H+	TSR3H	DDR2	DDR2	12.5
13	191.9	58.50	DDR2H	TDR2	TDR2H	TDR3H	DSR2H	TSR2	TSR2H	TSR2H	DDR1	DDR1	DDR1H	DDR1H	TSR3H	DDR2	DDR2	DDR2	DDR2	13
13.5	199.3	60.75	DDR2H	TDR2H	TDR2H	—	TSR2	TSR2H	TSR2H	DDR1	DDR1H	DDR1H	TSR3H	DDR2	DDR2	DDR2	DDR2	DDR2	DDR2	13.5
14	206.7	63.00	TDR2	TDR3	TDR3	—	DDR1	DDR1H	DDR1	DDR2	DDR2	DDR2	DDR2	DDR2	DDR2	DDR2H	DDR2H	DDR2H	DDR2H	14
14.5	214.1	65.25	TDR2H	TDR3	TDR3H	—	DDR1	DDR1H	DDR1H	DDR2	DDR2	DDR2	DDR2	DDR2H	DDR2H	DDR2H	DDR2H	DDR2	TDR2	14.5
15	221.5	67.50	TDR3	TDR3H	TDR3H	—	DDR1H	DDR1H	DDR2	DDR2	DDR2	DDR2	DDR2	DDR2H	DDR2H	DDR2H	TDR2	TDR2	TDR2	15
15.5	228.8	69.75	TDR3H	TDR3H	—	—	DDR2	DDR2	DDR2	DDR2	DDR2	DDR2H	DDR2H	DDR2H	TDR2	TDR2	TDR2H	TDR2H	TDR2H	15.5
16	236.2	72.00	TDR3H	TDR3H	—	—	DDR2	DDR2	DDR2	DDR2H	DDR2H	DDR2H	TDR2	TDR2	TDR2H	TDR2H	TDR3	TDR3	TDR3	16
16.5	243.6	74.25	—	—	—	—	DDR2H	DDR2	DDR2H	DDR2H	DDR2H	TDR2	TDR2H	TDR2H	TDR3	TDR3	TDR3	TDR3	TDR3	16.5
17	251.0	76.50	—	—	—	—	DDR2H	DDR2H	DDR2H	TDR2	TDR2H	TDR2H	TDR2H	TDR3	TDR3H	TDR3H	TDR3H	TDR3H	TDR3H	17
17.5	258.4	78.75	—	—	—	—	DDR2H	DDR2H	TDR2H	TDR2H	TDR3	TDR3	TDR3	TDR3	TDR3H	TDR3H	TDR3H	TDR3H	—	17.5
18	265.7	81.00	—	—	—	—	TDR2	TDR2H	TDR2H	TDR3	TDR3	TDR3	TDR3H	TDR3H	TDR3H	TDR3H	TDR3H	TDR3H	—	18

+ Denotes heavy end panels in end bays only.
++ Denotes heavy end panels in two end bays only.

'Mabey Universal' Bailey Bridge Decks for Saudi Highway-Bridge-Loading-1981 for Various Spans *(see Table 14.7)*

Table 14.7			
No of Bays	*Span Length* (m)	*Construction*	
		Single Lane Extra-Wide (4.20 m)	*Two Lane* (7.35 m)
4	18.00	SS++	DS++
4.5	20.25	SSRH++	DS++
5	22.50	SSRH++	DSR2H++
5.5	24.75	SSRH++	DSR2H++
6	27.00	SSRH++	DSR2H++
6.5	29.25	SSRH++	DSR2H++
7	31.50	SSRH++	DSR2H++
7.5	33.75	SSRH++	DSR2H++
8	36.00	DSR1++	TSR2H++
8.5	38.25	DSR1++	TSR2H++
9	40.50	DSR1++	TSR2H++
9.5	42.75	DSR1++	TSR2H++
10	45.00	DSR2H++	TSR3H++
10.5	47.25	DSR2H++	TSR3H++
11	49.50	DSR2H++	DDR2H++
11.5	51.75	DSR2H++	TDR2H++
12	54.00	TSR2H++	TDR2H++
12.5	56.25	TSR2H++	TDR2H++
13	58.50	TSR3H++	TDR2H++
13.5	60.75	TSR3H++	TDR3H++
14	63.00	DDR2H++	TDR3H++
14.5	65.25	DDR2H++	TDR3H++
15	67.50	DDR2H++	QDR3H++
15.5	69.75	DDR2H++	QDR3H++
16	72.00	DDR2H++	QDR4H++
16.5	74.25	TDR2H++	—
17	76.50	TDR2H++	—
17.5	78.75	TDR3H++	—
18	81.00	TDR3H++	—

Table 14.8

Contractor's Loadings
(Special Trucks)

13'-7" (4.14 m) Roadway · 15'-11" (4.85 m) Roadway

VEHICLE • MODEL • TYPE • GROSS WEIGHT (tons)

Bridge for vehicles with axles up to 40 tonnes · Bridge for Vehicles with axles up to 60 tonnes

Bays	Feet	Metres	Terex R25 Rear Dumper 46	Terex TS 14B Scraper 50	Cat 627B Scraper 59	Terex R35B Rear Dumper 65	Cat 769B Rear Dumper 65	Cat 631D Scraper 81	Cat 637 Scraper 80	Terex TS24 Mk5 Scraper 85	Terex R45 Rear Dumper 82	Terex R50 Rear Dumper 88	Cat 773B Rear Dumper 92	Terex TS32 Scraper 118	Cat 657B Scraper 118	Bays
4	59.1	18.00	DS	DS	DS	DS	DS	DS	DS	DS	DS+	DS+	DS+	DS+	DS+	4
4.5	66.4	20.25	DS	DS	DS	DS	DS	DS	DS	DS	DS+	TS	TS	DS+	DS+	4.5
5	73.8	22.50	DS	DS	DS	DS	DS	DS	DS	DS	TS	TS	TS	TS	TS	5
5.5	81.2	24.75	DS	DS	DS	DS	DS	DS	DS	DS	TS	TS	TS	TS	TS	5.5
6	88.6	27.00	DS	DS	DS	DS	DS	DS	DS	TS	TS	TS	TS	TS	TS	6
6.5	96.0	29.25				TS	TS	TS	TS	TS	TS	DSR2+	DSR2++	DSR2++	DSR2++	6.5
7	103.3	31.50	DS	DS	DS	TS	TS	TS	TS	TS	DSR2+	DSR2++	DSR2++	DSR2++	DSR2++	7
7.5	110.7	33.75	DS	DS	TS	TS	TS	TS	TS	TS	DSR2+	DSR2++	DSR2++	DSR2++	DSR2++	7.5
8	118.1	36.00	TS	TS	TS	TS	TS	TS	TS	DSR2+	DSR2+	DSR2++	DSR2++	DSR2H++	DSR2H++	8
8.5	125.5	38.25	TS	TS	TS	TS	TS	DSR2	DSR2	DSR2+	DSR2+	DSR2++	DSR2H++	DSR2++	DSR2++	8.5
9	132.9	40.50	TS	TS	TS	DSR2	DSR2	DSR2	DSR2+	DSR2+	DSR2+	DSR2H++	DSR2+	TSR2++	TSR2++	9
9.5	140.3	42.75	TS	TS	TS	DSR2	DSR2	DSR2+	DSR2+	DSR2++	DSR2++	DSR2H+	TSR2+	TSR2H++	TSR2H++	9.5
10	147.6	45.00	TS	TS	TS	DSR2	DSR2	DSR2+	DSR2+	DSR2++	DSR2H++	TSR2	TSR2+	TSR3+	TSR3+	10
10.5	155.0	47.25	TS	TS	DSR2	DSR2	DSR2	DSR2+	DSR2+	DSR2H++	TSR2	TSR2	TSR2H+	TSR3+	TSR3+	10.5
11	162.4	49.50	DSR2	DSR2	DSR2	DSR2	DSR2	DSR2+	DSR2H+	TSR2	TSR2	TSR2H+	TSR2H+	TSR3H++	TSR3H++	11
11.5	169.8	51.75	DSR2	DSR2	DSR2	DSR2H	DSR2H	DSR2H+	DSR2H+	TSR2	TSR2H	TSR2H+	TSR3	TSR3H++	TSR3H++	11.5
12	177.2	54.00	DSR2	DSR2	DSR2	DSR2H	DSR2H	TSR2	TSR2	TSR2	TSR2H+	TSR3	TSR3	DDR2	DDR2	12
12.5	184.5	56.25	DSR2	DSR2	DSR2		TSR2	TSR2	TSR2	TSR2H	TSR3	TSR3H	TSR3H	DDR2	DDR2	12.5
13	191.9	58.50	DSR2	DSR2H	DSR2H	TSR2	TSR2	TSR2H	TSR2H	TSR3	TSR3H	TSR3H	TSR3H	DDR2	DDR2	13
13.5	199.3	60.75	DSR2H	DSR2H	TSR2	TSR2H	TSR2H	TSR3	TSR3	TSR3H	TSR3H	DDR2	DDR2	DDR2H	DDR2H	13.5
14	206.7	63.00	TSR3	TSR3	TSR3	TSR3	TSR3	TSR3H	TSR3H	TSR3H	DDR2	DDR2	DDR2	DDR2H	DDR2H	14
14.5	214.1	65.25	TSR3	TSR3	TSR3	TSR3H	TSR3H	TSR3H	DDR2	DDR2	DDR2	DDR2	DDR2	TDR2	TDR2	14.5
15	221.5	67.50	TSR3	TSR3	TSR3	DDR2	DDR2	DDR2	DDR2	DDR2	DDR2	DDR2H	DDR2H	TDR2H	TDR2H	15
15.5	228.8	69.75	DDR2	DDR2	DDR2	DDR2	DDR2	DDR2	DDR2	DDR2	DDR2H	DDR2H	DDR2H	TDR2H	TDR2H	15.5
16	236.2	72.00	DDR2	DDR2	DDR2	DDR2	DDR2H	DDR2	DDR2	DDR2H	DDR2H	TDR2	TDR2H	TDR3	TDR3	16
16.5	243.6	74.25	DDR2	DDR2	DDR2	DDR2	DDR2H	DDR2H	DDR2H	DDR2H	TDR2H	TDR2H	TDR2H	TDR3H	TDR3H	16.5
17	251.0	76.50	DDR2	DDR2	DDR2	DDR2H	DDR2H	DDR2H	DDR2H	TDR2	TDR2H	TDR3	TDR3	TDR3H	TDR3H	17
17.5	258.4	78.75	DDR2	DDR2	DDR2	DDR2H	DDR2H	DDR2H	DDR2	TDR2H	TDR3	TDR3	TDR3	—	—	17.5
18	265.7	81.00	DDR2	DDR2	DDR2H	DDR2H	DDR2H	TDR2H	TDR2H	TDR3	TDR3	TDR3H	TDR3H	—	—	18

14.24 DIMENSIONS AND DETAILS AT SUPPORTS (MABEY SYSTEM)

See Fig. 14.13 and Table 14.9, and Figs 14.14, 14.15 and 14.16.

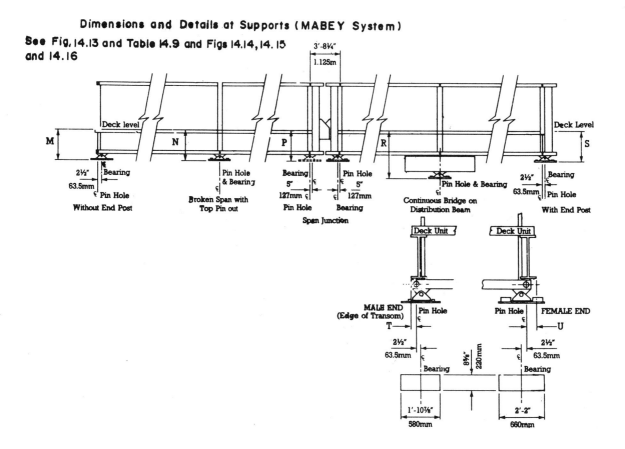

Dimensions and Details at Supports (MABEY System)

See Fig. 14.13 and Table 14.9 and Figs 14.14, 14.15 and 14.16

Fig. 14.13

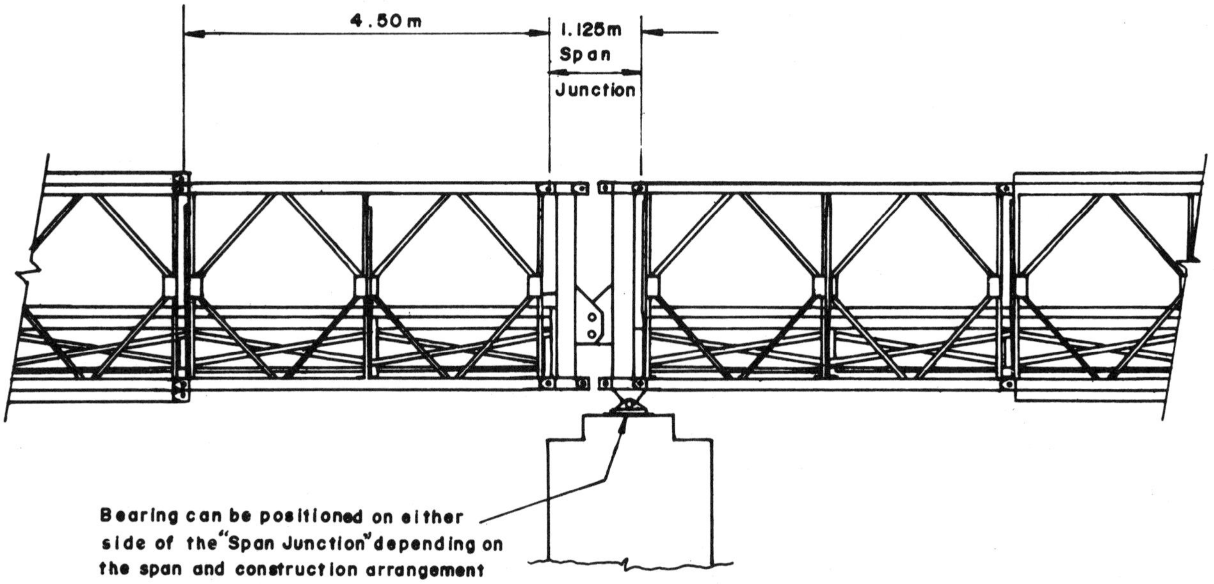

Bearing can be positioned on either
side of the "Span Junction" depending on
the span and construction arrangement

Fig. 14.14   Detail showing "Mabey" span junction arrangement

Table 14.9

	Roadway Width	Dimension → / Unit	M	N	P	R	S	T	U
HIGHWAY	Standard Single Lane	Impl.	2′10″	3′1⅝″	3′1⅝″	4′11⅝″	2′10″	2¼″	6′⅛″
		Met.	862	957	957	1515	862	58	156
	Extra Wide Single Lane	Impl.	2′10¼″	3′1⅞″	3′1⅞″	4′11¾″	2′10¼″	2¼″	6′⅛″
		Met.	867	962	962	1520	867	58	156
	Two Lane Carriageway	Impl.	3′6″	3′9¾″	3′9¾″	5′7¾″	3′6″	2⅝″	6′⅜″
		Met.	1067	1162	1162	1720	1067	67	161
	Three Lane Carriageway	Impl.	3′9⅜″	4′1⅛″	4′1⅛″	5′11$\frac{1}{16}$″	3′9⅜″	3¼″	6′⅞″
		Met.	1152	1247	1247	1805	1152	82	173
CONTRACTOR'S	13′–7″ 40 Tonne Single Axle 4.14m	Impl.	3′8″	3′11¾″	3′11¾″	5′9¾″	3′8″	2⅝″	6′⅜″
		Met.	1119	1214	1214	1772	1119	67	161
	15′–11″ 60 Tonne Single Axle 4.85m	Impl.	3′9½″	4′1⅛″	4′1⅛″	5′11⅛″	3′9½″	2⅝″	6′⅜″
		Met.	1154	1249	1249	1807	1154	68	161

Dimensions are subject to Manufacturing Tolerances.

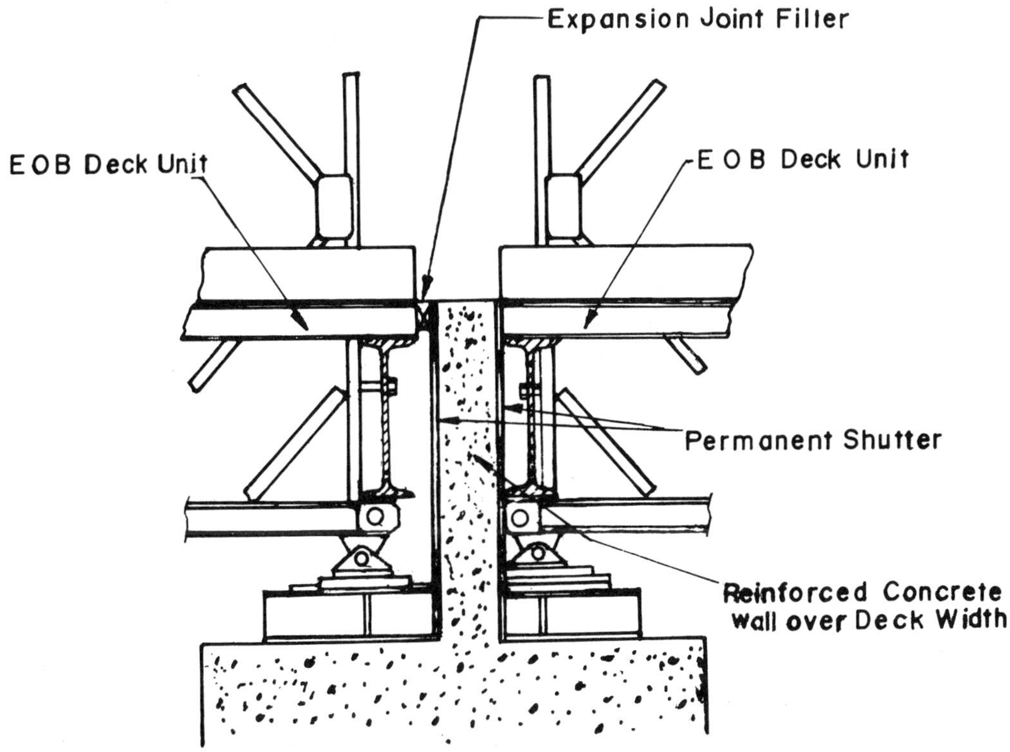

Fig. 14.15 Alternative deck junction details

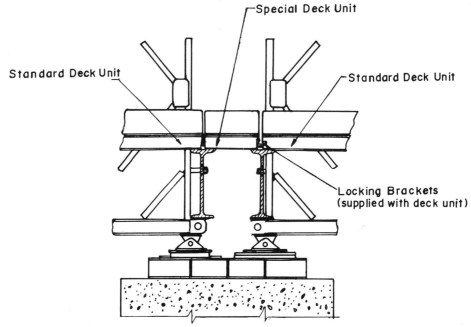

Fig. 14.16 Alternative deck junction details

14.25 CONTINUOUS SPANS

- In addition to simple spans, Mabey-Universal may be built as a multi span bridge over a number of supports.

- The simplest method is to construct the bridge with the girders continuous over the whole length of the bridge (*See* Fig. 14.17). This may lead to economy in the construction of the side girders since the maximum bending moment may well be less than in the equivalent simple span. There is an advantage inasmuch as the intermediate supports may be sited anywhere along the length of the bridge and are not restricted to being at exact multiples of 14'9" (4.5m), the panel length.

- There are, however, inherent problems in this system. Should an end span be shorter than the adjacent intermediate span, the ends of bridge (posts) will tend to lift off their bearings when a heavy load passes over the intermediate span. Also, if the nature of the ground is such that settlement of an intermediate support is likely, then the side girders may be subjected to stresses beyond the safety limit. In either of these cases, the correct solution is to build the bridge as a series of 'broken' spans (Fig. 14.18).

14.26 BROKEN SPAN

- There are two possible ways of producing a broken-span-structure with the Mabey Universal. If the degree of rotation over the intermediate support is small, but a discontinuity is required, the top panel chord pin may be removed. This would be done once

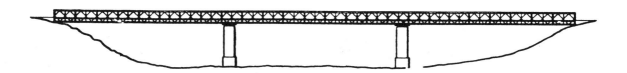

Fig. 14.17

BROKEN SPAN

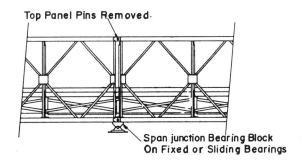

Top Panel Pins Removed.

Span junction Bearing Block
On Fixed or Sliding Bearings

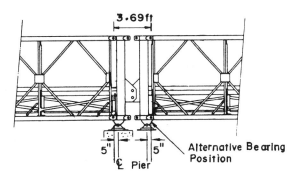

3·69ft

5"   5"   Alternative Bearing
Position

℄ Pier

SPAN JUNCTION EQUIPMENT

Fig. 14.18

the bridge has been launched and is in its final position (Fig. 14.17).

- If a higher degree of rotation is expected at the support and (or) a high shear force is to be transferred across from one span to the support as would be the case for a double-storey truss, then the 'span-junction-equipment' is introduced (Fig. 14.18).

- Short chord sections are introduced for continuity over the span junction length for the launching stage, and are later removed.

14.27 LAUNCHING AND ERECTION

A basic principle of the Universal Bridge is that it is designed to be completely erected on rollers on one side of the gap to be bridged, and then 'launched' across without requiring any temporary supports in the gap. This is achieved by building onto the front-end of the bridge a temporary skeleton structure, called the 'Launching Nose', which is constructed from the same standard components as the bridge. The 'nose' is built of such a length that when the whole structure is rolled forward, the tip of the nose lands on the rollers on the

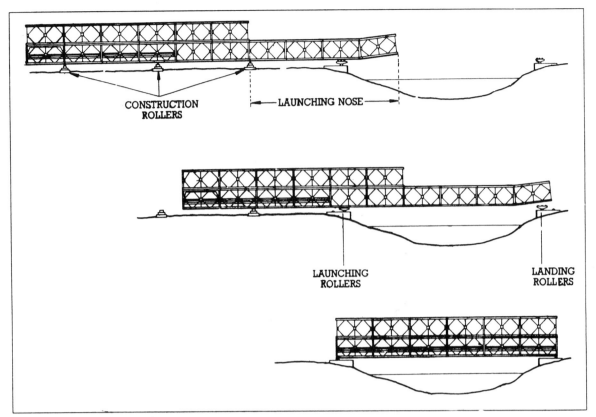

Fig. 14.19 Launching and erection

far bank before the centre of gravity passes the launching rollers. Once the bridge is in position across the gap, the launching nose is dismantled and the bridge jacked up off the rollers and lowered onto its permanent bearings on the abutments (piers).

Alternatively, where adequate cranage is available, the bridge can be lifted-in as a complete unit or built up 'in situ'.

14.28 'MABEY'-UNIVERSAL BRIDGE RAMP-UNITS . . . . 9.0 M (30 FT.)

See Figure 14.20.

14.29 'M2' BAILEY BRIDGE

(a) *Description*
— The M2 Bailey Bridge was developed by the U. S. Army Research and Development Laboratories. It has been well proven both as a tactical and as a line of communications bridge, and is capable of carrying heavy traffic loads. They are used for both temporary and permanent service. In emergencies they can be opened to traffic in 1 to 3 days.
— The M2 Bailey Bridge is an all-purpose prefabricated steel panel bridge deck, designed for porta-

bility and speed of erection under adverse conditions. Optimum spans are 12.2 to 61 m. Width is 3.809 m between steel kerbs and 4.343 m between trusses. The components are manufactured in fixtures to ensure accuracy and interchangeability. The heaviest component weighs 281 kg.
— The Bailey roadway is supported between two trusses or multiple truss girders. It consists of longitudinal runner planks over transverse planks (chess), laid over steel stringers supported by the floor beams (transoms) which rest on (and are clamped to) truss bottom chords. Steel kerbs secure the 'chess' to the stringers. The basic truss element is the 'panel' which is 3.048 m long, 1.448 m deep and 165.1 mm wide. Pin-connected end-to-end trusses of any length are formed. Where strength exceeding that of single trusses is needed, multiple-truss girders can be assembled, with either two or three panels side-by-side in single, double, or triple storey heights. End-ramps extend the deck 3.048 m onto the approach runway.
— Bailey truss panels, end posts, transoms, and ramps, are of low-alloy high tensile steel, having a yield point of 3515 kg/cm^2 and an ultimate strength of 4921.7 kg/cm^2.

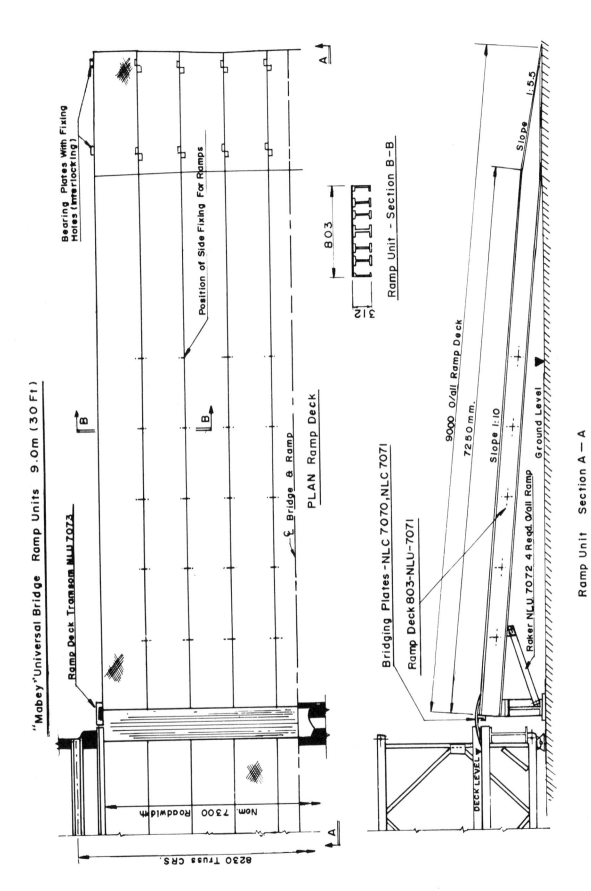

Fig. 14.20 Ramp arrangement

— The cantilever method of erection is accomplished without falsework. The bridge is assembled on stationary rollers, and then pushed or pulled across the gap. A skeleton launching nose is assembled from standard bridge components and fixed to the leading-end of the bridge. The nose precedes the leading-end of the bridge while the bridge proper, acting as a counterweight, enables the nose to reach and 'land' on rollers on the far bank. The bridge is then 'rolled' into position, the nose is removed, and the span is lowered onto its bearings. A 24.4 m double-truss bridge has been completed in less than 40 minutes in competitive trials!

— There are approximately 15 major components in an average Bailey Bridge, and a total of about 50 components; fittings, accessories, special items and tools are available.

— See Tables 14.10 and 14.11 ahead for details.

(b) *Specifications*

• Length of Single Span
  — 9.44 m min.
  — 61 m max.

• Width of Roadway: 3.809 m

(c) *Status*

Production is complete, and it is in service with the U. S. Army and with other armed forces.

(d) *Supplier*

Bailey Bridge Inc., P. O. Box 1186, San Luis Obispo, California, 93406, U. S. A.
Tel: (805) 543-8083, Fax: (805) 543-8983.

Table 14.10   U.S. 'M2' Bailey Bridge Selection Chart

'Span' in Feet (a)	'Class' Tons (b)	'Configuration' (c)
50	75/70	DS
60	65/65	DS
70	60/60	DS
80	50/55	DS
90	65/65	TS
100	50/55	TS
110	65/70	DD
120	45/55	DD
130	55/60	TD
140	45/55	TD
150	60/60	DT
160	55/55	DT
170	45/50	DT
180	45/55	TDR
190	45/50	TDR

NOTE:

(a) 'Spans' are clear distance between bearings, in feet.

(b) 'Class' is Military Class Loading in tons as specified in U.S. Army Technical Manual FM5-277, 1986 and TM5-312, 1968 (refer to 14.11 ahead).

(c) *Configurations*

DS	=	double-single
TS	=	triple-single
DD	=	double-double
TD	=	triple-double
DT	=	double-triple
TDR	=	triple-double chord-reinforced

U.S. ARMY VEHICLE — LOAD CLASSIFICATION DATA. (Table 14·II)

Table 14·II   Standard Class Hypothetical Vehicle Characteristics

HYPOTHETICAL VEHICLES FOR CLASSIFICATION OF ACTUAL VEHICLES AND BRIDGES.

WHEELED VEHICLES.

1	2	3	4	5	6	7	8	9
Class	Tracked Vehicles	Axle Loads And Spacing	Maximum Single Axle Load In Short Tons	Minimum Wheel Spacing And Tire Sizes Of Critical Axles.			Maximum Tire Load And Minimum Tyre Size	Class

N. B.
(1) SINGLE AXLE TIRE SIZES SHOWN IN COLUMNS 5,6 & 7 REFER TO THE MAXIMUM SINGLE AXLE LOADS GIVEN IN COLUMN 4.
(2) BOGIE AXLE TIRE SIZES SHOWN IN COLUMNS 5,6 & 7 REFER TO THE MAXIMUM BOGIE LOADS SHOWN ON THE DIAGRAM IN COLUMN 3.
(3) THE MAXIMUM TIRE PRESSURE FOR ALL TIRES SHOWN IN COLUMN 8 SHALL BE TAKEN AS 75 LBS./SQ. INCH.
(4) THE FIRST DIMENSION OF TIRE SIZE REFERS TO THE OVERALL WIDTH OF TIRE AND THE SECOND DIMENSION IS THE RIM DIAMETER OF THE TIRE.

SHORT TONS.

15
DIAGNOSTIC TESTING

15.1 INTRODUCTION

Apart from visual inspection (described in Chapter 5), certain diagnostic tests may have to be performed on the materials (concrete, steel, etc.) of the structure in order to ascertain the 'material-properties' and any onset of 'chemical attacks'. This information can have far reaching effect on the details of appropriate nature of repair and strengthening that may be called for. After detailed visual investigations, the diagnostic test results can assist the engineer to decide upon the appropriate rehabilitational steps. Diagnostic tests are somewhat like the 'clinical tests' meant to diagnose the diseases and defects in the body of the structure to enable the engineer (the doctor) to plan and administer the appropriate 'medication' (the repairs and strengthening) so as to 'cure' the structure (the patient) as best as possible within the prevailing constraints of ability and funds. This should not be confused with the full scale Load Tests on the structure which may become incumbent if the as-built drawings are unavailable and the engineer is unable to measure the physical dimensions and reinforcement details within the deteriorated structure and yet has to rate the bridge capacity. (Load Testing of some major bridge elements is described in Chapter 16.)
    Broadly speaking
— the visual inspection methods are useful for detecting visible cracking, scaling, wear and all the physically feelable and visible distresses;
— the Sonic, the Electrical and the various Chemical Diagnostic Tests are suited for corrosion detection;
— the Ultrasonic (pulse velocity) Diagnostic Tests are more suited for detecting cracks and deformities;
— the Thermographic and Radar-type Diagnostic Tests are suitable for corrosion detection as well as discovering delaminations within concrete (and also beneath the bituminous surfaces);
the Radiographic and Air-permeability Diagnostic Tests have only a limited application (to detect corrosion and voids in grout for instance).

15.2 AN INTRODUCTION TO SOME DIAGNOSTIC TESTING METHODS

15.2.1 General

A variety of nondestructive diagnostic testing methods are available for investigating different properties of concrete and chemical attacks on concrete and steel. Tests are aimed at assessment of strength and other properties and to locate and obtain comparative results indicating permeable regions, cracks or laminations and areas of lower integrity than the rest. It is essential to emphasize here that it is not necessary to carry out all the tests in each cases except the most relevant ones. Also, not much may be achieved by spending time and money on carrying out other than the most essential tests in many cases. In fact, in some cases, engineering judgements could help taking decisions faster!

15.2.2 Some Chemical Diagnostic Tests

    (a) For Carbonation Depth: The carbonation of concrete, on the surface, results in loss of alkaline protection of the cover over the steel against corrosion. Carbondioxide of atmosphere reacts with hydrated cement compounds causing reduction in pH and alkalinity (i.e. increase in acidity) in concrete, and the process is referred to as carbonation. The depth of carbonation is measured by applying on the freshly broken surface of concrete a one per cent solution of phenolphthalein. The concrete undergoes a colour change to purple red (red violet) when pH value is > 9.5. The colour of the concrete surface after the spray may be compared with the classified standard test results to indicate the areas of serious carbonation. (For details, see ahead, as also Chapter 4.) Where carbonation has taken place and acidity has been brought about, the pH reduces to below 9, concrete undergoes no colour change.

    (b) For Sulphate Attack and its Concentration: The concrete attacked by sulphate has a characteristic white stained appearance. The quantity of sulphate is

estimated by the precipitation of barium sulphate (sulphate confined by identification of calcium sulpho aluminate by microscopy). (For details *see* ahead, as also Chapter 4.)

(c) For Chloride Content: The chloride content in concrete is measured in laboratory by Mohr's method using potassium chromate as indicator in a neutral medium or by Volhard's Volumetric Titration method in acidic medium. The presence of water-soluble chlorides in concrete beyond the permissible limit (0.20 per cent by weight of chloride ion in concrete mix) is considered as a corrosion hazard in concrete structures. (For details, *see* ahead, as also Chapter 4.)

15.2.3 Some Non-Destructive Diagnostic Tests (NDT)

(a) Schmidt Hammer and Other Tests for Concrete Strength: This is used to measure hardness at concrete surface, which can be related to its strength. The instrument used is very handy. The pull-out methods and penetration resistance techniques are also adopted for estimation of strength of concrete and assessment of its overall quality. (For details refer to the author's book: *'Concrete for Construction — Facts and Practice'*, Tata McGraw-Hill, New Delhi, India.)

(b) Magnetic Methods for Reinforcement Detection: These are used to determine the position of reinforcement with reference to the surface of concrete, and thus the adequacy or otherwise of the cover over the reinforcement can be assessed. 'Pachometers' detect position of reinforcement and measure the depth of cover. A battery is used to generate a magnetic field which gets distorted where there is steel in the vicinity of the probe and the difference of distortion is the function of the mass of the steel and its distance from the probe. Several portable battery operated 'cover-meters' are available which can measure the cover with an accuracy of 15 mm up to a depth of about 75 mm. Special calibration curves must be established when some of the constituents of the concrete like pozzolana or sand contain magnetic particles. (*See* ahead for details.)

(c) Radar Technique for Detecting Cracks and Deterioration in Concrete: A high frequency pulsed radar can be used to detect deterioration in concrete decks. The echos produced from the pavement surface and from the interface with the bridge deck concrete in case of bituminous surfaced bridge decks, are very distinct so that thickness can be measured accurately. Short duration pulses of radio-frequency energy are directed into the deck portion and is reflected from any interface and the output is displayed on an oscilloscope. The 'interface' can be any discontinuity or differing dielectric, such as cracks in concrete. A permanent record can be stored on magnetic tape and the unit is normally mounted on a vehicle and data is collected as the vehicle moves slowly along and across the deck. (This method is not commonly used in practice.)

(d) Radiography for Detecting Defects in Steel and Grout, etc.: Radiographic techniques are applied to the prestressing cables to detect the defects in them and to examine the quality of grout within the ducts. Most applications involve transmission of wave-energy. The emerging radiation is detected by photographic emulsion or by a radiation-detector. The former is called radiography and the latter radiometry. The back-scatter techniques, based on reflected intensity of X-rays, can be used to detect the voids in grout and for testing strands or wires that are broken or are out of position. However, small amounts of corrosion will not be detected and the technique is suitable only for isolated cables without any other obstruction in the path of the wave. (This method is not commonly used in practice.)

(e) Thermography for Detecting Delaminations and Cracks: Infra-red thermography is the method of detecting delamination in concrete bridge decks and columns exposed directly to sun. The method works on the principle that discontinuity within the concrete, such as a delamination, interrupts the 'heat transfer' through concrete. The differences in surface temperature are measured by sensitive infra-red detection systems which consist of infra-red signal, control-unit and a display-screen. The images are recorded on photographic plates or video tapes. The equipment can be truck-mounted, permitting a lane width to be scanned by a single pass. The main disadvantage of thermography is that while a positive result is valid, a negative result may not always be reliable because it relates to results under conditions prevailing at the time of tests. Nevertheless, the method itself has a considerable promise as a rapid screening tool for determining whether a more detailed investigation is required. (It is still in development stages.)

(f) Ultrasonic Pulse Velocity Measurement for Detecting Defects in Concrete: The quality of concrete can be assessed by passing though concrete ultrasonic pulse and measuring its velocity. The measured values can be affected somewhat by surface texture, moisture content, temperature and specimen size, but are significantly affected by reinforcement, deformities and cracks in concrete and density of concrete.

Correlations with concrete strength are difficult to make since there will be influence of types and proportions of mix constituents and maturity. However, correlation with control specimens gives a 'relative' picture. (For more details, *see* the author's book: *'Concrete for*

Construction — Facts and Practice', Tata McGraw-Hill, New Delhi, India.)

(g) Pull-out Test for Finding Compressive Strength of Hardened Concrete: It is possible to assess the compressive strength of hardened concrete by correlating it to the pulling force required to pull an embedded metal device inserted in concrete. (*See* ahead for details.)

(h) Coring for Detecting Crack Penetration, Delaminations and Concrete Strength: Cores of concrete from the structure are drilled with the help of a coring machine. The core is then analyzed in a laboratory for various properties including its strength (Portable Core drilling machines are available commercially, and cores can be taken bottom-up from a slab, for instance, although with some inconvenience.)

(i) Endoscopic Examination: Optical fibre (flexible) viewing tubes are inserted into holes drilled in the concerned bridge component or the cable-duct in prestressed concrete. Light is transmitted in from an external source, enabling a visual inspection. Endoscopes are available with attachments for a camera or a TV-Monitor and are used for detailed examination of parts of the bridge structure which cannot otherwise be seen. They are useful in detecting voids in the grout and concrete, and corrosion in steel, etc. (*See* ahead for details.)

(j) Electrical Device for Corrosion Detection: The electrode (half-cell) potential of reinforcement embedded in concrete provides a measure of the corrosion-risk, and indicates if electrochemical reaction has taken place on the electrode surface. The electrical potential difference between the steel and electrode (concrete) is measured with copper/copper-sulphate half-cell or copper calomel electrode or silver chloride electrode (*see* Fig. 15.1). The test by the 'pathfinder' equipment is a refined version of copper/copper-sulphate electrode for better scanning. However, this method does not give

information on the 'rate of corrosion' and also it gives only the probability of activity of corrosion. (*See* ahead for more details.)

15.3 DETAILS OF THE MORE USUAL DIAGNOSTIC TEST METHODS

15.3.1 General

The philosophy of the special Bridge Inspection is to combine a visual assessment of the structure with appropriate test methods to obtain sufficient information on the 'condition of the structure'. Location and selection of representative samples are important to get accurate conclusions for the entire structural element considered. The extent of tests must be sufficient for determining the right repair strategy, and for giving a good estimate of the total area requiring repair.

Personnel performing the special bridge inspection should be experienced and competent to know
— how to carry out the available methods of testing in practice, including how to operate the equipment;
— how to select the right type of test-method and the right test-locations for different types of damage;
— how to interpret the results of the measurements.

The test methods can be divided into three categories
— the non-destructive 'survey' methods, which are suitable for *mapping* the damage on large areas of the structure;
— the *detailed* non-destructive and destructive sampling and measurements on small areas;
— the *laboratory analysis*, which, when applied on the samples, provides detailed and precise information about specific locations in the structure.

Normally, the combination of all three categories can lead to reliable conclusions on the mechanisms of

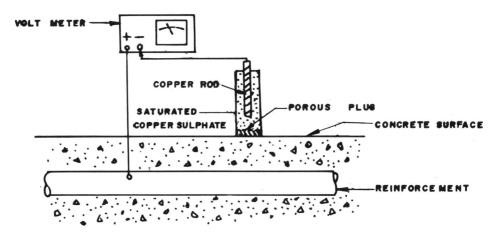

Fig. 15.1 Electrical potential measurement of reinforcement

deterioration, the cause, and the extent of damage. However, in most cases, the first two categories of tests are satisfactory enough to conclude upon many types of distress.

On site, a special inspection always starts with a hands-on type of visual inspection in order to get an overview of the structure and the damage. This concludes an assessment of the design, construction and maintenance of the structure and the effect of the surroundings, since the initial cause of damage is often related to these.

When the visual inspection is completed, a hypothesis for the cause of damage is made, and based upon this, test methods and locations of test areas are selected.

Visual registrations are then recorded on sketches, showing plan and elevation views. The sketches must include identification and orientation of the elements in question.

When indicating positions, vertical distances are normally measured from ground level. Horizontal distances must be measured from a well defined point. (e.g. the 'south-west' corner of a column, etc.)

When all planned tests are completed, the visual registrations and test results must be evaluated to see if they form a sufficient basis for concluding the cause, extent, and possible development of the damage. Otherwise, supplemental tests must be chosen and performed.

If the test results do not confirm the hypothesis of cause of damage, the hypothesis must be revised. It may be necessary to perform supplementary tests to confirm the revised hypothesis.

15.3.2 Investigation of Concrete Structures

1. *Mapping of Damage*

Mapping is normally used for providing an overview of damage and for registering the extent of the damage. This is necessary for the rehabilitation design.

For investigating structural cracks, the specific crack pattern must be mapped, indicating the type (bending, shear, etc.) and extent of each represented section. For non-structural crack patterns, the areas must be mapped and the assumed cause noted.

Spalling, scaling and rust/rust-stains should be indicated by area. Splitting-cracks due to corrosion must be mapped.

When mapping specific cracks, characteristic widths are measured with a 'scale' and the location indicated as a supplement to the visual inspection of the surfaces.

Regarding both structural cracks and the non-structural (i.e. non-load-induced) cracks, the risk of corrosion

in the future should be evaluated. If risk of corrosion is possible, it should be noted in the report.

2. *Break-ups from Concrete in the Structure*

Break-ups (e.g. cores) taken from concrete in the structure provide in most cases a necessary supplement

— to register the general condition of the concrete
— to register the type of reinforcement, dimension, position and possible corrosion activity
— to provide a reference level for Electrochemical potential (ECP) measurements, and to calibrate the measured corrosion activity indicated by ECP measurements

Break-ups are also necessary in case the concrete surface in question is not visible because of the wearing course or in similar cases

Break-ups are a common part to inspections of concrete structures and are usually performed with the help of power tools. The backside of exposed rebars can be inspected by the use of a dental mirror.

When performing break-ups, take care not to weaken the structure to a harmful extent.

The structure should always be re-established immediately after making break-ups.

3. *Cover-Meter Measurements*

The cover-meter, also called Pachometer, is used to locate the reinforcement in the concrete and to measure the depth of the concrete-cover. The cover-meter is often used to locate the rebars before starting other investigations.

The cover-meter measurement is based on changes in the magnetic field lines/eddy-current. The presence of nearby magnetic rebars will cause changes, which can be measured by passing the measuring-head over the surface above the rebars.

This measuring-head is a rectangular encapsulated unit containing the search coil. As the coil windings are directional, the head should always be used with it's longitudinal axis parallel to the expected line of the reinforcing bars. A 'lead' from the head is plugged into the battery-operated cover-meter.

The method is suitable in general. Tests have shown that the inaccuracy increases from 5–10 per cent at approx. 35 mm depth to about 15–25 per cent at 60–70 mm depth.

4. *Electrical (electro-chemical) Potential Measurements for Corrosion (Pathfinder, ECP Test)*

4.1 The **purpose** of the potential measurements is to map the electrochemical potentials in order to locate areas with risk of corrosion.

The potentials are measured either by the pathfinder equipment or by an ordinary multimeter (voltmeter) and a reference electrode.

In the field the following steps 1 to 5 have to be followed.

Step 1 *For exposing a rebar for the electrical connection*

— normally go for stirrups; The most convenient area is where the cover is minimum; try locate the areas with the smallest cover

— on bridge decks with an asphaltic overlay, the connection can be made on the edge beams

Step 2 *For checking the circuit of the reinforcement*

— on columns, make a contact with the reinforcement of another column to check the circuit by using the multimeter; The potential difference must be zero to have the circuit required—if that is the case, use the same connection during the whole measuring

If the difference is not zero, then first check the connection to the rebar. If this connection is good, the internal connection of the reinforcement is not sufficient and you have to make contact with every column; Look for joints in the bridge deck and check columns on both sides of the joints

— on decks, use the above described procedure

Step 3 *Make a 'measuring grid' (columns and rows) on each part to be measured*

— when making survey measurements on large areas, a mesh size of 500 mm × 500 mm may be chosen.

Prior to making the grid, survey-measurements at (more or less) random locations may help locating the areas to be mapped.

When making measurements in areas where corrosion is likely to occur (selected as a result of condition survey, experience, or other tests), the mesh size must be 250 mm × 250 mm or less.

The grid-size, its location, and orientation, must be marked on sketches of the structure.

— If the measurements are started in the upper left corner of the grid and made one horizontal row at a time, the printout from the Pathfinder equipment will have the same orientation as the grid.

In many cases, e.g. on columns, it is more convenient to measure in vertical rows. In this case, start the lower left corner of the grid and make the measurements of a vertical row from the bottom and upwards. Proceed to the right to the next row (i.e. counterclockwise on columns).

In this case the printout must be rotated 90° counterclockwise to get the correct orientation.

Step 4 *Check the stability of the 'potential' measurements*

— wet a single measuring point,

— place the electrode, and note the potential and time,

— wait until the potential is stable. Note the potential and time. This time difference is the necessary time required between wetting and measuring. In very dry concrete, it is normally necessary to make continuous wetting for a longer period. It means that one person is constantly wetting while the other person is doing the measuring.

Step 5 *Commence the measurements*

4.2 *Evaluation of Corrosion*: It should be noted that the 'potential measurement' itself does not lead to a final assessment of the condition. Supplementary testing has to be carried out. As a first guide to an evaluation of the reliability of the measured potential values, the measurements are normally divided into phases:

Phase 1: Immediately after completing the measurements (x), the measurements are printed out on the printer and then evaluated according to a scale based on experience, e.g.

Group (1):	x above 0 mV	No Corrosion
Group (2):	x between 0 and − 200 mV	Corrosion in very early phase, No signs of corrosion on bars.
Group (3):	x between -200 to − 300 mV	Corrosion in a phase where the first sign of corrosion will be visible on the bars.
Group (4):	x more than − 300 mV	Corrosion in a phase where the corrosion is notorious, with evident corrosion signs on the bars and in some cases on the concrete too.

Hint! Use colours to distinguish between the groups, e.g.

Group (1) — white
Group (2) — green
Group (3) — yellow
Group (4) — red

In this way, the most critical areas are easily pinpointed.

NOTE: Areas which are already suffering from evident corrosion attack (the surface of the concrete for instance has severe cracking, spalling and/or it sounds hollow) *cannot* be evaluated according to these limits. These areas are always placed in Group (4).

Phase 2: Make ''break-ups'' of concrete to confirm the first evaluation and to understand the approximate reduction of cross-sections if any.

Note that the potential measurement is meant only for the detection of areas with corrosion activity. The reduction in cross-sections cannot be assessed by ECP-measurements.

Breaks-ups must be carried out for each group. As a rule of thumb, the break-ups are located in the highest 'negative' areas in each group.

If the first evaluation confirms the results, further break-ups have to be carried out only as required for confirmation. When the correlation between the potential values and the actual corrosion condition has been established through the break-ups, the ECP-measurements can be used to assess the size of the damaged areas as a basis for rehabilitation design.

If the first elevation does not confirm the results (if e.g. severe corrosion is found in Group (3), limits must be changed for the four groups accordingly.

These new limits must be confirmed by new break-ups.

The above mentioned limits (and the potentials mentioned in the following) are based on measurements with an Ag/AgCl electrode. If using another electrode, limits have to corrected according to Table 15.1.

Table 15.1		
Type of Electrode	Cu/CuSO₄	Calomel
Constant*	$-80\,mV$	$-75\,mV$

Type of Electrode uses $Cu/CuSO_4$ and Calomel columns.

* When using an electrode other than Ag/AgCl, add the constant mentioned to the measured value for a comparison with measurement taken with a Ag/AgCl electrode.

4.3 'Potential' Levels in Different Situations — Principles: The potential levels in the following examples must only be regarded as guidelines. Local conditions may influence the levels.

(a) Presence of Oxygen (Fig. 15.2): The potentials may be low below ground/sea level, even if no corrosion takes place.

When corrosion takes place around ground/sea level where sufficient oxygen is available, the potential is expected to be around -300 mV above ground/sea level and a little lower below ground/sea level. Corrosion may take place far below ground/sea level, even if no oxygen is available, because the oxygen-consuming cathode-formation process takes place above ground/sea level. In this case the potential may be around -200 mV above ground/sea level and much lower below.

(b) Moisture Content (Fig. 15.3): If no corrosion is going on, the potentials are expected to be around +100 mV in dry concrete. If chloride-initiated corrosion is going on, the potentials are expected to be around -300 mV in (relatively) dry concrete. In wet concrete, the potentials will be approximately 100 mV lower.

(c) Cause of Corrosion (Fig. 15.4): The figure shows the 'range' for potentials for CO_2 and Cl^- initiated corrosions, respectively.

(d) Degree of Deterioration (Fig. 15.5): 'Local corrosion' which may not cause spalling of concrete-cover because of incomplete rusting, shows lower potentials than 'general corrosion' which is associated with complete rust-formation and consequent high expansion pressure.

In general, the potentials decrease with increasing temperature.

5. Carbonation Test

This test is performed by applying an 'indicator' solution to concrete surfaces just fractured. The colour of the solution will change with corresponding changes in pH

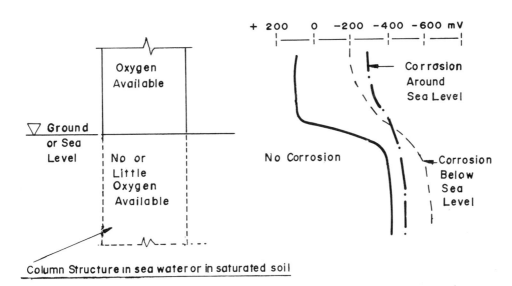

Fig. 15.2   Presence of oxygen

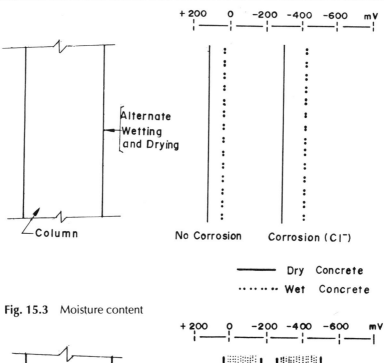

Fig. 15.3 Moisture content

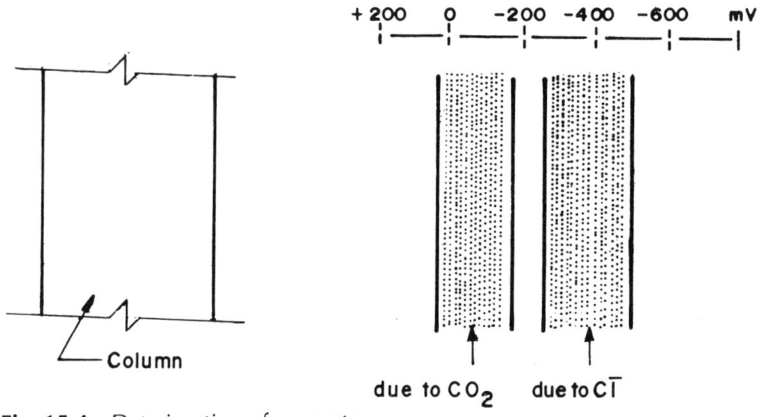

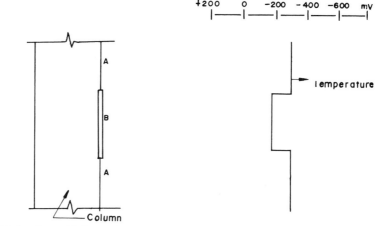

Fig. 15.4 Deterioration of concrete

Fig. 15.5 Degree of deterioration

of the concrete. The carbonation depth is then measured by means of a scale.

Two different indicator solutions are available in the laboratory:

— *Phenolphthalein (1% solution)*—after its spray-application, if the colour of the concrete surface immediately turns red-violet, it indicates a pH 9.5, but if pH has fallen lower, i.e. acidity has set-in, then carbonation has been caused and such carbonated portions will remain colourless. The depth of the colour-less portion will be the depth of carbonation, maximum carbonation being at the surface;

— *Rainbow Indicator* — this solution, when applied, will produce a series of colours corresponding to various pH-levels, again distinguishing the acidity levels of the fractured concrete along its depth.

6. *Relative Humidity Test (RH-Test)*

This test measures the relative humidity in the pores of the concrete (usually in the outer 100 mm). A 16 mm hole is drilled into the area to be tested and a plastic sleeve, which has a thin membrane closing its outer end, is inserted into the hole and left in for approximately one hour. This provides enough time for picking up the prevailing relative humidity and temperature. The test is conducted by breaking the closed end of the sleeve with the probe of the measuring device, and recording the values.

7. *Rapid Chloride Test (RCT)*

The Rapid Chloride Test, RCT, provides a fast method of determining the 'acid-soluble' amount of chlorides in concrete on-site.

Pulverized concrete obtained by hammer drilling of hardened concrete is mixed with a chloride extraction liquid and shaken for 5 minutes. The amount of acid-soluble chlorides, expressed as weight per cent of concrete weight, is determined directly by means of a calibrated chloride-sensitive-electrode connected to the RCT-electrometer.

Taking more samples from different depths in the same location — usually in steps of 30 mm — the 'chloride-profile' is determined by testing pulverized concrete from each depth-interval.

Examining the profile, the probable source of the chlorides and mechanism of penetration can be detected (for instance curing water, saline soil, sea water, freed chlorides from aggregates, air-borne chlorides, etc.).

The RCT measures 'bounded' as well as 'free' chlorides, because the concrete sample is dissolved in an acid that frees the bounded chlorides.

An assessment of the content of 'free' chlorides alone can be made by carrying out the measurements in distilled water instead of in the acid. The 'free' chlorides will rapidly be dissolved in the water, while the 'bounded' will remain 'bounded .

8. *Concrete 'Strength' Test*

In general, the concrete strength can be measured *in situ* by following methods

    (i) Drilling out 'cores' and later crushing them for compressive strength in the laboratory

   (ii) Lok-Test (for compressive strength)

  (iii) Bond-Test (for tensile strength)

  (iv) Capo-Test (for compressive strength)

The **Capo-Test** is much quicker than drilling and testing cores. And in most cases the reliability of the method is satisfactory as an *in situ* test. (For details, *see* ahead.)

Lok-Test is only used to measure the strength of new concrete. The inserted expanding element used in the Capo-Test is replaced here by an element left cast in the concrete. The test method and equipment is in principle identical to that of the Capo-Test.

Bond-Test is only used for measuring the concrete tensile strength. Normally the method is primarily used for testing the bond strength of a newly cast concrete layer on the existing concrete. Further, it can be used for testing the tensile strength of the existing concrete when a certain strength is required before the new concrete layer and/or waterproofing can be placed. (It can also be used for testing the bond of a membrane to a concrete surface.)

Capo-Test: The theoretical background of the Capo-test is simply that the compression strength is correlated to the force necessary to pull-out a bolt from the concrete, if the fracture shape is a cone with a specific angle.

This correlation is independent of aggregate type and strength, as long as the correct fracture shape is achieved.

To ensure the correct geometry of the fracture, a special bolt and a circularly applied counterpressure are employed. The bolt is expanded in a recess milled at the bottom of a 25 mm hole in the concrete and pulled out by means of a small hydraulic jack. Reading the hydraulic pressure gives the pull-force, from which the compression strength of the concrete can be found by means of a calibration chart. (It is a relatively new test method, but dependable, provided at least three satisfactory tests are carried out in the area under question.)

Special attention must be paid to

— selection of representative areas of concrete in question

— milling of the recess with sharp edges (by keeping the miller at a right angle to the surface all the time)

— greasing of the insert before inserting

— Assembling and tightening the various parts of the expansion-bolt and jack in the right sequence

— Fastening of the insert without rotating it

If the failure surface is not conical, the measurement is not valid and a new test must be made.

The test measures the strength in a very small area. The presence of coarse aggregates or minor deficiencies in the concrete at the test location may affect the measured strength.

To compensate for this, at least three tests are made.

15.3.3 Investigation of Post-Tensioned Steel

1. Visual Inspection

A visual inspection is the first step in obtaining an estimate of any possible damage to the post-tensioned reinforcement. The visual inspection will cover all exposed surfaces, looking for rust stains and any cracks relating to the prestressing cables — be it due to their profile-effect or at their anchorages, etc.

In case of *box girders*, an inspection of the boxes from within must be carried out too. When carrying out this inspection, the drainage system should also be checked.

Special attention should be paid to the condition of *construction joints* and *cold joints*, through which water penetrates the concrete more easily. In girders constructed by the 'span' and 'segment' construction methods, some cables are 'coupled' at the construction joints where the concrete may not be very dense, *see* Fig. 15.6. In superstructrues made of precast segments, the number of construction joints is high and a number of cables are anchored at these joints. Concrete may sometimes not be very densely packed at these anchorages. Cables in the top of the cross-section can be affected by the penetrating water from the wearing course in case no waterproofing is made. It must be noted that high strength prestressed steel may develop a serious type of corrosion due to the high stress level it can also become brittle (brittle-fracture and stress-corrosion).

Further investigations (described as follows) may be required for certain areas.

2. Visual Investigation of Cables

The following investigation methods may be used for inspection of the post-tensioned steel concerning tensioning, corrosion, grouting, and evaluation of future durability. Investigating post tensioned steel requires special care in order not to damage the steel. A small damage may lead to fracture of the tendon.

(i) Check for grout efficacy: The purpose of this investigation is to check if the cabel ducts have been completely filled with grout.

A simple method is to puncture the vent-pipes of duct with a round chisel and check if they are fully injected. This method is simple, quick (if the vent-pipes are located), inexpensive and reliable. Alternatively, 'contact-drilling' may be performed to the cable ducts. But procedure involves extreme care.

Random tests should be performed when inspection of most of the vent-pipes is possible.

If the tests indicate lack of injection in the inspected ducts, the inspection must be supplemented by contact drilling at other parts of the same ducts to investigate the extent of the voids and possible corrosion of the tendons within.

(ii) Break-ups/strand exposure: Visual inspections may be obtained by break-ups or contact drilling at various positions. This investigation method requires accuracy in locating the ducts, otherwise difficultly will arise during the break-up and/or contact drilling. The location of the ducts is normally based both on 'as built'

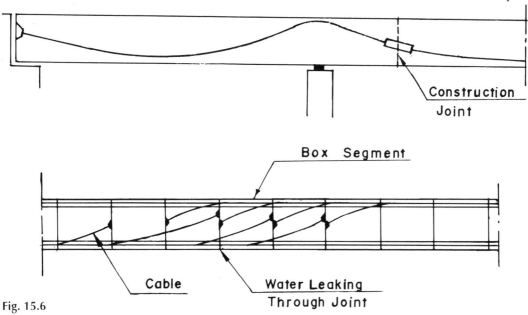

Fig. 15.6

drawings and cover-meter tests. Experience shows that drilling deeper than 0.35–0.45 m is usually unsuccessful because of the likehood of obtaining good contact with the ducts is reduced significantly. All drilling must be performed carefully in order not to damage the tendons.

(iii) Endoscopy: If the cable ducts are found not well-grouted, contact drilling up to the ducts makes it possible to inspect the tendons by means of an 'endoscope'. The endoscope equipment includes a lighting-source and a fibre optic cable to transfer the light to the endoscope. A system of lenses enables the endoscope to be used as a 'monocular'. A camera can also be mounted on the endoscope to make photo-documentation.

Generally speaking, the method is appropriate and may also be used for inspection of inaccessible small voids, honeycombs, cracks, slots, etc.

3. *Volumetric Measurement Check for Possible Voids in Cable Ducts*

This investigation method enables an 'estimate' of the total volume of voids in cable ducts.

The principle of the method is that a known volume of air at a known pressure is connected to the unknown volume (at the pressure of 1 bar). By measuring the drop of air pressure when the connection is made (by opening a valve), the unknown volume can be calculated.

When the cross-section of the cable duct is known (excluding the cross-section of the strands), the length that is uninjected can be calculated, although not very accurately in case the ducts had got slightly squashed before or during concreting. The reliability of the results is further limited because the calculations assume that the duct is either fully injected or uninjected. Water in the ducts will also affect the measurement.

The reliability also depends on the air tightness of the duct and of the connections to the equipment. This is checked by watching the pressure-meter. The pressure should drop immediately after the valve is opened and then stay at the same lower level. If the pressure drops further, there is leakage in the system.

15.3.4 Inspection of 'Steel' Structures

1. *Visual Inspection*

The inspection is normally concentrated on important structural parts, such as joints subject to static and dynamic loads.

Inspection of the joints should be made before any cleaning begins, as any deposits will often highlight possible crack formations in the welds. After the initial inspection, the welds should be cleaned thoroughly. During the inspection, attention must be paid to all crack-like irregularities since many types of surface irregularities look like cracks (particularly on painted surfaces).

Attention must be given to the bolts/rivets/welds at the edges of the joint-plates when inspecting structures connected by the use of bolts or rivets. Loose connections may imply member failure. Rust precipitation at connections or at unintended eccentric position often indicates such damage.

2. *'Magnetic Particle Flow' Test for Detecting Cracks at or Near Surface*

The test method is used to detect surface cracks in welds and in steel surfaces whenever the cracks are less than 1.0–1.5 mm below the steel surface. The test is specially suitable for fillet-welds where radiography and ultrasonic test may not be easy to perform.

The test procedure is as follows

(a) Clean and dry the test area
(b) Apply contrast colour (e.g. Castrol 710) in a thin uniform layer to the test area
(c) Focus the magnet by the adjustable legs on the test area and turn it on
(d) Thoroughly shake the spray-can containing the magnetic particles (e.g. Suprainor 4 Black) and apply to the area
(e) Register any black lines interrupting the magnetic field lines. Such lines indicate surface irregularities on or just below the surface
(f) Rotate the magnet by 90° and repeat the procedure
(g) Small test areas can be demagnified by turning off the magnet and slowly pulling the area between the poles. When testing larger areas, it is easier to shake off the magnet while it is switched on

3. *Ultrasonic Thickness-Gauge for Testing 'Remaining-Thickness'*

The thickness-gauge is used to determine the 'remaining' thickness of corroded metallic items. The gauge uses dual transducers to measure the thickness of corroded, pitted, scaled, granular materials from one side only. Thicknesses between 0.50 mm and 200 mm can usually be determined within ± 2 per cent accuracy.

The transducer shall be placed on the surface of the test item. The thrown ultrasonic waves will then be reflected by the opposite surface. The thickness is shown digitally.

For uneven or uncorroded surfaces, it may be necessary to grind the surface at the test positions to make proper contact. Possible rust on the opposite surface does not disturb the reflections. If steel plates are laminated, the measured thickness will only be the depth of the first layer.

Before starting the measurements, a 'contact-liquid' is applied to the test-locations. Further, the equipment must be calibrated. For common steel-alloys, the calibration is performed by means of test-blocks. For unknown alloys (or if one is not sure), the calibration is performed by adjusting the sound-velocity-setting of the equipment until the equipment shows the same thickness as can be measured by a slide-caliper at a free edge.

The method is generally dependable, and is specially useful for measuring possible corrosion in structural elements where access is possible from one side only.

4. Coating Thickness Measurements

Coating thickness measured by the use of the Elcometer 246 is based on the assumption that the inductance between an electromagnet and a metallic surface varies with the thickness of a non-magnetic interface coating. Change in the inductance is electronically transmitted to a digital measurement of the coating thickness.

The coating must be non-metallic with a thickness preferably less than 1 mm. The Elcometer can be used on paint, plastic, vitreous, enamel, galvanizing and hard chrome to name a few.

The reliability of the readings depends on the calibration, the test item and the geometry of the item. With the exception of the thin coatings (less than 0.005 mm), the accuracy is usually within ± 3 per cent if the Elcometer is calibrated at the thickness to be measured. Otherwise, the precision is ± 5 per cent. The minimum thickness that can be measured without special calibration is 0.003 mm.

The choice of the test areas may be based on one of the following: (i) Random, (ii) Filed observations, (iii) Advance knowledge of possible problem area (corners, edges, and inaccessible areas).

5. Dye-Penetration Test

It is the most popular field test method for detecting surface cracks. After wetting the surface with the dye spray penetrant, the penetrant is wiped off from the surface, then a developer (normally a contrasting colour) is applied to the surface. If there is a crack, the dye colour appears on the developer.

Technique of Acoustic Emission has lately been introduced for detecting delayed cracks. This is used 72 hours after completion of welding and can be used in the field for locating and detecting cracks in existing beams that have vehicular collision damage.

6. Radiographic Test

It is based on the use of X-ray radiation to take pictures of internal cracks including shape and size. This test does not show the depth of defects and is difficult to use in field locations.

7.

See Chapter 5 for details regarding various factors that can cause deterioration in steel and the phenomena of Fracture and Fatigue, Brittle Fracture and Stress Corrosion.

8. Methods for Non-Destructive Testing of Welds

See Table 15.2 for some details.

Table 15.2 Reference Guide to Major Methods for Non-Destructive Testing of Welds

Inspection Method	Equipment Required	Enables Detection of	Advantages	Limitations	Remarks
Visual	Magnifying glass Weld-Size Gauge Pocket rule Straight edge Workmanship-standards	Surface flaws—cracks, porosity, unfilled craters, slag inclusions Warpage under-welding, over-welding, poorly formed beads, misalignments, improper fitup	Low cost Can be applied while work is in process, permitting correction of faults Gives indication of incorrect procedures	Applicable to surface defects only Provides no permanent record	Should always be primary method of inspection, no matter what other techniques are required Is the only 'productive' type of inspection Is the necessary function of everyone who in any way contributes to the making of the weld

(Table continued on next page.)

Inspection Method	Equipment Required	Enables Detection of	Advantages	Limitations	Remarks
Radiographic	Commercial X-ray or gamma units, made especially for inspecting welds, castings and forgings Film and proccesing facilities Fluoroscopic viewing equipment	Interior macroscopic flaws—cracks, porosity, blow holes, non-metallic inclusions, incomplete roof penetrations, undercutting, icicles, and burnthrough	When the indications are recorded on film, it gives a permanent record When viewed on a fluoroscopic screen, a low-cost method of internal inspection	Requires skill in choosing angles of exposure, operating equipment, and interpreting indications Requires safety precautions. Not generally suitable for fillet weld inspection	X-ray inspection is required by many codes and specifications Useful in qualification of welders and welding processes Due to cost, its use should be limited to those areas where other methods will not provide the assurance required
Magnetic Particle	Special commercial equipment Magnetic powders—dry or wet form; may be fluorescent for viewing under ultraviolet light	Excellent for detecting surface discontinuities—especially surface cracks	Simpler to use than radiographic inspection Permits controlled sensitivity Relatively low-cost method	Applicable to ferromagnetic materials only Requires skill in interpretation of indications and recognition of irrelevant patterns Difficult to use on rough surfaces	Elongated defect parallel to the magnetic field may not give pattern; for this reason the field should be applied from two directions at or near right angles to each other
Liquid Penetrant	Commercial Kits, containing fluorescent or dye penetrants and developers Application equipment for the developer A source of ultraviolet light—if fluorescent method is used	Surface cracks not readily visible to the unaided eye Excellent for locating leaks in weldments	Applicable to magnetic and non-magnetic materials Easy to use Low cost	Only surface defects are detectable Cannot be used effectively on hot assemblies	In thin-walled vessels, will reveal leaks not ordinarily located by usual air tests Irrelevant surface conditions (smoke, slag) may give misleading indications
Ultrasonic	Special commercial equipment, either of the pulse-echo or transmission type Standard reference patterns for interpretations of RF or video patterns	Surface and sub-surface flaws including those too small to be detected by other methods Especially for detecting subsurface lamination-like defects	Very sensitive Permits probing of joints inaccessible to radiography	Requires high degree of skill in interpreting pulse-echo patterns Permanent record is not readily obtained	Pulse-echo equipment is highly developed for weld inspection purposes The transmission-type equipment simplifies pattern interpretation where it is applicable

Reproduced at the courtesy of the LINCOLN ELECTRIC Co., U. S. A.

16

LOAD TESTING

16.1 FULL SCALE LOAD TESTING OF BRIDGE ELEMENTS

As pointed out in the previous chapter, a full-scale load testing of, say, a bridge deck, may become necessary if its load rating cannot be ascertained through detailed structural investigative computations (assisted by relevant diagnostic tests) perhaps because the as-built drawings may not be available or structural integrity may be doubtful or may be it is a contract condition to test-load a particular representative element of the bridge.

The typical full-scale test may range from a simple proof test coupled with a visual inspection, to sophisticated test loading vehicles and instrumentation to measure the strains and deflections! Loading may be applied by means of fully loaded vehicles of known axle weights and spacings, simulating the design truck loading. Other loading mediums such as concrete blocks of known weights, water tanks, or hydraulic jacks, can also be used. Loads are applied in increments and the structural response is monitored to arrive at an elastic load limit before the loading is discontinued.

Load tests can be expensive, and, on larger bridges, they require considerable planning, and demand the use of sophisticated equipment. Testing in remote locations can present additional difficulties. Load testing can be justified where the effect of defects and/or deterioration on load capacity cannot be determined by analysis alone. However, a decision to carry out full scale load testing should not be undertaken without a serious thought.

A recently completed Research Project ('*Comparison of Measured and Computed Ultimate Strengths of Four Highway Bridges*'—Highway Research Record, NO: 382, HRB, NCR, 1972, U. S. A.) on improving the Bridge Load Capacity Estimates by correlation with Test Data, contains a summary of some 150 references pertaining to load testing of bridges. The report contains a detailed discussion of factors which enhance the load capacity of existing bridges. These factors include the effect of lateral load distribution, unintended composite action,

unintended continuity, and several other contributing factors.

Load testing may be done for a whole bridge deck in a span, for a foundation caisson (preferably prior to its plugging), for a foundation pile, or any other structural item, depending on the requirement. The first three items will be discussed here.

16.2 LOAD TESTING A BRIDGE 'SUPERSTRUCTURE'

The load test is intended to check and establish the serviceability and working strength of the superstructure for the limit state of 'deflection and local damage' in a particular span (vis-a-vis a critical section, for its critical bending or shear, etc.).

Test Load and Computation of Deflections Theoretically: This 'Test Load' is the equivalent static load equal to 'the impacted working design Live Load' (applied in addition to all dead loads) placed for maximum effect (bending moment, shear, deflection, etc., as the case may be) and applied either through an appropriate test-loading-truck or through loading platforms resting on tyre-contact area plates simulating the actual contact areas of wheels. (If any of the final dead loads, e.g. wearing course, parapets, etc., are not yet in position on the structure, equivalent compensating loads should be placed for simulation. Also, the effective contact area under a wheel may be calculated on the basis of tyre pressure of 5.50 kg/cm^2, taking the contact dimension of a wheel in the direction of traffic as 25 cm normally.)

Under the above mentioned test load, placed longitudinally and transversely for the maximum desired effect (e.g., for maximum bending moment at midspan section in the extreme girder in a girder-slab type structure, or for maximum bending moment at midspan section of a unit-width strip under the longitudinal centre line of the extreme truck (closest to the kerb) in a slab type superstructure, etc.), calculate theoretically the deflections at various critical points, assuming instantaneous value for the concrete modulus of elasticity and considering the section property only of the structural deck (ignoring the contribution of wearing course, parapets, etc., to the section properties).

The above calculations should preferably be done by a suitable grid analysis. It may also be done (approximately) manually, using reaction factors for different girders (by Courbon's Method or Little and Morice's Method, or using the AASHTO factors, as applicable) in girder-slab and box type superstructures, and by using appropriate dispersion-width formulae in slab-type superstructures.

Load Application and Testing: After placing the equivalent compensating loads for any dead loads that may have not been placed up to the time of load testing, fix deflection gauges (supported on unyielding supports) under the structure, just touching its soffit, at the pre-marked critical 'points'. Apply the test load in five equal increments (including the weight of platforms in the first increment), allowing about two hours in between completion of one load increment and commencement of the next. About an hour after completing a load increment, note the deflections and any crack patterns together with maximum crack widths.

After measuring the deflections, etc., one hour after applying the fifth, i.e., the final load increment, keep the test load maintained for about 24 hours, and again measure the deflections and crack widths (if any).

Remove the test load in five equal decrements, allowing about two hours in between completion of one load decrement and commencement of the next. Note the deflections and crack widths (if any) about one hour after the end of each load decrement.

Note the deflections and crack widths (if any) 24 hours after removal of the test load in order to see whether the recovery is complete or there is any residual deflection, etc.

Assessment of Results

If

- Increase and decrease in deflections at a point follows a linear relation, with actual deflections not exceeding the theoretically estimated ones by more than about 5 per cent;
- The maximum crack width in reinforced concrete does not exceed about 0.25 mm to 0.30 mm (but no cracks in case of prestressed concrete);
- The cracks, if any (in reinforced concrete), close upon removal of test load but may not completely disappear;
- The deflections at the end of the 24 hours of maintaining the test load do not exceed the corresponding values at the start of this 24 hours period by about 2.5 per cent;
- The residual deflection at any point 24 hours after complete unloading does not exceed 5 per cent of the maximum observed deflection at that point (i.e., recovery: 95 per cent or more);

. . . **then** the superstructure may be accepted for service.

Precautions: There shall be sufficient instruments for accurately measuring small order deflections of the superstructure (horizontal, vertical and rational movements in box sections and curved decks).

Strong scaffolding, capable of supporting both the superstructure weight as well as the test load, must be erected and placed in position prior to testing, leaving a gap under the deck-soffit which is sufficient for the deck deflections. Adequate precautions shall be taken to safeguard the personnel at and in the vicinity of the test area.

If any severe cracking/spalling, etc., appears at any stage of test, the test shall be discontinued, load removed and matter immediately reported to the designer, in detail.

Some Important Considerations: The bridge deck hogs or sags owing to its top and bottom surfaces being at different temperatures. This is strain induced deformation. In simply supported cases, no moments are induced by this, however. Any change in the temperatures of the deck surfaces during the load test will therefore affect the observed deflection readings. The true values of deflection due to test load alone are, therefore, the recorded values of deflection algebraically adjusted by the temperature induced deflections.

Where day temperatures are high, the deck may be loaded and unloaded at night in order to 'minimise' temperature induced deflections. One additional gauge must, however, be positioned in a nearby similar span which is unaffected by the test load, in order to check the magnitude of temperature deflections! In its location in this dummy span, this additional gauge must be placed at the same location as some critical gauge in the loaded span, e.g., the gauge under the maximum predicted deflection point.

The actually recorded deflections in the span being loaded represent the 'algebraic' summation of

(a) The (downward) deflections due to the test load

(b) 'Varying-in-magnitude' and 'now-upward-and -now-downward' deflections due to differing top and bottom surface temperatures during the test

The true downward deflection at any particular instant due to the test load alone is the measured downward deflection 'minus the downward deflection due to temperature' or 'plus the upward deflection due to temperature' at that instant at that location, depending on whether the temperature-deflection is downward or upward. In the load test of a bridge deck of, for instance, a 30 m span, the deflections due to the test load will generally be much greater than those due to temperature variation. Effect (b) will then be drowned by effect (a). By the same token, both these effects may be of somewhat similar magnitudes if the span is short! If this is not understood carefully, the resulting serrations in the load-deflection graph may misguide into feeling of creep problems!

If the readings are taken during night time, and in case the span is small, then, even if the values of temperature induced deflections (as recorded by the dummy gauge) are adjusted in the 'load versus deflection' graph of the corresponding gauge of the loaded span, this graph may still not look neatly linear! This may be partially explained by the fact that the dummy gauge does not reflect the true temperature induced deflections in the loaded span because the soffit of the loaded span may be lit continuously by strong arc lights whereas the dummy gauge may be read merely by torchlight or an isolated light. The arc lights are an additional heat-source in the loaded span which would in fact tend to sag this deck (bottom fibre hotter than top fibre). Hence the dummy gauge would not be truely representing the temperature-deflections of such arc-lit loaded span—point to note.

• In many test load cases it is observed that the measured load response of the deck differs significantly from the theoretically calculated behaviour. The decks are in fact much stiffer than assumed in the theoretical computations, and hence the actual behaviour, as measured, tends to show lower deflections. There can be several possible reasons for this:

(a) In the theoretical analysis the bridge deck may be assumed to be 'ideally' supported. However, the width over which the deck in fact actually 'bears on the seating arrangements' may be large relative to the span so that the 'rotation at the supports will be restricted'. The deck is, therefore , subject to rotational restraint at its supports! The magnitude of this restraint lies somewhere between the 'simply supported' and 'encastre' conditions. And if the span is short (say 7 to 10 m), this effect will reduce the span moment not insignificantly!

(b) In the theoretical analysis the deck is generally assumed to act as a thin plate loaded transversely. Since there can be considerable friction between the soffit and its seating, some of the load could perhaps be carried by 'arch-action'. This effect would, however, be of secondary importance if the span is not short.

(c) Conventional design practice assumes that the parapet does not contribute to the structural strength of the deck. The concrete parapet in actual case is, however, more or less integral with the deck! It must therefore contribute some stiffening effect to the edge of the slab, and thus to the whole deck section. Even the wearing course (if of concrete) can stiffen the deck.

(d) The theoretical analysis assumes that concrete strength is the minimum as specified. The average strength of concrete in the actual structure could however be significantly greater.

(e) Periodic random 'releases' (periodic 'jerks') of the aforementioned 'rotational restraints' and the 'possible horizontal friction', coupled with the unpredictable up or down temperature deflection (particularly if the span is small), could account for sudden and sometimes erratic jumps in the load-deflection graph.

Acceptance Criteria for the load-test performance sometimes may not appear satisfied (particularly if the span is small), causing 'creep and residual deflection' fear! This, however, may be discounted in case of such spans, because

— the jumps in the load versus deflection curves under maintained test load and, to some extent, after removal of the load, are a matter of scale. The absolute values of the 'change' in deflection could be minuscule;

— these jumps, under maintained test load condition, may be due essentially to the deck hogging or sagging in response to temperature variation if the span is small. They are not then primarily caused by creep. Creep would only cause increase in deflection under sustained load, not an oscillating deflection (now more, now less);

— when deflections are small, even a small change in their magnitudes will appear as a large percentage change. The erratic jumps (i.e. the 'periodic jerks' referred to above) must not be lost sight of before coming to alarming conclusions!

Effect of Assumptions in Grillage Analysis: Initially a grillage analysis (by computer) may be made assuming the deck to be pin supported at nodes over the bearings, with no edge stiffenings.

Modified grillage analysis may then be made attempting to take account of factors (a) and (c) above. This maybe done by

(i) Increasing the flexural and torsional stiffness of the grillage members above the bearings (support diaphragms) by an arbitrary factor.

(ii) Assuming a fraction of a parapet section to act structurally with the deck, thus enhancing the flexural and torsional stiffness of the longitudinal edge members of the grillage.

With regard to the choice of factor in (i) above, multiplication of diaphram stiffness by 10, may be found to lead to 10 to about 40 per cent reduction in maximum deflection depending on whether the span is long or short. Multiplication by a factor of 100, however, only causes a further 5 to 10 per cent decrease in estimated deflection values! These results tend to justify the assumptions made in (i) and (ii) above. All this explains why the actual deflections, as observed, are generally lesser than those calculated simply and theoretically.

16.3 LOAD TESTNG A 'CAISSON' (A FOUNDATION 'WELL')

This is generally done before the well is 'plugged'. The 'test load' is calculated so as to balance the assumed 'ultimate soil resistance' against the 'bearing area' of the 'cutting edge' together with an assumed skin friction on the outer surface of the well-steinng till about 2 m below the surrounding ground level, taking due account of the buoyant weight of the submerged part of the well and dry weight of the dry part of the well. The calculated 'test load' is then applied in five equal increments and the 'sinking' of the well observed regularly. Before commencing the loading, the well is temporarily plugged with sand up to half the curb height to prevent 'squeezing out' of soil from under the bearing area. If the sump was large, the well initially will sink faster. If the actual effective sinking during testing does not exceed about 0.005 inch per ton of 'test load' (unless the well is on rock), with the total 'test load' maintained for 24 hours, and if upon removal of the test-load there is no appreciable rebound, the well may be accepted for service, remembering that the well is not yet concrete-plugged! (Rebound may occur only if the subsoil is predominantly cohesive. The amount of such rebound will be much less if the testing is done after one to three months of constructing the caisson. In this time the thixotropic strength gain around cohesive soil surroundings matures because time is needed to dissipate the surrounding adhesive cohesion. Indeed, in cohesive soils, the sinking of caissons (or driving of piles) demands lesser effort if sinking (driving) is resumed after a few days 'rest'.)

It should be remembered that in the actual design of the well-. foundation, it is prudent to either ignore the skin friction on the well surface (or at best consider only half of it as dependable). The 'well', tested as above, will then be safer, because all the load balancing (computation of base-pressures, etc.) will be done on the basis of the 'safe bearing capacity' of substrata and increased well-size to suit.

16.4 LOAD TESTING A 'PILE'

16.4.1 General

In areas in which no reliable experiences are available about the behaviour of formerly established pile foundations (if any), every major project requires a better knowledge about their bearing capacity. This can be obtained by making a loading test on a pile which has been installed in the same way as the other piles will be. These field tests have to be made well in advance of the construction in order to permit working out a design which fits the test results.

A loading test is usually made on single piles, by applying to the pile head a direct load on a platform, or by jacking down against a dead load on a platform, or by jacking against a crossbeam held down by anchor piles at each end.

Piles in granular soils may be load-tested after several days after the pile is driven, by when load test arrangements also have been made. Piles in cohesive soils, however, should be tested after a significant time elapses (about 30 to 90 days) so that the thixotropic strength gain is substantially complete. In any soil, sufficient time should elapse to allow partial dissipation of residual compression and frictional stresses under the shoe and around the shaft.

16.4.2 Load testing of a pile will be discussed here in terms of general detail. In an actual Project, follow the applicable codal provisions (in respect of settlement under test-load, etc. as may be legally binding on the contract.

16.4.3 Where feasible, the immediate area around the test-pile shall be excavated to the proposed pile cut-off elevation. The test-pile shall be cut off or built up to the proper grade as necessary to permit construction of the load-application apparatus, placement of the necessary testing and instrumentation equipment and observation of the instrumentation. Where necessary, the unsupported length of the test pile shall be braced to prevent buckling without influencing the test results.

If the hydraulic jack pump is to be left unattended at any time during the test, it shall be equipped with an automatic regulator to hold the load constant as pile settlement occurs. Calibration report shall be furnished for all testing equipment for which calibration is required, and shall show the temperature at which the calibration was done.

16.4.4 Schematic Arrangements for Axial-Compression Load-Testing of Piles

These are shown in Figs 16.1, 16.2 and 16.3, as various alternative arrangements, and the one appropriate to a particular site may be selected.

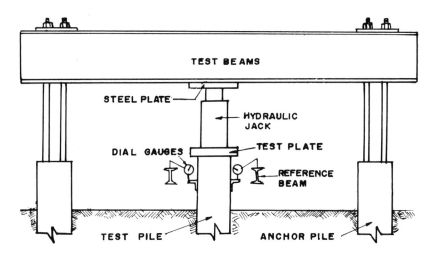

Fig. 16.1 Schematic set-up for applying loads to pile, using hydraulic-jack acting against anchored reaction frame

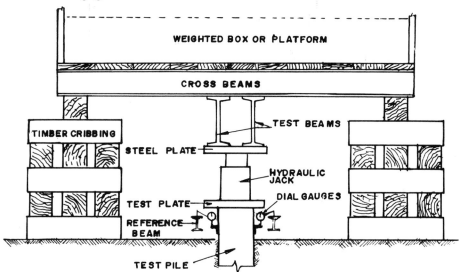

Fig. 16.2 Schematic set-up for applying load to pile using hydraulic-jack acting against weighted box or platform

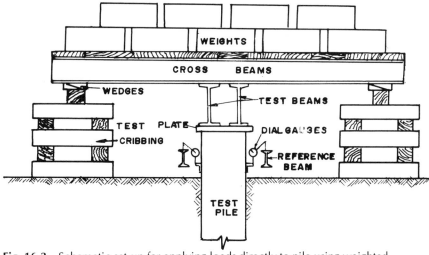

Fig. 16.3 Schematic set-up for applying loads directly to pile using weighted platform

16.4.5 Apparatus for Measuring Movement

(a) All reference beams and wires shall be independently supported with supports firmly embedded in the ground at a clear distance of on less than 8 ft (2.5 m) from the test-pile or pile-group and as far as practical from the anchor pile or cribbing. Reference beams shall be sufficiently stiff to support the instrumentation such that excessive variations in readings do not occur and should be cross-connected to provide additional rigidity. If steel reference beams are used, one end of each beam shall be free to move horizontally as the beam length changes with temperature variations.

(b) Dial gauges shall have at least a 2" (50 mm) travel; longer gauge stems or sufficient gauge blocks shall be provided to allow for greater travel if anticipated.

The gauges shall have a precision of at least 0.01" (0.25 mm). Smooth bearing surfaces (such as glass), perpendicular to the direction of gauge-stem travel, shall be provided for the gauge stems. Scales used to measure pile movements shall read to 1/64th of an inch or to 0.01" (0.25 mm).

(c) Dial Gauges — two parallel reference beams, one on each side of the test-pile, shall be oriented in a direction that permits placing their supports as far as practicable from anchor piles or cribbing. A minimum of two dial gauges shall be mounted on the reference beams, approximately equidistant from the center of and on opposite sides of the test-pile, with stems parallel to the longitudinal axis of the pile and bearing on lugs firmly attached to the sides of the pile below the test plate. Alternatively, the two dial gauges shall be mounted on opposite sides of the test-pile, below the test plate, with stems parallel to the longitudinal axis of

the pile and bearing on lugs firmly attached to the reference beams.

For test on individual batter-piles, the dial gauges shall be mounted along a line perpendicular to the direction of batter.

16.4.6 Some Factors that can Influence Interpretation of Test Results

(a) Potential residual loads in the pile which could influence the interpreted distribution of load at the pile tip and along the pile shaft.

(b) Possible interaction of friction loads from test pile with upward friction transferred to the soil from anchor piles obtaining part or all of their support in soil at levels above the tip level of the test pile.

(c) Changes in pore water pressure in the soil caused by pile driving, construction fill, and other construction operations which may influence the test results for frictional support in relatively impervious soil such as clay and silt.

(d) Differences between conditions at the time of testing and after final construction, such as changes in the grade or groundwater level.

(e) Potential loss of soil supporting the test pile from such actions as excavation and scour.

(f) Possible differences in the performance of 'a pile in a group' or of a 'pile group' from that of a single isolated test pile.

(g) Effect (on long-term pile performance) of factors such as creep, environmental effects on pile material, negative friction loads not previously accounted for, and strength losses.

(h) Requirement that all conditions for non-tested piles be basically identical to those for test pile, including such things as subsurface conditions, pile type, length, size and stiffness, and pile installation methods and equipment, so that application or 'extrapolation' of the test results to such other piles is valid.

16.4.7 'Pile Load Testing' and 'Procedures'

(a) **General**

Initially a few representative piles, placed slightly away from the 'working piles' (i.e. the piles which will be used for the bridge foundations), must be test loaded to 'ultimate' or at least to 'twice' the design load value (working load value) so as to ascertain the true design load capacity.

The actual number of such piles, and their locations vis-a-vis the location of various foundations in the bridge, will depend on the magnitude of the piling work, the uniformity or otherwise of the subsoil characteristics, etc. In certain projects, the client may directly specify the number (1 in 200 is not unusual). Once these ultimate load tests have been carried out and working load (i.e. design load) capacity of the pile established, the design work is completed and the construction work taken up.

During this actual project construction work, it is customary (and indeed essential) to test-load approximately 1 to 2 per cent of the 'working piles'. Such tests are called 'Working Load' Tests or 'Routine' Tests, and in these the piles are test-loaded only to 1.25 times their design load value.

(b) **Basic Procedure for 'Ultimate Load Test' on a Pile**

- Select the appropriate pile-loading-methodology from the various schematic arrangements shown earlier. Arrange these details, rehearse the procedure, and train the crew.

- Load the pile in increments of 0.25 X, up to 2 X (where X = design load value of a pile, in tons), giving 2 hours time between two successive load increments. Note the pile-settlement at the beginning and at the end of each load increment. Leave the 2 X load on the pile constantly for 24 hours and if then the total net (effective) settlement does not exceed 0.01 inch per ton of Test Load, or 1 inch, whichever lesser, the design load capacity (i.e. working load capacity) of the pile can be taken X Tons. However, if the pile appears to be able to take more load without exceeding the settlement limits, apply more load — a smaller increment of, say, 0.1 X per increment, and note the total net (effective) settlement after 12 hours. That Test load, at which this settlement reaches 0.01 inch per ton of its value, or 1 inch, whichever is lesser, shall be taken as the ultimate load capacity of the pile. Half of such Test Load can then be taken as the design (i.e. working) load capacity of the pile. Now remove this Test Load in four equal decrements, giving one hour gap between ending one decrement and commencing the next, and note the corresponding settlements also. Note if any permanent settlement has taken place 24 hours after complete unloading. (A load-settlement curve can be drawn for the loading and unloading cycles just completed.)

NOTE: If the pile settlement exceeds the above limitation either before the test load reaches the 2 X value, or in less than 24 hours after the load reaches the 2 X value, then the pile is assumed to have an ultimate capacity of less than 2 X. A fresh pile and a fresh ultimate load test (to less than 2 X value) will be called for, unless the capacity can be fixed by engineering judgment.

(c) **Basic Procedure for the 'Routine' i.e. 'Working Load' Test on a Pile**

The procedure here is similar to that described above for the Ultimate Load Test but the 'Test Load' here is only 1.25 X and the settlement criterion is 'either 0.005 inch per ton of this Test Load, or ¼ inch, whichever lesser'. The pile on which a Working Load Test has been done, can still be used as a 'working pile' in the bridge foundations, unlike the pile on which Ultimate Load Test is done (which is no more usable as a regular working pile).

17
BRIDGE DISASTERS —
Some Reflections

When a doctor makes a mistake — he 'buries' it two metres underground. When a lawyer makes a mistake— the man is 'hidden' behind the bars. But when an engineer makes a mistake — there it is in the 'open' for everyone to see!

Daring new ventures have a glamour that tends to blind both engineers and the public alike, to the costs and dangers involved. Such was the enthusiasm for the first railway bridge over the Niagara Falls, the first comet aeroplane and today perhaps the Channel Tunnel and the Concorde. A handful of 'novel plans' of bridges have ended in disaster during or after construction. To err is but human and even a great engineer makes gross errors, unwittingly though. The Tay bridge in the U. K. designed by the much maligned Sir Thomas Bouch, was first hailed as an engineering masterpiece because of its novel features. But 19 months after its construction 'down went the bridge to disaster land' with its high girders buffeted by a gale bringing about the downfall of the bridge. Not only did thousands of guineas float away, but also a passenger train plunged into the river in that black night of 28th December, 1879 taking a toll of 75 human lives.

Society over-reacts to such accidents, to say the least. In spite of a Tay or a Tacoma, there is no denying the fact that bridge disasters that inevitably happen, are too few and far between. If the entire backdrop is viewed with equanimity by an unbiased observer, it will be obvious that the successes in bridge engineering, which tend to go unnoticed, outweigh the much publicised failures. Yet we live in an age when instant public adulation or indignation is aroused by mass media coverage. A bridge which stands successfully for a hundred years or more has no news value at all. One that falls is the only one that is newsworthy. If the consequences of a bridge failure are disastrous, the bridge

and its builders earn the dubious distinction of hittings the headlines. An inquest is ordered post-haste into the inglorious death of the bride. Often it turns into a legal inferno with responsibilities to be pinned somewhere. Here are the brief details of two bridge disaster enquiries in Australia which should be illuminating:

"... The King's Bridge Royal Commission lasted for 71 sitting days. There were 45 witnesses and 237 exhibits. The parties were represented by 17 barristers.

"... The West Gate Royal Commission lasted for 80 sitting days. There were 52 witnesses and 319 exhibits. The parties were represented by 25 barristers."*

With so many barristers around after a disaster, when in each case the immediate cause of the collapse is reasonably obvious, it is clear that bridge engineers should better have a good training and experience in law as well.

The keen interest that a bridge failure creates has its reasons, but society shows a deep-rooted psychological attitude towards safety which is almost an irrational and emotional evaluation. Engineers are charged with the social responsibility of building the bridges 'safe'. It is but natural that a failure to live up to this responsibility will bring social condemnation. However, the actual situations are not so simple and well defined. What does one really mean by 'safe'.

"In everyday usage the work 'safe' means 'no danger' This is so deep-rooted a feeling that the emotional value of safety is obsolete. We do not want to know about any risk of insecurity, and danger, except as an Act of God or a blow delivered by fate. If insecurity does not arise from any of these causes, but is instead due to some form of human action, we feel that it ought to be punished".**

In Babylonian times, the punishment prescribed (in the edict of Hammurabi) for the building failure was death to the builder, if it caused death to owner of the building. Today, loss of life in a bridge failure earns no less social indignation. Although the bridge builder may

---

*Barber, E., 'Engineers, lawyers and the failure of two bridges', *Journal of Australian Institution of Engineers*, Jan – Feb., 1973.
**Dicke, D., 'Remarks on Introductory Report, Theme Ib', Final Report, *Tenth Congress IABSE*, 1976, p. 9.

not hang, he does surely get hauled up. The heat and glare of public outcry and an enquiry using pathological methods to fix legal responsibilities tends to create an environment in which the attitudes of bureaucrats and lawyers may predominate. A bridge builder, ill-equipped to counter these people (by training and apti-tude) may, in some cases, suffer worse public condemnation than he deserves. The story how Sir Thomas Bouch, the nineteenth century British bridge builder, was crucified by malignant publicity following the Tay Bridge disaster enquiry is recalled here as a ready example:

> "From contemporary accounts of the incident and from the enquiry proceedings, there can be no doubt as to the sense of shock felt by the engineering profession and the general public. The much publicised minority report of the Court ... reflected a widespread feeling that a scape-goat should be found. The retiring and rather introverted Bouch soon found himself filling this role."*.

Bouch's Tay bridge collapsed almost certainly due to an obvious single reason. He failed to estimate the probable wind load intensity. The 13 spans of the Tay bridge, that were higher and longer than the rest, failed, as the wrought iron diagonal bracing of the cast iron columns were too weak. The remaining 72 spans stood. Bouch's design approach for wind load effects was, nevertheless, quite in accordance with the widely accepted British practice of the day. A committee of eminent engineers retained by his clients endorsed the wind load values adopted by Bouch. The contemporary design approach was highly deficient. By hind-sight he could be condemned for not looking beyond his nose and lacking the intuitive judgement that is the hallmark of a great bridge builder. But even in the unkindest views of an engineer, it was but a genuine mistake that Bouch made, nothing more and nothing less.

The court of enquiry also concluded very clearly that 'the fall of the bridge was occasioned by the insuffi-ciency of the cross bracing and its fastenings to sustain the force of the gale'. But this was not to be the end of the story. The chairman of the Court of Enquiry, Mr. H. C. Rothers, the Board of Trade's Wreck Commissioner, was not an engineer and he had different ideas. Influ-enced by the climate of witch-hunt that prevailed, he was more inclined to turn the enquiry into an inquisi-tion of Bouch. The majority report did not satisfy his intentions. Bouch had to be painted as a villian who had utter disregard for the safety of his fellowmen. So Rothers chose to write a separate report condemning Bouch squarely. He minced no words and concluded that 'The bridge was badly designed, badly constructed and badly maintained; its downfall was due to inherent defects in the structure which would, sooner or later,

have brought it down. Sir Thomas Bouch was mainly to blame'.

That Bouch had been very unjustly blamed is borne out by cold facts. Seventy-two spans of the bridge did not collapse at all. Most of their main girders were used in the rebuilt bridge and still carries trains after almost a century! The diagonal braces did not fail due to bad materials. The weakest brace was found to be only 19 per cent weaker than the strongest one. But the truth of the matter was to be submerged in the publicity of the minority report of Rothers the bureaucrat, which pushed the majority report into oblivion. His colleagues in the court of Enquiry refused to sign the report of the Chairman. It mattered little as the majority report was never read by many. Unfortunately though, there were very good reasons for the success of Rothers in twisting the truth. The conclusions of the majority report were clear, but as befits the 'weight' of an enquiry report, these were hidden in verbose and massive irrelevancies. Rothers did not beat about the bush in his shorter report and as he was not an engineer he wrote in more readable and more quotable style to hold the public attention he desired and offer the scapegoat everybody wanted. Perhaps the bridge builder should learn from this example to pay a little more attention to his ability in advocacy and the art of brevity and presentation in written or spoken words. Otherwise he will never be heard by the public when he badly needs to be heard.

Although, the above example relates to an incident in the nineteenth century, the public reaction to a bridge failure may run essentially the same course even today. Imagining any risk of failure of the bridge under-foot remains as much a taboo in the modern human as it was in his Babylonian predecessor. Today it is not a question of means but of ends. Actually the acceptable risk of failure is an interdisciplinary economic decision in the given environment of social aspirations and objec-tives. If responsible members of the society at large seem to be totally unaware of the calculated risk of failure and its consequences, it is simply because they have never been informed.

A bridge failure hurts the self-esteem of the profes-sion and, smarting under the pressure of uninformed public criticism, one may easily lose one's bearings. Hasty non-optimal safety level decisions for future works may be taken. For example, a tornado, a flood, or an earthquake of 100 year-return period may occur today and take its toll of bridges that could never be designed for them from very rational considerations of safety and economy. A thoughtless doubling of wind or earthquake loads to avert such failures in future bridges may be still worse because scarce resources of the

---

*Sibly, P. G. et al.: 'Structural Accidents and Their Causes', *Proc. Inst. Civ. Engrs.*, Part 1, May 1977, p. 196.

society may thus be squandered away by aiming at such irrational safety levels. The following comment of Gerwick in an FIP Congress is relevant here:

"Safety and risk standards vary; man can hardly worry about the 100-year earthquake if he has no roof over his head today".

Bridges do collapse and perhaps much oftener than we hear or read about them. Only when the consequences of failure assume the dimensions of a disaster, the incidents become public knowledge and get investigated. We are generally not informed about the cases of near misses or near disasters. A more complete record and airing of information regarding bridge failures would have helped engineers to learn as much from them as from more assiduously reported successes. But talk of failure has, by tradition, been a taboo among engineers too. The implicit risk of failure has been carefully hidden from view in traditional codes of practice. The pathology of failures has not been discussed as widely as it should have been. In recent years, a rational definition of safety philosophy on probabilistic principles, using the concept of an acceptable risk of failure, however, has been attempted.

The most sensational failures in history of bridges are those resulting from an inherent weakness in design, a flaw which seems very glaring by hindsight but escapes notice due to contemporary thinking. Often this results from an inapplicable extrapolation of the available inadequate experience to predict the behaviour of new materials or new structural arrangements. When such adventures end in calamities after many years of earlier successes on smaller scale of jobs, besides the direct losses, much wider implications are involved. Engineers have to reevaluate their entire methodology and introduce correctives after a thorough search for what went wrong. The exercise requires a degree of application and introspection that is not as easy.

To this category of bridge failures belong the many historical bridge disasters, e.g., the Dee (1847, U. K.), Tay (1879, U. K.), Quebec (1907, Canada), Tacoma Narrows (1940, U. S. A.), West Gate (1970, Australia). All these bridge disasters have earned wide publicity and proved as turning points in bridge engineering concepts. But for Tacoma Narrows, all others caused unfortunate loss of human lives. The Dee disaster killed 5, Tay 75, Quebec 74 and West Gate 34.

The failure of Dee bridge provides a classic example to demonstrate how the nineteenth century engineers, basking in the confidence gained by their early successes, overreached conceptual capabilities and walked unwittingly into disaster. These were the exciting days when railway bridge spans were growing longer and cast and wrought iron had substituted for timber and

stone. The ambitions of bridge builders soared sky high. But their knowledge of structural behaviour lagged much behind.

Looking back at the earliest days of iron bridges, the design of the day appeared to be guided 'more by a carpenter's intuition than mathematical logic . . . Engineers then came to see cast iron as an improved form of masonry as well as a stronger timber, and only much later developed new shapes of section suited to the properties of iron as a material by its own right.' It is at this advanced stage of development, that the Dee bridge (girders of 30 m span) was designed in the offices of Stephenson in 1854. Robert Stephenson, son of George Stephenson, was the most celebrated bridge builder of his days and had covered himself with glory for his tubular girders of Conway and Britannia bridges. Six months after the completion of the bridge, the girders failed on the May. 24, 1847 under the weight of a train crossing the bridge and some extra ballast. With the bridge sagging under him, the driver accelerated and took the engine across. The less fortunate coaches separated from the engine and fell into the river. Five lives were lost and the shock of the first iron railway bridge collapse in Britain enraged the public mind.

Today, it does not require an expert to find the glaring error in designs which led inevitably to disaster. For Dee bridge, they used a cast iron girder in three pieces bolted together and "trussed" up with tensioned wrought iron ties. The trussed cast iron girder was no innovation. The form of construction had been used for about sixty structures in the preceding two decades for larger span cast iron girders. Only the Dee girder of 30 m span was longer than any attempted previously. All would have been well if the engineers had any rational conceptual understanding of the interaction of the highly redundant 'trussing' and girder. Their intuitive thinking that the interaction benefited the carrying capacity of the girder was all wrong. Actually the ties, instead of helping, could only hinder safety as considerable bending and compression could be transmitted to the girder. Why did so many distinguished engineers fail to realise that for a simply supported beam it is the depth at midspan that matters and was inadequate here, while the rods were as helpful as the bootlaces with which an idiot may try to lift himself off the ground?

Yet it is but a fact of life that this glaring error in the basic concept of a structure system held its sway for a good twenty years before disaster struck at Dee and even afterwards they were prone to brush the failure aside rather as a mischance than a blunder. The learned engineers who sat in a public enquiry of the failure made eloquent discourses on such erudite but irrelevant areas as vibration, impact, fatigue and workmanship.

Stephenson's was too big a name to be dragged into as having committed a mistake. So, beating about the bush provided a good cover-up for Stephenson as also their complete lack of comprehension of what really went wrong. Of course, blind over- reaction followed in the wake of the disaster, put an end to the construction system for future bridges and all those built earlier, using the system, were strengthened or retired prematurely. This was a costly experience indeed.

The Dee disaster and its aftermath have important messages for modern engineers. An engineer has to have a thorough conceptual understanding of the structural system. A simple comprehension of a complex structure-system ensures more safety than tons of mathematics. This spirit of enquiry into the fundamentals has to be a continuing process and an integral part of the mental make-up of bridge designers. Lawyers like precedents. For engineers, they have no such permanent values. Any theory, concept or system in building is to be judged by a fundamental enquiry, irrespective of however many leading men have accepted them for however long. The Dee bridge enquiry report failed to help engineers unlearn the errors made by their eminent predecessors.

About thirty years after Dee, came the Tay bridge disaster. The bridge failure came at about the fag end of the last century. The 'brief period of iron bridges', marking the transition from stone and timber to steel, began and ended within the span of a single generation. The rapid expansion of railways in this period had suddenly pushed the scale of bridge building to far greater heights than ever before. Bridge engineers had to cope with many bridges to build, bigger spans to attempt, a new material like iron, and last but not least a tight purse! The fast pace of change in the scale of his activities made the likelihood of an engineer making mistakes greater in adventurous designs based on inadequate data and subjective experience. The period of iron bridges was, therefore, marked by many recorded failures. A new high in notoriety of iron railway bridges was reached by the collapse of the Ashtabula bridge in Ohio, U. S. A. On the December 29, 1876, the Pacific Express crossing the bridge plunged into the river and the loss of 91 lives that resulted mark it out as one of the worst disasters in the history of recorded failures. Disaster of almost similar proportions struck at Tay only 3 years later.

In Tay bridge, Bouch made an error in failing to recognise the gross inadequacies of the British experience with wind pressure. There might have been little in British bridge building history at the time to make him wary of the dangers of wind induced failure, but reports of wind failures from abroad, specially in the U. S. A., might have helped his judgement if he had cared to pay heed to them. It is easy to blame Bouch by hindsight, but the contemporary attitude of British engineers showed a blissful ignorance of the dangers that wind posed to economically designed bridge girders. The attitude shocked Clarke, an American engineer, who read a paper on the subject in the Institution of Civil Engineers just before Tay Bridge was opened. The lesson that Tay disaster teaches is the need for collation of information available from round the world regarding failures and preventive practices adopted. But the professional did not bother much about this duty in the days of Bouch. Had the contemporary American experience with wind been recorded and given the publicity it merited, the Tay bridge collapse could perhaps have been avoided. The current awareness of the profession in this vital area also leaves much to be desired—even in these days of information explosion.

Much has changed from the past century to the present in bridge technology. The nineteenth century had belonged to pioneers in bridge building and their bold imagination. The saga of Saint Louis bridge in America and its builder Eads will be recalled as typifying the pioneering spirit of the bridge builders of the past century. James Buchanan Eads had designed a variety of salvage equipments in 1842 and 'iron-clad' gunboats during the civil war, but never a bridge, before he took up the challenge in 1865 to build a bridge with 'spans no less 150 m and clearance of 15 m.' The first major steel bridge of America was built by Eads in 1874. The nineteenth century saw many other such giants like a Stephenson or a Roebling. Yet in the past century, bridge engineering had lacked the firmer scientific footing to be found later. Engineers were then busy grappling with new materials and some big gaps in the scientific understanding of structural behaviour. They had to rely largely on their intuitive judgement and, as a result, were likely to make bigger mistakes more frequently. The large crop of railway bridge collapses bears this out.

The twentieth century heralded a new era in bridge building concepts, with significant improvements in materials and methods. Structural steel had come to stay by the beginning of this century, and reinforced concrete, a versatile material of construction to which this century belongs, was waiting to make its grand entry. What changed the scene more than anything else, was the rapid developments in the theory of structures. Bridge builders gained rapidly increasing new confidence and reliance on the sophisticated mathematical tools of analysis to predict the behaviour of structures with a high degree of redundancy. The advent of the digital computer in recent years and the accent on research efforts have stimulated many new developments in this direction. The analysis of complex structures has become almost commonplace. The bridge engineer today is a specialist, educated and trained in the science of bridge building.

Sensational bridge failures have occurred in this century also. Some have failed during construction and some prematurely after a few years of completion. On 29th August, 1907, the Quebec bridge collapsed killing 74 people. The second Quebec disaster on September 11, 1916, took 13 lives. The Tacoma Narrows suspension bridge collapsed due to aerodynamic oscillation on the November 7, 1940. Two-204 m spans of a bridge over the Mississipi in Illinois fell into the river due to a tornado on the July 29, 1944. In the second Narrows bridge in Vancouver the steel truss collapsed during construction in June 1958 killing 18 people. Then there is the series of collapses of big steel box girders of the seventies. At the Milford Haven disaster of June 1970, 4 were killed; at Koblenz on the 10th November, 1971, the steel box girder collapsed during cantilever erection, killing 13 people. In the West Gate bridge, Melbourne, the box girder collapse took a toll of 34 lives. Needless to say, this sordid list may run very much longer if we project data of all bridge failures round the world. A free discussion may hurt some egos and open old wounds. It would, however, bring no disrepute to our profession to know more precisely about the 'frequency of our failures.' We do need the data for bridge failures, not only for an objective assessment of our achievements but also and much more so for a realistic appraisal of our future aspirations in relation to safety aims.

Looking back to the much publicized bridge failures of the twentieth century, the Quebec one seems to be in the forefront. The Quebec disaster came about three decades after the Tay or the Ashtabula (Ohio) bridge disaster, and it was of similar proportions. It seems an interesting but odd coincidence, as pointed out by Sibly et al., that the major bridge disasters followed the approximate interval of three decades.

"Dee bridge 1847, Tay bridge 1879, Quebec bridge 1907, Tacoma Narrows bridge 1940, are approximately thirty years apart. Projecting this nearer to present time one arrives at the series of accidents of box girder bridges Circa 1970 (West Gate, 1970)."*

It may be interesting to watch out for Circa 2000. But none of these bridge disasters could, for sure, be brushed aside as design incompetence. At the same time they also underline the simple fact that engineers of the twentieth century, in spite of their improved knowledge of structural analysis, are not immune to the mistakes in basic understanding of the structural behaviour. The

mistake they committed in Quebec, West Gate or Tacoma, is very much of the same category as the one that caused the downfall of Dee bridge.

Here are the candid comments of Sibly on the single and most damaging gross error that ailed the design of Quebec bridge:

"The Quebec Railway Bridge was designed as the greatest cantilever span of its age, 550 m from pier to pier. The unfinished arm collapsed in August 1907 when the main lower chord buckled, following failure of lattice components in its web. The chord had been designed according to the Gordon—Rankine formula, which was the currently accepted procedure, but unfortunately the engineers overlooked the fact that this expression was based on experimental tests on solid columns rather than on the built-up sections they were using. Furthermore, they made a risky extrapolation from the existing data to ascertain permissible stress values for the new components, and, during construction, failed to reproduce the axial loading condition shown on the contract drawings.

These three factors, none of which appeared unacceptable at the time, all contributed to cause failure by a combined local and overall buckling mechanism. Again, unthinkingly, the engineers had pushed the existing data too far, forgetting the underlying limitations."**

The analysis of Quebec bridge disaster also throws light on another salient aspect of the twentieth century bridge building. A large modern bridge project is really an assembly of human endeavours. Getting an eminent designer to design a bridge does not automatically ensure success of the project. Theodore Cooper, the consultant of Quebec bridge, was the doyen of American bridge builders of the day. Nobody could doubt his competence to rival or excel Benjamin Baker's great cantilever bridge over the Fourth in Scotland. But Cooper, who had checked the design calculations made by contractor, overlooked some glaring errors in them and contributed little by getting well researched information to aid his intuitive judgement. The actual design effort was based on small-bridge experience. Cooper only lent his big name to generate a false sense of security. The gross error in design concept was further aggravated by a confusion of responsibilities between the consultant and the client's chief engineer. 'Cooper rarely visited the site but advised, frequently.' At least the unfortunate loss of human lives could have been averted, had the ample forewarning of the failure been read by a trained eye. The buckling of compression chords was noted in July and also in early August. Construction continued ignoring this. Only more buckles set off some alarm. Work was stopped temporarily and Cooper's assistant at site travelled to New York to tell him that the situa-

* Sibly, P. G. et. al., 'Structural Accidents and Their Causes', *Proceedings, Institution of Civil Engineers*, Part 1, 1977, 62, May 1977, p. 208.
** Sibly P. G., 'Open Discussion' on "The Relevance of History by Pugsley, Mainstone and Sutherland", *The Structural Engineer*, September 1975, Vol. 53, p. 392.

tion was serious. This was on August 27 and the site management could not wait for more than a couple of days. Perhaps the political pressure to stick to a target date of completion drowned all murmers of the conscience. On the August 29 they resumed work without waiting for Cooper's advice and the tragic death of 75 workmen resulted from the folly. Cooper's telegram ordering all works to stop reached the site after it was all over.

The failure of Tacoma Narrows suspension bridge is of a different kind altogether. The bridge failed due to wind effects, but not in a gale or a tornado. The wind blew across the bridge only at a moderate speed of about 65 km/h when the elegant 850 m suspension bridge span fluttered and failed. It was but evident that an increase in design wind load, would not have saved the bridge. Whether the reasons for the collapse could have been accurately pin pointed had Tacoma perished in a tornado, is anybody's guess. It was fortunate for the engineering posterity that it did not. The movements of the bridge were dramatically recorded in an amateur's film. The bridge withstood the simple bending oscillations; which appeared early. After the bridge was opened in July, 1940, these oscillations provided some sort of tourist attraction. But the joyride on the 'galloping Gerite', as it came to be called, was soon ended when on 7 November, 1940, torsional oscillations developed and gained in amplitude in steady cross wind. Vertical hangers cracked and the slender suspended structure heaved, heaped, and collapsed. Fortunately no lives were lost. But Mosseiff, an eminent expert in suspension bridges, who had designed the bridge, languished and died within a year.

Mosseiff could not comprehend the problem of 'bridge flutter' and, as a matter of fact, none of his illustrious predecessors could either. Of course, if one searches the long history of suspension bridges, early warnings of failure by oscillation in high wind may not be very difficult to dig out. But the growth of sophistication in theory of suspension bridges and the brilliant progress in building larger and more economical, more slender, and more ambitious suspension structures, which had resulted from the efforts of the great bridge builders, continued to turn a blind eye to the problem. It was only after the Tacoma disaster that massive research, experimentation, and theoretical analysis, led to a thorough understanding of the phenomenon and solution of the aerodynamic instability problem. It may be easy to be wise after the event and conclude that Mosseiff could avoid the disaster by a critical study of the history of past failures. But the question that looms large is whether such errors are really avoidable.

Failure during construction of a series of steel box girders in more recent years have worried the engineering community. The striking features of these failure incidents are the close conformity in design, places of occurrence in widely separated parts of the world and coincidence in points of time, *Circa* 1970. Although no two failures were really alike, the basic mode of failure was instability of the thin plates in compression. That such minor details as geometric imperfections or so-called secondary stresses could pose primary dangers, was the lesson learnt from this.

If the engineers had to learn these lessons from failures, the West Gate disaster taught them in no uncertain manner. In West Gate, they decided to avoid a welded structure, perhaps as an overreaction to the fate of the King's Bridge which failed due to careless and incompetent welding. Of course, this could hardly save the bridge as the constructors chose to ignore the specific danger of instability of thin plates by ignorance or accident. The box girder span of 112 m was in a simply supported condition of erection when it failed due to buckling of the top flange, killing 34 men. The particular action which precipitated the collapse was the premature removal of 37 top flange splice bolts and consequent loss of restraint of the flange. The top flange was already distorted and instability in such ill-restrained conditions was but inevitable. The confusion in responsibility and gaps in communication plagued the team effort necessary and created the climate for the ultimate calamity.

At Koblenz the genesis of the disaster was clear enough, uncluttered by other contributory causes. The vulnerable point was a transverse welded splice in the bottom flange. There was nothing wrong with welding or the workmanship. Only geometric imperfections (the plate was out of straight by 2 mm (!) and the splice in the stiffener was eccentric) caused critical second order effects not foreseen by the designer. So the bottom flange buckled and the bridge deck collapsed during construction on November 10, 1971 killing 13 people.

These unfortunate incidents leading to accidents should bring home the point more clearly than ever before that our greater capability in computation and sophisticated analysis do not prevent mistakes being committed in the basic concepts and practices. All our mathematical modelling starts from an initial idealisation of the prototype and things can go wrong in this basic conceptual stage. The structure is unaware and completely unimpressed by the volume of massive computational efforts. It is, therefore, more important to think than to merely calculate. The lessons to be learnt from the collapse of steel box girders are best described in the following words of Leonhardt:

"In general, I wish to say again, that during the last 30 years there was too much emphasis on theory and too little consideration of structural detailing and of seeing the realities like geometrical imperfections and other avoidable facts. The dramatic failures and accidents, which have occurred recently, like the collapses of big steel box girder bridges or of large cooling towers, were more or less caused by a blind faith in theories. The large steel box girder bridges, which failed or almost failed, had no sufficient safety against buckling, because most engineers all over the world believed in the linear buckling theory of webs or of stiffened plates, based on plane plates and straight stiffeners. In reality, there is no plane steel plate and no straight stiffener; there are always imperfections. If we consider these imperfections, then classical buckling theories are no longer valid and we better apply stress analysis with the second order theory, introducing the effect of deformations due to these imperfections. It is bad that such big collapses had to occur to teach us this lesson."

Going into the genesis of these failures one can see the original sin of the engineer. It is the instinctive desire of the great bridge builders to build spans larger or attempt forms of construction more elegant and more economical than ever before. The likelihood of making errors, of course, increases inevitably in this process. To desist from using new methods and materials seems to be a wise way of reducing the risk of errors and consequent failures, but, had this false wisdom prevailed on the profession, 'we would still be crossing ravines on bridges made of vines.' Such aberrations in the reaction of the engineer would have been still worse and would have cost the society dearly compared with the failures discussed above. Thanks to the dynamism of our profession, failures have not proved deterrents to 'novel plans' of bridges. 'Bridge builders are perhaps the most unrepent sinners and hard boiled optimists.' Failures have always inspired them to find the means to more glorious success in more innovative ways.

For the bridge engineer at cross-roads after a disaster and social condemnation, it pays to be as much cautious about the overreactions to a failure-incident as to avoid the errors which led to the failure. A competent engineering analysis from purely pathological point of view and an uninhibited autopsy of the accident are the only ways to help the understanding rise from the ashes of disaster. Thus the failure of Tacoma did not condemn suspension bridges. Instead, simple aerodynamic stability measures found to beat the bridge flutter problem have paved the way for bigger and more elegant suspension structures. King's bridge failure due to brittle fracture did not mean that welding of steel box girders was banned. It only underlined the need for the

right welding technique to be made available at most bridges. The series of box girder failures of the seventies could not lead to the conclusion that there was anything fundamentally wrong with the thin plated structures. The second order analysis of instability of thin plates due to geometric imperfections only earned the design attention, previously lacking.

Bridges that show no unique features in design and construction methods fail due to gross errors of various other types, many of which are easily avoidable. For example, a large number of bridge accidents has occurred throughout the world, in all projects, large and small, during the construction or erection stages. The reasons and consequences of such incidents tend to be underplayed unless death or injury of construction personnel is involved. The risks of failure of a falsework or an erection method undertaken may be quite inconsistent with the aims of optimisation of construction cost. The entire climate may be vitiated by a lack of cohesion in team effort. More often than not, all engineering effort is concentrated on proportioning permanent structures with erudite computations of materials strength and stresses in support of the minutest details. Such mundane matters as falsework, erection, etc., are left to lower echelons of engineering without any participation or sharing in the responsibility by the design consultant or top levels of management. This climate of confused priorities and responsibilities inevitably leads to gross errors and finally to failure.

Bridges, designed and built successfully, sometimes do fail to stand nature's fury. Although knowledge about the probable wind, earthquake, and flood, as well as their effects, has progressed in recent years, there remain serious gaps to be bridged. For example, estimation of flood discharge, river behaviour and scour are very basic problems in bridge engineering for which rational solutions are not always easy to find and inaccurate design predictions can lead to failure. From a compilation of facts relating to 143 bridge failures, Smith concluded that—

"The most noticeable feature emerging from this study is the large number of failures due to flood. In number they constitute almost half the total, because a single river in spate can wash away the foundations of a whole series of bridges lying in its course."*

To Indian engineers, more familiar with frequent flood havocs and the human misery they cause, the risk of a flood-induced bridge-failure always looms large in planning and designing a bridge. Yet, an ethical question involved in fixing an optimal safety level deserves to be restated here. It is easy to increase the design flood from 50-year to 100-year return period, even adding another third to the value to play safe, or double the

---

* Smith, D. W., 'Bridge failures', *Proc. Inst. Civ. Engineers*, Part 1, 60, August 1976, p. 378.

design scour-depth to play yet safer. Although the wide social mandate given to engineers may hide the irrational cost of such decisions, one should pause and think hard about the additional financial investment involved which may mean denial of other special demands, equally pressing or more basic. Our responsibility is to strike a balance between safety and economy and to take a calculated risk compatible with the socio-economic objectives, not gamble. The same logic should apply to an analysis of bridge failure even due to wind, earthquake, or a ship collision.

In this context, it is equally important to avoid hasty revisions in design load values, induction of a new design specification or even the formulation of a new half cooked code of practice in the wake of a failure incident. All this might amount to making the code more complex and uncalled for increases in safety margins all around. It is worth remembering that simplicity, rationality and general applicability to fundamental and broad areas of experience are the prime functional values of a code of practice. What stands between success and failure of a bridge even today, is not the quality of code or design calculations but that of man — the engineer who represents a conglomerate of technological endeavours. Sophisticated analysis of structure, supported by massive computer output, may inspire visions of proportionate increase of perfection in engineering efforts, very much greater than our predecessors were capable of, but an analysis of failure events

around us would prove such dreams to be entirely false. 'Few, if any of the failures became inevitable merely through the inexactness of available methods of computing stresses and strains. It would be a dangerous illusion to suppose that greater perfection in such methods can of itself reduce the risk of accident. Sound design is achieved above all by the wisdom and judgement with which the designer applies his results', not by mere computations.

In the words of Pugsley: "A profession that never has accidents is unlikely to be serving the country efficiently". While bridge failures cannot be treated as a prerogative of incompetence of engineer, it would be irresponsible on the part of the engineer to brush them aside as mere mischance and hide them away. Instead it is necessary to take a close look at the past mistakes, analyse them fully and learn the right lessons from them. This is no easy task and sooner the engineers and practical researchers devote attention to it the better for the future of the profession.

A careful enquiry probes extensively into many possible sources of trouble, and sometimes reveals points deserving criticism, not all of which contributed to the actual failure. A failure report misses some of its impact if it gives a mere undifferentiated list of criticism; nor is a blanket condemnation of much value. *Great courage and judgement are demanded of the enquirer, as he has a task which requires freedom from bias while at the same time demanding a definite opinion.*

Appendix 1
MAINTENANCE OF TIMBER BRIDGES

1. Causes of Defects in the Timber in the Bridge
— 'decay' due to fungul growth under damp conditions
— insect attack (by Termites and Ants)
— attack by marine borers e.g. *shipworms, Toredos, and crustaceans (Limnoria)*
— 'checking' (splitting of grains) due to shrinkage
— 'splitting' due to imposed tension
— fire

2. Definition of Various Defects
See Fig. A1.1.

Decay is the disintegration of wood substances by fungi.

Check is the lengthwise separation of wood that usually extends across the ring of annual growth and commonly results from stresses set up during seasoning.

Split is the lengthwise separation of the wood due to the tearing apart of the wood cells.

Shake is the separation along the grain, the greater part of which occurs between the rings of annual growth.

Cross Break (Crack) is the separation of wood cells across the grain' such breaks may be due to internal strains resulting from unequal longitudinal shrinkage or to external forces.

Holes could extend partially or entirely through a piece and be from a knot, insects, marine borers, grub worms, and so on.

3. Possible Damages in Timber due to Various Defects
(a) *Damage due to Decay*: Surface moulds, stains and decay are caused by fungi which require water, oxygen, and a favorable temperature (50 to 90° F) for growth. Timber which has a moisture content below 20 per cent or is fully soaked with water is safe from decay.

The fungi produce thousands of seed-like structures called 'spores' which are carried by the wind and grow on the timber through chemical cell enzymes. A sign of early decay is discolouration or a mushroom type growth. As the decay advances, the surface of the timber grows spongy, weak and highly absorbent and has a hollow sound upon hammering. The decay eventually destroys the timber.

At the joints in timber stringers decay often starts along the contact surface of a field cut between stringers,

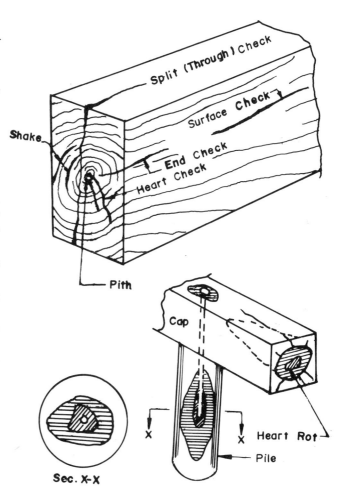

Advance-Decay

Incipient Decay

Fig. A1.1 Defects

and spreads into the center of each member. A member may appear to be in good condition although it may be hollow inside. In the later stages, crushing is visible along the contact surfaces. Accumulated debris between stringers over the caps and sills retains moisture and accelerates general decay in the stringers. When dirt or moisture comes into contact with the end of stringer, the decay often works into the cuts, rotting them hollow in a short time. All of these points should be examined with an increment borer.

Poor drainage in Laminated timber decks, (e.g. because of plugged drains) causes 'spore' infected water to splash over the wearing surface and it eventually penetrates into the laminated timber deck.

When timber is neither always wet nor extremely dry for a long period of time, it will decay. A favourable place for such a condition is at the ground line. The earth should be dug out from around piles to a depth of a foot or so and the timber should be probed thoroughly. Timber continuously submerged will not decay except in the splash zone. Even though it is in salt water it may be attacked by toredos or other than marine borers. When any indication of the toredo action is discovered, the pile post should be carefully checked for marine borer distraction. The infestation under these circumstances will occur at slightly above the point where the timber is continuously submerged. toredos will not attack the outside of the pile below the mudline.

(b) Damage due to Shrinkage: Shrinkage of damp timber leads to splitting cracks (called 'checks').

Shrinkage in the laminated floors separates the laminations, breaks the bond between the stringers and flooring, and permits movement which may break up the surfacing. Shrinkage of the stringers results unequal bearing and consequent damage.

(c) Damage due to Insects: Timber is damaged by beetles, carpenter-ants and termites. Termites do not show any outward evidence of their presence until the damages is well advanced. Termites look like ants and live partly in the ground and must have a permanent water supply. Dark and poorly ventilated locations, where the wood is in moist environment, provide ideal conditions for *termite attack*.

(d) Damage due to Marine Borers: A major concern near salt water is to protect the structure against marine borers, the main point of attack is usually between the high-tide level and the mudline. The most damaging to timber are shipworms, toredos, and the crustaceans (limnoria).

(e) Damage from Fire, Chemicals, Overload, Under-design, Poor Detailing, etc.

4. Protection and Repair against Various Damages
These can be done in following ways.

(a) Protection

(i) Pressure treatment with preservatives, specially of cuts, holes, and exposed surface areas.

(ii) Design for fast rain run-off, keeping wood dry and avoiding water traps.

(iii) Heavy treatment with coal-tar creosote or pentachlorophenol in heavy oil solvent (penta) can be used for deep penetration. For marine piles, the following treatments are used: Dual treatment with salt and creosote, Creosote and Waterborne salts.

(iv) For positive and improved uniform preservative penetrations, the incising method is recommended. It is a process of making a systematic pattern of punctures in the face of timber.

Preservative: Remedial treatment for the decay in the timber with internally applied non volatile, liquid preservatives such as vapam (sodium methyl dithiocarbamate) and Chloropicrin (Tricroloro-nitro-methane) are effective. It is suggested that the additional holes be drilled systematically through the decay and beyond the damaged areas so as to assure adequate spread of the preservative into the incipient decay areas.

In marine environment: soluble and insoluble salts, creosote-oil and creosote-coal-tar-solution injected preservatives have been used to protect against marine borers. The piles in the marine environments can be treated as follows.

For areas where the *toredo* (shipworm) and *pholad* attacks are known and where the *limnoria tripunotata* attacks are not prevalent, creosote or creosote-coal tar solution treatment will probably provide an adequate protection.

For areas where *toredo* and *limnoria tripunctata* attacks are known and where *pholad* attack is infrequent, high retentions of chromated copper arsenate (CCA) treatment will provide an adequate protection.

For areas where *limnoria tripunctata* and *pholad* attacks are known or expected, a dual treatment (creosote and CCA) provides a maximum protection.

(b) Repair

(i) by bolting, clamping and splicing. See Figs A1.2, A1.3, and A1.4.

(ii) by protective armour: The metallic or concrete armour jacket usually in sheeted form, could be used as an effective protection against marine borer attack. The armour should not be punctured and should extend from the high water elevation to far below the stream bed. Since this method is inconvenient and perhaps expensive, it is not widely used. *See* Fig. A1.5.

(iii) by post tensioning: In a longitudinally laminated timber deck, a transverse post-tensioning system will

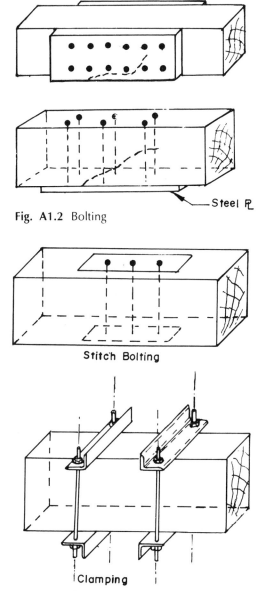

Fig. A1.2 Bolting

Stitch Bolting

Clamping

Fig. A1.3 Clamping

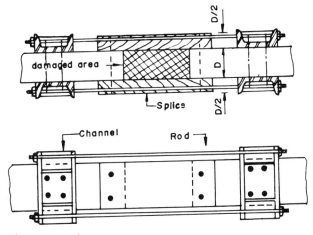

Fig. A1.4 Splicing

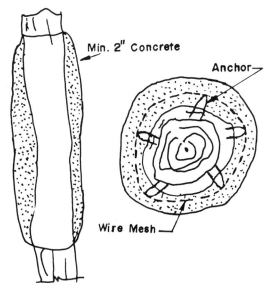

Fig. A1.5 Concrete encasement

develop friction through compression in the deck to promote load-sharing of the laminated deck. It will close the deck tight and result in reducing the deflection and preventing passage of moisture.

After removal of asphalt wearing surface, an epoxy coated high strength bar is installed at top and below the deck at each post-stressing point along the bridge. The bar is enclosed in tight fitting PVC pipe for additional protection. The PVC pipe is connected to steel sleeve of the steel anchorage plate. Post tensioning is applied, using jacks at each point to provide uniform contact pressure between steel and wood. After post tensioning, the deck should be resurfaced.

(iv) by replacement of major structural part: If the existing bridge shows severe deteriorations on the superstructure members, but minor defects in the substructure, only replacement of the superstructure may be required with reinforcing of the substructure members.

In a timber bridge, the bare plank bridge deck has poor skid resistance. When it is wet and the surface is bleeding the creosote, the bridge skid resistance decreases. In order to provide a smooth riding surface and improve skid resistance, asphalt overlay is used, but there are problems of bond between the asphalt and the deck.

Asphalt cracks over the joints between planks. For replacement of the timber plank deck, it is suggested to use treated glulaminated bridge materials to eliminate differential deck deflections and to reduce the dead loads.

(v) by replacement of only some members: Replacement of a member is required when its load carrying

capacity is found to be considerably reduced otherwise the member could be strengthened by adding reinforcements or by repairing it. If the structure has several more years of the remaining life, the replacement should comprise of pressure treated timbers and the adjacent member should be carefully inspect for defects. if a portion of the member shows defect, partial replacement of the member, preferably two feet beyond the defected area, should be made. It is recommended that the area surrounding the defect and the newly exposed cut off area should receive field preservative treatment.

For a highly stressed short span member, it is suggested to replace entire member instead of splicing or any partial repair (Difficulty in designing an effective splice.) For a long span stringer or member, splicing or partial repair may be an economical alternative.

5. Preventive Maintenance for Timber Bridges

(i) Field Preservative Treatment: The field treatment of the existing highway bridge would extend the life of the structure. Treatment should be applied at regular intervals the major structural component such as stringers, piles, caps and the exposed surface of the untreated part of the timber with collision damage, delamination, check and split to protect them from infection. Also, the localized areas should be treated to prevent decay from excessive moisture accumulation in the untreated parts of the timber.

Liquid preservatives are applied with hand brush and sprayer and the heavier grease or emulsion may be applied with proper equipment to the existing member during rehabilitation.

(ii) Moisture Control: Moisture control is required at location where the timber is subjected to frequent wetting causing decay. Proper and effective bridge surface drainage is the key to good maintenance. A plugged drain will cause the spore-infected water to leak through the deck and the joint between the stringers and the top of the caps or leak into the laminated deck. Also improper location of the drainage can cause certain part of the timber structure, to be vulnerable to decay through dampness and fungal attack ('spore' formation by fungi leads to decay, as explained earlier) the waterway must be cleared of debris because the accumulation of debris against the piles will promote decay. Protection of piles by metal or concrete jacketing should be looked into.

6. Some Useful References for Additional Reading:

— McGee, Dennis, 'The Timber Bridge Inspection Program in Washington State', FHWA Olympia, Washington, Jan. 1975.
— 'Timber Structure Standard Plans', FHWA.
— 'Glulam Bridge Systems: Plans and Details', The American Institute of Timber Construction, 1975.
— 'Transverse Post-Tensioning of Longitudinally Laminated Timber Bridge Decks', Ontario Ministry of Transportation and Communications.
— Bruesch, L. : 'Forest Service Timber Bridge Specification', ASCE Journal of Structural Division Vol. 108, No. ST 12 December 1982, pp 2737–2746.
— AREA Manual, Chapter 7: Timber Structures, American Railway Engineering Association.

Appendix 2
SPECIFICATIONS FOR CERTAIN REPAIR ITEMS

1. REPAIR MATERIALS

MATERIAL NAME : (A) *Class I Concrete (Non-Vibrated)*	
Materials:	Note:
94 lbs Cement Type I 218 lbs Fine Aggregate 325 lbs Coarse Aggregate 6.2 gals Water 3 Tablespoons Air Entraining Agent	1. Accelerators shall not be used. 2. Coarse aggregate is Grade 57 (3/4" and smaller). 3. For a wheelbarrow load (approx. 2 cu. ft.), use 1 and a 1/2 tablespoons air entraining agent. 4. Volume of water will vary slightly. 5. Chloride free materials shall be used. 6. Use only air entraining agents on approved list.

MATERIAL NAME: (B) *Class II Concrete (Non-Vibrated)*	
Materials:	Note:
94 lbs Cement Type I 182 lbs Fine Aggregate 297 lbs Coarse Aggregate 5.7 gals Water 3 Tablespoons Air Entraining Agent Slump 3"–5"	1. Accelerators shall not be used. 2. Coarse aggregate is Grade 57 (3/4" and smaller). 3. For a wheelbarrow load (approx. 2 cu. ft.), use 1 and a 1/2 tablespoons air entraining agent. 4. Volume of water will vary slightly. 5. Chloride free materials shall be used. 6. Use only air entraining agents on approved list.

MATERIAL NAME: (C) *Class III Concrete (Seal/Plug)*	
Materials:	Note:
94 lbs Cement Type I 166 lbs Fine Aggregate 263 lbs Coarse Aggregate 5.6 gals Water 3 Tablespoons Air Entraining Agent Slump 7"–9"	1. This is a one-bag tremie mix and is air-entrained, non-vibrated, with water-reducing, retardant admixture. 2. A one-bag mix is approximately two wheelbarrow loads. For one wheelbarrow load (approx. 2 cu. ft.), use 1 and a 1/2 tablespoons air entraining agent; Water-reducing retardant used per manufacturer's instructions. 3. Rock shall be coarse aggregate, Grade 57 (9), 3/4" or smaller 4. Volume of water will vary slightly. 5. Chloride free materials shall be used. 6. Use only admixtures on the approved list. 7. The above mix may be used for pumped concrete. Variations may be required, consult with the engineer.

MATERIAL NAME: (D) *Class IV Concrete (Vibrated)*

Materials:

Note:

94 lbs Cement Type I (AASHTO)
155 lbs Fine Aggregate
251 lbs Coarse Aggregate
4.6 gal Water
3 Tablespoons Air Entraining Agent
Slump 0"–3 and 1/2"

1. Rock shall be coarse aggregate, Grade 57 (3/4" or smaller).
2. Volume of water will vary slightly.
3. Chloride free materials shall be used.
4. For a wheelbarrow load (approx. 2 cu.ft.,), use 1 and a 1/2 tablespoons of air entraining agent.
5. Water-reducing retardant as per manufacturer's recommendations.
6. Use only admixtures on the approved list.

MATERIAL NAME: (E) *Latex Modified Portland Cement Concrete*

Materials:	By Weight		By Volume	
Portland Cement	94	lbs	1	cu ft
Sand	235	lbs (dry wt)	2.25	cu ft
Coarse Aggregate*	178.3	lbs (dry wt)	2.25	cu ft
Latex Additive	37.1	lbs	4½	gallons
Water**	4–25	lbs	½–3	gallons

*Coarse aggregate shall be No. 14, No. 16 or No. 16S
**The amount of water required per batch will depend on the moisture content of the sand.

MATERIAL NAME: (F) *Portland Cement Grout Filler*

Materials:

Note:

94 lbs Cement Type I (AASHTO)
436 lbs Fine Aggregate
5.4 gal Water
3 tablespoons Air Entraining Agent

1. Accelerators shall not be used.
2. Volume of water will vary slightly.
3. Chloride free materials shall be used.
4. Use only air entraining agents on approved list.

MATERIAL NAME: (G) *Non-Shrink Grout*

Non-shrink grout shall be mixed in the following portions by volume:

(i)   One part Portland Cement
(ii) One part fine aggregate
(iii) Non-shrink admixture proportioned as recommended by the manufacturer
      (for aluminum powder, 1/2 to 3/4 ounce by weight to one sack of cement)

The dry mix shall be first thoroughly mixed to form a uniform mixture. Only enough water shall be added to give a mealy-appearing adhesive mix.

The grout shall be hammered, tamped and rammed into place to produce complete contact between the elements. The material shall be generally placed near the center and built-out towards the edges by compacting as required above.

MATERIAL NAME: (H) *Epoxy Grout Filler*

Epoxy bonding compound: A two component material that is moisture insensitive for application both above and below water. It must be capable of adhering to wet concrete, steel and fiberglass forms.

Sand: Kiln dried Silica sand.

Machine mix the two component epoxy bonding compounds in strict accordance with the manufacturer's instructions. Mix in the ratio of 1 part epoxy bonding compound with a maximum of three parts sand. No more epoxy grout filler shall be mixed than can be placed in 20 minutes. Any excess on hand after this time shall be discarded.

MATERIAL NAME: (I) *Sand for Epoxy Grout Filler*

Kiln dried Silica sand meeting the following gradation requirements:

Standard Sand			Alternate Sand	
U. S. Sieve	Passing		U. S. Sieve	Passing
No. 4	100%		No. 4	100%
No. 16	90–100%		No. 10	90–100%
No. 30	30–50%		No. 20	0–5%
No. 50	0–10%		No. 40	0%
No. 100	0–5%			

ITEM NAME: (J) *Painting Structural Steel—Inorganic Zinc*

Description: Sandblast cleaning and repainting spot areas of deteriorated inorganic zinc paint system.

Application: Spot paint deteriorated areas on structures painted with inorganic zinc system. Applicable to both coastal and inland bridges.

Materials:
Sandblast Material --- M. S. #561-2
Zinc-Rich Primer ----APL
Intermediate Coat ---APL
Finish Coat ---------- APL

Construction Method:
General: Personnel performing this activity must be thoroughly trained with respect to required surface preparation and in the proper techniques for application of coatings.

1. All rusted areas are to be spot blasted to a 'near-white' condition.
2. Coat the blast-cleaned areas by brush or spray with a single coat (3.0 to 5.0 dry mils) of zinc-rich primer.
3. Twelve to twenty-four hours after application of the primer coat, apply finish coat to a minimum dry film thickness of 3.0 mils. Bridges located in coastal areas require an intermediate coat prior to application of the finish coat.

NOTE: Precautions must be taken not to damage the paint system outside the deteriorated area being repaired. This may be accomplished by shielding if necessary, however, any damaged areas from over-blast must be repaired. All mechanical components and electrical equipment must be protected. Blast material in gears and any coating on contractors must be removed.

ITEM NAME: (K) *Painting Structural Steel—Oil Base Paint*

Description: Sandblast cleaning and repainting spot areas of deteriorated oil base paint system.

Application: Spot repair of structural steel which has an existing oil base paint coating.

Construction Method:
1. All rusted areas are to be spot blasted to a 'commercial' blast or wire brushed.
2. The cleaned areas shall then be painted by brush or spray with two coats (min. 2.5 mils dry film thickness per coat).
3. A minimum of 48 hours shall elapse between first and second coat of primer.
4. After second coat of primer has cured 48 hours, apply by brush or spray one coat (0.5 mils min. dry film thickness).

MATERIAL NAME: (L) *Sandblast Material*

Sand: Silica sand of a maximum particle size no larger than that passing through a 16 mesh screen, U. S. sieve series.

Grit: Crushed grit made of cast iron, malleable iron, steel, or synthetic grits other than sand, of a maximum particle size no larger than that passing through a 16 mesh screen, U. S. sieve series.

Shot: Shot made of cast iron, malleable iron, steel, or synthetic shot of a maximum size no larger than that passing through a 16 mesh screen, U. S. sieve series.

NOTE: Type of sandblasting material to be furnished shall be as specified on the purchase order.

2. ADMIXTURES

Concrete Admixtures

1. Description: This item shall govern the materials used, methods of tests and construction methods for the use of admixtures in concrete.

2. General: An 'air-entraining admixture' is defined as a material which, when added to a concrete mixture in the correct quantity, will entrain uniformly dispersed microscopic air.

A 'water-reducing retarding admixture' is defined as a material which, when added to a concrete mixture in the correct quantity, will reduce the quantity of mixing water required to produce concrete of a given consistency and will retard the initial set of the concrete.

A 'water-reducing admixture' is defined as a material which, when added to a concrete mixture in the correct quantity, will reduce the quantity of mixing water required to produce concrete of a given consistency.

An 'accelerating admixture' is defined as an admixture that accelerates the setting time and the early strength development of concrete.

'High-range water reducing admixture' is defined as a material which when added to a concrete mix in the correct quantity will produce the results required herein.

3. Retarding and Water Reducing Admixtures: The admixture shall meet the requirements of Type A, Type D or Type F admixture as specified in ASTM Designation: C494, modified as follows:

(i) The water-reducing retarder shall retard the initial set of the concrete a minimum of 2 hours and a maximum of 4 hours, at a specified dosage rate, at an ambient temperature of 90° F.

(ii) The cement used in any series of tests shall be either the cement proposed for the specific work or a "reference" Type I cement from one mill.

(iii) Unless otherwise noted on the plans, the minimum relative durability factor shall be 80.

(iv) All concrete being tested will contain entrained air. The air entraining admixture used in the reference and test concrete shall be Neutralized Vinsol Resin.

High-range water reducing admixtures when compared to a standard mix design in accordance with ATSM Designation: C 494, in addition to (ii), (iii) and (iv) above shall produce the following:

(a) It shall reduce the required water by a minimum of 15 per cent.

(b) It shall increase the seven day compressive strength of the concrete by a minimum of 25 per cent.

(c) It shall not retard the initial set of concrete by more than one and one-half hours, when tested in accordance with ASTM Designation: C403.

(d) It shall contain no chlorides, air entraining agents or urea.

4. Air Entraining Admixture: The admixture shall meet the requirements of ASTM Designation: C260 modified as follows:

(i) The cement used in any series of tests shall be either the cement proposed for specific work or a "reference" Type I cement from one mill.

(ii) Unless otherwise noted on the plans, the minimum relative durability factor shall be 80.

The air entraining admixture used in the reference concrete shall be Neutralized Vinsol Resin.

5. Accelerating Admixture: The admixture shall meet the requirements for a Type C admixture as specified in ASTM C 494 modified as follows:

The accelerating admixture will contain no chlorides, and shall be used in the liquid form only.

Accelerators will be used only to meet special requirements and will require the approval of the Engineer on each specific project.

6. Approval of Admixtures: The manufacturer shall certify that the material to be furnished meets the requirements of this Item, and of ASTM Designation: C 260 or C 494, as modified herein, and shall furnish test reports from an approval laboratory, as defined by the Item. "Definition of Terms" having prior approval of the Materials and Tests Engineer. At the time of original request for approval of admixtures, the manufacturer shall state in writing the chloride content of the admixtures. No admixture to which chlorides have been added during manufacture will be permitted to be used.

In addition to the above, the manufacturer shall furnish the following:

(i) At the time of original request for approval of the admixture, the manufacturer shall supply a five gallon sample of the material to the Materials and Test Engineer of the client.

(ii) Each six months after approval of the material, the manufacturer shall furnish a notarized certification indicating that the material originally approved has not been changed or altered in any way. Any change in formulation of an admixture shall require retesting, and shall be approved by Materials and Tests Engineer prior to use.

(iii) Prior to approval, an Infrared Spectro-photometry scan, solids content, pH value, and unit weight will be submitted for further identification. A range of values will be given to cover the normal manu-facturing range. A change in formulation discovered by any of the tests prescribed herein, or other means, and not reported and retested, may be cause to permanently bar the manufacturer from furnishing admixtures for Departmental projects.

A list of pretested and approved admixtures will be maintained by the Materials and Tests Engineer. This list shall specify the approved dosage rates to be used except as modified herein.

The Department reserves the right to perform any or all of the tests required by ASTM Designation: C 260 and C 494 as a check on the tests reported by the manufacturer. In case of any variance the Departmental tests will govern.

7. Construction Use of Admixtures: When used in construction in conformance with the basic reference specifications and this specification, the Contractor will be allowed to use any admixture which has been approved. The Contractor shall submit to the Engineer one copy of the invoice showing the admixture or admixtures to be used on the project. Prior to using an admixture in the work, trial mixes shall be made and tested in the field using the materials and equipment to be used on the project.

Mix designs from previous or concurrent jobs may be used without trial batches if it is shown that no substantial change in any of the proposed ingredients has been made.

For air entraining admixtures the dosage shown on the approved list is the dosage utilized in approval tests of the admixture. This dosage *must* be adjusted as necessary to produce the required air content in the concrete within the specified tolerances.

For water reducing, water reducing-retarding, and high range water reducing admixtures, dosage rates different from those on the list of approved admixtures may be used provided they are established on the basis of trial mixes.

An approved retarding admixture (for normal hot weather concreting) may not perform satisfactorily for extended retardation, in which case its use will not be permitted.

All accelerating admixtures dosages will be based on trial mixes and approved by the Engineer.

All admixtures used shall be in the liquid state with the exception of high-range water reducers which may be in the powder form. All liquid admixtures shall be dispensed separately but at the same time as the mixing water except that liquid or powdered high-range water reducers shall be introduced into the concrete at the job site. No admixture shall be dispensed on dry aggregates. Each admixture shall be dispensed separately.

Admixtures shall be agitated as required to prevent separation or sedimentation of solids. Air agitation of Neutralized Vinsol Resin will not be permitted.

Air entraining agents shall be charged into the mixer at the beginning of the batch. Retarding or water reducing admixtures, except for high-range water reducers, shall be charged into the mixer during the last one-third of the batch.

Accelerating admixtures will be used only on the express approval of the Engineer. Accelerating admixtures will not be permitted in bridge decks, direct-traffic culvert slabs at any time, nor when Type II cement is specified.

The high-range water reducing admixtures (liquid or powder form) shall be incorporated in the mix entirely at the job site as determined by adequate trial batches, and approved by the Engineer. Normally powdered admixtures will be measured by the weight basis

and dispensed at the job site by method approved by the Engineer. The rotation of the mixer shall be sufficient to thoroughly mix the admixture into the concrete. The volume of water in the liquid high-range water reducer shall be taken into account in determining the water/cement ratio of the mix. If during the placement of concrete, an appreciable slump loss is noted, additional high-range water reducing admixture will be required to bring the concrete back into the desired consistency. The addition of water other than in the water reducer will not be permitted at the job site.

When high-range water reducing admixtures are used with Ready-Mixed concrete, the capacity of the truck mixer may be reduced for each batch by 25 per cent of the rated capacity to assure proper mixing of the admixture.

When deemed necessary by the Engineer, the Contractor shall furnish additional quantities of the admixture being used for further testing. Further use of the admixture will not be allowed until the results of such tests are known and the material meets the requirements of this item.

For individual placements of concrete of 25 cubic yards or more and for all Ready-Mix concrete, the admixture shall be measured and dispensed by a readily adjustable dispenser. When set to a predetermined volume the dispenser shall fill to the preset amount and hold it positively without leakage until the operator releases the content into the mixing water, by some positive means. Unless otherwise shown on the plans, completely automatic dispensing will not be required, except for use with a fully automatic plant.

The calibrated container shall be a measuring reservoir of the type where the level of the admixture is visible at all times. A strip gauge with one ounce increments for air-entraining admixtures and ten ounce increments for water reducing and/or retarding admixtures, shall be attached securely to the measuring apparatus. This strip shall be a material possessing weather-resistant qualities. The accuracy required for these systems shall be plus or minus three per cent. The equipment shall visibly show the total amount to be dispensed for ready check by the Engineer.

For contract work, with individual placements of less than 25 cubic yards and with the concrete batched on the job site, the Engineer may waive the requirements for mechanical dispensing equipment.

8. Measurement and Payment: No additional compensation will be made for the materials, equipment or methods required by this item, but shall be considered subsidiary to the various items included in the contract.

3. SHOTCRETE

Pneumatically Placed Concrete

1. Description: This item shall govern for furnishing and placing of 'Pneumatically Placed Concrete' for riprap, concrete channel or canal lining, encasement of designated structural steel members, the repair of deteriorated or damaged concrete and for other miscellaneous work as shown on the plans.

2. Materials: The cement, water, and aggregate shall conform to the requirements of the Item, 'Concrete for Structures'. Fine aggregate shall conform to the requirements of Table 2, Grade 1, and coarse aggregate shall conform to the requirements of Table 1, Grade 7, unless noted otherwise on the plans.

Air entraining admixtures, retarders and water reducing admixtures, if used, shall comply with the Item, 'Concrete Admixtures'.

Bar reinforcement and wire fabric reinforcement shall conform to the requirements of the Item, 'Reinforcing Steel'.

Expansion joint material shall conform to the requirements of the Item, 'Concrete Structures'.

Steel drive pins, studs or expansion bolts used for the attachment of reinforcing for repair of deteriorated or damaged concrete with pneumatically placed concrete, shall have a minimum diameter of one-eighth of an inch and a minimum length of 2 inches. Size and location of drive pins or studs and method of attachment of reinforcing shall be as specified herein or as detailed on the plans.

The equipment used for driving the pins or studs shall be of the type which uses an explosive for the driving force, and shall be capable of inserting the stud or pin to the required depth without damage to the surrounding concrete.

Expansion hook bolts (1/4 inch diameter), shall be placed in a drilled hole of the size and depth recommended by the manufacturer. The Engineer may require that a test be made of the driving equipment for steel drive pins and check the pull out quality of the expansion bolts, prior to approving their use.

3. Proportioning and Mixing: The Contractor shall submit a mix design for approval of the Engineer. The basic mix design shall conform to the following (see Table 1).

The cement and aggregates shall be measured by volume with enough water added to bring the materials to the desired consistency. Test panels will be required prior to approval of the mix design. The concrete will be applied to a plywood panel and shall be a minimum size of 18″ × 18″ × 3″ in depth. The panel will be shot with the same air pressure and nozzle tip to be used for the

Table 1

Type	*Minimum of One Part Cement To	28 Day Compressive Strength Cores
I	4 Parts Aggregate	3600 p.s.i. Min.
II	5 Parts Aggregate	3000 p.s.i. Min.
III	7 Parts Aggregate	2000 p.s.i. Min.

* The Contractor may use a design containing more cement than required by this specification, when approved by the Engineer.

production work. The panel will be cured in the same manner required for the particular usage required by the contract.

Three 2 inch diameter cores will be taken from each panel and tested in compression at 7 days. The average strength of the cores shall be a minimum of 70 per cent of the strengths required in Table 1 herein .

The Engineer may require additional test panels during the progress of the work if there is any change in materials, equipment or nozzle operator.

Mixing and application may be done by either by the dry mix or wet mix process. The materials shall be thoroughly and uniformly mixed using a mixer designed for use with pneumatic application. It may be either a paddle type or drum type mixer. Transit mix concrete may be used for the wet process.

All mixing and placing equipment shall be cleaned at regular intervals and be kept in acceptable working condition. The nozzle liner, water and air injection system should be inspected daily and replaced daily and replaced when the parts are worn.

4. Construction Methods

(i) Reinforcement: All reinforcement to be embedded in pneumatically placed concrete shall be clean and free loose mill scale, rust, oil, or other coatings which might prevent adequate bond.

Reinforcement shall be secured rigidly in the position indicated on the drawings. The clear distance between reinforcing bars shall be least 2-1/2 inches.

Minimum clear distance between forms and reinforcement and for cover shall be as shown on the plans. Space shall be provided for splicing bars in the approved manner.

For repair of structures, welded wire fabric shall be held securely approximately three-fourths of an inch out from the surface to be covered Adjacent sheets shall lap at least 6 inches and sheets shall be fastened together securely by tying at intervals not to exceed 18 inches. In placing the wire fabric, steel drive pins shall be driven to a penetration of not less than 1 inch or 1/4 inch hook bolts installed in accordance with the manufacturer's recommendation into the face of the designated portions to be covered or repaired. The wire fabric shall be fastened securely to each pin or bolt. Any pin that does not reach the desired depth or hook bolt that does not anchor properly in its hole may remain in place but must be supplemented by an additional pin or bolt installation. The welded wire fabric shall have a minimum of 1 inch cover to the finished concrete surface.

For the encasement of designated portions of steel structures, the welded wire fabric shall be bent to a template to conform as nearly as possible to the outlines of the steel members to be encased. Holes not less than 1/2 inch nor more than one inch in diameter, shall be provided in the webs of the members as near as practicable to the flanges for the purpose of attaching the reinforcing fabric. These holes shall be spaced approximately 3 feet on centers. The welded wire fabric shall be held securely approximately three-fourths of an inch out from the surfaces of the members to be encased. Adjacent sheets shall lap at least 6 inches and sheets shall be fastened together securely by tying at intervals not to exceed 18 inches. In placing the wire fabric, three-eighths of an inch round rods shall be fastened to the structural steel through the holes provided in the webs of the members to be encased and the fabric shall be tied securely outside to rods. Ties shall be spaced approximately 12 inches on centers. The formed fabric shall conform, in so far as possible, to the shape of the structural member.

(ii) Existing Structural Steel: Pneumatically placed concrete for encasing structural steel shall be Type II unless otherwise designated on the plans. All steel and concrete surfaces shall be cleaned thoroughly of all paint, rust, loose mill scale, grease, and such other foreign materials which are likely to prevent adequate bond between the surface to be encased or covered. Concrete surfaces to be covered with pneumatically placed concrete shall be kept damp a day in advance of application and shall be thoroughly cleaned and washed with water and filtered compressed air just prior to the application.

The encasement concrete shall be given a wood float finish and water cured for 4 days.

(iii) Repair of Existing Concrete: A Type I mixture shall be used for structure repair. All deteriorated or loose concrete shall be removed from the areas designated to be repaired or restored within the limits specified on the plans or designated by the Engineer. Concrete adjacent to a crack shall be removed in such a manner as to leave the existing reinforcing steel throughout the repair area as intact as possible. Concrete and reinforcing steel surfaces which will be in contact with pneumatically placed concrete shall be sandblasted clean, then the surface cleaned of loose material with filtered compressed air.

Exposed areas are to be sprayed with water, followed with another spraying after ten minutes, then not later than ten minutes after the second water spray, the repair area or the cavity will be filled with pneumatically placed concrete, tying in the wire mesh as required. A steel edged screed shall be used to cut surface to original lines. A steel trowel shall be used for final finish.

For curing, the repair area shall have a piece of wet cotton mat taped into place over the repaired area followed with a covering of 4 mil min. sheet plastic also taped into place. The sheet plastic shall be larger than the mat and shall be continuously taped at the edges with 3 inch min. width tape (air duct tape or better) to completely enclose the mat and hold in the moisture. After four days or longer, the mat and cover may be removed.

After the curing period the patches will be tested by striking with a hammer to check for soundness and bond to existing concrete.

(iv) Riprap and Ditch Lining: Pneumatically placed concrete for riprap and for channel or canal linings shall be the type designated on the plans. The concrete shall be placed within the limits specified on the plans or as designated by the Engineer. The surface shall be given a wood float finish or a gun finish as directed by the Engineer. Curing of riprap and/or ditch lining shall be by either Type I or II membrane in accordance with the Item, 'Membrane Curing'.

(v) Operating Requirements for the Dry Mix Process: The compressor or blower used to supply air shall be capable of delivering a sufficient volume of oil free air, at a pressure range of 30 to 85 psi as required by the size of the nozzle employed. Required capacity of compressor and operating pressures are shown in Table 2 for the various nozzle sizes. Steady pressure must be maintained throughout the placing process. The water pump shall be of sufficient size and capacity to deliver the water to the nozzle at a pressure of not less than 15 psi in excess of the required air pressure.

The values shown in Table 2 are based on a hose length of 150 ft. with the nozzle not more than 25 ft. above the delivery equipment. Operating pressures shall be increased approximately 5 psi for each additional 50 ft. of hose and approximately 5 psi for each 25 ft. the nozzle is raised.

Table 2 Compressor Capacities

Compressor Capacity cu ft/min.	Hose Dia (inches)	Max. Size of Nozzle Tip (inches)	Operating Air Pressure Available (p.s.i.)
250	1	¾	40
315	1 ¼	1	45
365	1 ½	1 ¼	55
500	1 ⅝	1 ½	65
600	1 ¾	1 ⅝	75
750	2	1 ¾	85

(vi) Operating Requirements for the Wet Mix Process: The pump shall operate so that the line pressure is between 100 psi and 300 psi for delivery hoses with 1-1/2" to 3" diameters. The mixing equipment shall be capable of thoroughly mixing the materials in sufficient quantity to maintain continuous placement. When transit mix concrete is used, this equipment shall conform to the Item, 'Ready-Mix Plants'.

The use of the wet process will not be permitted for the repair of deteriorated or damaged concrete.

(vii) Rebound: Rebound material shall not be used.

(viii) Construction Joints: Particular care shall be given to the formation of construction joints. Unless otherwise noted on the plans, all joints subject to compressive stress or over existing construction joints shall be square butt joints. Tapered joints will be permitted at other locations except the outside 1 inch shall be perpendicular to the surface.

(ix) Placing of Pneumatically Placed Concrete: Proper consistency shall be controlled at the nozzle valve by the operator for the dry mix process and a low water-cement ratio must be maintained. The consistency of the mix and the water shall be controlled by the mixer pump or by the transit mix truck when used for the wet mix process. The mix shall be sufficiently wet to adhere properly and sufficiently dry so that it will not sag or fall from vertical or inclined surfaces or separate in horizontal work.

When encasing structural steel members or covering portions of structures the concrete may be applied in one coat: however, if the concrete, after being placed, shows any tendency to sag, it shall be applied in two or more coats. Pneumatically placed concrete for overhead work shall be placed in two or more coats as may be necessary to insure proper bond and to eliminate sag. In covering vertical surfaces, placing of the concrete shall begin at the bottom and be completed at the top.

The nozzle shall be held at such distance (2 or 4 feet) and position that the stream of flowing concrete shall impinge as nearly as possible at right angles to the surface being covered. Any deposit of loose sand shall be removed prior to placing any original or succeeding layers of pneumatically placed concrete. Should any deposit of loose sand be covered with pneumatically placed concrete, the concrete shall be removed and replaced with a new coat of pneumatically placed concrete after the receiving surface has been properly cleaned.

Before channel or canal lining or riprap is placed, the earth canal or channel slopes shall have been compacted uniformly and thoroughly and brought to a uniform moist condition. The subgrade for lining shall be excavated and fine graded to the required section. The use of forms for lining will not be required. The surfaces of pneumatically placed concrete for both channel lining and riprap shall be finished accurately by hand floating methods before the concrete has attained its initial set.

The original surface and each surface which is permitted to harden before applying succeeding layers shall be washed with water and air blast, or a stiff hose stream, and loosened material removed. Sand which rebounds and does not fall clear of the work or which collects on horizontal surfaces shall be blown off from time to time to avoid leaving sand pockets. Concrete shall not be applied to a surface containing frost or ice. Where standing or running water is encountered it shall be removed before applying the concrete. No work shall be done without the permission of the Engineer when the temperature is lower than 35° F. After placing, the concrete shall be protected from freezing or quick drying.

(x) Workmen: Only experienced foremen, gunmen, nozzlemen, and rodmen shall be employed and satisfactory written evidence of such experience shall be furnished the Engineer or his representative upon demand.

5. Measurement: Measurement of pneumatically placed concrete for encasement of structural steel members will be by the square foot, in place, of the actual contact area.

Measurement of pneumatically placed concrete for repair and restoration of concrete structures, will be by the cubic foot, in place, using the surface area times the average depth of the patch.

4. EXTENSION OF STRUCTURE

Extending Concrete Structures

1. Description: This item shall govern for concrete for extending structures and for preparation of the existing structures for extending or widening, including the materials used; the removal of portions of the existing structure, preparation of exposed surfaces of steel and concrete for bonding new construction to old; and the construction of the proposed extensions, all as indicated on the plans.

2. *Materials*: All materials shall conform to the requirements of the pertinent items, viz.,
— concrete
— expansion joint material
— reinforcing steel

3. *Construction Methods*: The work shall be performed in accordance with the provisions of the Item, 'Concrete Structures', and in conformance with the requirements herein.

The Contractor shall verify all pertinent dimensions of the existing structure, prior to ordering materials required for the extensions.

Portions of the old structure shall be removed to the lines and dimensions shown on the plans, and these materials shall be disposed of as shown on the plans or as directed by the Engineer. Unless otherwise noted on the plans, metal railing shall be removed in such a manner that it will not be damaged, stacked neatly on the right-of-way at convenient loading points which will not interfere with traffic or construction and will remain the property of the Department. Any portion of the existing structure, outside of the limits designated for removal, damaged during the operations of the Contractor shall be restored to its original condition at his entire expense. Explosives shall not be used in the removal of portions of the existing structure unless approved by the Engineer, in writing.

When the headwalls, wingwalls and apron are specified on the plans to be reused in the extended structure, the portion to be reused shall be severed from the old structure to the lines and details shown on the plans. The headwall unit shall be moved to the new location specified, by methods approved by the Engineer, and the extension concrete and reinforcement placed according to the plan details. Any portion of the headwall unit damaged by the moving operation shall be restored to its original condition at the expense of the Contractor.

Unless otherwise noted on the plans, a demolition ball, other swinging weight or impact tool, will be permitted on those portions of the structure not immediately adjacent to the 'break' line of the concrete.

The concrete shall be severed at the 'break' line by pneumatic tools and may be followed by the use of the demolition ball, or other methods acceptable to the Engineer. The final removal of concrete at the 'break' line shall be with pneumatic tools. Damaged concrete shall be treated as specified above. Bridge slabs (other than multiple box culverts) shall first be sawed along the break line one-half of an inch deep prior to beginning the removal of concrete.

Except when otherwise provided on the plans, new reinforcing bars shall be spliced to exposed bars in the old structure by lap splices in accordance with Table 1 of the Item, 'Reinforcing Steel'. When welded splices are permitted by the plans, they shall conform with the Item, 'Structural Welding'. For lap splices (not welded), new reinforcing steel need not be tied to existing steel where spacing and/or elevation does not match that of existing steel provided the proper lap length is attained.

Dowels, if required by the plans, shall be installed by grouting reinforcing bars to a minimum length of 12 inches into the old structure. Holes for dowel bars shall be cleaned of all loose material, wetted and filled with a 1:3 mix grout or other approved materials, as may be specified on the plans, immediately prior to placing of dowel bars.

Concrete surfaces which will be in contact with new construction shall be roughened and cleaned prior to placing of forms. These surfaces shall be dampened and coated with mortar just prior to placing fresh concrete.

Roadway slabs shall be finished in accordance with standard practice except that when an overlay is required, the slabs shall be given a reasonably smooth surface finish by longitudinal or transverse screeding without any straight edge requirements.

The widened portion of bridges and direct traffic culverts shall not be opened to construction traffic or to the traveling public until authorized by the Engineer in accordance with the following:

Authorization may be given after the last slab concrete has been in place at least 21 days for light construction traffic not to exceed a three-quarter ton vehicle.

Authorization for normal construction traffic, and when necessary to the traveling public, may be given after the last slab concrete has been in place 30 days. Construction vehicles which exceed the legal load limit will not be authorized to cross structures. Vehicles used exclusively to transport ready-mixed concrete, and which have three axles, will be permitted only if the gross weight does not exceed 51, 000 pounds. Speed of such vehicles shall be limited to no more than 10 miles per hour.

Where a detour is not readily available or economically feasible to use, an occasional crossing of a structure with overweight equipment may be permitted for relocating equipment only but not for hauling material, providing a structural analysis of the structure using the exact equipment in question, indicates no damage will result. The structural analysis must be approved by the Engineer. Temporary matting and/or other requirements may be imposed by the Engineer if an occasional crossing is permitted.

4. Measurement: The quantities of concrete of the various classifications which will constitute the completed and accepted structure or structures in place will be measured by the cubic yard, each, square foot, square yard, or linear foot as the case may be. *Measurement will be as follows:*

(i) General

(a) All concrete quantities will be based on the dimensions shown on the plans or those established in writing by the Engineer. Diafram concrete, when required, will be included in the slab measurement.

(b) In determining quantities, no deductions will be made for chamfers less than two inches, embedded portions of structural steel or prestressed concrete beams, piling, anchor bolts, reinforcing steel, drains, weep holes, junction boxes, electrical or telephone conduit, conduit and/or voids for prestressed tendons or for embedded portions of light fixtures.

(c) For Pan Girder Spans, a quantity will be included for the screed setting required to provide proper camber in the roadway surface after form removal.

(d) For Slabs on Steel and Prestressed Beams, a quantity for the haunch between the slab and beams will be included when required. No measurement will be made during construction for variation in the amount of haunch concrete due to deviation from design camber in the beams.

(e) For Slabs on Panels or Tee-Beams, the combination of span length, theoretical camber in beams, computed deflections, and plan vertical curve will be taken into account in determining the quantity for the slab.

Additional concrete which may be required by an adjustment of the profile grade line during construction, to insure proper slab thickness, will not be measured for payment.

5. BANK PROTECTION

1. Description

The work under this section shall consist of furnishing all materials and constructing bank protection in accordance with the details shown on the plans and the requirements of these specifications. Bank protection shall be dumped riprap, grouted riprap, wire tied riprap, riprap in wire baskets or gabions, and other types of bank protection and shall be constructed at the locations and as shown on the project plans.

2. Materials

2.1 Rock

• *General*: Rock shall be sound and durable, free from clay or shale seams, cracks or other structural defects and shall have a specific gravity of at least 2.40. Rock used to construct dumped riprap shall be angular in shape. Rock used to construct other types of bank protection may be rounded stones or boulders. Rock shall have a least dimension not less than one-third of its greatest dimension and a gradation in reasonable conformity with that shown herein for the various types of bank protection. Control of the gradation will be by visual inspection.

No source of rock is designated. It shall be the contractor's responsibility to negotiate for the material, obtain the right-of-way and pay all royalties and damages.

The source from which the stone will be obtained shall be selected well in advance of the time when the stone will be required in the work. The acceptability of the stone will be determined by the Engineer. If testing is required, suitable samples of stone shall be taken in the presence of the Engineer at least 25 days in advance of the time when the use of the stone is expected to begin. The approval of some rock fragments from a particular quarry site shall not be construed as constituting the approval of all rock fragments taken from that quarry.

• *Grouted Riprap*: Gradation of the rock for grouted riprap shall be as specified in the special provisions.

• *Wire Tied Riprap*: Rock for wire tied riprap shall be well graded with at least 95 per cent, by weight, exceeding the least dimension of the wire mesh opening. The maximum size stone, measured normal to the mat, shall not exceed the mat thickness.

• *Dumped Riprap*: Gradation of the rock for dumped riprap shall be as specified in the special provisions.

The contractor shall provide two samples of rock of at least five tons each, meeting the gradation specified above. The sample at the construction site may be a part of the finished riprap covering. The other sample shall be provided at the quarry. These samples shall be used as a frequent reference for judging the gradation of the riprap supplied. Any difference of opinion between the Engineer and the contractor shall be resolved by dumping and checking the gradation of two random truck loads of stone. Mechanical equipment, a sorting site, and labor needed to assist in checking gradation shall be provided by the contractor at no additional cost to the Department.

• *Gabions*: Rock for gabions shall be well graded, varying in size from four to eight inches.

• *Riprap (Slope Mattress)*: Rock for slope mattress shall be well graded with 70 per cent, by weight, exceeding four inches. The maximum dimension of a single stone shall not exceed the least dimension of the gabion.

Broken concrete may be used upon approval of the Engineer.

- *Rail Bank Protection*: Rock used to construct rail bank protection (Type 1-Type 6) shall be as specified in the plans.

2.2 Metal Items

- *Wire Fabric*: Welded wire fabric shall be galvanized and shall conform to the requirements of AASHTO M 55, except that the minimum weight of the zinc coating shall be 0.15 of an ounce per square foot of actual surface.

Woven wire fabric shall conform to the requirements of ASTM A 116, except that the minimum weight of zinc coating shall be 0.3 of an ounce per square foot or ASTM A 584, Class 1.

Wire fabric shall be of the gauge, spacing, pattern, and dimensions shown on the plans. The selvedge on each sheet of mesh shall be galvanized steel wire two gauges heavier than that used in the body of the mesh.

- *Miscellaneous Fittings and Hardware*: Miscellaneous fittings and hardware shall be of the type and size provided by the manufacturer of the major item to which they apply for the use indicated on the project plans and shall be galvanized in accordance with the requirements of AASHTO M 232.

- *Tie Wires:* Tie wires shall be of good commercial quality and the gauge shall be as shown on the project plans, except that the minimum weight of the zinc coating shall be 0.30 of an ounce per square foot. At the option of the contractor, hog rings may be used on gabions in lieu of tie wires.

- *Steel Cable:* Steel cable shall be zinc-coated structural wire rope conforming to the requirements of ASTM A 475. Utilities Grade, Type 1 for the diameter shown on the plans.

- *Railroad Rail:* Railroad rails may be new or used. If used rails are furnished, they shall be free from rust and equal to at least 95 per cent of the original section.
Soil Anchor Stakes: Soil anchor stakes shall be steel and of the length called for on the plans. When not specified to be railroad rails, the following items may be used; crane rails with a weight of at least 40 pounds per yard; two inch diameter steel pipe conforming to the requirements of ASTM A 120, or 3" × 3/8" structural steel angles conforming to the requirements of ASTM A 36 or better. Used rails, pipe or angles may be used provided the material is not rusted or damaged to the extent that the strength of the item is reduced to less than 90 per cent of a new item of the same type and nominal size.

2.3 Bedding Material
Bedding material shall consist of granular material having a maximum dimension of two inches and shall be free of clay or organic material.

2.4 Grout
Grout shall consist of one part portland cement, three parts fine aggregate and one-fifth part hydrated lime, by volume. These materials shall be thoroughly dry mixed and sufficient water shall be added to provide a mixture of thick workable consistency.

Portland cement, fine aggregate and water, shall conform to the standard requirements. Hydrated lime shall conform to the requirements of ASTM C 207, Type N.

Grout that has been mixed more than one hour shall not be used. Retempering of grout will not be permitted.

2.5 Filter Fabric
Filter fabric shall conform to the requirements of AASHTO M 287.

2.6 Sacked Concrete
Sacked concrete shall be utility concrete conforming to the standard requirements. Except that the minimum cement content shall be 376 pounds per cubic yard; the slump shall be from three to five inches; and the aggregate shall conform to the following gradation:

Sieve Size	Per cent Passing
2 inch	100
1/4 inch	45-89
No. 200	0-12

Sacks for sacked concrete riprap shall be made of at least ten ounce burlap and shall be approximately 19 1/2 inches x 36 inches measured inside the seams when the sack is laid flat, with an approximate capacity of 1.25 cubic feet. Sound, reclaimed sacks may be used.

3. Construction Requirements

3.1 General
Areas on which bank protection is to be constructed shall be cleared, grubbed, and excavated or backfilled in accordance with the requirements of the appropriate sections of Division II to produce a ground surface in reasonable conformance with the lines and grades shown on the project plans or established by the Engineer.

Placement through water will not be permitted unless otherwise approved by the Engineer.

3.2 Filter Fabric
When filter fabric is required it shall be placed in the manner and at the locations shown on the project plans. The surface to receive the fabric shall be free of obstructions, depressions and debris. The filter shall be loosely laid and not placed in a stretched condition. The strips shall be placed to provide a minimum 24 inches of overlap for each joint. On horizontal joints, the uphill strip shall overlap the downhill strip. On vertical joints, the upstream strip shall overlap the downstream strip.

The fabric shall be protected at all times during construction from extensive exposure to sunlight.

When the maximum size of the rock to be placed on filter fabric exceeds 18 inches, the fabric shall be protected during the placement of the rock by a layer of bedding material. The bedding material shall be spread uniformly on the fabric to a depth of four inches and shall be free of mounds, dips or windrows. Compaction of the bedding material will not be required.

Rock shall be carefully placed on the bedding material and filter fabric in such a manner as not to damage the fabric. If, in the opinion of the Engineer, the fabric is damaged or displaced to the extent that it cannot function as intended, he will order the contractor to remove the rock, regrade the area if necessary, and replace the filter fabric.

3.3 Dumped Riprap

The rock shall be placed for its specified thickness in one operation and in a manner which will produce a reasonably well graded mass with a minimum practicable amount of voids with the larger rock evenly distributed throughout the mass.

No method of placing the rock that will cause segregation will be allowed. Hand placing or rearranging of individual rock by mechanical equipment may be necessary to obtain the specified results.

3.4 Wire-tied Riprap

After installation of the lower portion of the wire mesh, rock shall be placed in accordance with the requirements of Subsection 913-3.03. After placement of the rock, the upper portion of the wire mesh shall be placed, laced, and tied in accordance with the details shown on the project plans.

3.5 Grouted Riprap

Rock for grouted riprap shall be placed in accordance with the requirements of Subsection 913-3.03. The stones shall be thoroughly moistened and any excess of fines shall be sluiced to the underside of the stone blanket before grouting.

The grout may be delivered to the place of final deposit by any means that will insure uniformity and prevent segregation of the grout. If penetration of grout is not obtained by gravity flow into the interstices, the grout shall be spaded or rodded into the interstices to completely fill the voids in the stone blanket. Pressure grouting shall not unseat the stones; and during placing by this method, the grout shall be spaded or rodded into the voids. Penetration of the grout shall be to the depth specified on the project plans. When a rough surface is specified, stone shall be brushed until from one-fourth to one-half of the depth of the maximum size stone is exposed. For a smooth surface, grout shall fill the interstices to within a 1/2 of the surface.

Grout shall not be placed when the descending air temperature falls below 40 degrees F, nor until the ascending air temperature rises above 35 degrees F. Temperatures shall be taken in the shade away from artificial heat.

Curing of the grout shall be in accordance with the standard requirements applicable here.

At the option of the contractor, shotcrete conforming to the standard requirements may be furnished in lieu of grout.

3.6 Slope Mattress Riprap

The mattress bed shall be excavated to the width, line and grade as staked by the Engineer. The mattress shall be founded on this bed and laid to the lines and dimensions required.

Excavation for toe or cut-off walls shall be made to the neat lines of the wall.

Mattresses shall be fabricated in such a manner that the sides, ends, lid and diaphragms can be assembled at the construction site into rectangular units of the specified sizes. Mattresses are to be of single unit construction, the base, ends and sides either to be woven into a single unit or one edge of these members connected to the base section of the unit in such a manner that strength and flexibility at the point of connection is at least equal to that of the mesh.

All perimeter edges of the mattresses are to be securely selvedged or bound so that the joints formed by tying the selvedges have at least the same strength as the body of the mesh.

Mattresses shall be placed to conform with the details shown on the project plans. Stone shall be placed in close contact within the unit so that maximum fill is obtained. The units may be filled by machine with sufficient hand work to accomplish the requirements of this specification.

Slope mattresses shall be filled with at least two layers of stone. Broken concrete may be placed in the bottom layer with approval of the Engineer. Before the mattress units are filled the longitudinal and lateral edge surfaces of adjoining units shall be tightly connected by means of wire ties placed every four inches or by a spiral tie having a complete loop every four inches. The lid edges of each unit shall be connected in a similar manner to adjacent units. The slope mattress shall be anchored as shown on the project plans. Each anchor stake shall be fastened to the cover mesh with a tie wire.

3.7 Gabions

The gabion bed shall be excavated to the width, line and grade as staked by the Engineer. The gabions shall be founded on this bed and laid to the lines and dimensions required.

Excavation for toe or cut-off walls shall be made to the neat lines of the wall.

Gabions shall be fabricated in such a manner that the sides, ends, lid and diaphragms can be assembled at the construction site into rectangular units of the specified sizes. Gabions are to be of single unit construction, the base, ends and sides either to be woven into a single unit or one edge of these members connected to the base section of the unit in such a manner that strength and flexibility at the point of connection is at least equal to that of the mesh.

Where the length of the gabion exceeds its horizontal width the gabion is to be equally divided by diaphragms, of the same mesh and gauge as the body of the gabions, into cells whose length does not exceed the horizontal width. The gabion shall be furnished with the necessary diaphragms secured in proper position on the base section in such a manner that no additional tying at the juncture will be necessary.

All perimeter edges of gabions are to be securely selvedged or bound so that the joints formed by tying the selvedges have at least the same strength as the body of the mesh.

Gabions shall be placed to conform with the project plan details. Stone shall be placed in close contact in the unit so that maximum fill is obtained. The units may be filled by machine with sufficient hand work to accomplish requirements of this specification. The exposed face or faces shall be hand-placed using selected stones to prevent bulging of the gabion cell and to improve appearance. Each cell shall be filled in three lifts. Two connecting tie wires shall be placed as shown on the project plans between each lift in each cell. Care shall be taken to protect the vertical panels and diaphragms from being bent during filling operations.

The last lift of stone in each cell shall be level with the top of the gabion in order to properly close the lid and provide an even surface for the next course.

All gabion units shall be tied together each to its neighbor along all contacting edges in order to form a continuous connecting structure.

Empty gabions stacked on filled gabions shall be laced to the filled gabion at the front, side and back.

3.8 Sacked Concrete Riprap

The sacks shall be filled with concrete, loosely packed so as to leave room for folding or tying at the top. Approximately one cubic foot of concrete shall be placed in each sack. Immediately after filling, the sacks shall be placed according to the details shown on the project plans and lightly trampled to cause them to conform with the earth face and with adjacent sacks in place.

The first two courses shall provide a foundation of double thickness. The first foundation course shall

consist of a double row of stretchers laid level and adjacent to each other in a neatly trimmed trench. The trench shall be cut back into the slope a sufficient distance to enable proper subsequent placement of the riprap. The second course shall consist of a row of headers placed directly above the double row of stretchers. The third and remaining courses shall consist of stretchers and shall be placed in such a manner that joints in succeeding courses are staggered.

All dirt and debris shall be removed from the top of the sacks before the next course is laid thereon. Stretchers shall be placed so that the folded ends will not be adjacent. Headers shall be placed with the folds toward the earth face. Not more than four vertical courses of sacks shall be placed in any tier until initial set has taken place in the first course of any such tier.

When there will not be proper bearing or bond for the concrete because of delays in placing succeeding layers of sacks or because of the work having been hampered by storms, or mud, or for any cause, a small trench shall be excavated back of the row of sacks already in place, and the trench shall be filled with fresh concrete before the next layer of sacks is laid. The size of the trench and the concrete used for this purpose shall be approved by the Engineer. The Engineer may require header courses at any level to provide additional stability to the riprap.

Sacked concrete riprap shall be cured by being covered with a blanket of wet earth or by being sprinkled with a fine spray of water every two hours during the daytime for a period of four days.

3.9 Rail Bank Protection (Type 1-Type 6)

Excavation, where required for rock fill, shall be performed in reasonably close conformity to the lines and grades established or shown on the plans.

Rails shall be driven at the locations and to the minimum penetrations shown on the plans. Driving equipment shall be capable of developing sufficient energy to drive the rails to the specified minimum penetration and be approved by the Engineer.

If hard material is encountered during driving before minimum penetration is reached and it has been demonstrated to the satisfaction of the Engineer that additional attempts at driving would result in damage to the rails, the Engineer may order additional work to be performed, such as jetting or drilling, in order that minimum penetration may be obtained or he may order the minimum penetration to be reduced as required by the conditions encountered.

Wire fabric shall be securely fastened to the rails, placed in the trenches and laid on the slopes. The rock backfill shall then be carefully placed so as not to displace the wire fabric or rails. The wire fabric shall entirely enclose the rock backfill.

The completed rock fill shall be backfilled as necessary and the waste material disposed of as directed.

4. *Method of Measurement*

Riprap, except gabions and sacked concrete will be measured by the cubic yard of protection constructed by computing the surface area measured parallel to the protection surface and the total thickness of the riprap measured normal to the protection surface.

Riprap (gabions) will be measured by the cubic yard by computing the volume of the rock filled wire baskets used.

Riprap (sacked concrete) will be measured by the cubic yard of concrete placed in the completed work. The measurement will be based on mixer volumes.

Rail bank protection (Type 1-Type 6) will be measured by the linear foot. Measurement will be made from top of rail to top of rail (longest rail where rails of two or more lengths are used) and the distance measured will be from end rail to end rail. Where two parallel rows of vertical rails are used, the measurement for payment will be the average of the distance along the two rows.

6. EPOXY BONDING COMPOUND

Epoxy Bonding Compound

1. Description: Epoxy bonding compound shall include the furnishing and placing of an epoxy bonding compound on existing concrete and/or existing reinforcement steel before placing new concrete or mortar at the locations prescribed on the plans or determined by the engineer.

2. Materials: Epoxy bonding compound shall be a two component epoxy-resin bonding system for application to Portland cement concrete and shall conform to the requirements of ASTM C-881. The system type, grade and class shall depend on the conditions of the intended use. The color shall be clear or gray to match the color of the adjacent concrete.

The epoxy bonding compound shall be furnished as two separate components. Each container shall be clearly labeled and the following information shown:

    (i) Specification number and type

    (ii) Component designation (A or B)

    (iii) Manufacturer's batch number (a batch shall consist of a single charge of all components in a mixing chamber)

    (iv) Expiration date (shelf life for separate components in original containers)

    (v) Mixing ratio and directions (by volume of weight as designated by the manufacturer)

    (vi) Potential hazards and precautions shall be displayed in accordance with the Federal Hazardous Products Labeling Act.

3. Methods of Construction: All surfaces to be bonded shall be thoroughly blast cleaned with no. 40 boiler slag grit or no. 2 sandblast sand or abraded by approved mechanical means.

After blast cleaning and just prior to the application of epoxy, the surface shall be vacuum cleaned.

The two components of the epoxy bonding compound are furnished in separate containers. It is essential that the entire contents of both containers be thoroughly blended together.

The epoxy bonding compound shall be mixed and applied in strict accordance with the manufacturer's recommendations.

4. Quantity and Payment: No specific payment will be made for epoxy bonding compound and all costs thereof shall be included in the unit price bid for the Class B concrete of patching concrete deck or other appropriate item.

7. CRACK REPAIRS

Pressure Injection

1. Description: Pressure injection shall include the furnishing and placing of a suitable epoxy to seal the cracks at the approximate crack location shown on the plans or as directed by the engineer at the time of construction.

2. Materials: The epoxy resin shall meet the requirements of ASTM C-881 and the type and grade shall be submitted for the Engineer's approval after the Contractor's careful analysis of the area to be injected.

The system shall be 100 per cent solids with component 'A' being completely reactive with component 'B'. The system shall be suitable for use on damp surfaces but shall not be used where the temperature is 50 degrees or below.

All materials furnished shall be shipped in strong substantial containers. The containers shall be clearly identified as 'Part A — contains epoxy resin' and 'Part B — contains curing agent'. They shall also be plainly marked with the following information:

1. Name of product
2. Mixing proportions and instructions
3. Name and address of manufacturer
4. Lot number and batch number
5. Date of manufacturer
6. Quantity

The expiration date of acceptance of this material shall be one year after the date of manufacture. Any unauthorized tampering of breaking of the seals on the containers between the time of sampling and delivery to the job site will be cause for rejection of the material.

Component 'A' shall be the condensation product of Bisphenol 'A' and Epichlorohydrin. It shall contain no non-reactive diluents and shall be designated as a low viscosity resin suitable for injection.

3. *Methods of Construction*: The actual location and total quantity of the pressure injection of cracks may be adjusted by field inspection by the Contractor at the time of construction subject to the approval of the Engineer.

The Contractor shall be required to have present a manufacturer's technical representative for the duration of the injection process. Also, the detailed methods of repairs and the injection procedure should be submitted for the Engineer's approval.

The epoxy injection equipment shall be a positive displacement pump system. The system shall have a suitable mixing chamber where the epoxy components are accurately metered and thoroughly mixed immediately prior to injection. A clear, legible, and accurate pressure gauge shall be located in the supply line adjacent to the mixing chamber.

The equipment shall also be capable of providing a continuous and uninterrupted pressure head to continually force the injection epoxy into the cracks. Epoxy flow shall be capable of being fully controlled by operator controls at the mixing chamber.

All working personnel shall be familiar with the equipment, materials and procedures to be used during the operation. Extra (backup) equipment to assure the continuous injection of epoxy, in the event of primary equipment failure, shall be required.

All materials and equipment, including backup equipment, shall be at the work site before injection is begun. All equipment shall be in proper calibration and in good working order as determined by, and to the full satisfaction of, the Engineer. Epoxy shall be injected only by the use of automatic mechanical pumping, metering, and mixing equipment as described above. Pressure pot systems and hand held caulking guns, or grease guns, shall not be allowed.

The two components shall be mixed in accordance with the manufacturer's recommendations. The ratio of the components shall be maintained within a tolerance of five per cent.

Before applying the surface seal, all loose matter shall be removed. Surfaces shall be cleaned and prepared as per the seal manufacturer's specifications.

Any solvent used for cleaning shall be non-chlorinated. Acceptable solvents are mineral spirits, methyl ethyl ketone, acetone, low boiling naptha, xylene, or any other non-chlorinated solvent.

Prior to injection of the epoxy in the crack, a surface seal material shall be applied to the face of the crack. Surface seal material should be capable of use on vertical as well as horizontal surfaces.

Openings in the surface seal shall be established along the crack. The distance between entry ports shall not be less than the thickness of the concrete member being repaired.

Port Spacing and Installation
Sites for ports shall be drilled to a width and depth sufficient to assure a snug fit of the port. These holes shall be cleaned to remove any dust or debris left by the drilling operation. Care shall be exercised to assure that oil or other contaminants shall not be introduced into the air feed hoses, or deposited on any air blown surfaces.

Injection Procedure
No epoxy injection, or surface sealing, shall be done when the concrete temperature or ambient temperature is, or is expected to fall, below 50 degrees F during the 24 hours following the time of epoxy injection. Application or injection of epoxy at temperatures lower than that allowed by this specification, shall be done with the written approval of the Director, Materials Bureau.

The injection of the adhesive into each crack shall begin at the entry port at the lowest elevation. Injection shall continue at the first port until the epoxy adhesive begins to flow out of the port at the next highest elevation. The first port shall be plugged and injection started at the second port until the adhesive flows from the next port. This sequence shall be followed until the entire crack is repaired.

Before injection of any crack, the automatic mixing and metering pump shall be activated and a small amount (about one pint) of injection epoxy shall be mixed and wasted into a disposable container. The Engineer shall observe this trial operation, and be fully satisfied that the equipment is working properly. If the equipment is not working properly it shall be immediately repaired to full working condition or replaced with the backup equipment. If the backup equipment is used, additional and fully operable equipment shall be provided to replace it in case of malfunctions.

The feed line from the mixing equipment shall be securely held or properly attached to the end most port (bottom-most for vertical cracks). The operator shall then initiate epoxy injection and flow shall be allowed at ten (10) to forty (40) psi. The injection procedure shall be monitored to assure the epoxy flow does not cease before the injected epoxy exudes from the adjacent port. When epoxy flows from the adjacent port, injection shall be stopped, the feed line removed from the port, and the port sealed. The feed line shall then be attached to the next port and the procedure repeated until the last port is sealed. If the epoxy flow stops before epoxy appears at the adjacent port, the feed line shall be moved to the adjacent port and the port just used shall be sealed.

When the epoxy supply in the mixing equipment is about to be exhausted, it shall be replenished. Each epoxy component shall be thoroughly stirred before adding it to its respective storage tank in the mixing equipment. No discontinuity of epoxy flow through the feed lines of each or both components shall be allowed. In this manner, a continuous injection operation shall result.

In the event of leakage from the crack, injection will be stopped until the leak is sealed. Any work stoppage of fifteen (15) minutes, or longer, will necessitate cleaning of the mixing chamber and any equipment in contact with mixed epoxy.

After the injection process has been completed and the epoxy allowed to fully cure, the injection ports, surface seal and any spillage shall be removed from all surfaces easily visible to the public. Ports may be cut, or knocked off, while the surface seal and any spillage shall be ground off flush with the original surface, using a hand grinder. Any damage to the concrete during the cleanup procedure shall be repaired in a manner satisfactory to the engineer, at no additional cost to the state.

Payment for pressure injection will be made for the quantity as above determined, measured in linear feet, at the unit price bid in the proposal for the item pressure injection, which price shall be full compensation for the satisfactory complete repair of the cracks, including removal and disposal of unsound concrete, all saw cutting, chipping, cleaning, sand-blasting, drilling, furnishing and installing the ports, furnishing equipment, furnishing detailed methods of epoxy pressure injections, tests and retests required as a result of substandard work of material, furnishing, installing, finishing, and curing epoxy materials and all else necessary therefor and incidental thereto.

The Contractor shall submit to the Engineer a certification from the manufacturer that the epoxy furnished is the manufacturer's recommended epoxy for pressure injection of cracks and that the method of application proposed by the Contractor is acceptable.

The injected epoxy shall have penetrated a minimum of 90 per cent of the visible crack. The state may take cores of the repaired concrete to determine the length of penetration. If the penetration is less than 90 per cent of the visible crack, the crack from which the core was taken will be deemed not to have been repaired.

4. Quantity and Payment: The quantity of pressure injection for which payment will be made will be the total length of cracks, measured along the surface of the concrete, actually repaired in accordance with these specifications at the locations shown on the plans or as directed by the Engineer.

8. REPAIRING SPALLED CONCRETE

1. Description: Repairing spalled concrete shall include all work required to repair deteriorated concrete surfaces of abutments, wingwalls and where indicated, all in accordance with the plans and specifications or as directed by the Engineer. This work consists of the removal and disposal of all loose and disintegrated concrete, saw-cutting, the preparation of the surface, cleaning or replacement of existing reinforcement steel, field epoxy coating exposed existing reinforcement, the application of an epoxy bonding compound and placing of Class B concrete

2. Materials

Concrete: Concrete for repair shall be Class B, air-entrained, conforming with the applicable requirements of division 4, section 1 of the specifications except that the slump shall not exceed 2" + 1/2". The maximum size of course aggregate shall be 3/8 inch.

The Contractor shall be responsible for furnishing the design proportions of cement, fine aggregate, coarse aggregate, water, air-entraining admixture and water-reducing admixture which will produce a workable concrete mix meeting the above requirements for class B concrete.

At the Contractor's option, superplasticizer may be added to the concrete mix at the job site. The slump of the concrete mix, as delivered, shall conform to the requirements above and shall not exceed 8 inches after the addition of the superplasticizer.

3. Methods of Construction: The lateral limits of each area to be removed and replaced will be delineated by the Engineer and suitably marked. The outline of each such area shall first be saw-cut where required to a depth of one (1) inch with a power saw capable of making straight cuts.

All deteriorated of unsound concrete shall be removed to sound concrete or as directed by the Engineer. Minimum depth of removal shall be two (2) inches. Concrete may be removed by means of approved hand held pneumatic chipping hammers, not exceeding 30 pounds. Pneumatic tools should not be placed in direct contact with reinforcement steel. Extreme care shall be taken when reinforcement steel or anchor bolts are uncovered so as not to damage the steel or the anchor bolts or their bond in the surrounding sound concrete. Reinforcement steel or other embedded items damaged during concrete removal shall be repaired by the Contractor, as directed by the Engineer, at no cost to the state. If reinforcement bars are exposed, the removal shall continue until at least 3/4 of the bar's circumference is exposed. If unsound concrete is encountered at or below the mid-depth of reinforcement bars, removal shall extend to at least 3/4 inch beyond the bars. The limits of the cavity for deep, narrow repairs (depth equals or exceeds twice the width) shall be undercut to lock the repair in place.

After removal of unsound concrete the surfaces of the remaining concrete shall be cleaned of all loose concrete, dust, and other foreign material. Embedded items shall be cleaned of all loose adhering concrete, rust, and scale. Adhering concrete shall be removed from exposed reinforcement steel and the surfaces of the exposed bars shall be cleaned and field epoxy Supplementary Specifications.

Deteriorated reinforcement shall be replaced with new reinforcement as directed by the Engineer.

Adequate provisions must be made to prevent pieces of broken concrete from falling below.

Details of platforms, catches or other methods of collecting materials resulting from the preparation of spall areas for repair shall be of the Contractor's design and shall be submitted to the Engineer for review before commencing concrete removal. Materials collected shall not be allowed to accumulate but shall be promptly removed and disposal of away from the site.

After completion of the removal and cleaning operations and immediately prior to the placement of the concrete repair material, the surfaces shall receive a coating of the bonding compound applied in accordance with the manufacturer's recommendations and as directed by the Engineer.

The pot-life of the epoxy bonding compound, mixing period, maximum time lapse between mixing compound and placing of the new concrete, are all dependent on the temperature, humidity, and wind conditions. The contractor shall acquaint himself with such information as recommended by the manufacturer and shall schedule his operations accordingly.

Concrete shall be placed immediately while the compound is still tacky. If the bonding compound dries before the new concrete is placed, another coating of epoxy bonding compound shall be applied, as herein described, at no additional cost. After the new concrete is in place, normal finishing operations shall be completed.

All details of design, such as v-grooves, chamfers, joints, etc., in the existing abutments and pier shall be duplicated in the repair work under this section.

Placing of concrete, forming, removal of forms, finishing and curing of the new concrete shall conform with the applicable requirements.

Whatever means of placing the concrete is employed, it shall be such that the material will completely fill the space to be replaced, be thoroughly compacted, and free of air pockets.

When repairs are to be made of fast-setting mortar, the manufacturer's recommendations shall be strictly followed in all procedures relating to preparation and placement of the fast-setting mortar, based on prevailing climatic and job conditions. The Contractor shall arrange to have a technical representative of the manufacturer at the site during the initial phase of the spall repair work to acquaint the Contractor's forces with the proper methods of preparation and application.

4. Quantity and Payment: The quantity of repairing spalled concrete for which payment will be made, will be the total area of repair actually performed as prescribed regardless of depth.

Individual area measurements will be computed to the nearest half square foot. Individual areas measuring less than one square foot will be considered as one (1.0) square foot for the purpose of payment. The Engineer's measurements will be final and conclusive.

Payment for repairing spalled concrete will be made for the quantity as above determined, measured in square feet, at the price per square foot bid for the item repairing spalled concrete in the proposal, which prices shall include the cost of sawcutting, catching, removing and disposing of concrete; Cleaning remaining concrete surfaces; Protecting and repairing uncovered reinforcement steel; Additional reinforcement as required; furnishing, mixing and applying epoxy bonding compound; Furnishing, forming, placing, finishing and curing concrete; all materials, labour, equipment and all else necessary therefor and incidental thereto.

9. CORING AND GROUTING

1. Description: The work shall consist of coring holes into concrete grouting the holes and properly positioning reinforcing bars, or anchor bolts, into the grouted holes where shown on the plans.

For purposes of this specification, the terms reinforcing bars and anchor bolts shall be considered identical.

2. Materials: The grout material shall be a nonmetallic, non-shrink grout conforming to the standard requirements.

3. Methods of Construction: Equipment used for coring holes shall be approved by the engineer prior to use.

Coring holes into a structural slab or into any structural concrete element shall be done by means of a core drill.

Coring holes into any other structural element shall be done by methods satisfactory to the engineer, unless a specific method is noted elsewhere in the contract documents.

Coring with a lubricant shall not be permitted. Water will not be considered a lubricant. Coring methods shall not cause spalling or other damage to the concrete. Concrete spalled or otherwise damaged by the contractor's operations shall be repaired in a manner approved by, and to the satisfaction of the engineer. Such repair shall be done at the expense of the contractor. Holes shall be surface dry and shall have had all foreign and loose material removed immediately prior to grout placement. Coring through steel parts or reinforcing steel shall be permitted only where shown on the plans and as approved by the engineer.

Grout shall be mixed and placed in strict accordance with the manufacturer's instructions unless otherwise modified herein.

No grout placement shall be permitted when the ambient air temperature is 50 degrees F or below during the working day. Grout shall be inserted to a depth sufficient to insure complete filling of the hole after insertion of the reinforcing bar.

Grout material shall be thoroughly brushed into all surfaces of the hole immediately prior to the actual placement of the grout.

Reinforcing bars shall be clean and dry prior to insertion into the grouted hole. They shall be inserted full depth into the hole and shall be manipulated to insure complete coverage by the grout. After insertion of the reinforcing bar, all excess grout shall be struck off flush with the concrete face. Should the grout fail to fill the hole after reinforcing bar insertion, additional grout shall be added to the hole to allow a flush strike-off.

If the reinforcing bar is inserted in a hole which has an axis predominantly horizontal to the ground surface, care shall be taken to prevent the reinforcing bar from changing position prior to the setting of the grout, and to prevent grout from running down the face of the concrete. These precautions shall be done in a manner satisfactory to the Engineer.

4. Quantity and Payment: The quantity of coring and grouting for which payment will be made will be the length of holes, measured in linear feet, into which grout and reinforcing bars have been inserted where shown on the plans.

Payment for coring and grouting will be made for the quantity as above determined, measured in linear feet, at the unit price bid in the proposal for the item coring and grouting, which price shall include the cost of all labour, materials and equipment necessary to do the work, except that reinforcing bars and anchor bolts shall be paid for under their appropriate items.

10. PIER REPAIR ABOVE AND BELOW WATER

Description
Repairs to pier bases above EL. 'X' shall consist of the following:

1. Removal and disposal of unsound and deteriorated concrete from the pier bases of piers EL. and 'X', as shown on the plans, and the replacement thererof with class B concrete.

2. Removal and disposal of unsound and deteriorated concrete from the nose area of all piers and the replacement thereof with class B concrete.

3. The application of epoxy waterproofing to the tops of all piers to the limits shown on the plans.

All of the above work takes place above elevation 'X' and shall be performed in accordance with the plans and as directed by the engineer.

Materials
Class B concrete materials shall meet the standard requirements specified for such concrete. Class B concrete shall be air-entrained. The requirements for air-entrained concrete, air-entraining additives, and air-entraining admixtures shall be as specified under article . . .

Welded steel wire fabric shall meet the requirements of ASTM A 185.

Expansion type anchor bolts and shields shall meet the requirements specified under federal specifications FF-S-325, Group VIII, Type 1.

Epoxy bonding compound shall meet the requirements specified in AASHTO M325-Class 1.

Epoxy waterproofing shall meet the standard requirements specified under article . . . . Epoxy waterproofing shall be grey to match the color of the adjacent concrete.

Methods of Construction
Methods of construction for removal and disposal of unsound and deteriorated concrete shall meet the requirements specified under article . . . as amended elsewhere herein, and as hereinafter specified.

Where the term 'chip to sound matrix' or words of similar intent are indicated on the plans, the level or levels of sound matrix will be determined by the Engineer, based on observation of actual in place conditions, but generally on the basis that concrete which is difficult to remove by use of a pneumatic hammer equipped with a spade is sound concrete. Removal of existing concrete shall be performed in a manner as not to damage sound matrix. Existing reinforcement bars uncovered by chipping shall be preserved, in-place and in correct position. Such bars shall be thoroughly cleaned with wire brushes before being embedded in new concrete.

All surfaces to be bonded shall be thoroughly blast cleaned with No. 40 boiler slag grit or No. 2 sandblast sand or abraded by approved mechanical means.

After blast cleaning and just prior to the application of epoxy, the surface shall be vacuum cleaned.

The two components of the epoxy bonding compound are furnished in separate containers. It is essential that the entire contents of both containers be thoroughly blended together. A paddle attached to a slow speed electric drill is recommended for mixing. For small batches of one gallon or less, thorough hand stirring may be satisfactory.

The Contractor shall schedule his concreting operations so that concrete shall be placed while the epoxy bonding compound has not set and is still tacky. If in the opinion of the engineer, the bonding compound has begun to set, no concrete shall be placed until a new film of bonding compound has been applied to the required areas, at no cost to the state.

Quantity and Payment

The quantity of repairs to pier bases above EL. 'X' for which payment will be made will be the volume of concrete actually placed within the limits shown on the plans and as directed by the engineer.

Payment for repairs to pier bases above EL 'X' will be made for the quantity as determined above, measured in cubic feet, at the price per cubic foot bid for the item repairs to pier bases above EL 'X' in the proposal, which price shall include all the work specified herein and as shown on the plans and the cost of furnishing all labour, materials, equipment, including removal and disposal of unsound and deteriorated concrete, welded steel wire fabric, anchor bolts and shields, epoxy bonding compound, epoxy waterproofing and all else necessary and incidental thereto.

- REPAIRS TO PIER BASES AT EL. 'X' AND BELOW (— CONCRETE)

Description

Repairs to pier bases at EL. 'X' and below (—concrete) applies to all piers and shall include removal and disposal of unsound and deteriorated concrete of pier bases at EL 'X' and below and the replacement thereof with pumped concrete fill, as indicated on the plans and in accordance with the specifications.

Materials

- *Pumped Concrete Fill*: Pumped concrete fill shall have a minimum, ultimate compressive strength of 3500 p.s.i. at the age of 28 days and shall not be leaner than one part cement to five parts of aggregate. The Contractor shall be responsible for, at his own cost, preparation and testing of trial mixes, which mixes and test results shall be approved by the engineer before placing pumped concrete fill.

Concrete materials shall meet the requirements specified under Article 4.1.2. Cement shall be Type II. The maximum size of coarse aggregate shall be 3/4 inch. The minimum size of coarse aggregate shall be 3/8 inch. The ratio of weight of coarse to fine aggregate in the mix shall not be less than 1.0.

Methods of Construction

- *Removals*: Methods of construction for removal and disposal of unsound and deteriorated concrete above water line shall meet, the requirements specified under repairs to pier bases above EL. 'X' and as hereinafter specified. Underwater sections shall be chipped with a pneumatic hammer equipped with a spade, using an extended shaft on the spade. Where the workmen operate from floats, the floats shall be secured in position so that the workmen can apply pressure between the spade and the surface being chipped by taking leverage against the floats.

- *Pumped Concrete Fill*: Methods of construction for pumping concrete fill shall meet the requirements specified under article . . . and as specified herein.

The concrete shall be placed under an air pressure sufficient to assure a smooth and continuous flow. The discharge end of the hose shall be inserted into the form until top of concrete footing is reached before any concrete is pumped into place. As the level of the concrete rises in the form, the hose shall be slowly withdrawn, care being taken that the end of the pipe or hose be continuously immersed in the newly placed concrete. Kinking the hose or other sources of sudden surges of flow shall be avoided insofar as possible. In general, concrete shall be placed in a smooth, continuous operation, at a rate which will avoid damage to the forms and permit the concrete to flow along the length of the form to the adjacent hose, and in such manner as to avoid segregation of the sand and cement particles. The form shall be filled and the concrete allowed to settle for one hour, after which the form shall be refilled. The form shall be completely filled when the concreting operation is complete. All concrete in place shall be free of voids, and streaks, honey-combs, or other evidence of segregation or porous construction. A manifold shall be provided on the hose leading from the concrete pump and a sufficient number of hoses shall lead from the manifold into the form so that any one section of form between baffles will be uniformly filled from hoses spaced at intervals not to exceed 8 feet. Running the concrete along the form a distance in excess of 5 feet will not be permitted, nor shall the differential head of the concrete fill in any form section between baffles exceed 2 feet at any time during the concreting operation. Additional concrete pumps may be used in lieu of a manifold.

When appropriate during the placing of the concrete, the forms shall be vibrated lightly on the outside to assure the complete filling of the form. The vibrating shall be done using approved equipment.

Only approved mixing and pumping equipment shall be used in the preparation and handling of pumped concrete. All oil or other rust inhibitors shall be removed from the mixing drum stirring mechanisms and other portions of the equipment in contact with concrete before the mixers are used. All materials shall be accurately measured by volume or weight as they are fed into the mixer.

Top of concrete fill shall be finished with trowelled mortar, so pitched and to such level, that water will not pond atop the protection facing.

Quantity and Payment: The quantity of repairs to pier bases at EL. 'X' and below (– Concrete) for which payment will be made will be the volume of pumped concrete fill actually placed within the limits indicated on the plans or as revised by authority of the Engineer.

Payment for repairs to pier bases at EL. 'X' and below (– Concrete) will be made for the quantity as above determined, measured in cubic feet, at the price per cubic foot bid in the proposal for the item repairs to pier bases at EL. X and below – concrete, which price shall include the performance of all the work specified herein and indicated on the plans, and the cost of furnishing all labor, materials and equipment and all else necessary therefor and incidental thereto.

11. 'BEARING' REPLACEMENT

New Expansion Bearings

Description: New expansion bearings shall include the removal and disposal of portions of the existing expansion bearings, the cleaning and painting of all fixed bearings and expansion bearings that are to remain, the furnishing and installing of the new expansion bearings, including the jacking of existing girders, all as indicated on the plans and in accordance with the specifications.

Materials: Structural steel components for new expansion bearings shall meet the requirements specified under current ASTM designation A588 and shall be hot-dip galvanized after fabrication in accordance with ASTM designation A123.

Bronze washers shall meet the requirements of current ASTM designation bloc, alloy No. 510 or Federal Specification QQ-B-637, Alloy 464.

Bedding materials for masonry plates shall be furnished and placed according to the requirements of AASHTO specifications for highway bridges.

Jacks used for the jacking operations shall have the rated capacity shown clearly on the manufacturer's name plate attached to each jack.

Methods of Construction: Methods of construction for removal and disposal of portions of existing bearings shall meet the requirements specified under article... as amended elsewhere herein and as indicated on the plans.

Methods of construction including fabrication, erection and welding of new expansion bearings shall meet the applicable requirements specified under Article and as specified herein.

The plans indicate a suggested scheme for jacking operations and supporting the existing structure. The Contractor may use the method indicated or use alternate methods. The method of jacking operations and the method for temporarily supporting the existing structure shall be of the Contractor's choosing but will be subject to approval by the Engineer. The Contractor shall submit detailed drawings showing all elements of the proposed jacking operations and temporary supports he proposes to use, including design calculations therefor, for approval by the Engineer. No materials shall be ordered and no work shall be performed until written approval by the Engineer has been obtained. Approval by the Engineer will not in any way relieve the Contractor of his responsibility for the safety and adequacy of the jacking and supports systems and operations.

The Contractor will not be permitted to jack the ends of a girder after the deck joint at that location has been sealed.

Jacking of the girders will only be permitted between the hours of twelve midnight and 5 a.m.

The Contractor shall supply all lighting and equipment that is required to safely perform the jacking operations and all related work as a night operation.

No vehicular traffic or pedestrians will be permitted on the structure during the actual jacking operations or while the dead load of the span is still being supported by the jacks. As a means to accomplish this, the Contractor shall set-up and maintain the detour shown on the plans. The detour will only be permitted between the hours of twelve mid-night and 5 a.m.

However traffic on detour will be permitted from the end of the working day on Friday until twelve midnight of the following Monday.

The detour need only be in effect during the actual jacking operations and until the jacks have been relieved of all load by transferring the full dead load of that particular span to the temporary supports that make up part of the jacking frame.

The detour shall be completely removed by 5 a.m., or sooner if no longer required, and all signs covered or removed so that normal traffic flow may resume.

It shall be understood by the Contractor that the detour will only be permitted during the jacking operations of the Girders. The time length of each detour shall be a short as possible so as to minimize any inconvenience to the driving public.

The amount of jacking movement required to release and remove each bearing shall be the very minimum required. The Contractor shall submit to the engineer for approval, at least 30 days before the start of jacking operations, the method he intends to use and the procedure he intends to follow in removing the bearing.

The removal of the bearing shall not start until the load from the jacks has been transferred to the temporary supports and the temporary supports secured.

All fixed bearings, all expansion bearings that are to remain and the upper portion of expansion bearings that is to remain, shall be cleaned and painted two coats of zinc dust-zinc oxide paint primer. Surface preparation shall be in accordance with Steel Structures Painting Council SSPC-SP6 "Commercial blast Cleaning". The first coat of zinc dust-zinc oxide paint primer must be applied within four hours after the surface preparation has been completed.

The Contractor's attention is directed to the necessary of providing means to measure constantly the possible yielding or deflection of the temporary supports and shall continuously observe this at all times. This precaution is of importance since the yielding of the support will induce permanent residual stresses in the structure during construction. Any yielding or deflection of the temporary supports shall be immediately corrected by the contractor. The Contractor shall immediately notify the Engineer of such condition. The contractor's attention is directed to the fact that it is of extreme importance to safeguard the temporary supports. It will be the sole responsibility of the contractor to provide and maintain adequate protection for the temporary supports for the duration of the contract. In the event the temporary supports fail and any damage is done to the structure, the contractor shall correct this damage at his own expense. When no longer required, as determined by the engineer, all materials used for the temporary supports shall become the property of the Contractor and shall be disposed of by with clear of the site.

In the performance of the work under this item, the Contractor will be held responsible for damage to existing construction to remain in place. The Contractor shall use extreme care and exercise every precaution necessary to prevent damage or injury thereto. All damage as a result of the Contractor's operations shall be repaired or replaced by him at his own expense to the satisfaction of the engineer at no cost to the state.

Quantity and Payment: The quantity of new expansion bearings for which payment will be made will be a lump sum covering all the work indicated on the plans and in accordance with the specifications.

Payment for new expansion bearings will be made at the lump sum price bid in the proposal for the item new expansion bearings, which price shall include the performance of all the work specified herein and as shown on the plans and the cost of furnishing all labour, materials, equipment including structural steel, bolts, nuts and washers, jacks and jacking operations, temporary supports, removal and disposal of portions of existing expansion bearings, shims, welding, galvanizing, cleaning and paining and all else necessary and incidental thereto.

12. BEARING REPAIR

Description: Strapping of bearings shall include the furnishing and installing galvanized steel angles to strap bearings together, as indicated on the plans and in accordance with the specifications.

Materials: Structural steel for angles shall meet the requirements specified under article . . . for ASTM designation A36 Steel and shall be hot-dip galvanized in accordance with ASTM Designation A123.

Methods of Construction: Methods of construction including fabrication, erection and welding of steel angles to base plates shall meet the applicable requirements specified.

No painting of galvanized steel angles will be required. Surfaces of galvanized steel angles damaged after welding operations shall be regalvanized in the field in accordance with the requirements specified under article.

Quantity and Payment: The quantity of strapping of bearings for which payment will be a lump sum covering all the work indicated on the plans and in accordance with the specifications.

Payment for strapping of bearings will be made at the lump sum price bid in the proposal for the item strapping of bearings, which price shall include the performance of all the work specified herein and indicated on the plans, and the cost of furnishing all labour, materials and equipment including structural steel, galvanizing, welding, and all else necessary therefor and incidental thereto.

13. TEMPORARY SHIELD OVER RAILWAY TRACK

1. Description: Temporary shield shall include the furnishing, erecting and subsequent removal of a temporary protective shield over the railroad tracks constructed to provide a safe working platform during construction of the bridge and to preclude the dropping of any materials that may endanger the safety of the railroad and its operating train services.

2. Materials: Structural steel, unless otherwise indicated, of all types, sizes, shapes and configurations shown on the plans, shall conform to the requirements of ASTM A36.

Plywood shall conform to the requirements of Commercial Standard CS 35 type hardwood exterior technical Type 1.

Timber and lumber to include posts, bridging filler, nailers and all other indicated uses and terms shall be select structural grade 1800 psi, southern pine or equal.

Fasteners shall conform to the following: structural steel fastened with nut, bolts, washers or bent U-bolts shall conform to the requirements of ASTM A307 or as specified in article . . . except as otherwise indicated on the plans.

Fasteners for attachment of wood to wood members and plywood to wood construction shall be of required size commercially available standard common galvanized nails; where structural steel is attached to wood construction fasteners shall conform to the requirements of ASTM A307 as specified above for structural steel.

All fasteners shall be galvanized in conformance with the applicable requirements of ASTM A123 or ASTM A153.

Roofing paper shall be 15 pound felt conforming to the requirements of ASTM D3158; adhesive and sealant shall be of cold or hot applied and of the Type recommended by the product manufacturer designed for rapid and full adhesion between roofing paper layers and wood substrates.

3. Methods of Construction: The contractor shall submit detailed drawings showing all elements of the system he proposes to use and design calculations therefor, for approval by the engineer and shall construct the shield in accordance with the approved working drawings.

The Contractor shall construct the temporary shield during periods of track outages as specified in division I of these specifications. The contractor shall not commence further progressive schedules of work, until the protective shield has been completed and temporary grounding measures have been implemented.

All fasteners shall be drawn up tight or set flush and should any type fastener become loosened for any reason it or they shall be maintained at all times during construction.

Walkways of plywood construction shall be protected and sealed, without the use of nails, and all roofing shall be fastened in place using asphalt adhesives.

All work shall provide rigidly secured in place construction, fitted to the bridge skew and to provide safe working conditions without any endangerment to the railroad operations, property or other facilities.

Upon completion of the bridge construction and permanent barricade and when directed by the engineer, all temporary construction attached to or suspended or cantilevered from the bridge, shall be removed and be disposed of off site. Removal shall be carefully conducted in consideration of all factors of track outages and safety as specified herein for construction of the shield.

4. Quantity and Payment: The quantity of temporary shield for which payment will be made will not be measured but all costs for the work of this section shall be included in the lump sum price for the work shown on the plans.

Payment for furnishing, erecting and removing of the temporary shield will be made for the quantity as above determined, at the lump sum price bid for the item temporary shield in the proposal, which price shall include all costs for labour, materials, tools, equipment for the temporary protective shield, complete as shown on the plans and specified herein, and the subsequent removal and disposal when directed by the engineer.

14. TIEBACKS

1. Description: The work shall consist of designing, furnishing and installing a permanent, prestressed anchor tieback system at the locations indicated on the plans, or where ordered by the Engineer.

The work will require the in-place testing of the tieback system.

2. Materials: Tieback tendons shall be fabricated from single or multiple elements of the following:

(a) Steel bars conforming to ASTM Designation A-722, 'Uncoated High-Strength Steel Bars for Prestressed Concrete'.

(b) Seven-wire strand conforming to ASTM Designation A-416, 'Uncoated Seven-Wire Stress-Relieved Strand for Prestressed Concrete'.

(c) Wires conforming to ASTM Designation A-421, 'Uncoated Stress-Relieved Wire for Prestressed Concrete'.

(d) Compact seven-wire strands conforming to ASTM designation A- 779-80, 'Uncoated Seven-Wire Compacted, Stress-Relieved Steel Strand for Prestressed Concrete'.

The tension members shall be of such size that the design load does not excess 60 per cent of the guaranteed ultimate tensile strength of the tension member.

Couplers for tension members shall be capable of developing 100 per cent of the guaranteed ultimate tensile strength of the tension member.

Anchorages shall be capable of developing 95 per cent of the guaranteed ultimate tensile strength of the anchor material when tested in an unbonded state.

Bearing plates shall be fabricated from mild steel and be capable of developing 95 per cent of the guaranteed minimum ultimate tensile strength of the prestressing steel.

All end hardware shall meet the requirements of ACI 318.

Grease used to coat the stressing length shall be compounded to provide corrosion inhibiting and lubrication properties. Acceptable greases for the stressing length shall be:

1. Exxon Rust Ban 326
2. Chevron Polyurea EP Grease, #2 grade or equal

Grease other than 1 or 2 above shall be submitted to the Engineer for his approval and shall not be allowed to exceed the maximum allowable quantity of certain substances as shown in the following table.

Substance	Maximum Allowable Quantity—PPM	Test Method
Chlorides	10	ASTM D-512
Nitrates	10	ASTM D-992
Sulfides	10	APHA-"test methods; sulfides in water."

Centralizers shall consist of plastic steel or any material not detrimental to the prestressing steel. Wood centralizers shall not be used. Centralizers shall be capable of positioning the tendon in the middle of the drilled holes and of providing no less than 0.5 inches of grout cover along the bonded length.

Spacers shall be used to separate elements of multi-element tendons. They shall be fabricated from material which is nondetrimental to the prestressing steel.

A combination centralizer—spacer can be used. Sheathing shall be a smooth plastic tube having a minimum wall thickness of 0.02 inches and shall encapsulate the total stressing length of the anchor.

The grout to be used for anchorage shall consist of a pumpable mixture of types I, II, or III portland cement, sand and water meeting the requirements of ASTM C 150.

Chemical additives to control bleed or retard set may be used with the anchor grout. Expansive additives will not be allowed. Additives if used, shall be mixed in accordance with the manufacturer's recommendations. Epoxy resin will not be allowed as a substitute for cement grout.

Shop drawings shall be submitted to the engineer for written approval.

Each tieback shall have a minimum unbonded length of 15 feet (4.58 m). The contract drawings indicate the unbonded length required for each tier of tiebacks. The tieback shall be installed at an angle varying between 10 degree and 30 degree from the horizontal.

The minimum total tieback lengths are indicated on the contract drawings. In no case shall the anchor length be less than 15 feet.

3. Methods of Construction: The holes for the anchors may be either driven or drilled. Core drilling, rotary drilling, auger drilling or percussion drilling may be used. If the hole will not stand open, casing shall be installed as required to maintain a clean and open hole. The hole diameter shall not be less than three inches if pressure grouting is used in the bond length and four inches.

If pressure grouting is not used (pressure grouting is defined as grouting with a pressure greater than 60 p.s.i.). The diameter of the drill bid shall not be smaller than the specified hole diameter minus 1/8 inch. The hole shall extend a minimum of two feet beyond the specified bar length. The holes shall be drilled to the inclination specified on the contract plans within a plus or minus three-degree tolerance.

The tendon shall be installed in the casing or hole drilled for the anchor. Care shall be taken to ensure that the tendon's corrosion protection is not damaged during handling or installation. The tendon in the bond length shall be installed in such a way as to ensure that it has a minimum of one-half inch grout cover. The bond length of the tendon shall be degreased prior to installation by using acetone, MEK, or MIBK. No residue shall be left on the tendon. Other substances may be used subject to approval by the engineer.

If artesian or flowing water is encountered in the drilled hole, pressure shall be maintained on the consolidation grout until the grout has initially set.

After a hole is drilled to the final depth and watertightness, if required, is attained, the tendon shall be inserted. Anchor tendons shall not be subjected to sharp bends. Centralizers provided at a maximum of 10 feet center to center spacing throughout the bond length shall be used to ensure that the tendons do not contact the wall of the drill hole and no less than 0.5 inches of grout cover along the bond length is achieved.

The grouting operation shall be performed after the tendon is inserted. Grout shall always be injected at the lowest point of the anchor.

The grouting equipment shall be capable of continuous mixing and shall produce grout free of lumps. The grout pump shall be equipped with a grout pressure gage capable of measuring 150 p.s.i.

The annular space between the sheathing and the drilled hole shall be filled with grout for its entire length.

A pipe or trumpet integral with the bearing plate, shall extend from the anchor plate a sufficient distance to encapsulate the front portion of the sheath. The trumpet void shall be filled with grease or grout. On completion of the work, the anchorage shall be encased as shown on the contract drawings.

- *TESTS*

Anchor testing and stressing: each anchor shall be tested. The maximum test load shall not exceed 80 percent of the guaranteed ultimate tensile strength of the tendon. The first two anchors installed at each specified design load capacity and 15 per cent of the remaining anchors (locations to be chosen by the Engineer) shall be performance tested. All remaining anchors shall be proof tested.

(i) Performance Test

Performance tests shall be made by incrementally loading and unloading the anchor in accordance with the following schedule. At each load increment the movement of the tendon shall be recorded to the nearest 0.001 inch with respect to an independent fixed reference point. The jack load shall be monitored with a calibrated load cell. The load cell shall be calibrated by an independent testing laboratory within 14 days of start of testing the anchors. The Contractor shall provide the Engineer with the calibration curve before start of testing.

At the completion of the test the anchor load shall be reduced to 0.80 P and transferred to the permanent stressing anchorage.

Each load increment shall be held until movement ceases or for two minutes, whichever is longer. Loading and unloading rates (tons per minute) shall be submitted by the Contractor for approval.

Cycle	Load
1	0.00 P
	0.25 P
2	AL
	0.25 P
	0.50 P
	0.25 P
3	AL
	0.25 P
	0.50 P
	0.75 P
	0.50 P
	0.25 P
4	AL
	0.25 P
	0.50 P
	0.75 P
	1.00 P
	0.75 P
	0.50 P
	0.25 P
5	AL
	0.25 P
	0.50 P
	0.75 P
	1.00 P
	1.25 P
	1.33 P (Hold 50 minutes for creep test)
	Adjust to transfer load of 0.8 P

P = Design Load for the anchor
AL = Alignment Load

The creep test shall consist of holding the 1.33 P load for 50 minutes. While the load is maintained constant, anchor movement (total movement) referenced to a fixed point, shall be recorded at 0, 1/2, 1, 5, 10, 30 and 50 minutes.

Two copies of all test data shall be submitted to the engineer.

The Engineer will review all performance tests to determine if the anchor is acceptable. An anchor will be accepted if:

(a) The total elastic movement obtained at the design load exceeds 80 per cent of the theoretical elastic elongation of the stressing length and is less than the theoretical elastic elongation of the stressing length plus 50 per cent of the bond length.

(b) The creep movement does not exceed 0.080 inches during the time increment between 5 and 50 minutes regardless of tendon length and load.

For anchors that the Engineer finds unacceptable, the contractor shall submit a written proposal containing a suggested course of action. The action to be taken will be subject to written approval by the engineer.

(ii) Proof Tests

The proof tests shall be performed by incrementally loading and unloading the anchor in accordance with the following schedule. At each increment, the movement of the tendon shall be recorded to the nearest 0.001 inches with respect to an independent fixed reference point. The jack load shall be monitored with a load cell.

Load
0
0.25 P
0.50 P
0.75 P
1.00 P
1.25 P
1.33 P (Hold for creep test)
Adjust to transfer load of 0.8 P

Two copies of all test data shall be submitted to the engineer.

Acceptance criteria for an anchor which has been proof tested shall be the same as the performance test.

The creep test shall hold the 1.33 P load for 5 minutes. With the load held constant, anchor movement (total movement) shall be recorded at 0 seconds, 30 seconds, 2 minutes, and 5 minute intervals. If the movement between the 30 seconds and the 5 minutes reading is 0.080 inches or more, the load shall be maintained for an additional 45 minutes and the movement measured. If the additional movement exceeds 0.080 inches, the anchor shall be rejected. All movements shall be measured in relation to a fixed reference point.

4. Quantity and Payment: The unit price bid shall include the cost of furnishing all labour, equipment, and material required to complete the work.

The quantity to be paid for under this item shall be the number of anchors successfully installed, tested and accepted.

15. SOME OF THE AASHTO STANDARDS WHICH MAY BE UTILISED IN WRITING THE SPECIFICATIONS

AASHTO	NAME
T 23	Making and Curing Concrete Compressive Test Specimens in the Field
T 141	Sampling Fresh Concrete
T 2	Sampling Stone, Gravel, and Sand for Use as Highway Materials
T 127	Sampling Hydraulic Cement
T 26	Quality of Water to be Used in Concrete
T 22	Compressive Strength of Cylindrical Concrete Specimens
T 119	Slump of Portland Cement Concrete
M 85	Portland Cement
M 6	Fine Aggregate for Portland Cement Concrete
T 104	Soundness of Aggregate by Use of Sodium Sulphate or Magnesium Sulcate
T 112	Clay Lumps and Friable Particles in Aggregate
T 21	Organic Impurities in Sands for Concrete
T 176	Plastic Fines in Graded Aggregates and Soils by Use of the Sand Equivalent Test
T 27	Sieve Analysis of Fine and Coarse Aggregate
M 80	Coarse Aggregate for Portland Cement Concrete
T 96	Resistance to Abrasion of Small Size Coarse Aggregate by Use of the Los Angeles Machine
M 31	Deformed and Plain Billet-Steel Bars for Concrete Reinforcement
M 55	Steel Welded Wire, Fabric, Plain, for Concrete Reinforcement
M 194	Chemical Admixtures for Concrete
T 126	Making and Curing Concrete Test Specimens in the Laboratory
M 205	Molds for Forming Concrete Test Cylinders Vertically
M 121	Creosote Primer Used in Roofing, Damp-proofing, and Waterproofing
M 148	Liquid Membrane-Forming Compounds for Curing Concrete
M 235	Epoxy-Resin Adhesives

Appendix 3
METRIC UNITS AND CONVERSION FACTORS

Subject	Imperial Unit	S.I. Unit	Symbol	Conversion Factor
Length	mile	kilometre	km	1 mile = 1.609 km
	yard	metre	m	1 yd = 0.914 m
	foot	metre or millimetre	m or mm	1 ft = 0.305 m = 304.8 mm
	inch	millimetre	mm	1 in = 25.400 mm
Area	square mile	square kilometre	km^2	$1\ mile^2 = 2.590\ km^2$
	acre	square kilometre or hectare	km^2 or ha	$1\ acre = 0.004\ km^2$ = 0.405 ha
	square yard	square metre	m^2	$1\ yd^2 = 0.836\ m^2$
	square foot	square metre	m^2	$1\ ft^2 = 0.093\ m^2$
	square inch	square millimetre	mm^2	$1\ in^2 = 645.16\ mm^2$
Volume	cubic yard	cubic metre	m	$1\ yd^3 = 0.765\ m^3$
	cubic foot	cubic metre	m^3	$1\ ft^3 = 0.0283\ m^3$
	cubic inch	cubic millimetre	mm^3	$1\ in^3 = 16387.1\ mm^3$
Capacity	gallon	litre	litre	1 gal = 4.546 litre
Mass of materials	ton	tonne	tonne	1 ton = 1.016 tonne
	hundred weight	kilogramme	kg	1 cwt = 50.802 kg
	pound	kilogramme	kg	1 lb = 0.454 kg
	ounce	gramme	g	1 oz = 28.350 g
Density	pound per cubic foot	kilogramme per cubic metre	kg/m	$1\ lb/ft^3 = 16.019\ kg/m^3$
	pound per cubic yard			$1\ lb/yd^3 = 0.593\ kg/m^3$

Subject	Imperial Unit	S.I. Unit	Symbol	Conversion Factor
Force	pound force	newton	N	1 lbf = 4.448 N
	ton force	kilonewton	kN	1 tonf = 9.964 kN
	pound force per foot	newton per metre	N/m	1 lb/ft = 14.591 N/m
	ton force per foot	kilonewton per metre	kN/m	1 ton /ft = 32.690 kN/m
Pressure	pound force per square foot	newton per square metre	N/m^2	$1\ lbf/ft^2 = 47.880\ N/m^2$
	pound force per square inch	newton per square millimetre	N/mm^2	$1\ lbf/in^2 = 0.006\ 89\ N/mm^2$
	ton force per square foot	kilonewton per square metre	kN/m^2	$1\ ton/ft^2 = 107.250\ kN/m^2$
	ton force per square inch	kilonewton per square millimetre	kN/mm^2	$1\ tonf/in^2 = 0.0154\ kN/mm^2$
Stress	pound force per square inch	newton per square millimetre	N/mm^2	$1\ lbf/in^2 = 0.006\ 89\ N/mm^2$
	ton force per square inch	newton per square millimetre	N/mm^2	$1\ tonf/in^2 = 15.444\ N/mm^2$
	ton force per square foot	kilonewton per square metre	kN/m^2	$1\ tonf/ft^2 = 107.250\ kN/m^2$
Modulus of elasticity	pound force per square inch	newton per square millimetre	N/mm^2	$1\ lbf/in^2 = 0.006\ 89\ N/mm^2$
Bending	pound force inch	newton millimetre or newton metre	N mm or N m	1 lbf in = 112.985 N mm = 0.113 N m
	pound force foot	newton metre	N m	1 lbf ft = 1.356 N m
	ton force foot	kilonewton metre	kN m	1 tonf ft = 3 037 kN m
Section modulus	in^3	mm^3	mm^3	$1\ in^3 = 16\ 386\ mm^3$
Second moment of area	in^4	mm^4	mm^4	$1\ in^4 = 416\ 210\ mm^4$

Decimals of an Inch
(For each 64th of an inch With Millimetre Equivalents)

Fraction	1/64ths	Decimal	Millimetres (Approx.)	Fraction	1/64ths	Decimal	Millimetres (Approx.)
. . .	1	.015625	0.397	. . .	33	.515625	13.097
1/32	2	.03125	0.794	17/32	34	.53125	13.494
. . .	3	.046875	1.191	. . .	35	.546875	13.891
1/16	4	.0625	1.588	9/16	36	.5625	14.288
. . .	5	.78125	1.984	. . .	37	.578125	14.684
3/32	6	.9375	2.381	19/32	38	.59375	15.081
. . .	7	.109375	2.778	. . .	39	.609375	15.478
1/8	8	.125	3.175	5/8	40	.625	15.875
. . .	9	.140625	3.572	. . .	41	.640625	16.272
5/32	10	.15625	3.969	21/32	42	.65625	16.669
. . .	11	.171875	4.366	. . .	43	.671875	17.066
3/16	12	.1875	4.763	11/16	44	.6875	17.463
. . .	13	.203125	5.159	. . .	45	.703125	17.895
7/32	14	.21875	5.556	23/32	46	.71875	18.256
. . .	15	.234375	5.953	. . .	47	.734375	18.653
1/4	16	.250	6.350	3/4	48	.750	19.050
. . .	17	.265625	6.747	. . .	49	.765625	19.447
9/32	18	.28125	7.144	25/32	50	.78125	19.844
. . .	19	.296875	7.541	. . .	51	.796875	20.241
5/16	20	.3125	7.938	13/16	52	.8125	20.638
. . .	21	.328125	8.334	. . .	53	.828125	21.034
11/32	22	.34375	8.731	27/32	54	.84375	21.431
. . .	23	.359375	9.128	. . .	55	.859375	21.828
3/8	24	.375	9.525	7/8	56	.875	22.225
. . .	25	.390625	9.922	. . .	57	.890625	22.622
13/32	26	.40625	10.319	29/32	58	.90625	23.019
. . .	27	.421875	10.716	. . .	59	.921875	23.416
7/16	28	.4375	11.113	15/16	60	.9375	23.813
. . .	29	.453125	11.509	. . .	61	.953125	24.209
15/32	30	.46875	11.906	31/32	62	.96875	24.606
. . .	31	.484375	12.303	. . .	63	.984375	25.003
1/2	32	.500	12.700	1	64	1.000	25.400

Appendix 4
SOME GENERAL DATA

SI PREFIXES

Multiplication Factor								Prefix	Symbol
1	000	000	000	000	000	000	$= 10^{18}$	exa	E
	1	000	000	000	000	000	$= 10^{15}$	peta	P
		1	000	000	000	000	$= 10^{12}$	tera	T
			1	000	000	000	$= 10^{9}$	giga	G
				1	000	000	$= 10^{6}$	mega	M
					1	000	$= 10^{3}$	kilo	k
						100	$= 10^{2}$	hecto[b]	h
						10	$= 10^{1}$	deka[b]	da
						0.1	$= 10^{-1}$	deci[b]	d
						0.01	$= 10^{-2}$	centi[b]	c
						0.001	$= 10^{-3}$	milli	m
					0.000	001	$= 10^{-6}$	micro	μ
				0.000	000	001	$= 10^{-9}$	nano	n
			0.000	000	000	001	$= 10^{-12}$	pico	p
		0.000	000	000	000	001	$= 10^{-15}$	femto	f
	0.000	000	000	000	000	001	$= 10^{-18}$	atto	a

Temperature Conversion Tables

$$\deg C = \frac{\deg F - 32}{1.8}, \deg F = 1.8 \deg C + 32$$

deg C	deg F	deg F	deg C
− 40	− 40.0	− 40	− 40.0
− 35	− 31.0	− 30	− 34.4
− 30	− 22.0	− 20	− 28.8
− 25	− 13.0	− 10	− 23.3
− 20	− 0.0	0	− 17.7
− 15	+5.0	+10	− 12.2
− 10	14.0	20	− 6.6
− 5	23.0	30	− 1.1
0	32.0	40	+4.4
+5	41.0	50	10.0
10	50.0	60	15.5
15	59.0	70	21.1
20	68.0	80	26.6
25	77.0	90	32.2
30	86.0	100	37.7
35	95.0	110	43.2

deg C	deg F	deg F	deg C
40	104.0	120	48.8
45	113.0	130	54.4
50	122.0	140	60.0
55	131.0	150	65.5
60	140.0	160	71.1
65	149.0	170	76.6
70	158.0	180	82.2
75	167.0	190	87.7
80	176.0	200	93.3
		210	98.8
90	194.0	220	104.4
100	212.0		
120	248.0	240	115.5
140	284.0	260	126.6
160	320.0	280	137.7
180	356.0	300	148.9
200	392.0	400	204.4
300	572.0	500	260.0
400	752.0	600	315.6
500	932.0	800	426.7

An *approximate* method of converting temperatures which is correct to within 5 deg F, between 0 and 100 deg F, is—

$$\deg C = \frac{\deg F - 30}{20}$$

Shrinkage of Concrete
The shrinkage of concrete made with quartz aggregate due to ageing is approximately-

Time	Shrinkage (per cent)
After 28 days	0.025
After 3 months	0.035
After 12 months	0.050

Conversion Factors
Note that it is strict continental practice to use lower case letters always for units unless the unit derives from a name, e.g. Watt—W, Fahrenheit—F, and kilowatt—kW.

To Convert	Into	Multiply by	Reciprocal
Linear			
Inches	centimetres	2.54	0.3972
Inches	metres	0.0254	39.3701
Feet	centimetres	30.4799	0.0328
Yards	metres	0.9144	1.0936
Miles (5.280 ft)	kilometres	1.6093	0.62137
Chains	miles	0.0215	80.0
Furlongs	miles	0.125	8.0
Area			
Acres	square chains	10.0	0.1
Acres	hectares	0.4047	2.471
Acres	square miles	0.0016	640.0
Acres	square yares	4,840.0	0.0002
Square inches	square centimetres	6.4516	0.155
Square feet	square decimetres	9.2903	0.1076
Square yards	square metres	0.8361	1.196
Square chains	square metres	404.7	0.0025
Square chains	square yards	484.0	0.0021
Square feet	square metres	0.0929	10.764
Square yards	hectares	0.00008	11,960.0
Square inches	square metres	0.000645	1,550.0
Volume			
Cubic inches	cubic centimetres	16.387	0.061
Cubic feet	cubic metres	0.02832	35.32
Cubic feet	litres	28.317	0.0353
Cubic yards	cubic metres	0.7645	1.308
Cubic inches	cubic decimetres	0.01639	61.024
Miscellaneous			
British Thermal Units	calories	252.0	0.004
British Thermal Units	kilocalories	0.252	3.968
Btu per square foot	kilocalories per square metre	2.713	0.369
Atmospheres	pounds per square inch	14.7	0.0679
Feet of water	pounds per square inch	0.4335	2.3067
Metres of water	pounds per square inch	1.418	0.7031
Inches of mercury	pounds per square inch	0.4912	2.036
Millimetres of mercury	pounds per square inch	0.0193	51.71
Kilowatts	horsepower	1.34	0.746
Miles per hour	feet per second	1.4667	0.6818

To Convert	*Into*	*Multiply by*	*Reciprocal*
Weight			
Ounce (Avoirdupois)	drams	16	0.0625
Ounce (Av.)	grammes	28.35	0.0353
Ounce (Troy)	grammes	31.1035	0.03215
Pound (16 oz)	kilogrammes	0.45359	2.2046
Hundred weights (112 lb)	kilogrammes	50.8	0.0197
Tons (2,240 lb)	tonnes (= 1000 kg)	1.016	0.9842
Hundred weights	quintal (= 100 kg)	0.508	1.968
Pounds	tons	0.0004	2,240.0
Fluid			
Cubic feet	litres	28.317	0.0353
Callons (Imp.)	litres	4.546	0.22
Gallons	cubic feet	0.1605	6.232
Litres	cubic centimetres	1,000.0	0.001
Pints	litres	1.7598	0.568
Gallons (Imp.)	gallons (U.S.)	1.2	0.8327
Circular			
Degrees	radians	0.0175	57.29
Compound			
lb/lineal foot	kg/lineal metre	1.488	0.6711
ton/lineal foot	kg/lineal metre	3,333.0	0.0003
lb/in^2	kg/cm^2	0.0703	14.223
lb/in^2.	kg/mm^2	1.575	0.635
lb/ft^2	kg/m^2	4.883	0.2048
lb/in^2	ton/ft^2	0.0643	15.55
lb/ft^2	kg/m^3	16.019	0.9625
lb/yd^3	kg/m^3	0.5917	1.686
lb/gal	kg/l	0.0998	10.0166
ft-lb	kg-1	0.1382	7.233
ft-ton	tonne-m	0.3096	3.229
Horsepower	Force de Cheval	1.0139	0.9863
ft^3/sec	m^3/hr	101.94	0.098
ft^3/hr	l/sec	0.0787	127.2
lb^3/hp	kg/Force de Cheval	0.447	2.237
Grain/gal	g/l	0.0143	70.15

Figures Relating to Water

1 cubic foot of fresh water weighs 62.4 lb
1 cubic foot of sea water weighs 64.0 lb
1 British gallon of fresh water weighs 10.0 lb

1 United States gallon = 0.83 British gallon
1 litre of fresh water = 0.22 British gallon
1 cubic foot of fresh water = 6.24 British gallons
Head of fresh water in feet $\times 0.433$ = pressure (lb/in^2)
Head of sea water in feet $\times 0.444$ = pressure (lb/in^2)

Weights of Constructional Materials

Miscellaneous materials		kN/m³	lb/ft³
	Tarmacadam	22.6	144
	Macadam (waterbound)	25.1	160
	Snow : compact	2.4 to 8.0	15 to 50
	loose	0.8 to 1.9	5 to 12
	Vermiculite (aggregate)	0.8	5
	Terracotta	20.8	132
	Glass	26.7	170
	Cork: granular	1.2	7½
	compressed	3.8	24

	N/m²	lb/ft²
Clay floor tiles	575	12
Pavement lights	1.200	25
Damp-proof course	48	1
	N/m² per mm thickness	lb/ft² per in thickness
Felt (insulating)	1.9	1
Paving slabs (stone)	26.4	14
Granite sets	28.3	15
Asphalt	22.6	12
Rubber paving	15.1	8
Polyvinylchloride	19(av.)	10(av.)
Glass-fibre (forms)	1.9	1

Timber		kN/m³	lb/ft³
	General	7.9 (av.)	50 (av.)
	Douglas fir	4.7	30
	Yellow pine, spruce	4.7	30
	Pitch pine	6.6	42
	Larch, elm	5.5	35
	Oak (English)	7.1 to 9.4	45 to 60
	Teak	6.3 to 8.6	40 to 55
	Jarrah	9.4	60
	Greenheart	10.2 to 11.8	65 to 75
	Quebracho	12.6	80

	N/m² per mm	lb/ft² per in.
Wooden boarding and blocks		
Softwood	4.7	2½
Hardwood	7.5	4
Hardboard	10.4	5½
Chipboard	7.5	4
Plywood	6.1	3¼
Blockboard	4.7	2½
Fibreboard	2.8	1½
Wood-wool	5.7	3
Plasterboard	9.4	5
Weather boarding	3.8	2

Stone and other materials		kN/m³	lb/ft³
	Natural stone (solid)		
	Granite	25.1 to 28.7	163 to 183
	Limestone: Bath stone	20.4	130
	Marble	26.7	170
	Portland stone	22.0	140
	Sandstone	22.0 to 23.6	40 to 150
	Slate	28.3	180

	kN/m³	lb/ft³
Stone rubble (packed)	22.0	140
Quarry waste	14.1	90
Hardcore (consolidated)	18.9	120
All-in aggregate	19.6	125

	kN/m	lb/ft³
Iron: cast	70.7	450
wrought	75.4	480
ore: general	23.6	150
(crushed) Swedish	36.1	230
Steel (see also below)	77.0	490
Copper: cast	85.6	545
wrought	87.7	558
Brass	83.3	530
Bronze	87.7	558
Aluminium	27.2	173
Lead	111.0	707
Zinc (rolled)	70.0	446

Structural steelwork rivetted	Net weight of member + 10% for cleats, rivets, bolts, etc.
welded	+ 1¼% to 2½% for welds, etc.
Rolled sections: beams stanchions	+ 2½% +5% (extra for caps and bases)
Plate-webgirders	+ 10% for rivets or welds, stiffeners, etc.

		g/mm² per metre	lb/in² per foot		N/m	lb/ft
Metals, steel constructions, etc.	Steel bars	7.85 (7.85T/m³)	3.4	Steel stairs: industrial type 1m or 3ft wide	820	56
				Steel tubes: 50 mm or 2 in bore	45 to 60	3 to 4
				Gas piping: 20 mm or ¾ in	18	1¼
	Rail tracks standard gauge main line				kN/m of track	lb/ft of track
				Bull-head rails, chairs, transverse timber (softwood) sleepers, etc.	2.4	165
				Flat-bottom rails, transverse prestressed concrete sleepers. etc.	4.1	280
				Add for electric third rail	0.5	35
				Add for crushed stone ballast	25.5	1.750
				Overall average weight: rails, connections, sleepers, ballast, etc.	kN/m² 7.2	lb/ft² 150
				Bridge rails, longitudinal timber sleepers, etc.	kN/m of rail 1.1	lb/ft of rail 75

To convert values in kN to values in kg multiply by 102.

Normal Concrete	kN/m³	kg/m³
—plain	22	2250
—reinforced	24	2450
—prestressed reinforced	25	2550

COEFFICIENTS OF EXPANSION (α **per 100°**)

The coefficient of linear expansion (α) is the change in length, per unit of length, for a change of one degree of temperature. The coefficient of surface expansion is approximately two times the linear coefficient, and the coefficient of volume expansion, for solids, is approximately three times the linear coefficient.

A bar, free to move, will increase in length with an increase in temperature and will decrease in length with a decrease in temperature. The change in length will be $\alpha\,tl$ where α is the coefficient of linear expansion, t the change in temperature, and l the length. If the ends of a bar are fixed, a change in temperature (t) will cause a change in the unit stress of $E\,\alpha\,t$, and in the total stress of $AE\,\alpha\,t$, where A is the cross-sectional area of the bar and E the modulus of elasticity.

The following table gives the coefficient of linear expansion for *100°* or *100 times the value indicated above*.

EXAMPLE: A piece of medium steel is exactly 40 feet long at 60° F. Find the length at 90° F assuming the ends free to move.

Change of length $= \alpha tl = \dfrac{.00065 \times 30 \times 40}{100} = .0078$ ft

The length at 90° F is 40.0078 ft.

EXAMPLE: A piece of medium steel is exactly 40 feet long and the ends are fixed. If the temperature increases 30° F. what is the resulting change in the unit stress?

Change in unit stress $= E\,\alpha\,t =$

$\dfrac{29,000,000 \times .00065 \times 30}{100} = 5655$ lb. per sq. in.

(397.6 kg.cm²)

Coefficients of Expansion for 100 Degrees = 100 α

Materials	Linear Expansion		Materials	Linear Expansion	
	Centigrade	Fahrenheit		Centigrade	Fahrenheit
METALS AND ALLOYS			STONE AND MASONRY		
Aluminum, wrought	.00231	.00128	Ashlar masonry	.00063	.00035
Brass	.00188	.00104	Brick masonry	.00061	.00034
Bronze	.00181	.00101	Cement Portland	.00126	.00070
Copper	.00168	.00093	Concrete	.00099	.00055
Iron, cast, gray	.00106	.00059	Granite	.00080	.00044
Iron, wrought	.00120	.00067	Limestone	.00076	.00042
Iron, wire	.00124	.00069	Marble	.00081	.00045
Lead	.00286	.00159	Plaster	.00166	.00092
Magnesium, various alloys	.0029	.0016	Rubble masonry	.00063	.00035
Nickel	.00126	.00070	Sandstone	.00097	.00054
Steel, mild	.00117	.00065	Slate	.00080	.00044
Steel stainless, 18–8	.00178	.00099			
Zinc, rolled	.00311	.00173			
TIMBER			TIMBER		
Fir ⎫	.00037	.00021	Fir ⎫	.0058	.0032
Maple ⎬ parallel to fibre	.00064	.00036	Maple ⎬ perpendicular to	.0048	.0027
Oak ⎪	.00049	.00027	Oak ⎪ fibre	.0054	.0030
Pine ⎭	.00054	.00030	Pine ⎭	.0034	.0019

Expansion of Water
If Maximum Density = 1 unit at 4° C, then

C°	Volume	C°	Volume	C°	Volume	C°	Volume	C°	Volume	C°	Volume
0	1.000126	10	1.000257	30	1.004234	50	1.011877	70	1.022384	90	1.035829
4	1.000000	20	1.001732	40	1.007627	60	1.016954	80	1.029003	100	1.043116

Appendix 5
PROPERTIES OF GEOMETRIC SECTIONS AND SHAPES

PROPERTIES OF GEOMETRIC SECTIONS

RECTANGLE
Axis of moments on base

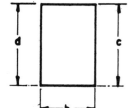

$$A = bd$$
$$c = d$$
$$I = \frac{bd^2}{3}$$
$$S = \frac{bd^2}{3}$$
$$r = \frac{d}{\sqrt{3}} = .577350\, d$$

RECTANGLE
Axis of moments on diagonal

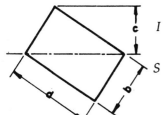

$$A = bd$$
$$c = \frac{bd}{\sqrt{b^2 + d^2}}$$
$$I = \frac{b^3 d^3}{6(b^2 + d^2)}$$
$$S = \frac{b^2 d^2}{6\sqrt{b^2 + d^2}}$$
$$= \frac{bd}{\sqrt{6(b^2 + d^2)}}$$

RECTANGLE
Axis of moments any line through centre of gravity

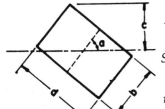

$$A = bd$$
$$c = \frac{b \sin a + d \cos a}{2}$$
$$I = \frac{bd(b^2 \sin^2 a + d^2 \cos^2 a)}{12}$$
$$S = \frac{bd(b^2 \sin^2 a + d^2 \cos^2 a)}{6(b \sin a + d \cos a)}$$
$$r = \sqrt{\frac{b^2 \sin^2 a + d^2 \cos^2 a}{12}}$$

HOLLOW RECTANGLE
Axis of moments through centre

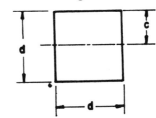

$$A = bd - b_1 d_1$$
$$c = \frac{d}{2}$$
$$I = \frac{bd^3 - b_1 d_1^3}{12}$$
$$S = \frac{bd^3 - b_1 d_1^3}{6d}$$
$$r = \sqrt{\frac{bd^3 - b_1 d_1^3}{12A}}$$
$$Z = \frac{bd^2}{4} - \frac{b_1 d_1^2}{4}$$

SQUARE
Axis of moments through centre

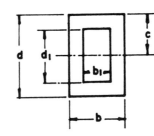

$$A = d^2$$
$$c = \frac{d}{2}$$
$$I = \frac{d^4}{12}$$
$$S = \frac{d^2}{6}$$
$$r = \frac{d}{\sqrt{12}} = .288675\, d$$
$$Z = \frac{d^3}{4}$$

SQUARE
Axis of moments on base

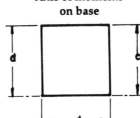

$$A = d^2$$
$$c = d$$
$$I = \frac{d^4}{3}$$
$$S = \frac{d^3}{3}$$
$$r = \frac{d}{\sqrt{3}} = .577350\, d$$

PROPERTIES OF GEOMETRIC SECTIONS (*Contd.*)

SQUARE
Axis of moments on diagonal

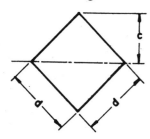

$A = d^2$

$c = \dfrac{d}{\sqrt{2}} = .707107\,d$

$I = \dfrac{d^4}{12}$

$S = \dfrac{d^2}{6\sqrt{2}} = .117851\,d^3$

$r = \dfrac{d}{\sqrt{12}} = .288675\,d$

$Z = \dfrac{2\,c^3}{3} = \dfrac{d^3}{3\sqrt{2}}$

$\quad = .235702\,d^3$

$A = b\,t + b_1\,t_1$

$c = \dfrac{\frac{1}{2}\,b\,t^2 + b_1\,t_1\,(d - \frac{1}{2}\,t_1)}{A}$

$I = \dfrac{b\,t^3}{12} + b\,t\,y^2 + \dfrac{b_1\,t_1^{\,3}}{12} + b_1\,t_1\,y_1^{\,2}$

$S = \dfrac{1}{c} \qquad S_1 = \dfrac{1}{c_1}$

$r = \sqrt{\dfrac{1}{A}}$

$Z = \dfrac{A}{2}\left[d - \left(\dfrac{1 + t_1}{2}\right)\right]$

RECTANGLE
Axis of moments through centre

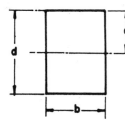

$A = b\,d$

$c = \dfrac{d}{2}$

$I = \dfrac{b\,d^3}{12}$

$S = \dfrac{b\,d^2}{6}$

$r = \dfrac{d}{\sqrt{12}} = .288675\,d$

$Z = \dfrac{b\,d^2}{4}$

TRIANGLE
Axis of moments through centre of gravity

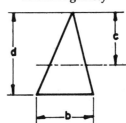

$A = \dfrac{b\,d}{2}$

$c = \dfrac{2\,d}{3}$

$I = \dfrac{b\,d^3}{36}$

$S = \dfrac{b\,d^2}{24}$

$r = \dfrac{d}{\sqrt{18}} = .235702\,d$

EQUAL RECTANGLES
Axis of moments through centre of gravity

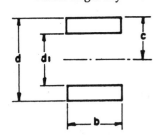

$A = b\,(d - d_1)$

$c = \dfrac{d}{2}$

$I = \dfrac{b\,(d^3 - d_1^{\,3})}{12}$

$S = \dfrac{b\,(d^3 - d_1^{\,3})}{6\,d}$

$r = \sqrt{\dfrac{d^3 - d_1^{\,3}}{12\,(d - d_1)}}$

$Z = \dfrac{b}{4}\,(d^2 - d_1^{\,2})$

TRIANGLE
Axis of moments on base

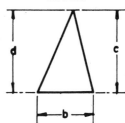

$A = \dfrac{b\,d}{2}$

$c = d$

$I = \dfrac{b\,d^3}{12}$

$S = \dfrac{b\,d^2}{12}$

$r = \dfrac{d}{\sqrt{6}} = .408248\,d$

UNEQUAL RECTANGLES
Axis of moments through centre of gravity

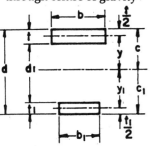

TRAPEZOID
Axis of moments through centre of gravity

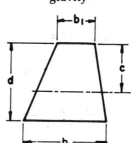

$A = \dfrac{d\,(b + b_1)}{2}$

$c = \dfrac{d\,(2\,b + b_1)}{3\,(b + b_1)}$

$I = \dfrac{d^3\,(b^2 + 4bb_1 + b_1^{\,2})}{36\,(b + b_1)}$

$S = \dfrac{d^2\,(b^2 + 4bb_1 + b_1^{\,2})}{12\,(2\,b + b_1)}$

$r = \dfrac{d}{6\,(b + b_1)}\,\sqrt{2\,(b^2 + 4bb_1 + b_1^{\,2})}$

PROPERTIES OF GEOMETRIC SECTIONS (*Contd.*)

CIRCLE
Axis of moments through centre

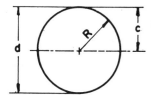

$$A = \frac{\pi d^2}{4} = \pi R^2 = .785398\, d^2$$
$$= 3.141593\, R^2$$

$$c = \frac{d}{2} = R$$

$$I = \frac{\pi d^4}{64} = \frac{\pi R^4}{4} = .049087\, d^4$$
$$= .785398\, R^4$$

$$S = \frac{\pi d^3}{32} = \frac{\pi R^3}{4} = .098175\, d^3$$
$$= .785398\, R^3$$

$$r = \frac{d}{4} = \frac{R}{2}$$

$$Z = \frac{d^3}{6}$$

HOLLOW CIRCLE
Axis of moments through centre

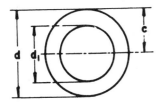

$$A = \frac{\pi (d^2 - d_1^2)}{4} = .785398\,(d^2 - d_1^2)$$

$$c = \frac{d}{2}$$

$$I = \frac{\pi (d^4 - d_1^4)}{64} = .049087\,(d^4 - d_1^4)$$

$$S = \frac{\pi (d^4 - d_1^4)}{32\, d} = .098175\,\frac{d^4 - d_1^4}{d}$$

$$r = \frac{\sqrt{d^2 - d_1^2}}{4}$$

$$Z = \frac{d^3}{6} - \frac{d_1^3}{6}$$

HALF CIRCLE
Axis of moments through centre of gravity

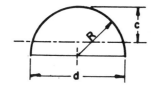

$$A = \frac{\pi R^2}{2} = 1.570796\, R^2$$

$$c = R\left(1 - \frac{4}{3\,\pi}\right) = .575587\, R$$

$$I = R^4\left(\frac{\pi}{8} - \frac{8}{9\,\pi}\right) = 1.09757\, R^4$$

$$S = R^3/24\,\frac{(9\,\pi^2 - 64)}{(3\,\pi - 4)} = .190687\, R^3$$

$$r = R\,\frac{\sqrt{9\,\pi^2 - 64}}{6\,\pi} = .264336\, R$$

PROPERTIES OF THE CIRCLE

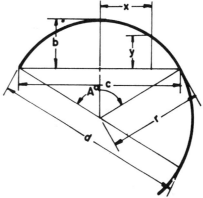

Circumference $= 6.28318\, r = 3.14159\, d$

Diameter $= 0.31831$ circumference

Area $= 3.14159\, r^2$

Arc $\quad a = \dfrac{\pi r A°}{180°} = 0.017453\, r\, A°$

Angle $\quad A° = \dfrac{180°\, a}{\pi r} = 57.29578\,\dfrac{a}{r}$

Radius $\quad r = \dfrac{4\, b^2 + c^2}{8\, b}$

Chord $\quad c = 2\sqrt{2\, b\, r - b^2} = 2\, r \sin\dfrac{A}{2}$

Rise $\quad b = r - ½\sqrt{4\, r^2 - c^2} = \dfrac{c}{2}\tan\dfrac{A}{4}$

$$= 2\, r \sin^2\frac{A}{4} = r + y - \sqrt{r^2 - x^2}$$

$$y = b - r + \sqrt{r^2 - x^2}$$

$$x = \sqrt{r^2 - (r + y - b)^2}$$

Diameter of circle of equal periphery as square
$= 1.27324$ side of square

Side of square of equal periphery as circle
$= 0.78540$ diameter of circle

Diameter of circle circumscribed about square
$= 1.41421$ side of square

Side of square inscribed in circle
$= 0.70711$ diameter of circle

PROPERTIES OF THE CIRCLE

CIRCULAR SECTOR

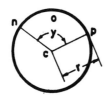

r = radius of circle y = angle ncp in degrees

Area of sector $ncpo$ = ½ (length of arc $nop \times r$)

= Area of circle $\times \dfrac{y}{360}$

= $0.0087266 \times r^2 \times y$

CIRCULAR SEGMENT

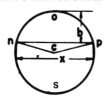

r = radius of circle x = chord b = rise

Area of segment nop

= Area of sector $ncpo$ − Area of triangle ncp

= $\dfrac{(\text{length of arc } nop \times r) - x\,(r-b)}{2}$

Area of segment nsp

= Area of circle − Area of segment nop

VALUES FOR FUNCTIONS OF π

π = 3.14159265359, log = 0.4971499

π^2 = 9.8696044, log = 0.9942997 $\dfrac{1}{\pi}$ = 0.3183099,

log = $\overline{1}.5028501$

$\sqrt{\dfrac{1}{\pi}}$ = 0.5641896, log = $\overline{1}.7514251$

π^3 = 31.0062767, log = 1.4914496 $\dfrac{1}{\pi^2}$ = 0.1013212,

log = $\overline{1}.0057003 \dfrac{\pi}{180}$ = 0.0174533, log = $\overline{2}.2418774$

$\sqrt{\pi}$ = 1.7724539, log = 0.2485749 $\dfrac{1}{\pi^2}$ = 0.0322515,

log = $\overline{2}.5085504$

$\dfrac{180}{\pi}$ = 57.2957795, log = 1.7581226

NOTE: Logs of fractions such as $\overline{1}.5028501$ and $\overline{2}.5085500$ may also be written 9.5028501—10 and 8.5085500—10 respectively.

PROPERTIES OF GEOMETRIC SECTIONS (*Contd.*)

PARABOLA

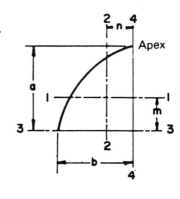

$A = \dfrac{4}{3} a b$

$m = \dfrac{2}{5} a$

$I_1 = \dfrac{16}{175} = a^3 b$

$I_2 = \dfrac{4}{15} = a b^3$

$I_3 = \dfrac{32}{105} = a^3 b$

HALF PARABOLA

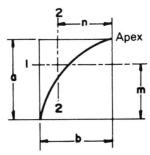

$A = \dfrac{2}{3} a b$

$m = \dfrac{2}{5} a$

$n = \dfrac{3}{8} b$

$I_1 = \dfrac{8}{175} = a^3 b$

$I_2 = \dfrac{19}{480} = a b^3$

$I_3 = \dfrac{16}{105} = a^3 b$

$I_4 = \dfrac{2}{15} = a b^3$

COMPLEMENT OF HALF PARABOLA

$A = \dfrac{1}{3} a b$

$m = \dfrac{7}{10} a$

$n = \dfrac{3}{4} b$

$I_1 = \dfrac{37}{2100} a^3 b$

$I_2 = \dfrac{1}{80} a b^3$

PARABOLIC FILLET IN RIGHT ANGLE

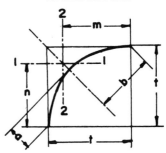

$a = \dfrac{t}{2\sqrt{2}}$

$b = \dfrac{t}{\sqrt{2}}$

$A = \dfrac{1}{6} t^2$

$m = n = \dfrac{4}{5} t$

$I_1 = I_2 = \dfrac{11}{2100} t^4$

PROPERTIES OF GEOMETRIC SECTIONS (*Contd.*)

* HALF ELLIPSE

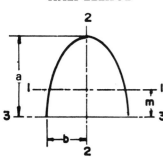

$$A = \frac{1}{2}\pi a b$$

$$m = \frac{4a}{3\pi}$$

$$I_1 = a^3 b \left(\frac{\pi}{8} - \frac{8}{9\pi} \right)$$

$$I_2 = \frac{1}{8}\pi a b^3$$

$$I_3 = \frac{1}{8}\pi a^3 b$$

* QUARTER ELLIPSE

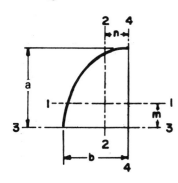

$$A = \frac{1}{4}\pi a b$$

$$m = \frac{4a}{3\pi}$$

$$n = \frac{4b}{3\pi}$$

$$I_1 = a^3 b \left(\frac{\pi}{16} - \frac{4}{9\pi} \right)$$

$$I_2 = a b^3 \left(\frac{\pi}{16} - \frac{4}{9\pi} \right)$$

$$I_3 = \frac{1}{16}\pi a^3 b$$

$$I_4 = \frac{1}{16}\pi a b^3$$

* ELLIPTIC COMPLEMENT

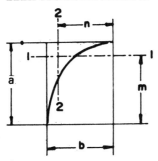

$$A = a b \left(1 - \frac{\pi}{4} \right)$$

$$m = \frac{a}{6 \left(1 - \frac{\pi}{4} \right)}$$

$$n = \frac{b}{6 \left(1 - \frac{\pi}{4} \right)}$$

$$I_1 = a^3 b \left(\frac{1}{3} - \frac{\pi}{16} - \frac{1}{36 \left(1 - \frac{\pi}{4} \right)} \right)$$

$$I_2 = a b^3 \left(\frac{1}{3} - \frac{\pi}{16} - \frac{1}{36 \left(1 - \frac{\pi}{4} \right)} \right)$$

PROPERTIES OF PARABOLA AND ELLIPSE

PARABOLA

When H ÷ B = 0.1 or less, approximate ½ perimeter = $\sqrt{B^2 + 4/3\, H^2}$ or use formulae for circular arcs.

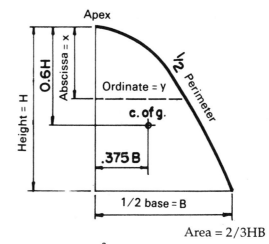

Area = 2/3HB

Perimeter $P = B^2 \div H$

$x = y^2 \div P$

$x = \sqrt{x\, P}$

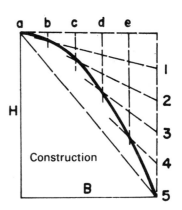

Construction

---

* To obtain properties of half circle, quarter circle and circular complement substitute $a = b = R$.

PROPERTIES OF PARABOLA AND ELLIPSE
(*Contd.*)

ELLIPSE

$(x^2 \div H^2) + (y^2 \times B^2) = 1$

$x = (H \div B) \sqrt{B^2 - y^2}$

$y = (B \div H) \sqrt{H^2 - x^2}$

Approximate 1/4 perimter =

$\dfrac{\pi}{4} \sqrt{2(H^2 + B^2)}$

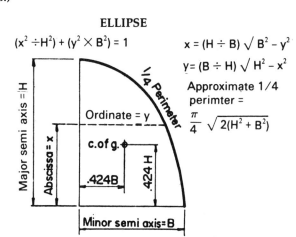

Area=.7854 Dd

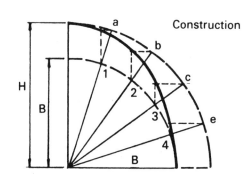

Construction

AREA BETWEEN PARABOLIC CURVE AND SECANT

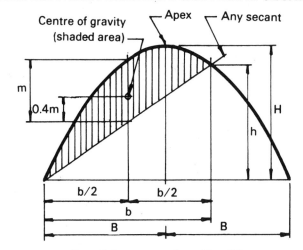

$h = Hb\left(\dfrac{2B - b}{B^2}\right)$

$m = \dfrac{Hb^2}{4B^2}$

Shaded area = 2/3 bm

$= \dfrac{Hb^3}{6B^2}$

Length *b* may vary from O to 2B.

BRACING FORMULAS

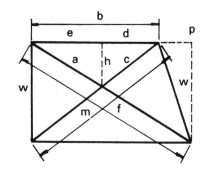

Given	To Find	Formula
bpw	f	$\sqrt{(b + p)^2 + w^2}$
bw	m	$\sqrt{b^2 + w^2}$
bp	d	$b^2 \div (2b + p)$
bp	e	$b(b + p) \div (2b + p)$
bfp	a	$bf \div (2b + p)$
bmp	c	$bm \div (2b + p)$
bpw	h	$bw \div (2b + p)$
afw	h	$aw + f$
cmw	g	$cw + m$

BRACING FORMULAS (*Contd.*)

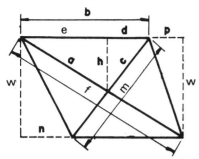

Given	To Find	Formula
bpw	f	$\sqrt{(b+p)^2 + w^2}$
bnw	m	$\sqrt{(b-n)^2 + w^2}$
bnp	d	$b(b-n) \div (2b+p-n)$
bnp	e	$b(b+p) \div (2b+p-n)$
bfnp	a	$bf \div (2b+p-n)$
bmnp	c	$bm \div (2b+p-n)$
bnpw	h	$bw \div (2b+p-n)$
afw	h	$aw \div f$
cmw	h	$cw \div m$

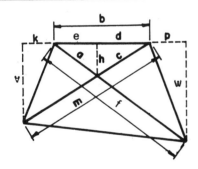

Given	To Find	Formula
bpw	f	$\sqrt{(b+p)^2 + w^2}$
bkv	m	$\sqrt{(b+k)^2 + v^2}$
bkpvw	d	$bw\,(b+k) \div [\,v(b+p) + w(b+k)\,]$
bkpvw	e	$bv(b+p) \div [v(b+p) + w(b+k)]$
bfkpvw	a	$fbv \div [v(b+p) + w(b+k)]$
bkmpvw	c	$bmw \div [v(b+p) + w(b+k)]$
bkpvw	h	$bvw \div [v(b+p) + w(b+k)]$
afw	h	$aw \div f$
cmv	h	$cv \div m$

PARALLEL BRACING

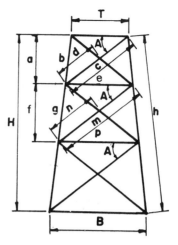

$k = (\log B - \log T) \div$ no. of panels. Constant k plus the logarithm of any line equals the log of the corresponding line in the next panel below.

$a = TH \div (T + e + p)$

$b = Th \div (T + e + p)$

$c = \sqrt{(\tfrac{1}{2}T + \tfrac{1}{2}e)^2 + a^2}$

$d = ce \div (T + e)$

$$\log e = k + \log T$$
$$\log f = k + \log a$$
$$\log g = k + \log b$$
$$\log m = k + \log c$$
$$\log n = k + \log d$$
$$\log p = k + \log e$$

The above method can be used for any number of panels.

In the formulas for "*a*" and "*b*" the sum in parentheses, which in the case shown is $(T + e + p)$, is always composed of all the horizontal distances except the base.

PROPERTIES OF GEOMETRIC SECTIONS AND STRUCTURAL SHAPES

REGULAR POLYGON

Axis of moments through centre

$n = $ Number of sides

$$\phi = \frac{180°}{n}$$

$$a = 2\sqrt{R^2 - R_1^2}$$

$$R = \frac{a}{2 \sin \phi}$$

$$R_1 = \frac{a}{2 \tan \phi}$$

$$A = \frac{1}{4} na^2 \cot \phi = \frac{1}{2} nR^2 \sin 2\phi = nR_1^2 \tan \phi$$

PROPERTIES OF GEOMETRIC SECTIONS AND STRUCTURAL SHAPES (*Contd.*)

$$I_1 = I_2 = \frac{A(6R^2 - a^2)}{24} = \frac{A(12R_1^2 - a^2)}{48}$$

$$r_1 = r_2 = \sqrt{\frac{6R^2 - a^2}{24}} = \sqrt{\frac{12R_1^2 - a^2}{48}}$$

ANGLE

Axis of moments through centre of gravity

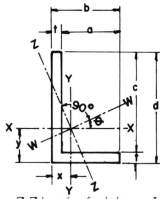

Z-Z is axis of minimum I

$$\tan 2\theta = \frac{2K}{I_y - I_x}$$

$$A = t(b + c) \quad x = \frac{b^2 + ct}{2(b + c)} \quad y = \frac{d^2 + at}{2(b + c)}$$

K = Product of Inertia about X–X and Y–Y

$$= \mp \frac{abcdt}{4(b + c)}$$

$$I_x = \frac{1}{3}[t(d - y)^3 + by^3 - a(y - t)^3]$$

$$I_y = \frac{1}{3}[t(b - x)^3 + dx^3 - c(x - t)^3]$$

$$I_z = I_x \sin^2\theta + I_y \cos^2\theta + K \sin^2\theta$$

$$I_w = I_x \cos^2\theta + I_y \sin^2\theta - K \sin^2\theta$$

K is negative when heel of angle, with respect to c. g., is in 1st or 3rd quadrant, positive when in 2nd or 4th quadrant.

BEAMS AND CHANNELS

Transverse force oblique through centre of gravity

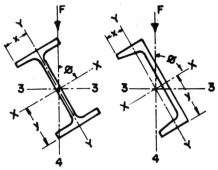

$$I_3 = I_x \sin^2\phi + I_y \cos^2\phi$$

$$I_4 = I_x \cos^2\phi + I_y \sin^2\phi$$

$$f_b = M\left(\frac{y}{I_x}\sin\phi + \frac{x}{I_y}\cos\phi\right)$$

where M is bending moment due to force F.

Appendix 6
MATHEMATICAL DATA

$\pi = \dfrac{355}{113}$ (approx.) $= \dfrac{22}{7}$ (approx.)

$= 3.141592654$ (approx.)

One radian $= \dfrac{180°}{\pi} = 57.3°$ (approx.)

$= 57.2957795°$ (approx.)

Length of arc subtended by an angle of one radian
= radius of arc.

One degree Fahrenheit = 5/9 degree Centigrade or
Celsius.

Temperature of $t°F = 5/9 (t - 32)°C$.
Temperature of $t°C = (1.8\,t + 32)°F$.

Base of Naperian logarithms, $e = \dfrac{193}{71}$ (approx.)

$= \dfrac{2721}{1001}$ (approx.) $= 2.718281828$ (approx.)

To convert common into Naperian logarithms, multiply
by $\dfrac{76}{33}$ (approx.) $= \dfrac{3919}{1702}$ (approx.) $= 2.302585093$

(approx.)

Nominal value of $g = 9.80665$ kg/sec^2 = 32.174 ft/sec^2.
Diameter of inscribed circle of a triangle:

$$D = \dfrac{2b\sqrt{a^2 - \left(\dfrac{a^2 + b^2 - c^2}{2b}\right)^2}}{a + b + c}$$

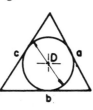

For isosceles triangle, $a = c$:

$$D = \dfrac{b\sqrt{4a^2 - b^2}}{2a + b}$$

Solution of Triangles

Applicable to any triangle ABC in which $AB = c$; $BC = a$;

$AC = b$; $\dfrac{\sin A}{a} = \dfrac{\sin B}{b} = \dfrac{\sin C}{c}$

Area $= \dfrac{bc \sin A}{2} = \dfrac{ac \sin B}{2} = \dfrac{ab \sin C}{2}$

$= \sqrt{s(s-a)(s-b)(s-c)}$, where $s = \frac{1}{2}(a + b + c)$

$$\sin \dfrac{A}{2} = \sqrt{\dfrac{(s-b)(s-c)}{bc}} \qquad \cos A = \dfrac{b^2 + c^2 - a^2}{2bc}$$

THE GREEK ALPHABETS

Letters from the Greek alphabet are frequently used as
symbols in engineering problems.

Alpha	A	α	Iota	I	ι	Rho	P	ρ
Beta	B	β	Kappa	K	κ	Sigma	Σ	σ
Gamma	Γ	γ	Lambda	Λ	λ	Tau	T	τ
Delta	Δ	δ	Mu	M	μ	Upsilon	Υ	υ
Epsilon	E	ε	Nu	N	ν	Phi	Φ	φ
Zeta	Z	ζ	Xi	Ξ	ξ	Chi	X	χ
Eta	H	η	Omicron	O	o	Psi	Ψ	ψ
Theta	Θ	θ	Pi	Π	π	Omega	Ω	ω

ALGEBRAICAL FORMULAE

Products and Factors

$$(x + a)(x + b) = x^2 + (a + b)x + ab$$
$$(x + a)(x - a) = x^2 - a^2$$
$$(x \pm a)(x^2 \pm ax + a^2) = x^3 \pm a^3$$
$$(x \pm a)^2 = x^2 \pm 2ax + a^2$$
$$(x \pm a)^3 = x^3 \pm 3x^2 a + 3xa^2 \pm a^3$$

General Solutions to a Quadratic Equation

If $\qquad ax^2 + bx + c = 0$

$$x = \dfrac{-b \pm \sqrt{b^2 - 4ac}}{2a}$$

Indices

$$a^m \times a^n = a^{m+n}$$
$$a^m \div a^n = a^{m-n}$$
$$(a^m)^n = a^{mn}$$
$$\log(ab) = \log a + \log b$$
$$\log(a/b) = \log a - \log b$$
$$\log(a^n) = n \log a$$
$$\log \sqrt[n]{a} = \log a \times 1/n$$

STANDARD FORMS OF DIFFERENTIALS AND INTEGRALS

Differential $\dfrac{dy}{dx}$	Function $\longleftarrow y \longrightarrow$	Integral $\int y\,dx$
1	x	$\frac{1}{2}x^2 + C$
0	a	$ax + C$
1	$x \pm a$	$\frac{1}{2}x^2 \pm ax + C$
$2x$	x^2	$\frac{1}{2}x^3 + C$
nx^{n-1}	x^n	$\dfrac{1}{n+1}x^{n+1} + C$
$-x^{-2}$	x^{-1}	(except when $n = -1$) $\log_e x + C$
$\dfrac{du}{dx} \pm \dfrac{dv}{dx} \pm \dfrac{dw}{dx}$	$u \pm v \pm w$	$\int u\,dx \pm \int v\,dx \pm \int w\,dx$
$u\dfrac{dv}{dx} + v\dfrac{du}{dx}$	uv	No general form
$\dfrac{v\dfrac{du}{dx} - u\dfrac{dv}{dx}}{v^2}$	u/v	No general form
$\dfrac{du}{dx}$	u	$ux - \int x\,du + C$
e^x	e^x	$e^x + C$
x^{-1}	$\log_e x$	$x(\log_e x - 1) + C$
$0.4343\,x^{-1}$	$\log_{10} x$	$0.4343\,x(\log_e x - 1) + C$
$a^x \log_e a$	a^x	$\dfrac{a^x}{\log_e a} + C$
$\cos x$	$\sin x$	$-\cos x + C$
$-\sin x$	$\cos x$	$\sin x + C$
$\sec^2 x$	$\tan x$	$-\log_e \cos x + C$
$\cosh x$	$\sinh x$	$\cosh x + C$
$\sinh x$	$\cosh x$	$\sinh x + C$
$\operatorname{sech}^2 x$	$\tanh x$	$\log_e \cosh x + C$
$-\dfrac{1}{(x+a)^2}$	$\dfrac{1}{x+a}$	$\log_e (x+a) + C$

TRIGONOMETRIC FUNCTIONS

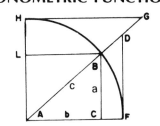

Radius $AF = 1$
$$= \sin^2 A + \cos^2 A = \sin A \operatorname{cosec} A$$
$$= \cos A \sec A = \tan A \cot A$$

Sine $A = \dfrac{\cos A}{\cot A} = \dfrac{1}{\operatorname{cosec} A} = \cos A \tan A$
$$= \sqrt{1 - \cos^2 A} = BC$$

Cosine $A = \dfrac{\sin A}{\tan A} = \dfrac{1}{\sec A} = \sin A \cot A$
$$= \sqrt{1 - \sin^2 A} = AC$$

Tangent $A = \dfrac{\sin A}{\cos A} = \dfrac{1}{\cot A} = \sin A \sec A = FD$

Cotangent $A = \dfrac{\cos A}{\sin A} = \dfrac{1}{\tan A} = \cos A \operatorname{cosec} A = HG$

Secant $A = \dfrac{\tan A}{\sin A} = \dfrac{1}{\cos A} = AD$

Cosecant $A = \dfrac{\cot A}{\cos A} = \dfrac{1}{\sin A} = AG$

$$\sin^2 \theta + \cos^2 \theta = 1 \quad \sec^2 \theta - \tan^2 \theta = 1 \quad \operatorname{cosec}^2 \theta - \cot^2 \theta = 1$$
$$\sin(\theta + \phi) = \sin\theta \cos\phi + \cos\theta \sin\phi$$
$$\cos(\theta + \phi) = \cos\theta \cos\phi - \sin\theta \cos\phi$$
$$\sin(\theta - \phi) = \sin\theta \cos\phi - \cos\theta \sin\phi$$
$$\cos(\theta - \phi) = \cos\theta \cos\phi + \sin\theta \sin\phi$$
$$\tan(\theta + \phi) = \dfrac{\tan\theta + \tan\phi}{1 - \tan\theta \tan\phi} \quad \tan(\theta - \phi) = \dfrac{\tan\theta - \tan\phi}{1 + \tan\theta \tan\phi}$$
$$\sin\theta + \sin\phi = 2\sin\tfrac{1}{2}(\theta + \phi)\cos\tfrac{1}{2}(\theta - \phi)$$
$$\sin\theta - \sin\phi = 2\cos\tfrac{1}{2}(\theta + \phi)\sin\tfrac{1}{2}(\theta - \phi)$$
$$\cos\theta + \cos\phi = 2\cos\tfrac{1}{2}(\theta + \phi)\cos\tfrac{1}{2}(\theta - \phi)$$
$$\cos\theta - \cos\phi = -2\sin\tfrac{1}{2}(\theta + \phi)\sin\tfrac{1}{2}(\theta - \phi)$$
$$\sin\theta \sin\phi = \tfrac{1}{2}[\cos(\theta - \phi) - \cos(\theta + \phi)]$$
$$\sin\theta \cos\phi = \tfrac{1}{2}[\sin(\theta + \phi) + \sin(\theta - \phi)]$$
$$\cos\theta \cos\phi = \tfrac{1}{2}[\cos(\theta + \phi) + \cos(\theta - \phi)]$$

$\text{Sin } A = \cos A, \tan A = 2\sin\dfrac{A}{2}\cdot\cos\dfrac{A}{2}$
$$= \sqrt{\tfrac{1}{2}(1 - \cos 2A)} = \sqrt{1 - \cos^2 A}$$

$\text{Cos } A = \dfrac{\sin A}{\tan A} = 2\cos^2\dfrac{A}{2} - 1 = 1 - 2\sin^2\dfrac{A}{2} = \sqrt{1 - \sin^2 A}$

$\text{Tan } A = \dfrac{\sin A}{\tan A} = \dfrac{\sin 2A}{1 + \cos 2A} = \sqrt{\dfrac{1}{\cos^2 A} - 1}$

$\text{Sin } \dfrac{A}{2} = \sqrt{\dfrac{1 - \cos A}{2}}$ \qquad $\text{Sin } 2A = 2\sin A.\cos A$

$\text{Cos } \dfrac{A}{2} = \sqrt{\dfrac{1 + \cos A}{2}}$ \qquad $\text{Cos } 2A = 2\cos^2 A - 1$

$\text{Tan } \dfrac{A}{2} = \dfrac{1 - \cos A}{\sin A}$ \qquad $\text{Tan } 2A = \dfrac{2\tan A}{1 - \tan^2 A}$

RIGHT ANGLED TRIANGLES

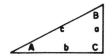

$$a^2 = c^2 - b^2$$
$$b^2 = c^2 - a^2$$
$$c^2 = a^2 + b^2$$

Known	Required					
	A	B	a	b	c	Area
a, b	$\tan A = \dfrac{a}{b}$	$\tan B = \dfrac{b}{a}$			$\sqrt{a^2 + b^2}$	$\dfrac{ab}{2}$
a, c	$\sin A = \dfrac{a}{c}$	$\cos B = \dfrac{a}{c}$		$\sqrt{c^2 - a^2}$		$\dfrac{a\sqrt{c^2 - a^2}}{2}$
A, a		$90° - A$		$a \cot A$	$\dfrac{a}{\sin A}$	$\dfrac{a^2 \cot A}{2}$
A, b		$90° - A$	$b \tan A$		$\dfrac{b}{\cos A}$	$\dfrac{b^2 \tan A}{2}$
A, c		$90° - A$	$c \sin A$	$c \cos A$		$\dfrac{c^2 \sin 2A}{4}$

OBLIQUE ANGLED TRIANGLES

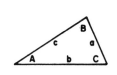

$$s = \frac{a + b + c}{2}$$

$$K = \sqrt{\frac{(s - a)\,(s - b)\,(s - c)}{s}}$$

$$a^2 = b^2 + c^2 - 2bc \cos A$$
$$b^2 = a^2 + c^2 - 2ac \cos B$$
$$c^2 = a^2 + b^2 - 2ab \cos C$$

Known	Required					
	A	B	C	b	c	Area
a, b, c	$\tan \dfrac{1}{2} A = \dfrac{K}{s - a}$	$\tan \dfrac{1}{2} B = \dfrac{K}{s - b}$	$\tan \dfrac{1}{2} C = \dfrac{K}{s - c}$			$\sqrt{s(s - a)\,(s - b)\,(s - c)}$
a, A, B			$180° - (A + B)$	$\dfrac{a \sin B}{\sin A}$	$\dfrac{a \sin C}{\sin A}$	
a, b, A		$\sin B = \dfrac{b \sin A}{a}$			$\dfrac{b \sin C}{\sin B}$	
a, b, C	$\tan A = \dfrac{a \sin C}{b - a \cos C}$				$\sqrt{a^2 + b^2 - 2ab \cos C}$	$\dfrac{ab \sin C}{2}$

TRIGONOMETRICAL RATIOS

Angle Degrees	Sine	Tangent	Cotangent	Cosine	
1	0.0175	0.0175	57.2900	0.9998	89
2	0.0349	0.0349	28.6363	0.9994	88
3	0.0523	0.0524	19.0811	0.9986	87
4	0.0698	0.0699	14.3007	0.9976	86
5	0.0872	0.0875	11.4301	0.9962	85
6	0.1045	0.1051	9.5144	0.9945	84
7	0.1219	0.1228	8.1443	0.9925	83
8	0.1392	0.1405	7.1154	0.9903	82
9	0.1564	0.1584	6.3138	0.9877	81
10	0.1736	0.1763	5.6713	0.9848	80
11	0.1908	0.1944	5.1446	0.9816	79
12	0.2079	0.2126	4.7046	0.9781	78
13	0.2250	0.2309	4.3315	0.9744	77
14	0.2419	0.2493	4.0108	0.9703	76
15	0.2588	0.2679	3.7321	0.9659	75
16	0.2756	0.2867	3.4874	0.9613	74
17	0.2924	0.3057	3.2709	0.9563	73
18	0.3090	0.3249	3.0777	0.9511	72
19	0.3256	0.3443	2.9042	0.9455	71
20	0.3420	0.3640	2.7475	0.9397	70
21	0.3584	0.3839	2.6051	0.9336	69
22	0.3746	0.4040	2.4751	0.9272	68
23	0.3907	0.4245	2.3559	0.9205	67
24	0.4067	0.4452	2.2460	0.9135	66
25	0.4226	0.4663	2.1445	0.9063	65
26	0.4384	0.4877	2.0503	0.8988	64
27	0.4540	0.5095	1.9626	0.8910	63
28	0.4695	0.5317	1.8807	0.8829	62
29	0.4848	0.5543	1.8040	0.8746	61
30	0.5000	0.5774	1.7321	0.8660	60
31	0.5150	0.6009	1.6643	0.8572	59
32	0.5299	0.6249	1.6003	0.8480	58
33	0.5446	0.6494	1.5399	0.8387	57
34	0.5592	0.6745	1.4826	0.8290	56
35	0.5736	0.7002	1.4281	0.8192	55
36	0.5878	0.7265	1.3764	0.8090	54
37	0.6018	0.7536	1.3270	0.7986	53
38	0.6157	0.7813	1.2799	0.7880	52
39	0.6293	0.8098	1.2349	0.7771	51
40	0.6428	0.8391	1.1918	0.7660	50
41	0.6561	0.8693	1.1504	0.7547	49
42	0.6691	0.9004	1.1106	0.7431	48
43	0.6820	0.9325	1.0724	0.7314	47
44	0.6947	0.9657	1.0355	0.7193	46
45	0.7071	1.0000	1.0000	0.7071	45
	COSINE	COTANGENT	TANGENT	SINE	ANGLE DEGREES

ADDITIONAL READING

1. RAINA, V. K., 'Concrete for Construction — Facts and Practice', Tata McGraw-Hill, New Delhi.
2. RAINA, V. K., 'Concrete Bridge Practice — Construction, Maintenance and Rehabilitation', Tata McGraw-Hill, New Delhi.
3. RAINA, V. K., 'Concrete Bridge Practice — Analysis, Design and Economics', Tata McGraw-Hill, New Delhi.
4. WOOD and WYATT, 'Reinforced Concrete Repairs Integrated with Cathodic Protection'.
5. AASHTO 'Guide Specifications for Strength-Evaluation of Existing Steel and Concrete Bridges — 1989.
6. *U. S. Transportation Research Board*
   - Minkarah, I., 'Behavior and Repair of Deteriorated Reinforced Concrete Beams' Transportation Research Record 821, pp 73–78.
   - 'Bridge Deck Repair' Research Results Digest 85, March 1976.
   - 'Bridges on Secondary Highways and Local Roads—Rehabilitation and Replacement' NCHRP Report 222, 1980.
   - Biswas, M., 'Bridge Replacement with Precast Concrete Panel', TRB Special Report 148, 1974, pp 136–148.
   - Buth, Eugene, 'Use of Prestressed, Precast Concrete Panels in Highway Bridge Construction', Special Report 148, 1974, pp 122–35.
   - Nicholson, J. P., 'New Approach Cathodic Protection of Bridge Deck and Concrete Structure', TR Record 762, 1980, pp 13–17.
   - Ellis, W., 'Cathodic Protection of Concrete Structures', Draft Report of NCHRP Project 12–19, Sept. 1980.
   - Manning, D., 'Decision Criteria for Rehabilitation of Concrete Bridge Decks', TR Record 762, pp 1–8, 1980.
   - 'Durability of Concrete Bridge Decks', NCHRP Synthesis of Highway Practice 57, 1979.
   - Manning, David G., 'Effects of Traffic-Induced Vibrations on Bridge Deck Repairs', NCHRP Report 86, Dec. 1981.
   - Boyd, W. K., 'Influence of Bridge Deck Repairs on Corrosion of Reinforcing Steel', Draft Report of NCHRP Project 12–16, Jan. 1979.
   - 'Rehabilitation and Replacement of Bridges on Secondary Highways and Local Roads', NCHRP Report 243, Dec. 1981.
   - Kliethermess, John, 'Repair of Spalling Bridge Deck' Highway Research Record 400, pp 83–91.
   - Darter, Michael I., 'Patching Continuously Reinforced Concrete Pavements', TR Record 800, pp 12–17.
   - Fromm, H. J., 'Successful Application of Cathodic Protection to a Concrete Bridge Deck', TR Record 762.
   - 'Use of Polymers in Highway Concrete', NCHRP Report 190.
   - 'Waterproof Membrane for Protection of Concrete Bridge Deck', NCHRP Report 165, 1976.
   - 'Rapid-Setting Materials for Patching of Concrete', NCHRP Practice 45.
   - 'Concrete Sealer for Protection of Bridge Structures', NCHRP Report 244, TRB, Dec. 1981.
   - 'Evaluation of Methods of Replacement of Deteriorated Concrete in Structure', NCHRP Report 1.
7. AASHTO, 'Manual for Maintenance Inspection of Bridge, 1978', AASHTO, Washington D. C.
8. *Federal Highway Administration (U. S., F. H. W. A.)*
   - 'Bridge Inspector's Training Manual'.
   - 'Federal-Aid Highway Program Manual Vol. 6—Engineering and Traffic Operation'.
   - 'Third Annual Report to Congress, Highway Bridges Replacement and Rehabilitation Program', Bridge Division, March 1982.
   - 'Recording and Coding Guide for the Structural Inventory and Appraisal of Nation's Bridges'.
   - 'FCP Annual Progress Report-Cost Effective Rigid Concrete Construction and Rehabilitation in Adverse Environments'.
   - 'Extending the Service of Life on Existing Bridges by Increasing Their Load Carrying Capacity', Report No: FHWA-RD-78-133.
9. 'Rehabilitation and Replacement of Bridges on Secondary Highways and Local Roads', NCHRP Report 243, TRB, Dec. 1981.
10. MISHLER, H. W. and B. N. LEIS, 'Evaluation of Repair Techniques for Damaged Steel Member', Draft copy of Phase I Final Report, May 1980.
11. HEINS, C. P. and H. KATO, 'Load Distribution of Cracked Girders', ASCE Journal of Structural Division Vol. 108, No. ST8, Aug. 1982.
12. 'Manual of Concrete Practice', ACI Detroit, Mich., U. S. A.
13. 'Guide to Durable Concrete', ACI Committee 201, ACI Journal, Dec. 1977, pp 573–609.
14. 'Standard Practice for Concrete Highway Bridge Deck Construction', ACI Committee 345, Concrete International, Sept. 1981.
15. 'Guide for Repairs of Concrete Superstructure', Report No. 546, ACI 1980.
16. 'Extending the Service of Life of Existing Bridges by Increasing Their Load Carrying Capacity', FHWA/RD-78-133, FHWA June 1978.

17. 'Design and Construction of Welded Bridge Members and Connection', FHWA Training Course Material, Sept. 1980.

18. 'Manual of Concrete Practice', *Part 1* Material and General Properties of Concrete; *Part 4* Bridge, Substructure, Sanitary and Other Special Structures; Structural Property; *Part 5* Masonry; Precast Concrete; Special Process, American Concrete Institute, Detroit, MI 48219.

19. 'Area Manual for Railway Engineering', Chapter 8, Concrete Structure and Foundation, *Part 13* Shotcrete; *Part 14* Repairing and Solidifying Masonry Structure, American Railway Engineering Association, Washington D. C.

20. DUSSECK, IAN, 'Strengthening of Bridge Beams and Similar Structure by Means of Epoxy-Resin Bonded External Reinforcement', Transpt. Research Record 785, pp 21–24.

21. SHANAFELT, G. O., 'Damage Evaluation and Repair Methods for Prestressed Concrete Bridge Members', NCHRP Report 226, 1980.

22. CLARK, R. and T. PAULAY, 'Reinforced Concrete Structures', John Wiley & Sons, Inc., 1975.

23. CORDON, WILLIAM 'Repairs of Concrete, Concrete Construction Handbook' Chapter 48, McGraw-Hill, N.Y.

24. JOHNSON, SIDNEY M., 'Deterioration, Maintenance, and Repairs of Structures', McGraw-Hill, N. Y., 1965.

25. 'Cement and Concrete Technical Data', Kaiser Cement, Oakland, CA.

26. 'Concrete Information', Portland Cement Association, Skokie, Ill.

27. *FHWA Publications*
    - 'Federal-Aid Highway Program Manual', Vol. 6, Chapter 7, Section 2, Subsection 7, 1976.
    - 'Cost Effective Rigid Concrete Construction and Rehabilitation in Adverse Environment', 4th Annual Report, 1982.
    - 'Extending the Service Life of Existing Bridge by Increasing Their Load Carrying Capacity', FHWA-RD-78-133, June 1978.
    - 'Evaluation of Super-Water Reducers for Highway Application', FHWA/RD-80/132, March 1981.
    - 'Connections for Modular Precast Concrete Bridge Details', FHWA/RD-82/106, August 1983.
    - 'Evaluation of Portland Cement Concrete for Permanent Bridge Deck Repair', FHWA/RD-74/5, 1974.
    - 'Internally Sealed Concrete', FHWA/DP-49-2, Jan. 1982.
    - 'Latex Modified Concrete Bridge Deck Overlays: Field Performance Analysis', FHWA/OH-79/004, 1979.
    - 'Polymer Impregnation of New Concrete Bridge Deck Surfaces', FHWA Implementation Package 78–5, Jan. 1978.
    - 'Polymer Concrete Patching Manual', FHWA/IP-82-10, June 1982.
    - 'Polymer Concrete Overlays Interim User Manual', FHWA-TS-78-225 and TS-78-218.
    - 'Styrene-Butadiene Latex Modifiers for Bridge Deck Overlay Concrete', FHWA/RD-78-35, April 1978.
    - 'Transverse Cracking of Asphalt Pavement', FHWA-TS-82-206, July 1982.
    - 'Waterproofing Membranes for Bridge Deck Rehabilitation', FHWA/NY-77-59-1.
    - 'Transverse Cracking of Asphalt Pavement', FHWA-TS-82-206, July 1982.

28. Maine Dept. of Transportation: 'Bridge Design Manual'.

29. Mass. Dept. of Public Works: 'Bridge Manual'.

30. Minnesota Dept. of Transportation: 'Bridge Design Manual', 'Bridge Details', 'Bridge Standard Plans Manual'.

31. New Mexico Highway Department: 'Bridge Manual'.

32. N. Y. State Dept. of Transportation: 'Standard Details for Highway Bridges', (Accounting and Fiscal Service Bureau, NY), State DOT, 1220 Washington Ave. State Campus, Albany, NY 12232.

33. Ohio Dept. of Transportation: 'Ohio Design Regulation', 'Ohio Supplement to AASHTO', Bridge and Structure Design, 25 South Front St. P. O. Box 899, Columbus, CH 43216.

34. Penn. Dept. of Transportation: 'Design Manual Part 4—Structural Design', Publication and Sales, P. O. Box 134 Middle Town PA 17057.

35. UNDERWOOD, TOM 'Preliminary Design of Highway and Railroad Strucutures', Columbus Engineering Ltd, Columbus, Ohio, 1966.

36. Washington State Dept. of Transportation: 'Bridge Design Manual', Bureau of Bridge Design, Highway Administration Bldg. Olympia, Washington, 98504.

37. Wisconsin Dept. of Transportation, 'Bridge Design Manual', Bridge Development Engineer, 4802 Sheboygan Ave. P. O. Box 7916, Madison, WI 53707.

38. BRANSON, D. E., 'Deformation of Concrete Structures', McGraw-Hill, N. Y., NY 1977.

39. American Concrete Institute: 'Manual of Concrete Practice—Bridge Analysis and Design', (American Concrete Institute, P. O. Box 19150, Redford Station, Detroit, MI 48219).

40. Florida Dept. of Transportation: 'Manual for Bridge Maintenance Planning and Repair Methods'.

41. New Jersey Department of Transportation: 'Bridge Design Manual, 1983'.

42. Transportation Research Board (U. S. A.): 'Bridge Engineering Vols 1 and 2', TRB Record 664 and 665, (TRB: 2101 Constitution Ave. N. W. Washington D. C. 20418).

43. Calif. Dept. of Transportation: 'Bridge Design Manual volume 1, 2 and 3', 'Project Development Procedure'.

44. UHLIG, HERBERT H., 'Corrosion and Corrosion-Control', John Wiley & Sons Inc.

45. 'Handbook of Corrosion Protection for Steel Pile Structures in Marine Environments', American Iron and Steel Institute, (AISI) Washington, D. C., 1981.

46. DOLAN, E. D., 'Corrosion Control of Steel Structure, Maintenance of Marine Structure', The Institution of Civil Engineers, London, England 1978, pp 217–229.

47. *U. S. FHWA Technical Advisory*
    - ' Integral No-joint Structure and Required Provision for Movement—T 5140. 13', Jan. 28, 1980.
    - ' Expansion Devices for Bridge—T 5140. 15' March 26, 1980.
    - ' Bridge Deck Joint Rehabilitation (Retrofit)—T 5140.16' March 26, 1980.
    - FHWA Notice N-5140.11 and 12 'Development of Watertight Bridge Deck Joint Seals'.
    - Kozlov, George, 'Performed Elastomeric Joint Sealer for Bridges', FHWA/NJ-80/003 Nov. 1978.
    - Azevedo, Vemon, 'Evaluation of Watertight Bridge Expansion Joints' Kentucky DOT Research Report, UKTRP-81-21, July 1981.
    - ' Evaluation of Various Bridge Deck Joint Sealing System', Final Report Research Project 72F-128, Michigan Transportation Commission, July 1979.
    - ' Bridge Deck Joint Sealing System-Evaluation and Performance Specification', NCHRP Report 204, TRB, June 1979.

48. 'Foundations of Earth Structures', Design Manual 7.2, Naval Facility Engineering Command, May 1982.

49. ANDERSEN, PAUL, 'Substructure Analysis and Design', Second Edition, Ronald Press, NY 1956.

50. ELIAS, VICTOR and, PAUL SWANSON, 'Cautions of Reinforced Earth in Residual Soils', Paper Presented: 1983 TRB Annual Meeting.

51. FHWA Memorandum dated Nov. 25, 1981, "Doublewal —Interlocking Precast Retaining Wall System'.

52. 'Doublewal', Technical Manual Doublewal Corpn.

53. 'VSL Rock and Soil Stabilization System', Technical Manual VSL Corpn.

54. 'Concrete Bridge Details', Portland Cement Association.

55. FHWA Technical Advisory T 5140.13, 'Integral, No-joint Structures and Required Provision for Movement', Jan. 28, 1980.

56. GREIMANN, L., A. WOLDE-TINSAE and P. YANG, 'Skewed Bridge with Integral Abutments', Engineering Research Institute, Iowa State University, Dec. 1982.

57. JORGENSON, JAMES L., 'Behavior of Abutment Piles in an Integral Abutment Bridge', Paper Presented 1983: TRB Annual Meeting, Jan. 1983.

58. 'Underwater Inspection and Repair of Bridge Substructures', NCHRP Practice 88, TRB.

59. 'Superplasticized Concrete for Rehabilitation of Bridge Decks and Highway Pavements', US DOT/ RSPA/ DPB-50/81/31, Jan 1981.

60. FURR, HOWARD L., 'Bridge Slab Concrete Placed Adjacent to Moving Live Loads', Report FHWA/ TX-81/1+266-IF, Texas Transportation Institute, 1981.

61. GANGARAO, HOTA V. S., 'Feasibility Study of Steel Grid Decks for Bridge Floors', West Virginia University, July 1980.

62. 'Gunite', Pressure Concrete Construction Co. Florence, AL 35630.

63. 'Solving Corrosion Problems of Bridge Surfaces Could Save Billions', US General Accounting Office, Jan. 1979.

64. 'Polymer Concrete Overlay on S4-51 Bridge Deck', Oklahoma DOT Research and Development Division, March 1982.

65. MONT, R. M., EUGENE FASULLO, DANIEL M. HAHN, 'Replacement of the Upper Deck of the George Washington Bridge', Annals of the NY Academy of Sciences, Vol. 352, Dec. 1980, pp 143–146.

66. 'Pre-Cast Slabs for Composite Bridge Decking', US Steel State of the Art Report, Nov. 1981.

67. REILICH, H. D, 'Golden Gate Bridge: Deck Investigation', ASCE Journal of Technical Council, Vol. 107, April 1981, pp 101–116.

68. SEIM, CHARLES, 'Studies Leading to the Golden Gate Bridge Deck Replacement Project', Proceedings of Bridge Maintenance and Rehabilitation Conference, W. VA College of Engineering, Morgantown, WV, 1980, pp 727–744.

69. STEWART, C. F., 'Consideration for Repairing Salt-Damaged Bridge Decks', ACT Journal Vol. 72 1975, pp 685–698.

70. TIMMER, DONALD H. "A Study of Concrete Filled Steel Grid Bridge Decks in Ohio" Proceeding of Bridge Maintenance and Rehabilitation Conference pp 442–475, 1980, Morgantown, WV.

71. 'Technical Data', E-Poxy Industry, Inc. Ravena, NY.

72. RYELL, J. M. OWENS and D. G. MANNING, 'Durable Highway Structures—Ontario's Approach', Ontario Ministry of Transportation, Dec. 1982.

73. FONDRIEST, F. F. and M. J. SNYDEL, "Paving Practice for Wearing Surface on Orthotropic Steel Deck", Jan. 1968 and "Addendum" Oct., 1971 American Iron and Steel Institute.

74. 'Adhesive/Coating/Sealant Construction', Adhesive Engineering Co. San Carlos, CA.

75. 'Benjamin Franklin Bridge Design Engineering Report', Delaware River Port Authority, 1983.

76. 'Optimized Section for Major Prestressed Concrete Bridge Girder', FHWA/DR-82/005, FHWA, Feb. 1982.

77. 'Recommended Practice for Segmental Construction in Prestressed Concrete', PCI Journal Vol. 20, No. 2, March-April, pp 22–41, 1975.

78. 'Precast Segmental Box Girder Bridge Manual', Post Tensioning Institute and PCI, 1978.

79. PODOLY, WALTER Jr., 'An overview of Precast Prestressed Segmental Bridges', PCI Journal, Vol. 24, No. 1, Jan.–Feb. 1979, pp 56–87.

80. MULLER, JEAN, 'Construction of Long Key Bridge', PCI Journal, Vol. 25, No. 6, Nov. –Dec. 1980, pp 97–111.

81. BARKER, JAMES, 'Construction Techniques for Segmental Concrete Bridges', PCI Journal, Vol. 25, No. 4, July–Aug 1980, pp 66–86.

82. FHWA Structural Engineering Series No. 6 'Prestressed Concrete Segmental Bridge', FHWA Bridge Division, Aug. 1979.

83. KULKA, F., S. J. THOMAS, and T. Y. LIN, 'Feasibility of Standard Sections for Segmental Prestressed Concrete

Box Girder Bridges', FHWA/RD-82/024 FHWA, July 1982.

84. 'Prestressed Concrete Segmental Bridges in the United States as of September 30, 1982', FHWA Bridge Division.

85. PCI Technical Bulletin, Number 1, 1983

86. MARTIN, L. D. and A. E. OSBORN, 'Connection for Modular Precast Concrete Bridge Decks', FHWA/RD-82/106 FHWA, August 1983.

87. 'Recommendations for Prestressed Rock and Soil Anchors', Post-Tensioning Institute.

88. 'Tiebacks' FHWA/RD-82/046 and 047, FHWA, July 1982.

89. WEATHERBY DAVID E. and JAMES W. SIGOURNEY, 'Tiebacks and their Application', ASCE National Geotechnical Conference, Jan. 6–8, 1981.

90. REEVES, RONALD, B., 'Control of Landslides with Permanent Tiebacks'.

91. 'Soil Mechanics, Design Manual 7.1' and 'Foundation and Earth Structures Design Manual 7.2', U. S. Dept. of Navy, Naval Facilities, Engineering Command, May, 1982.

92. 'Development and Use of Prestressed Steel Flexural Member', ASCE Journal of Structural Division Vol. 94 No. ST9, Sept. 1968, pp 2033.

93. 'Chemical Grouts for Soil', FHWA/RD-77-51, FHWA June 1977, 'Design and Control of Chemical Grouting'.

94. 'Guideline for the Hydraulic Design of Culverts', AASHTO Highway Drainage Guideline Series, 1975.

95. 'Handbook of Steel Drainage and Highway Construction Products', American Iron and Steel Institute.

96. *FHWA Hydraulic Engineering Circular*
   • 'Hydraulic Charts for the Selection of Highway Culverts', HEC No. 5, Dec. 1965.
   • 'Capacity Chart for the Hydraulic Design of Highway Culverts', HEC No. 10, Nov. 1972.
   • 'Hydraulic Design of I, proved Inlets for Culverts', HEC No. 13, August 1972.
   • 'Hydraulic Design of Energy Dissipators for Culverts and Channels', HEC No. 14, Dec. 1975.
   • 'Culvert Design System', FHWA-TS-80-245, FHWA, Jan. 1980.
   • 'Improved Inlet for Culvert—Example Structural Plans', FHWA Technical Advisory T 5140.6, Jan. 6, 1979.

97. BRICE, J. C. and J. C., BLODGETT, 'Countermeasures for Hydraulic Problems at Bridge, *Vol. I* Analysis and Assessment, (FHWA/RD-78-162) and *Vol. II* Case Histories for Sites (FHWA/RD- 78-163), FHWA, Sept. 1978.

98. BROWN, S. A., R. S. MCQUIVEY and T. N. KEEFER, 'Stream Channel Degradation and Aggradation Analysis of Impacts to Highway Crossings, FHWA/RD-80-159, FHWA, March 1981.

99. HOPKINS, G. R., R. W. VANCE and B. KASRAIE, 'Scour Around Bridge Pier', FHWA/RD-79-103, FHWA, Feb. 1980.

100. KEEPFER, T. N., 'Stream Channel Degradation and Aggradation: Causes and Consequence to Highways', FHWA/RD-80-038, FHWA, June 1980.

101. SCHNEIDER, V. R. and K. V. WILSON, 'Hydraulic Design of Bridges with Risk Analysis', March 1980.

102. Scour at Bridge Waterway', NCHRP Practice 5, TRB, 1970.

103. 'Hydraulic Analysis for Location and Design and Bridges', AASHTO, 1982.

104. 'Bank and Shore Protection in California Highway Practice', Calif. Dept. of Transportation, Nov., 1970.

105. 'New Life for Deteriorated Piles', Intrusion-Prepakt Inc. Cleveland, Ohio.

106. 'Pile Restoration and Preservation System', Symons Corp. Des Plaines, Ill.

107. 'Grouting Soil-Stabilization Water Control', Intrusion-Prepakt Inc. Cleveland, Ohio.

108. 'Ground Modification', Hayward Baker Co., Odenton, MD.

109. HEINZ, RONEY, 'Protection of Piles in Wood, Concrete, Steel', Civil Engineering—ASCE, Dec. 1975.

110. 'Design of Pile Foundation', NCHRP Practice 42, TRB, 1977.

111. 'Steel Construction Manual', NY State DOT.

112. WADE, BILL C., 'Retrofit of Pin Connected Truss Bridge', Proceedings of Bridge Maintenance and Rehabilitation, Conference 1980, W. Va. Univ.

113. KEANE, JOHN D., 'Protective Coating for Highway Structural Steel', NCHRP Report 74, TRB, 1969.

114. FISHER, JOHN W., 'Detection and Repair of Fatigue Damage in Welded Highway Bridge', NCHRP Report 206, TRB, 1979.

115. FISHER, JOHN W., 'Fatigue Behavior and Full Scale Welded Bridge Attachment', NCHRP Project 12-15(3), Draft Copy of Final Report, TRB.

116. SWEENEY, RAP, 'Some Examples of Detection and Repair of Fatigue Damage in Railway Bridge Member', TR Record 676, pp 8–14, TRB.

117. SWEENEY, RAP, 'Importance of Redundancy in Bridge Fracture Control' TR Record 665, TRB.

118. HSIONG, WEI, 'Repair of Popular Street Complex Bridge', TR Record 664, pp 110–119, TRB.

119. AASHTO, 'Guide for Selecting, Locating, and Designing Traffic Barriers', 1977.

120. AASHTO, 'Highway Design and Operational Practices Related to Highway Safety' Second edn, 1974.

121. 'Upgrading Safety Performance in Retrofitting Traffic Railing Systems', FHWA-RD-77-40, 1976.

122. 'Location, Selection and Maintenance of Highway Traffic Barriers', NCHRP Report 118, 1971, TRB.

123. FORTUNIEWICZ, J. S., J. E. BRYDEN and R. G. PHILLIPS, 'Crash Tests of Portable Concrete Median Barrier for Maintenance Zones', NY State DOT Research Report 102, 1982.

124. 'Implementing Highway Safety Improvements', ASCE, 1980, NY.

125. Technical Information from Energy Absorption Systems, Inc. Chicago, Ill.

126. 'Design Manual', Fitch Inertial Barrier System Inc., Boston, Mass.

127. GIBEILY, JOSEPH, and GEORGE A. HARPER, 'Findings, Ideas, Conclusions and Recommendations Develop from Bridge Design Review-in-depth Program', Washington, D. C. : FHWA Bridge Division, March 1973.

128. F. I. P. Guide to Good Practice: 'Inspection and Maintenance of Reinforce and Prestressed Concrete Structures', 1986.

129. F. I. P. Guide to Good Practice: 'Repair and Strengthening of Concrete Structures', 1989.

130. JOHNSON, S. M., 'Deterioration, Maintenance, and Repair of Structures, McGraw-Hill, New York, 1965.

131. MANSUR, M. A. and K. C. G. ONG, 'Epoxy repaired Beams', Concrete International, Oct. 1985.

132. VIRGINIA, F., 'Filling the Cracks', Civil Engineering, Oct. 1986.

133. WAGH, V. P., 'Bridge-Beam Repair', Concrete International , Oct. 1986.

134. PLECNIK, J. M. and R. W. GAUL, 'Epoxy Penetration', Concrete International, Feb. 1986.

135. OECD Road Research: 'Bridge Inspection', OECD, Paris, 1976.

136. OECD Road Research: 'Evaluation of Load Capacity of Bridges', OECD, Paris, 1979.

137. OECD Road Research: 'Bridge Maintenance', OECD, Paris, 1981.

138. OECD Road Transport Research: 'Bridge Rehabilitation and Strengthening', OECD, 1983.

139. OECD Road Transport Research: 'Durability of Concrete Bridges', OECD, 1989.

140. Transportation and Research Board: 'Underwater Inspection and Repair of Bridge Structures Programme—Synthesis of Highway Practice 88, Washington D. C., 1981.

141. ACI committee 224, 'Causes, Evaluation and Repair of Cracks in Concrete Structures', ACI Journal, May–June 1984.

142. PODOLNY Jr., W., 'The Causes of Cracking in Post Tensioned Concrete Box Girder Bridges and Retrofit Procedure', PCI Journal, Mar.–Apr. 1985.

143. PCI Committee on Quality Control, 'Fabrication and Shipment Cracks in Prestressed Hollow-Core Slabs and Double Tees', PCI Journal, Jan.–Feb. 1983.

144. PCI Committee on Quality Control, 'Fabrication and Shipment Cracks in Precast or Prestressed Beams and Columns', PCI Journal, May–June, 1985.